Challenges and Advances in Computational Chemistry and Physics

Volume 21

Series editor

Jerzy Leszczynski
Department of Chemistry and Biochemistry
Jackson State University Chemistry, Jackson, Mississippi, USA

This book series provides reviews on the most recent developments in computational chemistry and physics. It covers both the method developments and their applications. Each volume consists of chapters devoted to the one research area. The series highlights the most notable advances in applications of the computational methods. The volumes include nanotechnology, material sciences, molecular biology, structures and bonding in molecular complexes, and atmospheric chemistry. The authors are recruited from among the most prominent researchers in their research areas. As computational chemistry and physics is one of the most rapidly advancing scientific areas such timely overviews are desired by chemists, physicists, molecular biologists and material scientists. The books are intended for graduate students and researchers.

More information about this series at http://www.springer.com/series/6918

Jean-Louis Rivail · Manuel Ruiz-Lopez
Xavier Assfeld
Editors

Quantum Modeling of Complex Molecular Systems

Springer

Editors
Jean-Louis Rivail
UMR 7565—SRSMC
Université de Lorraine
Nancy-Vandoeuvre
France

Xavier Assfeld
UMR 7565—SRSMC
Université de Lorraine
Nancy-Vandoeuvre
France

Manuel Ruiz-Lopez
UMR 7565—SRSMC
Université de Lorraine
Nancy-Vandoeuvre
France

Challenges and Advances in Computational Chemistry and Physics
ISBN 978-3-319-21625-6 ISBN 978-3-319-21626-3 (eBook)
DOI 10.1007/978-3-319-21626-3

Library of Congress Control Number: 2015947402

Springer Cham Heidelberg New York Dordrecht London

Printed on acid-free paper

Springer International Publishing AG Switzerland is part of Springer Science+Business Media (www.springer.com)

Preface

The extraordinary developments of scientific computers during the past 50 years, together with the advances of theoretical chemistry have made possible, for the time being, the computations with great accuracy of the properties of molecular systems comprising about a hundred atoms. However, many systems of chemical or biochemical interest are much larger than allowed by the present limits of computational chemistry. For example, the apparently simple system made of a molecule dissolved in a solvent may exhibit properties quite different from those observed or computed in the isolated state. Similarly, the properties at the interface of a liquid, or a solid and a gas are expected to be quite different from those predicted for a not interacting molecule. The case of biochemical systems is a very typical example since a large number of systems of interest are made of thousands of atoms and if, from a chemical point of view, one feels allowed to focus on the chemical or physical part of interest, the interactions with the rest of the system cannot be discarded.

Historically, accounting for the solvent effects in quantum chemical computations became possible, in a rather simplified representation of the solvent by a dielectric continuum, more than 40 years ago.[1] In the meantime, the development of classical force fields opened the way to molecular mechanics i.e. the classical computation of intermolecular interaction energies and consequently the prediction of thermodynamic properties of these systems, thanks to statistical mechanical tools.

Finally, the coupling of an elaborated quantum description of a chemically important subsystem with the rest of the system described at a lower level such as the classical mechanical one or even the continuum allowed the development of multiscale models for the treatment of the complex systems considered above. These methodologies represent a real scientific advance which has been recently acknowledged by the Nobel prize awarded to Martin Karplus, Michael Levitt, and Arieh Warshel.

---

[1] D Rinaldi and J-L Rivail, Polarisabilités moléculaires et effet diélectrique de milieu à l'état liquide. Etude théorique de la molécule d'eau et de ses dimères. *Theor. Chim. Acta*, 1973, **32**, 57–70.

The present book aims to cover various aspects of the state-of-the-art of computational treatment of complex molecular systems. This includes the development of accurate force fields, including induced electronic polarization and the applications of lower level models, in particular hybrid Quantum Mechanical/Molecular Mechanical (QM/MM) approaches to various problems such as solutions in which the solute–solvent and solvent–solvent interactions can be quite important, in particular in the case of aqueous solutions. The case of molecules at water–air or water–organic liquid interfaces is a special example of the necessity of taking into account the presence of a very large number of ancillary molecules. The situation is rather very similar when one of the phases is solid. Finally, the emergence of reliable theoretical approaches of the innumerable chemical problems encountered in the understanding of living processes is a very promising subject. It generally involves very large molecules such as enzymes or nucleic acids, in which the chemically significant fraction only can be treated at an accurate level and its links with the rest of the macromolecule raise some specific methodological problems.

In this book various aspects of this field in great development are addressed, both from the point of view of the methods and their illustrative applications.

Jean-Louis Rivail
Manuel Ruiz-Lopez
Xavier Assfeld

Contents

Chapter 1
Addressing the Issues of Non-isotropy and Non-additivity in the Development of Quantum Chemistry-Grounded Polarizable Molecular Mechanics

Nohad Gresh, Krystel El Hage, Elodie Goldwaser, Benoit de Courcy, Robin Chaudret, David Perahia, Christophe Narth, Louis Lagardère, Filippo Lipparini and Jean-Philip Piquemal

Abstract We review two essential features of the intermolecular interaction energies (ΔE) computed in the context of quantum chemistry (QC): non-isotropy and non-additivity. Energy-decomposition analyses show the extent to which each comes into play in the separate ΔE contributions, namely electrostatic, short-range repulsion, polarization, charge-transfer and dispersion. Such contributions have their counterparts in anisotropic, polarizable molecular mechanics (APMM), and each of these should display the same features as in QC. We review examples to evaluate the performances of APMM in this respect. They bear on the complexes of one or several ligands with metal cations, and on multiply H-bonded complexes. We also comment on the involvement of polarization, a key contributor to non-additivity, in the issues of multipole transferability and conjugation. In the last

---

N. Gresh (✉) · K.E. Hage · E. Goldwaser · B. de Courcy
Chemistry and Biology Nucleo(S)Tides and Immunology for Therapy (CBNIT), UMR 8601 CNRS, UFR Biomédicale, Paris, France
e-mail: nohad.gresh@parisdescartes.fr

N. Gresh · E. Goldwaser · B. de Courcy · R. Chaudret · C. Narth · L. Lagardère · F. Lipparini · J.-P. Piquemal (✉)
Laboratoire de Chimie Théorique, Sorbonne Universités, UPMC, UMR7616 CNRS, Paris, France

K.E. Hage
Faculté des Sciences, Centre d'Analyses et de Recherche, UR EGFEM, Saint Joseph University of Beirut, Beirut 1107 2050, Lebanon

D. Perahia
Laboratoire de Biologie et de Pharmacologie Appliquées (LPBA), UMR 8113 CNRS, Ecole Normale Supérieure de Cachan, Paris, France

Present Address
K.E. Hage
Department of Chemistry, University of Basel, Klingelbergstrasse 80, 4056 Basel, Switzerland

J.-L. Rivail et al. (eds.), *Quantum Modeling of Complex Molecular Systems*, Challenges and Advances in Computational Chemistry and Physics 21,
DOI 10.1007/978-3-319-21626-3_1

section we provide recent examples of APMM validations by QC, which relate to interactions taking place in the recognition sites of kinases and metalloproteins. We conclude by mentioning prospects of extensive applications of APMM.

Abbreviations

AMOEBA	Atomic multipoles optimized energetics for biological applications
APMM	Anisotropic polarizable molecular mechanics
ATP	Adenosine triphosphate
aug-cc-pVTZ	Augmented correlation-consistent polarized valence triple dzeta
BO	Born-Oppenheimer
CDP	Claverie, Dreyfus, Pullman
CEP	Coreless effective potential
CP	Car-Parrinello
CSOV	Constrained space orbital variation
CTU	Carboxythiourea
dd-Cosmo	Domain-decomposition conductor screening model
DIIS	Direct inversion of iterative space
EFP	Effective fragment potential
ELF	Electron localization function
EM	Energy-minimization
EVG	Elvitegravir
FAK	Focal adhesion kinase
FEP	Free energy perturbation
GEM	Gaussian electrostatic model
GPU	Graphics processor unit
GS	Garmer and Stevens
HF	Hartree-Fock
IEHT	Iterative extended Huckel theory
I-NoLLS	Interactive non linear least squares
INT	Integrase
KM	Kitaura-Morokuma
LC	Langlet-Claverie
LMO	Localized molecular orbitals
MC	Monte-Carlo
MD	Molecular dynamics
MEP	Molecular electrostatic potential
MO	Molecular orbitals
NA	Nucleic acids
NRP1	Neuropilin-1
OPEP	Optimized partitioning of electrostatic properties
PBC	Periodic boundary conditions
PME	Particle Mesh Ewald
PMI	Phosphomannose isomerase
PMM	Polarizable molecular mechanics

QC	Quantum chemistry
QM/MM	Quantum mechanics/molecular mechanics
RVS	Reduced variational space
SAPT	Symmetry-adapted perturbation theory
SIBFA	Sum of interactions between fragments Ab initio computed
SOD	Superoxide dismutase
vdW	van der Waals

1.1 Introduction

The field of molecular simulation is undergoing major strides forward. It is spearheaded by ab initio quantum chemistry (QC), whose realm of applications is being extended to complexes of increasing size, nowadays totaling a few hundreds of atoms, and are amenable as well to increasingly long simulation times as is the case for Car-Parrinello (CP) [1, 2] or Born-Oppenheimer (BO) [3, 4] molecular dynamics (MD) approaches. Such extensions are enabled by advances both in informatics and in the QC codes themselves. The first kind of advances results from progress both in computer hardware, as exemplified by the advent of Graphics Processor Units (GPU) [5–6] and in computer software, as enabled by parallelism on Open MP/MPI software [7]. The second advances are due to promising developments in linear-scaling and divide-and-conquer approaches [8–13]. Despite such advances, simulations of very large molecular complexes and/or on very long-time scale MD or of computer-intensive Monte-Carlo (MC) simulations are likely to remain out of reach of high-level QC for a long foreseeable time in a diversity of domains. These encompass drug design, protein folding, or material science. An appealing alternative consists into the so-called QM/MM (Quantum mechanics/molecular mechanics) approaches, in which the core of the recognition site is treated quantum-mechanically while the periphery is treated by classical MM. Multiscale approaches are a generalization of QM/MM [14–22]. However, such approaches could not be used in the absence of a single ‘privileged’ recognition site, as is the case for multidomain sites in, e.g., protein-nucleic acid (NA), protein-protein or NA-NA recognition. There are also examples of enlarged or accessible recognition sites, such that long-range polarization effects could blur the distinction between QM and MM zones. This can occur with the complexes of inhibitors to the Zn-metalloenzyme phosphomannose isomerase, in which networks of polarizable water molecules can connect together the inhibitor in the recognition site to farther sites on N- and/or C-terminal sites [23]. ‘Classical’ MM/MD potentials are on the other hand fully able nowadays to handle very large molecular complexes over time-scales which even for proteins can approach the microsecond if not the millisecond time. They have rendered prominent service to the field and will most probably continue to do so for a long-time to come. Nevertheless their limitations are duly recognized, notably the absence of non-additivity and the lack of appropriate directionality. Such shortcomings will not be commented in this review. An ideal situation would consist into a QC-grounded MM potential formulated and calibrated so as to reproduce each individual QC contribution. Transferability can be compromised if the aim of a given MM potential were limited to reproduce the total QC intermolecular interaction, ΔE(QC) forfeiting that on its individual contributions. It is also clear that a term-to-term agreement with the ΔE (QC) contributions should carry out beyond the training set used in the calibration. Striving at reproducing QC results has been the object of several endeavors, past and present. These have been surveyed in several recent reviews [24–29]. Summarizing, we recall that the main efforts bore on the inclusion of polarization: dipole polarization [24–29], charge equalization [30–32], or Drude models [33, 34]. It is fitting to recall that the very first inclusion of polarization for the modeling of biological molecules were from the mid-sixties [35–37] and mid-seventies [14]. A subset of 'polarizable' MM/MD potentials have also targeted first-order electrostatics, resorting to distributed multipoles derived from ab QC calculations on the fragments [23, 24, 26, 29, 38–50]. This is the case of the SIBFA (Sum of Interactions Between Fragments Ab initio computed [38, 39] procedure, on which this review focuses, and which was among the prime ones in this regard. Further refinements have also borne on the two short-range contributions, repulsion in first-order and charge-transfer in second-order. This appears to restrict the subset of polarizable potentials to SIBFA, ORIENT [40, 41], and the Effective Fragment Potential (EFP) [42–45].

This review is organized as follows. We use here italics to underline the sought-for features. We will first summarize the features of the SIBFA potential, in light of the requirements for *transferability*. We will then present the results of validation tests regarding first *anisotropy*, then *non-additivity*. Their synergistic impact will be illustrated with the complexes of halobenzene derivatives with a guanine-cytosine base-pair in the recognition site of the HIV-1 nucleocapsid. This will be extended to a case problem in which both come into play and should enable to address in synergy two additional issues, namely *multipole transferability* and *conjugation*. We will next present results from recent studies from two of our Laboratories. The first study relates to the organization of highly structured waters in a bimetallic Zn/Cu enzyme, superoxide dismutase (SOD). The two other relate to inhibitor or ligand binding to a tyrosine kinase, Focal Adhesion Kinase (FAK), on the one hand, and to a Zn-metalloenzyme, phosphomannose isomerase (PMI) on the other hand. These will help to highlight the importance of the second-order contributions, and possibly non-additivity, on molecular recognition.

1.1.1 Procedure

In the SIBFA procedure, the total intermolecular interaction energy between molecules or molecular fragments is computed as a sum of five contributions:

$$\Delta E_{tot} = E_{MTP} + E_{rep} + E_{pol} + E_{ct} + E_{disp}$$

which are the electrostatic multipolar, the short-range repulsion, the polarization, the charge-transfer, and the dispersion contributions.

E_{MTP} is computed as a sum of multipole-multipole interactions, encompassing six terms from monopole-monopole till quadrupole-quadupole. The multipoles are located on the atoms and mid-points of the chemical bonds, and are derived from the ab initio QC molecular orbitals (MO) of the fragment. We resort to a procedure pioneered in 1970 by Claverie, Dreyfus and Pullman (CDP) at the Institut de Biologie Physico-Chimique in Paris [51–53]. It was initially applied to computations of the Molecular Electrostatic Potential (MEP) around biologically important molecules, such as the nucleic acid bases [54–57], DNA and RNA [58–60], proteins [61], phospholipids [62] and ionophores [63, 64]. The CDP method was first applied to the computations of intermolecular interaction energies in 1979 in the context of a polarizable potential [65] and, in 1980-1983, for a series of molecular recognition problems [66–71] using forerunners of the SIBFA procedure. Current applications have since resorted to a variant of the CDP procedure due to Vigné-Maeder and Claverie [72]. For completeness we mention that the derivation of distributed multipoles was also done in the context of IEHT wave-functions by Rein in 1973 [73, 74], and, regarding ab initio QC MO's, in the early eighties by Stone et al. [75, 76], Sokalski et al. [77, 78], and Karlstrom et al. [47, 79]. The more recent OPEP procedure enables to derive both distributed multipoles and polarizabilities from such MO's [80, 81].

Subsequent improvements to E_{MTP} have consisted into including an explicit 'penetration' term, E_{pen}. This is an overlap-dependent term: it translates the fact that at short intermolecular distances, there is a lesser shielding of the nuclear charge of a given atom by its own electronic density, due to its 'penetration' by the corresponding density of the incoming atom, and conversely. This results into an actual increase of the electron-nucleus attraction of the interacting pair. The first formulations of E_{pen} were in 2000–2001 in the context of EFP [82, 83] and in 2003 in the context of SIBFA [84]. The latter has been used systematically ever since 2005 with this procedure. Other formulations have also been recently put forth [85–87]. The formulation by Piquemal et al. recently underwent a promising extension enabling to use it in conjunction with the Particle Mesh Ewald (PME) procedure [194]. This will pave the way for efficient MD simulations on very large systems in the context of the SIBFA or AMOEBA procedures.

The other contribution of predominantly electrostatic nature is polarization. It translates the gain in energy upon rearrangement of the electronic distribution of a given molecular fragment due to the electrostatic field generated on it by all the other interacting fragments. E_{pol} on any 'polarizable' center is a function of the electrostatic field it undergoes and of its polarizability. The field is computed with the same distributed multipoles as E_{MTP} and is screened by a Gaussian function, S, of the distance between the center and the interacting polarizing partner, modulated by the mean of the effective radii of the interacting pair. The polarizability can be used: -either as a scalar, the magnitude of which is derived from experimental measurements. This was done in the earlier versions of SIBFA and in most contemporary polarizable potentials which use the induced dipole approach [27–29]; -or as a QC-derived tensor. This is the case of but a handful of polarizable potentials, namely EFP, SIBFA, and ORIENT. In fact, both SIBFA and EFP resort to the same procedure to derive the polarizabilities, specifically a method published in 1989 by Garmer and Stevens (GS) [88], locating them on the centroids of the Boys localized molecular orbitals (LMO): these are the barycenters of the chemical bonds and the 'tips' of the saturated lone pairs. The two first papers reporting the use of the GS polarizabilities, namely using EFP and SIBFA, were actually published in the same 1994 issue of the ACS Symposium series [89, 90].

Thus given any rigid molecule or molecular fragment, we derive once and for all, in a consistent fashion both distributed multipoles and polarizabilities. These are stored in the SIBFA library as one file, along with the information on the internal geometry and connectivities and types of atoms. Each fragment-specific file can be extracted and concatenated with others whenever the fragment is needed to assemble a large, flexible molecule, or a biomolecule, such as a protein or a nucleic acid.

SIBFA embodies two overlap-dependent contributions.

E_{rep} in first-order is formulated under the form of a sum of bond-bond, bond-lone pair, and lone pair-lone pair interactions. This was inspired by an earlier proposal by Murrell et al. [91] following which the exchange-repulsion is proportional to the square of the intermolecular overlap between localized orbitals. For each pair of interacting orbitals, it could be formulated as S^2/R^n, S denoting their overlap and R the distance between their centroids, n taking the values of 1, 2 and possibly beyond. In the context of SIBFA, representations of the orbitals are the chemical bonds and the lone-pairs.

The formulation of E_{ct} in second order is based on another work by Murrell et al. [92], in which it is formulated as the integral of an overlap transition density convoluted with the electrostatic potential it undergoes. Starting from it, E_{ct} was derived as a function of the overlap between representations of the localized orbitals of the electron-donor fragment and of the virtual orbitals of the electron-acceptor fragment [93, 94]. On account of their greater mutual proximities, they are restricted to the sole lone-pairs of the donor, and to the sole B-H bonds of the acceptor, where B denotes any heavy atom. The different contributions are calculated for all pairs of such orbitals, and embody dependencies upon the electrostatic potentials V separately undergone by the electron-donor and by the electron acceptor in the large molecular complex. These potentials also intervene in the denominator of E_{ct}. In it the ionization potential, I, of the electron donor is modulated by the (predominantly positive) potential it undergoes; while the electron affinity of the electron acceptor, A, is modulated by the (predominantly negative) potential it undergoes. In the whole range of energy-relevant distances, this prevents the I-A differences from being negative (and therefore E_{ct} from becoming positive) whenever I has a smaller magnitude than A, as is the case for most di- or multivalent cations. We also note that those Vs, along with the permanent multipoles, embody the contributions due to the induced dipoles, whence an indirect coupling between the polarization and charge-transfer contributions.

The last contribution, E_{disp}, uses a formulation by Creuzet et al. [95]. It is computed as a sum of atom-atom terms with $1/R^6$, $1/R^8$, and $1/R^{10}$ dependencies. It is augmented by an explicit 'exchange-dispersion' term ($E_{exch\text{-}disp}$) and by contributions from the lone pairs at their centroid positions.

Two points could be noted at this stage. Firstly, all five SIBFA contributions, not only E_{rep} and E_{ct}, embody dependencies upon the overlap: E_{pen} for E_{MTP}; a Gaussian screening factor of the distance between each interacting pair for E_{pol}; and $E_{exch\text{-}disp}$ for E_{disp}. Secondly, with the exception of E_{MTP}, all contributions embody dependencies upon explicit representations of the lone pairs.

We denote throughout by ΔE the energies in the absence of E_{disp}, and by E_1 and E_2 the summed first- and second-order contributions, namely $E_1 = E_{MTP} + E_{rep}$ and $E_2 = E_{pol} + E_{ct}$, respectively.

On account of its very separability, critical for the refinement of the SIBFA potential is the availability of energy-decomposition analyses. The Reduced Variational Space (RVS) procedure by Stevens and Fink [96] was particularly instrumental. On the one hand, it affords an operational and dependable separation of E_2 into E_{pol} and E_{ct}. On the other hand, it lends itself to analyses of complexes having more than just two interacting partners. This enable to compute the separate components of E_1 and E_2 in multi-molecular complexes, evaluate the extent of their non-additivities, and how well these could be accordingly retrieved by their SIBFA counterparts. There are also other energy-decomposition analyses, such as the Constrained Space Orbital Variation (CSOV) [97] and the Symmetry-Adapted Perturbation Theory (SAPT) [98], which can both be done at the correlated level. We have resorted to the former upon dealing with open-shell cations [99] and for correlated computations [100] and are presently resorting to the latter in analyses of the interactions involving nucleic acid bases (Gresh et al. submitted).

1.2 Aspects of Anisotropy and Non-additivity in Molecular Complexes

A. ***Anisotropy***

MM potentials should be able to account for the fine angular features of ΔE(QC) upon performing in- and out-of-plane variations of the approach of two interacting atoms at fixed equilibrium distance: that is, the departure from the assumption of atom-centered spherical symmetry. This might not be warranted by simple point-charge electrostatics and atom-atom Lennard-Jones-like formulas for E_{rep} and E_{disp}, unless neighboring atoms restricted the available space.

(1) *Linear water dimer*. The earliest example we considered dating back from the mid-eighties is the linear water dimer at equilibrium distance (d_{OO} = 2.95 A) [93, 94]. We monitored the evolutions, as a function of the theta angle, of ΔE

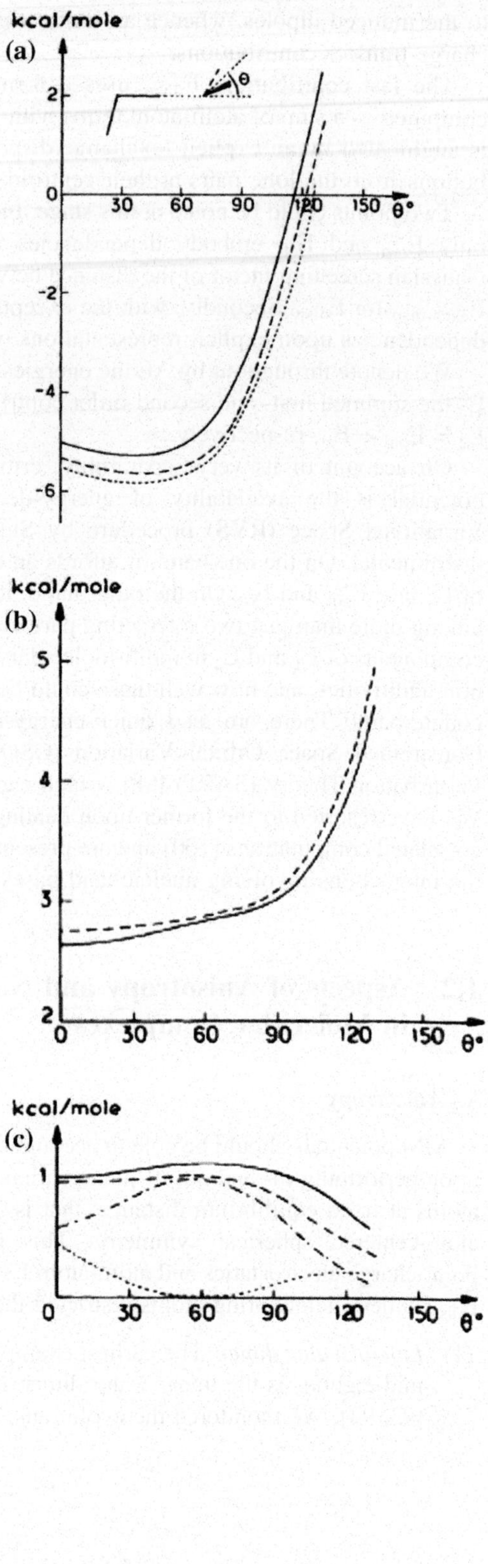

Fig. 1.1 Linear water dimer. Compared evolutions of: **a** ΔE(SIBFA), ΔE(HF), and E_{MTP} (*full*, *dashed*, and *dotted lines*, respectively; **b** E_{exch}(HF), and E_{rep}(SIBFA) (*full* and *dashed lines*); and **c** E_{ct}(HF) and E_{ct}(SIBFA) (*full* and *dashed lines*). The *dotted* and *dashed-dotted lines* represent the contributions to E_{ct}(SIBFA) from each of the two individual sp^3 lone pairs of the donor oxygen. Reprinted with permission from Gresh et al. [94]. Copyright 1986 Wiley

(SIBFA) and its E_{MTP}, E_{rep} and E_{ct} contributions (Fig. 1.1a–c). For such evolutions the H and O atoms of the incoming OH bond of the electron-acceptor monomer predominantly sense the electrostatic potential around the O atom of the electron donor monomer and their overlaps with its sp^3 lone-pairs and there is little, if any, overlap with its two OH bonds. A very close parallelism was observed not only between the total ΔE(SIBFA) and ΔE (HF) energies, but also between the corresponding individual contributions. This early demonstration already lends credence to the representation of electrostatics with QC multipoles and to that of the short-range contributions with lone-pairs. The behavior of E_{ct} was instructive. Its relative shallow behavior matching that of E_{ct}(HF) was found to result from strongly opposed trends from the two sp^3 lone-pairs. The contribution from the first lone-pair increases regularly until for theta = 60° the direction of the incoming HO bond aligns with it, while that of the second lone-pair decreases in concert. Past 60°, the two trends are reversed. The overall shape of ΔE appears dictated by that of E_{MTP}. This appears consistent with the proposal by Buckingham and Fowler [101]. Nevertheless the present example as well as several subsequent others (vide infra) clearly show instances where the other contributions do have inherently anisotropic characters of their own and do bear on the overall angular dependencies of ΔE.

(2) *Cation-ligand complexes.* Zn(II) is a 'soft' divalent cation. This translates into large values of the charge-transfer and of the dispersion contributions [102, 103], and to an enhanced propensity to bind to 'softer' ligands than oxygen, such as nitrogen and in particular sulfur. Its divalent charge gives rise to very large values of E_{MTP} and E_{pol} thus to a propensity to bind to oxygen ligands as well. In proteins and NAs, Zn binds in a versatile fashion to N, O, and S ligands, can adopt a diversity of coordination numbers ranging from 4 to 6, and can exert a structural as well as a catalytic role. It is thus most important, but it could also be particularly challenging, to correctly represent it with PMM potentials. This was addressed in 1995 [104]. By parallel QC and SIBFA computations, we probed by Zn(II) the in- and out-of-plane angularities of ΔE and its individual contributions in a diversity of complexes with N, O, and S ligands. Marked directionalities were found for both E_{rep} and E_{ct} for in-plane variations of the theta angle around the X–O–Zn bond upon binding to hydroxy (X = H), formate and formamide (X = C), and around the C–S–Zn bond of methanethiolate. These could complement or oppose the directionalities of E_{MTP}. Lesser directionalities of E_{rep} and E_{ct} were found on the other hand for out-of-plane variations on a cone at a fixed theta angle. E_{pol} was found to display a generally shallower behavior than the other contributions, although it is a major contributor to non-additivity.

(3) *Hydrogen bonding to an anionic ligand.* Water acting as proton acceptor can be also used to probe the electron-rich sites of a bound ligand. An illustrative example is that of its monodentate complex with formate, upon performing in-plane variations of the C–O–H angle, at a fixed O–H distance (Fig. 1.2). E_{exch}(HF) has a marked angular character with a maximum at 120°, which

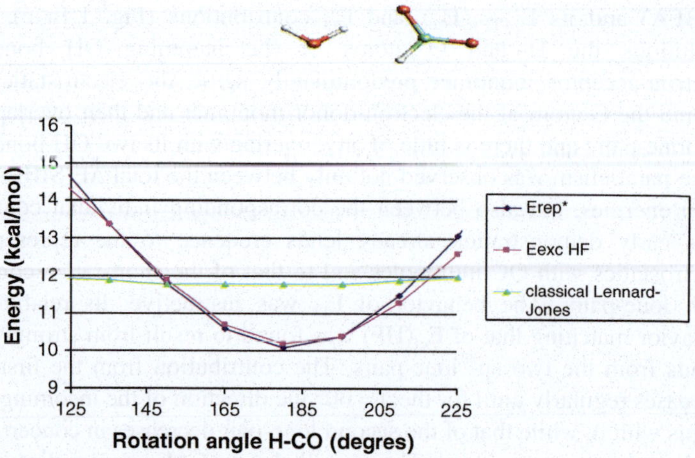

Fig. 1.2 Monodentate formate-water Compared evolutions as a function of the theta angle of E_{rep}(SIBFA) E_{exch}(HF) and a $1/R^{12}$ atom-atom repulsion. Reprinted with permission from Piquemal et al. [105]. Copyright 2007 American Chemical Society

corresponds to a maximum overlap with one sp^2 oxygen lone-pair. Its behavior is paralleled by E_{rep}(SIBFA), while by contrast a $1/R^{12}$ atom-atom repulsion expression gives rise to a shallow behavior [105].

(4) *Directionality in stacked complexes.* Stacking interactions are a major determinant of NA stability, but act also to stabilize a diversity of ordered structures. We have considered a representative stacked dimer of formamide and rotated step-wise the second monomer with respect to the first around the z-axis [105]. The curve of E_{MTP} augmented with E_{pen} is virtually superimposable over that of E_C (Fig. 1.3a. The curve of E_{rep} matches well that of E_{exch} with only a small indentation at the most repulsive point (Fig. 1.3b. The parallelism of ΔE (SIBFA) with ΔE(HF) is evidenced in Fig. 1.3c. Extensions are presently underway to NA bases, and are carried out at both uncorrelated and correlated levels, and they appear conclusive as well (Gresh et al. J Phys Chem B, 2015, 119, 9477). Such results suggest that the anisotropy of ΔE(QC) in stacked complexes can be reliably accounted for in the context of the SIBFA procedure.

(5) *Directionality in halobenzene complexes.* About thirty-five per cent of therapeutic drugs embody at least one halogen atom [106–108]. There is an outstanding feature of the CX bond in halobenzenes (X = F, Cl, Br, I) which was discovered thanks to quantum chemistry [109–111]. It is the presence of a 'sigma-hole', namely a zone of electron depletion along the bond, concomitant with a zone of electron buildup around a cone circumscribing the bond. Such features increase along the F < Cl < Br < I sequence. To account for electron depletion in the context of classical MM, a fictitious charge of δ+ has been located prolonging the CX bond, with the charge of X accordingly modified by δ -. The magnitude of δ and the δ-X distance can be optimized in order to fit the QC-computed interaction energy of an electron-rich ligand such as water

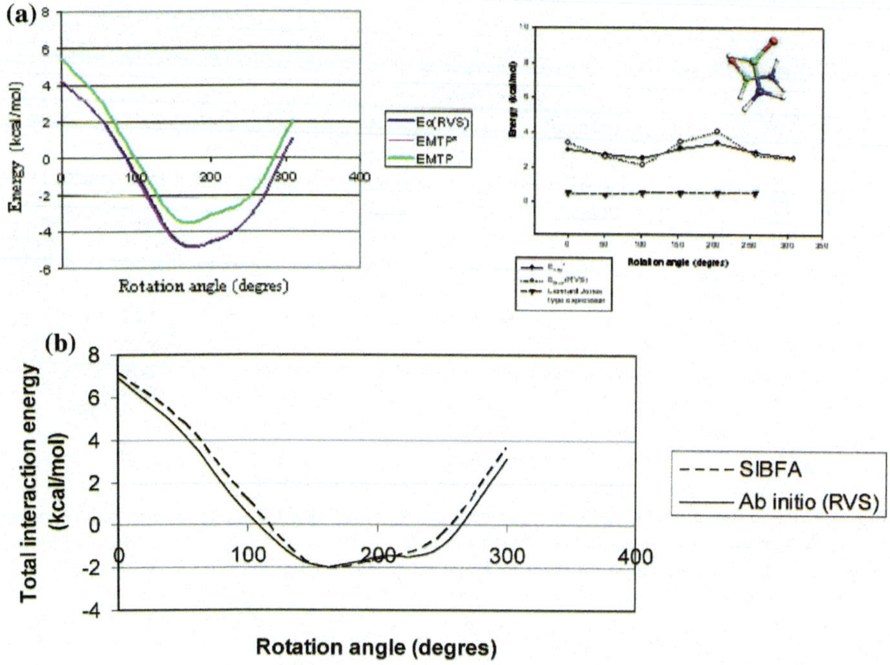

Fig. 1.3 Stacked formamide dimer Compared evolutions of **a** E_{MTP} and $E_C(HF)$, **b** $E_{rep}(SIBFA)$ and $E_{exch}(HF)$, $\Delta E(SIBFA)$ and $\Delta E(HF)$. Reprinted with permission from Piquemal et al. [105]. Copyright 2007 American Chemical Society

or formamide approaching the CX bond through its O atom [112–114]. While this representation can be successful to translate the binding of electron-rich ligands along the CX bond, it leaves the outcome unresolved regarding approaches along the electron-rich cone. In the prospect of future simulations with APMM potentials in drug design or material science, it was thus critical to evaluate if the E_{MTP} contribution, without any extra fitting and/or fictitious center, could reproduce the angular features of halogen bonding. We have thus resorted to a divalent cation, Mg(II), as a probe of the CX bond (X = F, Cl, Br) [115] in halobenzenes. A doubly charged species was deliberately chosen to exacerbate the impact of the sigma hole on the angularity of ΔE. The Mg-X distance, d, was first optimized for an approach at theta (C–X–Mg) of 180°. This was followed by in-plane variations of theta and by out-of-plane of the phi angle describing rotations of Mg(II) on a cone at fixed d and optimized theta (see Fig. 1.4). RVS energy-decomposition showed the E_C to be the essential contribution to the angularity of $\Delta E(HF)$. Would its representation going as far as distributed quadrupoles enable E_{MTP} to match E_C? This is evaluated in Fig. 1.5a, b. Figure 1.5 confirms this to be the case all throughout for all three halogens: for both Cl and Br, there is a pronounced maximum at 180° and two accented minima: in the 105–120° and 240–255° regions for Cl

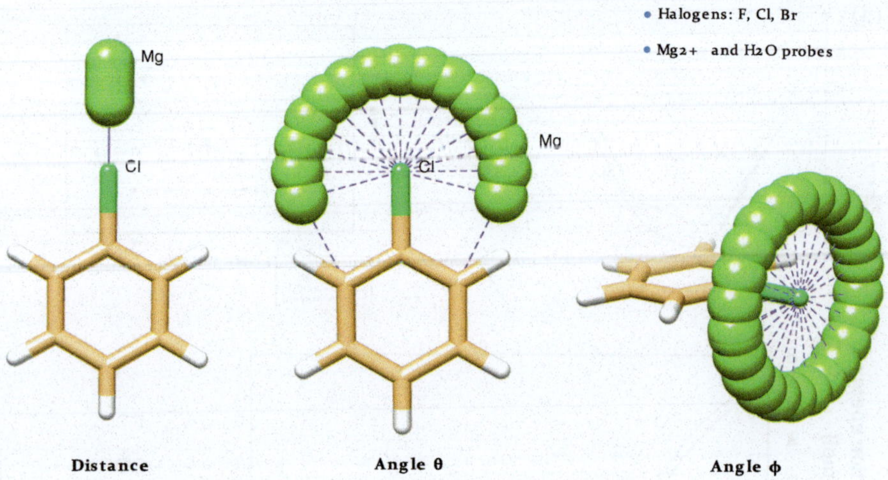

Fig. 1.4 Representation of the variations undergone by an Mg(II) probe around halobenzenes **a** in-plane variations of d with theta = 180, **b** in-plane variations of theta, **c** out-of-plane variations of phi. Reprinted with permission from El Hage et al. [115]. Copyright 2013 Wiley

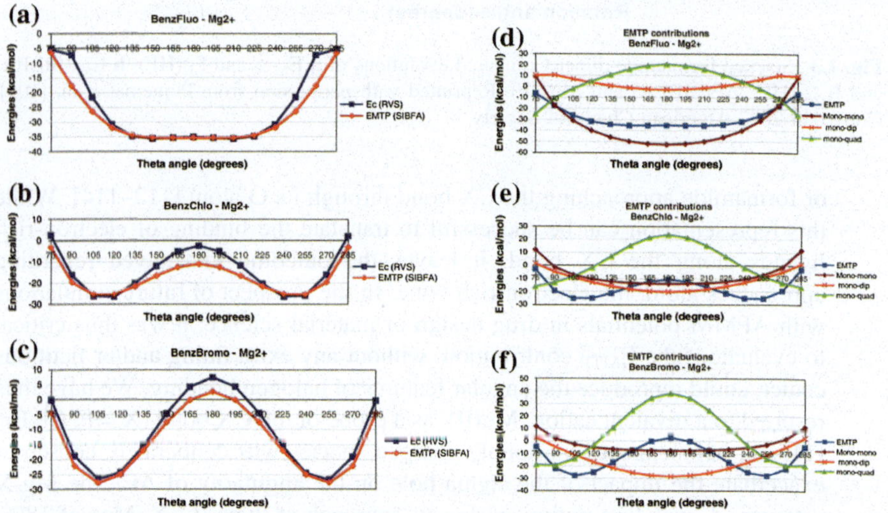

Fig. 1.5 **a–c** Compared evolutions of E_C(HF) and E_{MTP} for in-plane variations of the theta angle for Mg(II) binding to the C–F (a) C–Mg (b), and C–Br (c) bonds of halobenzene, **d–f** in-plane theta evolutions of the charge-charge charge-dipole and charge-quadrupole components of E_{MTP} for Mg(II) binding to the C–F (d), C–Mg (e), and C–Br (f) bonds of halobenzene. Reprinted with permission from El Hage et al. [115]. Copyright 2013 Wiley

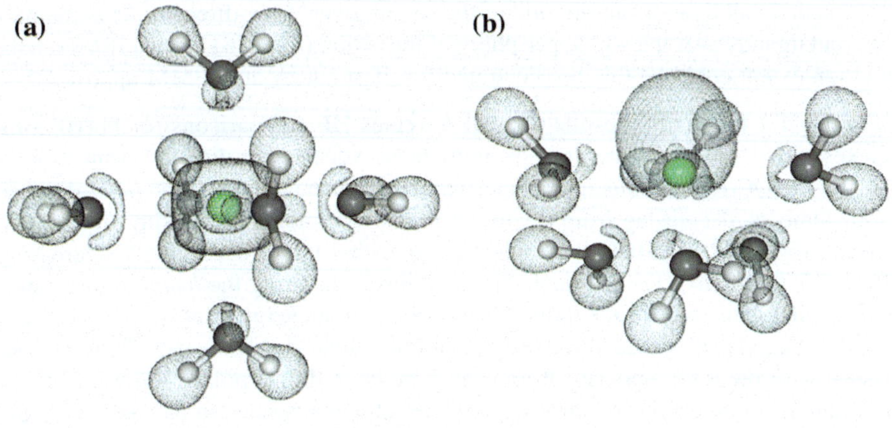

Fig. 1.6 Representation of the tetraligated complex of Pb(II) with six water molecules in **a** holo- and **b** hemi-directed arrangements. Reprinted with permission from Devereux et al. [119]. Copyright 2011 American Chemical Society

and more accented ones at 105° and 255° for Br. In marked contrast, the curve is for F shallow over the whole 135–225° region. It is also possible to unravel the origin of both the angularities of the Cl and Br curves, and of the shallowness of the F curve. Thus Fig. 1.6a–c separate each E_{MTP} curve into its charge-charge (CC), charge-dipole (CD), and charge-quadrupole (CQ) components. The angularities of the Cl and Br curves is clearly dominated by CQ overcoming the opposed preferences of both CC and CD. Possibly just because of its featurelessness, the F curve is revealing. It shows that the flat behavior of E_{MTP} is due to a near-exact compensation between the antagonizing angular features of CQ on the one hand and of CC and CD on the other hand. This is an important a posteriori and not taken for granted demonstration that the three components are correctly balanced within E_{MTP}.
For the out-of-plane variations of phi, E_{MTP} again displayed parallel features to E_C. Thus for both Cl and Br, one maximum for found at 180º, and two minimum-energy regions at 60–120° and 240–300°. For F, the curve was virtually flat throughout.
We have performed a similar evaluation now with a water probe approaching through either one H atom or through it O atom. For all three halogens, and for theta as well as for phi variations, E_{MTP} could invariably match the angular features of E_C.

(6) *Could there be departures from spherical symmetry around some metal cations?* In the polyligated complexes of metal cations, the ligand cations could justifiably be anticipated to bind around a 'crown' surrounding the cation. On account of the spherical symmetry of the cation, this is largely borne out experimentally and computationally in the vast majority of cases, but could there be exceptions? A notable exception is indeed provided by the Pb(II) cation, which in some instances is prone to prefer '*hemi*-directed' arrangements,

namely all ligands on one side of the cation, over '*holo*-directed' ones, namely all ligands over the whole periphery of the cation (Fig. 1.6) [116–118]. To what an extent could be this be accounted for, if at all, by an APMM approach?

Table 1.1 reports the results of SIBFA versus HF comparisons for Pb(II) complexes with 4, 5, and 6 water ligands in the *hemi*- versus *holo*-directed arrangements [119]. The QC/RVS results indicate a small preference in favor of the *hemi*-directed arrangements, decreasing from 6.1 to 4.8 to 1.6 kcal/mol upon passing from 4 to 6 of out 160–200. ΔE(SIBFA) is able to match this trend, accordingly decreasing from 3.6, to 2.8, to −1.5 kcal/mol. The preferences favoring the *hemi*-arrangements can in SIBFA be traced back to the Pb(II) polarization energy, that is, in its absence, such arrangements would have lesser stabilities than the *holo*-ones. This is consistent with the RVS analyses. It translates the large dipole polarizability of Pb(II), biasing the arrangements towards the *hemi*-arrangements since in the *holo*-ones, the summed electrostatic field polarizing the cation has near-zero values. It is thus fitting to observe that while for the mono-ligated complexes of a cation such as Zn (II) Epol had a limited in-plane directional character, in the polyligated complexes of Pb(II), it is the dominant factor in favor of directionality.

B. ***Non-additivity***.

(1) *Sign and magnitude of* δE_{nadd}. Non-additivity is a critical feature of ΔE. That, is, in multimolecular complexes, its magnitude differs from the sum of its magnitudes in all bimolecular (pair-wise) complexes considered separately. Non-additivity could either increase or decrease the stability of the complexes, resulting into cooperativity or anticooperativity, respectively: these two cases are mostly encountered in multiply hydrogen-bonded complexes and in polycoordinated complexes of metal cations, respectively. The onset of non-additivity, δE_{nadd}, was realized in the earlier stages of development of polarizable potentials. However, to our knowledge until the early nineties there have been surprisingly few, if any at all, attempts to quantify δE_{nadd} on model complexes by QC computations and the separate weights of the ΔE(QC) contributions, let alone evaluate if PMM were able to reliably match δE_{nadd}. This could be traced back to the fact that with the sole exception of the RVS method, all available energy decomposition procedures are limited to bimolecular complexes. Thanks to this procedure, we could consider several multimolecular complexes and address such points. Such QC/SIBFA comparisons date back from the mid-to late nineties and have borne on: (a) mono-, bi-dentate, and through-water complexes of Zn(II) with formate and first- and second-shell waters [120]; polycoordinated complexes of Zn(II) with water [104], and water, hydroxy, methanethiol and/or methanethiolate ligands [121]; (b) polyligated complexes of Zn(II) with the end side-chains of proteins and some Zn-ligating groups of metalloprotein inhibitors [122]; (c) several representative water oligomers [105, 123–125] and d) the complexes of N-methylformamide in an array of linear H-bonded complexes as encountered in models of multilayered β-sheets [126]. The most important conclusions were:

Table 1.1 Polyligated complexes of Pb(II) with n = 4–6 waters Comparisons of the QC and SIBFA intermolecular interaction energies (in kcal/mol) and their contributions

	$[Pb(H_2O)_4]^{2+}$				$[Pb(H_2O)_5]^{2+}$				$[Pb(H_2O)_6]^{2+}$			
	RVS		SIBFA		RVS		SIBFA		RVS		SIBFA	
	Hemi	*Holo*	*Hemi*	*Holo*	*Hemi*	*Holo*	*Hemi*	*Holo*	*Hemi*	*Holo*	*Hemi*	*Holo*
Coulomb /E_{MTP}	−179.7	−171.4	−181.1	−172.8	−206.4	−198.8	−210.0	−201.9	−229.0	−224.3	−235.9	−229.2
Repulsion /E_{rep}	103.5	87.5	103.1	85.1	106.1	91.9	106.9	91.1	108.2	94.5	111.8	95.3
Epol. RVS /E_{pol}*	−70.2	−61.2	−64.9	−59.7	−71.2	−63.9	−67.6	−63.2	−70.4	−64.6	−67.7	−64.8
Epol. VR /E_{pol}	−65.3	−54.2	−67.6	−55.0	−64.3	−54.5	−66.1	−56.1	−61.1	−53.1	−63.3	−55.9
Pb-polarization	−7.4	−1.6	−5.3	0.0	−6.1	−1.2	−4.5	−0.1	−6.0	−0.6	−4.7	0.0
Charge-transfer	−18.3	−15.8	−17.0	−16.4	−18.5	−16.3	−18.1	−17.5	−18.2	−16.7	−19.0	−18.2
ΔEtotal	−159.7	−153.8	−162.7	−159.1	−182.5	−177.7	−187.3	−184.5	−200.1	−199.4	−206.4	−207.9

(a) With all ligands in the first coordination shell, polycoordinated complexes of divalent metal cations are anticooperative. E_{ct} has a greater anticooperativity than E_{pol}, undergoing a smaller increase in magnitude than E_{pol} upon increasing the coordination number n. A different outcome however occurs in through-water complexes, where for both contributions, a balance between cooperativity and anticooperativity can occur. Both E_{pol}(SIBFA) and E_{ct}(SIBFA) can match the behavior of their RVS counterparts upon increasing n, regarding their magnitudes, extent of non-additivities, and dependence upon the number of ligands in first second shells. As an illustration, we give in Table 1.2 a comparison between the QC and SIBFA computations for the complex of Zn(II) with six water molecules. The three considered complexes have either 0, 1, or 2 second-shell water molecules. There is a close correspondence between the SIBFA and QC contributions. As noted originally concerning the best-bound $[Zn(H_2O)_6]^{2+}$ complex [104, 121], while the number of ligand has increased by 6, E_{pol} and E_{ct} with both RVS and QC have increased in magnitude by factors of only 2.6 and 2 with respect to their corresponding values in the monoligated $[Zn\text{-}H_2O]^{2+}$ complex at the optimized Zn-O distance of 2.10 A found in the hexamer. This, and several other related examples [120, 121, 127] give a clear indication of anticooperativity and of its control by both SIBFA E_{pol} and E_{ct}. This could have been expectable regarding E_{pol}, but was not granted for E_{ct}: it shows that the formulation of E_{ct} embodying dependencies upon the electrostatic potential and field undergone by the electron donor as well as by the electron acceptor can reliably ensure for this critical feature.

(b) Multiply H-bonded complexes are generally cooperative. E_{pol} is the dominant, but not unique, contributor to δE_{nadd}. Cooperativity is optimized in cyclic structures if each molecule can act simultaneously as an H-bond donor with one neighbor and as an H-bond acceptor with the other. Anticooperativity was also noted with one cyclic water tetramer [124], in which one water molecule acts as an H-bond acceptor from both neighbors.

Table 1.2 Polycoordinated complex of Zn(II) with six water molecules Values (kcal/mol) of the QC and SIBFA intermolecular interaction energies and of their individual contributions

	$[Zn\text{-}(H_2O)_6]^2$		$[Zn\text{-}(H_2O)_5\text{–}H_2O]^{2+}$		$[Zn\text{-}(H_2O)_4\text{–}(H_2O)_2]^{2+}$	
	RVS	SIBFA	RVS	SIBFA	RVS	SIBFA
El	−209.1	−218.2	−191.9	−198.9	−174.8	−179.6
E_{pol}	−81.3	−78.9	−92.4	−89.5	−101.8	−99.3
Ect	−20.6	−14.6	−24.5	−18.6	−28.7	−22.5
ΔE	−311.0	−311.7	−308.8	−307.0	−305.3	−301.4
E_{corr}/E_{disp}	−34.3	−35.9	−32.2	−35.3	−32.1	−37.3
ΔE_{tot}	−345.3	−347.7	−341.0	−342.3	−337.4	−338.7

Reprinted with permission from Tiraboschi et al. [121]. Copyright 2000 Wiley

Again, both E_{pol} and E_{ct}(SIBFA) could closely match their QC counterparts. In large oligomers (n = 12 and beyond) having an ice cube-like structure, cooperativity results into a significant compression of the structures, several H-bonded distances shortening by up to 0.2 Å. The dominant magnitude of E_{MTP} becomes strongly opposed by E_{rep}, and thus E_1 attains a much smaller magnitude than E_{pol} and in some cases even than E_{ct}. The fact that in such complexes E_2 could have a significantly larger weight than E_1 does not appear to have its pendant with *any* other competing PMM potential, yet is fully supported by RVS computations [105, 123].

(c) It is also possible to evaluate, within E_{pol}, the contribution to δE_{nadd} of the iterative calculations of the induced dipoles. The polarizing field is a vector quantity and the polarization energy of a given center is proportional in first approximation to the square of the field it undergoes. Non-additivity thus appears already when the field is computed with the permanent SIBFA multipoles and its magnitude is close to the corresponding magnitude of E_{pol}(RVS). This is consistent with the fact that in the RVS approach each monomer is considered in turn and relaxed in the presence in the frozen MO's of all the others. At the outcome of the iterative calculation of E_{pol} taking into account the additional contributions of the induced dipoles, E_{pol} has values that are now close to those of E_{pol}(QC) calculated in a fully variational manner. We have found that in both cooperative and anticooperative complexes at equilibrium, the contribution to δE_{nadd} of the iterative procedure was generally close to 30 %, and consistently enforced the preexisting cooperativity or anticooperativity.

(2) *On the need for off-centered polarizabilities*. The importance of having off-centered polarizabilities, specifically on the tips of saturated lone-pairs, was shown in a study that bore on three model water oligomers [125]. They were denoted as bifurcated, transverse H-bonded, and longitudinal chains of helical shape (Fig. 1.7). For all three complexes, the SIBFA calculations

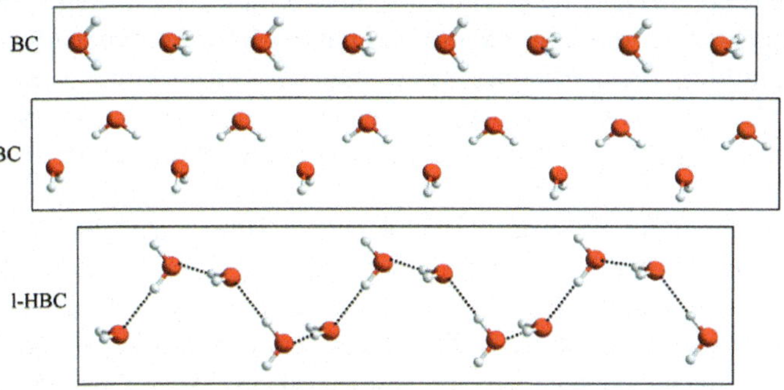

Fig. 1.7 Representation of three water oligomers in transverse bifurcated and linear H-bonding arrangements. Reprinted with permission from Piquemal et al. [125]. Copyright 2007 American Chemical Society

enabled to reliably reproduce both the magnitude of E_{pol}(QC) and the values of the total dipole moments. This is due to the fact that in such complexes polarizable centers on the lone-pairs can be closer to an incoming molecule than the bearer atom, and thus sense a more intense field than it. This constitutes an illustrative example of the impact of non-anisotropy on non-additivity. This is a reverse situation to the one encountered in the Pb(II) oligohydrates, for which it was the non-additive behavior of Pb(II) polarization that resulted into non-isotropy.

(3) *Impact of quadrupolar polarizabilities.* E_{pol} stems predominantly from the dipolar polarizabilities. There are cases, however, in which the polarization energy of some atoms can be increased due to their quadrupolar polarizabilities, whereby quadrupoles can be induced by the gradient of the field. This was observed with the Cu(I) cation [128]. We have compared several Cu(I) di- and oligo- planar ligated complexes in which the ligands are at equal distances from Cu(I) and on opposite sites. In such arrangements the field at the cation position is null, and so would E_{pol}(Cu) if it were limited to the sole Cu(I) dipole polarizability. On the other hand the field gradients are non-null, resulting into a significant Cu(I) quadrupolar polarization, the inclusion of which enabled an improved agreement with ΔE(QC). Only few attempts to include quadrupolar polarizabilities in PMM potentials have been reported so far [129].

(4) *Handling intramolecular polarization and the issue of multipole transferability.* A large flexible molecule is assembled from its molecular fragments as follows: X and Y denoting two heavy atoms, two successive fragments, Fa and Fb, connect at the level of their X–H and H–Y bonds to create a junctional bond X–Y. The multipoles on the two H atoms and at the centers of the two X–H and H–Y bonds disappear and give rise to multipoles on X, Y, and the mid-point of the X–Y bond, according to proportionality rules related to their distances from these three centers. This preserves the net charge of the assembled molecule, but not that of the individual fragments. This is of no consequence for the calculation of E_{MTP}, since the interactions of the charges along the junction bond connecting F_a and F_b are not computed anyway in SIBFA between the two fragments, although they are indeed with all the other fragments: the 'missing' interactions have no impact upon performing torsions around the junction bond, since they are located along this very bond. A different situation arises for the computation of E_{pol}, because F_a would be polarized by the field of F_b now bearing a non-net charge (0, −1, 1 for neutral, anionic or cationic), and conversely. This would again be immaterial were it not for the presence of other fragments in a larger molecule. The summed field polarizing F_a would be that of a non-net charge before being squared, so that due to non-additivity the residual non-net charge could lead to uncontrolled over- or underestimations of E_{pol}. These would be further amplified by the fact that F_a would be polarized by a centroid at a very close distance from it, namely half the length of the XY bond. Therefore we resorted to an alternative representation. Both HX and HY bonds are shrunk, each H atom being superimposed over its 'bearer' X or Y atom. All polarizing centers thus retain

their net initial charge and their shortest distances from the neighboring fragments have lengths never smaller than those of XY bonds. Such a procedure should be more immune to short-distance artifacts. It has been evaluated by several comparisons with QC in a diversity of cases. We have thus compared the conformational energies, δE_{conf}, of ten alanine tetrapeptides put forth by Beachy et al. [130] to evaluate the accuracies of MM force-fields. In our published study [131] the ten conformers differed by their sole torsion angles, i.e. with rigid constitutive fragments, pending completion of the SIBFA stretching and bending harmonic potentials. Inclusion of E_{pol} enabled the relative SIBFA δE_{conf} values to closely reproduce the corresponding QC values for all ten conformers. Refinements presently underway bear on the representation of the π lone pairs of the prime constituent of the peptide backbone, the N-methylformamide moiety, to further improve the accuracy of the intra-molecular short-range contributions E_{rep} and E_{ct}; this leads back to the issue of non-isotropy. We have also investigated the conformational dependency of: Na^+ binding to conformers of glycine and the glycine zwitterion [132]; Zn(II) binding to mercaptocarboxamides [133], which constitute the Zn-binding moiety of some Zn-metalloenzyme inhibitors; and Zn(II) binding to tetra-anionic triphosphate, which is the backbone of ATP [134]. The comparisons with the QC results constituted revealing tests of the procedure, requesting the simultaneous and consistent computation of *both* intra- and intermolecular polarization and charge-transfer effects. They enabled to address the issue of multipole transferability [135]. Namely, a limitation of the use of distributed multipoles in intramolecular studies would reside in the variation of their intensities, thus lack of transferability, following conformational changes. The results of our above-mentioned studies on the other hand showed that, exactly as in intermolecular interactions between fragments unconnected by covalent bonds, E_{pol} enables to account for the consequences of conformational changes on *both* δE_{conf} *and* the intermolecular interactions with additional molecules. This occurs provided that: -the permanent multipoles of the constitutive unconnected fragments are the ones used to compute first-order electrostatics; -and the H atoms are carried back on their 'bearer' X and Y atoms of the XY junction bond. Very recent tests have borne on the solvation energies of four inhibitors of the PMI Zn-metalloenzyme, which embody a dianionic phosphate and a monoanionic Zn-binding hydroxamate. Shells with 64 waters were considered and up to 26 complexes in different conformations. Despite the large magnitudes of the interaction energies, the large number of interacting partners, and the diversity of binding modes, close agreements between SIBFA and QC calculations at both HF and correlated levels were demonstrated (Gresh et al. to be submitted).

(5) *Could the first-order exchange repulsion contribute to non-additivity?* The short-range exchange-repulsion between two molecules is due to reorthogonalization of their MO's upon complex formation. This leads to a destabilization of the total energy of the complex in first-order counteracting the Coulomb contribution. In a multimolecular complex, the energy cost due to

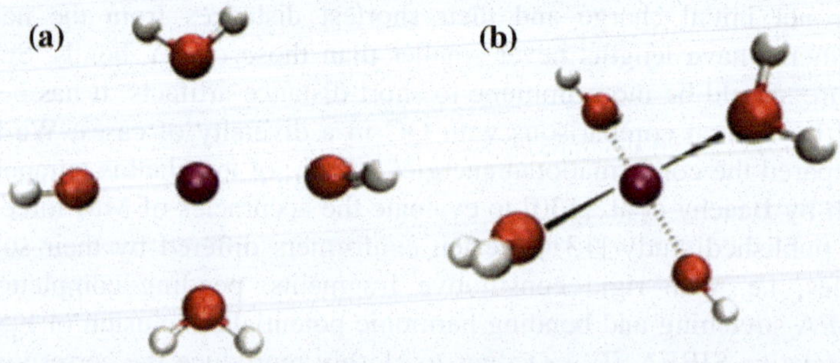

Fig. 1.8 Representation of the $[Hg(H_2O)_4]^{2+}$ complex in it's **a** square planar and **b** tetrahedral arrangements. Reprinted with permission from Chaudret et al. [136]. Copyright 2011 Wiley

simultaneous reorthogonalization can be expected to differ from that due to the summed separate pair-wise reorthogonalizations. The resulting non-additivity of E_{exch} was found to be very small, except for the polyligated complexes of some metal cations. In the context of SIBFA, it was formulated under the form of a three-center exponential involving the cation and all pairs of ligating heteroatoms, calibrated beforehand on model diligated cations, and then validated by QC comparisons on tetra- and hexaligated complexes [136]. The impact of the non-additivity of E_{exch}/E_{rep} is illustrated in the case of one metal cation for which it is the most pronounced, a heavy toxic metal, Hg(II). We have thus considered the tetrahydrated $[Hg(H_2O)_4]^{2+}$ complexes in competing square-planar versus tetrahedral arrangements (Fig. 1.8). The results are reported in Table 1.3. The three-body repulsion $\Delta E_{three\text{-}body}$ is 3 kcal/mol larger in the square planar arrangement than in the tetrahedral one. For both arrangements its values computed by SIBFA were close to the QC ones. Its explicit inclusion in SIBFA enabled to account for the preference of Hg(II) in favor of a tetrahedral arrangement. The energy difference between the two arrangements might appear small, but could nevertheless be sufficient to bear on the structural preferences found at the outcome of large-scale MD or MC simulations.

Table 1.3 Complexes of Hg (II) with four water molecules in square-planar and tetrahedral arrangements Comparisons of the QC and SIBFA intermolecular interaction energies (in kcal/mol) and their contributions

$[Hg(H_2O)_4]^{2+}$	Tetrahedral		Square planar	
	Ab initio	SIBFA	Ab initio	SIBFA
E_{rep}	78.0	81.8	132.3	126.9
E^c	−194.2	−196.0	−226.9	−226.6
E_1	−116.1	−114.3	−94.6	−99.7
E_2	−89.3	−86.8	−101.7	−102.8
ΔE	−205.5	−201.1	−196.3	−202.5
$\Delta E_{three\text{-}body}$	1.7	1.6	4.8	4.8
$E_{rep\ modified}$	78.0	83.3	132.3	131.7
$\Delta E_{modified}$	−205.5	−199.6	−196.3	−197.7

Reprinted with permission from Chaudret et al. [136]. Copyright 2011 Wiley

1.3 Applications

(1) *Conformational study of a polyconjugated drug. Interplay of cooperativity, anisotropy, multipole transferability, and conjugation.*

Several drugs which target proteins involved in disease embody conjugated or aromatic groups connected together by conjugated or partly conjugated bonds. This is the case of an inhibitor of neuropilin-1 (NRP1), a protein which when overexpressed is involved in diseases such as cancer or macular degeneracy [137]. A ligand, denoted as Lig-47, was identified in one of our Laboratories after experimental high-throughput screening, and shown to be endowed with a sub-micromolar activity on several cell lines [138]. Its structure is represented in Fig. 1.9. It is made out of the following chemical groups: benzimidazole, methyl benzene, carboxythiourea (CTU), and benzene-substituted dioxane. It is a highly conjugated drug, and the connections between its four groups all involve unsatured atoms having sp or sp^2 hybridization. The three-dimensional structure of NRP1 is known from high-resolution X-ray crystallography [139, 140], but not that of its complex with Lig-47. As a prerequisite to APMM MD or MC studies on the NRP1-Lig-47 complex, it is essential to control the ligand conformational flexibility, which is mostly governed by torsions around conjugated bonds. Overestimating it would result into too 'floppy' a ligand and the onset of a manifold of unlikely candidate protein-binding poses. Conversely, underestimating it would give rise to an unrealistically stiff ligand, unlikely to favorably bind its receptor. It was thus imperative to ensure if the SIBFA δE_{conf} calculations were reliable when compared to the QC ones. The work published in [141], focused essentially on the central CTU unity, and it proceeded along several successive steps:

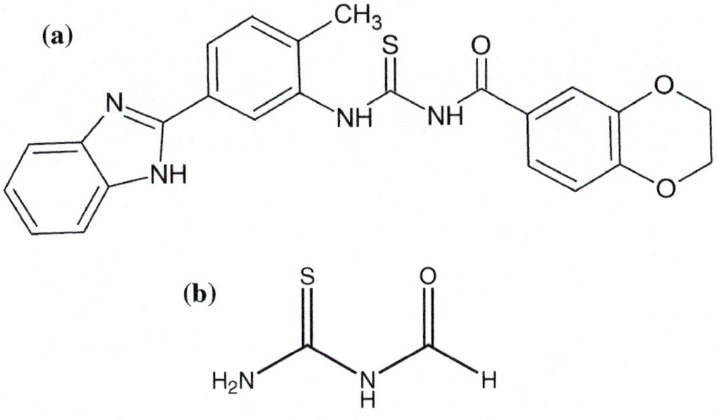

Fig. 1.9 Representation of **a** the molecular structure of Lig-47 and **b** of the structure of its central carboxythiourea (CTU) moiety. Reprinted from Goldwaser et al. [141]

(a) The distributed multipoles and the polarizabilities were derived from the MO's of the constitutive fragments of Lig-47, namely benzimidazole, methyl-benzene, CTU in an extended conformation, and benzene-connected dioxane. The location of the sp^2 lone-pairs of CTU was determined on the basis of ELF [142] analyses using the TopMod package [143], and is shown in Fig. 1.10;
(b) For all four constitutive fragments, we calibrated the increments or decrements of the effective van der Waals (vdW) radii of the atoms along their lone-pair directions by probing them with an incoming water probe approaching through one of its H atoms, in order for E_{rep}(SIBFA) to match E_{exch}(RVS) over the range of energy-relevant distances;
(c) Having set the increment of the vdW radii, CTU was split into four pseudo-fragments. The first and the third one are an sp^2 amine, the second one is thioaldehyde, and the fourth one is aldehyde. In this process fictitious connecting H atoms are created which are given null multipoles, the centroids of each of the two connecting bonds making up a junction are superimposed at the mid-point of the bond and given half the multipoles and polarizabilities of the original bond centroid prior to splitting. CTU thus retains its net initial charge of 0. 15° step-wise variations are performed around the four junction bonds, and the values of one- and two-fold rotational barriers V_0 are calibrated so that δE_{conf}(SIBFA) reproduces δE_{conf}(QC). CTU is then integrated in the entire Lig-47 ligand. The torsional variations are redone to evaluate the transferability of V_0, for which only limited changes were found necessary. The values of V_0

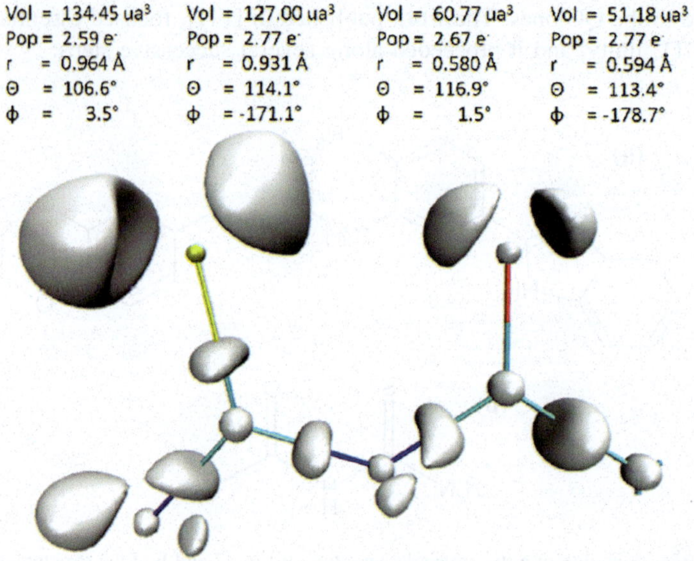

Fig. 1.10 Representation of the ELF contours around the CTU locating the positions of the sp^2 lone-pairs of O and S atoms. Reprinted from Goldwaser et al. J. Mol. Mod. 2014, 20, 2472

for the rotations around the junctional bonds connecting CTU to methylbenzene and dioxane, and methylbenzene to benzimidazole were similarly calibrated with respect to QC. In all cases, the curves of δE_{conf}(SIBFA) were found to reproduce very satisfactorily the corresponding QC ones.

Steps a-c are virtually exclusively fitting steps, but which accuracy is to be expected upon passing to a real validation step? This was carried out as follows. We selected all conformations which, for rotations around all junction bonds, corresponded to local minima or to the global one. For each of them, we performed energy-minimization with the 'Merlin' minimizer [144] on all torsional angles simultaneously. We were thus able to characterize up to 20 distinct conformations. The four most relevant ones are represented in Fig. 1.11. The most stable one, denoted 1b, has an intramolecular H-bond between the NH bond of the first thioamide group of CTU and the carbonyl bond of its formamide group. In Fig. 1.12 are plotted the SIBFA and QC δE_{conf} curves. The following analyses will be limited to the HF level. Calculation at the correlated level and with E_{disp} contribution in the SIBFA calculations led to the same conclusions as those to follow. While 1b is confirmed by the QC calculations to be the lowest-energy minimum, its relative stability is clearly underestimated by δE_{conf}(SIBFA) compared to δE_{conf}(QC). The overall R^2 coefficient is 0.88. Could this be improved? CTU had been used as one single building block in an extended conformation enabling to account for the impact of conjugation on the multipoles and polarizabilities. This disables, however, the computation of E_{pol} between the four pseudo-junctions. Firstly, the intramolecular polarization of CTU, which took place during the variational HF procedure, is inherently embodied. Secondly, since the sub-fragments bear non-net charges, and for the reasons mentioned above, attempts to include E_{pol} can be easily anticipated to severely overestimate it. We thus sought for an alternative representation, in which CTU is built from two separate fragments, thioamide and

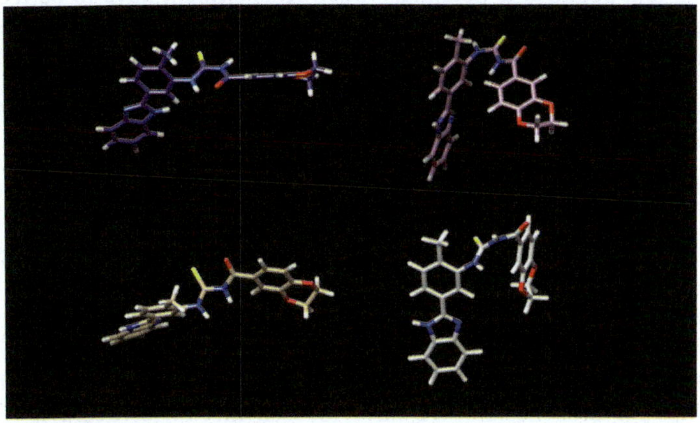

Fig. 1.11 Three-dimensional structures of four representative energy-minimized conformations of Lig-47. Reprinted from Goldwaser et al. J. Mol. Mod. 2014, 20, 2472

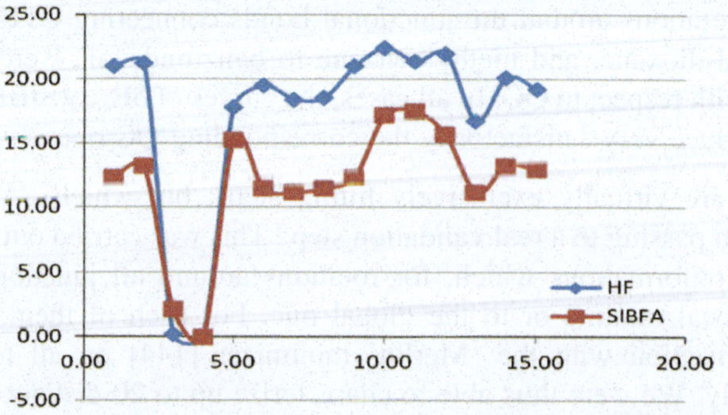

Fig. 1.12 Compared evolutions of the relative QC/HF and SIBFA conformational energies as a function of the conformer number Representation 'a' is used to construct CTU. Reprinted from Goldwaser et al. J. Mol. Mod. 2014, 20, 2472

formamide. This enables for their mutual polarization simultaneously with the other Lig-47 fragments, although at the cost of the loss of conjugation between them. The two fragments were again split as before into two pseudo sp^2 amine, one pseudo thioaldehyde and a pseudo aldehyde fragment. The one- and two-fold V_0 torsional barriers were refit as above, and the values of δE_{conf} for the twenty minima were recomputed. The corresponding δE_{conf}(SIBFA) curve is plotted in Fig. 1.13a along with the QC one, now showing a very close superposition and a R^2 coefficient of 0.96. Figure 1.13b, c represent the corresponding evolutions of δE_1 and δE_2. It is clearly seen that the evolution of δE_{conf}(SIBFA) is governed by E_2 and not by E_1. While Epol can be shown to be a leading determinant in several cases of molecular recognition, the finding that it plays *the leading role* in shaping the conformational energies of some conjugated molecules has no precedent. Several papers have mentioned [145, 146] that in order to treat polyconjugated compounds, introduction of couplings between successive torsion angles was necessary. Those studies were done using non-polarizable potentials. Such results might have to be reconsidered since such couplings might have been accounted for by polarization effects which are non-additive and long-range.

(2) *Binding of halobenzene derivatives to a recognition subsite of the HIV-1 integrase Joint involvement of anisotropy and non-additivity*

The HIV-1 integrase (INT) is a viral enzyme responsible for the integration of the viral DNA genome into the genome of the host cell. It is the target for the development of novel antiviral drugs. Three of these are currently used in therapy, and all three have a halobenzene ring: raltegravir, dolutegravir, and elvitegravir [147–150]. The high-resolution structure of their complexes with the DNA-bound integrase of Moloney foamy virus has been solved by X-ray crystallography [151]. This INT has very close homologies to the HIV-1 one, enabling inferences from structure-activity relationships for drug design. We have focused first on elvitegravir (EVG), the chlorofluorobenzene ring of which interacts which a guanine-cytosine (GC) base-pair in an INT recognition subsite (Fig. 1.14a, b). Figure 1.14b shows

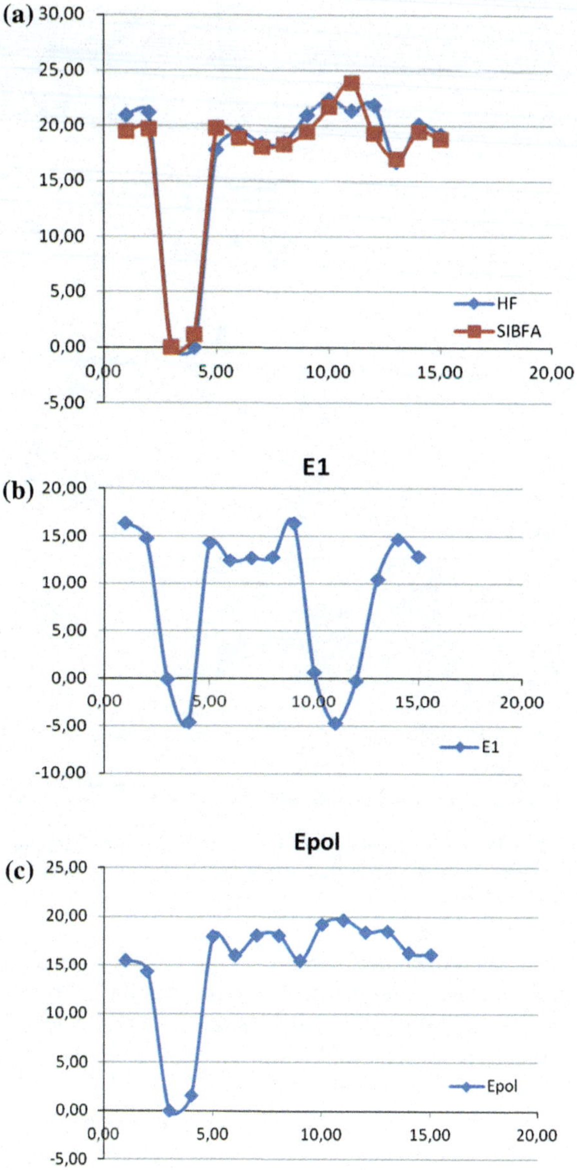

Fig. 1.13 **a** Compared evolutions of the relative QC/HF and SIBFA conformational energies as a function of the conformer number Representation 'b' is used to construct CTU, **b** corresponding evolutions of δE_1, **c** corresponding evolutions of δE_{pol}. Reprinted from Goldwaser et al. J. Mol. Mod. 2014, 20, 2472

that on the one hand, the electron-deficient prolongation of the C-Cl bond points towards the electron-rich ring of guanine, and that on the other hand, an area of its electron-rich cone can interact favorably with an electron-deficient region around the extracyclic N_4H_2 zone of cytosine. The double-faceted, 'Janus-like' property of the CX bond in halobenzene could be leveraged to enhance its affinity for each of these two regions separately, upon resorting to electron-withdrawing and to electron-donating groups, respectively [152]. A simultaneous enhancement for both

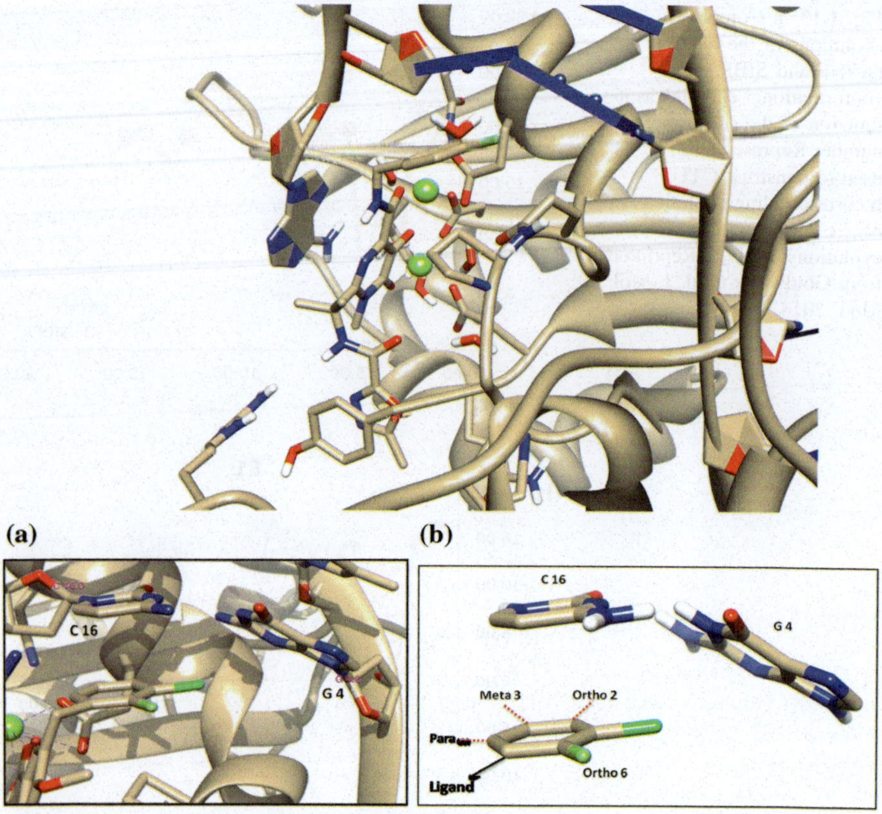

Fig. 1.14 **a** Representation of the three-dimensional structure of the Foamy virus integrase/DNA complex with elvitegravir, **b** close-up on the interaction of the halobenzene ring with the G-C subsite. Reprinted with permission from El Hage et al. [152]. Copyright 2011 Wiley

sites could even be sought for. Figure 1.15 gives the structures of several substituted halobenzene candidates. Their interaction energies with the G-C base-pair were optimized by QC calculations at the B97-D level, and relative energy balances were done which took into account their desolvation energies. Several derivatives were found endowed with more favorable energy balances than the parent fluorochlorobenzene ring of EVG, including one, compound H, having both an electron-donating and an electron-withdrawing ring. While the search for improved first-order electrostatics due to anisotropy was indeed justified by the energy decomposition analyses as well by QC-derived contours of Molecular Electrostatic Potential (MEP) [152], an unanticipated result concerned the role of non-additivity now coming into play on two counts: (a) the intermolecular interaction energies in the trimeric complexes could in some cases significantly differ from the sum of the pairwise interactions in the three dimeric complexes. While this has been documented in previous work on multiply H-bonded complexes, it was unprecedented in the case of stacking interactions; (b) the energy gains resulting from di- and

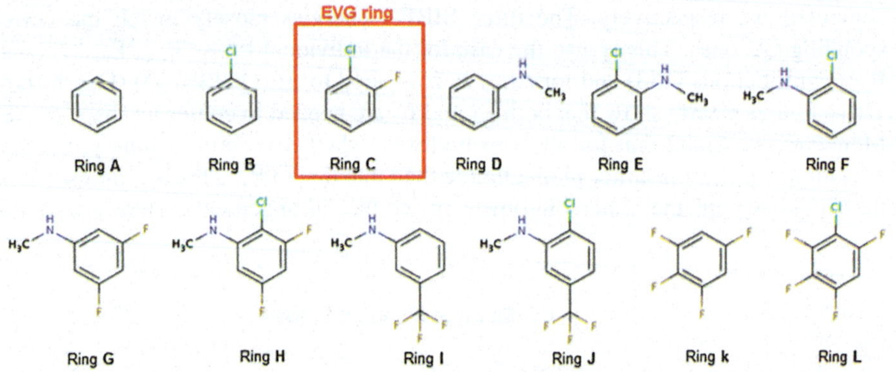

Fig. 1.15 Molecular structures of halobenzene derivatives substituted with electron-donors or electron attractors or with both. Reprinted with permission from El Hage et al. [153]. Copyright 2014 American Chemical Society

polysubstitutions could differ from the summed gains resulting from monosubstitutions. To what an extent could APMM account for such QC results [153]? E_{MTP} was shown above to account faithfully for the anisotropy of E_C in halobenzene binding, but is non-additivity still in control for the considered trimeric complexes, which could either involve cooperativity and anticooperativity? The values of the nonadditivities are compared in Table 1.4 for the QC and SIBFA computations. δE_{nadd}(SIBFA) reproduces correctly the trends of δE_{nadd}(QC), whether cooperative or anticooperative. There is a close agreement between E_{pol}(RVS) and E_{pol*}(SIBFA) prior to iterating on the induced dipoles, as also noted from previous publications. E_{pol}(KM) resulting from the Kitaura-Morokuma procedure [154] has greater δE_{nadd} values than found from SIBFA after iterations on the induced dipoles, but this might be caused in part by the non-orthogonalization of the MO's by this procedure. We have considered a total of 18 complexes of compounds selected from Fig. 1.15 with the G–C pair as well as with G and C separately, for which we monitored the evolutions of E_1, E_2, and their ΔE sums in Fig. 1.16a, b for the SIBFA and QC

Table 1.4 Values of the binding non-additivities of a series of halobenzene derivatives for the G-C base pair of the HIV-1 biding subsite

Non-additivity	ΔE		E_{POL}		E_{POL} *	
(kcal/mol)	SIBFA	RVS	SIBFA	Morokuma	SIBFA	RVS
Ring A	0.3	0.1	0.2	0.3	0.3	0.3
Ring D	−1.1	−1.8	−1.2	−1.6	−0.7	−0.8
Ring E	−0.9	−1.3	−1.0	−1.2	−0.4	−0.5
Ring G	0.1	0.3	0.2	0.1	0.3	0.3
Ring H	0.7	0.6	0.4	0.4	0.4	0.6
Ring L	0.9	0.8	0.8	0.7	0.8	0.7

Reprinted with permission from El Hage et al. 115. Copyright 2014 American Chemical Society computations, respectively. The three SIBFA energies closely match the corresponding QC ones. This is also the case for the individual E_{MTP}, E_{rep}, E_{pol}, E_{ct} and E_{disp} contributions [153] and for ΔE_{tot} as compared to ΔE(QC/B97-D) (Fig. 1.16c). These figures clearly show that both E_1 and E_2 are needed to confer its shape to ΔE. Moreover, we found that for all 18 complexes, whether ternary or binary, the stabilization due to E_2 is larger in magnitude than that due to E_1. This is reminiscent of the situation with the water oligomers in ice-like arrangements. Here again, the

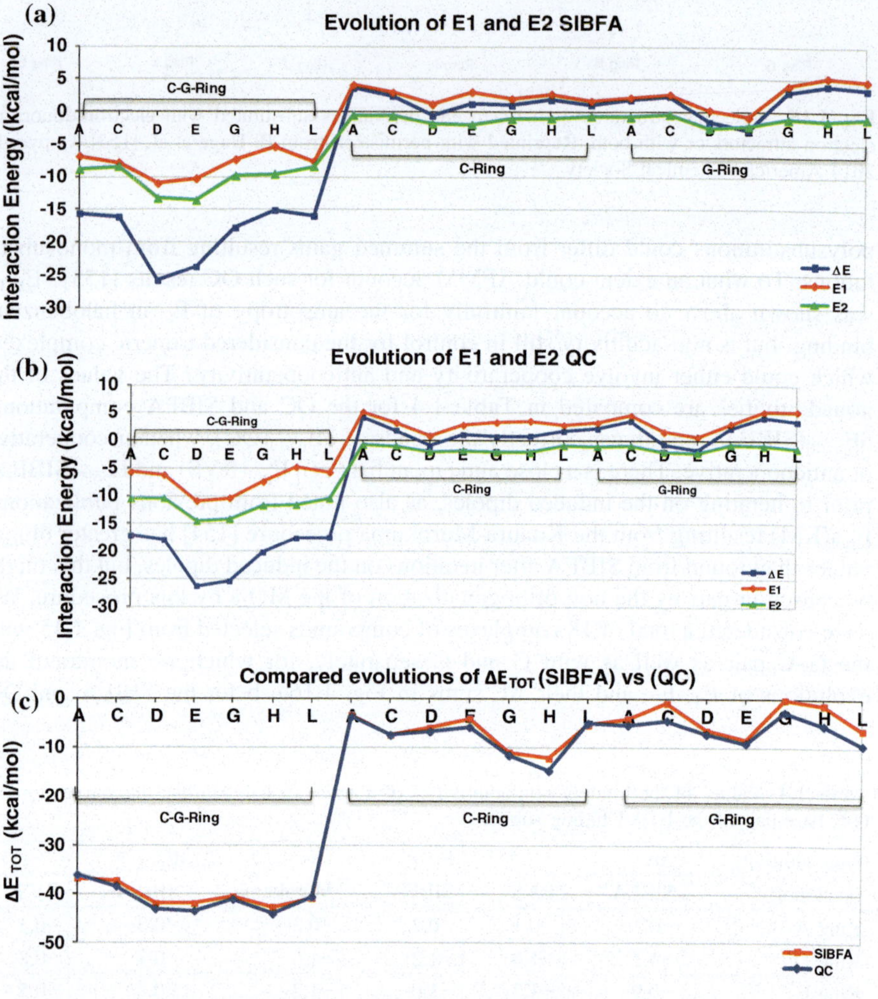

Fig. 1.16 **a** Evolutions of the SIBFA E_1 E_2 and $E_1 + E_2$ intermolecular interaction energies along a series of halobenzene derivatives, **b** Evolutions of the QC E1 E^2 and $E_1 + {}_{E2}$ intermolecular interaction energies along a series of halobenzene derivatives, **c** Compared evolutions of ΔE_{tot}(SIBFA) and ΔE(DFT/QC) along a series of halobenzene derivatives. Reprinted with permission from El Hage et al. 115. Copyright 2014 American Chemical Society

dominant magnitude of E_{MTP} and E_C are counteracted by those of E_{rep} and E_{exch}. The present results indicate that ab initio QC-derived multipoles and polarizabilities can be necessary to control *both* non-isotropy and non-additivity. The latter feature also concerns the gains or losses in binding energies from polysubstitutions relative to the summed monosubstitutions, the impact of which on the MO's can be reflected by the distributed QC multipoles and polarizabilities whence they are derived.

(3) *Structural waters in and around protein recognition sites. Impact of the second-order contributions.* Discrete water molecules, whether individually or in arrays, are considered to be an integral part of protein or NA structures. We summarize below recent findings aiming to unravel the impact of polarization and charge-transfer on the stabilization energies they confer. These studies bore on superoxide dismutase (SOD), a bimetallic Zn/Cu metalloenzyme, and on the complexes of inhibitors with the FAK tyrosine kinase and with the PMI Zn-metalloenzyme.

(a) *Superoxide dismutase.* SOD is a bimetallic enzyme catalyzing the dismutation of $O2^-$ into dioxygen and hydrogen peroxide [155]. It is essential for the survival of cells, but even more so for the cancer cells. SOD is thus an emerging target for the design of novel anticancer strategies [156, 157] including photodynamic therapy which can involve Ru(II)-based compounds [158]. Its high-resolution structure has been solved by X-ray crystallography [159, 160] showing the presence of several structural waters in close vicinity to the 85/Zn/Cu binding site. This is an incentive to evaluate whether an APMM approach could single out some privileged water network(s), and the possible extent of its/their overlap with the one found experimentally. As a first step toward such an evaluation [161], we started from the X-ray structure retaining eleven well-defined structural waters, and completed the solvation with up to 296 waters. We then performed short-duration MD runs at temperatures in the 10–300 K range with a simplified SIBFA potential and constrained the waters in a 22 Å sphere centered around Cu(I) using a quadratic potential. We selected six snapshots, denoted *a-f*. For each we then retained the 64 waters closest to Cu(I), and performed energy-minimization (EM) on their positions, on the side-chain conformations of the SOD residues making up, or neighboring, the Zn/Cu binding site, and on the positions of the two cations. Model binding sites were then extracted at the outcome of EM for validation by parallel SIBFA and QC computations. These encompassed selected main-chains and/or the side chains of 25 residues, the two metal cations and the 28 closest waters, totaling 301 atoms. Single-point computations were done for validation by parallel QC and SIBFA computations.
Figure 1.17a gives a representation of the most stable of the six energy-minimized structure, namely *c*. *c* is the structure for which the water networks have the greatest overlap with the one determined by X-ray crystallography used as a starting structure, despite the 'scrambling' it underwent by the initial MD steps. Figure 1.17b shows a close-up on the water network. There is a dense array of waters connecting two ionic residues, Glu131 and Arg141, which

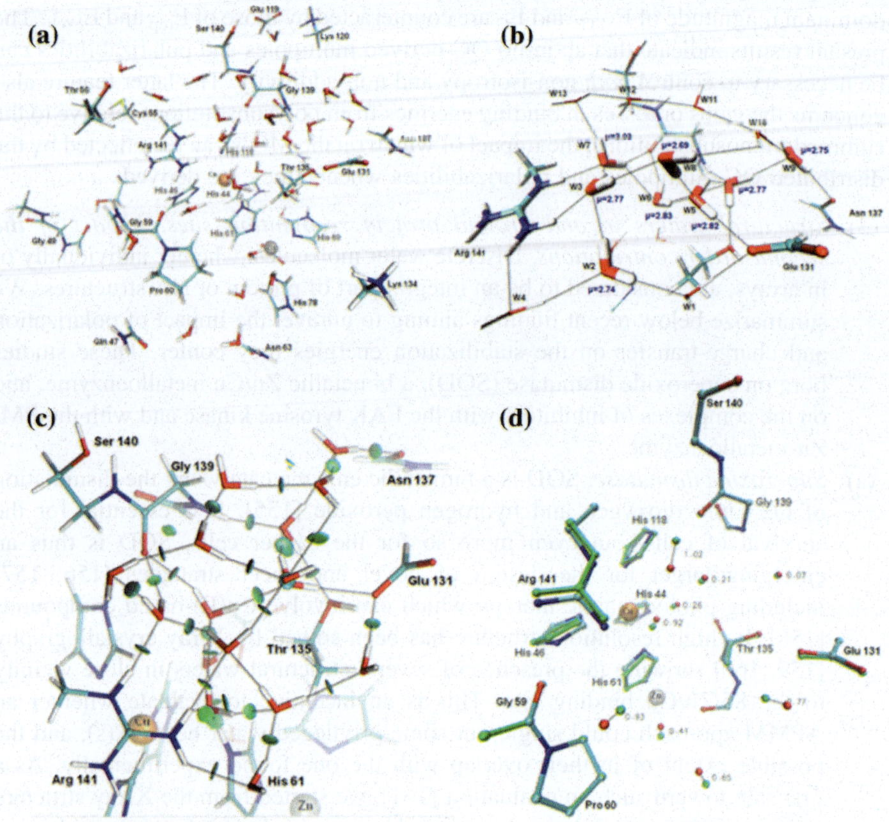

Fig. 1.17 **a** Organization of the most stable network of 28 waters around the bimetallic site of SOD, **b** representation of the water networks and values of the highest dipole moments of structural waters. Reprinted with permission from Gresh et al., J. Comput Chem., 2014, 35, 2096. Copyright 2011 Wiley

are located beneath the bimetallic site and are at about 10 Å distance from one another. These residues are connected, in fact, by several intermeshed water networks, which are also channeled in other vicinal regions, such as between the side-chains of residues Thr56 and Asn137. These two residues are linked together by a five-water near-linear array, the first and the last of which interact with the dense Glu131-Arg141 connecting water network. Six waters have dipole moments (μ) greater than 2.70 Debye, which is the value found for water oligomers in ice using SIBFA [123]. High values of the dipole moment were also found for discrete waters in the recognition site of FAK kinase which similarly mediated the interactions between ionic sites. This is mentioned below. Parallel SIBFA and QC computations were performed on complexes *a-f* in the presence and in the absence of the water networks. This enabled to compute the stabilization brought by the water networks. Figure 1.18a displays the evolutions of ΔE(SIBFA) and ΔE(HF) in the six complexes. The evolution of ΔE(HF) is very closely matched by that of ΔE(SIBFA), the relative errors being <2 %. As was the case for substituted halobenzene binding to the G–C base pair of HIV-1 integrase, both E_1 and E_2 terms are necessary to confer its proper shape to ΔE. The superimposition of ΔE(QC/B97D) and ΔE_{tot}(SIBFA) curves is even better at the correlated level (Fig. 1.18b). One means to validate the high values of the dipole moments found for water is to compare the total values of μ in the six 28-water clusters extracted from complexes *a-f* (retaining a net charge of 0 ensures that the dipole moment is translation-independent). Figure 1.19 compares the evolutions of μ as derived from both QC and SIBFA in the six clusters, with a close and clear correspondence. The R^2 factor is of 0.99. It is also seen that the most stable complex, *c*, is the one for which μ has the highest value, attesting to the importance of polarization and non-additivity in its preferential stabilization.

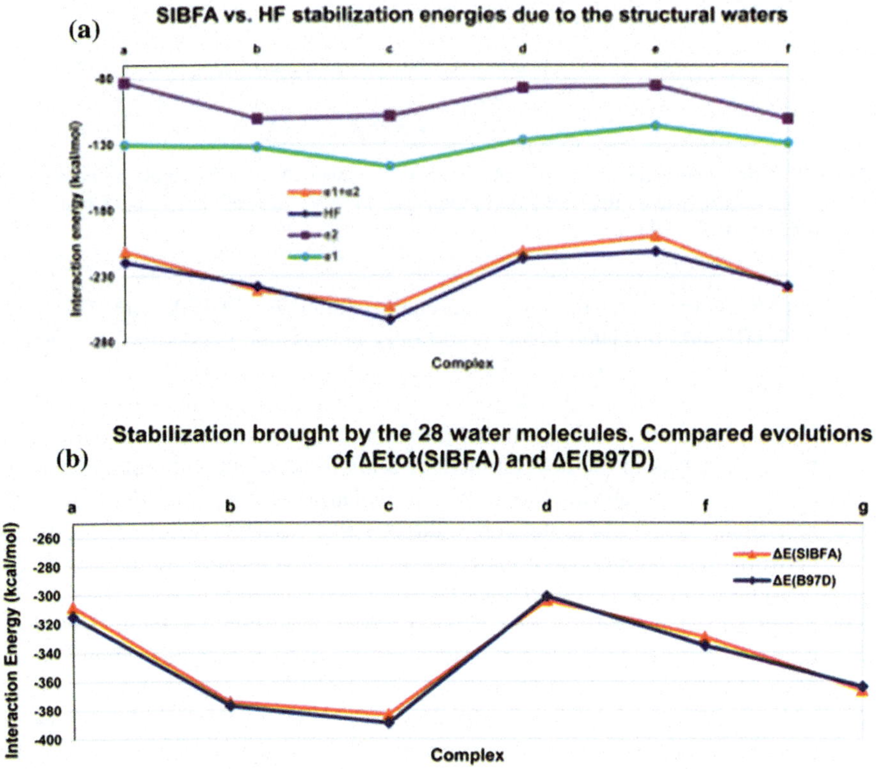

Fig. 1.18 Stabilization brought by the 28-water networks in six distinct arrangements **a** Evolutions of ΔE(QC) and of ΔE(SIBFA) and its separate E_1 and E_2 terms **b** Evolutions of ΔE(QC/B97D) and ΔE_{tot}(SIBFA). Reprinted with permission from Gresh et al., J. Comput Chem., 2014, 35, 2096. Copyright 2011 Wiley

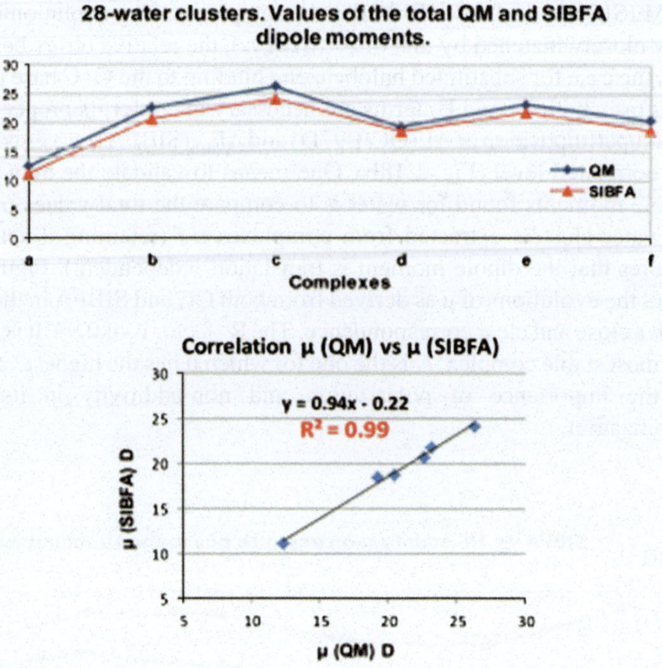

Fig. 1.19 Compared variations of the QC and SIBFA values of the total dipole moments in six distinct 28-water networks. Reprinted with permission from Gresh et al., J. Comput Chem., 2014, 35, 2096. Copyright 2011 Wiley

We stress again that the close agreements found for SIBFA with ΔE(QC) at both HF and correlated levels could only be possible thanks to the separable nature of the potential, a proper balance of first- and second-order contributions, and control of non-additivity in large complexes. They are encouraging in the perspective of long-time MD simulations, enabled by very important recent advances in the development of a highly efficient and scalable code by one of our Laboratories [162]. We thus plan to monitor the lifetimes of the individual waters in the networks, the possible existence of other, competing networks, the possibility of their mutual interconversions, and the channeling of solutes toward the bimetallic site.

(b) *Focal Adhesion Kinase (FAK)*. Kinases presently account for about 40 % of the targets for drug design [163]. They are a class of enzymes which draft a phosphate group from ATP to hydroxylated amino-acids, namely tyrosine, serine and threonine. This results into cellular activation but also if overexpressed into pathologies such as cancer, arthrosis and neurodegenerative diseases [164–166]. FAK is a tyrosine kinase which subsequent to autophosphorylation, can trigger a cascade of protein-protein interactions resulting into signal transmission to the cellular nucleus to trigger cell division and motility. Its three-dimensional structure has been resolved by X-ray diffraction showing the ATP-binding site in a hinge between the N- and C-terminal lobes (Fig. 1.20) [167]. Five inhibitors in the pyrrolopyrimidine were designed, synthesized and tested by the Novartis company [168]. They all have in common a benzene substituting the five-membered ring nitrogen. The benzene is substituted by a carboxylate group at the *ortho* or *meta* position. In the latter, there are 0, 1, or 2 methylene groups interposed. They are represented in Fig. 1.21, along with their IC_{50} values in μM and using the notations of the original paper. Compound *16i* with an *ortho* carboxylate substituent is the least active (1.6 μM). Compounds *17g, 17h, 17i* all have a *meta* carboxylate substituent and have the same submicromolar affinity (0.04 μM), i.e. a two-order-of-magnitude enhancement in affinity. A further gain results from the replacement of benzene of *17i* by pyridine, with compound *32* now nanomolar (IC_{50} = 0.004 μM). Thus, apparently modest structural changes can result into very large (thousand-fold) changes in the binding affinities. Could APMM procedures shed light on the factors governing such large changes [169]? Energy-minimizations were performed in which the solvation free energy was computed by a Continuum reaction field procedure, which was designed by Langlet, Claverie and their coworkers [170]. We denote this contribution as

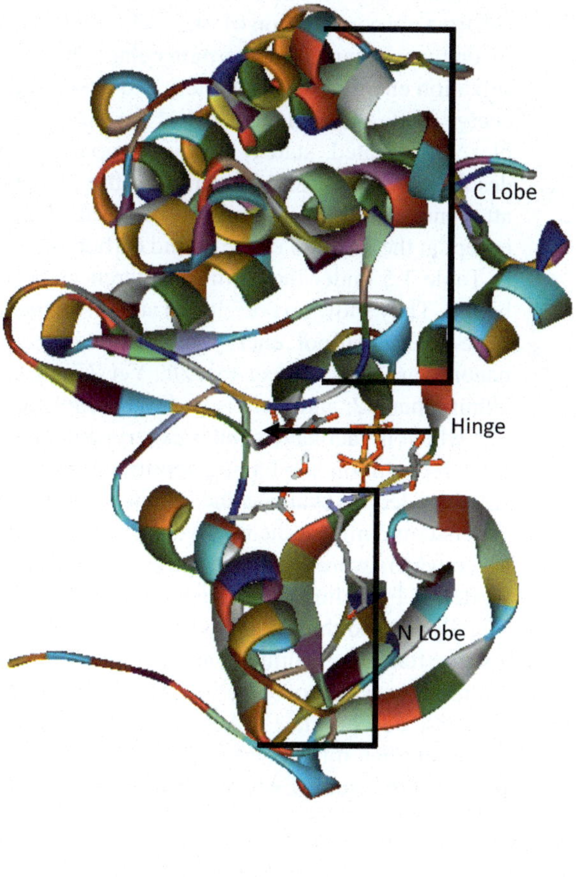

Fig. 1.20 Representation of the three-dimensional structure of FAK kinase. Reprinted with permission from de Courcy et al. J. Am. Chem. Soc. 2010, 132, 3312. Copyright 2010 American Chemical Society

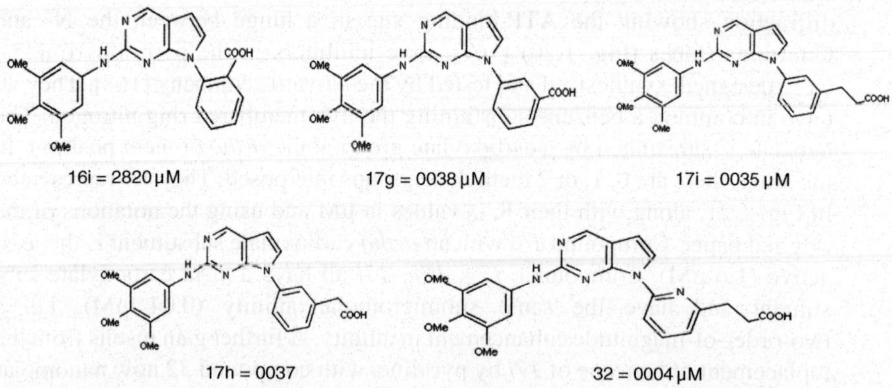

Fig. 1.21 Representation of the molecular structures of the five pyrolopyrimidine inhibitors. . Reprinted with permission from de Courcy et al. J. Am. Chem. Soc. 2010, 132, 3312. Copyright 2010 American Chemical Society

ΔG_{solv}(LC). The solvent is represented by a 'bulk' which responds to the electrostatic potential generated by the solute on its van der Waals by creating fictitious charges, which interact with the solute potential to give rise to the electrostatic contribution of ΔG_{solv}(LC). The energy balances take into account: (a) on the one hand: the intermolecular ligand-protein interaction energy and the solvation energy of the complex; and (b) on the other hand, the conformational energy costs of the ligand and of the protein upon passing from their uncomplexed, solvated states to the complex, along with their corresponding desolvation energies. An overlay of the five complexes in the recognition site after energy-minimization is given in Fig. 1.22, showing an extensive overlap except at the position of the ligand carboxylates. The energy balances are given in Table 1.5 under the form of differences between a) and b) for each contribution. They show no correlation at all with the experimental results. Thus, e.g., there is 7 kcal/mol energy difference disfavoring compound *32*, which is nanomolar, with respect to *17h*. Yet *17h* has a ten-fold lesser experimental affinity than *32*; and there is an 11 kcal/mol energy difference between *17g* and *17h* which in fact have similar experimental affinities. Separate QC tests on the FAK recognition site having confirmed the accuracy of the SIBFA procedure, we were led to evaluate the extent to which a limited number of 'discrete' waters could impact the relative energy balances. For that purpose, five, six, or seven waters were located in the complexes between each ligand and a reduced model of FAK limited to the recognition site shown in Fig. 1.22. They were initially located thanks to a procedure [171] which minimizes with a simplified energy function a limited number of discrete waters around the accessible hydrophilic sites of the solute. These positions were reoptimized by a Generalized Simulated Annealing [172, 173] procedure with the SIBFA potential, then minimized with Merlin. The final resulting positions were then ported to the entire FAK and EM redone again in the presence of ΔG_{solv}(LC).

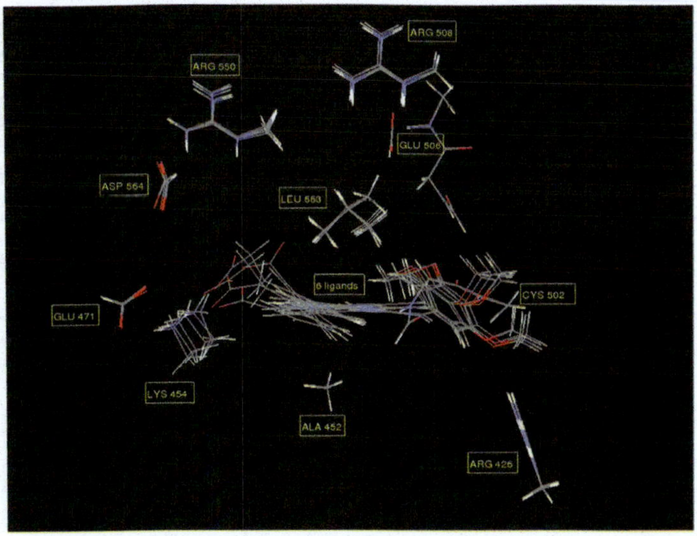

Fig. 1.22 Overlay of the inhibitors in the FAK recognition site. Reprinted with permission from de Courcy et al. J. Am. Chem. Soc. 2010, 132, 3312. Copyright 2010 American Chemical Society

Table 1.5 Relative energy balances for the binding of the five pyrrolopyrimidine inhibitors to FAK in the absence of the structural waters

	16i	17g	17h	17g	32
E_1	35.6	22.5	−4.3	13.0	18.7
E_{pol}	−12.1	9.5	7.7	−6.5	−8.1
E_{ct}	−1.8	−1.2	−0.1	−1.4	−1.3
E_{disp}	−52.7	−48.8	−45.7	−51.2	−50.0
ΔE_{tot}	−30.9	−18.0	−42.3	−46.1	−40.7
ΔG_{solv}	14.8	4.8	22.4	25.8	24.8
$\Delta E_{tot} + \Delta G_{solv}$	−19.4	−17.5	−28.5	−22.0	−21.5
$\delta(\Delta E_{tot} + \Delta G_{solv})$	2.1	4.0	−7.0	−0.5	0.0

Reprinted with permission from de Courcy et al. [169]. Copyright 2010 American Chemical Society

Figure 1.23 gives a simplified representation limited to the complex of the end carboxylate of 32, the ionic sites of FAK, and five discrete waters (denoted as complexes 'c_w' in Ref. [169]). These waters can clearly snugly fit in the structure. They can either mediate the interactions between the ligand and PMI as occurs with residues Glu471, Arg550 and Asp564, or complement them, as occurs with residue Lys454. Several waters have much stronger dipole moments than ice, so that the polarization energy contribution could be expected to be a key contributor to ΔE. This led us to perform a parallel SIBFA/QC(RVS) analysis of the intermolecular interactions on complexes

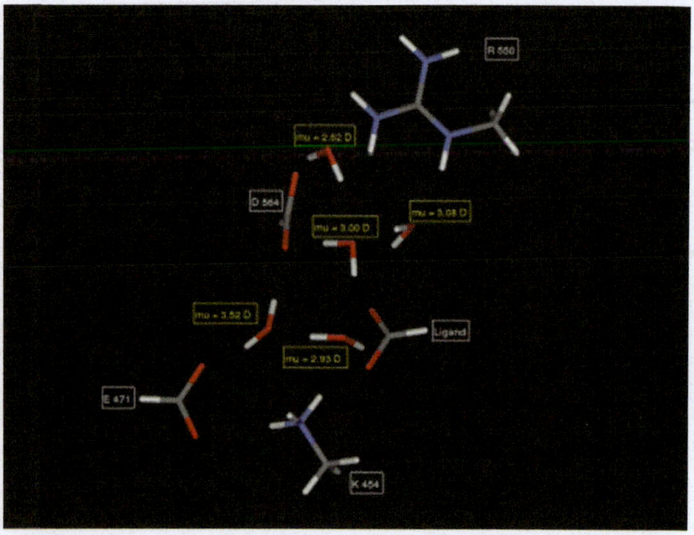

Fig. 1.23 Representation of the complexes between the carboxylate group of the inhibitor and the ionic groups of the recognition site of FAK together with the discrete waters. Reprinted with permission from de Courcy et al. J. Am. Chem. Soc. 2010, 132, 3312. Copyright 2010 American Chemical Society

c and c_w, without and with the discrete waters. The analyses done for the two extreme compounds, nanomolar *32*, and micromolar *16i*, are reported in Table 1.6. In the absence of the waters, E_1 favors by 17 kcal/mol *16i* over *32*, while E_2 favors by 3–4 kcal/mol *32* over *16i*. A remarkable reversal in the magnitudes of the preferences takes place in the presence of the discrete waters. E_1 now favors by only 4–5 kcal/mol *16i* over *32*, but E_2 favors *32* over *16i* by a very significantly augmented preference, namely 17 kcal/mol. The resulting preference of 13–14 kcal/mol in terms of ΔE(SIBFA) and ΔE(RVS) is thus the one imposed by E_2. The persistent agreements between all SIBFA and RVS individual contributions is noteworthy for all four complexes. The final energy balances in the presence of the five structural waters are given in Table 1.7. All five ligands are now ranked at least qualitatively along the correct sequence ranking first the nanomolar compound *32*, then the three submicromolar compounds *17g, 17h, 17i,* and then the micromolar compound *16i*. The same conclusions hold with 6 and 7 discrete waters. The critical role of E_{pol} in such balances is noteworthy. The need for E_{pol} for a correct ranking of affinities had been previously shown in a study with the AMOEBA potential that bore on the complexes of trypsin with benzamidine derivatives [174], although its role was possibly not as extreme as in the present study. As an extension of the present work, we plan to resort to a massively parallel version of the TINKER software on which the SIBFA potential is ported to perform long-duration MD. These should enable us to monitor the life-times of the discrete waters and their rates

Table 1.6 Energy decomposition for the binding of the carboxylate of the micro- and the nanomolar compounds to the ionic sites of FAK without and with the structural waters

	E_{MTP} /E_C		E_{rep} /E_{exch}		E_1	
	SIBFA	RVS	SIBFA	RVS	SIBFA	RVS
16i	−206.0	−203.6	19.4	18.5	−186.6	−185.1
32	−194.0	−190.8	24.0	22.5	−169.9	−168.3
16i + 5 w	−348.1	−344.2	142.1	138.0	−206.0	−206.2
32 + 5 w	−352.4	−350.2	152.6	148.7	−199.8	−201.6
	E_{pol}		E_{ct}		$E_2 = E_{pol} + E_{ct}$	
	SIBFA	RVS	SIBFA	RVS	SIBFA	RVS
16i	−21.5	−23.3	−2.6	−2.6	−24.2	−25.9
32	−24.6	−26.7	−2.5	−3.0	−27.1	−29.7
16i + 5 w	−48.5	−51.3	−18.4	−16.6	−66.9	−67.9
32 + 5 w	−63.1	−66.0	−21.0	−19.6	−84.0	−85.6
	ΔE(SIBFA) /ΔE(RVS)					
	SIBFA			RVS		
16i	−210.8			−210.0		
32	−197.0			−196.9		

Reprinted with permission from de Courcy et al. [169]. Copyright 2010 American Chemical Society

Table 1.7 Relative energy balances for the binding of the five pyrrolopyrimidine inhibitors to FAK in the presence of the structural waters

	16i	17g	17h	17g	32
E_1	47.1	34.3	57.8	60.3	61.4
E_{pol}	0.2	7.2	−14.1	−26.4	−28.6
E_{ct}	9.7	5.8	2.2	8.1	8.1
E_{disp}	−41.5	−49.8	−55.2	−53.5	−53.1
ΔE_{tot}	15.6	−2.5	−9.5	−11.6	−12.1
ΔG_{solv}	32.4	45.7	54.0	53.1	51.8
$\Delta E_{tot} + \Delta G_{solv}$	−1.8	−3.7	−6.1	−11.8	−14.1
$\delta(\Delta E_{tot} + \Delta G_{solv})$	12.3	10.3	8.0	2.3	0.0

Reprinted with permission from de Courcy et al. [169]. Copyright 2010 American Chemical Society

of exchange with the solvent, and possibly the path for interconversion between 'DFG-in' and'DFG-out' conformations [175].

(c). *Phosphomannose isomerase (PMI).* PMI is a Zn-metalloenzyme which catalyzes the reversible isomerization of fructose-phosphate into mannose-phosphate [176]. It is responsible for several infectious and parasitic diseases but there is no clinically-useful inhibitor against it [177–179]. A hydroxamate inhibitor, denoted 5PAH was designed, synthesized and tested in the Laboratory of Bioorganic and Bioinorganic Chemistry at Orsay, France, and shown to display a submicromolar inhibitory potency. By contrast, an analogue with formate replacing hydroxamate was devoid of potency [180].

Based on the X-ray crystal structure of Zn-bound PMI [181, Fig. 1.24], we were able to account for these experimental results and derived a structural model locating the hydroxamate in the Zn-binding pocket and the phosphate at the entrance of the cavity where it binds simultaneously to two cationic residues, Arg304 and Lys310 [182]. In a next step, four ligands in the sugar family were considered (Fig. 1.25) [23]. The first three ones (*1–3*) have a dianionic phosphate. The first (1) is β-D-mannopyranose 6-phosphate (β-6-MP1). The fourth (*4*) is an analog of β-6-MP1 but has a malonate with two monoanionic carboxylates replacing the phosphate. Only compound *1* in the phosphate series displayed a measurable PMI binding affinity. The malonate derivative *4* displayed a ten-fold larger binding affinity than it. This could constitute a step toward the design of therapeutically relevant drugs, because malonate is more resistant to enzymatic hydrolysis than phosphate, and is also more easily transported. The SIBFA calculated have enabled to account for the greater affinity fort PMI of *1* than *2* and *3*. However the energy balances with the sole ΔG_{solv}(LC) terms failed to account for the greater affinity of the malonate derivative. This led us, as in the case of FAK, to solvate their

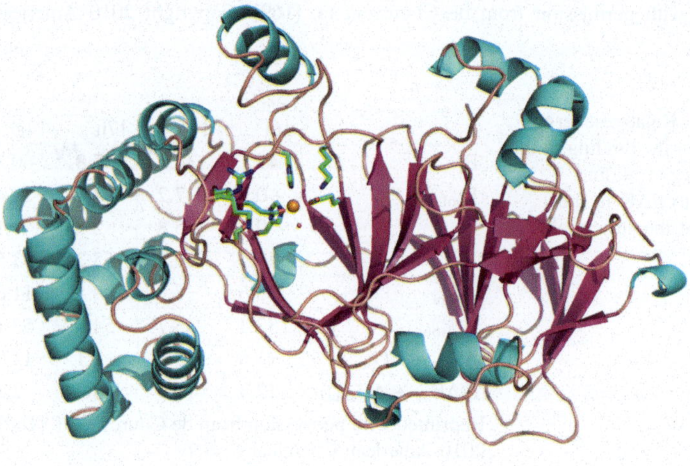

Fig. 1.24 Three-dimensional structure of Zn(II)-bound PMI. Reprinted with permission from Gresh et al. J. Phys. Chem. B. 2011, 115, 8304. Copyright 2011 American Chemical Society

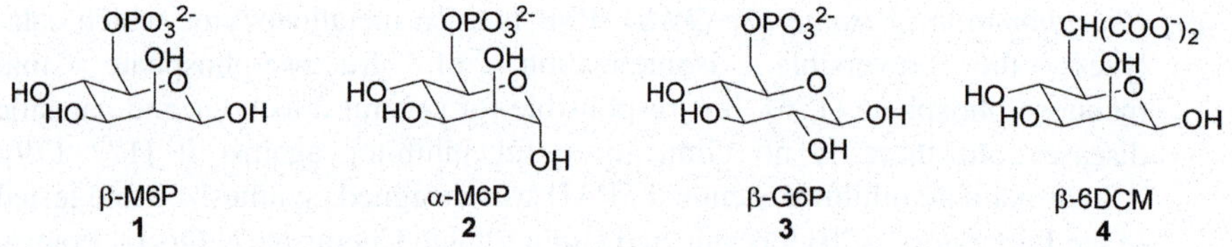

Fig. 1.25 Representation of the structures of four PMI ligands. Reprinted with permission from Gresh et al. J. Phys. Chem. B. 2011, 115, 8304. Copyright 2011 American Chemical Society

complexes with discrete waters. Nine waters were optimized in the PMI-*1* and PMI-*4* complexes. The structure of the latter is shown in Fig. 1.26. Three networks are found. The first network bridges one O of the most accessible carboxylate of the ligand with Trp18 and Glu48 residues. These interactions are extended by ionic interactions with residues Lys100 and Glu294. The second network bridges the other anionic O with Asp17. The third network bridges the sole accessible anionic O of the second carboxylate with Arg304 and Asp300. Residues Asp17, Glu48, and Asp300 are at about 18 Å distance from one another. We thus observe an extension of the recognition site to PMI residues that do not interact directly with the ligand, and such an extension is mediated by the polarizable water molecules.

In the complex of I, the phosphate fits more snugly than malonate between Arg304 and Lys310. This on the one hand gives rise to stronger electrostatic interactions with these two residues, but at the price of a lesser accessibility to the structural waters. The differential stability due to the nine waters can be illustrated by comparing the interaction energies of the two ligands in the recognition site. Each site has been extracted from the energy-minimized ligand-PMI complex, and

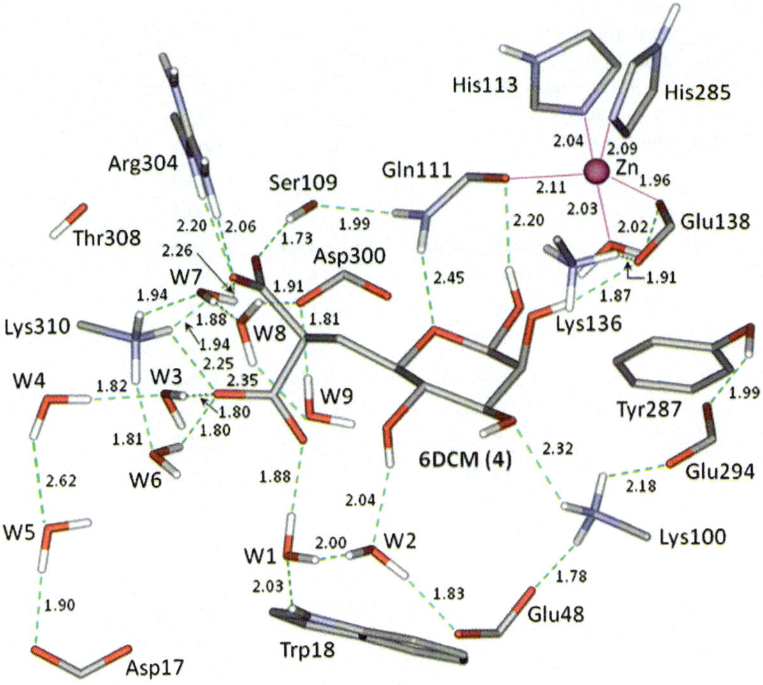

Fig. 1.26 Representation of the optimized complex of the malonate derivative with PMI and nine water molecules. Reprinted with permission from Gresh et al. J. Phys. Chem. B. 2011, 115, 8304. Copyright 2011 American Chemical Society

single-point computations were done with and without the nine structural waters. The SIBFA interaction energies are reported in Table 1.8 and compared to the QC ones. For each contribution and for each ligand, we also report the values of δ(a-b), namely the gain due to the nine waters. The gain in E_1 is by only −1.6 kcal/mol more favorable for the malonate than for the phosphate ligand. Such relative gains increase very significantly with the second-order contributions, E_{pol} and E_{ct}, passing to −7.7 and −5.2 kcal/mol, respectively. Thus the networks totaling nine waters would stabilize by −14.7 kcal/mole the malonate ligand over the phosphate one. This value is very close to the corresponding QC(HF) value of −15.7 kcal/mol. Inclusion of dispersion/correlation does not alter the outcome. The comparative energy balances done on the complete ligand-PMI gave rise to the same conclusion: there is a distinct preference in favor of the malonate over the phosphate derivative, but it is only enabled by the networks of structural waters. However, as noted in [23], more exhaustive sampling of the energy surface, along with long-time MD and accounting for entropy effects, are needed for a more quantitative evaluation. This will be further discussed in the last section of this review.

Table 1.8 Comparisons of the weights of the different energy contributions to the stabilization of the PMI-1 and PMI-4 complexes with and without the structural waters

	a		b	
	1	4	1	4
E_{MTP}	−1453.8	−1437.0	−1272.8	−1223.8
E_{rep}	411.6	433.4	312.4	303.6
E_1	−1042.1	−1003.5	−960.4	−920.2
δE_1(a-b)	**−81.7**	**−83.3**		
E_{pol}	−176.0	−190.4	−152.1	−158.8
δE_{pol}(a-b)	**−23.9**	**−31.6**		
E_{ct}	−59.9	−68.3	−43.9	−47.1
δE_{ct}(a-b)	**−16.0**	**−21.2**		
ΔE(SIBFA)	−1278.1	−1262.2	−1156.4	−1126.1
δE(a-b)	**−121.7**	**−136.1**		
ΔE(HF)[1]	−1268.2	−1243.0	−1145.2	−1104.3
δE(a-b)[1]	**−123.0**	**−138.7**		
ΔE(HF)[2]	−1278.9	−1253.1	−1148.9	−1107.3
δE(a-b)[2]	**−130.0**	**−145.8**		
E_{disp}	−139.8	−138.7	−103.1	−97.4
$ΔE_{tot}$(SIBFA)	−1417.9	−1400.9	−1259.5	−1223.5

[1]CEP 4-31G(2d) basis
[2]aug-cc-pVTZ(-f) basis

1.4 Conclusions and Perspectives

There are numerous fields of application of computational chemistry where next-generation QC-derived anisotropic polarizable molecular mechanics/dynamics could very significantly extent the realm of ab initio quantum chemistry. These encompass, e.g., drug design, material science, and supramolecular chemistry. APMM should be able to handle systems with sizes larger by at least four orders of magnitude than QC, and/or enable simulations times also larger by similar orders. But a prerequisite to very large scale applications is an objective evaluation of its expectable accuracy. In this respect, a distinctive asset of the SIBFA procedure, which appears to this day to be shared by very few other potentials [42–45], is the separability of ΔE into five distinct contributions, each of which is formulated and calibrated on the basis of its ab initio QC counterpart, and subsequently extensively tested against it.

We have reviewed in this paper the inherent non-isotropy and non-additivity features of several of these contributions, and their impact on overall structure and energetics. The SIBFA procedure has lent itself to numerous confrontations against QC, more so than any other competing method. It could be adequate to conclude to shortly develop on three points: its refinements and enrichments, its integration into highly optimized softwares, and the realm of its applications.

(a) **Refinements.**

- *Regarding electrostatics.* The distributed multipoles and polarizabilies used to construct the SIBFA library of fragments were derived from QC fragment calculations using the CEP 4-31G(2d) basis set by Stevens et al. [183, 184]. Accordingly, most validation studies resorted to QC computations using this basis as well. Upon comparing the evolutions of intermolecular interaction energies for a series of different complexes, we found invariably the values of DE(HF) with this base to very closely parallel those with larger basis sets [185, 186], the most extended one being aug-cc-pVTZ(-f) (Gresh et al. to be submitted). This attests to the high reliability of this basis set, and justifies its use to construct the SIBFA library of fragments. There are several cases, however, where it could be preferable to resort to very extended basis for calibration and validation purposes. Accordingly, we are assembling a new library of fragments with multipoles and polarizabilities now derived from uncorrelated as well as correlated aug-cc-pVTZ(-f) calculations. The parametrization phase can be automatized thanks to the I-NoLLS algorithm [187, 188]. It was recently reported in the context of SIBFA [189]. Once the 'general' parameters are set, the calibration of individual atom types or the introduction of new atoms, such as metal cations, becomes straightforward. This was done recently in the Li^+ – Cs^+ alkali cation series [100].
- *Regarding short-range.* E_{rep}, E_{ct}, and $E_{disp\text{-}exch}$ have dependencies upon the location of the lone-pair tips. Such locations can be derived from QC

analyses such as Boys' localization procedure [190] or ELF [142], as analyzed for a series of ligands by Chaudret et al. [191], whence an additional filiation of SIBFA to QC.

(b) **Enrichments**. Multi-scale approaches such as QM/MM, pioneered in 1976 [14] constitute nowadays an emerging field of computational chemistry. Steps toward merging SIBFA and the Gaussian Electrostatic Model (GEM) [192] have been completed [193], and a complete integration of the two approaches is underway. Owing to their polarizable nature, APMM approaches are well suited to a merging with QM approaches. This was completed recently concerning AMOEBA (Piquemal et al. submitted) and could be pursued with SIBFA.

(c) **Integration into massively parallel codes**. There could be very important perspectives for the use of the SIBFA potential on a much larger scale than before. It is presently being integrated in the newly developed Tinker-HP software. It will therefore benefit from novel algorithmic developments to speed up polarizable molecular dynamics. For example, the bottleneck of the polarization energy and associated derivative evaluation on parallel computers has been overcome by the use of new iterative techniques such as the Jacobi/DIIS approach offering good scaling on hundreds and even thousands processors with gains in time going up to three orders of magnitude upon using advanced MD predictor-corrector algorithms [162]. SIBFA should also benefit from the high performance Smooth Particle Mesh periodic boundary condition implementation for electrostatics, including the short-range penetration correction [194] and for the polarization energy that uses newly introduced solvers and benefits from a Nlog(N) scalability [195]. Overall, all derivatives and torques have been coded and production simulation runs could be anticipated to start this year. Moreover, another asset will be the availability of the newly developed domain decomposition Cosmo (dd-Cosmo) continuum solvation model that is now available in direct connection with polarizable molecular dynamics [196]. To conclude on the technical part, new parametrization strategies have been defined with automatic parametrization using the INOLLS software [187, 188]. Such an approach should greatly reduce the time effort required for the definition of new parameter sets [189]. A previous version of the SIBFA software had been earlier (1999–2005) deposited at the Computational Chemistry List (CCL). A version of the Tinker-HP code integrating the SIBFA potential and its gradients is destined to a release in the forthcoming year.

(d) **Prospective applications**.

– *Ligand-macromolecule complexes*. One of the most attractive fields of APMM applications is ligand-protein complexes. There have been published applications regarding kinases [169] and metalloproteins [23, 161, 197–201]. It could be rewarding to adapt Free Energy Perturbation (FEP)

methods [202] or non-equilibrium MD [203] to such targets, and particularly metalloproteins, on account of the demonstrated reliable handling of metal cation-ligand interactions. Along these lines, we note that an AMOEBA application was recently coauthored by one of us, which bore on the binding to MMP-13 of four dicarboxamide inhibitors [202]. This was the first ever reported FEP study on a metalloprotein using polarizable potentials. Although the ligand structural changes bore on sites distinct from the Zn-binding site, it is expectable that changes directly affecting Zn (II)-binding are amenable to prospective SIBFA FEP calculations.
- *Supramolecular chemistry and material science*. As reviewed above, SIBFA has been adapted to a diversity of metal cations. These encompass the following: alkali Li(I)-Cs(I) [100], alkaline-earth Mg(II) and Ca(II) [103], transition metals Cu(I) [128], Cu(II) [99], Zn(II) and Cd(II) [103, 104, 127], heavy metals Pb(II) [119] and Hg(II) [136], and lanthanides and actinides [204]. On the one hand, this should enable to investigate their binding, stationary or transitory, to a diversity of proteins and NA's. On the other hand, this should enable to address the issue of preferential entrapment of one cation over that of others by a supramolecular host, and possibly the design of cation-selective hosts for extraction or detoxification. The computation of free energies of binding could be based on FEP, non-equilibrium MD, or possibly by computing the contribution of vibrational entropy [205, 206]. Finally, there is a little charted domain of application of APMM approaches, which relates to surfaces and nanostructures. Accessing the values of their distributed multipoles and polarizabilities, along with the handling of Periodic Boundary Conditions (PBC) in SIBFA, and following validations against QC in model systems, could pave the way for numerous studies on adsorption events.
- *Modeling of nucleic acids*. The application of APMM to nucleic acids constitutes another virtually uncharted ground [see, e.g., Refs. 207–211]. Yet NA's are an ideal domain of application for such approaches, considering the polyanionic nature of the sugar-phosphate backbone and the strongly polar and polarizable nature of the bases. Several issues can be mentioned: the structure, organization, and dynamics of the water networks in the groove, the dynamics of binding of metal cations to the backbone, the groove, and/or the water networks; the amplitude of stacking energies of successive base-pairs, from which the sequence-dependent conformation of NA's depend; and the conformational properties of the phosphodiester backbone needing to handle properly polarization and anomeric effects.

Acknowledgments We wish to thank Dr. Michael Devereux (University of Basel, Switzerland) for his recent contributions on Pb(II) modeling, and for enabling to derive INollS-optimized SIBFA parameters in conjunction with the aug-cc-pVTZ(-f) basis set. Several colleagues are thanked for rewarding collaborations in the last few years: Tom Darden (Open Eyes, Santa Fe, USA), G. Andres Cisneros (University of Detroit, USA), Aude Marjolin (Quantum Theory Project, Pittsburgh, USA), Marie Ledecq (Union Chimiques Belges, Brussels, Belgium), Johan Wouters

(Universites Notre-Dame de la Paix, Namur, Belgium), Dr. Dorothée Berthomieu (Dorothée Berthomieu, Institut Charles-Gerhardt, Université de Montpellier, France) and Markus Meuwly (University of Basel, Switzerland). We would also like to mention, regarding the experimental work, Prof. Christiane Garbay (Université ParisDescartes, Paris, France) and Prof. Laurent Salmon (Institut de Chimie Moléculaire d'Orsay, France).

We wish to thank the Grand Equipement National de Calcul Intensif (GENCI): Institut du Développement et des Ressources en Informatique Scientifique (IDRIS), Centre Informatique de l'Enseignement Supérieur (CINES), France, project No. x2009-075009), and the Centre de Ressources Informatiques de Haute Normandie (CRIHAN, Rouen, France), project 1998053. This work was supported in part by the French state funds managed by CALSIMLAB and the ANR within the Investissements d'Avenir program under reference ANR-11-IDEX-0004-02. F.L. We also wish to thank the Association Philippe Jabre (Beirut, Lebanon) for financing the PhD Thesis of Krystel El Hage.

References

1. Car R, Parrinello M (1985) Phys Rev 55:2471
2. Kuhne TD, Krack M, Mohamed FR, Parrinello M (2007) Phys Rev Letts 98:066401
3. Schlegel HB, Iyengar S, Li X, Millam JM, Voth GA, Scuseria GE, Frisch M (2002) J Chem Phys 117:8694
4. Tongraar A, Liedl KR, Rode BM (1998) J Phys Chem A 102:10340
5. Titov AV, Ufimtsev IS, Luehr N, Martinez TJ (2013) J Chem Theory Comput 9:213
6. Ufimtsev IS, Luehr N, Martinez TJ (2011) J Phys Chem Lett 2:1789
7. Schmidt MW, Baldridge KK, Boatz JA, Elbert ST, Gordon MS, Jensen JH, Koseki S, Matsunaga N, Nguyen KA, Su S, Windus TL, Dupuis M, Montgomery JA (1993) J Comput Chem 14:1347
8. York DM, Lee T-S, Yang W (1998) Phys Rev 80:5011
9. Scuseria G (1999) J Phys Chem A 103:4782
10. Nam K, Gao J, York DM (2005) J Chem Theory Comput 1:2
11. Yang W, Lee T-S (1996) J Chem Phys 103:5674
12. Monard G, Bernal-Uruchurtu MI, van der Waart A, Merz KM, Ruiz-Lopez M (1998) J Phys Chem A 109:3425
13. VandeVondele J, Borstnik U, Hutter J (2012) J Chem Theory Comput 8:3565
14. Warshel A, Levitt M (1976) J Mol Biol 103:227
15. Maseras F, Morokuma K (1995) J Comput Chem 16:1170
16. Senn HM, Thiel W (2009) Angewandte Chem 48:1198
17. Rod TH, Ryde U (2005) Phys Rev Lett 94:138
18. Monari A, Rivail J-L, Assfeld X (2013) Acc Chem Res 46:596
19. Ferre N, Assfeld X, Rivail J-L (2002) J Comput Chem 23:610
20. Very T, Despax S, Hebraud P, Monari A, Assfeld X (2012) Phys Chem Chem Phys 14:12496
21. Antonczak S, Monard G, Ruiz-Lopez M, Rivail J-L (1998) J Am Chem Soc 120:8834
22. Ryde U, Olsen L, Nilsson K (2002) J Comput Chem 23:1058
23. Gresh N, de Courcy B, Piquemal JP, Foret J, Courtiol-Legourd S, Salmon L (2011) J Phys Chem B 115:8304
24. Gresh N (2006) Curr Pharm Des 12:2121
25. Dehez F, Angyan JG, Soteras Gutierrez IS, Luque FJ, Schulten K, Chipot C (2007) J Chem Theory Comput 3:1914
26. Gresh N, Cisneros AG, Darden TA, Piquemal J-P (2007) J Chem Theory Comput 3:1960
27. Warshel A, Kato M, Pisliakov AV (2007) J Chem Theory Comput 3:2034

28. Cieplak P, Dupradeau F-Y, Duan Y, Wang J (2009) J Phys: Condens Matter 21:33102
29. Ponder JW, Wu C, Ren PY, Pande VS, Chodera JD, Mobley DL, Schnieders MJ, Haque I, Lambrecht DS, DiStasio RA Jr, Head-Gordon M, Clark GNI, Johnson ME, Head-Gordon T (2010) J Phys Chem B 114:2549
30. Rick SW, Stuart SJ, Berne BJ (1994) J Chem Phys 101:6141
31. Stern HA, Rittner F, Berne BJ, Friesner RA (2001) J Chem Phys 115:2237
32. Chelli R, Procacci P (2002) J Chem Phys 117:9175
33. Lamoureux G, MacKerell AD Jr, Roux B (2003) J Chem Phys 119:5185
34. Anisimov V, Vorobyov IV, Roux B, MacKerell AD Jr (2007) J Chem Theory Comput 3:1927
35. Pullman B, Claverie P, Caillet J (1967) Proc Natl Acad Sci USA 57:1663
36. Rein R, Claverie P, Pollack M (1968) Int J Quantum Chem 2:1129
37. Claverie P, Rein R (1969) Int J Quantum Chem 3:537
38. Gresh N, Claverie P, Pullman A (1984) Theor Chim Acta 66:1
39. Gresh N, Pullman A, Claverie P (1985) Theor Chim Acta 67:11
40. Price SL, Stone AJ (1984) Mol Phys 51:569
41. Price SL, Stone AJ (1992) J Chem Soc Faraday Soc 88:1755
42. Day P, Jensen JH, Gordon MS, Webb SP, Stevens WJ, Krauss M, Garmer DR, Basch H, Cohen D (1996) J Chem Phys 105:1968
43. Slipchenko LV, Gordon MS (2007) J Comput Chem 28:276
44. Gordon MS, Fedorov DG, Pruitt SR, Slipchenko LV (2012) Chem Rev 112:632
45. Gordon MS, Smith QA, Xu P, Slipchenko LV (2013) Annu Rev Phys Chem 64:553
46. Millot CJ, Soetens JC, Martins Costa NTC, Hodges MP, Stone AJ (1998) J Phys Chem A 102:754
47. Karlstrom G, Linse P, Wallqvist A, Jonsson B (1983) J Am Chem Soc 105:3777
48. Hermida-Ramon JM, Brdarski S, Karlstrom G, Berg U (2003) J Comput Chem 24:161
49. Ren PY, Ponder JW (2003) J Phys Chem B 107:5933
50. Shi Y, Xia Z, Zhang J, Best R, Wu C, Ponder JW, Ren PY (2012) J Chem Theory Comput 9:4046
51. Dreyfus M (1970) PhD Thesis, University of Paris
52. Claverie P (1973) PhD Thesis, Paris CNRS library number AO 8214
53. Claverie P Localization and delocalization in quantum chemistry. In: Chalvet O, Daudel R, Diner S, Malrieu JP (eds) Reidel, Dordrecht Vol II, p 127
54. Pullman A, Zarkzewska C, Perahia D (1979) Int J Quantum Chem 16:395
55. Pullman A, Perahia D (1979) Theoret Chim Acta 50:351
56. Perahia D, Pullman A, Pullman B (1979) Theoret Chim Acta 51:349
57. Pullman B, Perahia D, Cauchy D (1979) Nucleic Acids Res 10:3821
58. Lavery R, Pullman A, Pullman B, de Oliveria M (1980) Nucleic Acids Res 8:5095
59. Lavery R, Pullman B, Corbin S (1981) Nucleic Acids Res 9:6539
60. Lavery R, Pullman B, Zakrzewska K (1982) Biophys Chem 15:343
61. Lavery R, Pullman A, Wen YK (1982) Int J Quantum Chem 24:353
62. Zarkzewska C, Pullman A (1982) Int J Quantum Chem Quantum Biol Symp 9:411
63. Etchebest C, Lavery R, Pullman A (1982) Studia Biophys 90:7
64. Gresh N, Pullman A (1983) Int J Quantum Chem Quantum Biol Symp 10:215
65. Gresh N, Claverie P, Pullman A (1979) Int J Quantum Chem Symp 13:243
66. Gresh N (1980) Biochim Biophys Acta 597:345
67. Gresh N, Pullman B (1980) Biochim Biophys Acta 625:356
68. Gresh N, Etchebest C, de la Luz Rojas O, Pullman A (1981) Int J Quantum Chem Quantum Biol Symp 8:109
69. Gresh N, Pullman A (1982) Int J Quantum Chem 22:709
70. Gresh N, Pullman B (1984) Mol Pharmacol 25:452
71. Gresh N (1985) Biopolymers 24:1527
72. Vigné-Maeder F, Claverie P (1988) J Chem Phys 88:4934
73. Rein R, Rabinowitz JR, Swissler TJ (1972) J Theor Biol 34:215

74. Rein R (1973) Adv Quantum Chem 7:335
75. Stone AJ (1981) Chem Phys Lett 83:233
76. Stone AJ, Alderton M (1985) Mol Phys 56:1047
77. Sokalski WA, Poirier RA (1983) Chem Phys Lett 98:86
78. Sokalski WA, Sawaryn AJ (1987) Chem Phys 87:526
79. Andersson M, Karlstrom G (1985) J Phys Chem 89:4957
80. Angyan JG, Chipot C, Dehez F, Hattig C, Jansen G, Millot C (2003) J Comput Chem 24:997
81. Angyan JG, Chipot C (2005) New J Chem 29:411
82. Freitag M, Gordon MS, Jensen JH, Stevens WJ (2000) J Chem Phys 112:7300
83. Gordon MS, Freitag MA, Bandyopadhyay A, Jensen JH, Kairys V, Stevens WJ (2001) J Phys Chem A 105:293
84. Piquemal J-P, Gresh N, Giessner-Prettre C (2003) J Phys Chem A 107:10353
85. Slipchenko LV, Gordon MS (2006) J Comput Chem 28:276
86. Wang B, Truhlar DG (2010) J Chem Theory Comput 6:3330
87. Stone AJ (2011) J Phys Chem A 115:7017
88. Garmer DR, Stevens WJ (1989) J Phys Chem 93:8263
89. Gresh N, Leboeuf M, Salahub DR (1994) Modeling the hydrogen bond American Chemical Society Symposia No 569 82 D A Smith Ed
90. Jensen JH, Day PN, Gordon MS, Basch H, Cohen D, Garmer DR, Krauss M, Stevens WJ Ibid p 139
91. Murrell JN, Teixeira-Dias JJC (1970) Mol Phys 19:521
92. Murrell JN, Randic M, Williams DR (1966) Proc R Soc Lond Ser A 284:566
93. Gresh N, Claverie P, Pullman A (1982) Int J Quantum Chem 22:199
94. Gresh N, Claverie P, Pullman A (1986) Int J Quantum Chem 29:101
95. Creuzet S, Langlet J, Gresh N (1991) J Chim Phys Phys Chim Biol 88:2399
96. Stevens WJ, Fink W (1987) Chem PhysLett 139:15
97. Bagus PS, Hermann K, Bauschlicher C (1984) J Chem Phys 80:4378
98. Jeziorski B, Moszynski R, Szalewicz K (1984) Chem Rev 94:1887
99. Piquemal J-P, Hubbard BW, Fey N, Deeth R, Gresh N, Giessner-Prettre C (2003) J Comput Chem 24:1963
100. Dudev T, Devereux M, Meuwly M, Lim C, Piquemal J-P, Gresh N (2015) J Comput Chem 36:285
101. Buckingham AD, Fowler PW (1983) J Chem Phys 79:6426
102. Garmer DR, Gresh N (1994) J Am Chem Soc 116:3556
103. Gresh N, Garmer DR (1996) J Comput Chem 17:1481
104. Gresh N (1995) J Comput Chem 16:856
105. Piquemal J-P, Chevreau H, Gresh N (2007) J Chem Theory Comput 3:824
106. Lu Y, Shi T, Wang Y, Yang H, Yan X, Luo X, Jiang H, Zhu W (2009) J Med Chem 52:2854
107. Hernandes MZ, Cavalcanti SM, Moreira DR, de Azevedo WF, Jr Leite AC (2010) Curr Drug Targets 11:303
108. Matter H, Nazaré M, Güssregen S, Will DW, Schreuder H, Bauer A, Urmann M, Ritter K, Wagner M, Wehner V (2009) Angew Chem Int Ed 48:2911
109. Clark T, Hennemann M, Murray JS, Politzer P (2007) J Mol Mod 13:291
110. Murray JS, Lane P, Politzer P (2009) J Mol Mod 15:723
111. Politzer P, Murray JS, Clark T (2010) Phys Chem Chem Phys 12:7748
112. Ibrahim MA (2011) J Comput Chem 32:2564
113. Jorgensen WL, Schyman P (2012) J Chem Theory Comput doi:101021/ct300180w
114. Kolar M, Hobza P (2012) J Chem Theory Comput 8:1325
115. El Hage K, Piquemal J-P, Hobaika Z, Maroun RG, Gresh N (2013) J Comput Chem 34:1125
116. Gourlaouen C, Piquemal J-P, Parisel O (2006) J Chem Phys 124:174311
117. Gourlaouen C, Parisel O (2007) Angew Chem Int Ed 46:553
118. van Severen M-C, Piquemal J-P, Parisel O (2009) Chem Phys Lett 478:17
119. Devereux M, van Severen M-C, Parisel O, Piquemal J-P, Gresh N (2011) J Chem Theory Comput 7:138

120. Tiraboschi G, Roques BP, Gresh N (1999) J Comput Chem 20:1379
121. Tiraboschi G, Gresh N, Giessner-Prettre C, Pedersen LG, Deerfield DW (2000) J Comput Chem 21:1011
122. Garmer DR, Gresh N, Roques BP (1998) Proteins Struct Funct Genet 31:42
123. Gresh N (1997) J Phys Chem A 101:8690
124. Masella M, Gresh N, Flament JP (1998) J Chem Soc, Faraday Trans 94:2745
125. Piquemal J-P, Chelli R, Procacci P, Gresh N (2007) J Phys Chem A 111:8170
126. Guo H, Gresh N, Roques BP, Salahub DR (2000) J Phys Chem B 104:9746
127. Gresh N, Piquemal J-P, Krauss M (2005) J Comput Chem 26:1113
128. Gresh N, Policar C, Giessner-Prettre C (2002) J Phys Chem A 106:5660
129. Ohrn A, Karlstrom G (2007) J Chem Theory Comput 3:1993
130. Beachy MD, Chasman D, Murphy RB, Halgren TA, Friesner RA (1997) J Am Chem Soc 119:5908
131. Gresh N, Kafafi SA, Truchon J-F, Salahub DR (2004) J Comput Chem 25:823
132. Rogalewicz F, Gresh N, Ohanessian G (2000) J Comput Chem 21:963
133. Tiraboschi G, Fournie-Zaluski MC, Roques BP, Gresh N (2001) J Comput Chem 22:1038
134. Gresh N, Shi G-B (2004) J Comput Chem 25:160
135. Faerman CH, Price SL (1990) J Am Chem Soc 112:4915
136. Chaudret R, Gresh N, Parisel O, Piquemal J-P (2011) J Comput Chem 31:2949
137. Barr MP, Byrne AM, Duffy AM, Condron CM, Devocelle M, Harriott P, Bouchier-Hayes DJ, Harmey JH (2005) Br J Cancer 92:328
138. Borriello L, Montès M, Lepelletier Y, Leforban B, Liu W-Q, Demange L, Delhomme B, Pavoni S, Jarray R, Boucher JL, Dufour S, Hermine O, Garbay C, Hadj-Slimane R, Raynaud F (2014) Cancer Lett 349:120
139. van der Kooi C, Jusino MA, Perman B, Neau DB, Bellamy HD, Leahy D (2007) Proc Natl Acad Sci USA 104:6152
140. Jarvis A, Allerston CK, Jia H, Herzog B, Garza-Garcia A, Winfield N, Ellard K, Aqil R, Lynch R, Chapman C, Hartzoulakis B, Nally J, Stewart M, Cheng L, Menon M, Tickner M, Djordjevic S, Driscoll PC, Zachary I, Selwood DL (2010) J Med Chem 53:2215
141. Goldwaser E, de Courcy B, Demange L, Garbay C, Raynaud F, Hadj-Slimane R, Piquemal J-P, Gresh N (2014) J Mod Mol 20:2472
142. Silvi B, Savin A (1994) Nature 371:683
143. Piquemal J-P, Pilme J, Parisel O, Gerard H, Fourre I, Berges J, Gourlaouen C, De La Lande A, Van Severen M-C, Silvi B (2008) Int J Quantum Chem 108:1951
144. Evangelakis GA, Rizos JP, Lagaris IE, Demetropoulos IN (1987) Comput Phys Commun 46:401
145. Zheng J, Yu T, Papajak E, Alecu IM, Mielke S, Truhlar DG (2011) Phys Chem Chem Phys 13:10885
146. Tafipolsky M, Schmid R (2005) J Comput Chem 26:1579
147. Marchand C, Maddali K, Metifiot M, Pommier Y (2009) Curr Top Med Chem 9:1016
148. Klibanov OM (2009) Curr Opin Invest Drugs 10:190
149. Satom M, Motomura T, Aramaki H, Matsuda T, Yamashita M, Ito Y, Kawakami H, Matsuzaki Y, Watanabe W, Yamataka K et al (2006) J Med Chem 49:1506
150. Schafer JJ, Squires KE (2010) Ann Pharmacother 44:145
151. Hare S, Gupta S, Valkov E, Engelman A, Cherepanov P (2010) Nature 464:232
152. El Hage K, Piquemal J-P, Hobaika Z, Maroun RG, Gresh N (2015) J Comput Chem 36:210
153. El Hage K, Piquemal J-P, Hobaika Z, Maroun RG, Gresh N (2014) J Phys Chem A 118:9772
154. Kitaura K, Morokuma K (1976) Int J Quantum Chem 10:325
155. McCord JM, Fridovich I (1969) J Biol Chem 244:6049
156. Somwar R, Erdjumen-Bromage H, Larsson E, Shum D, Lockwood WW, Yang G, Sander C, Ouerfelli O, Tempst PJ, Djaballah H, Varmus HE (2011) Proc Natl Acad Sci USA 108:16375
157. Glasauer A, Sena LA, Diebold LP, Mazar AP, Chandell NS (2014) J Clin Invest 124:117

158. Bijeire L, Elias B, Souchard J-P, Gicquel E, Moucheon C, Kirsch-de Mesmaeker A, Vicendo P (2006) Biochem 45:6160
159. Hough MA, Hasnain SS (2003) Structure 11:937
160. Tainer JA, Getzoff ED, Richardson JS, Richardson DC (1983) Nature 306:284
161. Gresh N, El Hage KE, Perahia D, Piquemal J-P, Berthomieu C, Berthomieu D (2014) J Comput Chem 35:2096
162. Lipparini F, Lagardere L, Stamm B, Cances E, Schnieders M, Ren P, Maday Y, Piquemal J-P (2014) J Chem Theory Comput 10:1638
163. Krause DS, Van Etten RA (2005) N Engl J Med 353:172
164. Nikolopoulos S, Giancotti FG (2005) Cell Cycle 4:e131
165. McLean GW, Carragher NO, Avizienyte VG, Frame MC (2005) Nat Cancer 5:505
166. Cohen LA, Guan JL (2005) Curr Cancer Drug Targets 5:629
167. Lietha D, Cai X, Ceccarelli DFJ, Li Y, Schaller MD, Eck M (2007) J Cell 129:1177
168. Choi HS et al (2006) Bioorgan Med Chem Lett 16:2689
169. de Courcy B, Piquemal J-P, Garbay C, Gresh N (2010) J Am Chem Soc 132:3312
170. Langlet J, Claverie P, Caillet J, Pullman A (1998) J Phys Chem 92:1617
171. Claverie P, Daudey J-P, Langlet J, Pullman B, Piazzola D (1978) J Phys Chem 8:405
172. Mundim KC, Tsallis C (1996) Int J Quantum Chem 58:373
173. Moret MA, Pascutti PG, Bisch PM, Mundim KC (1998) J Comput Chem 19:647
174. Jiao D, Golubkov PA, Darden TA, Ren P (2008) Proc Natl Acad Sci USA 105:6290
175. Lietha D, Eck M (2008) J PLOS one. doi:10.1371/journalpone0003800
176. Gracy RW, Noltmann EA (1968) J Biol Chem 243:3161
177. Patterson JH, Waller RF, Jeevarajah D, Billman-Jacobe H, McConville M (2003) J Biochem J 372:77
178. Garami A, Ilg T (2001) J Biol Chem 276:6566
179. Shinabarger D, Berry A, May TB, Rothmel R, Fialho A, Chakrabarty AM (1991) J Biol Chem 266:2080
180. Roux C, Lee JH, Jeffery CJ, Salmon L (2004) Biochemistry 43:2926
181. Cleasby A, Wonacott A, Skarzynski T, Hubbard RE, Davies GJ, Proudfoot AEI, Bernard AR, Payton MA, Wells TNC (1996) Nat Struct Biol 3:470
182. Roux C, Gresh N, Perera LE, Piquemal J-P, Salmon L (2007) J Comput Chem 28:938
183. Stevens WJ, Basch H, Krauss M (1984) J Chem Phys 81:6026
184. Stevens WJ, Krauss M, Basch H, Jasien PG (1992) Can J Chem 70:612
185. Dunning TH (1989) J Chem Phys 90:1007
186. Feller D (1996) J Comput Chem 17:1571
187. Law MM, Hutson JM (1997) Comp Phys Commun 102:252
188. Devereux M, Meuwly M (2010) J Chem Inf Model 50:349
189. Devereux M, Gresh N, Piquemal J-P, Meuwly M (2014) J Comput Chem 35:1577
190. Foster JM, Boys SF (1960) Rev Mod Phys 32:300
191. Chaudret R, Gresh N, Cisneros GA, Scemama A, Piquemal J-P (2013) Can J Chem 91:804
192. Piquemal J P, Cisneros GA, Reinhardt P, Gresh N, Darden TA (2006) J Chem Phys 124:104101
193. Chaudret R, Gresh N, Narth C, Lagardere L, Darden TA, Cisneros GA, Piquemal J-P (2014) J Phys Chem A 118:7598
194. Wang Q, Rackers JA, He C, Qi R, Narth C, Lagardere L, Gresh N, Ponder JW, Piquemal JP, Ren PY (2015) J Chem Theory Comput 11:2609
195. Lagardère L, Lipparini F, Polack E, Stamm B, Cances E, Schnieders M, Ren PY, Maday Y, Piquemal JP (2015) J Chem Theory Comput 11:2589
196. Lipparini F, Lagardère L, Raynaud C, Stamm B, Cances E, Schnieders M, Ren PY, Mennucci B, Maday Y, Piquemal JP (2015) J Chem Theory Comput 11:623
197. Antony J, Gresh N, Olsen L, Hemmingsen L, Schofield CJ, Bauer R (2002) J Comput Chem 23:1281
198. Antony J, Piquemal J-P, Gresh N (2005) J Comput Chem 26:1131
199. Gresh N, Derreumaux P (2003) J Phys Chem B 107:4862

200. Miller Jenkins LM, Hara T, Durell SR, Hayashi R, Inman JK, Piquemal J-P, Gresh N, Appella E (2007) J Am Chem Soc 129:11067
201. Gresh N, Audiffren N, Piquemal J-P, de Ruyck J, Ledecq M, Wouters J (2010) J Phys Chem B 114:4884
202. Zhang J, Yang W, Piquemal J-P, Ren P (2012) J Chem Theory Comput 8:1314
203. Nicolini P, Frezzato D, Gellini C, Bizzarri M, Chelli R (2013) J Comput Chem 34:1561
204. Marjolin A, Gourlaouen C, Clavaguéra C, Ren P, Wu J, Gresh N, Dognon J-P, Piquemal J-P (2012) Theor Chem Acc 131:1199
205. Muddana H, Gilson MK (2012) J Chem Theory Comput 8:2023
206. Grimme S (2012) Chem Eur J 18:9955
207. Gresh N, Sponer J (1999) J Phys Chem B 103:1415
208. Gresh N, Sponer JE, Spackova N, Leszczynski J, Sponer J (2003) J Phys Chem B 107:8669
209. Ghosh D, Kosenkov D, Vanovschi V, Williams CR, Herbert JM, Gordon MS, Schmidt MW, Slipchenko LV, Krylov A (2010) J Phys Chem A 114:12739
210. Smith QA, Gordon MS, Slipchenko LV (2011) J Phys Chem A 115:11269
211. Savelyev A, MacKerell AD (2014) J Comput Chem 16:1219

Chapter 2
Proton Transfer in Aqueous Solution: Exploring the Boundaries of Adaptive QM/MM

T. Jiang, J.M. Boereboom, C. Michel, P. Fleurat-Lessard and R.E. Bulo

Abstract In this chapter, we review the current state-of-the-art in quantum mechanical/molecular mechanical (QM/MM) simulations of reactions in aqueous solutions, and we discuss how proton transfer poses new challenges for its successful application. In the QM/MM description of an aqueous reaction, solvent molecules in the QM region are diffusive and need to be either constrained within the region, or their description (QM versus MM) needs to be updated as they diffuse away. The latter approach is known as adaptive QM/MM. We review several constrained and adaptive QM/MM methods, and classify them in a consistent manner. Most of the adaptive methods employ a transition region, where every solvent molecule can continuously change character (from QM to MM, and vice versa), temporarily becoming partially QM and partially MM. Where a conventional QM/MM scheme partitions a system into a set of QM and a set of MM atoms, an adaptive method employs multiple QM/MM partitions, to describe the fractional QM character. We distinguish two classes of adaptive methods: Discontinuous and continuous. The former methods use at most two QM/MM partitions, and cannot completely avoid discontinuities in the energy and the forces. The more recent continuous adaptive methods employ a larger number of QM/MM partitions for a given configuration. Comparing the performance of the methods for the description of solution chemistry, we find that in certain cases the low-cost constrained methods are sufficiently accurate. For more demanding purposes, the continuous adaptive schemes provide a

T. Jiang · J.M. Boereboom · R.E. Bulo
Inorganic Chemistry and Catalysis Group, Debye Institute for Nanomaterials Science, Utrecht University, Universiteitsweg 99, 3584CG Utrecht, The Netherlands
e-mail: R.E.Bulo@uu.nl

C. Michel · P. Fleurat-Lessard (✉)
Laboratoire de Chimie de L'ENS de Lyon, Université de Lyon, 46 Allée d'Italie, 69364 Lyon Cedex 7, France
e-mail: Paul.Fleurat-Lessard@ens-lyon.fr

P. Fleurat-Lessard
Université de Bourgogne, Institut de Chimie Moléculaire de L'Université de Bourgogne (ICMUB), 9 Avenue Alain Savary, 21078 Dijon, France

J.-L. Rivail et al. (eds.), *Quantum Modeling of Complex Molecular Systems*, Challenges and Advances in Computational Chemistry and Physics 21,
DOI 10.1007/978-3-319-21626-3_2

good balance between dynamical and structural accuracy. Finally, we challenge the adaptive approach by applying it to the difficult topic of proton transfer and diffusion. We present new results, using a well-behaved continuous adaptive method (DAS) to describe an alkaline aqueous solution of methanol. Comparison with fully QM and fully MM simulations shows that the main discrepancies are rooted in the presence of a QM/MM boundary, and not in the adaptive scheme. An anomalous confinement of the hydroxide ion to the QM part of the system stems from the mismatch between QM and MM potentials, which affects the free diffusion of the ion. We also observe an increased water density inside the QM region, which originates from the different chemical potentials of the QM and MM water molecules. The high density results in locally enhanced proton transfer rates.

2.1 Introduction

Proton transfer is one of the unifying elements that (almost) all aqueous chemical reactions have in common. The computational study of aqueous reactions, already difficult due to the long distance effects in this highly structured solvent, [1, 2] is further complicated by this challenging phenomenon. Acid-base reactions involve proton transfer to or from a reactant compound, often across an extended chain of water molecules (Grotthuss mechanism) [3]. This means that in these processes many bonds break and form nearly simultaneously, causing the reaction to be less local than reactions in organic media [4]. Recent simulations on the solvated hydroxide ion have further shown that the proton transfer motion is strongly dependent on water wires linking the OH^- to the rest of the solution [5]. Global effects, like changes in the electric field or compression of the wires, can be the cause of proton transfer events [6]. If the wires compress, the proton can even hop several water molecules along, in a nearly concerted process.

Aqueous reactions such as those described above involve constant breaking and forming of bonds, and can therefore best be described using electronic structure methods, preferably including nuclear quantum effects [7, 8]. However, this approach is costly, due to the long time scales on which the reactions occur. A multi-scale quantum mechanical (QM)/molecular mechanical (MM) approach assumes a localized reactive region, and thereby makes a QM description of a reaction feasible. In most cases bond breaking or forming can indeed be described as local (with the possible exception of the proton transfer events discussed above), but even for local reactions a multi-scale description is not straightforward. In standard QM/MM approaches [9–15], the nature (QM or MM) of each particle is defined initially and remains fixed during the whole simulation. This can lead to problems in dealing with solvent molecules in solution chemistry, which are diffusive by nature. Let us consider a QM/MM molecular dynamics (MD) simulation of a solvated system. The initial QM particles comprise a solute (such as a sodium or a hydroxide ion) and its first solvation shell. However, after a few tens of picoseconds, the solvent molecules of the solvation shell are exchanged with bulk solvent molecules, leading to an incongruous simulation in which the first solvation shell contains MM solvent molecules, while QM solvent molecule have diffused into the MM bulk. To overcome this diffusivity problem two classes of methods have emerged. The most direct approach constrains the QM molecules to remain within a certain region (henceforth denoted by 'active region'), while the MM molecules are constrained to their respective part of space ('environment region'). A more involved approach allows the nature of the solvent molecules to change from QM to MM and vice versa as they move in and out of the active region. This approach is called adaptive QM/MM [16–18]. Among the adaptive methods, some lead to discontinuities in the forces exerted on the atoms, while more complex ones manage to keep a continuous description. This book chapter aims at giving a comprehensive overview of the main constrained and adaptive methods currently available.

The adaptive QM/MM methods have already provided insights into several processes in water [19–21]. Local proton transfer events have been studied with conventional QM/MM [22], as well as with adaptive QM/MM [23]. However, none of the previous applications considered more global proton transfer and diffusion into the aqueous solution. A hydroxide ion in water can in principle exit the predefined QM region that is usually centered around a reactive solute. In this book chapter we address the issues involved with the description of such proton diffusion events. More specifically, we will consider the solvation of a methanol molecule in alkaline water, with the aim to understand the values and limitations of adaptive QM/MM.

This chapter is organized as follows. In Sect. 2.2 the most dominant diffusive QM/MM methods aimed at aqueous reactions are reviewed, first the constrained, and then the adaptive methods. Their performances are compared, based on published results, in Sect. 2.3. In Sect. 2.4, one of the methods is applied to describe proton transfer in solution, and the performances of the method are discussed based on structural and dynamical data. Finally, in Sect. 2.5, the general conclusions of this work are briefly summarized.

2.2 Review of QM/MM Methods for the Simulation of Chemistry in Solution

In this section the available QM/MM methods designed for solution chemistry are reviewed, and the relevant terms are introduced. As we focus on describing a reactive process in solution at the atomistic level, multi-scale methods resorting to coarse-graining will not be described and we refer the interested readers to recent reviews [24–26]. Similarly, methods targeting gas phase reactions only are omitted [27, 28].

All the QM/MM methods dedicated to treat solution chemistry are based on the definition of two zones: (i) an active region that is generally defined as a sphere around the reactive center and (ii) an environment region that embeds the active region. This is illustrated in Fig. 2.1. The active region generally includes the reactive center (the orange disk in Fig. 2.1) and at least the first coordination sphere and is described at the QM level. QM molecules will always be shown in

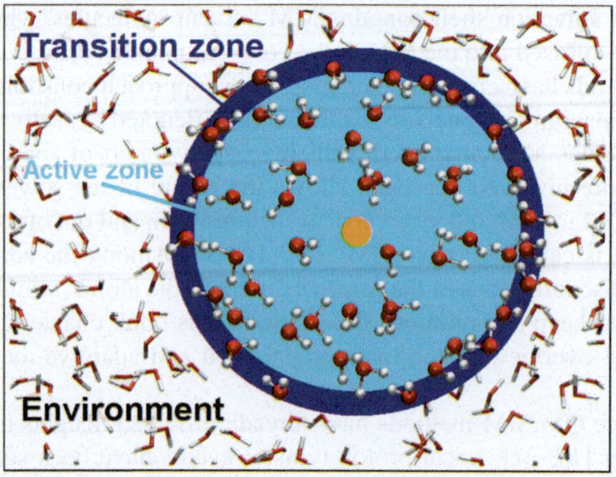

Fig. 2.1 Partitioning of a simulation box into an active zone (with light *blue* background), an environment zone (*white* background) and for some methods a transition zone (with *blue* background). The active center is symbolized with an *orange* disk. QM molecules are shown in *Ball&Stick* while MM water molecules are shown with *lines*

Ball&Stick in the Figures. The environment region includes the bulk solvent molecules described at the MM level, and will be depicted with lines. Each solvent molecule is either in the active zone, or in the transition region or in the environment zone. The QM/MM border does not cut through solvent bonds, even though methods that allow it have been developed [29].

As mentioned in the introduction, there are two main classes of methods. The first class constrains the QM molecules to remain inside the active region, while simultaneously keeping the MM molecules in the environment region (outside the active region). The solvent molecules cannot diffuse across the boundary between the active region and the environment. The most prominent examples of this approach will be discussed in Sect. 2.2.1. The second class of methods discussed does allow the solvent molecules to diffuse across the QM/MM boundary, changing the nature (QM vs. MM) of those molecules on the fly. This change can be instantaneous [16, 30–32] (Sect. 2.2.2: Discontinuous adaptive QM/MM), or smooth [33–35] (Sect. 2.2.3: Continuous adaptive QM/MM).

Including the continuity referred to above, there are three defining characteristics of the adaptive QM/MM methods that reveal much about their performance:

1. Continuity of the energy and/or the forces;
2. Energy conservation;
3. Momentum conservation.

The methods are classified according to these definitions in Table 2.1. By definition none of the discontinuous methods are energy conserving, but even among the continuous methods only some fall into this class. All categories have advantages as well as disadvantages, and these will be addressed in the following subsections.

Table 2.1 Classification of the various methods

	Continuous	Energy conserving	Momentum conserving
Abrupt [30]	✗	✗	✓
ONIOM-XS [30]	✗	✗	✓
LOTF [31]	✗	✗	✓
Hot-spot [16]	✗	✗	✗
BF [32]	✗	✗	✗
PAP [33]	✓	✓	✓
SAP [33]	✓	✓	✓
DAS [34]	✓	✗	✓
SCMP [35]	✓	✗	✓

They are classified according to three features: Continuity of the energy and/or forces, energy conservation, and momentum conservation

It is worth emphasizing that all constrained and adaptive QM/MM methods described in this chapter are fundamentally QM/MM simulations and will thus experience the same problems, all stemming from the fact that two different potentials are applied to describe the same system. The bulk of the problem originates in the interaction between QM and MM particles, and as a result, artifacts are found mainly at the QM/MM boundary. Several different schemes exist for the description of the interaction between QM and MM particles in conventional QM/MM [18, 36–38]: (i) mechanical embedding, (ii) electrostatic embedding, (iii) polarizable embedding, and (iv) flexible embedding [39, 40]. The most straightforward choice is mechanical embedding, in which one simply uses the MM (pair)-potential to describe the interaction between the QM and the MM molecules. Electrostatic embedding includes the partial charges of the MM atoms in the QM Hamiltonian, thereby describing part of the interaction using the QM description and allowing the MM environment to polarize the electrons in the QM region. Polarizable embedding is a less used option that employs a polarizable force-field in combination with electrostatic embedding, thereby allowing the QM region to in turn polarize the MM region. The mutual polarization then needs to be solved in an iterative manner. The last scheme, flexible embedding, is very recent, and allows the total charges of the active and environment regions to adapt during the course of the simulation, effectively allowing charge transfer between the two regions. In this chapter we will not go into the advantages and disadvantages of these options, but they have been recently reviewed and compared [18]. Suffice it to say that all of them have certain advantages and disadvantages, and none of them are exempt from structural and dynamical artifacts at the boundary. Nonetheless, all QM/MM methods discussed in this chapter can be straight-forwardly combined with either mechanical, electrostatic or polarizable embedding.

Another important decision in conventional QM/MM lies in the choice of the size of the QM region (the further away the boundary is from the reactive site, the better). The same holds for adaptive QM/MM simulations, and the possible options as to where to locate the QM/MM boundary (distance to the QM center, number of active solvent molecules, or density based) will be discussed at the end of this section.

2.2.1 Constrained QM/MM

In this section we will discuss two constrained QM/MM methods: the Flexible Inner Region Ensemble Separator (FIRES) method, [41] and the recent Boundary based on Exchange Symmetry Theory (BEST) method [42].

Flexible Inner Region Ensemble Separator (FIRES)
The FIRES method is an elegant example of the constrained approach. It prevents the solvent molecules from crossing the QM/MM boundary, but does not require the volume of the QM region to be determined in advance. The constraining potential is located on a sphere around the active region, and the radius of the sphere is defined by the position of the QM solvent molecule farthest away from the QM center. This is illustrated in Fig. 2.2. Since the radius of the sphere is not constant throughout the simulation, the density of the QM region can adjust and equilibrate to the underlying free energy surface. Such a simulation can be viewed as having the active region placed in a balloon immersed in the environment. FIRES calculations are as efficient as conventional QM/MM since the overhead of the computation of the forces keeping the two sets of molecules apart is minimal.

While the imposed constraint alters the dynamics with respect to an unconstrained system, the authors argue that if the constraint is infinitely steep, average values of a property $X(\mathbf{r})$ (i.e. the distance between two reactants, the number of

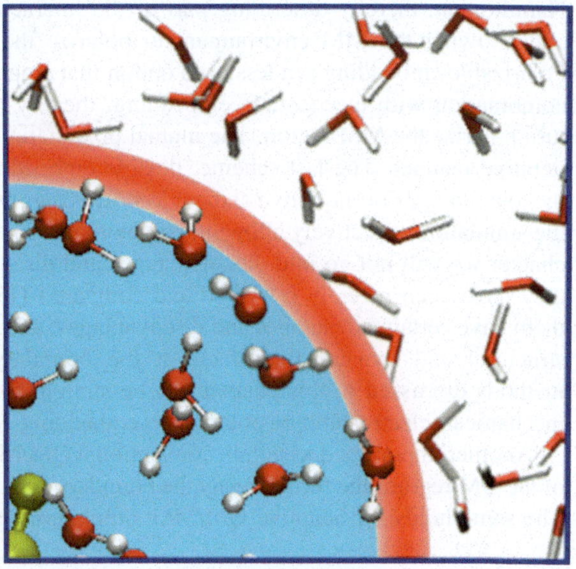

Fig. 2.2 FIRES method. The constraining potential is schematized as a *red* ring

hydrogen bonds to a molecule, etc.) remain unaltered by the constraint. If we define the partition function Z for a system consisting of a central reactant a and N_s equivalent solvent molecules,

$$Z = \frac{1}{N_s!}\int d\mathbf{r}_a \int e^{-V(\mathbf{r})/k_BT} d\mathbf{r}^{N_s}, \tag{2.1}$$

with $V(\mathbf{r})$ the potential energy of the system. We can define the configurational average of $X(\mathbf{r})$,

$$\langle X\rangle = \frac{\int d\mathbf{r}_a \int X(\mathbf{r}) e^{-V(\mathbf{r})/k_BT} d\mathbf{r}^{N_s}}{\int d\mathbf{r}_a \int e^{-V(\mathbf{r})/k_BT} d\mathbf{r}^{N_s}}, \tag{2.2}$$

In the following, we will prove that providing the constrained particles are identical and $V(\mathbf{r})$ and $X(\mathbf{r})$ are invariant under exchange of two particles, constraining two sets of particles to different regions of space will not affect the value of $\langle X\rangle$. If we define a central molecule (the reactant a), and then partition our system into a set S_n of n molecules and a set of S_m of $(m = N_s - n)$ molecules, we can rewrite the partition function as,

$$Z = \frac{1}{N_s!}\int d\mathbf{r}_a \int d\mathbf{r}^n \int d\mathbf{r}^m e^{-V(\mathbf{r})/k_BT}. \tag{2.3}$$

We can select a subset of configurations contributing to the integral, where the molecules in S_n are the n solvent molecules closest to the reactant a (which we would like to assign QM character). They occupy a flexible region $A(\mathbf{r})$ generally defined as the region of space containing the n solvent molecules closest to a. Then, the MM molecules in the set S_m occupy a further part of space $(E(\mathbf{r}))$.

$$Z' = \int d\mathbf{r}_a \frac{1}{n!}\int^{A(\mathbf{r})} d\mathbf{r}^n \frac{1}{m!}\int^{E(\mathbf{r})} d\mathbf{r}^m e^{-V(\mathbf{r})/k_BT}. \tag{2.4}$$

The contributions to the integral Z that we discarded in Z' are simply those where solvent molecules in S_n and in S_m exchanged position. Since all solvent molecules are identical, the discarded contributions to the integral Z are equivalent to the ones in Z' (Eq. 2.4), and the two expressions differ only by a multiplicative constant factor C: $Z' = Z/C$.

This separation is exact, and implementing the integral of Eq. (2.4) into Eq. (2.2) has the factors C cancel in the numerator and denominator, yielding an unchanged thermodynamic average $\langle X\rangle$, provided that the molecules in the active and environment region do not interchange [41].

To ensure that QM and MM molecules do not exit their regions, an infinitely steep constraint can be used. However, in practice, such a constraint would lead to divergent molecular dynamics. Therefore, a soft constraint is used, typically a simple half-harmonic potential that exerts a repulsive force on any MM molecules as they approach the active region $A(\mathbf{r})$. As this constraint is not infinitely steep, it might happen that a MM molecule sneaks into the active region $A(\mathbf{r})$ and thus becomes closer to the QM center than a QM molecule. As a result, the averages computed with Eq. (2.4) are no longer exact. The authors have shown that an aqueous system retains the correct structure in a test MM/MM simulation [41]. However, it has been shown that when two different potentials are used for the active zone and the environment, even equilibrium properties can be wrong at the boundary such as the radial distribution of oxygen atoms around the core QM water oxygen [17, 42].

Boundary Based on Exchange Symmetry Theory (BEST)

The soft harmonic constraint used in the FIRES method changes the partition function, resulting in slightly distorted equilibrium structures at the QM/MM boundary. The BEST method aims to solve this problem [42]. In this section, we illustrate the BEST method for the simplest example of only two particles i and j (belonging to S_n and S_m respectively) that violate the requirement in Eq. (2.4): the molecules in S_m cannot be closer to reactant a than any of the molecules in S_n. The partition function Z' in Eq. (2.4) can be rewritten as,

$$Z' = \frac{1}{n!m!}\int d\mathbf{r}_a \int^{A(\mathbf{r})} d\mathbf{r}_i \int^{E(\mathbf{r})} d\mathbf{r}_j \int^{A(\mathbf{r})} d\mathbf{r}^n_{rest} \int^{E(\mathbf{r})} d\mathbf{r}^m_{rest} e^{-V(\mathbf{r})/k_BT}. \tag{2.5}$$

Because FIRES uses a soft wall, additional configurations are explored, where particles i and j have swapped positions. Because the soft wall prevents these configurations from occurring too often, they have a relatively low probability. So they appear with a fractional weight $f(\mathbf{r}_i, \mathbf{r}_j)$ in the computed FIRES partition function Z'':

$$\begin{aligned} Z'' = &\frac{1}{n!m!}\int d\mathbf{r}_a \int^{A(\mathbf{r})} d\mathbf{r}_i \int^{E(\mathbf{r})} \mathbf{r}_j \int^{A(\mathbf{r})} d\mathbf{r}^n_{rest} \int^{E(\mathbf{r})} d\mathbf{r}^m_{rest} e^{-V(\mathbf{r})/k_BT} \\ &+ \frac{1}{n!m!}\int d\mathbf{r}_a \int^{E(\mathbf{r})} d\mathbf{r}_i \int^{A(\mathbf{r})} d\mathbf{r}_j f(\mathbf{r}_i, \mathbf{r}_j) \int^{A(\mathbf{r})} d\mathbf{r}^n_{rest} \int^{E(\mathbf{r})} d\mathbf{r}^m_{rest} e^{-V(\mathbf{r})/k_BT} \end{aligned} \tag{2.6}$$

The partition function Z'' thus differs from the exact one (Z) by something other than a constant. This leads to inexact average values (Eq. 2.1). The authors of BEST recognized that the partition function Z′ in Eq. (2.5) can be recovered by additional weighting of the first term of Eq. (2.6).

$$f(\mathbf{r}_i, \mathbf{r}_j) + f'(\mathbf{r}_i, \mathbf{r}_j) = 1 \tag{2.7}$$

$$Z' = \frac{1}{n!m!}\int d\mathbf{r}_a \int^{A(\mathbf{r})} d\mathbf{r}_i \int^{E(\mathbf{r})} d\mathbf{r}_j f'(\mathbf{r}_i,\mathbf{r}_j) \int^{A(\mathbf{r})} d\mathbf{r}^n_{rest} \int^{E(\mathbf{r})} d\mathbf{r}^m_{rest} e^{-V(\mathbf{r})/k_BT}$$
$$+\frac{1}{n!m!}\int d\mathbf{r}_a \int^{E(\mathbf{r})} d\mathbf{r}_i. \int^{A(\mathbf{r})} d\mathbf{r}_j f(\mathbf{r}_i,\mathbf{r}_j) \int^{A(\mathbf{r})} d\mathbf{r}^n_{rest} \int^{E(\mathbf{r})} d\mathbf{r}^m_{rest} e^{-V(\mathbf{r})/k_BT}. \tag{2.8}$$

In BEST, this formulation was extended to include all possible exchanges of solvent molecules between the QM and the MM sets. This approach does, however, become very laborious, and the recommended implementation assigns weights $f(\mathbf{r}_i, \mathbf{r}_j)$ to only the most frequent exchange contributions and neglects the rest.

2.2.2 *Discontinuous Adaptive QM/MM*

The list of discontinuous adaptive QM/MM methods discussed here counts five members: (1) Abrupt, [30] (2) ONIOM-XS, [30] (3) Hot Spot, [16] (4) Learn On The Fly (LOTF), [31] and (5) Buffered Force (BF) [32]. In QM/MM simulations for chemistry in solution, the molecular system is generally partitioned into a subset of QM solvent molecules S_{QM} and a subset of MM solvent molecules S_{MM} (see also Sect. 2.2.1). In order to account for the diffusivity of the solvent molecules, the QM/MM partition needs to be adjusted during the course of the simulation to always describe the solvent molecules in the active region as QM. As pointed out in the introduction to this section, none of the discontinuous methods are energy conserving. In addition, Hot Spot and BF do not conserve momentum, thereby breaking Newton's third law, which states that the force a molecule exerts on its neighbor is equal and opposite to the force the other molecule exerts in return. At first glance this seems an odd setup, and to clarify this, we address the different reasons why this option can be advantageous in the paragraphs below.

2.2.2.1 Abrupt

The simplest and most intuitive adaptive QM/MM simulation involves an abrupt switch of description as a solvent molecule crosses a given cut-off radius. This corresponds to a sudden change in the QM/MM partition, as the molecule is reassigned to a different set (QM or MM). For each configuration, an energy $V^{(a)}(\mathbf{r})$ is defined, which corresponds to the energy of the desired QM/MM partition with label a. In this chapter, this approach will be denoted by Abrupt.

Due to the difference between the QM and the MM potential, an instantaneous change of the partition is accompanied by a sudden change in the potential energy $V^{(a)}(\mathbf{r})$, and in the forces on the atoms. As a result, the total energy of a simulation is not conserved, which is demonstrated by an acceleration of the molecules at the

QM/MM boundary [17, 30]. Strong thermostats are required to correct for this acceleration. Although the energy is not conserved, at each configuration the forces on all atoms are defined as the negative gradient of $V^{(a)}(\mathbf{r})$. As a result, the total force on the system equals zero, and the momentum is conserved.

2.2.2.2 ONIOM-XS

This method aims to introduce a certain degree of continuity into the switch from one QM/MM partition to the next (which accompanies the change in the QM character of the molecules). ONIOM-XS is still classified as discontinuous, as some degree of discontinuity remains. The sudden changes occurring in Abrupt are diminished through introduction of a transition region between the active region and the environment, where the solvent molecules can semi-continuously switch between QM and MM character. As a consequence, the molecules in that region have partial QM and partial MM character. We define a transition region variable T as a set of two distances R_s and R_e ($T = \{R_s, R_e\}$, with $R_s < R_e$), denoting the start and the end of the transition region. The partial QM character of the solvent molecules in the transition region is reflected in a quantity $\lambda(\mathbf{r}_i)$, which is a switching function that depends on the position of the molecule i in the transition region. In ONIOM-XS, the switching function $\lambda(\mathbf{r}_i)$ is a polynomial function that equals 1 inside the QM zone (for $r_i < R_s$), 0 inside the environment (for $R_e < r_i$) and smoothly switch between these two values inside the transition zone:

$$\lambda(\mathbf{r}_i : T) = 6\left(x_i - \frac{1}{2}\right)^5 - 5\left(x_i - \frac{1}{2}\right)^3 + \frac{15}{8}\left(x_i - \frac{1}{2}\right) + \frac{1}{2}, \tag{2.9}$$

with $x_i = (r_i - R_s)/(R_e - R_s)$.

Another, more common choice for a continuous switching function in adaptive methods is,

$$\lambda(\mathbf{r}_i : T) = \begin{cases} 1 & \text{if } r_i < R_s, \\ \frac{(R_e - r_i)^2 (2r_i + R_e - 3R_s)}{(R_e - R_s)^3} & \text{if } R_s \le r_i \le R_e, \\ 0 & \text{if } R_e < r_i. \end{cases} \tag{2.10}$$

The switching function $\lambda(\mathbf{r}_i : T)$ is a function of the distance r_i between the center of the QM core and the position of (a representative atom in) the solvent molecule i. The value of $\lambda(\mathbf{r}_i : T)$ decreases smoothly from 1 to 0 if the molecule diffuses from the center of the QM core across the transition region (see Fig. 2.3).

As illustrated in Fig 2.4, at each time-step, two QM/MM partitions a and b are computed. In partition a, only molecules in the active region are included in the set $S_{QM}^{(a)}$. In partition b, the set $S_{QM}^{(b)}$ also contains the molecules in the transition region.

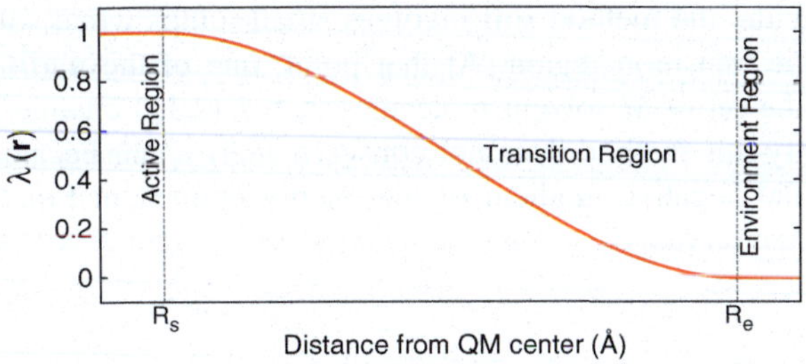

Fig. 2.3 Value of the switching function $\lambda(\mathbf{r})$ from Eq. (2.10) over the transition region

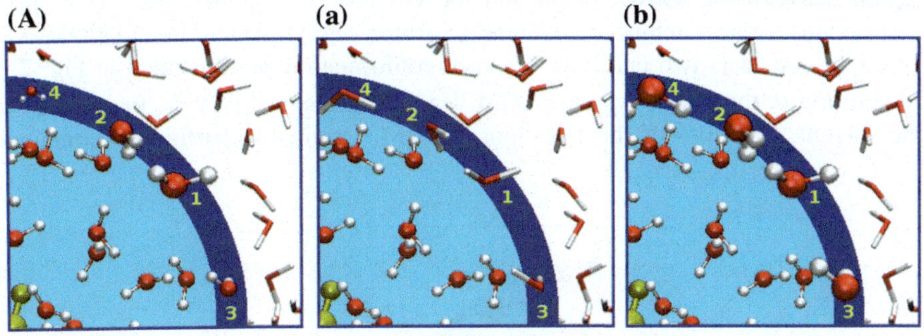

Fig. 2.4 Schematic illustration of ONIOM-XS method. (**A**) ONIOM-XS energy with (partially) QM solvent molecules in *Ball&Stick*, while molecules with *lines* are MM. Size of the *Ball&Stick* QM molecules in the transition zone illustrate their percentage of QM character; (**a**) Small QM/MM partition: $S_{QM}^{(a)}$ contains only solvent molecule in the active zone; (**b**) Large QM/MM partition: $S_{QM}^{(b)}$ contains solvent molecules in the active and transition regions

The energy is then defined as a weighted average of the two computed partition energies,

$$V^{XS} = \sigma^{(a)}(\mathbf{r})V^{(a)}(\mathbf{r}) + \sigma^{(b)}(\mathbf{r})V^{(b)}(\mathbf{r}). \tag{2.11}$$

The functions $\sigma^{(a)}(\mathbf{r})$ and $\sigma^{(b)}(\mathbf{r})$ are weights of the two computed partition energies, and $\sigma^{(a)}(\mathbf{r}) + \sigma^{(b)}(\mathbf{r}) = 1$. If the M molecules in the transition region define a set S_T, the value of $\sigma^{(b)}(\mathbf{r})$ is defined as the average over the QM character ($\lambda(\mathbf{r}_i : T)$ values) of all M molecules in S_T,

$$\sigma^{(b)}(\mathbf{r}) = \frac{1}{M}\sum_{i \in S_T}^{M} \lambda(\mathbf{r}_i : T). \tag{2.12}$$

When only one molecule is in the transition region, the change from QM to MM (and vice versa) is continuous. However, when the transition region contains more than one molecule, the method still involves small jumps when solvent molecules exit or enter the transition region. At that point, one of the partitions (*a* or *b*) is replaced, and the set of *M* solvent molecules in Eq. (2.12) changes. As a result of the discontinuity this method does not conserve energy, but because the forces on all atoms are the negative gradient of the energy defined in Eq. (2.11), the total momentum is conserved.

2.2.2.3 Hot Spot

The Hot Spot method, like ONIOM-XS, is based on the introduction of a transition region between the active region and the environment region. The "Hot Spot" corresponds to the combined active and transition region. Again, like ONIOM-XS, Hot Spot considers two partitions for each conformation, as illustrated in Fig. 2.5. In partition *0*, the whole system is MM, leading to the energy $V^{(0)}$. In partition *a*, the solvent molecules of the Hot Spot are QM. Originally, partition *a* contained

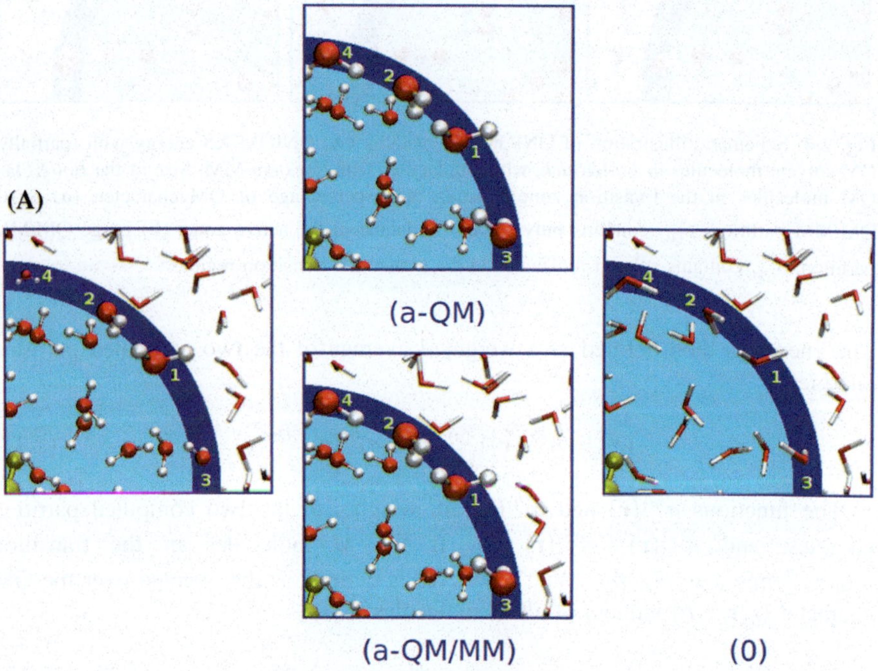

Fig. 2.5 Illustration of Hot Spot method. (**A**) Forces applied to each molecule. *Ball&Stick* molecules are assigned forces from the QM/MM calculations, while molecules with *lines* are assigned MM forces. Sizes of the *Ball&Stick* QM molecules in the transition zone illustrate their percentage of QM character. (**a**) QM or QM/MM simulation to get the force for the molecules in the active and transition zones; (**0**) MM calculation to get the forces for the molecules in the transition zone and in the environment

only these molecules, [16] but recent applications used a QM/MM energy for this partition: molecules in active region and the transition change are QM while the environment is MM [43]. The molecules in the active region feel the forces computed for partition a at the QM or QM/MM level. The molecules belonging to the environment region feel the forces obtained from the MM calculation. And last, the molecules in the transition region feel a weighted average of the two forces. The force exerted on a molecule i in Hot Spot writes:

$$\mathbf{F}_i^{HS}(\mathbf{r}) = -\lambda(\mathbf{r}_i : T)\frac{\partial V^a(\mathbf{r})}{\partial \mathbf{r}_i} - (1 - \lambda(\mathbf{r}_i : T))\frac{\partial V^{(0)}(\mathbf{r})}{\partial \mathbf{r}_i}. \tag{2.13}$$

In this approach, no unique energy can be defined that is consistent with the force on all atoms, and Newton's third law is violated. This happens because the force that a molecule exerts on a neighboring molecule is no longer equal to the negative force the second molecule exerts in return, since it comes from a different calculation. This type of approach has a continuity similar to an ONIOM-XS type simulation, at a reduced cost, since only one QM/MM calculation per conformation is required. Like ONIOM-XS, the forces in this method are still discontinuous, which results in an energy gain (or loss) during the simulation. As a consequence of the violation of Newton's third law, the total force acting on the system is not zero. This results in an additional energy flux into the system, so that the simulations require even stronger thermostats than Abrupt to maintain the desired temperature.

A very large part of the accelerations of the atoms in the Hot Spot simulations comes from the lack of momentum conservation. A way to reduce this effect to the level of ONIOM-XS, without requiring an additional QM/MM computation, is to fit a simple force field to the Hot Spot forces for the entire system on the fly (for each conformation). This results in a unique energy expression $V^{FF}(\mathbf{r})$, which has approximately the Hot Spot forces as its gradient. Then, the forces used for the molecular dynamics propagation of the system are taken as the negative gradient of the fitted potential $V^{FF}(\mathbf{r})$.

A method such as this has been developed under the name Learn On The Fly (LOTF) [31]. As $V^{FF}(\mathbf{r})$ is fitted at each time step, it is a discontinuous method that does not conserve the total energy. It behaves similarly to a Hot Spot simulation, but with the added advantage that it conserves momentum. Indeed, the forces on all atoms come from a unique potential $V^{FF}(\mathbf{r})$ so that the total force is zero.

2.2.2.4 Buffered Force (BF)

BF, like Hot Spot, obtains the forces assigned to different atoms from different calculations, as schematized in Fig. 2.6. As a consequence, no unique potential energy is defined, and Newton's third law is violated. The total momentum of the simulations will therefore not be conserved and the method requires an efficient thermostat to be used. There is, however, a major distinction between Hot Spot and

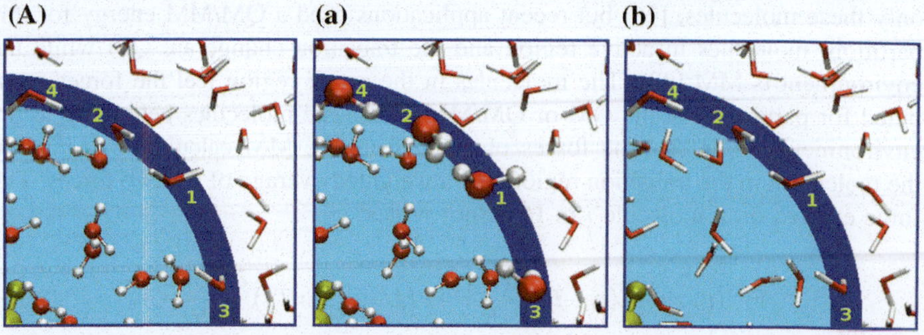

Fig. 2.6 Schematic illustration of BF method. (**A**) Forces applied to each molecule. *Ball&Stick* molecules are assigned forces from the QM/MM calculations, while molecules with *lines* are assigned MM forces; (**a**) Nature of the solvent molecules in the QM/MM simulation; (**b**) Nature of the solvent molecules in the MM calculation

BF, in that they are designed to solve different problems. One general problem with QM/MM, whether conventional or designed for diffusive systems, is that particles at the boundary interact with other particles of both QM and MM character, and may prefer one over the other. This may result in either a density increase, or a density depletion at the boundary, which would not be present in a fully QM simulation. This artifact may be reduced somewhat in the above-mentioned ONIOM-XS, Hot Spot, and LOTF, if the transition region is chosen sufficiently large, but in the BF method this issue is tackled in a more thorough manner.

Like Hot Spot and LOTF, BF computes one QM/MM energy $V^{(b)}(\mathbf{r})$ for a partition b with a large set of QM molecules ($S^{(b)}{}_{QM}$ in ONIOM-XS) and a fully MM energy $V^{(0)}(\mathbf{r})$ for the entire system. In a manner analogue to the Hot Spot method, the QM/MM forces are then assigned to all molecules in a small predefined active region ($S^{(a)}_{QM}$ in ONIOM-XS), while the MM forces are assigned to all the rest.

In BF, there are no molecules that have partial QM and partial MM character, exactly as in the Abrupt approach. In that sense, there is no transition region, but we can define a buffer region as $S^{(b)}{}_{QM}$ - $S^{(a)}{}_{QM}$. The choice of transition region thickness is carefully selected based on the error in the forces on the QM reactive center, and amounts to values below 1 Å, similar to the sizes used in other methods. The considerable advantage over the Abrupt approach is that in BF the QM molecules at the boundary feel forces from a calculation that describes all their neighbors QM, while the MM molecules feel forces from a calculation that describes all their neighbors MM. As a result, unwanted clustering or depletion is avoided.

During a molecular dynamics simulation most molecules exhibit Brownian motion. Therefore, close to the boundary, a wandering molecule could switch between QM and MM character very often. To avoid this effect, the authors of BF introduced hysteresis in the QM $\leftrightarrow$ MM change: Four boundaries are defined in total. A molecule going out of the active region will be assigned MM forces beyond the R_s^{out} radius whereas a molecule coming towards the active region will be

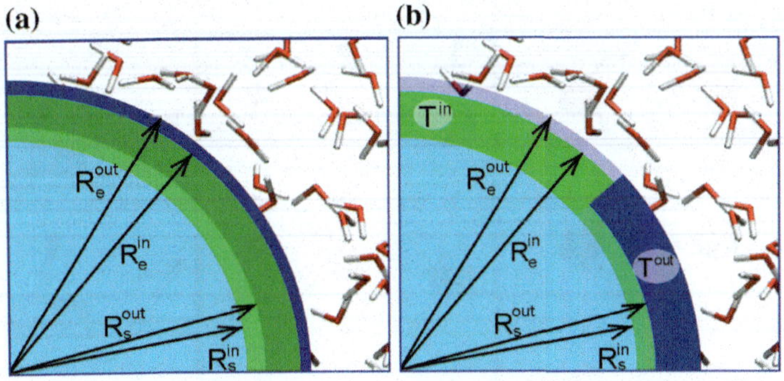

Fig. 2.7 Schematic hysteresis approach in the BF method. (**a**) Definition of the four boundaries; (**b**) Four boundaries seen as two buffer regions. T^{in} is shown as a *green* ring, T^{out} is shown in *blue*. For the sake of clarity, they are only partially shown

assigned QM forces below the R_s^{in} radius, with $R_s^{in} \leq R_s^{out}$. Similarly, R_e^{in} and R_e^{out} are introduced as the outer boundaries of the buffer region. These four boundaries can be grouped into two buffer regions defined by $T^{in} = \{R_s^{in}, R_e^{in}\}$ and $T^{out} = \{R_s^{out}, R_e^{out}\}$. This is schematized on Fig. 2.7.

2.2.3 *Continuous Adaptive QM/MM*

In the previous subsection, we presented several methods that aim to provide some continuity in the QM/MM transition of solvent molecules (ONIOM-XS, Hot Spot, LOTF). In this subsection, we discuss a set of more recent QM/MM methods that ensure a completely continuous transition: (1) Permuted Adaptive Partitioning (PAP), [33] (2) Sorted Adaptive Partitioning (SAP), [33] (3) Difference-based Adaptive Solvation (DAS), [34] and (4) Size-Consistent Multi-Partitioning (SCMP) [35]. Similar to the discontinuous methods from Sect. 2.2.2, a transition region is again introduced, in which the solvent molecules are assigned fractional QM (and MM) character based on their position. This is illustrated in Fig. 2.8a. As opposed to the discontinuous schemes that require only two different QM/MM partitions to be computed each time step, the continuous methods require calculation of a large set of partitions of the solvent molecules. This can be rationalized as follows: to assign a partial QM (and MM) character to a single molecule *i*, one should launch two calculations: one in which *i* is QM and one in which it is MM. This is illustrated in Fig. 2.8b, c for the second solvent molecule in the transition region. Then, an adaptive potential energy expression can be defined that is a weighted average of the energies for the two partitions (equivalent to Eq. (2.11) for ONIOM-XS). As all molecules in the transition region can have different QM characters, an adaptive potential energy expression that assigns the desired fractional QM characters to all

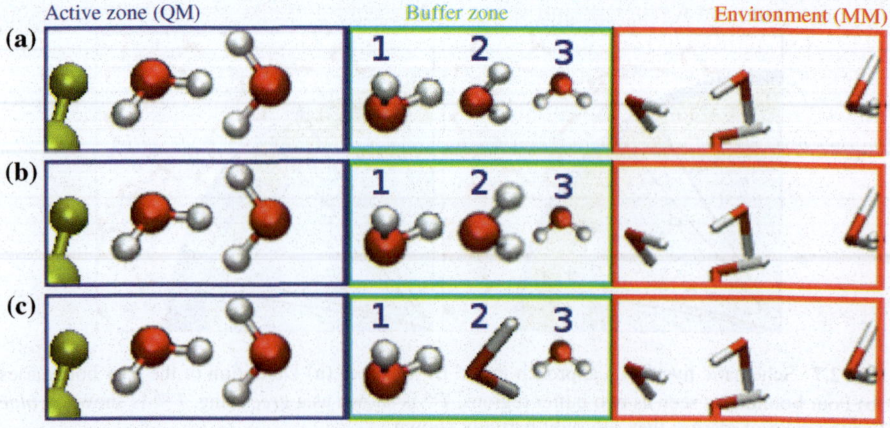

Fig. 2.8 Schematic illustration of adaptive methods. (**a**) Continuous evolution of the nature of solvent molecule from QM (shown as *Ball&Stick*) to MM (shown as *lines*). The percentage of QM nature is schematized by the size of the *Ball&Stick* drawing.(**b**) and (**c**) Two contributing partitions that determine the QM/MM character of water 2: (**b**) Water 2 is QM, (**c**) Water 2 is MM

these molecules needs to include contributions from many QM/MM partitions. All of the methods define this adaptive potential energy $V^{ad}(\mathbf{r})$ as,

$$V^{ad}(\mathbf{r}) = \sum_n \sigma^{(n)}(\mathbf{r}) V^{(n)}(\mathbf{r}). \tag{2.14}$$

For a system with M solvent molecules in the transition region, the maximum number of partitions that can meaningfully contribute to this partial character corresponds to the 2^M possible partitions of the M molecules into two sets. The behavior of the weight functions $\sigma^{(n)}(\mathbf{r})$ is such that when a solvent molecule m_i moves away from the QM center, the weight $\sigma_{(n)}(\mathbf{r})$ of a partition n that describes m_i QM decreases, becoming zero as m_i exits the transition region and enters the environment region.

As mentioned, all continuous methods have the same basic definition for the potential energy (Eq. 2.14). The most important difference between the continuous methods lies in the form of the weight-functions $\sigma^{(n)}(\mathbf{r})$, where the PAP method assigns them in perhaps the most intuitive manner. However, the PAP method also requires computation of a large number of computationally expensive QM/MM partitions for each conformation, since it includes all 2^M possible partitions, with M the number of solvent molecules in the transition region. SAP, DAS and SCMP are considerably more economic alternatives that include only the most important contributions, typically $M+1$.

There is, however, a less defining choice in how to perform a continuous adaptive molecular dynamics simulation. Molecular dynamics works on the premise that the forces (the negative gradient of the energy) are a good estimate for $\Delta V^{ad}(\mathbf{r})/\Delta\mathbf{r}$ over a time-step Δt. Only then can the correct kinetic energy for the next time step be predicted, and will the total energy be conserved. The forces are only a good estimate for $\Delta V^{ad}(\mathbf{r})/\Delta\mathbf{r}$ if the potential energy is continuous, which it is in all continuous methods. In adaptive approaches, however, there are two ways in which the forces on the atoms can be obtained from $V^{ad}(\mathbf{r})$, and the choice determines whether or not the simulation conserves energy.

(i) Energy conserving: The forces on the atoms are the negative gradients of the energy in Eq. (2.14) according to,

$$\mathbf{F}_i^{ad}(\mathbf{r}) = -\sum_n \sigma^{(n)}(\mathbf{r})\frac{\partial V^{(n)}(\mathbf{r})}{\partial \mathbf{r}_i} - \sum_n \frac{\partial \sigma^{(n)}(\mathbf{r})}{\partial \mathbf{r}_i} V^{(n)}(\mathbf{r}). \tag{2.15}$$

(ii) non-Energy conserving: The forces on the atoms are again the negative gradient of the energy in Eq. (2.14), but now the terms that describe the change of $\sigma^{(n)}(\mathbf{r})$ are neglected, resulting in the definition,

$$\mathbf{F}_i^{ad'}(\mathbf{r}) = -\sum_n \sigma_{\mathbf{r}}^{(n)} \frac{\partial V^{(n)}(\mathbf{r})}{\partial \mathbf{r}_i}. \tag{2.16}$$

The forces in Eq. (2.16) provide an inexact prediction for $\Delta V^{ad}(\mathbf{r})$, resulting in simulations that do not conserve energy.

Of the methods discussed below, PAP and SAP historically used the weighted energy from Eq. (2.15), and thus are energy conserving. DAS and SCMP are generally used in a non-energy conserving approach, directly computing the forces from Eq. (2.16).

As stated above, the four methods below differ mainly in the form of the weight functions, and as long as the weight functions have continuous derivatives (PAP, SAP, DAS), they can equally well be used in an energy conserving as in a non-energy conserving manner. While the former has the advantage of being formally more sound, it was demonstrated that it does not produce reasonable structures with either the DAS or the SAP weight functions [34]. As a result, the use of the approach of Eq. (2.16) is recommended for all practical purposes, as was recently confirmed by Lin et al. [18]. While the resulting simulations do not conserve energy, one can make use of the fact that the forces are completely continuous, and integrate the forces *a posteriori* along the trajectory, to obtain a quantity that is conserved. This so-called bookkeeping quantity can be used with Eq. (2.16) and any of the available weight functions, to test simulation parameters such as the size of the transition region or the selected time step [34, 44].

2.2.3.1 Permuted Adaptive Partitioning (PAP)

This method includes all possible partitions of the transition region solvent molecules into two sets (QM and MM) contributing to the total energy in Eq. (2.14). If the number of solvent molecules in the transition region equals M at a certain time-step, then the maximum number of non-zero PAP partitions equals 2^M.

To obtain the weight functions σ of these partitions, the switching function λ is required each time-step for all M transition region molecules according to Eq. (2.3).

Each partition n in Eq. (2.14) consists of a set of QM solvent molecules $S_{QM}^{(n)}$ and a set of MM solvent molecules $S_{MM}^{(n)}$, which together sum to the total number of solvent molecules in the system. We can now define QM and MM fade-out functions for each partition as,

$$O_{QM}^{(n)}(\mathbf{r}) = \prod_{i \in S_{QM}^{(n)}} \lambda(\mathbf{r}_i : T) \quad , \quad O_{MM}^{(n)}(\mathbf{r}) = \prod_{i \in S_{MM}^{(n)}} 1 - \lambda(\mathbf{r}_i : T). \tag{2.17}$$

The fade-out functions earn their name because $O_{QM}^{(n)}(\mathbf{r})$ becomes zero when a QM solvent molecule diffuses into the environment, while the fade-out function of the set of MM molecules becomes zero when a MM molecule penetrates the QM region. The partition weight functions $\sigma^{(n)}(\mathbf{r})$ are defined as products of the fade-out functions,

$$\sigma^{(n)}(\mathbf{r}) = O_{QM}^{(n)}(\mathbf{r}) \cdot O_{MM}^{(n)}(\mathbf{r}). \tag{2.18}$$

The sum of the weights in Eq. (2.18) over all contributing partitions equals one. The number of contributing partitions is often very large (the method scales as $O(2^M)$, see Fig. 2.9), but many of the weights are negligible, and in practice computation time is saved by computing only the partitions that have a non-negligible weight.

2.2.3.2 Sorted Adaptive Partitioning (SAP)

The SAP method drastically reduces the number of contributing partitions by constructing the weight functions $\sigma^{(n)}(\mathbf{r})$ in such a way that only $M + 1$ 'ordered' partitions contribute. The 'ordered' partitions are those in which all QM solvent molecules are closer to the QM core than the MM molecules (Fig. 2.10). This intrinsically means that each partition has a different number of QM solvent molecules. The main criterion for the weight functions $\sigma^{(n)}(\mathbf{r})$ in Eq. (2.14) is that when two solvent molecules are at similar distance from the QM core, then those partitions that describe one particle QM and the other particle MM have weights that smoothly approach zero. The expressions for the weight functions are fairly complex, and have been extensively discussed elsewhere [33, 34].

2.2.3.3 Difference-Based Adaptive Solvation (DAS)

Similar to SAP, in the DAS method only the 'ordered' partitions contribute. The weight functions used in DAS have a relatively simple form,

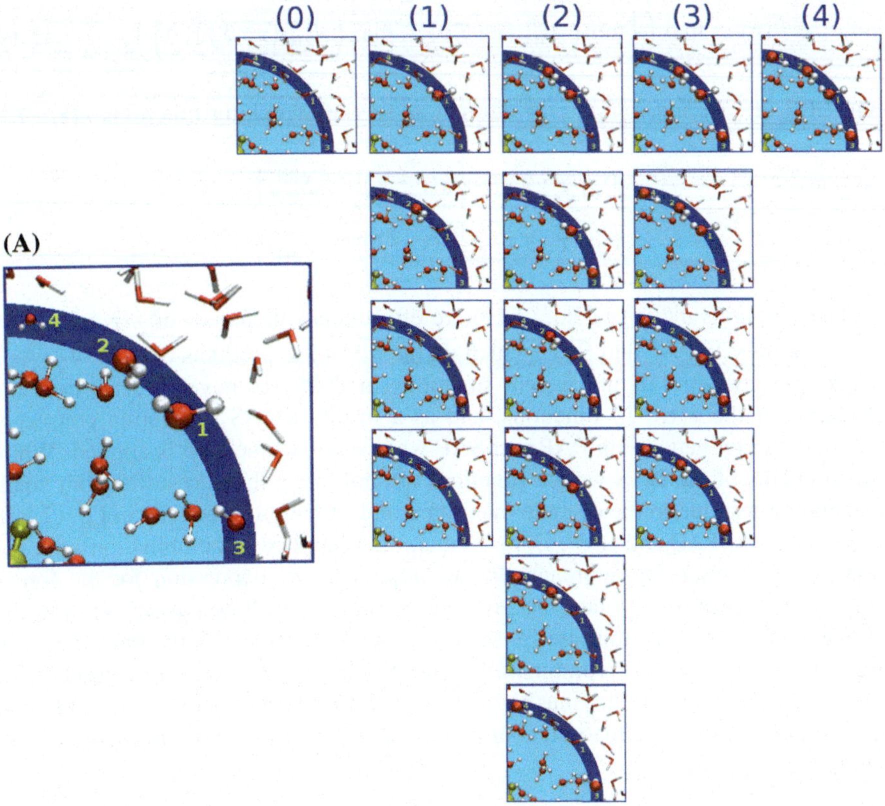

Fig. 2.9 Illustration of the PAP method for a methanol molecule (in the bottom left corner in *yellow*) solvated in water. QM molecules are shown in *Ball&Stick* while MM molecules are shown with *lines*. (**A**) QM characters of solvent molecules in a PAP simulation: In the transition zone, the size of a molecule indicates its proportion of QM character. On the *right*, all the computed partitions are shown, ordered by the number of QM solvent molecule, ranging from 0 to 4 in this example

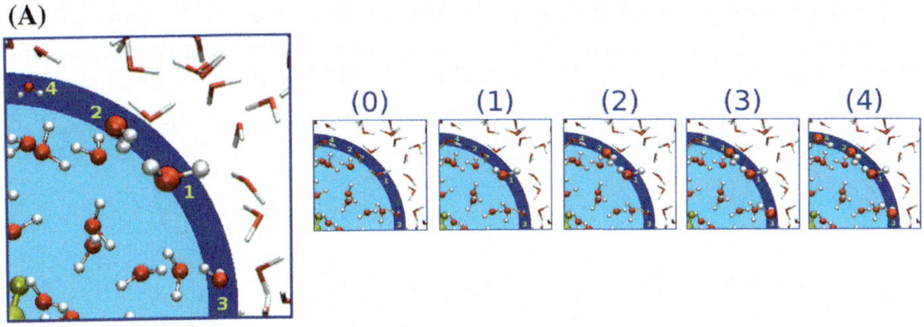

Fig. 2.10 SAP or DAS simulation (**A**) the percentage of QM nature is schematized by the size of the *Ball&Stick* drawing. In SAP, these weights are used to compute the energy, while in DAS they are used to compute the forces. On the *right*, the 'ordered' partitions are shown. As indicated in the text, they all have a different number of QM molecules, ranging from 0 to 4 in this example

$$\sigma^{(n)}(\mathbf{r}) = \max\left(\left\{\min\left(\Lambda_{MM}^{(n)}(\mathbf{r}:T)\right) - \max\left(\Lambda_{QM}^{(n)}(\mathbf{r}:T)\right), 0\right\}\right). \tag{2.19}$$

with $\Lambda_{MM}^{(n)}(\mathbf{r}:T)$ and $\Lambda_{QM}^{(n)}(\mathbf{r}:T)$ defined as sets of switching functions $\lambda(\mathbf{r}_i:T)$,

$$\begin{aligned} \Lambda_{QM}^{(n)}(\mathbf{r}:T) &= \left\{\lambda(\mathbf{r}_i:T) | i \in S_{QM}^{(n)}\right\}, \\ \Lambda_{MM}^{(n)}(\mathbf{r}:T) &= \left\{\lambda(\mathbf{r}_i:T) | i \in S_{MM}^{(n)}\right\}. \end{aligned} \tag{2.20}$$

The functions $\sigma^{(n)}(\mathbf{r})$ in Eq. (2.19) are effectively differences of $\lambda(\mathbf{r}_i:T)$ functions, contrary to PAP and SAP, which define $\sigma^{(n)}(\mathbf{r})$ as products of $\lambda(\mathbf{r}_i:T)$. As a result, the gradients of the weight functions in DAS are never greater than the gradients of the $\lambda(\mathbf{r}_i:T)$ functions, and as a result, a DAS simulation conserves the total energy better than PAP and SAP when a large time-step is applied. While the gradients of the DAS weight functions are not large, they do suffer from small discontinuities, due to the discrete maximum and minimum functions in Eq. (2.19).

Contrary to PAP and SAP, DAS is most often applied in combination with the non-Energy conserving approach. In this approach, the expression for the forces (Eq. (2.16)) does not contain the gradient of the weight functions $\sigma^n(\mathbf{r})$, and the above-mentioned discontinuities pose no problems. However, when the weights in Eq. (2.19) are used in combination with the energy conserving approach of Eq. (2.15), this discontinuity has to be dealt with to achieve energy conservation in the simulation. In that situation, continuous maximum and minimum functions can be used,

$$\begin{aligned} \max(A) &= \frac{1}{k}\ln\left(\sum_{a\in A} e^{ka}\right), \\ \min(A) &= 1 - \left(\frac{1}{k}\ln\left(\sum_{a\in A} e^{k(1-a)}\right)\right). \end{aligned} \tag{2.21}$$

where A is a set of fractional numbers between 0 and 1, and k is a large constant, typically $k = 250$. The function $\max(A)$ is a function that returns the maximum element out of the set A, and $\min(A)$ is a function that returns the minimum. When these continuous functions are applied in Eq. (2.19), the sum of the weights of the contributing partitions slightly deviates from normalization but the error is less than 0.01.

2.2.3.4 Size-Consistent Multi-partitioning (SCMP)

The Size-Consistent Multi Partitioning (SCMP) method is a scheme where the number of contributing partitions N is predefined by the user, and remains the same at every time-step. The partitions themselves are defined such that each partition treats the same number (N_{QM}) of molecules quantum mechanically (QM), as illustrated on Fig. 2.11. One can note here that SCMP includes only partitions that are in the same column in Fig. 2.9, while in SAP and DAS the contributing partitions are in the same row. Since each partition energy takes roughly the same time to calculate, this scheme is able to run efficiently in parallel.

In order to obtain the weight functions of SCMP, again, at each time-step λ switching functions are required for all molecules. The SCMP method defines four different transition regions labeled T_{out}^{QM}, T_{in}^{QM}, T_{out}^{MM}, and T_{in}^{MM}. Each molecule has thus four λ values associated to it. The analogue of the PAP fade-out functions in SCMP is

$$O_{QM}^{(n)}(\mathbf{r}) = \prod_{i \in S_{QM}^{(n)}} \lambda(\mathbf{r}_i : T_{out}^{QM}), \quad O_{MM}^{(n)}(\mathbf{r}) = \prod_{i \in S_{MM}^{(n)}} 1 - \lambda(\mathbf{r}_i : T_{out}^{MM}). \qquad (2.22)$$

In addition, the SCMP method introduces fade-in functions according to,

$$I_{QM}^{(n)}(\mathbf{r}) = 1 - \prod_{i \in S_{QM}^{(n)}} \lambda(\mathbf{r}_i : T_{in}^{QM}), \quad I_{MM}^{(n)}(\mathbf{r}) = 1 - \prod_{i \in S_{MM}^{(n)}} 1 - \lambda(\mathbf{r}_i : T_{in}^{MM}). \qquad (2.23)$$

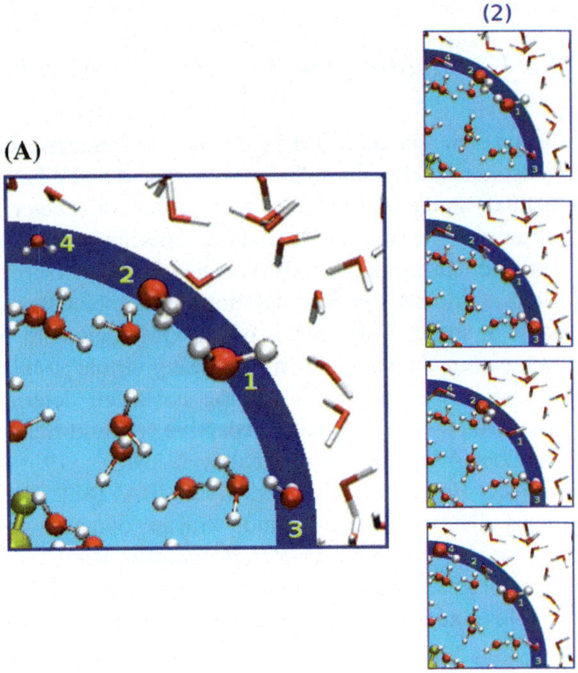

Fig. 2.11 Example of an SCMP simulation in which $N = 4$ partitions are considered with $N_{QM} = 2$ QM water molecules. (**A**) Forces computed for the molecules. The fraction of QM character is schematized by the size of the *Ball&Stick* drawing. On the *right*, the 'ordered' partitions are shown

Here, $I_{QM}^{(n)}(\mathbf{r})$ and $I_{MM}^{(n)}(\mathbf{r})$ are the fade-in functions of partition n for the QM and MM molecules, respectively. Finally, the weight function of partition n is defined as,

$$\upsilon^{(n)}(\mathbf{r}) = \frac{O_{QM}^{(n)}(\mathbf{r}) \cdot O_{MM}^{(n)}(\mathbf{r}) \cdot I_{QM}^{(n)}(\mathbf{r}) \cdot I_{MM}^{(n)}(\mathbf{r})}{\sum_n^N O_{QM}^{(n)}(\mathbf{r}) \cdot O_{MM}^{(n)}(\mathbf{r}) \cdot I_{QM}^{(n)}(\mathbf{r}) \cdot I_{MM}^{(n)}(\mathbf{r})}, \tag{2.24}$$

where the sum of all weight functions is normalized to 1.

At each time-step an updating algorithm is used that ensures that significant partitions (partitions where QM solvent molecules are close to the center) are contributing to the forces in Eq. (2.16). Although less intuitive, the fade-in functions are used because they ensure that for the most-compact partition (the partition where all the QM molecules are closer to the center than the MM molecules) either $I_{QM}^{(n)}(\mathbf{r}) = 0$ or $I_{MM}^{(n)}(\mathbf{r}) = 0$, leading to a zero weight for this partition. As a result, when one replaces the least significant partition with the most-compact one, the energy and the forces are continuous during the exchange. For the precise algorithm of SCMP simulations, see Ref. [35].

Exploiting the fact that the calculations run efficiently in parallel, the SCMP method allows one to consider a fairly large number of partitions (typically on the order of 48 instead of around 10 for DAS or SAP), leading to a very smooth change in QM character of molecules inside of the transition region.

2.2.4 *Choosing the Boundary Position in Adaptive QM/MM*

On top of the standard QM/MM issues, to define the shape and size of the active region is also a challenge in both the constrained and adaptive approaches. Up to now, all methods define a spherical active region with a center and a radius. Most works center the active zone on a particular atom. This approach can straightforwardly be extended to a spherical region around the center of mass of a set of atoms. In a recent contribution, Lin et al. showed that in simulations of biomolecules, it can be desirable to use a fixed point in space as the center of the QM region. In order to retain conservation of momentum, they assigned a very heavy virtual atom as the QM center [45].

The size of the active region has been determined in three different ways: Distance-based (Distance-Adaptive Multi-Scale: DAMS), number-based (Number-Adaptive Multi-Scale: NAMS) [19, 46, 47], and density-based (Density-Based Adaptive QM/MM: DBA) [48]. The size-selection methods can in theory all be used in combination with any of the above adaptive QM/MM approaches.

Distance and Number-Based Adaptive Multi-Scale (DAMS and NAMS) DAMS is the original and most often used approach, where the radius of the active sphere remains fixed during the simulation. Csányi and co-workers define the ideal fixed radius as the distance where the description of the solvent molecules no longer affects the forces on the central atom. The number of solvent molecules in the active region is adjusted on the fly.

Conversely, in NAMS the number of solvent molecules in the active zone is constant. The QM/MM boundary is adjusted to the density of solvent molecules inside the active region, and the radius of the active region is updated on the fly.

Density Based Adaptive QM/MM (DBA)
The DBA approach is an elegant and systematic way to adaptively choose the size of the active region. It has been applied in combination with the Abrupt method. In this approach, the QM/MM boundary is not an empirically selected distance from a QM region, but is based on the overlap of electron densities between a subset of molecules at the core of the active region, with the rest of the solvent molecules. To have a fast method, the molecular electron density of a molecule is pragmatically defined as the sum of the atomic densities. The electron density and the reduced density gradient are used to determine if the interaction between two molecules can be defined as covalent or not. In practice, this amounts to a maximum distance of around 3.9 Å between molecules. This is comparable to the commonly chosen distance of 4 Å in some distance-based schemes. An advantage of DBA over the distance-based scheme is that orientation plays a role in determining the QM character of the water molecule. As a result, the number of QM molecules can be slightly smaller than in DAMS with a comparable cut-off distance, making DBA computationally more favorable. The DBA approach has thus far been applied only in geometry optimizations, but extension to molecular dynamics (even coupling to a continuous adaptive method) would be relatively straightforward.

2.3 Method Comparison

While no general overview paper exists that compares all methods discussed above on the same grounds, several contributions have appeared that compare subsets of methods. In 2007, Heyden et al. compared the performance of Abrupt, Hot Spot, ONIOM-XS, PAP, and SAP for a system of argon in argon with two different classical (MM) potentials [33]. The observables compared were energy conservation, temperature conservation, and radial distribution functions. In 2013, the authors of this chapter compared the performance of Abrupt, SAP, DAS, BF, FIRES, and BF for a system of water in water with four permutations of two QM and two MM potentials [17]. They compared the results obtained with the various QM/MM simulations with a reference calculation, full QM. In 2015, Pezeshki et al. reviewed ONIOM-XS, PAP, SAP, DAS, Hot-Spot and BF [18]. However, they did not perform any new method comparison with respect to Ref. [33]. Below, we will summarize the results of these studies, separated into a section about dynamical performance (energy/temperature conservation, time correlation functions), a section about structural results, and a section about timings. The comparison is summarized in Table 2.2.

Table 2.2 Performance of the various methods. They are classified following the previous section

	$E^{(a)}$	Momentum$^{(b)}$	Structure$^{(c)}$	$N^{(d)}$
FIRES [41]	✗	✓	✗	1
BEST [42]	✗	✓	✓	1
Abrupt [30]	✗	✓	✓	1
ONIOM-XS [30]	✗	✓	✓	2
LOTF [31]	✗	✓	✗	1
Hot-spot [16]	✗	✗	✓	1
BF [32]	✗	✗	✓	1
PAP [33]	✓	✓	✗	$\simeq 2^M$
SAP [33]	✓	✓	✗	$M+1$
DAS [34]	✗	✓	✓	$M+1$
SCMP [35]	✗	✓	✓	User defined

The criteria of comparison are: (a) conservation of the total energy in a simulation in the microcanonical (NVE) ensemble (b) conservation of the total momentum (c) quality of radial distribution functions extracted from an MD simulation in the canonical ensemble (d) scaling in terms of number of QM/MM partitions, M being the number of molecules in the transition region

2.3.1 Dynamical Performance

The comparative study by Heyden et al. [33] revealed that regarding the energy conservation, the quality of the performance of the methods can be ordered PAP = SAP > ONIOM-XS > Abrupt > Hot Spot (with a simulation time-step of 0.1 fs) as shown in Fig. 2.12. This is in agreement with expectation, since the definition of the potential energy in PAP and SAP is designed to be continuous, so that energy conservation is possible—provided a small enough time-step is applied. While PAP and SAP are continuous and employ the energy conserving approach (Eq. 2.15), the ONIOM-XS potential energy (Eq. 2.11) is discontinuous, thereby sacrificing energy conservation. On the other hand, the energy in ONIOM-XS is considerably more continuous than the energy in the Abrupt approach, and indeed the latter performs worse in energy and temperature conservation. The Hot Spot method, besides its discontinuity, breaks Newton's third law, which further reduces the performance in energy conservation.

The second comparative study [17] listed only temperature conservation, ranking the tested methods as FIRES > DAS > Abrupt > BF. The ranking is again in agreement with expectation. The FIRES method applies forces that are exact derivatives of the potential energy, which only has minor discontinuities, and is therefore the best at total energy (and thus temperature) conservation. The three adaptive methods, DAS, Abrupt, and BF, are all combined with the 'non-energy conserving' approach. Still, DAS conserves the energy better since it is a continuous approach while Abrupt and BF have a completely discontinuous (sudden) energy definition. BF (like Hot Spot) furthermore violates the Newton's third law, which is why the temperature conservation is even worse.

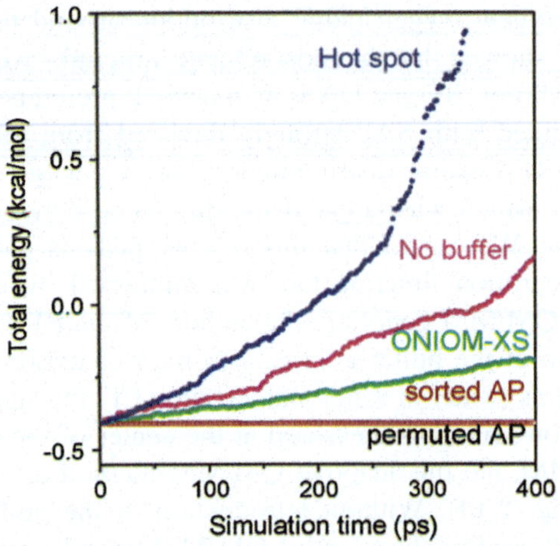

Fig. 2.12 Total energy during an MD simulation in the microcanonical ensemble. The system consists of 171 argon atoms in a cubic periodic box with a length of 20 Å. One atom is chosen to be the active center, and the radius of the active zone is 5 Å. Trajectories are computed using the hot spot (*blue*), ONIOM-XS (*green*), PAP (*red*, permuted AP), and SAP (*black*, sorted AP) methods with a transition region 0.5 Å thick. The 'No buffer' *pink line* corresponds to an Abrupt simulation within the nomenclature of this chapter. Total energy data from the simulation using the PAP method are directly underneath the data from the SAP simulation. Reproduced from Ref. [33] with permission

Reference [17] also computed the residence time of the water molecules in the first coordination shell of the central QM water. This dynamical quantity will clearly be affected by any errors in the equations of motion, and this can be seen from the ranking of the adaptive methods, which is the same as specified above for temperature conservation (DAS > Abrupt > BF). Being a constrained (non-diffusive) method, FIRES is not expected to properly reproduce a quantity that is determined by diffusion. Surprisingly, FIRES performs exceptionally well when the first solvation shell is buried deep enough in the active (QM) region, that is when the transition zone starts beyond the second solvation shell.

To sum up, only few adaptive methods conserve the energy. However, there is a very strong ranking in the quality of the dynamical performance: constrained > adaptive continuous > adaptive non-continuous.

2.3.2 *Recovery of Structures*

The results for structure recovery reveal a ranking that is almost opposite to the ranking found for dynamical performance. In Ref. [33], the radial distribution of the argon atoms with respect to the central argon atom were all very similar, and presumably of good quality, especially when a thermostat was applied in the simulations. The oxygen-oxygen radial distribution around the central QM water oxygen in Ref. [17] showed that the most reliable structures were obtained from the simulations using Abrupt, BF and DAS, with a small preference for BF (Fig. 2.13). The structures obtained with SAP strongly deviated from the reference full QM simulation. The FIRES radial distributions showed a large density peak at the QM/MM boundary, which was larger if the difference between the QM and MM potential was larger. We expect that this density peak is partially related to the problem with the partition function that was addressed by the BEST approach. Using electrostatic QM/MM coupling, the authors of the BF method demonstrated that the introduction of the buffer is crucial to recover reasonable structural properties. On two systems (water in water and an anion Cl^- in water), they compare the BF radial distribution functions measured at the center of the QM region with the full QM, the full MM and the adaptive QM/MM method referred to as Abrupt in this chapter (see Fig. 2.14). Without introduction of the buffer, the radial distribution is strongly affected by the adaptive QM/MM boundary with the apparition of a peak at the boundary. With the introduction of the buffer, the QM radial distribution is recovered in the active zone and the spurious peak disappears.

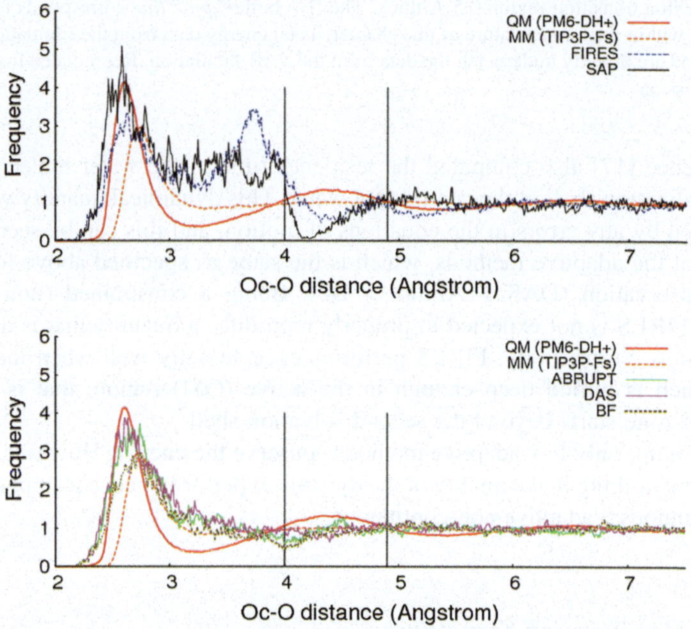

Fig. 2.13 Oc-O radial distribution function (RDF) (Oc is central oxygen) for water from QM/MM simulations (PM6-DH +/TIP3P-Fs) with an active region up to 4.0 Å (in FIRES simulation the active region is ≈4 Å). *Top* (FIRES, SAP) and *bottom* (Abrupt, BF and DAS). The *vertical lines* represent the inner and outer borders of the transition region. Adapted from Ref. [17] with permission

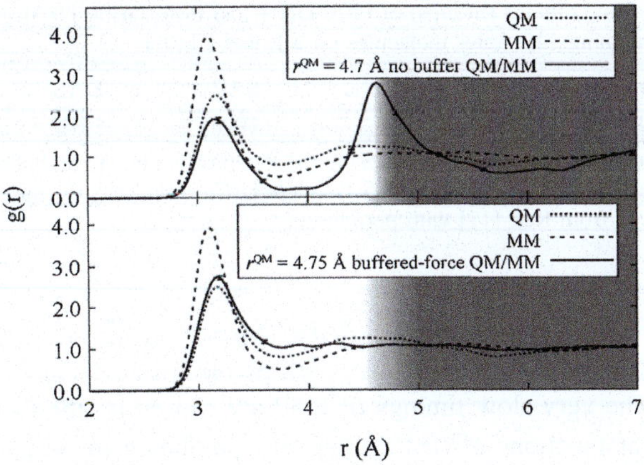

Fig. 2.14 Cl-O radial distribution function $g(r)$ for a chlorine anion solvated in water in a QM/MM (BLYP/fTIP3P) simulation. The Cl^- anion is the active center. Each panel shows reference full QM (*dotted*) and MM (*dashed*) results, as compared with an adaptive QM/MM simulation (*solid*). The top panel shows results of a Abrupt simulation, while the *bottom panel* shows the results of a BF simulation. The *unshaded* parts of the plots represent the QM regions, with the gradient in shading corresponding to the hysteresis range, a refinement used by the authors to reduce the fluctuations in the set of molecules that constitute the active and transition region. Reproduced from Ref. [32] with permission

2.3.3 *Timings*

In general, one can roughly rank the timings of the approaches discussed in Sect. 2.2 by the number of partitions that are computed each time-step. This would results in a speed ranking of constrained > discontinuous > continuous. The continuous methods compute many partitions each time-step, where PAP computes by far the most, and will therefore be the slowest among all. On the other hand, the viewpoint has been taken [19] that these partition-based calculations are trivially parallelizable, and as long as a large number of computing cores are available, more partitions do not lead to more wall time. At that point, the deciding issue becomes the size of the active and transition regions, the required time-step for the simulation, and for a minor part the overhead of the adaptive algorithm.

Most QM/MM methods are implemented in different program packages, which makes it difficult to compare their timings on the same ground. In Ref. [17] the timings for a subset of QM/MM methods were presented, all using FlexMD, [44] and are reproduced in Table 2.3. FlexMD is a QM/MM python library developed with the specific purpose of multi-scale simulations on aqueous systems in mind, and it is available within the ADF software [49]. The FlexMD simulations, using a relatively fast semi-empirical QM description (PM6-DH+), yield a speed ranking of Abrupt > FIRES > BF > DAS > SAP. The discontinuous adaptive methods compete with the constrained methods, while, regardless of parallelization of the partition calculations, the continuous adaptive methods are the slowest. In the

Table 2.3 Real Time timings (in h/ps) of PM6-DH + and PM6-DH +/TIP3P-Fs simulations of a periodic box containing 110 water molecules on a Linux Cluster

QM	FIRES	ABRUPT	SAP	DAS	BF
43.6	1.20	1.18	10.4	1.63	1.35

All simulations are run in serial, except DAS and SAP, which are un in parallel on 12 cores. For all simulations except SAP, a time-step of 0.5 fs is used. The time-step for the SAP simulation is 0.1 fs. Reproduced from Ref. [17] with permission

BF and DAS simulations, the size of the largest set $S_{QM}^{(n)}$ in the corresponding partition (n) is the same, and the slower performance of DAS can only be attributed to overhead. The very slow timings of SAP are caused by the fact that the steep partition weight functions $\sigma^{(n)}(\mathbf{r})$ require the simulation to be performed with a time-step of 0.1 fs, smaller than the 0.5 fs used for other methods in this example.

2.4 Proton Transfer at the Boundary of Adaptive QM/MM

The previous sections presented the state of art in adaptive QM/MM methods. Since their debut in 1996 with the Hot Spot method, they have largely reduced the demand on computational resources, while improving structural properties and energy conservation. These methods have already been successfully applied to simulate the solvation structure of ions in liquid water [41, 43, 50, 51] and even aqueous chemical reactions [19, 23]. The validation procedure is usually based on comparison with experiments [50] or with the corresponding fully QM simulation [19, 23, 41].

However, as stated in the introduction, the bulk of the aqueous reactions of interest involve proton transfer in some form. Even with advanced adaptive QM/MM methods it is still not trivial to perform simulations on proton transfer processes, due to their delocalized nature. Therefore, considering an example system in which proton transfer and diffusion occur can serve as a tough "crash test" for adaptive QM/MM, and should demonstrate the limitations of the methods. We specifically focus on diffusion and Grothuss shuttling, which are particularly non-local, and will affect thermodynamic reaction quantities. The impact of the introduction of an adaptive QM/MM boundary will be quantified through comparison with fully QM simulations on the same system. To underline the limits of the methodology, the behavior of the proton transfer events and their deviation from expectation will be thoroughly discussed.

The generally accepted notion is that the hydronium ion, once formed, exists in two preferred conformations: the dimer or Zundel ion ($O_2H_5^+$), or the Eigen cation ($O_4H_9^+$) [52]. In the dimer, a single step proton transfer between the contributing oxygen atoms is nearly barrierless, and occurs on a very short time-scale [53, 54]. This behavior is generally separated from the so-called forward hopping motion, where the proton moves to a new donor oxygen atom without hopping back on a short time-scale. This forward hopping motion is attributed to a rearrangement of the solvent shell around the hydronium ion (re-solvation) [55]. The hydroxide ion also requires resolvation to precede a forward hopping step. Molecular dynamics simulations indicate that the hydroxide ion in its stable state is involved in four hydrogen bonds to proton donors (hypercoordination) [8, 56] in agreement with experiments [57]. Only when one of the donated hydrogen bonds breaks while at the same time an acceptor hydrogen bond is in place [5], a proton transfer event can result in a stable new hydroxide ion.

The proton transfer reaction selected for study is the diffusion of a hydroxide ion in a solution of methanol in water ($pH \gg 7$). The model contains a methanol molecule dissolved in a cubic periodic water box of 15 Å side containing 112 water molecules and one (Na^+, OH^-) dissociated ion pair. This gives a total density of 1.03 kg/L and an approximate pH around 13.3.

Based on its satisfactory performance, as discussed in Sect. 2.3, the DAS method was selected for the adaptive QM/MM simulations. The center of the active region is defined by the position of the methanol oxygen atom O_{Me}, and the active region has a radius of 6.3 Å around it. The transition zone extends from 6.3 to 7.2 Å and ensures a smooth QM/MM transition for the water molecules. This setup is depicted in Fig. 2.15.

The QM and MM descriptions were selected based on two requirements: (1) The MM force field should allow proton shuttling in the environment region, and

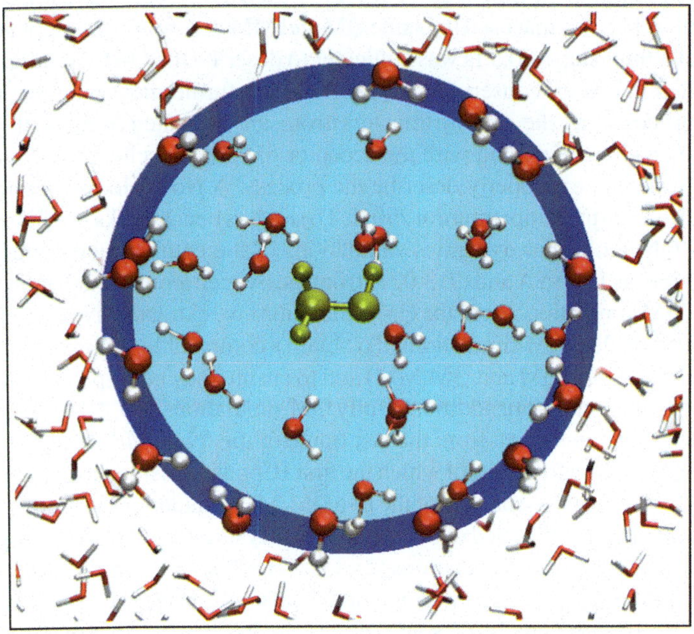

Fig. 2.15 Methanol solvated by an alkaline solution. Methanol is represented in *yellow*. The *light blue* disk (a sphere in 3D) corresponds to the active (QM) zone, the *blue* ring to the transition zone

(2) The difference between the QM and the MM description of water should not be too large, to minimize the artifacts at the QM/MM boundary. The reactive force field ReaxFF [58] is used as the MM level (as implemented in the ADF package), [49] with parameters specifically developed for proton transfer in water. In this way the OH^- ion is in principle allowed to diffuse in and out of the active region, while still retaining its reactive qualities. All QM calculations are done at the semi-empirical PM6-DH+ level of theory, [59, 60] as implemented in the MOPAC2012 program [61]. Earlier calculations [17] showed that the PM6-DH+ and ReaxFF descriptions of water are very similar, thereby minimizing the chance of artifacts due to differences between the potentials. PM6 offers a real improvement to AM1 or PM3 for the description of liquid water [59, 62]. This comes from the fact that the core-core interaction has been modified, analogously to the strategy proposed by the authors of PM3-MAIS or PM3-PIF [63, 64]. While able to describe proton transfer, the PM6-DH+ potential is not optimized for proton transfer processes [65], and is known to overestimate the methanol-water interaction [62]. As a result, our setup is perfectly capable of demonstrating the pitfalls of adaptive QM/MM, but it is not expected to provide meaningful physical insights into the selected process. The interaction between the QM and MM molecules is described with mechanical embedding, which means that the electron density is not polarized by the MM charges. However, the use of the charge equilibration scheme Qeq [66] in ReaxFF, ensures polarization of the QM region by the MM environment, even if this effect is purely classical.

In all simulations, a time-step of 0.5 fs is used. We base this choice on earlier contributions [17, 44] that used the same QM and MM descriptions as applied here. We found that a time-step of 0.5 fs was sufficient to describe the O-H vibrations in water, and resulted in energy conserving simulations. We furthermore note that the proton transfer process—as the proton transfers from an elongated bond with the donor oxygen to an elongated bond with the acceptor oxygen—occurs in 5–20 time-steps, which is sufficient to properly describe the process. A Nose-Hoover thermostat was employed to keep the temperature at 290 K. To serve as benchmarks, we first performed two fully QM molecular dynamics simulations of the entire system, with the initial OH^- position at (1) 8.85 Å and (2) 3.02 Å from the oxygen atom in methanol (O_{ME}), and one fully MM simulation. From the QM simulations we extracted five frames that were used to run five QM/MM simulations. To obtain a comparable amount of data for both types of simulations (QM and QM/MM) and to ensure they both spanned overlapping regions of phase space, four additional fully QM simulations ((3), (4), (5) and (6)) were started from positions taken from the last frame of the five QM/MM simulations. All simulations were run for 20 ps, of which the first 10 ps are considered equilibration, and excluded from analysis. Analysis of the five QM/MM simulations was performed over the entire 50 ps of equilibrated data available. Analysis of the QM data was performed over the trajectories from simulations (2), (3), (4), (5) and (6), also yielding a total of 50 ps of equilibrated data. The average time consumptions per time step for QM and QM/MM simulations are 24.82 s and 14.14 s, respectively. Note that the QM/MM simulations with DAS presented here are much slower than the ones in Sect. 2.3.3. This is simply because a bigger QM zone was used. Three different types of quantities were computed: (i) the probability to find the OH^- at a certain distance from the QM core, (ii) the radial distribution of water oxygen atoms around the hydroxide ion, and (iii) the rate of proton transfer, from the time-correlation of the OH^- oxygen identity. We found that the structural data was easily converged within the first 10 ps, but the relatively low mobility of the OH^- ion encourages some caution in interpreting the data on its distribution over the simulation space. To estimate the convergence of the distribution of the distance from the OH^- to the methanol, we compared the probabilities to find the OH^- ion somewhere in the solution over two different time intervals: Between 10 and 20 ps and between 10 and 19 ps. We did this for all the QM and QM/MM simulations. The sum of the absolute values of the difference was found to be approximately 1 % of the maximum probability, which was deemed so small as to have a negligible influence on the results.

2.4.1 *Influence of QM/MM Boundary on OH^- Diffusion*

We started the QM/MM simulations with the hydroxide far away from O_{Me}, at approximately 10 Å, which is well inside the environment region. The hydroxide ion is therefore initially described MM. At the start of all five QM/MM simulations, we observed fast consecutive proton transfer occurrences at the start of the simulations, involving the hydroxide ion and neighboring water molecules. An initial fast and frequent shuttling of the proton back and forth between two oxygen atoms is then suddenly followed by a large displacement of the OH^- ion, always in the direction of the central methanol molecule. Fig. 2.16c displays the distance between the OH^- ion and the methanol O_{Me} atom, which suddenly decreases within the first 2.5 ps of the simulation. The OH^- ion then finds itself inside the active region boundary, where it stays for the remainder of the simulation.

It is worth noting that the sodium cation Na^+ follows a Brownian diffusion and thus moves much more slowly. In all simulations, it stayed close to its initial location at ∼9 Å of the methanol molecule.

Of the six fully QM reference simulations, only one has a starting geometry with the OH^- ion located far away from the central methanol molecule. The evolution of the distance of the OH^- ion from the methanol molecule is depicted in Fig. 2.16a, and it reflects a very different behavior. The simulation starts with the OH^- ion located at 8.9 Å distance from the O_{Me} atom. Unlike in the QM/MM simulations, the OH^- ion did not approach the methanol molecule, but remains within a range of approximately 8 to 10 Å. This region corresponds to approximately 30 % of the total volume of the simulation box.

The deviating behavior of the OH^- ion in the QM/MM simulations has two separate aspects: (i) the fast migration of the OH^- ion towards the boundary of the QM/MM transition zone, and then (ii) the sudden jump into the QM zone of the OH^- ion. Since the mechanical embedding scheme is applied in the QM/MM simulations, it could be that a long range MM interaction causes the quick diffusion of the OH^- ion towards the central methanol molecule. However, the results of a

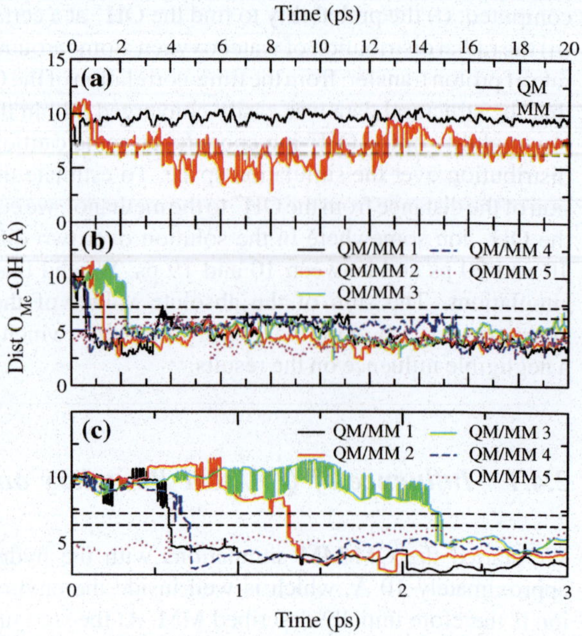

Fig. 2.16 Distance from the oxygen in methanol to the OH^- ion in (**a**) QM and MM simulations, and (**b**) QM/MM simulations. The first 3 ps of (**b**) is shown in more detail in (**c**)

fully MM simulation show that the MM potential does not steer the OH^- ion towards the MeOH molecule (Fig. 2.16a). The OH^- ion does move more freely through the solution than it does in the fully QM simulation, and like in the QM/MM simulation we observe many proton shuttling events (Figs. 2.16a). This is the direct result of a much higher proton transfer frequency in MM simulation. Including the proton shuttling events, we find proton transfer rates of 0.48^{-1} and 43 ps^{-1} in the QM simulation and MM simulations respectively. In our QM/MM simulations, the OH^- is initially positioned at the corner of the box, which is around 10 Å distance from MeOH. Due to the limited size of the box, the highly mobile MM OH^-can quickly migrate to the boundary of the transition region at 7.2 Å, We observed a few occasions in which the ion re-entered the transition region, but these were quickly followed by a proton shuttling event launching it back, and overall the OH^- ion stays trapped inside the active region. The apparent preference of the OH^- ion for the active region creates an unphysical pH gradient in the QM/MM solution, and increases the probability of a reaction with the solute present at the core of the active region.

We note that an artifact such as the one described above strongly depends on the choice of QM and MM potentials. The simplest solution to the above problems would be to develop force fields that are very similar to the QM description of proton transfer in water. However, we suspect that there will be a limit to the accuracy with which a simple force field can describe such a complex process. As an alternative solution, we suggest the introduction of a second active region, centered on the diffusive OH^- ion itself. In this manner, the OH^- ion is described at the QM level everywhere in the simulation box, and an un-physical preference for one region over another is avoided. Developing such a method is not be without its difficulties since the hydroxide anion does not always correspond to the same oxygen and proton.

2.4.2 *Reliability of QM/MM Equilibrated Structures*

We now focus on average structural differences between a QM and a QM/MM equilibrated system. We select three relevant structural indicators, namely the distance of the OH^- ion from the QM core, the radial distribution of water around water in the active region, and the radial distribution of water around the OH^- ions when it is located inside the active region.

The average position of the OH^- ion in QM/MM simulations is closer to the QM core than it is in the fully QM reference simulations, as shown in Fig. 2.17. In the QM reference simulations, the OH^- ion spends some time in the area that corresponds to the transition region in the QM/MM simulations. The difference between the two results is related to our earlier observation that in the QM/MM simulation the OH^- ion is quickly shuttled back into the active region whenever it enters the transition region.

The radial distribution of the oxygen atoms of the water molecules in the active region around one another in the active region is depicted in Fig. 2.18a. Overall, the water structure obtained in the QM/MM simulations agrees well with that of the QM reference. The first solvation peak in the QM/MM simulations is almost on top of that in the QM simulations, and the second solvation peak is a fraction further away. More significant is the increased density between the first and second solvation shell in the QM/MM simulations. This corresponds to a more compressed aqueous environment in the active region. Integration of the first solvation peak in the radial distribution function $g(r)$ yields an average water coordination number of 4.23 for the QM/MM water molecules in the active region, versus a coordination number of 3.43 in the same region in the QM simulation. This amounts to an

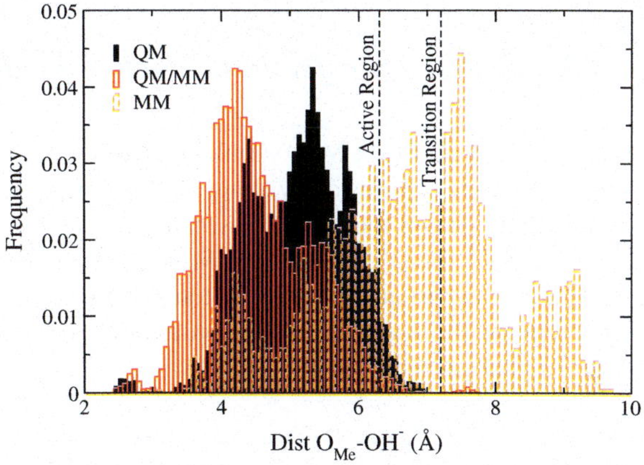

Fig. 2.17 Average histogram of distance from O_{Me} to OH^- after 10 ps of equilibration for both QM (*black solid*) and QM/MM (*red empty*) simulations, the histogram of the only MM simulation is also shown in *dashed orange*

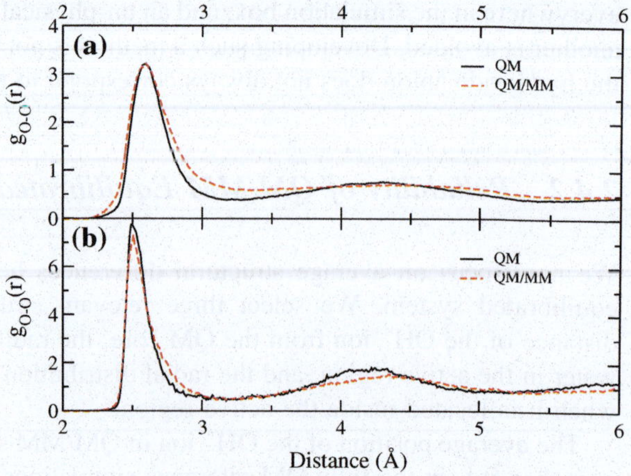

Fig. 2.18 Average oxygen-oxygen radial distribution around (**a**) all oxygen atoms in the range of 6.3 Å from O_{Me} in the QM and in the QM/MM simulations, and (**b**) all oxygen atoms around the OH^- ion in QM and QM/MM simulations

average increase in coordination with 0.80 water molecules. The integration of the radial distribution function over the entire QM region yields 43.21 water molecules for the QM/MM simulation, a 10.2 % increase over 39.22 water molecules in the same region in QM simulations. We attribute the higher QM/MM water density to the differences between the QM and the MM chemical potentials of water. The optimized geometries of a water dimer have optimized oxygen-oxygen bond lengths r_{O-O} of 2.79 Å with the QM potential, and 2.89 Å with the MM potential. This suggests that indeed the MM water prefers a lower density than the QM water, and therefore exerts a pressure on the QM region. The radial distribution of water oxygen atoms around the OH^- oxygen atom in the active region (Fig. 2.18b does not appear to be significantly affected by the increased density in the active region. Integration of the $g(r)$ for the distribution of oxygen atoms around the hydroxide ion reveals only a small increase in coordination, from 5.13 in the QM simulation to 5.46 in the QM/MM simulations. The overall altered environment of the OH^- ion may still affect its reactive behavior.

2.4.3 *Proton Transfer in the Active Region*

Both our QM and QM/MM simulations show that hydroxide migration occurs in step-wise hops followed by long resting periods. The average hopping rates (counting all proton transfer events, not only the forward-hopping motions) are 1.67 and 0.48 ps^{-1} for QM/MM and QM simulations respectively.

Another way to quantify the proton hopping rate is to compute the decorrelation with time of the location of the hydroxide ion at a certain oxygen atom. Let us define a characteristic function $h_i(t)$ which is 1 if a certain oxygen atom O_i can be identified as the hydroxide ion at time t, and zero when it is not. We can then evaluate the auto-correlation function of $h_i(t)$ as,

$$c(t) = \frac{1}{t_{max}} \int_{t'=0}^{t_{max}} \sum_{i \in O} h_i(t') h_i(t' + t) dt', \tag{2.25}$$

where the time-intervals $\{t', t' + t\}$ are extracted from all the QM (or QM/MM) trajectories after equilibration.

Decorrelation is not complete on the time-scales of our simulations ($c(t)$ in Fig. 2.19 does not decay to zero). Again, we observe a considerably faster decay (higher hopping-rate) in the QM/MM simulation than in the fully QM simulation. In both cases, the OH^- ion remains mainly in the active region.

The increased water density in the active region of the QM/MM simulation (discussed in Sect. 2.4.2) may explain the increased proton transfer rate observed in the active region of the QM/MM simulation. The generally accepted picture of proton transfer to an OH^- ion (shown in Fig. 2.20a) states that the OH^- ion is generally involved in hydrogen bonds to five neighboring water molecules, four of which are prospective proton donors. A water molecule, by contrast, has on average four hydrogen bonds. A forward hopping proton transfer event is preceded by a fluctuation of the solvent structure around the hydroxide ion, known as re-solvation. This involves cleavage of one of the proton donating hydrogen bonds, after which the hydroxide ion finds itself under more favorable conditions to become a water molecule. If a low-barrier proton transfer then occurs, the product state is accommodated by the solvation shell (presumably the solvation structure of the formed hydroxide ion simultaneously adjusts towards a coordination number of five), and the new conformation is stable. In the QM/MM simulations, the coordination number of the water molecules is significantly higher than in the QM reference simulations, shown in Fig. 2.20b. As a result, the water molecules adjacent to the hydroxide ions on average have more coordinated water molecules, and will therefore more often have a solvation structure that is akin to that of a

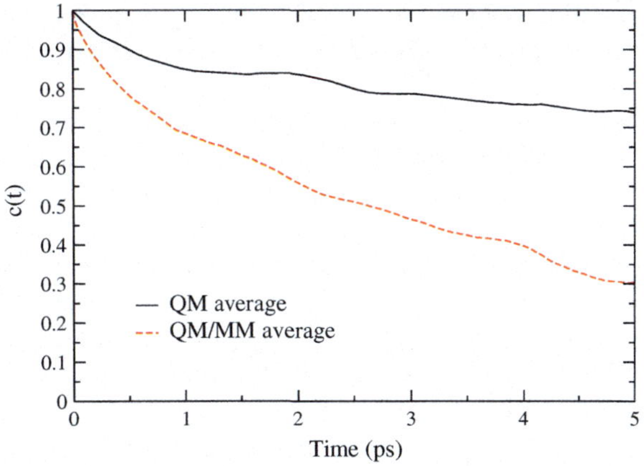

Fig. 2.19 Average auto-correlation function of OH^- for QM and QM/MM simulations

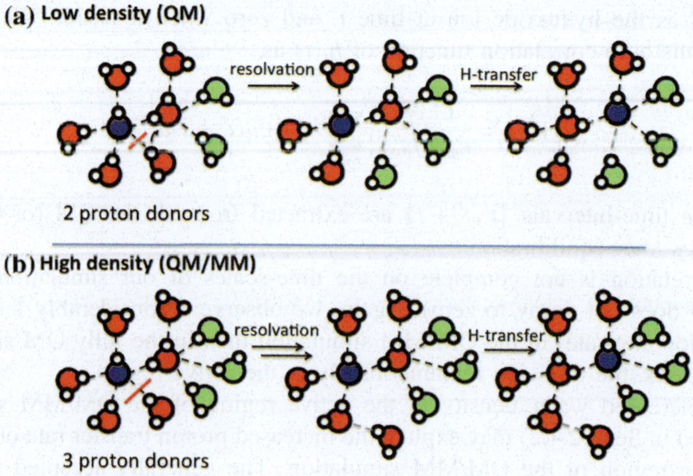

Fig. 2.20 Schematic representation of re-solvation accompanying a proton transfer event to a hydroxide ion (*blue*) at regular and high density. At high density many of the proton donors have sufficient coordinated water molecules (*green*) to stabilize the product conformation

hydroxide ion (5 hydrogen bonds). This can be viewed as favorable re-solvation to accommodate the product of a proton transfer event, thereby increasing the mobility of the hydroxide ion.

2.4.4 What Did We Learn?

To illustrate the capabilities and limitations of adaptive QM/MM methods, we focused on a highly challenging problem: the diffusion of a hydroxide ion from the bulk to the reactive center. We evaluated the performance of Difference-based Adaptive Solvation (DAS) as a representative adaptive QM/MM method in the description of proton diffusion, using an aqueous solution of NaOH and methanol as our model system. Even for our very small model system, the QM/MM simulation is faster than fully QM by approximately a factor of two. We find that performing a QM/MM simulation of a non-local process like proton transfer results in an amplification of the known artifacts induced by the QM/MM boundary.

The differences between the QM and the MM potential can result in strong artifacts in the diffusivity of a hydroxide ion, which in our simulations remained largely trapped in the active (QM) region. The nature of the artifacts in the diffusivity will strongly depend on the nature of the QM and MM potentials and their comparative description of the diffusion of a proton. As a possible solution we suggest the introduction of an extra active region that adapts its position to the location of the reactive OH^- ion.

Inside the active region of a QM/MM system, we found a significantly higher proton transfer rate with respect to the fully QM reference. We attribute this second discrepancy to a general problem—also found in conventional QM/MM—that the interaction between QM and MM molecules introduces artifacts. In our simulations, the water density inside the QM region was somewhat higher than the water density in the fully QM simulations (both were performed with fixed total volume in the NVT ensemble). We expect that this density difference stems from the chemical potential difference between QM and MM molecules, as a result of which the two regions exert a pressure on one another [67]. The resulting smaller volume of the QM region does not significantly affect the solvation shell of the OH^- ion, but the water molecules have a denser solvation structure. Because an OH^- ion generally prefers a higher coordination number than a water molecule, the water molecules that can serve as proton donors to a neighboring OH^- ion are better equipped for their future role as OH^- ion, resulting in the observed higher proton transfer rate.

Overall, we believe that the presented methods for a multi-scale description of reactions in water represent the future of simulations of aqueous reactions. However, our results show that for the successful description of proton transfer events, several challenges still lie in our path.

2.5 Conclusion

We have presented in some detail the state of the art in available methods for the QM/MM simulations of chemical reactions in solution. We distinguish three classes of methods, each with a different approach to the diffusion of solvent molecules across the QM/MM boundary. The first class, constrained QM/MM, prevents this diffusion, and as a consequence dynamical properties are correctly described only well within the QM region. In addition, recovery of the correct structural properties at the QM/MM boundary still poses a challenge. Nonetheless, the constrained methods have the non-negligible advantage of simplicity and low computation cost, equivalent to that of a conventional QM/MM simulation.

The second class, discontinuous adaptive QM/MM, adapts the description of a solvent molecule (QM versus MM) as it diffuses across the QM/MM boundary, and does this in a discontinuous manner. The straight-forward Abrupt method 'abruptly' modifies the description of a solvent molecule as it diffuses from the QM active region into the MM environment. This results in strong discontinuities in potential energy and forces, and in some cases involves structural artifacts at the QM/MM boundary. To reduce these effects, several strategies have been developed over the past 20 years. The basic feature is the transition region between the active and environment regions, and the partitioning of the system into QM and MM molecules in two different manners. If a weighted average of the resulting energies or forces is then applied, the molecules in the transition region can be assigned fractional QM (and MM) character. These methods have led to the first successful applications of the adaptive QM/MM methods for solution chemistry. While the dual partition approach can reduce discontinuities, it cannot remove them, and as a consequence the total energy is not conserved, and structural properties can be recovered only using a strong thermostat to compensate the anomalous heating of the system.

The third and more recent class contains the continuous adaptive methods. Like the second class they define a transition region, but they are not limited to the computation of only two different QM/MM partitions for each given configuration. By computing many partitions, these methods achieve a continuous description of the forces exerted on the solvent molecules as they cross the transition region. The accompanying increase in computational load is easily compensated by efficient parallelization of the numerous QM/MM computations that are required at each molecular dynamics time-step. The methods constitute a formal improvement, but they are not exempt from defects. While an energy conserving application is possible, it is generally sacrificed to obtain accurate molecular structures. In addition, the complex definitions of the energies and/or forces can result in fast changes, requiring a small time-step for proper integration.

A comparative performance of the three classes of QM/MM methods leads to the following conclusions. In terms of computation performance, a ranking can be approximately based on the number of required QM/MM calculations, which increases with the complexity of the method. For the constrained QM/MM methods only one calculation suffices, the discontinuous methods require at most two QM/MM calculations, and the continuous methods are the most demanding, with a computational cost at least proportional to the number of solvent molecules in the transition region. When ranking energy conservation, the continuous methods are at the top, closely followed by the constrained methods. Energy conservation suggests a good description of dynamical properties, but the constrained methods fail by construction when dynamical properties involving diffusion near the QM/MM boundary are desired. Finally, in reproducing the QM molecular structures, the discontinuous methods perform at least as well as the continuous ones. Overall, the best compromise between diffusion properties (solvation life-times) and structural properties (radial distribution) is provided by non-energy-conserving continuous adaptive methods such as DAS.

We exemplified the strengths and limitations of the DAS method on a challenging system by describing the diffusion of a hydroxide ion in water, in the presence of a methanol molecule. The active region is centered on the methanol molecule, and the hydroxide ion can freely diffuse in and out this active region. We used the semi-empirical Hamiltonian PM6-DH+ to describe the active region, while in the environment region all molecules are treated with the ReaxFF force field, which can describe a Grotthus shuttling mechanism for hydroxide diffusion. For the sake of simplicity, the mechanical embedding scheme was used. When comparing our DAS-QM/MM with fully QM and fully MM simulations, the main artifacts we observed resulted directly from the QM/MM interactions, not from the adaptive nature of the simulations. The difference between the QM and the MM description of proton transfer slows the diffusion of the hydroxide ion down once it enters the active region, thereby effectively trapping it inside. The hydroxide diffusion in the active region is still faster than in the fully QM simulation, which is a result of anomalous interactions at the QM/MM boundary.

In summary, the large choice of adaptive schemes available today allows QM/MM simulations of solution chemistry with the main artifacts similar to those of conventional QM/MM schemes. We expect that a careful choice of the QM/MM scheme combined with a QM region around each active center (here methanol *and* hydroxide anion) should cure the main sources of error.

References

1. Shelton DP (2000) Chem Phys Lett 325:513. doi:10.1016/S0009-2614(00)00734-X
2. Marcus Y (2009) Chem Rev 109(3):1346. doi:10.1021/cr8003828
3. von Grotthuss CJT(1806) Ann Chim (Paris) 58:54. Available at: http://gallica.bnf.fr/ark:/12148/bpt6k6573748h/f54.image.r=Annales%20de%20chimie%20Paris.langEN
4. Tielrooij KJ, Timmer RLA, Bakker HJ, Bonn M (2009) Phys Rev Lett 102:198303. doi:10.1103/PhysRevLett.102.198303
5. Hassanali A, Giberti F, Cuny J, Kühne TD, Parrinello M (2013) PNAS 110(34):13723. doi:10.1073/pnas.1306642110
6. Roberts ST, Mandal A, Tokmakoff A (2014) J Phys Chem B 118(28):8062. doi:10.1021/jp501145p
7. Giberti F, Hassanali AA, Ceriotti M, Parrinello M (2014) J Phys Chem B 118(46):13226. doi:10.1021/jp507752e
8. Tuckerman ME, Marx D, Parrinello M (2002) Nature 417(6892):925. doi:10.1038/nature00797
9. Warshel A, Levitt M (1976) J Mol Biol 103(2):227. doi:10.1016/0022-2836(76)90311-9
10. Thole BT, van Duijnen PT (1980) Theor Chim Acta 55(4):307. doi:10.1007/BF00549429
11. Field MJ, Bash PA, Karplus M (1990) J Comput Chem 11(6):700. doi:10.1002/jcc.540110605
12. Gao J (1995) Reviews in computational chemistry. In: Lipkowitz K, Boyd DB (eds) Wiley, pp 119–185. doi:10.1002/9780470125847.ch3
13. Sherwood P (2000) Modern methods and algorithms of quantum chemistry: proceedings. In: Grotendorst J (ed) NIC series, vol 3, 2nd edn. NIC-Directors, Jülich, pp 257–277
14. Carloni P, Rothlisberger U, Parrinello M (2002) Acc Chem Res 35(6):455. doi:10.1021/ar010018u
15. Magistrato A, DeGrado WF, Laio A, Rothlisberger U, VandeVondele J, Klein ML (2003) J Phys Chem B 107(17):4182. doi:10.1021/jp027032o
16. Kerdcharoen T, Liedl KR, Rode BM (1996) Chem Phys 211(1–3):313. doi:10.1016/0301-0104(96)00152-8
17. Bulo RE, Michel C, Fleurat-Lessard P, Sautet P (2013) J Chem Theory Comput 9(12):5567. doi:10.1021/ct4005596
18. Pezeshki S, Lin H (2015) Mol Simul 41(1–3):168. doi:10.1080/08927022.2014.911870
19. Park K, Gotz AW, Walker RC, Paesani F (2012) J Chem Theory Comput 8:2868. doi:10.1021/ct300331f
20. Mones L, Jones A, Goetz AW, Laino T, Walker RC, Leimkuhler B, Cányi G, Bernstein N (2015) J Comput Chem 36: 633. doi:10.1002/jcc.23839
21. Pezeshki S, Davis C, Heyden A, Lin H (2014) J Chem Theory Comput 10:4765. doi:10.1021/ct500553x
22. Tunon I, Martins-Costa MTC, Millot C, Ruiz-Lopez MF (1997) J Chem Phys 106(9):3633. doi:10.1063/1.473457

23. Várnai C, Bernstein N, Mones L, Csányi G (2013) J Phys Chem B 117(40):12202. doi:10.1021/jp405974b
24. Nielsen SO, Bulo RE, Moore PB, Ensing B (2010) Phys Chem Chem Phys 12:12401. doi:10.1039/c004111d
25. Potestio R, Fritsch S, Espanol P, Delgado-Buscalioni R, Kremer K, Everaers R, Donadio D (2013) Phys Rev Lett 110:108301/1. doi:10.1103/PhysRevLett.110.108301
26. Praprotnik M, Delle Site L (2013) Methods Mol Biol (NY, U.S.) 924:567. doi:10.1007/978-1-62703-017-5_21
27. Salazar MR (2005) J Phys Chem A 109(50):11515. doi:10.1021/jp053551q
28. Guthrie MG, Daigle AD, Salazar MR (2010) J Chem Theory Comput 6:18. doi:10.1021/ct900449q
29. Pezeshki S, Lin H (2011) J Chem Theory Comput 7:3625. doi:10.1021/ct2005209
30. Kerdcharoen T, Morokuma K (2002) Chem Phys Lett 355(3–4):257. doi:10.1016/S0009-2614(02)00210-5
31. Csányi G, Albaret T, Payne MC, De Vita A (2004) Phys Rev Lett 93:175503. doi:10.1103/PhysRevLett.93.175503
32. Bernstein N, Várnai C, Solt I, Winfield SA, Payne MC, Simon I, Fuxreiter M, Csányi G (2012) Phys Chem Chem Phys 14:646. doi:10.1039/c1cp22600b
33. Heyden A, Lin H, Truhlar DG (2007) J Phys Chem B 111(9):2231. doi:10.1021/jp0673617
34. Bulo RE, Ensing B, Sikkema J, Visscher L (2009) J Chem Theory Comput 5(9):2212. doi:10.1021/ct900148e
35. Watanabe HC, Kubar T, Elstner M (2014) J Chem Theory Comput. doi:10.1021/ct5005593
36. Bakowies D, Thiel W (1996) J Phys Chem 100(25):10580. doi:10.1021/jp9536514
37. Senn H, Thiel W (2007) Atomistic approaches in modern biology. In: Reiher M (ed) Topics in current chemistry, vol 268. Springer, Berlin, pp 173–290. doi:10.1007/128_2006_084
38. Senn HM, Thiel W (2009) Angew Chem Int Ed 48(7):1198. doi:10.1002/anie.200802019
39. Zhang Y, Lin H (2008) J Chem Theory Comput 4(3):414. doi:10.1021/ct700296x
40. Zhang Y, Lin H (2010) Theor Chem Acc 126(5–6):315. doi:10.1007/s00214-009-0704-z
41. Rowley CN, Roux B (2012) J Chem Theory Comput 8(10):3526. doi:10.1021/ct300091w
42. Shiga M, Masia M (2013) J Chem Phys 139(4):044120. doi:10.1063/1.4816629
43. Hofer TS, Pribil AB, Randolf BR, Rode BM (2005) J Am Chem Soc 127(41):14231. doi:10.1021/ja052700f
44. Fleurat-Lessard P, Michel C, Bulo RE (2012) J Chem Phys 137:074111/1. doi:10.1063/1.4739743
45. Pezeshki S, Davis C, Heyden A, Lin H (2014) J Chem Theory Comput 10:4765. doi:10.1021/ct500553x
46. Takenaka N, Kitamura Y, Koyano Y, Nagaoka M (2012) Chem Phys Lett 524:56. doi:10.1016/j.cplett.2011.12.053
47. Takenaka N, Kitamura Y, Koyano Y, Nagaoka M (2012) J Chem Phys 137:024501/1. doi:10.1063/1.4732307
48. Waller MP, Kumbhar S, Yang J (2014) ChemPhysChem 15:3218. doi:10.1002/cphc.201402105
49. Velde G, Bickelhaupt F, Baerends E, Fonseca-Guerra C, Van Gisbergen S, Snijders J, Ziegler T (2001) J Comput Chem 22:931. doi:10.1002/jcc.1056
50. Rode BM, Schwenk CF, Tongraar A (2004) J Mol Liq 110:105. doi:10.1016/j.molliq.2003.09.016
51. Hofer TS, Tran HT, Schwenk CF, Rode BM (2004) J Comput Chem 25(2):211. doi:10.1002/jcc.10374
52. Agmon N (1995) Chem Phys Lett 244:456. doi:10.1016/0009-2614(95)00905-J
53. Komatsuzaki T, Ohmine I (1994) Chem Phys 180(23):239. doi:10.1016/0301-0104(93)E0424-T
54. Newton MD, Ehrenson S (1971) J Am Chem Soc 93(20):4971. doi:10.1021/ja00749a001
55. Markovitch O, Chen H, Izvekov S, Paesani F, Voth GA, Agmon N (2008) J Phys Chem B 112(31):9456. doi:10.1021/jp804018y

56. Marx D, Chandra A, Tuckerman ME (2010) Chem Rev 110(4):2174. doi:10.1021/cr900233f
57. Botti A, Bruni F, Imberti S, Ricci M, Soper A (2005) J Mol Liq 117(1–3):81. doi:10.1016/j.molliq.2004.08.013
58. Rahaman O, van Duin ACT, Goddard WA, Doren DJ (2011) J Phys Chem B 115(2):249. doi:10.1021/jp108642r
59. Stewart J (2007) J Mol Model 13(12):1173. doi:10.1007/s00894-007-0233-4
60. Korth M (2010) J Chem Theory Comput 6(12):3808. doi:10.1021/ct100408b
61. Stewart JJP (2012) Mopac2012. Stewart computational chemistry, Version 7.263 W. http://OpenMOPAC.net
62. Marion A, Monard G, Ruiz-Lopez MF, Ingrosso F (2014) J Chem Phys 141(3):034106. doi:10.1063/1.4886655
63. Bernal-Uruchurtu MI, Ruiz-López MF (2000) Chem Phys Lett 330(1–2):118. doi:10.1016/S0009-2614(00)01062-9
64. Monard G, Bernal-Uruchurtu MI, van der Vaart A, Merz KM, Ruiz-López MF (2005) J Phys Chem A 109(15):3425. doi:10.1021/jp0459099
65. Wu X, Thiel W, Pezeshki S, Lin H (2013) J Chem Theory Comput 9(6):2672. doi:10.1021/ct400224n
66. Rappe AK, Goddard WA (1991) J Phys Chem 95(8):3358. doi:10.1021/j100161a070
67. Fritsch S, Poblete S, Junghans C, Ciccotti G, Delle Site L, Kremer K (2012) Phys Rev Lett 108:170602/1. doi:10.1103/PhysRevLett.108.170602

Chapter 3
Recent Progress in Adaptive-Partitioning QM/MM Methods for Born-Oppenheimer Molecular Dynamics

Soroosh Pezeshki and Hai Lin

Abstract Molecular dynamics simulations based on adaptive QM/MM methods feature on-the-fly reclassifications of atoms and molecular groups as either QM or MM without causing abrupt changes in the trajectory propagations, thus allowing QM subsystems to automatically change over time. Such treatments are not possible in the framework of conventional QM/MM, where the QM and MM partitions are predetermined and immutable throughout the simulation. The present contribution reviews the recent progress in the adaptive QM/MM algorithms developed by ourselves and our collaborators, namely the family of adaptive-partitioning (AP) schemes. Initially developed for the studies of solvated ions and molecules, AP methods have been extended to model large molecules, such as biopolymers, to monitor the exchange of solvent molecules between a protein active site and the bulk solvent, and to describe proton hopping in water via the Grotthuss mechanism.

Acronyms

AP	Adaptive-partitioning
BEST	Boundary-based-on-exchange-symmetry-theory
BZ	Buffer zone
CG	Coarse-grained
DAS	Difference-based adaptive solvation
EEMB	Electrostatically embedded many-body expansion
FIRES	Flexible inner region ensemble separator
MD	Molecular dynamics
MM	Molecular-mechanics
mPAP	Modified permuted adaptive-partitioning
NVE	Microcanonical ensemble
NVT	Canonical ensemble
ONIOM-XS	Our own *n*-layered integrated molecular orbital and molecular mechanics-exchange of solvent

---

S. Pezeshki · H. Lin (✉)
Department of Chemistry, University of Colorado Denver,
Denver, CO 80217-3364, USA
e-mail: hai.lin@ucdenver.edu

J.-L. Rivail et al. (eds.), *Quantum Modeling of Complex Molecular Systems*,
Challenges and Advances in Computational Chemistry and Physics 21,
DOI 10.1007/978-3-319-21626-3_3

QM	Quantum-mechanics
QM/MM	Combined quantum-mechanics/molecular mechanics
QM/MM-LPS	Combined quantum-mechanics/molecular mechanics with large primary-subsystem
PAP	Permuted adaptive-partitioning
PS	Primary subsystem
RC	Redistributed charge
RCD	Redistributed charge and dipole
SAP	Sorted adaptive-partitioning
SCMP	Size-consistent multi-partitioning
SS	Secondary subsystem

3.1 Introduction

Combined quantum-mechanics/molecular-mechanics (QM/MM) [1–36] methods are widely applied to the study of physical, chemical, and biological processes in complex environments. In a QM/MM setup, the entire system is divided into a primary subsystem (PS) treated at the QM level of theory and a usually much larger secondary subsystem (SS) described by MM force fields. The PS is also called the QM subsystem or active zone, and the SS the MM subsystem or environmental zone. Combined QM/MM models offer realistic and effective descriptions for reactions in complex environments by integrating accurate QM treatments for a small region of primary interest and computationally efficient MM treatments for the surroundings. Here, we use the term reaction for any changes in the electronic structure, including bond-breaking and bond-forming, electron excitation, charge transfer, etc. Numerous applications of QM/MM have been published, addressing a broad range of problems in biology, chemistry, physics, material science, and engineering [1–11, 13–15, 17–24, 26–34, 37–39].

QM/MM methods can be used for Born-Oppenheimer molecular dynamics (MD) simulations. (Applications of QM/MM in Car-Parrinello MD [40] have also been reported, [41–51] but they will not be discussed here.) There are two common approaches to carry out QM/MM dynamics simulations. The first approach is to perform MD simulations at the MM level and to extract snapshots from the trajectories, which are then used for QM/MM calculations such as geometry optimizations and free energy perturbation calculations. This approach is computationally efficient. However, for the results to be meaningful, the configurational space sampled by MM should overlap significantly with that sampled by QM/MM. This is, unfortunately, not always easy to judge. The second approach is to carry out the dynamics simulations directly at the QM/MM level. However this is much more expensive. Due to these high computational demands, the QM/MM dynamics simulations are limited to semi-empirical QM methods for large systems like enzymes if long simulation times are sought.

A conventional QM/MM setup designates atoms as members of either the QM or MM partition before a MD simulation starts. Furthermore, the partitions do not change during the simulation. This creates problems for the simulations of many systems, for which on-the-fly updates of the contents of PS are desired. For example, when modeling a Na^+ ion solvated in water, it is natural to include the ion and the solvent molecules in the ion's first solvation shell in the PS. However, the exchange of solvent molecules in the first solvation shell with the bulk solvent is expected. It is thus likely that, during the simulation, an MM water molecule replaces a QM water molecule in the first solvation shell. Now, one would like to treat this MM water molecule at the QM level, but that is not possible in the framework of conventional QM/MM. One workaround, the so-called QM/MM-LPS (QM/MM with large primary-subsystem), is to set up a sufficiently large PS that includes all water molecules that may enter the first solvation shell during the simulation. However, QM/MM-LPS calculations can be quite expensive, and sometimes even impractical, because it is not always possible to predict which water molecule will move into the first solvation shell. Other "difficult" cases include ion or molecule transport across a membrane, defect propagation in materials, adsorbate moving on catalytic surface, and solvent diffusion into/out of enzyme active sites. For those systems, it would be ideal if one could reclassify atoms as QM or MM on-the-fly as the trajectory is propagated. As such, the size of the PS could be kept small in long simulations.

The QM/MM methods that permit on-the-fly reclassification of atoms into PS or SS are called adaptive QM/MM methods, and they have drawn much attention in the past few years. Several adaptive algorithms [36, 38, 52–67] have been proposed, most of which aim to deal with the exchange of solvent molecules between the solvation shell of a given solute and the bulk solvent. The challenge in switching an atom from QM to MM (or vice versa) is the abrupt changes in potential energy and forces. One way to alleviate such abrupt changes is to insert a buffer zone between the PS and SS, which helps to smooth the transition of atoms between the PS and SS (Fig. 3.1). The buffer zone is often defined as a spherical shell with a radius between R_{min} and R_{max} centered at the solute. A molecule is considered to be a buffer-group if its distance to the active-zone center is within the range from R_{min} to R_{max}; the distance is usually measured using designated referencing atoms or centers of mass of certain atoms. The atoms in the buffer zone have dual QM and MM characters. Smoothing functions are applied so that potential energy and/or forces will vary gradually.

Of course, another way to deal with the possible exchange of solvent molecules between a solute's solvation shells and the bulk solvent is to simply prohibit the exchanges of atoms between the PS and SS. This can be achieved by applying restraining potentials to keep the QM atoms in the PS and the MM atoms in the SS. Examples in this category are the flexible inner region ensemble separator (FIRES) method by Rowley and Roux [68] and the boundary-based-on-exchange-symmetry-theory (BEST) by Shiga and Masia [69]. The formulation and implementation of those non-adaptive methods are usually less complicated than the adaptive schemes. The non-adaptive methods require that the region of interest is

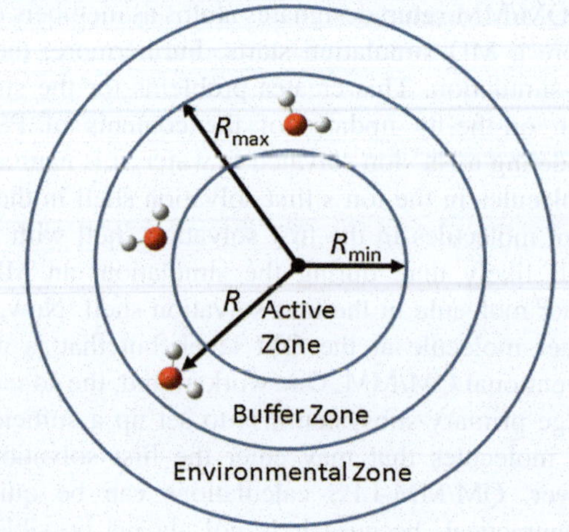

Fig. 3.1 Buffer zone between the active zone treated by QM and the environmental zone treated by MM. A molecule will usually be considered in the buffer zone if its distance R to the center of the active zone satisfies $R_{min} < R < R_{max}$, where R_{min} and R_{max} are two predefined thresholds, and the molecule will be called a buffer group. (Reprinted with permission from [66], Copyright 2014 American Chemical Society.)

sufficiently deep in the PS such that the impacts due to the restraints imposed at the boundary are insignificant. Those methods are not suitable to the studies of certain properties such as diffusion coefficients due to the enclosed PS.

There have been several excellent reviews on adaptive QM/MM methods in the literature [17, 36, 37]. In this contribution, we will provide a brief overview of the most recent progress in adaptive QM/MM algorithms, with the emphasis on the adaptive-partitioning (AP) QM/MM methods that have been developed by us and our collaborators. The scope of this review is limited to organic and biological systems, leaving out solid-state systems. At the end of this contribution, we will discuss a few key challenges in the future development of adaptive QM/MM methods.

3.2 Adaptive QM/MM Methods

3.2.1 Energy-Based Smoothing Approaches

Adaptive QM/MM methods can be largely divided into two categories according to whether the smooth functions are applied to the energy or to the forces. It is a natural idea to interpolate between the QM and MM energies when a buffer group is moving from the PS to SS (or vice versa), and the forces are computed as derivative from the energy. Schemes that adopt this idea are energy-based adaptive methods, and they include the "our own *n*-layered integrated molecular orbital and molecular mechanics-exchange of solvent" (ONIOM-XS) algorithm by Kerdcharoen and Morokuma, [53, 54] the permuted adaptive-partitioning (PAP) and sorted adaptive-partitioning (SAP) schemes by Heyden et al. [55] and their later extensions, [36, 60] as well as the difference-based adaptive solvation (DAS) method by Bulo et al. [56].

The ONIOM-XS method is the first energy-based adaptive QM/MM scheme. It requires only two QM/MM (ONIOM) calculations:

$$E^{\mathrm{ONIOM-XS}} = s\,E^{\mathrm{ONIOM}}(\mathrm{PS;SS+BZ}) + (1-s)E^{\mathrm{ONIOM}}(\mathrm{PS+BZ;SS}), \quad (3.1)$$

Here, $E^{\mathrm{ONIOM-XS}}$ is the interpolated energy, s smoothing function, E^{ONIOM}(PS; SS + BZ) the energy calculated with PS at the QM level while SS and buffer zone groups at the MM level, and E^{ONIOM}(PS + BZ;SS) the energy calculated with PS and the buffer zone groups at the QM level while SS and at the MM level. The smoothing functions used in the interpolation depend on the positions of the buffer groups related to the active-zone centre (this is also true for other adaptive QM/MM schemes). When there is only one buffer group, the ONIOM-XS method removes discontinuities in energy and forces. However, in the presence of more than one buffer-zone group, the energy and forces may still experience sudden changes.

The PAP and SAP schemes (to be discussed in more detail in Sect. 3.3) remove the discontinuities in energy and forces when more than one group is in the buffer zone. The PAP scheme expresses the potential in a many-body expansion manner. Each term in the expansion requires a set of QM/MM calculations where selected buffer groups are treated at the QM level. The calculation for the zero-order term treats only the active groups at the QM level, the calculations for the first-order terms include the active groups and one buffer group, the calculations for the second-order terms include the active groups and two buffer groups, … until all buffer groups are included. For a given order, all possible combinations of the buffer groups are considered.

The SAP scheme [55] reduces the computational demands of PAP by sorting the groups in the buffer zone in a canonical order based on their distances to the active-zone centre. This reduction in computational expense does not come free, as very complicated smoothing functions have to be used so that the energy and gradient stay constant when two buffer-zone groups switch ranks [55].

The DAS scheme [56] is similar to the SAP scheme in sorting the buffer-zone groups, but employs a different method in the construction of smoothing functions. By construction, the smoothing functions will approach 0 when two buffer-zone groups switch ranks (exactly be 0 at the point when two groups have the same distance to the active-zone centre). The smoothing functions in DAS are even more complicated than those in SAP, but they significantly reduce the variations in the forces when buffer groups switch ranks. Therefore, the DAS scheme has been found to be numerically more stable than the SAP scheme [56].

The advantage of the energy-based schemes is that a well-defined Hamiltonian exists. The PAP, SAP, and DAS schemes conserve energy and momentum in the propagation of trajectories in *NVE* simulations. It is recognized that conservations of energy and momentum are fundamental checks on the rigorousness of a given algorithm for MD simulations. However, the energy-based methods face a problem: the extra forces due to the smoothing functions. Those extra forces can lead to artifacts if they are not negligible when compared with the "real" forces acting on the atoms [66]. In principle, the extra forces can be removed or minimized by aligning the QM and MM potentials. However, that is very difficult to achieve for polyatomic molecules, especially polar molecules like water, because of the complicated dependence of the interaction energies on the relative orientations of one molecule with another. To our knowledge, there is no published simple and general yet effective way for aligning the potentials for the energy-based adaptive methods.

3.2.2 Force-Based Smoothing Approaches

The adaptive treatments in the second category interpolate the forces instead of the energy. In fact, the first adaptive QM/MM algorithm, the "hot-spot" method by Rode and coworkers, [52] was force-based. Other methods in this category include the "bookkeeping" algorithm by Bulo et al. [56], the buffered-force scheme by Bernstein and coworkers, [61] the size-consistent multi-partitioning (SCMP) method by Watanabe et al. [65] and more recently the modified permutated adaptive-partitioning (mPAP) treatment by Pezeshki et al. [66, 67].

In the hot-spot method, the forces acting on the buffer-group atoms are modulated by smoothing functions. More specifically, the force on the i-th buffer group f_i is interpolated between the QM force f_{QM} and MM force f_{MM} employing the smoothing function S_{m}:

$$f_i = S_{\mathrm{m}} f_{\mathrm{QM}} + (1 - S_{\mathrm{m}}) f_{\mathrm{MM}}, \tag{3.2}$$

The QM force f_{QM} is computed by QM calculations for the atoms in the PS and the buffer zone, and the MM force f_{MM} by force field calculations for the atoms in the buffer zone and SS. Note that only the forces acting on the buffer groups are interpolated, whereas the forces on the PS and SS are not treated. Clearly, such a treatment violates the Newton's Third Law of Motion and does not conserve momentum.

The buffered-force scheme [61] also does not conserve momentum. In fact, the forces are not interpolated at all in this method. Instead, the forces on the PS atoms are obtained by QM/MM calculations where the PS atoms are treated at the QM level and the buffer-zone atoms at the MM level, and the forces on the buffer group and SS atoms are obtained by pure-MM calculations for the entire system. Furthermore, the boundary between the buffer zone and the SS has to be constantly shifted to avoid successive border crossing by the atoms in the SS.

On the other hand, the bookkeeping [56] method, the SCMP scheme, [65] and the mPAP [66, 67] treatment do conserve momentum, offering better numerical stabilities, which is an advantage over the hot-spot and buffered-force methods.

The bookkeeping and the mPAP schemes are simplifications of the DAS and the PAP schemes, respectively. The simplification involves the deletion of the extra forces due to the smoothing functions. The SCMP method also neglects those extra forces. In contrast to the bookkeeping and mPAP methods, each of which employs one buffer zone, the SCMP scheme utilizes two different buffer zones for molecules moving into and out of the active zone, respectively. The SCMP treatment moreover requires that the numbers and types of the groups treated at the QM level in all calculations must be the same so that the zeros of the energy are constant. While maintaining the same zero of energy for all involved calculations offers certain convenience, the requirement may significantly limit the applications of the SCMP method. Note that the bookkeeping, mPAP, and SCMP methods smooth the forces on *all* atoms of the model system instead of just a selected subset of atoms. Removal of the extra forces due to smoothing functions does not violate Newton's Third Law of Motion, and thus the momentum is still conserved in all three methods.

Both the bookkeeping and SCMP schemes attempt to restore the energy of the system by adding a so-called bookkeeping term, which is an additional extra "pseudo-energy" term obtained by path integral along the propagated trajectory over the extra forces due to the smoothing function. However, it has been pointed out that because the extra forces on which the integration is carried out are not exact differentials, the integrated bookkeeping correction term is not a state function, and adding this term to the total energy does not lead to energy conservation [70, 71]. For example, it has been found that after the bookkeeping correction, the energy was path-dependent in test simulations [55, 56].

The force-based adaptive QM/MM methods do not have a well-defined energy parameter. They describe non-Hamiltonian systems, which can be coupled to thermostats so that *NVT* simulations can be carried out, even though the canonical ensemble may not be reproduced exactly. The major advantage offered by the force-based methods is the removal of the extra forces due to smoothing functions. It has been argued [64] that adaptive QM/MM schemes without a well-defined total energy can be used with free-energy techniques that require only forces or trajectories, e.g. umbrella integration, [72] constrained dynamics, [73] meta-dynamics, [74] and adaptive biasing force [75].

3.3 Adaptive-Partitioning QM/MM Methods

3.3.1 *Permuted and Sorted Adaptive-Partitioning QM/MM Methods*

As mentioned above, the PAP and SAP schemes remove the discontinuities in energy and forces when more than one group is in the buffer zone. The PAP scheme expresses the potential in a many-body expansion manner:

$$
\begin{aligned}
V = {} & V^{\mathrm{A}} + \sum_{i=1}^{N} P_i\left(V_i^{\mathrm{A}} - V^{\mathrm{A}}\right) + \sum_{i=1}^{N-1}\sum_{j=i+1}^{N} P_i P_j \left(V_{i,j}^{\mathrm{A}} - \left[V^{\mathrm{A}} + \sum_{r=i,j} \left(V_r^{\mathrm{A}} - V^{\mathrm{A}}\right) \right] \right) \\
& \sum_{i=1}^{N-2}\sum_{j=i+1}^{N-1}\sum_{k=j+1}^{N} P_i P_j P_k \left(V_{i,j,k}^{\mathrm{A}} - \left(V^{\mathrm{A}} + \sum_{r=i,j,k} \left(V_r^{\mathrm{A}} - V^{\mathrm{A}}\right) + \sum_{(p,q)=(i,j),(i,k),(j,k)} \left(V_{p,q}^{\mathrm{A}} - \left(V^{\mathrm{A}} + \sum_{r=(p,q)} \left(V_r^{\mathrm{A}} - V^{\mathrm{A}}\right) \right) \right) \right) \right) \\
& + \ldots
\end{aligned}
\tag{3.3}
$$

where V^{A} is the energy determined with the groups in the PS at the QM level, V_i^{A} with all PS groups and the i-th buffer-zone group at the QM level, $V_{i,j}^{\mathrm{A}}$ with all PS groups, the i-th buffer-zone group, and the j-th buffer-zone group at the QM level, … $V_{1,2\ldots N}^{\mathrm{A}}$ with all PS groups and all N buffer-zone groups at the QM level, and P_i is the smoothing function of the i-th buffer-zone group whose value varies between 0 and 1. All derivatives of the potential energy with respect to the coordinates vary smoothly up to the same order for which the smoothing functions P_i vary continuously. The equation can be rewritten in a more compact form: [60]

$$
V = V^{\mathrm{A}} \prod_{i=1}^{N} (1 - P_i) + \sum_{i=1}^{N} P_i V_i^{\mathrm{A}} \prod_{j \neq i}^{N} (1 - P_j) + \sum_{i=1}^{N-1}\sum_{j=i+1}^{N} P_i P_j V_{i,j}^{\mathrm{A}} \prod_{k \neq j \neq i}^{N} (1 - P_k) + \ldots
\tag{3.4}
$$

The complete evaluation of the PAP potential needs 2^N QM calculations to be performed. At first glance, this may seem overwhelming. However, the many-body contributions typically decrease rapidly, and their contributions are further scaled down by the product of weights P_i. Therefore, it is possible to truncate the series. Although the truncation introduces small (but controllable) discontinuities in the energy and derivatives, it greatly reduces the number of the QM calculations. In testing simulations, we found no noticeable discontinuities if the series was truncated at the 5th order. Although not yet fully tested, truncation at the 3rd order may be adequate if electrostatic embedding is adopted, as implied by the results of electrostatically embedded many-body expansion (EEMB) calculations [76–78]. Furthermore, all embedded-QM calculations are parallel in nature, facilitating parallel computations. Therefore, in principle, the wall time is only limited by the most expensive calculation, the calculation of $V_{1,2\ldots N}^{\mathrm{A}}$. One notable advantage of PAP is that it offers superior numerical stability as compared with all other adaptive schemes published so far.

The SAP scheme [55] reduces the computational demands of PAP by sorting the groups in the buffer zone in a canonical order based on their distances to the active-zone centre. SAP carries out only $N + 1$ calculations at the QM level, starting with the PS only and adding the buffer-zone groups one at a time according to the increasing distance. The SAP potential is given by the weighted sum

$$V = V^{\mathrm{A}} \prod_{j=1}^{N} (1 - \Phi_j) + \Phi_1 V_1^{\mathrm{A}} \prod_{j=2}^{N} (1 - \Phi_j) + \Phi_2 V_{1,2}^{\mathrm{A}} \prod_{j=3}^{N} (1 - \Phi_j) + \ldots \quad (3.5)$$

where Φ_i is the smoothing function. An important criteria of choosing the smoothing function Φ_i is that the energy and gradient must stay constant when two buffer-zone groups switch ranks [55]. Comparing Eqs. (3.4) and (3.5) implies that SAP may be considered as a truncated PAP scheme that retains only one (presumably the largest) contribution in the terms for a given order of expansion with the assistance of rather complicated smoothing functions [60].

3.3.2 *Molecular Fragments as Buffer Groups*

Just like other adaptive QM/MM methods, the original PAP scheme can only handle buffer groups that are whole ions (e.g. Na^+ ion) or molecules (e.g. water molecule). To simulate more complicated systems such as biopolymers, one may wish to treat molecular fragments as buffer groups. For example, for ion transport by channel proteins through membranes, it will be ideal to add and remove residues that solvate the ion as the ion moves through the pore. This requires dynamic relocations of the QM/MM boundary that pass through covalent bonds, for which molecular fragments are presented as buffer groups. To solve this problem, we [60] combined the PAP scheme with the redistributed charge (RC) and redistributed charge and dipole (RCD) schemes [79] that were developed to treat the QM/MM boundaries that pass through covalent bonds. The combinations led to the PAP-RC and PAP-RCD schemes.

In both the PAP-RC and PAP-RCD treatments, an automated procedure based on the molecular topology is implemented to on-the-fly relocate the QM/MM boundaries, reassign the link-atoms, recalculate the zeros of energy for the fragmental groups, and redistribute the MM point charges in the boundary region. The molecular fragments that are connected to each other by covalent bonds need special care, because they will merge into a super-group in the calculations if they are treated at the QM level of theory at the same time. Take for example a butanol molecule $CH_3CH_2CH_2CH_2O$, which is divided into three fragments: -CH_2OH as group 1, -CH_2CH_2- as group 2, and CH_3- as group 3. When only group 1 is treated at the QM level, the PS capped by a link atom is CH_3OH, and the zero of energy for group 1 is

$$E_0(\text{group 1}) = E(\mathrm{CH_3OH}). \quad (3.6)$$

Likewise, when only group 2 is treated at the QM level, the PS capped by link atoms is CH_3CH_3, and the zero of energy for group 2 is

$$E_0(\text{group 2}) = E(\mathrm{CH_3CH_3}). \quad (3.7)$$

When both groups 1 and 2 are treated at the QM level, the two groups will merge to form a super-group: -$CH_2CH_2CH_2OH$, and the zero of energy for the capped super-group is $E(CH_3CH_2CH_2OH)$. If we keep the zero of energy for group 1 unchanged, to be consistent, the zero of energy for group 2 should be computed as

$$E_0(\text{group 2}) = E(CH_3CH_2CH_2OH) - E(CH_3OH), \quad (3.8)$$

instead of being set to $E(CH_3CH_3)$. Despite the complication due to group merging, the PAP-RC and PAP-RCD schemes were demonstrated to conserve energy and momentum very well [60].

3.3.3 Deletion of Extra Forces Due to Smoothing Functions

As mentioned above, the extra forces due to the smoothing function are a ubiquitous problem in the energy-based adaptive algorithms, which include PAP and SAP. If those extra forces are non-negligible when compared with real forces, artifacts may occur in simulations. The extra forces depend on the gradient of the smoothing function and the difference between the energies $V^{A}_{i,j\ldots r}$ and V^{A}. Usually the thickness of the buffer zone is set to 1.0 Å or less. Because the smoothing function varies from 0 to 1 in such a short distance, the gradient of the smoothing function is in the order of 1/Å. One natural way to align the potential energies $V^{A}_{i,j\ldots r}$ and V^{A} is to let them equate with each other when the groups are separated infinitely away from each other. Such alignment may not be the best for a specific model system, but the treatment is simple and generally applicable. It is possible that $V^{A}_{i,j\ldots r}$ and V^{A} can differ by a few kcal/mol when the groups are in closer distances. Our rough estimations suggest that the magnitude of extra forces reaches 10–20 kcal/mol/Å for potentials aligned this way. The estimations were numerically verified by our recent test calculations of water entering and leaving the binding sites of the *E. coli* CLC chloride transport protein, where the extra forces were found to be about 10 kcal/mol/Å and occasionally spiked up to 60 kcal/mol/Å [66]. Those extra forces were transient in nature, but their impacts on the simulated structural and dynamical properties were noticeable. For example, as can be seen in Fig. 3.2, the extra forces caused water molecules to drift out of the binding site, in contrast to the reference conventional QM/MM-LPS calculations, where the number of water molecules in the binding pocket were largely maintained over the simulation duration; the average and standard deviations of numbers of water molecules in the binding sites are 5.4 ± 0.8 for QM/MM-LPS, 3.4 ± 1.0 for PAP, and 4.7 ± 0.6 for mPAP, respectively [66].

The mPAP scheme was proposed as a simple yet effective solution to the problem due to the extra forces. In principle, the extra forces can be minimized or even eliminated in favorable situations by carefully aligning the potentials.

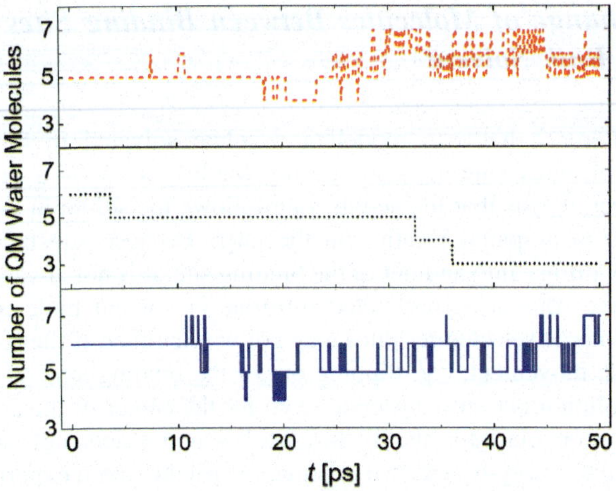

Fig. 3.2 Numbers of atoms in the intracellular binding site S_{int} of *E. coli* CLC Cl^- transporter by conventional QM/MM-LPS (*dashed*, *upper panel*), PAP (*dotted*, *middle panel*), and mPAP (*solid*, *lower panel*) dynamics simulations, respectively. The binding site is defined to be a sphere of 6.0 Å radius centered at the initial position of the Cl^- ion in the binding site at the beginning of the simulations. The model was built based on the crystal structure (PDB code 1OTS) [80]. (Re-plotted with permission based on the data from [66], Copyright 2014 American Chemical Society.)

However, in general, it is very complicated and difficult to align multi-dimensional potentials. Aligning the potentials along one coordinate may not necessarily be meaningful for another coordinate. The treatment in mPAP is to delete those extra forces in the PAP calculations. As stated above, the deletion of the extra forces does not affect the total momentum, which is still conserved. Removing the extra forces is equivalent to having external forces acting on the system to cancel out those extra forces. The system described by the mPAP algorithm is a non-Hamiltonian system. In practice, the system is coupled to a thermostat so that *NVT* simulations are performed. The mPAP method was applied to the same model system of the *E. coli* CLC chloride transport protein, where the water exchange between the bulk solvent and the binding sites of the protein were monitored. As revealed in Fig. 3.2, the results by mPAP are consistent with the referencing conventional QM/MM-LPS calculations, both maintaining the number of water molecules in the binding site despite fluctuations [66]. While not always an optimal solution, mPAP does offer a useful and convenient way to avoid the artifacts caused by the extra forces. Although the deletion of extra forces has only been demonstrated in the case of PAP, the same treatment can be applied to SAP.

3.3.4 Exchange of Molecules Between Binding Sites and Bulk Solvent

A notable difference between modeling a solute solvated in bulk solvent and monitoring molecules entering and leaving binding sites of a protein (like the example in Sect. 3.3) is that the active zone follows the solute in the former but is basically fixed in a spatial location in the later. Because substrates and solvent molecules may diffuse into and out of the binding site, it is not always convenient to center the active zone at a preselected substrate or solvent molecule. Placing the active-zone center at selected residues may not work, either, if the residues undergo large amplitude movement, e.g. flipping out of the binding site.

A simple solution is to use a pseudo atom for the center of the active zone [66]. The pseudo atom is placed in the binding pocket, and it does not interact with any real atoms. In the energy-based PAP scheme, the pseudo atom experiences the extra forces due to smoothing functions, and its position will be propagated just like a normal atom. By setting the mass to a sufficiently large value (e.g. 10^6 amu), the acceleration of the pseudo atom is negligible while not causing numerical difficulties, and its position remains essentially unchanged. Because the extra forces are deleted, in the force-based mPAP scheme, the pseudo atom does not experience any forces, and its position is strictly fixed. In either case, the pseudo atom serves as a convenient reference point to locate the active zone. As an example, Fig. 3.3 illustrates two water molecules diffusing into and out of the intracellular binding site S_{int} of the *E. coli* CLC chloride transport protein simulated by mPAP [66]. The model was built based on the crystal structure (PDB code 1OTS) [80]. In the mPAP dynamics simulations, the active zone was defined to be a sphere of 6.0 Å radius centered at the initial position of the Cl^- ion in the S_{int} binding site at the beginning of the simulations, which is denoted X_{int}. As indicated by the snapshots in panels (a–c), water molecule W1 moved from the buffer zone into the active zone after $t = 45$ ps, whereas W2 was in the environmental zone at $t = 43$ ps (and not shown in (a)) but diffused into the buffer zone after $t = 44$ ps. This is also evident from the distances of those two water molecules to the active-zone center X_{int} as functions of the simulation time, which are plotted in panel (d).

3.3.5 Proton Hopping via Grotthuss Mechanism

Simulations of hydrated proton in bulk water present a particular challenge for adaptive QM/MM methods. As opposed to ordinary ions like Cl^- and Na^+, the transfer of a hydrated proton requires the transfer of the structural feature of the hydrogen and covalent bonding network (rather than the given ion) via the Grotthuss mechanism [81]. Because the network of covalent and hydrogen bonds is dynamically reorganized, the identity of the hydrated proton can change, or "hop,"

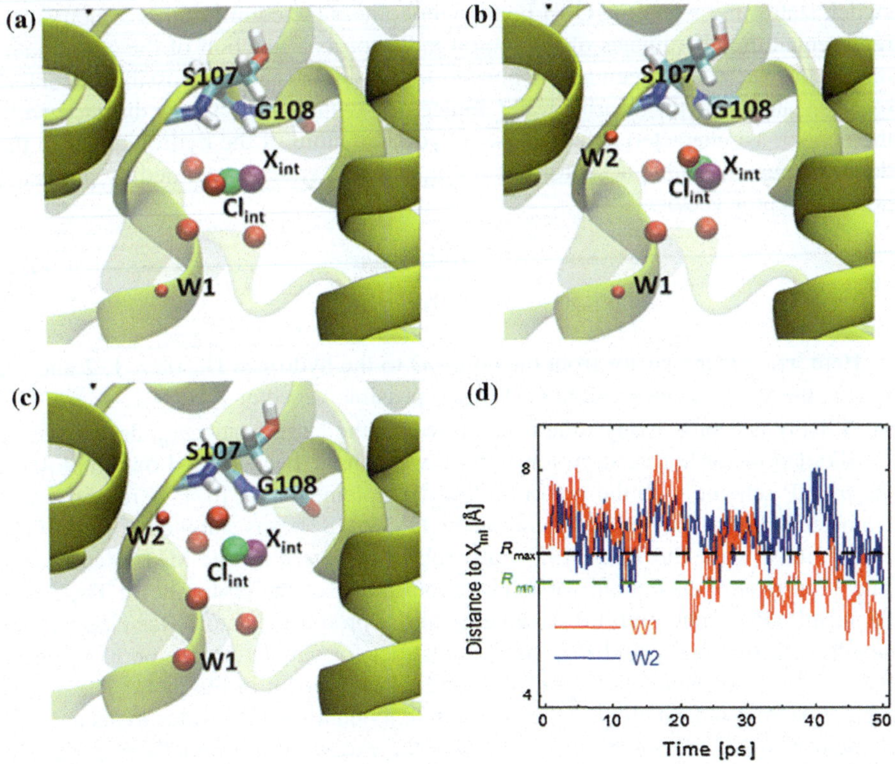

Fig. 3.3 Exchange of water molecules between bulk solvent and the intracellular binding site S_{int} of *E. coli* CLC Cl^- transporter by mPAP dynamics simulations. **a–c** Snapshots from the saved trajectory. The protein is shown as cartoon, with residues S107 and G108 in licorice. The Cl^- ion Cl_{int} and the pseudo atom X_{int} are indicated as large spheres, and water in the active and buffer zones as medium and small spheres, respectively. **d** Distances from X_{int} to W1 (*thick curve*) and from X_{int} to W2 (*thin curve*) over the simulation. The *dashed line* at R = 6.5 Å indicates the boundary between the environmental and buffer zones, and the dashed line at R = 6.0 Å between the buffer and active zones (Reprinted with permission from [66], Copyright 2014 American Chemical Society.)

quickly from one site to another. As a result, designating the active zone becomes a central problem for all adaptive QM/MM schemes.

We [67] solved this problem by introducing a proton indicator as the referencing point at which the active zone is centered. The position of the proton indicator represents the approximate location of the excess proton. In particular, it should superimpose with the donor (the oxygen atom of the hydronium ion) in the limit of Eigen ion $H_9O_4^+$ and with the shared hydrogen atom at the midpoint between the two oxygen atoms of the Zundel ion in the limit of the Zundel ion $H_5O_2^+$. When the structure varies between the two limits, the indicator's position should change accordingly in a smooth manner. Additional requirements include minimum perturbations to the proton indicator position when the proton donor and acceptor switch status and when the O–H bonds vibrate near the equilibrium geometry. We represented the coordinates of the proton indicator as a function of the coordinates of the donor, the hydrogen atoms bonded to the donor, and the possible acceptors (oxygen atoms of the nearby water molecules within a threshold distance from the donor). To characterize how close a hydrogen atom of the hydronium is to the donor versus a possible acceptor, a ratio ρ_{mj} is used, which is defined by projected donor-acceptor vectors

$$\rho_{mj} = \frac{\mathbf{r}_{\mathrm{DH}_m} \cdot \mathbf{r}_{\mathrm{DA}_j}}{\left|\mathbf{r}_{\mathrm{DA}_j}\right|^2} \tag{3.9}$$

Here, $\mathbf{r}_{\mathrm{DH}_m}$ is the vector from the donor D to the hydrogen H_m ($m = 1, 2$ and 3), $\mathbf{r}_{\mathrm{DA}_j}$ is the vector from the donor D to the possible acceptor A_j ($j = 1, 2 \ldots$). The most likely acceptor at any time is identified by the largest ratio ρ_{mj}, denoted as ρ.

Simulations of hydrated proton in bulk water have been carried out employing the mPAP scheme with the proton indicator implementation [67]. A recently reparameterized semi-empirical Hamiltonian OM3n [82] was adopted for the QM calculations, while the TIP3P [83] water model was used for the MM calculations. The atom type of each atom was updated on-the-fly as the proton hops. The small PS in mPAP (a sphere of 4.0 Å centered at the proton indicator) was close to the smallest PS that one could reasonably construct for simulating proton hopping in bulk water. A snapshot of the mPAP simulation is shown in Fig. 3.4a. The active zone, which is indicated by the dashed circle, contains 8 QM water molecules. In particular, W0 is the water molecule where the excess proton initially resided, and W1–W3 are the water molecules that formed the proton hopping path during the simulation. The snapshot shows the moment when excess proton is being transferred between W1 and W2. Figure 3.4b displays the radial distribution function for the pair of hydronium oxygen D and water oxygen O (upper panel) and the potential of mean force along the ratio ρ as the proton transfer reaction coordination (lower panel); both the conventional QM/MM-LPS (dashed curves) and by mPAP (solid curve) data are given. The conventional QM/MM-LPS data were taken from [82], where the PS was a sphere of 12 Å radius centered at the initial hydronium ion. The mPAP results were taken from [67]. Because a non-periodic droplet model was used in the QM/MM-LPS calculations whereas periodic boundary was employed in the mPAP calculations, the radial distribution functions have been scaled such that the heights of the first peak are 1 for easy comparisons. Both QM/MM-LPS and mPAP potential of mean forces indicated a small but finite barrier for proton transfer in bulk solvent, although the QM/MM-LPS barrier was higher (0.19 kcal/mol) than the mPAP barrier (0.04 kcal/mol). The results agree qualitatively with ab initio path integral calculations (barrier of 0.15 kcal/mol) that include the quantum fluctuations of all nuclei [84, 85]. Both QM/MM-LPS and mPAP suggest that the Zundel structure was the transition state at $\rho = 0.5$ and the Eigen structure the resting state at approximately $\rho = 0.45$.

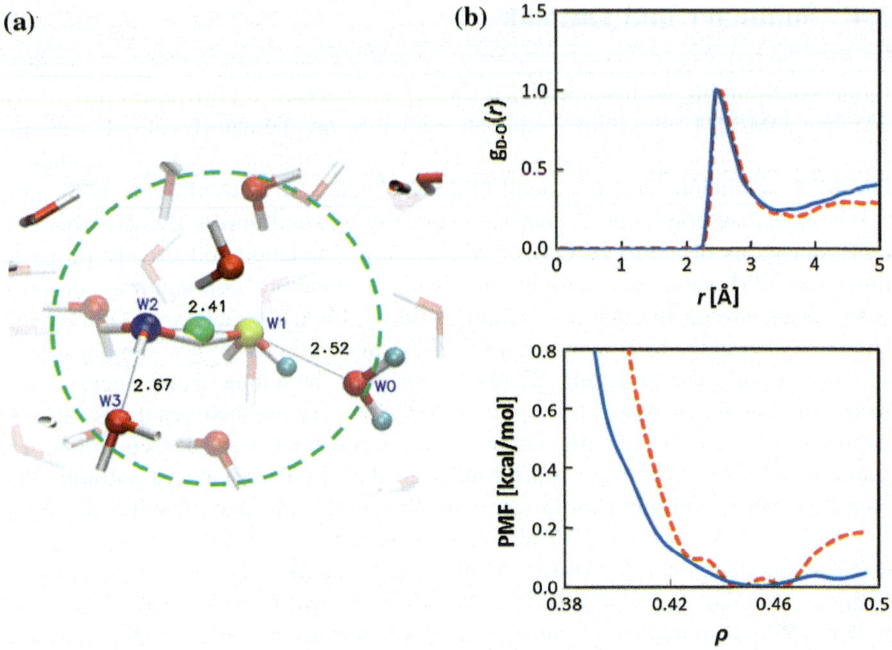

Fig. 3.4 **a** Snapshot of mPAP simulations of hydrated proton transfer in bulk water. The *dashed green circle* indicates the active zone. Water molecules are represented by *bonds*. The proton indicator and the O atoms of the active-zone water molecules are indicated by *large spheres*. The H atoms belonged to the original hydronium ion (W0) are shown as *small spheres*. W1–W3 are the water molecules involved in the proton transfer during the simulation. Distances are in Å. **b** Radial distribution function for the pair of hydronium oxygen D and water oxygen O (*upper panel*) and potential of mean force along the ratio ρ as the proton transfer reaction coordination (*lower panel*) by conventional QM/MM-LPS (*dashed curves*) and by mPAP (*solid curve*). See Eq. 3.9 for the definition of ρ

The simulation results demonstrated that the proton indicator serves well as the active-zone center, achieving smooth variations in the location and contents of the active zone. Despite a much smaller PS (a sphere of 4.0 Å radius), mPAP reproduced quite satisfactorily the results of conventional QM/MM-LPS model that utilized a large PS (a sphere of 12.0 Å radius). While we do not expect mPAP to reproduce all properties calculated by QM/MM-LPS due to different levels of theory for computing the interactions between the water molecules in the first solvation shell of the hydrated proton and the water molecules in the bulk, it is very encouraging to see that they agree reasonably with each other in the proton solvation structure and in the proton transfer dynamics.

3.4 Summary and Outlook

In this contribution, we have provided a succinct overview on the recent progress in adaptive QM/MM algorithms, with an emphasis on the family of AP schemes developed by ourselves and our collaborators. A significant advantage of adaptive QM/MM simulations is that a small QM subsystem can be used, with its location and contents are continuously and automatically updated on-the-fly. The dynamic reclassifications of atoms as QM or MM during simulations effectively turns the finite-size QM subsystem infinite and can in principle continuously describe behavior of interest through any length of time. This makes adaptive QM/MM an attractive alternative to conventional QM/MM in certain dynamics simulations.

Comparisons for different adaptive schemes, including both energy- and force-based methods, have been reported for model systems such as solutes solvated in bulk solvent in terms of structural properties, dynamical properties, and numerical stabilities [55, 66, 86]. The results confirmed that the energy-based methods that conserve energy and momentum were able to produce the same radial distribution functions in both *NVE* and *NVT* simulations, which is what one will expect at the thermodynamic limit [55]. However, the energy-based methods require properly aligned potentials, for which the extra forces due to smoothing functions are negligible [55]. If potential alignments are difficult to achieve or not pursued, force-based methods will be a more practical choice, as they avoid the artifacts caused by the extra forces [66]. The force-based methods do not have an explicitly defined energy, and it is not clear how well the canonical ensemble will be sampled, although the results produced by the force-based methods seem reasonable so far. Bulo et al. [86] observed that methods that conserve momentum overall outperformed methods that do not, the latter relying on massive thermostats to reduce numerical instabilities, and that the performance of a given method also partially depend on the specific QM and MM levels of theory that had been employed.

We envision that the number of applications of adaptive QM/MM will increase as these methods gain popularity in the future. The PAP method, which was originally proposed to model solutes solvated in bulk solvent, has been expanded to treat large molecules such as biopolymers, [60] to monitor molecules entering and leaving protein binding sites, [66] and to model proton hopping in bulk water [67]. Although other adaptive methods still focus on the studies of solvation structures for ions and molecules in bulk solvent (most notably in the works by Rode and coworkers [37]), investigations of chemical reactions in solutions have also recently been reported [62, 64].

We anticipate that further developments will continue to improve the adaptive QM/MM algorithms in various aspects. For energy-based adaptive methods, it is desirable to explore simple, general, and effective ways to align the potentials in order to minimize the extra forces due to smoothing functions. For force-based schemes, testing the rigorousness of the algorithms is highly recommended. For the AP family of algorithms, code parallelization that takes advantage of the parallel nature of the individual calculations of $V^{A}_{i,j...r}$ will greatly improve the computational efficiencies. Furthermore, it will be interesting to combine the adaptive schemes with the flexible-boundary treatments, [87–89] which describe on-the-fly partial charge transfer between the QM and MM subsystems during MD simulations. Such a marriage will lead to the so-called open-boundary QM/MM, [36, 88] which will allow on-the-fly exchanges of both atoms and partial charges at the same time between the QM and MM subsystems. The open-boundary treatment will transform QM/MM into truly seamless integrations of QM and MM descriptions. Finally, we note that adaptive algorithms have been proposed for on-the-fly forward and backward mapping between the atomistic and coarse-grained (CG) representations in MM simulations [58, 70, 90–95]. It will be intriguing, though very challenging, to develop three-tier QM/MM/CG models that feature adaptive treatments for both the QM/MM and MM/CG boundaries.

Acknowledgments This work is supported by National Science Foundation (CHE-0952337). This work used the Extreme Science and Engineering Discovery Environment (XSEDE) under grant CHE-140070, which is supported by National Science Foundation grant number ACI-1053575. HL thanks the Camille & Henry Dreyfus Foundation for support (TH-14-028). We are grateful to Adam Duster for critically reading our manuscript.

Disclosure statement: The authors declare no competing financial interest.

References

1. Warshel A, Levitt M (1976) Theoretical studies of enzymic reactions: Dielectric, electrostatic and steric stabilization of the carbonium ion in the reaction of lysozyme. J Mol Biol 103:227–249
2. Singh UC, Kollmann PA (1986) A combined ab initio quantum mechanical and molecular mechanical method for carrying out simulations on complex molecular systems: Applications to the CH_3Cl + Cl^- exchange reaction and gas phase protonation of polyethers. J Comput Chem 7:718–730
3. Field MJ, Bash PA, Karplus M (1990) A combined quantum mechanical and molecular mechanical potential for molecular dynamics simulations. J Comput Chem 11:700–733
4. Gao J (1996) Methods and applications of combined quantum mechanical and molecular mechanical potentials. Rev Comput Chem 7:119–185
5. Mordasini T, Thiel W (1998) Computational chemistry column. Combined quantum mechanical and molecular approaches. Chimia 52:288–291
6. Hillier IH (1999) Chemical reactivity studied by hybrid QM/MM methods. Theochem 463:45–52
7. Monard G, Merz KM Jr (1999) Combined quantum mechanical/molecular mechanical methodologies applied to biomolecular systems. Acc Chem Res 32:904–911
8. Hammes-Schiffer S (2000) Theoretical perspectives on proton-coupled electron transfer reactions. Acc Chem Res 34:273–281
9. Sherwood P (2000): Hybrid quantum mechanics/molecular mechanics approaches. In: Grotendorst J (ed) Modern methods and algorithms of quantum chemistry, vol 3. John von Neumann-Institute, Ho Chi Minh, pp 285–305
10. Gao J, Truhlar DG (2002) Quantum mechanical methods for enzyme kinetics. Annu Rev Phys Chem 53:467–505
11. Morokuma K (2002) New challenges in quantum chemistry: quests for accurate calculations for large molecular systems. Philos Trans R Soc Lond, Ser A 360:1149–1164

12. Riccardi D, Schaefer P, Yang Y, Yu H, Ghosh N, Prat-Resina X, König P, Li G, Xu D, Guo H et al (2006) Development of effective quantum mechanical/molecular mechanical (QM/MM) methods for complex biological process. J Phys Chem B 110:6458–6469
13. Lin H, Truhlar DG (2007) QM/MM: What have we learned, where are we, and where do we go from here? Theor Chem Acc 117:185–199
14. Senn HM, Thiel W (2007) QM/MM methods for biological systems. Top Curr Chem 268:173–290
15. Hu H, Yang W (2008) Free energies of chemical reactions in solution and in enzymes with ab initio quantum mechanics/molecular mechanics methods. Annu Rev Phys Chem 59:573–601
16. Sherwood P, Brooks BR, Sansom MSP (2008) Multiscale methods for macromolecular simulations. Curr Opin Struct Biol 18:630–640
17. Bernstein N, Kermode JR, Csányi G (2009) Hybrid atomistic simulation methods for materials systems. Rep Progr Phys 72:026501/1–25
18. Senn HM, Thiel W (2009) QM/MM methods for biomolecular systems. Angew Chem Int Ed 48:1198–1229
19. Acevedo O, Jorgensen WL (2010) Advances in quantum and molecular mechanical (QM/MM) simulations for organic and enzymatic reactions. Acc Chem Res 43:142–151
20. Sabin JR, Brändas E (eds) (2010) Combining quantum mechanics and molecular mechanics. Some recent progresses in QM/MM methods. Academic Press, New York
21. Yohsuke H, Masaru T (2010) Recent advances in jointed quantum mechanics and molecular mechanics calculations of biological macromolecules: schemes and applications coupled to ab initio calculations. J Phys: Condens Matter 22:413101/1–7
22. Ferrer S, Ruiz-Pernia J, Marti S, Moliner V, Tunon I, Bertran J, Andres J (2011) Hybrid schemes based on quantum mechanics/molecular mechanics simulations: Goals to success, problems, and perspectives. In Christov C (ed) Advances in protein chemistry and structural biology, vol 85: computational chemistry methods in structural biology. Advances in protein chemistry and structural biology, vol 85. Elsevier Academic Press Inc, New York, pp 81–142
23. Menikarachchi LC, Gascon JA (2010) QM/MM approaches in medicinal chemistry research. Curr Top Med Chem 10:46–54
24. Wallrapp FH, Guallar V (2011) Mixed quantum mechanics and molecular mechanics methods: looking inside proteins. Wiley Interdiscip Rev: Comput Mol Sci 1:315–322
25. Woodcock HL, Miller BT, Hodoscek M, Okur A, Larkin JD, Ponder JW, Brooks BR (2011) MSCALE: a general utility for multiscale modeling. J Chem Theory Comput 7:1208–1219
26. Chung LW, Hirao H, Li X, Morokuma K (2012) The ONIOM method: its foundation and applications to metalloenzymes and photobiology. Wiley Interdiscip Rev: Comput Mol Sci 2:327–350
27. Keil FJ (2012) Multiscale modelling in computational heterogeneous catalysis. In Kirchner B, Vrabec J (eds) Multiscale molecular methods in applied chemistry. Topics in current chemistry, vol 307. Springer, Berlin, pp 69–107
28. Lonsdale R, Harvey JN, Mulholland AJ (2012) A practical guide to modelling enzyme-catalysed reactions. Chem Soc Rev 41:3025–3038
29. Monari A, Rivail J-L, Assfeld X (2012) Theoretical modeling of large molecular systems. Advances in the local self consistent field method for mixed quantum mechanics/molecular mechanics calculations. Acc Chem Res 46:596–603
30. Site L, Holm C, Vegt NA (2012) Multiscale approaches and perspectives to modeling aqueous electrolytes and polyelectrolytes. In: Kirchner B, Vrabec J (eds) Multiscale molecular methods in applied chemistry. Topics in current chemistry, vol 307. Springer, Berlin, pp 251–294
31. Wu R, Cao Z, Zhang Y (2012) Computational simulations of zinc enzyme: challenges and recent advances. Progr Chem 24:1175–1184
32. Mennucci B (2013) Modeling environment effects on spectroscopies through QM/classical models. Phys Chem Chem Phys 15:6583–6594
33. Kussmann J, Beer M, Ochsenfeld C (2013) Linear-scaling self-consistent field methods for large molecules. Wiley Interdiscip Rev: Comput Mol Sci 3:614–636

34. Meier K, Choutko A, Dolenc J, Eichenberger AP, Riniker S, van Gunsteren WF (2013) Multi-resolution simulation of biomolecular systems: a review of methodological issues. Angew Chem Int Ed 52:2820–2834
35. van der Kamp MW, Mulholland AJ (2013) Combined quantum mechanics/molecular mechanics (QM/MM) methods in computational enzymology. Biochemistry 52:2708–2728
36. Pezeshki S, Lin H (2014) Recent developments in QM/MM methods towards open-boundary multi-scale simulations. Mol Simul 41:168–189
37. Rode BM, Schwenk CF, Hofer TS, Randolf BR (2005) Coordination and ligand exchange dynamics of solvated metal ions. Coord Chem Rev 249:2993–3006
38. Yockel S, Schatz G (2012) Dynamic QM/MM: a hybrid approach to simulating gas-liquid interactions. In: Kirchner B, Vrabec J (eds) Multiscale molecular methods in applied chemistry. Topics in current chemistry, vol 307. Springer, Berlin, pp 43–67
39. Allen C, McCann BW, Acevedo O (2014) Ionic liquid effects on nucleophilic aromatic substitution reactions from QM/MM simulations. J Phys Chem B
40. Car R, Parrinello M (1985) Unified approach for molecular dynamics and density-functional theory. Phys Rev Lett 55:2471–2474
41. Woo TK, Cavallo L, Ziegler T (1998) Implementation of the IMOMM methodology for performing combined QM/MM molecular dynamics simulations and frequency calculations. Theor Chem Acc 100:307–313
42. Eichinger M, Tavan P, Hutter J, Parrinello M (1999) A hybrid method for solutes in complex solvents: density functional theory combined with empirical force fields. J Chem Phys 110:10452–10467
43. Woo TK, Margl PM, Deng L, Cavallo L, Ziegler T (1999) Combined QM/MM and ab initio molecular dynamics modeling of homogeneous catalysis. ACS Symp Ser 721:173–186
44. Röthlisberger U, Carloni P, Doclo K, Parrinello M (2000) A comparative study of galactose oxidase and active site analogs based on QM/MM Car-Parrinello simulations. J Biol Inorg Chem 5:236–250
45. Woo TK, Blöchl PE, Ziegler T (2000) Towards solvation simulations with a combined ab initio molecular dynamics and molecular mechanics approach. Theochem 506:313–334
46. Colombo MC, Guidoni L, Laio A, Magistrato A, Maurer P, Piana S, Rohrig U, Spiegel K, Sulpizi M, VandeVondele J et al (2002) Hybrid QM/MM Car-Parrinello simulations of catalytic and enzymatic reactions. Chimia 56:13–19
47. Laio A, VandeVondele J, Röthlisberger U (2002) A Hamiltonian electrostatic coupling scheme for hybrid Car-Parrinello molecular dynamics simulations. J Chem Phys 116:6941–6947
48. Woo TK, Margl P, Blochl PE, Ziegler T (2002) Sampling phase space by a combined QM/MM ab initio Car-Parrinello molecular dynamics method with different (multiple) time steps in the quantum mechanical (QM) and molecular mechanical (MM) domains. J Phys Chem A 106:1173–1182
49. Sulpizi M, Laio A, VandeVondele J, Cattaneo A, Röthlisberger U, Carloni P (2003) Reaction mechanism of caspases: insights from QM/MM Ca-Parrinello simulations. Proteins: Struct, Function, Genet 52:212–224
50. Raugei S, Cascella M, Carloni P (2004) A proficient enzyme: insights on the mechanism of orotidine monophosphate decarboxylase from computer simulations. J Am Chem Soc 126:15730–15737
51. Laino T, Mohamed F, Laio A, Parrinello M (2006) An efficient linear-scaling electrostatic coupling for treating periodic boundary conditions in QM/MM simulations. J Chem Theory Comput 2:1370–1378
52. Kerdcharoen T, Liedl KR, Rode BM (1996) A QM/MM simulation method applied to the solution of Li^+ in liquid ammonia. Chem Phys 211:313–323
53. Kerdcharoen T, Morokuma K (2002) ONIOM-XS: an extension of the ONIOM method for molecular simulation in condensed phase. Chem Phys Lett 355:257–262
54. Kerdcharoen T, Morokuma K (2003) Combined quantum mechanics and molecular mechanics simulation of Ca^{2+}/ammonia solution based on the ONIOM-XS method: Octahedral coordination and implication to biology. J Chem Phys 118:8856–8862

55. Heyden A, Lin H, Truhlar DG (2007) Adaptive partitioning in combined quantum mechanical and molecular mechanical calculations of potential energy functions for multiscale simulations. J Phys Chem B 111:2231–2241
56. Bulo RE, Ensing B, Sikkema J, Visscher L (2009) Toward a practical method for adaptive QM/MM simulations. J Chem Theory Comput 5:2212–2221
57. Guthrie MG, Daigle AD, Salazar MR (2009) Properties of a method for performing adaptive, multilevel QM simulations of complex chemical reactions in the gas-phase. J Chem Theory Comput 6:18–25
58. Nielsen SO, Bulo RE, Moore PB, Ensing B (2010) Recent progress in adaptive multiscale molecular dynamics simulations of soft matter. Phys Chem Chem Phys 12:12401–12414
59. Poma AB, Delle Site L (2010) Classical to path-integral adaptive resolution in molecular simulation: towards a smooth quantum-classical coupling. Phys Rev Lett 104:250201/1–4
60. Pezeshki S, Lin H (2011) Adaptive-partitioning redistributed charge and dipole schemes for QM/MM dynamics simulations: on-the-fly relocation of boundaries that pass through covalent bonds. J Chem Theory Comput 7:3625–3634
61. Bernstein N, Varnai C, Solt I, Winfield SA, Payne MC, Simon I, Fuxreiter M, Csanyi G (2012) QM/MM simulation of liquid water with an adaptive quantum region. Phys Chem Chem Phys 14:646–656
62. Park K, Gotz AW, Walker RC, Paesani F (2012) Application of adaptive QM/MM methods to molecular dynamics simulations of aqueous systems. J Chem Theory Comput 8:2868–2877
63. Takenaka N, Kitamura Y, Koyano Y, Nagaoka M (2012) The number-adaptive multiscale QM/MM molecular dynamics simulation: application to liquid water. Chem Phys Lett 524:56–61
64. Várnai C, Bernstein N, Mones L, Csányi G (2013) Tests of an adaptive QM/MM calculation on free energy profiles of chemical reactions in solution. J Phys Chem B 117:12202–12211
65. Watanabe HC, Kubař T, Elstner M (2014) Size-consistent multipartitioning QM/MM: a stable and efficient adaptive QM/MM method. J Chem Theory Comput 10:4242–4252
66. Pezeshki S, Davis C, Heyden A, Lin H (2014) Adaptive-partitioning QM/MM dynamics simulations: 3. Solvent molecules entering and leaving protein binding sites. J Chem Theory Comput 10:4765–4776
67. Pezeshki S, Lin H (2015) Adaptive-partitioning QM/MM for molecular dynamics simulations: 4. Proton hopping in bulk water. J Chem Theory Comput 11:2398–2411
68. Rowley CN, Roux B (2012) The solvation structure of Na^+ and K^+ in liquid water determined from high level ab Initio molecular dynamics simulations. J Chem Theory Comput 8:3526–3535
69. Shiga M, Masia M (2013) Boundary based on exchange symmetry theory for multilevel simulations. I. Basic theory. J Chem Phys 139:044120/1–8
70. Heyden A, Truhlar DG (2008) Conservative algorithm for an adaptive change of resolution in mixed atomistic/coarse-grained multiscale simulations. J Chem Theory Comput 4:217–221
71. Praprotnik M, Poblete S, Delle Site L, Kremer K (2011) Comment on "Adaptive multiscale molecular dynamics of macromolecular fluids". Phys Rev Lett 107:099801/1–2
72. Kästner J, Thiel W (2005) Bridging the gap between thermodynamic integration and umbrella sampling provides a novel analysis method: "umbrella integration". J Chem Phys 123:144104/1–5
73. Carter EA, Ciccotti G, Hynes JT, Kapral R (1989) Constrained reaction coordinate dynamics for the simulation of rare events. Chem Phys Lett 156:472–477
74. Laio A, Parrinello M (2002) Escaping free-energy minima. Proc Natl Acad Sci USA 99:12562–12566
75. Darve E, Pohorille A (2001) Calculating free energies using average force. J Chem Phys 115:9169–9183
76. Dahlke EE, Truhlar DG (2007) Electrostatically embedded many-body expansion for large systems, with applications to water clusters. J Chem Theory Comput 3:46–53
77. Dahlke EE, Truhlar DG (2007) Electrostatically Embedded many-body correlation energy, with applications to the calculation of accurate second-order Moller-Plesset Perturbation Theory energies for large water clusters. J Chem Theory Comput 3:1342–1348

78. Speetzen ED, Leverentz HR, Lin H, Donald GT (2011) Electrostatically embedded many-body expansion for large systems. In: Manby FR (ed) Accurate condensed-phase quantum chemistry. Computation in chemistry. CRC Press, Taylor & Francis, New York, pp 105–127
79. Lin H, Truhlar DG (2005) Redistributed charge and dipole schemes for combined quantum mechanical and molecular mechanical calculations. J Phys Chem A 109:3991–4004
80. Dutzler R, Campbell EB, MacKinnon R (2003) Gating the selectivity filter in ClC chloride channels. Science 300:108–112
81. Agmon N (1995) The Grotthuss mechanism. Chem Phys Lett 244:456–462
82. Wu X, Thiel W, Pezeshki S, Lin H (2013) Specific reaction path Hamiltonian for proton transfer in water: reparameterized semiempirical models. J Chem Theory Comput 9:2672–2686
83. Jorgensen WL, Chandrasekhar J, Madura JD, Impey RW, Klein ML (1983) Comparison of simple potential functions for simulating liquid water. J Chem Phys 79:926–935
84. Marx D, Tuckerman ME, Hutter J, Parrinello M (1999) The nature of the hydrated excess proton in water. Nature 397:601–604
85. Marx D, Tuckerman ME, Parrinello M (2000) Solvated excess protons in water: quantum effects on the hydration structure. J Phys: Condens Matter 12:A153–A159
86. Bulo RE, Michel C, Fleurat-Lessard P, Sautet P (2013) Multiscale modeling of chemistry in water: are we there yet? J Chem Theory Comput 9:5567–5577
87. Zhang Y, Lin H (2008) Flexible-boundary quantum-mechanical/molecular-mechanical calculations: partial charge transfer between the quantum-mechanical and molecular-mechanical subsystems. J Chem Theory Comput 4:414–425
88. Zhang Y, Lin H (2010) Flexible-boundary QM/MM calculations: II. Partial charge transfer across the QM/MM boundary that passes through a covalent bond. Theoret Chem Acc 126:315–322
89. Pezeshki S, Lin H (2014) Molecular dynamics simulations of ion solvation by flexible-boundary QM/MM: on-the-fly partial charge transfer between QM and MM subsystems. J Comput Chem 35:1778–1788
90. Abrams CF (2005) Concurrent dual-resolution Monte Carlo simulation of liquid methane. J Chem Phys 123:234101/234101–234113
91. Praprotnik M, Delle Site L, Kremer K (2005) Adaptive resolution molecular-dynamics simulation: changing the degrees of freedom on the fly. J Chem Phys 123:224106/1–14
92. Ensing B, Nielsen SO, Moore PB, Klein ML, Parrinello M (2007) Energy conservation in adaptive hybrid atomistic/coarse-grain molecular dynamics. J Chem Theory Comput 3:1100–1105
93. Praprotnik M, Matysiak S, Site LD, Kremer K, Clementi C (2007) Adaptive resolution simulation of liquid water. J Phys: Condens Matter 19:292201/1–10
94. Wang H, Schütte C, Delle Site L (2012) Adaptive resolution simulation (AdResS): a smooth thermodynamic and structural transition from atomistic to coarse grained resolution and vice versa in a grand canonical fashion. J Chem Theory Comput 8:2878–2887
95. Wang H, Hartmann C, Schütte C, Delle Site L (2013) Grand-canonical-like molecular-dynamics simulations by using an adaptive-resolution technique. Phys Rev X 3:011018/1–16

Chapter 4
Probing Proton Transfer Reactions in Molecular Dynamics—A Crucial Prerequisite for QM/MM Simulations Using Dissociative Models

Thomas S. Hofer

Abstract The classical modeling of proton transfer reactions in chemical simulations requires the application of reactive force field formulations. A common feature of these dissociative potential models of aqueous systems is the possibility to transfer protons between water, oxonium and hydroxide ions. Since molecules undergo a change of their composition as the simulation progresses, the respective topology defining which atoms form a molecular unit at a given configuration becomes time-dependent. Knowledge of this variable topology is a key prerequisite to apply dissociative models in the framework of hybrid quantum mechanical/molecular mechanical (QM/MM) simulation studies. In order to effectively execute QM/MM simulations, the simulation software has to be able to independently monitor all occurring bond formation and cleavage events and automatically adjust the respective topology information, thereby discriminating between short-time fluctuations and sustained proton transfer events. The properties of a simple yet effective automated topology update criterion developed for excess protons are presented and its performance for hydroxide containing solutions and systems containing excess protons is compared. Furthermore, the influence of deuteration of the different systems is discussed. The data clearly demonstrates that it is possible to apply a global setting for the automated topology update of both proton and proton-hole migration.

List of Abbreviations

CPMD	Car–Parrinello molecular dynamics
DFT	Density functional theory
QM	Quantum mechanical
MM	Molecular mechanical
QM/MM	Quantum mechanical/molecular mechanical

T.S. Hofer (✉)
Theoretical Chemistry Division, Institute of General, Inorganic and Theoretical Chemistry, University of Innsbruck, Innrain 80-82, A-6020 Innsbruck, Austria
e-mail: T.Hofer@uibk.ac.at

J.-L. Rivail et al. (eds.), *Quantum Modeling of Complex Molecular Systems*, Challenges and Advances in Computational Chemistry and Physics 21,
DOI 10.1007/978-3-319-21626-3_4

MD	Molecular dynamics
PT	Proton transfer
MS-EVB	Multistate empirical valence bond
HPC	High-performance computing
ATU	Automated topology update
PTM	Proton transfer maps
PH	Proton-hole
NQE	Nuclear quantum effects
sOSS2	Scaled Ojamäe-Shavitt-Singer
PIMD	Path-integral molecular dynamics

4.1 Introduction

One of the most prominent technical advancements of the last decades was the continuous improvement of computational capacities. Computer simulations of chemical systems [1, 30, 35, 45, 49, 76, 91] are just one example of scientific applications, that benefited significantly from this still continuing trend. Computational studies of liquids, being central to a broad range of chemical phenomena and virtually all of biochemistry, evolved from applications using simplified model systems to describe the interactions to state-of-the-art investigations taking the quantum mechanical character of the electronic subsystem and in some cases even that of the nuclei into account. For instance Car–Parrinello molecular dynamics (CPMD) [18] simulations enable the treatment of systems containing a few hundred atoms at the level of density functional theory (DFT) [48, 81]. On the other hand simulation protocols based on Feynmans path integral representation of quantum mechanics [12, 16, 17, 22, 27, 44, 67] enable investigations of the quantum character of the nuclei, in particular that of hydrogen atoms which are known to exhibit a pronounced quantum character in many cases. Further advanced algorithms such as quantum thermostats [19, 20] enable a great reduction of the computational effort in PIMD simulations [20], while the development of GPU implemented quantum chemical algorithms [6, 96–99, 105] enjoyed increased interest in recent years.

Hybrid simulation techniques [5, 28, 31, 53, 80, 103] proved as another family of approaches with important significance in chemical sciences. Since quantum mechanical (QM) based computation techniques [21, 34, 86] are prohibitively expensive when applied to large systems and simplified molecular mechanical (MM, also referred to as force fields) potential models [24, 45, 49, 72] are in many cases not sufficiently accurate to investigate a chemical process, the advantages of both techniques are combined into a single computational framework. These hybrid quantum mechanical/molecular mechanical (QM/MM) approaches partition the system into a high- and a low-level zone. While the chemical most relevant region of the system is treated accurately via a suitable QM technique, efficient MM potentials are considered sufficiently accurate to describe the interactions in the remaining part of the system. The awarding of the Nobel price in chemistry to Martin Karplus, Michael Levitt and Arieh Warshel in 2013 for the development of multiscale models for complex chemical systems highlights the strong impact of this simulation approach (along many other techniques developed by the Nobel laureates). Initially being developed to address questions in biochemical research [53, 80], where active sites of enzymatic systems are included into the quantum mechanical zone, these techniques enjoy widespread application to numerous chemical systems including liquids and solutions [32, 36–39, 74, 75, 82, 92, 93, 107].

Since liquids combine the high mobility of its constituents comparable to the gaseous state with densities similar as observed in solids, liquid systems are the most difficult state of matter to describe (among the three mentioned states solid, liquid and gas being relevant for chemistry). Despite the requirement of an accurate representation of the potential energy surface of such systems, the pronounced dynamical properties represent a further obstacle to describe the liquid phase. Based on the concepts of statistical thermodynamics [61, 68, 79, 91] such systems can only be represented via an ensemble, accounting for all possible configurations of the systems according to their respective probability of occurrence. QM/MM simulations in the framework of the molecular dynamics (MD) technique [1, 30, 45, 49, 76, 91] describing the time evolution of chemical systems have proven as an excellent tool to investigate solvated species such as hydrated ionic [37–39, 74, 75, 85, 94] and molecular solutes [3, 36, 107] as well as coordination complexes in solution [7, 55, 62, 63, 88].

A further challenging yet important aspect in simulations of aqueous systems is the accurate description of proton transfer (PT) reactions in acidic or basic medium [10, 11, 47, 59, 102] Simple force field employing a spring-like harmonic potential to mimic chemical bonds are not capable of describing the cleavage or formation of covalent bonds. Particle pairs defined as bond in the initial configuration will remain bonded throughout the simulation, while on the other hand new bonds cannot be formed even if it was chemically favourable. Improved simulation techniques such as multistate empirical valence bond (MS-EVB) [13, 66, 77, 78, 95, 100, 101, 111] and dissociative force field formulations [8, 9, 41, 50, 65, 71, 83, 84, 109] do enable a description of bond formation and cleavage. While the first approach employs a superposition of educt and product states [104] which can also be realised via simple force field expressions, the latter require that no distinction between intra- or intermolecular interactions is made. A typical example to realise such a formalism is the bond length/bond order relationship used in the ReaxFF potential [14, 26, 51].

Due to the importance of excess protons and hydroxide ions in aqueous chemistry, a number of dissociative potential models for aqueous solutions have been developed. Although dissociation events have not been explicitly aimed for in the model derived by Stillinger and coworkers [71, 83, 84], its functional form in principle enables the dissociation of water molecules. More advanced potential models for the treatment of proton transfer reactions in aqueous solutions have been reported by Billeter and van Gunsteren [8, 9], Ojamäe et al. [50, 65], Hofmann et al. [41] as well as Wolf and Groenhof [109].

A further force field formulation for the description of (de-)protonated aqueous solutions has been reported by Garofalini and coworkers [54, 57, 58, 106], which enables proton transfer between water, oxonium and hydroxide in aqueous solution in addition to reproducing structure, vibrational frequencies and other properties of liquid water to high accuracy. Furthermore, it has been shown that due to its functional form this model can be applied in a straightforward way in the low-level region of hybrid QM/MM simulations [40]. Such a treatment is for instance required when investigating the hydration of a hydrolysing solute [15, 88] such as Lewis acids and bases, since the associated hydrolysis products (H_3O^+ or OH^-) may migrate freely into the low-level zone [15]. Alternatively, such an approach offers the possibility to prepare solutions at different pH values. The associate excess protons or hydroxide ions may then neutralise the above mentioned hydrolysis products or induce chemical reactions at a solute located in the QM zone. However, since hybrid computations are conceptually more difficult compared to purely MM-based simulations, proton transfer reactions occurring throughout the simulation and the associated time-dependent topologies introduce further challenges when applying dissociative force fields in mixed quantum mechanical/molecular mechanical studies [15, 40].

4.2 Monitoring of Proton Transfer Events

Since any kind of reactive potential model enables the change of molecules as the simulation progresses, the respective topology (i.e. which atoms form a molecule at a given configuration) is not predetermined as in simulations using non-reactive force field formulations (*c.f.* Fig. 4.1a). In a dissociative description as realised in the Garofalini model oxygen and hydrogen atoms are considered as independent species forming molecules according to the respective potential energy surface and therefore, this topology information is not required to execute molecular simulations. All analyses linked to the time-dependent topology can be conveniently addressed after the simulation has finished in a post-processing step.

This situation is dramatically changed in hybrid simulations [40]: Typically, all species within a given radius of a predetermined QM center (*e.g.* a selected atom or the center-of-mass of the solute species) are included in the quantum mechanical treatment. Use of a particle based cutoff scheme (P-sum, *c.f.* Fig. 4.1b) risks that solvent molecules may be fractured, resulting in unphysical spin-states of the system. This type of QM setup typically leads to a variety of convergence problems, thus effectively preventing the successful execution of the simulation. While a number of approaches such as capping approaches [2, 29, 43, 112] or frontier orbital techniques [4, 33, 69, 87] exist to cap fractured bonds in QM/MM simulations, the small size and the associated high mobility of solvent molecules make these methods less attractive. For these reasons, a molecular based cutoff scheme

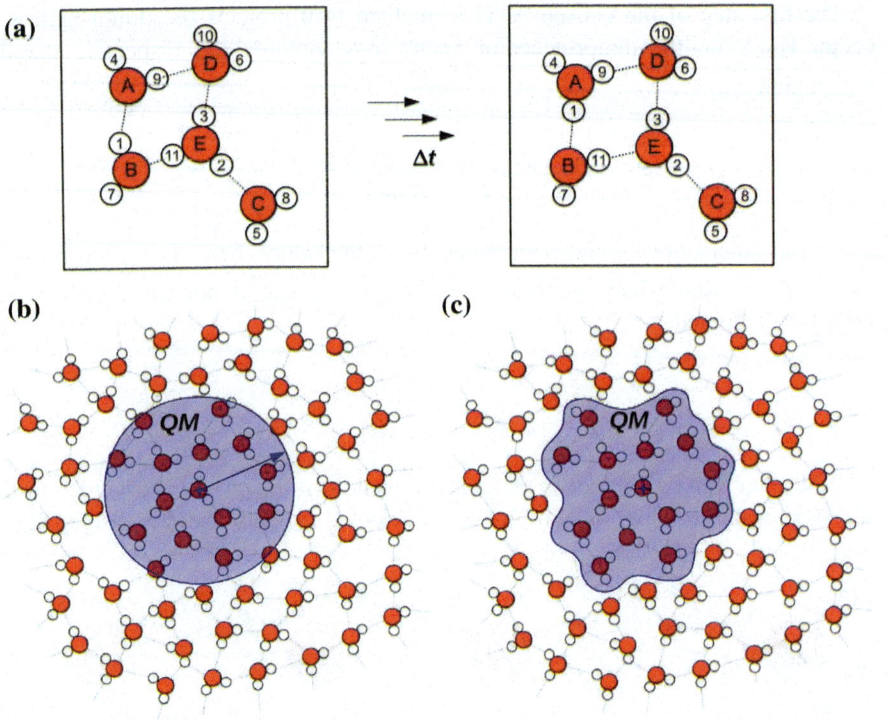

Fig. 4.1 **a** Application of a dissociative force field in molecular simulations enables changes of the molecular topology. Since oxygen and hydrogen atoms are independent species forming molecules, ions, etc. along the simulation, no topology information is required to execute purely classical simulation. **b** A particle-based assignment of atoms to the QM and MM regions results in a fragmentation of molecules, leading to unphysical spin-states effectively preventing the execution of a QM/MM simulation. **c** A molecule-based as assignment of atoms to the QM and MM regions requires knowledge of the actual chemical topology and thus, a monitoring of all occurring PT events

(M-sum, *c.f.* Fig. 4.1c) including all atoms of a molecule into the QM region based for instance on the respective centres of mass is the only viable option to achieve successful simulations. This on the other hand requires knowledge of the exact chemical topology for each configuration. Since in the dissociative description proton transfer events may occur, it is required that the time-dependence of the associated topology has to be monitored as the simulation progress. Because a user-based interference would be highly impractical in a simulation study requiring at least tens of thousands of individual configurations, typically executed on a high-performance computing (HPC) facility on a 24/7 basis, an automated topology update (ATU) criterion is required, enabling the simulation software to track proton transfer events independently and to adjust the chemical topology accordingly.

The first step of the chosen ATU formalism [40] projects the donor-hydrogen vector $\mathbf{r}_{DH}$ onto the donor-acceptor vector $\mathbf{r}_{DA}$ yielding the respective parallel component $\mathbf{r}_{DH}^{//}$.

$$\mathbf{r}_{DH}^{//} = \frac{(\mathbf{r}_{DH}^{T}\mathbf{r}_{DA})\mathbf{r}_{DA}}{\|\mathbf{r}_{DA}\|^2} \tag{4.1}$$

where the superscript ^{T} denotes the transpose of the vector. Next the ratio ρ between the projected donor-hydrogen distance $\|\mathbf{r}_{DH}^{//}\|$ and the donor-acceptor distance $\|\mathbf{r}_{DA}\|$ it computed:

$$\rho = \frac{\|\mathbf{r}_{DH}^{//}\|}{\|\mathbf{r}_{DA}\|} \tag{4.2}$$

Whenever this ratio exceeds the preset threshold $\bar{\rho}$ the proton is assigned to the acceptor species (cf. Fig. 4.2a, b). It should be noted that although the assignment of

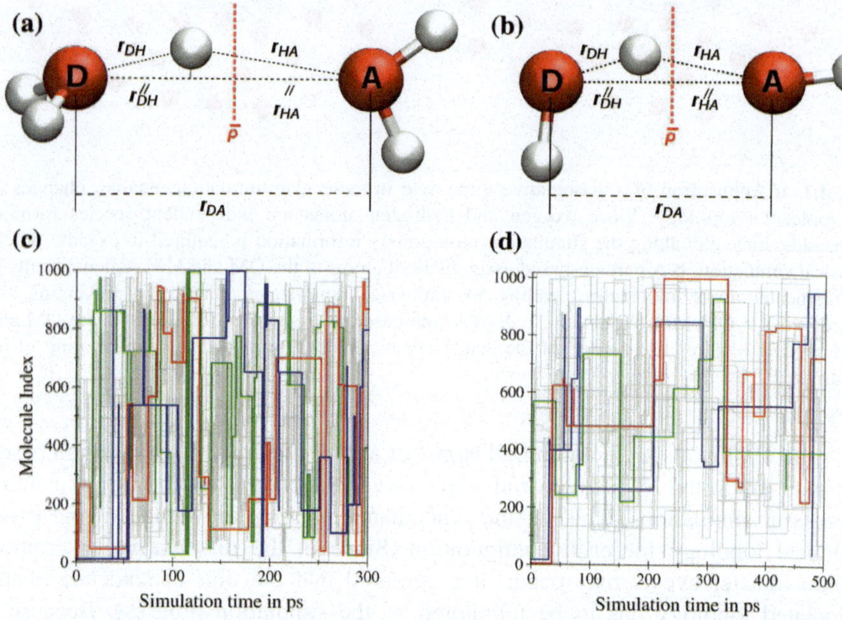

Fig. 4.2 Update criterion for a proton transfer of **a** an excess proton and **b** a proton-hole. As soon as the ratio ρ between the projected donor-hydrogen distance $\|\mathbf{r}_{DH}^{//}\|$ and the donor–acceptor distance $\|\mathbf{r}_{DA}\|$ exceeds the preset threshold $\bar{\rho}$, the proton transfer event is registered as a successful exchange. Registered proton transfer events observed in MD simulations of **c** 1M HCl and **d** 1M NaOH obtained for the recommended threshold ratio 0.585. *Horizontal lines* correspond to periods where water molecules carry an excess proton forming a H_3O^+ ion (HCl) or a proton-hole corresponding to a hydroxide ion OH^- (NaOH). *Vertical lines* correspond to proton transfer events to (HCl) or from (NaOH) the next molecule

the proton from the donor to the acceptor species results in a discontinuous change of the topology, the underlying potential energy surface described by the Garofalini model remains continuous.

A simulation study performed for an aqueous excess proton and 1M HCl [40] has shown that the outlined criterion is simple, time-independent and selective. Simplicity implies that this H-transfer criterion depends only on the position of the transferring hydrogen and the involved donor- and acceptor-atoms, but not on properties of other atoms (such as angle and dihedral data with respect to other hydrogen atoms bound to the donor or the acceptor). Enforcing such a simplicity in the update criterion results in a general formalism that is not restricted to H-transfer reactions between water molecules but can be applied to other systems as well as for example the transfer of a proton between a hydrohalogenide and a halide:

$$HBr + Cl^{-} \rightarrow Br^{-} + HCl \tag{4.3}$$

Ruling out a time-dependent criterion (such as the proton has to remain bound to the acceptor atom for a given time period $\bar{\tau}$) enables the application of the ATU in studies that do not propagate the system in time as it is the case in Monte-Carlo simulations and geometry optimisations.

Despite its simplicity the proposed criterion can be applied to discriminate between short-time fluctuations and sustained proton transfer events as demonstrated in Fig. 4.3 for an exemplary proton transfer event between the donor species and an acceptor molecule labeled 1 and 2, respectively. Figure 4.3a depicts the associated (unprojected) donor-hydrogen and hydrogen-acceptor distances. It can be seen that prior to the sustained proton transfer reaction occurring approximately at 2.1 ps these distances show short periods of overlap, which have to be considered as short-time fluctuations but not as a genuine transfer event. If a threshold ratio $\bar{\rho}$ of 0.5 is used, implying that the proton has to travel halfway from the donor to the acceptor to be considered transferred, all short-time fluctuation are registered as proton transfer immediately followed by the respective re-transfer. Mapping the respective topology changes to a graph indicating which of the two molecules carries the proton at a given instant leads to eight spikes prior to the actual transfer (cf. Fig. 4.3b). Increase of $\bar{\rho}$ to 0.525 (i.e. the proton has to travel 52.5 % of the donor-acceptor distance) eliminates a large fraction of these short-time fluctuations (cf. Fig. 4.3c) and a threshold of 0.55 is sufficient to only register the sustained proton transfer event (cf. Fig. 4.3d).

It can be seen that the number of registered proton transfer events depends strongly on the chosen value for $\bar{\rho}$ and it was of particular interest to investigate, whether a distinct value for this threshold criterion is to be preferred. To achieve this lengthy molecular dynamic simulations of an excess proton (300 ps) and hydroxide (0.5 ns) in aqueous solutions as well as 1M HCl (300 ps) and NaOH (0.5 ns) have been performed without monitoring the actual molecular topology. As outlined above this is perfectly possible, since the dissociative model treats intra- and intermolecular interactions with the same functional form [57, 106]. (The respective simulation protocols of this studies are summarised in the appendix).

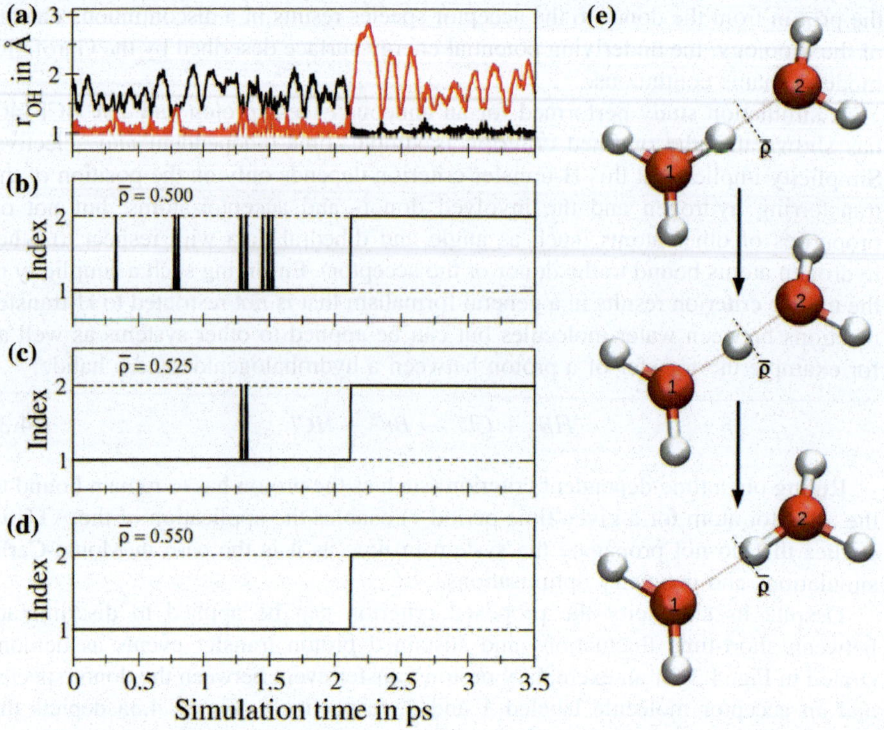

Fig. 4.3 **a** Donor-hydrogen (*red*) and donor-acceptor distance (*black*) of an exemplary proton transfer event between the donor species 1 and an acceptor molecule 2. Proton transfer maps determined using a threshold ratio $\bar{\rho}$ of **b** 0.500, **c** 0.525 and **d** 0.550. **e** Sketch of a sustained proton transfer passing the threshold $\bar{\rho}$

In the next step the obtained trajectories have been probed to monitor all occurring proton transfer events, which are then depicted in form of proton transfer maps (PTMs, cf. Fig. 4.2c, d). In these depictions horizontal lines correspond to time periods where a water molecule is carrying an excess proton forming a H_3O^+ ion in case of acidified systems or a proton-hole (PH) corresponding to an OH^- ion in simulations of basic systems. In contrast each vertical line represents a proton transfer event. In case an excess proton is monitored in the PTM, it is transferred to the next molecule in the graph, which then transform into an H_3O^+ ion. On the other hand when monitoring proton-hole migration, the direction of the transfer is reversed [50]: now the proton is transferred from the next molecule, which in turn becomes a hydroxide ion. This distinction is crucial when discussing the analysis of diffusive properties of protons and proton-holes discussed below.

With PTM data available it is straightforward to count the number of registered proton transfer events per excess proton/hydroxide N_{PT} as a function of the employed threshold ratio $\bar{\rho}$. Normalising N_{PT} by the simulation time t_{sim} yields the average proton hopping rate h, which is depicted in Fig. 4.4 for the excess proton

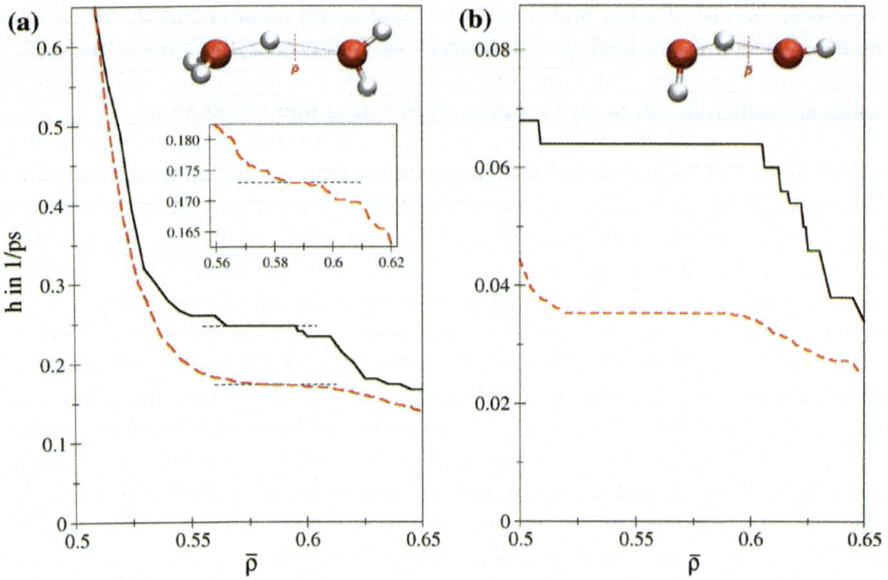

Fig. 4.4 Average proton hopping rate h per picosecond as a function of the threshold parameter $\bar{\rho}$ obtained for a simulation study of **a** an excess proton in aqueous solution (*solid black*) and 1 M HCl (*dashed red*) and **b** an aqueous excess hydroxide (*solid black*) and 1 M NaOH (*dashed red*)

and hydroxide in aqueous solution acting as model systems for infinite dilution as well as 1 M HCl and NaOH solutions. It can be seen that for all cases a similar pattern is observed, showing a steep decrease close to a value of 0.5, which levels out at different values of $\bar{\rho}$ for the different systems. In all cases a plateau of different width can be identified, after which the average hopping rate is further decreased. The observed dependence of h with respect to the $\bar{\rho}$ can be explained when compared to the data shown in Fig. 4.3 discussed above: In case of a too low threshold value short-time fluctuations contribute to h. As $\bar{\rho}$ approaches higher values only sustained exchange events are registered which explains the observed leveling of h. In case of 1M HCl the plot does not appear to form a plateau, but forms a saddle point. Close inspection of the respective graph reveals a small plateau in the region 0.585–0.59, however. Since this interval lies within the stationary regions observed for all other systems depicted in Fig. 4.4, it appears as a natural choice for the update criterion to automatically adjust the chemical topology in simulations studies, with 0.585 being the recommended value since it marks the lower bound of the interval. Simulations performed for an aqueous excess proton have shown that this recommendation for $\bar{\rho}$ is also valid for elevated (323 K) as well as reduced (283 K) temperatures [40].

In case the threshold is increased beyond this interval, the hopping rate decreases even further, albeit not at the same rate as observed at the beginning of the graph near a $\bar{\rho}$-value of 0.5. This decrease can be interpreted that in this region a number of sustained and thus, important exchange events are missed by the analysis and therefore, the use of a too high threshold value is not recommended. As expected higher concentrations lead to a reduction of h for both the proton as well as proton-hole migration compared to the excess proton and hydroxide in aqueous solutions, which act as model systems representing infinite dilution.

A further striking detail observed in Fig. 4.4 is the fact that the number of PT events in case of hydroxide is substantially lower than in case of the excess proton. The same applies to the case of 1 M NaOH when compared to 1 M HCl. In addition the observed leveling in the dependence of h with respect to the $\bar{\rho}$ is much broader in case of hydroxide containing systems. While it is known from experiment [52] as well as simulations [50], that the diffusion and thus the number of exchange events is higher in case of the excess proton compared to the proton-hole migration, the neglect of nuclear quantum effects in classical molecular dynamics simulations may be responsible for the observed differences. While it has been concluded that nuclear quantum effects (NQEs) play a minor role in case of proton transfer reaction of excess protons [60, 90], showing no sign of tunneling and only a very minor influence on the potential energy surface due to zero-point contributions [59, 110], a different situation is observed in case of proton-hole migration. In this case the influence of NQEs was shown to be significant [60, 90].

Therefore, the simulations have also been carried out replacing all hydrogen atoms in the system with deuterium. Since hydrogen and deuterium are chemically similar in that they exhibit the same potential both in electronic structure theory as well as in classical potential models, the deuterated systems can be treated with the same interaction potential and threshold ratio $\bar{\rho}$ as their hydrogenated counterparts. However, due to the increased mass of deuterium nuclear quantum effects are in general considered to be strongly diminished. Table 4.1 compares the respective data obtained for the different systems with simulation results obtained after deuteration. In case of the diluted systems the average proton hopping rate is noticeably reduced upon deuteration, while for concentrated solutions very similar h-values are obtained.

The observed hopping rates of 0.25 ps^{-1} obtained for the aqueous excess proton is well within the range of 0.16–0.25 ps^{-1} obtained with MS-EVB models [78, 111]. Explicit consideration of either the electronic structure via CPMD simulation or nuclear quantum effects via path integral molecular dynamics leads to an increase of the hopping rate to 0.5 ps^{-1} [78, 89]. Application of a scaled Ojamäe-Shavitt-Singer (sOSS2) potential yields an average time between proton transfer events of 0.72 ps [50] corresponding to a h value of 1.38 ps^{-1}, which is higher than the experimental estimate of 0.75 ps^{-1} [110] based on the average lifetime of a proton transfer determined via NMR measurements [56].

Table 4.1 Comparison of the proton hopping rate h in ps^{-1} and the diffusion coefficient D in $Å^2$/ps obtained from molecular dynamics simulations using the dissociative model

	H^+/H_2O	D^+/D_2O	1 M HCl	1 M DCl	OH^-/H_2O	OD^-/D_2O	1 M NaOH	1 M NaOD
h	0.25	0.15	0.17	0.16	0.064	0.032	0.035	0.032
D	0.38	0.35	0.29	0.30	0.28	0.23	0.22	0.21

For proton transport between water and hydroxide Ufimetsev et al. determined a hopping rate of 0.19 ps^{-1} [100] using an MS-EVB model. Lee and coworkers report an average time span between PT events of 2.52 ps [50] using the sOSS2 potential model, which corrsponds to a h-value of 0.40 ps^{-1}. These data are significantly larger than 0.064 ps^{-1} obtained via the Garofalini model. Neither of these investigations takes the influence of nuclear quantum effects into account, which highlights the potential benefit when performing simulations also for deuterated systems. Such an approach eliminates the uncertainty introduced by a classical treatment, thus enabling a direct comparison of the various parametrisations.

4.3 Locator Approach to Analyse Proton and Proton–Hole Diffusion

In addition to the evaluation of the proton hopping rate h the data provided by the proton transfer maps act as prerequisite for the evaluation of proton diffusion. Wright [110] distinguished two stages of the diffusion of an excess proton in liquid water: As long as it remains bound to its donor, the diffusion is limited by the heavy oxygen atom. Transfer of one of the protons to an adjacent water molecule gives rise to the diffusive properties of the proton, after which the diffusion is again determined by the oxygen atom of the acceptor species. Based on this discussion it is important to note that proton diffusion is different from the diffusion of a single hydrogen core and hence, the diffusion should rather be expressed via a generalised metaparticle [40]. This particle referred to as locator L links subsequent proton transfer events into a single diffusive entity by following the respective donor atom and upon transfer of a proton to the next molecule, migrates along an interpolated path from the donor to the acceptor atom (cf. Fig. 4.5). This interpolation is carried out by least-square fitting of a quadratic polynomial to the donor-hydrogen and hydrogen acceptor vectors $\mathbf{r}_{DH}$ and $\mathbf{r}_{HA}$ under the constraint that the interpolated graph visits the coordinates of the donor and acceptor atoms. Since the donor-hydrogen-acceptor configuration is in general close to linear in proton-transfer reactions, the contribution of the quadratic term is very minor. To determine the position of the locator particle on the interpolated graph a smoothing function [46] S_{PT} continuously moving from zero to one is employed:

$$S_{PT} = 6(x - 0.5)^5 - 5(x - 0.5)^3 + \frac{15}{8}(x - 0.5) + 0.5 \tag{4.4}$$

with the helper variable x being:

$$x = \frac{\|\mathbf{r}_{DH}^{//}\| - r_{\mathrm{on}}}{r_{\mathrm{off}} - r_{\mathrm{on}}} \tag{4.5}$$

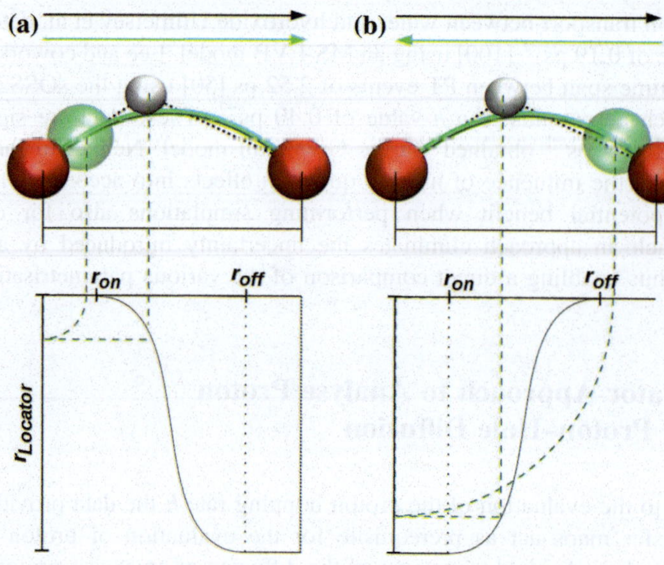

Fig. 4.5 Interpolation scheme to determine the position of the locator particle *L* (*green*) in case of **a** proton and **b** proton-hole migration. The interpolation is only applied in the interval r_{on} to r_{off} to ensure that any oscillation about the equilibrium position of the OH bond does not interfere with the determination of the diffusion coefficient of *L*. While in case of a proton transfer the locator moves in the same direction, the direction is opposed in case of proton-hole migration

As can be expected it has been shown that this interpolation should only be applied once a proton is significantly displaced from its equilibrium position, which is realised by restricting the interpolation to the interval between the radii r_{on} and r_{off} (cf. Fig. 4.5). It was shown that this restriction is a crucial prerequisite to ensures that oscillations of the proton about its equilibrium position do not influence the determination of the diffusion coefficient. A value of approx. ±1.0 Å for the two bounds of r_{on} and r_{off} was found to be the minimum value to yield consistent diffusion coefficients [40]. Too large values for r_{on} and r_{off} turning the interpolation into a step function (i.e. the locator is instantaneously transferred from the donor to the acceptor bypassing the interpolation) were found to only have a little influence on the resulting diffusion coefficient [40].

After the interpolation procedure the previous acceptor atom now acts as donor and is followed until the next proton transfer reaction takes place. The trajectory of the locator particle is thus not a measure for the diffusion of an individual proton but corresponds to the diffusion of the excess charge throughout the system. Application of the Einstein relation thus enables the determination of the diffusion coefficient of *L*:

$$D = \frac{1}{2d} \lim_{t \to \infty} \frac{\left\langle (\mathbf{r}_t - \mathbf{r}_0)^2 \right\rangle}{t} \tag{4.6}$$

In case of the basic systems the monitoring of the diffusion of the proton-hole is essentially similar (cf. Fig. 4.5b). Again a locator particle is employed to connect donor and acceptor species, however, in this case the donor atom has to be followed instead of the acceptor. To achieve this the interpolation has to be reversed: upon transfer of the proton the locator particle migrates in the opposite direction, because the donor species will carry the proton-hole after completion of the transfer reaction. Reversal of the interpolation can be realised in a straightforward by inverting the smoothing function resulting in S_{PH}:

$$\begin{aligned} S_{PH} = 1 - S_{PT} = 0.5 - 6(x - 0.5)^5 + 5(x - 0.5)^3 \\ - \frac{15}{8}(x - 0.5) \end{aligned} \tag{4.7}$$

A sketch of different stages of the interpolation in case of a proton and proton-hole transfer is depicted in Fig. 4.6. It can be seen that for a H-transfer the locator particle migrates into the same direction as the proton, whereas the directions are opposite in case of the proton-hole.

The respective diffusion coefficients obtained for the different systems have been included in Table 4.1 as well. It can be seen that despite the lack of nuclear quantum effects, the order of D obtained for dilute systems is $H_3O^+ > OH^- > H_2O$ (the latter being reported as 0.245 Å^2/ps for the Garofalini model [57]). This order agrees well with experimental [52] and theoretical [50] data. The increased concentration in case of the 1 M HCl and NaOH solutions leads to an overall decrease of the diffusivity as expected.

Although the observed diffusion coefficient for the excess proton observed as 0.38 Å^2/ps is lower than the experimental value of 0.93 Å^2/ps [70, 73], it is well within the range of 0.29–0.77 Å^2/ps determined by various investigations employing the MS-EVB [13, 66, 95, 101, 111] approach and simulations performed with the sOSS2 dissociative potential yielding a value of 0.762 Å^2/ps for 0.26 M HCl [50].

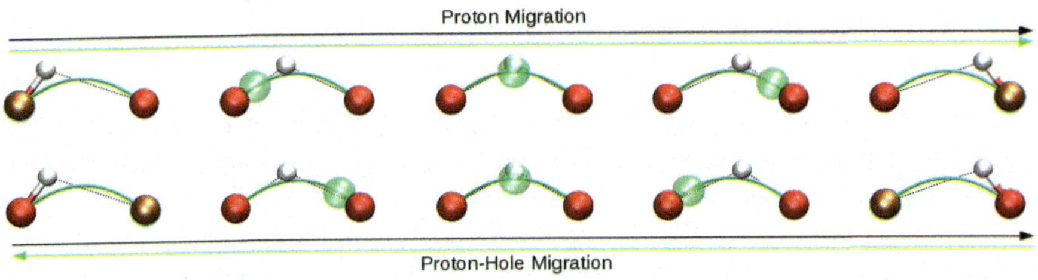

Fig. 4.6 Locator particle (*green*) at different stages of the migration of a proton (*top*) and a proton-hole (*bottom*), respectively. While for the proton transfer the locator particle migrates into the same direction as the proton, the directions are opposite in case of proton-hole migration

The diffusion coefficient determined for the proton-hole migration of 0.28 Å^2/ps agress well with data obtained for a charge-ring and an MS-EVB model for aqueous hydroxide reported as 0.17 and 0.31 Å^2/ps [100], respectively. Simulation data obtained with the modified sOSS2 potential yields a value 0.456 Å^2/ps [50] being closest to the experimentally determined value of 0.53 Å^2/ps [52].

As expected the influence of deuteration on the diffusive properties is strongly dependent on the pH value of the systems. For excess protons a noticeable decrease of D is observed, the associated kinetic isotope effect of 1.31 being in good agreement with the experimental value of 1.39 [59].

Also in case of hydroxide containing systems a decrease in D is observed, however, the difference is much smaller. Considering the low number of registered PT events and the fact that NQEs are strongly diminished in deuterated systems, the results clearly emphasize the importance of NQE for PT reactions in case of OH^-.

A widely employed approach to investigate nuclear quantum effects in molecular simulations is the path-integral molecular dynamics (PIMD) framework [12, 16, 17, 22, 27, 44, 67], propagating a number of interacting replicas of the system in time. (Use of only one replica corresponding to standard molecular dynamics, which is thus considered as classical limit.) However, it has been shown that this framework can only deliver approximate time-correlation functions, formally being valid only in the case ($\lim t \rightarrow 0$) [23]. This on the other hand is contradictory to the evaluation of the diffusion coefficient, which is only valid in the long-time limit $\lim t \rightarrow \infty$. The question whether the neglect of nuclear quantum effects or the use of long-time correlation functions to analyse PIMD simulations results in a larger error is strongly dependent on the system in question and the property to be analysed. To the best of knowledge no data on the diffuion coefficient obtained from PIMD simulations for proton hole migration is available as of yet. In this context purely classical simulations performed with and without deuteration of the system provide a valueable first reference for future investigations of proton transfer phenomena. Because the influence of NQEs can be negelected in case of deuterated systems, it provides an ideal framework to not only compare different theoretical approaches but also validate the different potential models against experiments performed with and without deuterated solutions.

4.4 Concluding Remarks

The observed differences between the outlined potential models as well as with respect to experiment highlight the diverse and challenging nature proton transfer reactions pose in modern research more than 200 years [59] after the influential publication of de Grotthus [25] laying the foundation of todays perception of proton migration via the well-known Grotthus mechanism. Nevertheless, the presented data clearly demonstrates that the outlined automated topology update as well as the determination of proton diffusion via a locator particle approach yields data in good agreement with other simulation techniques. Initially only applied to systems containing excess protons, the systematic comparison with hydroxide containing systems confirms that these techniques provide simple yet effective means to enable an independent analysis of proton transfer reactions by the simulation software. Based on the presented data it can be concluded that irrespective of the involved species (i.e. H_3O^+ or OH^-) the use of a common threshold ratio $\bar{\rho}$ is possible with 0.585 marking the recommended value. Since a monitoring of proton transfer events and the associated time-dependent topology is a crucial prerequisite for the application of dissociative water potentials in QM/MM simulation studies, the possibility to employ a single threshold $\bar{\rho}$ for proton transfer reactions between water, H_3O^+ and OH^- is of particular advantage. For instance hydrolysing solutes located in the QM region can be solvated with an appropriately adjusted solution in the MM region having a pH value that enables neutralisation of hydrolysis products generated along the simulation. It remains to be seen, however, whether this technique is applicable in the same way in simulation studies explicitly taking nuclear quantum effects into account.

Furthermore, the data obtained in case of deuterated systems was shown to be of high importance when estimating the influence of nuclear quantum effects, which highlights the importance to study of proton transfer phenomena with and without deuteration.

Considering the consistently increasing capacities of computational facilities, it can be expected that simulation studies will provide increasingly accurate and consistent data of proton transfer phenomena, which due to their extremely fast dynamical properties occurring on the picosecond scale are challenging targets for experimental and theoretical investigations alike.

Acknowledgments This work was supported by the Austrian Ministry of Science BMWF as part of the Konjunkturpaket II of the Focal Point Scientific Computing at the University of Innsbruck.

Appendix: Simulation Protocol

Starting structures for all simulations have been prepared by pre–equilibration of a number of simulations systems with an adequate number of water molecules employing periodic, cubic simulation cells for 100 000 MD steps. The velocity-Verlet algorithm has been employed to integrate the equations of motion with a time step of 0.2 fs. The Nose Hoover thermostat [42, 64] was employed to maintain constant temperature of 298.15 K. To account for long-range Coulombic interactions, the Wolf summation technique [108] with a cutoff distance of 10.0 Å was applied as defined in the Garofalini model [57, 106]. The proton transfer update criterion ρ was set to the recommended value of 0.585 [40].

Next, random water molecules have been adjusted to form H_3O^+ and OH^- ions or have been replaced by Na^+ and Cl^- ions as required. 1 M HCl and NaOH systems have been generated by replacing random water molecules with Na^+ or Cl^- and by adding/deleting hydrogen atoms to/from another randomly chosen water molecule, yielding net stochiometries of 17 HCl and 18 NaOH molecules per 1000 water molecules. For deuterated systems each hydrogen atoms was replaced by deuterium.

The solutions have then been equilibrated for at least 1 000 000 MD steps (200 ps). Data sampling has been performed for 1 500 000 and 2 500 000 MD steps in case of acidified and basic systems, respectively, corresponding to sampling times of 300 and 500 ps. To tightly monitor all occurring proton transfer events data sampling was performed every tenth MD step, leading to trajectory file sizes of approximately 60–110 GB per system, respectively.

References

1. Allen MP, Tildesley DJ (1990) Computer simulation of liquids. Oxford Science Publications, Oxford
2. Amara P, Field MJ (2003) Evaluation of an ab initio quantum mechancial/molecular mechanical hybrid-potential link-atom method. Theor Chem Acc 109:43
3. Anglada JM, Martins-Costa M, Ruiz-López MF, Francisco JS (2014) Spectroscopic signatures of ozone at the airwater interface and photochemistry implications. Proc Natl Acad Sci USA 111:11, 618
4. Assfeld X, Rivail J (1996) Quantum chemical computations on parts of large molecules: the ab initio local self consistent field method. Chem Phys Lett 263:100
5. Bakowies D, Thiel W (1996) Hybrid models for combined quantum mechanical and molecular mechanical approaches. J Phys Chem 100(25):10, 580
6. Bao J, Feng X, Yu J (2011) GPU triggered revolution in computational chemistry. Acta Phys Chim Sin 27(9):2019
7. Bhattacharjee A, Weiss AKH, Artero V, Field MJ, Hofer TS (2014) Electronic structure and hydration of tetramine cobalt hydride complexes. J Phys Chem B 118:5551
8. Billeter SR, van Gunsteren WF (1997) A modular molecular dynamics/quantum dynamics program for non-adiabatic proton transfers in solution. Comp Phys Commun 107:61
9. Billeter SR, van Gunsteren WF (1998) Protonizable water model for quantum dynamical simulations. J Phys Chem A 102:4669
10. Borgis D, Hynes JT (1991) Molecular dynamics simulation for a model nonadiabatic proton transfer reaction in solution. J Chem Phys 350:3619
11. Borgis D, Hynes JT (1993) Dynamical theory of proton tunneling transfer rates in solution: general formulation. Chem Phys 170:315
12. Braams BJ, Manolopoulos DE (2006) On the short-time limit of ring polymer molecular dynamics. J Chem Phys 125:124, 105
13. Brancato G, Tuckerman ME (2005) A polarizable multistate empirical valence bond model for proton transport in aqueous solution. J Chem Phys 122:224, 507
14. Buehler MJ, van Duin ACT, Goddard WA III (2006) Multiparadigm modeling of dynamical crack propagation in silicon using a reactive force field. Phys Rev Lett 96:095, 505
15. Canaval LR, Lutz OM, Weiss AKH, Huck CW, Hofer TS (2014) Electronic structure and hydration of tetramine cobalt hydride complexes. Inorg Chem 53:11, 861
16. Cao J, Voth GA (1994a) The formulation of quantum statistical mechanics based on the feynman path centroid density. i. equilibrium properties. J Chem Phys 100:5093
17. Cao J, Voth GA (1994b) The formulation of quantum statistical mechanics based on the feynman path centroid density. ii. dynamical properties. J Chem Phys 100:5106
18. Car R, Parrinello M (1985) Unified approach for molecular-dynamics and density functional theory. Phys Rev Lett 55(22):2471

19. Ceriotti M (2010) A novel framework for enhanced molecular dynamics based on the generalized langevin equation. PhD thesis, Swiss Federal Institute of Technology, Zürich, Switzerland
20. Ceriotti M, Bussi G, Parrinello M (2011) Accelerating the convergence of path integral dynamics with a generalized langevin equation. J Chem Phys 134:084, 104
21. Cook DB (2005) Handbook of computational chemistry. Dover Publications, New York
22. Craig IR, Manolopoulos DE (2004a) Quantum statistics and classical mechanics: Real time correlation functions from ring polymer molecular dynamics. J Chem Phys 121:3368
23. Craig IR, Manolopoulos DE (2004b) Quantum statistics and classical mechanics: real time correlation functions from ring polymer molecular dynamics. J Chem Phys 121:3368
24. Cramer CJ (2002) Essentials of computational chemistry. Wiley, West Sussex
25. de Grotthuss CJT (1806) Mémoire–sur la décomposition de l'eau et des corps qu'elle tient en dissolution à l'aide the l'électricité galvanique. Ann Chim (Paris) LVIII:54
26. van Duin ACT, Dasgupta S, Lorant F, Goddard WA (2001) ReaxFF: a reactive force field for hydrocarbons. J Phys Chem A 105(41):9396
27. Feynman RP, Hibbs AR, Styer DF (2010) Quantum mechanics and path integrals, Emended edn. Dover Publications, New York
28. Field MJ, Bash PA, Karplus M (1990a) A combined quantum mechanical and molecular mechanical potential for molecular dynamics simulations. J Comput Chem 11(6):700
29. Field MJ, Bash PA, Karplus M (1990b) A combined quantum mechanical and molecular mechanical potential for molecular dynamics simulations. J Comput Chem 11(6):700
30. Frenkel D, Smit B (2002) Understanding Molecular Simulation. Academic press, San Diego – London
31. Gao J (1993) Potential of mean force for the isomerization of dmf in aqueous solution: a monte carlo QM/MM simulation study. J Am Chem Soc 115:2930
32. Gao J (1996) Hybrid quantum and molecular mechanical simulations: an alternative avenue to solvent effects in organic chemistry. Acc Chem Res 29:298
33. Gao J, Amara P, Alhambra C, Field MJ (1998) A generalized hybrid orbital (gho) method for the treatment of boundary atoms in combined QM/MM calculations. J Phys Chem A 102 (24):4714
34. Helgaker T, Jørgensen P, Olsen J (2000) Molecular electronic structure theory. Wiley, Chichester
35. Hockney RW, Eastwood JW (1989) Computer simulation using particles. Taylor & Francis, New York
36. Hofer TS (2014) Perspectives for hybrid ab initio/molecular mechanical simulations of solutions: from complex chemistry to proton-transfer reactions and interfaces. Pure & Appl Chem 86:105
37. Hofer TS, Pribil AB, Randolf BR, Rode BM (2010a) Ab initio quantum mechanical charge field molecular dynamics: A nonparametrized first-principle approach to liquids and solutions. Adv Quant Chem 59:213
38. Hofer TS, Rode BM, Pribil AB, Randolf BR (2010b) Simulations of liquids and solutions based on quantum mechanical forces. Adv Inorg Chem 62:143
39. Hofer TS, Weiss AKH, Randolf BR, Rode BM (2011) Hydration of highly charged ions. Chem Phys Lett 512(4–6):139. doi:10.1016/j.cplett.2011.05.060
40. Hofer TS, Hitzenberger M, Randolf BR (2012) Combining a dissociative water model with a hybrid QM/MM approach—a simulation strategy for the study of proton transfer reactions in solution. J Chem Theory Comput 8:3586
41. Hofmann DW, Kuleshova L, DAguanno B (2007) A new reactive potential for the molecular dynamics simulation of liquid water. Chem Phys Lett 448(13):138
42. Hoover WG (1983) Nonequilibrium molecular dynamics. Annu Rev Phys Chem 34:103
43. Ihrig AC, Schiffmann C, Sebastiani D (2011) Specific quantum mechanical/molecular mechanical capping-potentials for biomolecular functional groups. J Chem Phys 135:214, 107
44. Jang S, Voth GA (1999) A derivation of centroid molecular dynamics and other approximate time evolution methods for path integral centroid variables. J Chem Phys 111:2371

45. Jensen F (2006) Introduction to computational chemistry, 2nd edn. John Wiley & Sons Ltd., Chichester
46. Kerdcharoen T, Morokuma K (2002) Oniom-xs: an extension of the oniom method for molecular simulation in condensed phase. Chem Phys Lett 335:257
47. Knight C, Voth GA (2012) The curious case of the hydrated proton. Acc Chem Res 45:101
48. Koch W, Holthausen MC (2002) A chemist's guide to density functional theory, 2nd Edn. Wiley–VCH, Weinheim
49. Leach AR (2001) Molecular modelling, 2nd edn. Prentice-Hall, Essex
50. Lee SH, Rasaiah JC (2011) Proton transfer and the mobilities of the H^+ and OH^- ions from studies of a dissociating model for water. J Chem Phys 135:124, 505
51. Liang T, Shin YK, Cheng Y, Yilmaz DE, Vishnu KG, Verners O, Zou C, Phillpot SR, Sinnott SB, van Duin AC (2013) Reactive potentials for advanced atomistic simulations. Ann Rev Mater Res 43:109
52. Light TS, Licht S, Bevilacqua AC, Morash KR (2008) Reactive potentials for advanced atomistic simulations. Electrochem Solid State Lett 8:E16
53. Lin H, Truhlar DG (2007) QM/MM: what have we learned, where are we, and where do we go from here? Theor Chem Acc 117:185
54. Lockwood GK, Garofalini SH (2013) Lifetimes of excess protons in water using a dissociative water potential. J Phys Chem B 117:4089
55. Lutz OMD, Messner CB, Hofer TS, Glätzle M, Huck CW, Bonn GK, Rode BM (2013) Combined ab initio computational and infrared spectroscopic study of the cis- and trans-bis (glycinato)copper(ii) complexes in aqueous environment. J Phys Chem Lett 4:1502
56. Luz Z, Meiboom S (1964) The activation energies of proton transfer reactions in water. J Am Chem Soc 86:4768
57. Mahadevan TS, Garofalini SH (2007) Dissociative water potential for molecular dynamics simulations. J Phys Chem B 111:8919
58. Mahadevan TS, Garofalini SH (2009) Dissociative chemisorption of water onto silica surfaces and formation of hydronium ions. J Phys Chem C 112:1507
59. Marx D (2006) Proton transfer 200 years after von Grotthuss: insights from ab initio simulations. Chem Phys Chem 7:1848
60. Marx D, Tuckerman ME, Hutter J, Parrinello M (1999) The nature of the hydrated excess proton in water. Nature 397:601
61. McQuarrie DA (1976) Statistical mechanics. Harper & Row, New York
62. Messner CB, Bonn GK, Hofer TS (2015) QM/MM md simulations of la(iii)-phosphopeptide complexes. Mol BioSyst 11:232
63. Moin S, Hofer TS (2014) Hydration of porphyrin and mg-porphyrin: ab initio quantum mechanical charge field molecular dynamics simulations. Mol BioSyst 10:117
64. Nose S (1984) A unified formulation of the constant temperature molecular-dynamics methods. J Chem Phys 81:511
65. Ojamäe L, Shavitt I, Singer S (1998) Potential models for simulations of the solvated proton in water. J Chem Phys 109(13):5547. doi:10.1063/1.477173
66. Park K, Lin W, Paesani F (2012) A refined MS-EVB model for proton transport in aqueous environments. J Phys Chem B 116:343
67. Parrinello M, Rahman A (1984) Study of an F center in molten kcl. J Chem Phys 80:860
68. Penrose O (2005) Foundations of statistical mechanics: a deductive treatment. Dover Publications, New York
69. Philipp D, Friesner RA (1999) Mixed ab initio qm/mm modeling using frozen orbitals and tests with alanine dipeptide and tetrapeptide. J Comput Chem 20(14):1468
70. Pines E, Huppert D, Agmon N (1988) Geminate recombination in excited-state proton-transfer reactions—numerical solution of the debye-smoluchowski equation with backreaction and comparison with experimental results. J Chem Phys 88:5620
71. Rahman A, Stillinger FH, Lemberg HL (1975) Study of a central force model for liquid water by molecular dynamics. J Chem Phys 63:5223

72. Ramachandran KI, Deepa G, Namboori K (2008) Computational chemistry and molecular modeling: principles and applications. Springer, Berlin
73. Roberts NK, Northey HL (1974) Proton and deuteron mobility in normal and heavy water solutions of electrolytes. J Chem Soc Faraday Trans 70:253
74. Rode BM, Schwenk CF, Hofer TS, Randolf BR (2005) The combination of quantum chemistry and statistical simulations: a most powerful tool to access structure and dynamics of liquid systems. Coord Chem Rev 249:2993
75. Rode BM, Hofer TS, Randolf BR, Schwenk CF, Xenides D, Vchirawongkwin V (2006) Ab initio quantum mechanical charge field molecular dynamics—a qm/mm md procedure for accurate simulations of ions and complexes. Theor Chem Acc 115:77bbl
76. Sadus RJ (1999) Molecular simulation of fluids: theory, algorithms, and object-orientation. Elsevier, Amsterdam
77. Sagnella DE, Tuckerman ME (1998) An empirical bvalence bond model for proton transfer in water. J Chem Phys 108:2072
78. Schmitt UW, Voth GA (1998) Multistate-empirical valence bond model for proton transport in water. J Phys Chem B 102:5547
79. Schwabl F (2010) Statistical mechanics, 2nd edn. Springer, Berlin
80. Senn HM, Thiel W (2007) QM/MM studies of enzymes. Curr Opin Chem Biol 11:182
81. Sholl DS, Steckel JA (2009) Density functional theory—a practical introduction. Wiley, Hoboken
82. Staib A, Borgis D (1995) Molecular dynamics simulation of an excess charge in water using mobile gaussian orbitals. J Chem Phys 103:2642
83. Stillinger FH (1975) Theory and molecular models for water. Adv Chem Phys 31:1
84. Stillinger FH, Rahman A (1978) Revised central force potentials for water. J Chem Phys 68:666
85. Strnad M, Martins-Costa MTC, Millot C, Tuñón I, Ruiz-López MF, Rivail JL (1997) Molecular dynamics simulations of elementary chemical processes in liquid water using combined density functional and molecular mechanics potentials. ii. Charge separation processes. J Chem Phys 106:3643
86. Szabo A, Ostlund NS (1996) Modern quantum chemistry. Dover Publications, New York
87. Thèry V, Rinaldi D, Rivail J, Maigret B, Ferenczy GG (1994) Quantum mechanical computations on very large molecular systems: The local self-consistent field method. J Comput Chem 15:269
88. Tirler AO, Hofer TS (2014) Structure and dynamics of the uranyl tricarbonate complex in aqueous solution: Insights from quantum mechanical charge field molecular dynamics. J Phys Chem B 118:12, 938
89. Tuckerman M, Laasonen K, Sprik M, Parrinello M (1995) Ab initio molecular dynamics simulation of the solvation and transport of hydronium and hydroxyl ions in water. J Chem Phys 103:150
90. Tuckerman M, Marx D, Klein M, Parrinello M (1997) On the quantum nature of the shared proton in hydrogen bonds. Science 275:817
91. Tuckerman ME (2010) Statistical mechanics: theory and molecular simulation. Oxford University Press, New York
92. Tuñón I, Martins-Costa MTC, Millot C, Ruiz-López MF (1995) Coupled density functional/molecular mechanics monte carlo simulations of ions in water. The bromide ion. Chem Phys Lett 241:450
93. Tuñón I, Martins-Costa MTC, Millot C, Ruiz-López MF, Rivail JL (1996) A coupled density functional-molecular mechanics monte carlo simulation method: the water molecule in liquid water. J Comput Chem 17:19
94. Tuñón I, Martins-Costa MTC, Millot C, Ruiz-López MF (1997) Molecular dynamics simulations of elementary chemical processes in liquid water using combined density functional and molecular mechanics potentials. i. proton transfer in strongly H-bonded complexes. J Chem Phys 106:3633

95. Čuma M, Schmitt UW, Voth GA (2001) A multi-state empirical valence bond model for weak acid dissociation in aqueous solution. J Phys Chem A 105:2814
96. Ufimtsev IS, Martinez TJ (2008a) Graphical processing units for quantum chemistry. Comput Sci Eng 10:26
97. Ufimtsev IS, Martinez TJ (2008b) Quantum chemistry on graphical processing units. 1. Strategies for two-electron integral evaluation. J Chem Theory Comput 4.222
98. Ufimtsev IS, Martinez TJ (2009a) Quantum chemistry on graphical processing units. 2. Direct self-consistent-field implementation. J Chem Theory Comput 5:1004
99. Ufimtsev IS, Martinez TJ (2009b) Quantum chemistry on graphical processing units. 3. Analytical energy gradients, geometry optimization, and first principles molecular dynamics. J Chem Theory Comput 5:2619
100. Ufimtsev IS, Kalinichev AG, Todd MJ, Kirkpatrick RJ (2009) A multistate empirical valence bond model for solvation and transport simulations of OH in aqueous solutions. Phys Chem Chem Phys 11:9420
101. Vuilleumier R, Borgis D (1999) Transport and spectroscopy of the hydrated proton: a molecular dynamics study. J Chem Phys 111:4251
102. Vuilleumier R, Borgis D (2012) Proton conduction: hopping along hydrogen bonds. Nat Chem 4:432
103. Warshel A, Levitt M (1976) Theoretical studies of enzymic reactions: Dielectric, electrostatic and steric stabilization of the carbenium ion in the reaction of lysozyme. J Mol Biol 103:227
104. Warshel A, Weiss RM (1980) An empirical valence bond approach for comparing reactions in solutions and in enzymes. J Am Chem Soc 102:6218
105. Watson M, Olivares-Amaya R, Edgar RG, Aspuru-Guzik A (2010) Accelerating correlated quantum chemistry calculations using graphical processing units. Comput Sci Eng 12:40
106. Webb MB, Garofalini SH, Scherer GW (2009) Use of a dissociative potential to simulate hydration of na^{+} and cl^{-} ions. J Phys Chem B 113:9886
107. Weiss AK, Hofer TS (2012) Exploiting the capabilities of quantum chemical simulations to characterise the hydration of molecular compounds. RSC Adv 3:1606
108. Wolf DPK, Phillpot S, Eggebrecht J (1999) Exact method for the simulation of coulombic systems by spherically turncated pairwise r summation. J Chem Phys 110(17):8254
109. Wolf M, Groenhof G (2014) Explicit proton transfer in classical molecular dynamics simualtions. J Comput Chem 35:657
110. Wraight CA (2006) Chance and design–proton transfer in water, channels and bioenergetic proteins. Biochim Biophys Acta 1757:886
111. Wu Y, Chen H, Wang F, Paesani F, Voth GA (2008) An improved multistate empirical valence bond model for aqueous proton solvation and transport. J Phys Chem B 112:467
112. Zhang Y, Lee T, Yang W (1999) A pseudobond approach to combining quantum mechanical and molecular mechanical methods. J Chem Phys 110(46)

Chapter 5
Accelerating QM/MM Calculations by Using the Mean Field Approximation

M. Elena Martín, M. Luz Sánchez, Aurora Muñoz-Losa, Ignacio Fdez. Galván and Manuel A. Aguilar

Abstract It is well known that solvents can modify the frequency and intensity of the solute spectral bands, the thermodynamics and kinetics of chemical reactions, the strength of molecular interactions or the fate of solute excited states. The theoretical study of solvent effects is quite complicated since the presence of the solvent introduces additional difficulties with respect to the study of analogous problems in gas phase. The mean field approximation (MFA) is used for many of the most employed solvent effect theories as it permits to reduce the computational cost associated to the study of processes in solution. In this chapter we revise the performance of ASEP/MD, a quantum mechanics/molecular mechanics method developed in our laboratory that makes use of this approximation. It permits to combine state of the art calculations of the solute electron distribution with a detailed, microscopic, description of the solvent. As examples of application of the method we study solvent effects on the absorption spectra of some molecules involved in photoisomerization processes of biological systems.

5.1 Introduction

There are many situations in which the electron distribution of molecules suffers important changes; some examples are chemical reactions, where bonds are formed or broken, or electron excitations, where large charge redistribution takes place. It is well known that a classical description (through force fields) does not reproduce adequately the charge flows that accompany these processes and the use of quantum

---

M.E. Martín · M.L. Sánchez · A. Muñoz-Losa · M.A. Aguilar (✉)
Área de Química Física, University of Extremadura, Avda. Elvas s/n, Edif. José Mª Vigera Lobo 3ª planta, 06006 Badajoz, Spain
e-mail: maguilar@unex.es

I. Fdez. Galván
Department of Chemistry–Ångström, The Theoretical Chemistry Programme, Uppsala University, PO Box 518, 751 20 Uppsala, Sweden

J.-L. Rivail et al. (eds.), *Quantum Modeling of Complex Molecular Systems*, Challenges and Advances in Computational Chemistry and Physics 21, DOI 10.1007/978-3-319-21626-3_5 mechanics becomes compulsory. Luckily, in most cases, the changes in the electron distributions are limited to only a small part of the system, usually, the active center or the chromophore. In these situations the use of focused methods [1] is especially useful. In focused methods the whole system is partitioned into two parts, the part of interest or focused part and the surroundings. In general, the part of interest of the system is quantum-mechanically described while the rest of the system is classically described. Examples of focused methods are dielectric continuum methods [1–4] or quantum mechanics/molecular mechanics (QM/MM) methods [5–7].

A characteristic of liquids is that they have many thermally accessible conformations. Consequently, in the study of the properties of these systems we must resort to some of the techniques developed by statistical mechanics. The presence of a surrounding medium can have important effects on the geometry, charge distribution, reactivity and spectroscopic properties of solutes. Different solvent configurations can yield slightly different solute properties; consequently the value of any molecular property must be calculated by averaging over a large enough set of conformations or configurations.

These two factors, the great number of thermally accessible configurations and the need to describe the charge distribution changes through quantum mechanics, taken together, increase dramatically the computational cost associated to the theoretical study of chemical reactions and electron transitions of molecules in solution. Throughout the years several strategies and approximations have been proposed to tackle this problem, in this chapter we will center in the study of one of the most useful: the mean field approximation (MFA) [8–10]. It is worth noting that the MFA is always applied in conjugation with focused methods. This approximation permits to dramatically improve the computational efficiency, which explains its widely extended use for molecules and biomolecules in solution. The importance of this approximation is evidenced by the fact that it is employed by many of the most used methods for the study of solvent effects on chemical or biochemical systems. The different quantum versions of dielectric continuum models [1–4], the methods based on Langevin dipoles [11] or more elaborate methods such as RISM-SCF [12–14], Mean-Field QM/MM [15], statistically-mechanically averaged solvent density [16] or ASEP/MD [17–19] are representative examples of this.

In the following sections we will try to clarify the theoretical fundament of MFA and we will discuss some applications.

5.2 Methods and Details

From a conceptual point of view the MFA is very simple [20–22]: it replaces the energy obtained by averaging over a set of configurations with the energy of an average configuration. The same procedure is applied in calculating any other property. Note that in the first case we need to calculate the energy of every configuration, while with the MFA we calculate only one energy value. Obviously, in this latter case the problem is how to determine and represent the average configuration. The important point to keep in mind when we use the MFA is that the effect of the classical subsystem, which can adopt many different configurations, on the quantum subsystem is introduced in an averaged way.

ASEP/MD, acronym for averaged solvent electrostatic potential from molecular dynamics, is a QM/MM method oriented towards the study of solvent effects that makes use of the mean field approximation. It permits to combine state of the art quantum calculations of the solute electronic wavefunction with a microscopic description of the solvent. Its main features are: (1) in optimizing the geometry and electronic structure of the solute the liquid structure is kept fixed. In the same way, when the liquid phase space is explored it is assumed that the geometry and charge distribution of the solute do not change. (2) The solute wavefunction and the liquid structure around it are optimized using an iterative procedure where a quantum calculation follows a molecular dynamics simulation. (3) The perturbation generated by the solvent on the solute enters into the molecular Hamiltonian in an averaged way. (4) The location of the critical points (minima, transition states, etc.) on the free energy surfaces is performed using a modification of the free energy gradient method.

We pass now to detail the method. Let us suppose the total system is formed by one solute molecule and N solvent molecules in a volume V. As usual in QM/MM methods the total Hamiltonian of the system is defined as:

$$\hat{H} = \hat{H}_{\mathrm{QM}} + \hat{H}_{\mathrm{class}} + \hat{H}_{\mathrm{int}} \tag{5.1}$$

corresponding to the quantum region, $\hat{H}_{\mathrm{QM}}$, the classical region, $\hat{H}_{\mathrm{class}}$, and the interaction between them, $\hat{H}_{\mathrm{int}}$. The quantum region is formed by the solute molecule and the classical one by the N solvent molecules. This distinction is important only during the quantum calculation, in the MD simulation all the molecules (solute +solvent) are classically treated.

The energy and wavefunction of the QM region are obtained by solving the effective Schrödinger equation:

$$\left(\hat{H}_{\mathrm{QM}} + \left\langle \hat{H}_{\mathrm{int}} \right\rangle\right)|\Psi\rangle = \bar{E}|\Psi\rangle \tag{5.2}$$

Here the mean field interaction Hamiltonian, $\hat{H}_{int}$, is defined as [20–22]:

$$\left\langle \hat{H}_{\mathrm{int}} \right\rangle = \int dr \cdot \hat{\rho} \cdot \langle V_S(r) \rangle \tag{5.3}$$

where $\hat{\rho}$ is the charge density operator of the quantum mechanical region (the solute) and $\langle V_S(r) \rangle$, which is named ASEP, is the average electrostatic potential generated by the solvent molecules at the position r. The brackets denote a statistical average over the configurational space of the classical subsystem.

In ASEP/MD, the information necessary to calculate $\langle V_S \rangle$ is obtained from a classical molecular dynamics simulation.

From a computational point of view it is convenient to split the interaction term into two components associated to the electrostatic and van der Waals contributions:

$$\hat{H}_{\text{int}} = \hat{H}_{\text{int}}^{\text{elect}} + \hat{H}_{\text{int}}^{\text{vdW}} \tag{5.4}$$

In general, it is assumed that $\hat{H}_{int}^{vdW}$ has only a small effect on the solute wavefunction and therefore it is usual to represent it through a classical potential that depends only on the nuclear coordinates but not on the electron ones. If this is the case, and for a given configuration of the classical subsystem, the $\hat{H}_{int}^{vdW}$ term can simply be added to the final value of the energy as a constant.

Equations (5.2) and (5.3) show how the classical region perturbs the quantum one and originates the solute polarization. Obviously, the classical subsystem depends in turn on the charge distribution of the quantum subsystem. As a consequence, Eq. (5.2) becomes an implicit non-linear expression that needs to be solved iteratively. At the end of this procedure, when convergence is reached, the solute charge distribution and the solvent structure around it become mutually equilibrated.

Scheme 5.1 can help to clarify the different steps of the ASEP/MD method. It begins by solving the Schrödinger equation of the isolated quantum subsystem. From this we obtain the electron distribution or any other derived quantity like the atomic point charges, etc., that are used to represent the solute in a classical molecular dynamics simulation. The ASEP calculation implies an average of the different configurations that the classical subsystem can adopt. Next, the ASEP is introduced into the molecular Hamiltonian. By solving the associated Schrödinger Eq. (5.2) we get the wavefunction of the quantum subsystem perturbed by the classical one. This perturbation modifies both the geometry and electron distribution of the quantum subsystem that consequently becomes polarized. The new charge distribution is used to recalculate the ASEP that is again introduced into Eq. (5.2). The procedure is repeated until convergence is attained, which typically occurs in a few cycles. At the end of this iterative procedure we get the energy, geometry and wavefunction of the quantum region and the structure of the classical region.

It is worth of note that most of the configurations generated in a MD simulation are not statistically independent, i.e., they do not provide additional information from a statistical point of view [23, 24]. So, in order to decrease the statistical correlation between the selected configurations it is important to include just those configurations separated by a time interval larger than the relaxation time. The length of this interval varies depending on the system under study, although intervals of at least 0.05 ps are usually needed. More important than including many configurations in the ASEP calculation is that they span a long enough simulation time (at least several hundreds of ps). In general, converged results are obtained with 500–1000 statistically independent configurations.

Another important point to clarify is what is the most adequate way of introducing ASEP into the Hamiltonian. There are several possibilities: numerically, as a set of point charges, using multipole expansions, etc. In general, we use a point charge representation, as it provides a compromise between accuracy and computational cost. The problem with this option is that the number of charges increases rapidly with the size of the system and with the number of configurations included. So, for instance, for a simulation with 500 molecules with five atoms per molecule and using 1000 configurations in the ASEP calculation we have 5 × 500 × 1000 = 2500000 charges. In order to keep a tractable number of charges we include explicitly only those charges associated to molecules that belong to the first solvation shell of the solute. The effect of the remaining solvent molecules is described by using potential-fitted charges. A further charge number reduction method permits to reduce their number to only a few thousands. More details about the calculation and representation of the ASEP can be found elsewhere [17–19, 25–27].

For optimizing the geometry of the quantum system we use a variant [28] of the free energy gradient method [29–31] that permits the determination of critical points on free energy surfaces (FES). The FES is defined as the energy associated with the time average of the forces acting on the solute molecule. Let A be the Helmholtz free energy of a system. The force felt by the solute molecule reads

$$\langle F(R)\rangle = -\frac{\partial A}{\partial R} = -\left\langle\frac{\partial E}{\partial R}\right\rangle = -\frac{\partial E_{QM}}{\partial R} - \left\langle\frac{\partial E_{int}}{\partial R}\right\rangle \tag{5.5}$$

R being the nuclear coordinates of the solute, E the energy obtained as the solution of the Schrödinger Eq. (5.2), $E_{QM} = \langle\Psi|H_{QM}|\Psi\rangle$, $E_{int} = \langle\Psi|H_{int}|\Psi\rangle$ and where we have assumed that E_{class} does not explicitly depend on the solute nuclear coordinates R and that the geometry of the quantum part is keep fixed during the MD simulation. The brackets denote a configurational average. Note that E incorporates both intramolecular, E_{QM}, and intermolecular, E_{int}, contributions.

In the same way the Hessian reads:

$$\langle G(R,R')\rangle = \left\langle\frac{\partial^2 E}{\partial R \partial R'}\right\rangle - \beta\left\langle\frac{\partial E}{\partial R}\left(\frac{\partial E}{\partial R'}\right)^T\right\rangle + \beta\left\langle\frac{\partial E}{\partial R}\right\rangle\left\langle\frac{\partial E}{\partial R'}^T\right\rangle \tag{5.6}$$

where the superscript T stands for *transpose* and $\beta = 1/k_BT$. The last two terms in Eq. (5.6) are related to the thermal fluctuations of the force.

Turning to the expression of the force, it is convenient to split the interaction term into two components associated with the electrostatic and van der Waals contributions:

$$\langle F(R)\rangle = -\frac{\partial E_{QM}}{\partial R} - \left\langle\frac{\partial E_{int}^{elect}}{\partial R}\right\rangle - \left\langle\frac{\partial E_{int}^{vdW}}{\partial R}\right\rangle \tag{5.7}$$

At this point one can introduce the MFA by replacing the average derivative of E_{int}^{elect} with the derivative of the average value. The force now reads [28]:

$$\langle F(R)\rangle = -\frac{\partial E_{QM}}{\partial R} - \frac{\partial \bar{E}_{int}^{elect}}{\partial R} - \left\langle \frac{\partial E_{int}^{vdW}}{\partial R} \right\rangle \tag{5.8}$$

and, analogously, the Hessian reads:

$$\langle G(R,R')\rangle = -\frac{\partial^2 E_{QM}}{\partial R \partial R'} - \frac{\partial^2 \bar{E}_{int}^{elect}}{\partial R \partial R'} - \left\langle \frac{\partial^2 E_{int}^{vdW}}{\partial R \partial R'} \right\rangle \tag{5.9}$$

where, in agreement with the mean field approximation, the force fluctuation term has been neglected. Once the gradient and Hessian values are known, we can use any of the usual optimization methods, RFO [32] for instance, to get the optimized geometry and couple it with Scheme 5.1.

A point that deserves attention is the contribution of the electronic solvent polarization to the solvatochromic shifts. Most force fields use atomic point charges that include the effect of electron polarization in an implicit way. The main problem with the use of this implicit description of the electronic polarization is that it gives a vanishing contribution when one compares situations where the solvent structure is fixed, for instance when the Franck-Condon principle is applied in vertical transitions. In these situations it is convenient to have a model to compute explicitly

Scheme 5.1 Scheme of the MFA iterative procedure

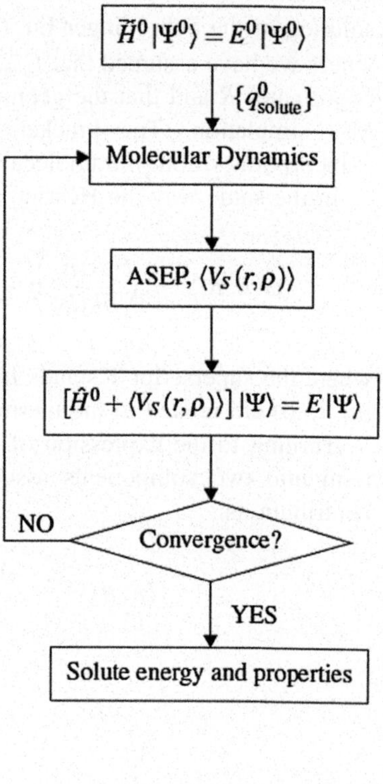

the contribution of the electronic polarization component. To this end, we assign a molecular polarizability to every solvent molecule, and simultaneously replace the effective solvent charge distribution used in the MD calculation with the gas phase values of the solvent molecule. This point is important, because otherwise the solvent polarization is accounted twice. The dipole moment induced on each solvent molecule is a function of the dipole moments induced on the rest of the molecules and of the solute charge distribution, and hence the electrostatic equation has to be solved self-consistently.

The total energy of the system (quantum solute + polarizable solvent) is obtained as:

$$U = U_{qq} + U_{pq} + U_{pp} + U_{\rho q} + U_{\rho p} + U_{dist}^{solute} + U_{dist}^{solvent} \tag{5.10}$$

Here, q refers to the permanent charges of solvent molecules, p to the solvent induced dipoles, and ρ is the solute charge density. The last two terms in this equation are the distortion energies of the solute and solvent molecules respectively, i.e., the energy spent in polarizing them. In the case of the solute it reads:

$$U_{dist}^{solute} = \langle \Psi | H_{QM} | \Psi \rangle - \langle \Psi^0 | H_{QM} | \Psi^0 \rangle \tag{5.11}$$

where Ψ and Ψ^0 are the in solution and in vacuo solute wave-functions, respectively. Explicit expressions for the different contributions can be found elsewhere [26].

The final expression for the total energy of the system is:

$$U = U_{qq} + \frac{1}{2}U_{pq} + +U_{\rho q} + \frac{1}{2}U_{\rho p} + U_{dist}^{solute} \tag{5.12}$$

The firsts two terms of this expression are strictly classical, while the last three involve the solute charge distribution and are calculated quantum-mechanically.

Finally, we address the application of the ASEP/MD methodology to the study of electronic transitions. Here, we can consider two situations depending on the description, implicit or explicit, of the solvent electronic polarization. If one uses an implicit description of this component then it is only necessary to perform the calculation of the different excited states in presence of the solvent charge distribution obtained during the ASEP/MD procedure. If, on the contrary, we explicitly include the contribution of this component then it is necessary to perform an additional self-consistent process. Using the solvent structure and solute geometry obtained in the first self-consistent process, we couple the quantum mechanical solute and the electron polarization of the solvent. The process finishes when the solute charge distribution and the solvent induced dipole moments become mutually equilibrated.

Once the solvation energy, Eq. (5.12), has been calculated for the ground and excited states, the solvent shift, δ, can be obtained as the difference

$$\delta = U_{ex} - U_g = \frac{1}{2}\delta_{pq} + \delta_{\rho q} + \frac{1}{2}\delta_{\rho p} + \delta_{dist}^{solute} \quad (5.13)$$

The term δ_{qq} cancels out because, in vertical transitions where the Franck-Condon approximation is applicable, the U_{qq} term takes the same value in both the ground and the excited state.

A problem with this procedure is that the solvent perturbs each electronic state of the solute molecule in a different way, i.e., each state is an eigenfunction of a different Hamiltonian; consequently, the different states are not mutually orthogonal. This fact complicates the calculation of oscillator strength and other transition properties of the system.

5.3 Examples

In this section we present results of the application of the ASEP/MD method to the study of solvent effects on electron transitions in p-coumaric acid derivatives. This system was chosen because it is an example of electron excitations that promote internal rotation around formal double bonds. Internal rotations are characterized by large flows of charge and, consequently, important solvent effects. Furthermore, they are involved in very interesting phenomena, as are dual fluorescence in push-pull chromophores [33–35] or *cis-trans* photoisomerization reactions [36–38]. An adequate description of the excited states involved in these processes demand state of the art quantum calculations including both static and dynamic electron correlation contributions. Furthermore, in many cases the solute is stabilized by hydrogen bonds, consequently, it is compulsory to use a microscopic description of the solvent in order to account for specific interactions. In these conditions ASEP/MD becomes a good alternative to other methods and it can help to shed light on these processes [39–42].

5.3.1 p-Coumaric Acid in Different Protonation States

p-coumaric acid (pCA) has been used as a model of the photoactive yellow protein (PYP) chromophore. This protein is related with the negative phototaxis of *Halorhodospira halophila* under blue light irradiation [38]. After irradiation (446 nm), the protein enters a photocycle where the primary event is the isomerization of the chromophore's double bond on a subpicosecond time scale, very similar to retinal photoisomerization in the visual process [43–46]. A first evidence of the medium effect on the p-coumaric acid chromophore is found in the absorption spectrum. In gas phase, the p-coumaric monoanion (pCA^-) absorbs at 430 nm [47, 48] whereas in water solution the absorption maximum is located at around 310 nm and it varies with the pH [48, 49]. It is important to note that pCA has two hydrogen atoms susceptible of deprotonation being the carboxylic hydrogen more acid than the phenolic one. Thus, in passing from acidic to neutral condition, the chromophore will first lose the carboxylic hydrogen yielding the carboxylate anion rather than the phenolate. Nevertheless, in gas phase, the phenolate anion is found to be more stable than the carboxylate due to the delocalization the negative charge along the structure. At basic pH the dianionic form is obtained.

In this section we compare the absorption spectra of p-coumaric acid as a function of its protonation state. During the ASEP/MD runs, quantum calculations were performed with the Gaussian 09 package [50]. The final SA-CASSCF and CASPT2 calculations were done with Molcas 7.4 [51]. All MD simulations were performed using Moldy [52] or Gromacs [53] and assuming rigid molecules. Lennard-Jones parameters and solvent atomic charges were taken from the OPLS-AA (optimized potentials for liquid simulations, all atoms) force field [54, 55]. Solute atomic charges were obtained from the quantum calculations with the CHELPG (charges from electrostatic potential in a grid) method [56, 57].

We started our study with the neutral form of the *trans*-pcoumaric acid (pCA). Four conformes were studied. Two of them correspond to the *cis* or *trans* disposition of the central vinyl double bond and the carboxylic double bond of the acid terminal group, s-*cis* and s-*trans*. For each of these species, the hydrogen of the phenolic group can be disposed in syn or anti position relative to the central double bond. These four species are displayed in Scheme 5.2 and are identified as I (s-*cis*-anti), II (s-*cis*-syn), III (s-*trans*-anti) and IV (s-*trans*-syn). In gas phase, anti rotamers are preferred to the syn ones, in solution the opposite is verified and the syn conformations are slightly more stable than anti.

Regarding the absorption spectrum, the most probable transition in gas phase is a ($\pi \rightarrow \pi^*$) absorption to the S_2 state with an oscillator strength of around 0.5. This is

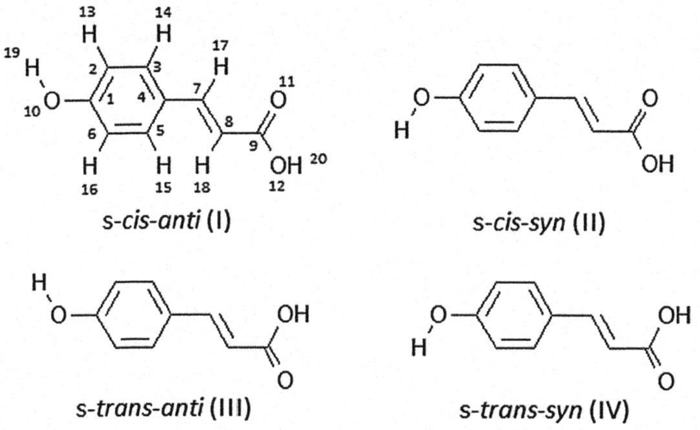

Scheme 5.2 Conformers for the neutral form of the *trans*-p-coumaric acid

a HOMO→LUMO transition involving a gap of 4.5 and 4.6 eV for the s-*cis* and s-*trans* isomers, respectively. The transition to the S_1 state corresponds to a HOMO→LUMO+1 transition with an energy gap of around 4.23 eV for all the forms.

In water, the most probable transition is still that corresponding to the S_2 state, with an oscillator strength of 0.4. This transition results in around 4.30 eV, which overestimates the experimental value in about 0.3 eV. However, if the solvent electronic polarization is explicitly considered we get a value of 4.00 eV, in perfect agreement with the experiment. It is also observed that the $S_0 \rightarrow S_1$ transition is hardly affected by the solvent interaction, being the vertical transition energies in gas phase and in solution practically coincident. The different behavior of S_2 and S_1 comes from the distinct charge distribution of these states. Whereas the transition to S_2 implies an electron density displacement from the phenolic ring to the alkyl fragment and a considerable increase in the dipole moment, S_1 has a charge distribution similar to S_0 and consequently they interact with the solvent in a similar way. Due to the larger stabilization of S_2 with respect to S_1 in water solution the relative stability order of the two states is reversed.

The situation is somewhat more complicated for the monoanionic form of p-coumaric acid (pCA^-) as it can appear (see Scheme 5.3) in two forms: phenolate and carboxylate, with very different charge distributions. So, whereas in the carboxylate the negative charge is localized on the carboxylate end, in the phenolate anion, the negative charge is spread along the whole molecule. In fact, in phenolate, there is an equilibrium between a quinolic structure with the negative charge localized at the COOH fragment, and a nonquinolic structure with the negative charge at the phenolate oxygen. In gas phase, the ground state of the phenolate form is clearly more stable than the carboxylate species due to its larger charge delocalization. Interaction with solvent molecules modifies the relative stability of the different isomers, and in water solution, the carboxylate becomes now the most stable form.

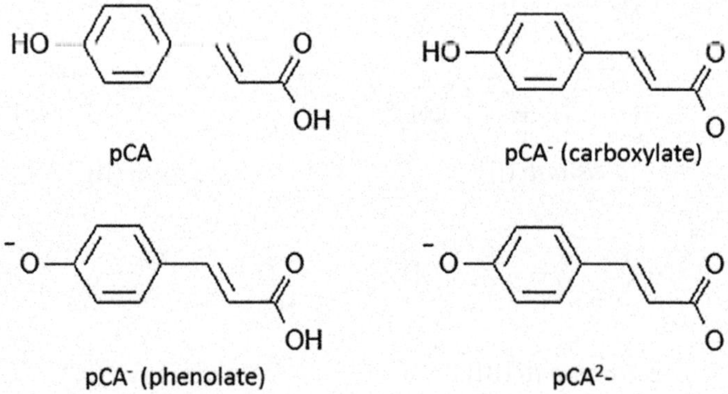

Scheme 5.3 Neutral, monoanionic and dianionic forms of the *p*-coumaric acid

An additional complication comes from the fact that for both carboxylate and phenolate monoanions, the gas-phase ($\pi \rightarrow \pi^*$) low-lying excited states are found in the detachment continuum [58–60], that is, their electron detachment energies are below the first vertical excitation energy, and therefore, these excited states are metastable. In any case, there is a very good agreement between the calculated CASPT2/CASSCF(12,24) gas phase excitation energy for phenolate (2.89 eV) and the experimental data (2.88 eV).

The most probable transition in gas phase is that leading to the S_1 state, with an oscillator strength of 1.0. In water solution, the bright state remains S_1, with the same nature as in gas phase. The absorption band appears displaced toward larger energies (3.17 or 3.09 eV depending whether the solvent is explicitly polarized or not); consequently, there is a blue solvent shift of around 0.25 eV.

The theoretical electronic absorption spectrum of the carboxylate anion is quite complex due to the large number of excited states lower in energy than the bright one. In fact, the transition with the larger oscillator strength (0.50) is $S_0 \rightarrow S_4$. This corresponds to a ($\pi \rightarrow \pi^*$) transition involving the HOMO and LUMO orbitals. We find a difference of 2.1 eV between the gas-phase transition energies of carboxylate (5.06) and phenolate (2.89). In water solution, the bright state is the S_2 excited state. It is a H→L transition, and it has an oscillator strength slightly larger than that found in gas phase. This state is more effectively solvated and more stabilized than the ground state, the transition energy is 4.76 eV and a red solvent shift of 0.3 eV is found. Explicit inclusion of the electronic solvent polarization slightly increases the transition energy until 4.81 eV.

The double anionic form of the trans-p-coumaric acid (pCA^{2-}) is unstable in gas phase and would suffer spontaneous autoionization. However, the interaction with the solvent increases the ionization potential permitting the existence of pCA^{2-} in water solution. In gas phase, the bright state is the S_2 state that corresponds to a H→L transition, whereas the S_1 state implies the transition H→L+1. In solution, the bright excited state is stabilized and becomes the first excited state as the charge displacement involved in this transition (−0.35 e in solution) is favored by the solvent. The calculated transition energy in solution (3.75 and 3.73 eV for non-polarizable and polarizable solvent, respectively) is in very good agreement with the experimental data recently published by Boggio-Pasqua and Groenhof [61] where the *trans*-p-coumaric acid in aqueous solution at pH > 10 showed an absorption maximum at around 3.71 eV.

The variation of the transitions energies with the protonation state is displayed in Fig. 5.1. For all the studied species, the bright state is a $\pi \rightarrow \pi^*$ transition involving a charge displacement along the system. For the neutral form, the transition involves an increase of the dipole moment of the excited state, a fact that leads to a larger stabilization of the excited state in water solution and consequently a bathochromic shift of the absorption maximum in solution. Phenolate and carboxylate monoanions show different behavior with respect to solvation. On the one hand, the phenolate monoanion in gas phase shows a displacement of the charge from the phenolic oxygen to the rest of the system during the transition, and this is enhanced in water solution. As the displacement involves a delocalization of the negative

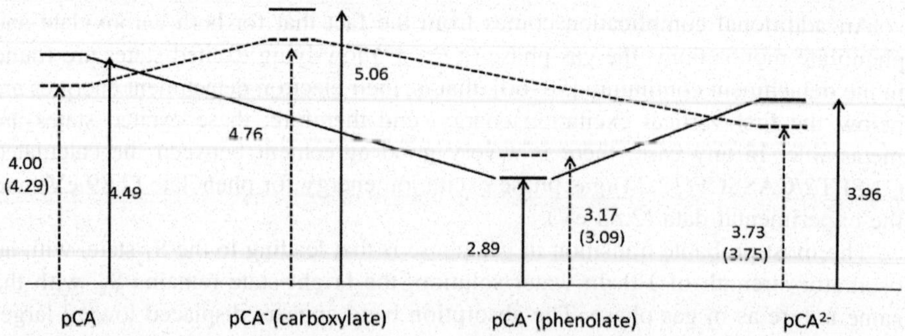

Fig. 5.1 Variation of the transition energies (in eV) for the *trans*-p-coumaric acid with the protonation state in gas phase (*full lines*) and in water solution (*dashed lines*). For the in-solution transition energies, the effect of the electronic solvent polarization has been included (values in parentheses correspond to those obtained without this contribution)

charge, the excited state is worse stabilized than the ground state, and a final solvent blue shift is achieved. On the other hand, in the carboxylate monoanion the negative charge moves from the COO^- toward the phenolic ring during the excitation. As the solvent hinders this displacement the excited state charge becomes more localized in solution, and consequently more stabilized than the ground state and a solvent red shift is found. Finally, for the dianionic species the electronic transition to the bright state involves a charge displacement from the phenolic part toward the rest of the system and most of the negative charge is concentrated in the carboxyl end. This displacement is enhanced in water solution; as a consecuence, the charge is more localized and more effectively solvated in the excited state than the ground state, and a final solvent red shift is found. In sum, except for the phenolate anion where a blue shift is found in the rest of cases a blue shift is registered.

As for the effect of the electronic solvent polarization in the transition energies and solvent shift values two cases can be distinguished. In neutral species, the polarization contribution depends on the state nature. That is, it is relevant in those cases in which there exists a significant charge displacement between the ground and the excited state. On the contrary, in ionic forms (mono- and dianionic species), the larger contribution to the solute-solvent interaction energy comes from the charge–potential term and, in general, solvent polarization has only a minor influence.

Finally, we would like to draw the attention to some attempts to mimic the solvent environment effect by including a limited number of solvent molecules in the quantum calculation. Putschögl et al. [62], for instance, studied the pCA^{2-} dianion surrounded by eight water molecules. They found that the solvent stabilization was not enough to cause the inversion between S_1 and S_2 states in solution, something that occurs when bulk solvation is accounted. In our case, the S_1 bright state is more stabilized than the ground state and consequently a red solvent shift of around 0.22 and 0.23 eV (nonpolarizable and polarizable solvent, respectively) is obtained in water solution. This fact evidences the importance that the bulk solvent contribution has in this system.

5.3.2 p-Coumaric Acid Derivatives

In this section we focus on the analysis of the solvent effect and coumaryl tail on the absorption spectrum of some p-coumaric derivatives: acid (pCA$^-$), thioacid (pCTA$^-$), methyl ester (pCMe$^-$) and methyl thioester (pCTMe$^-$), see Scheme 5.4. The comparison of the behavior of these systems permits to analyze the modifications introduced by the substitution of a sulfur by an oxygen atom and the influence of the methyl group. As we will show the presence of the sulfur modulates the solvent effect, as a consequence the first two excited states become practically degenerated for pCA- and pCMe- but moderately well separated for pCTA- and pCTMe-.

In gas phase and for the four derivatives (see Table 5.1) the bright state is the first excited state with a $\pi - \pi^*$ character and oscillator strength close to one. The second excited state is a $n - \pi^*$ state that involves the phenolic oxygen lone pair and the third one corresponded to a $\pi - \pi_2^*$ transition. A good agreement was found between the calculated transition energies and the available experimental data. So, the absorption maximum for the methyl ester derivative is experimentally found at 2.88 eV [63] whereas our CASPT2//CASSCF(14,12) value is 2.94. Explicit solvent electronic polarization was not considered, as its contribution is negligible for monoanionic derivatives. For the thiophenyl ester (pCT$^-$) the transition is experimentally found at 2.70 eV [64]. This value is a good reference for pCTMe$^-$ where we found a transition energy of 2.73 eV as the electron conjugation does not extend

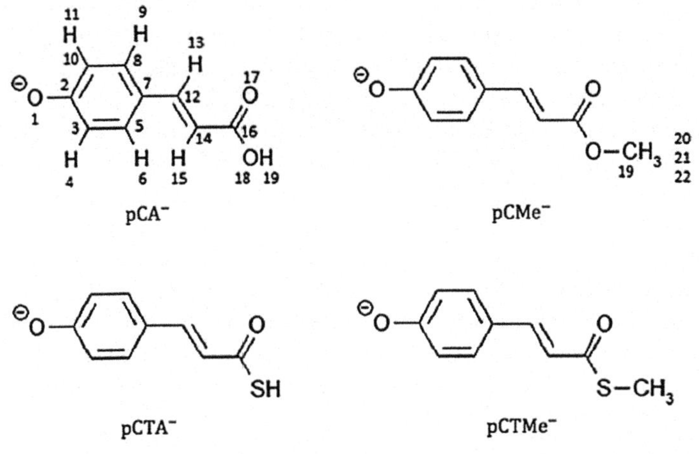

Scheme 5.4 Acid (pCA$^-$), thioacid (pCMe$^-$), methyl ester (pCTA$^-$) and methyl thioester (pCTMe$^-$) derivatives of the anionic p-coumaric acid

Table 5.1 Solute-solvent Interaction energies (kcal/mol) in the ground state and excited state

	GS					$\pi\rightarrow\pi_1^*$				
	O	Ph	C = C	COXY	Total	O	Ph	C = C	COXY	Total
pCA-	−158.4	16.8	−6.4	−24.6	−172.6	−139.8	34.5	−17.6	−27.6	−150.5
pCMe-	−149.1	14.2	−10.7	−10.3	−155.9	−131.8	31.5	−21.8	−13.3	−135.4
pCTA-	−147.5	9.4	2.2	−15.4	−151.3	−129.0	24.6	−7.3	−18.2	−129.9
pCTMe-	−148.5	12.5	−4.2	−5.6	−145.8	−131.4	25.8	−13.5	−8.1	−127.2

beyond the sulfur atom. Our results also agree with the values published by Zuev et al. [65], 2.98 eV, using a similar level of calculation (SS-CASPT2/ANO-RCC-VTZP) and by Gromov et al. [66], 2.89 eV, using CC2/SV(P). The TDDFT method overestimates excitation energies. So, Muguruza González et al. [67] reported values of 3.40 eV for the vertical energies of $pCTMe^-$ and Sergi et al. [68] a value of 3.24 eV for pCA^-. According to these values it seems evident the poor performance of TDDFT in describing charge-transfer excited states.

The replacement of oxygen by sulfur results in a red shift of the first absorption band of around 0.21 eV. However, the substitution of the terminal hydrogen for the methyl group does not modify the band position. In the four derivatives there is a flux of charge from the phenolic part toward the rest of the molecule. This flux is larger in pCA^- and $pCMe^-$, ≈0.22 e, than in $pCTA^-$ and $pCTMe^-$, ≈0.13 e. Consequently, the delocalization of the charge is larger in the excited state of the oxo derivatives that becomes more stable than the thio derivatives.

The solvent has important effects on the geometry of the four models analyzed: bond lengths are notably different in gas phase and in solution. So, for instance, the phenolic oxygen bond length increases from 1.23 to 1.28 Å when one passes from gas phase to water solution. In addition, there is a certain loss of the quinolic character displayed in gas phase, single bonds becoming now longer and double bonds shorter. However, and contrary to what could be expected, the carboxylic double bond length is not modified by the solvent, probably because of the steric hindrance and the low charge on this group (C16O17).

In the ground state of the four studied molecules the solvent originates a flux of negative charge toward the phenolic moiety. However, during the excitation the flux goes from the phenolic part (0.32 e) to the central double bond (0.24 e) and the terminal moieties (0.08 e). As a consequence the negative charge in the excited states is smoothed out along the molecule. The solvent penalizes the charge delocalization in the two excited states and originates a blue shift. The destabilization is larger for the $n\rightarrow\pi^*$ state, which becomes the third excited state. The $\pi\rightarrow\pi^*$ state is also destabilizated with respect to the $\pi\rightarrow\pi_2^*$ state and more in pCA^- and $pCMe^-$ than in $pCTA^-$ and $pCTMe^-$; consequently, while in the first case the S1 and S2 states become practically degenerated, in $pCTA^-$ and $pCTMe^-$ there is a gap of around 0.35 eV.

A quantity that can be useful in the analysis of solvent effects is the group contribution to the solute-solvent interaction energy (Table 5.1). The molecules are divided in portions and their relative contribution to the interaction energy and solvent shift are analyzed. In the present case we divided the molecules in four parts: the phenolic oxygen, the phenyl group, the central double bond and the terminal part (acid, thio-acid, ester or thio-ester depending on the case). The larger contribution to the total interaction energy comes from the phenolic oxygen that is the group that carries most of the negative charge. However, the contributions of the remaining groups are far from negligible. Electronic transition results in a decrease of the solute-solvent interaction energy. This is mainly due to the decrease of circa 35–40 kcal/mol in the contribution of the phenolic part of the molecule (Ph-O) as consequence of the charge flux from the phenolic part toward the rest of the molecule. The phenyl and carbonyl groups present a reduced number of water molecules placed in their neighborhood compared to those existing around the phenolic oxygen, and consequently the solvent does not stabilize the transferred charge as effectively.

It is interesting to compare ASEP/MD results with those obtained by other authors using different methodologies. So, Gromov et al. [66] calculated a CC2/aug-*cc*-pVDZ value of 2.96 eV for the transition energy of $pCTMe^{-}$ with two water molecules placed close to the phenolic oxygen (the part of the molecule with the largest interaction energy). This value is in reasonable agreement with the experimental value of 3.22 eV published by Naseem et al. [69]. Nevertheless, those authors report a theoretical solvent shift of 0.05 eV as they found the absorption band in gas phase at 2.91 eV. This value is clearly underestimated when compared with the experiment. Assuming the gas phase experimental value for pCT^{-} (2.70 eV) as suitable value as well for $pCTMe^{-}$, the experimental solvent shift can be estimated in 0.52 eV. Furthermore, the solvent effect on the bond length variations reported by these authors represents only one-third of that obtained with ASEP/MD. The situation is even worse when the solvent is described using continuum methods as they fail in reproducing the correct trend, which is a blue solvent shift. For instance, Wang [70]. performed a CPCM/TD-B3LYP study of $pCTMe^{-}$ in different media. In water, the absorption energy was estimated in 3.03 eV being this value lower than the one computed in gas phase (3.17 eV). Consequently a final red shift was obtained. In this case, the failure in predicting the solvent shift is due to the neglecting by continuum models of specific solvent interaction such as hydrogen bonding. Therefore, we must conclude that the representation of the solvent through a few solvent molecules or making use of continuum methods does not accurately account for the solvation effects on the studied systems. In the first case because solvation is a global property hardly represented just through a few solvent molecules and in the second because specific interactions between solute and solvent are missing.

5.4 Conclusions

Microscopic solvent effect theories imply an extensive sampling of the configurational space of the solute–solvent system. Furthermore, most of processes of chemical interest involve large charge redistribution and its study requires the use of high-level quantum-mechanical methods with the consequent increase in the computational cost. The mean field approximation provides a way of reducing the number of quantum calculations and, consequently, it permits to reduce the computational cost associated with the inclusion of solvent effects. In this chapter the theoretical basis of this approximation have been analyzed. We have paid special attention to the ASEP/MD method that implements this approximation in QM/MM methods.

The main characteristics of ASEP/MD are: (1) A reduced number of quantum calculations, that permits to increase the description level of the solute electronic structure which, in fact, can be described at the same level as in gas phase studies. (2) Since the solvent is described through MM force fields, there exists a great flexibility to include both bulk and specific interactions into the calculations. (3) At the end of the procedure the solute wavefunction and the solvent structure become mutually equilibrated, i.e., the solute is polarized by the solvent and the solvent structure is in equilibrium with the polarized solute charge distribution. (4) Finally, the method permits to perform in an efficient way optimizations on free energy surfaces.

In this chapter we have presented some applications of the ASEP/MD method to the study of electron transitions that promote internal rotation around formal double bonds. More specifically we have addressed the solvent effects on the electronic spectrum of the p-coumaric acid. Important differences in facing the solvent were verified depending on the protonation state of the acid and the nature of the terminal group.

References

1. Tomasi J, Mennucci B, Cammi R (2005) Quantum mechanical continuum solvation methods. Chem Rev 105:2999–3093
2. Cramer CJ, Truhlar DG (1999) Chem Rev 99:2161
3. Rivail JL, Rinaldi D (1976) Chem Phys 18:233–242
4. Ruiz-López MF (2008) In: solvation effects on molecules and biomolecules: computational methods and applications. In: Canuto S (ed) Springer series: Challenges and advances in computational chemistry and physics, Springer
5. Warshel A, Levitt M (1976) J Mol Biol 103:227–249
6. Singh UC, Kollman PA (1986) J Comput Chem 7:718–730
7. Field M J, Bash PA, Karplus M (1990) J Comput Chem 11(6):700–733
8. Sánchez ML, Martín ME, Galván IF, Olivares del Valle FJ, Aguilar MA (2002) J Phys Chem B 106:4813

9. Martín ME, Sánchez ML, Corchado JC, Muñoz-Losa A, Galván IF, Olivares del Valle FJ, Aguilar MA (2011) Theor Chem Acc 128:783–793
10. Yamamoto T (2008) J Chem Phys 129:244104
11. Warshel A (1991) Computer modelling of chemical reactions in enzymes and solutions. Wiley Interscience Publication, New York
12. Ten-no S, Hirata F, Kato S (1993) Chem Phys Lett 214:391
13. Sato H, Hirata F, Kato S (1996) J Chem Phys 105:1546
14. Hirata F (ed) (2003) Molecular theory of solvation (understanding chemical reactivity). Springer, Berlin
15. Nakano H, Yamamoto T (2013) J Chem Theory Comput 9:188–203
16. Kaminski JW, Gusarov S, Kovalenko A, Wesolowski TA (2010) J Phys Chem A 114:6082
17. Sánchez ML, Aguilar MA (1997) Olivares del Valle FJ. J Comput Chem 18:313
18. Sánchez ML, Martín ME, Aguilar MA (2000) Olivares del Valle FJ J Comput Chem 21:705
19. Galván IF, Sánchez ML, Martín ME, Olivares del Valle FJ (2003) Aguilar MA Comput Phys Commun 155:244
20. Tapia O (1991) In: Maksic ZB (ed) Theoretical treatment of large molecules and their interactions, vol 4. Springer, Berlin, p 435
21. Angyán JG (1992) J Math Chem 10:93
22. Sánchez ML, Aguilar MA, Olivares del Valle FJ (1997) J Comput Chem 18:313
23. Canuto S, Coutinho K (1997) Avd Quantum Chem 28:89
24. Coutinho K, Oliveira MJ et al (1998) Int J Quantum Chem 66:249
25. Martín ME, Sánchez ML, Olivares del Valle FJ, Aguilar MA (2002) J Chem Phys 116:1613
26. Martín ME, Muñoz Losa A, Galván IF, Aguilar MA (2003) J Chem Phys 118:255
27. Galván IF, Martín ME, Aguilar MA (2004) J Comput Chem 25:1227
28. Galván IF, Sánchez ML, Martín ME, Olivares del Valle FJ, Aguilar MA (2003) J Chem Phys 118:255
29. Okuyama-Yoshida N, Nagaoka M, Yamabe T (1998) Int J Quantum Chem 70:95
30. Okuyama-Yoshida N, Kataoka K, Nagaoka M, Yamabe T (2000) J Chem Phys 113:3519
31. Hirao H, Nagae Y, Nagaoka M (2001) Chem Phys Lett 348:350
32. Banerjee A, Adams N, Simons J, Shepard R (1985) J Phys Chem 89:52
33. Lippert E, Lüder W, Moll F, Nägele W, Boos H, Prigge H, Seibold-Blankenstein I (1961) Angew Chem 73:695–706
34. Rotkiewicz K, Grellmann KH, Grabowski ZR (1973) Chem Phys Lett 19:315–318
35. Grabowski ZR, Rotkiewicz K, Rettig W (2003) Chem Rev 103:3899–4032
36. Kukura P, McCamant DW, Yoon S, Wandschneider DB, Mathies RA (2005) Science 310:1006
37. Muñoz Losa A, Martin ME, Galván IF, Sánchez ML, Aguilar MA (2011) J Chem Theory Comput 7:4050–4059
38. Meyer TE (1985) Biochim Biophys Acta 806:175
39. Muñoz-Losa A, Galván IF, Aguilar MA, Martín ME (2007) J Phys Chem B 111:9864–9870
40. Muñoz-Losa A, Martín ME, Galván I, Sánchez ML, Aguilar MA (2011) J Chem Theory Comput 7:4050–4059
41. Muñoz-Losa A, Galván IF, Aguilar MA, Martín ME (2013) J Chem Theory Comput 9:1548–1556
42. García-Prieto FF, Galván IF, Muñoz-Losa A, Aguilar MA, Martín ME (2013) J Chem Theory Comput 9:4481–4494
43. Kort R, Vonk H, Xu X, Hoff WD, Crielaard W, Hellingwerf K (1996) J FEBS Lett 382:73
44. Xie A, Hoff WD, Kroon AR, Hellingwerf K (1996) J Biochem 35:14671
45. Unno M, Kumauchi M, Sasaki J, Tokunaga F, Yamaguchi S (2000) J Am Chem Soc 122:4233
46. Genik UK, Soltis SM, Kuhn P, Canestrelli IL, Getzoff ED (1998) Nature 392:206
47. Rocha-Rinza T, Christiansen O, Rajput J, Gopalan A, Rahbek DB, Andersen LH, Bochenkova AV, Granovsky AA, Bravaya KB, Nemukhim AV, Chistiansen KL, Nielsen MB (2009) J Phys Chem A 113:9442

48. Nielsen IB, Boyé-Péronne S, El Ghazaly MOA, Kristensen MB, Nielsen SB, Andersen LH (2005) Biophys J 89:2597
49. Putschögl M, Zirak P, Penzkofer A (2008) Chem Phys 343:107
50. Frisch MJ, Trucks GW, Schlegel HB, Scuseria GE, Robb MA, Cheeseman JR, Zakrzewski VG, Montgomery JA Jr, Stratmann RE, Burant JC, Dapprich S, Millam JM, Daniels AD, Kudin KN, Strain MC, Farkas O, Tomasi J, Barone V, Cossi M, Cammi R, Mennucci B, Pomelli C, Adamo C, Clifford S, Ochterski J, Petersson GA, Ayala PY, Cui Q, Morokuma K, Malick DK, Rabuck AD, Raghavachari K, Foresman JB, Cioslowski J, Ortiz JV, Stefanov BB, Liu G, Liashenko A, Piskorz P, Komaromi I, Gomperts R, Martin RL, Fox DJ, Keith T, Al-Laham MA, Peng CY, Nanayakkara A, Gonzalez C, Challacombe M, Gill PMW, Johnson BG, Chen W, Wong MW, Andres JL, Head-Gordon M, Replogle ES, Pople JA (1998) Gaussian 98, revision A11.3. Gaussian, Inc., Pittsburgh
51. Andersson K, Barysz M, Bernhardsson A, Blomberg MRA, Carissan Y, Cooper DL, Cossi M, Fleig T, Fu¨lscher MP, Gagliardi L, de Graaf C, Hess BA, Karlström G, Lindh R, Malmqvist P-Å, Neogrády P, Olsen J, Roos BO, Schimmelpfennig B, Schütz M, Seijo L, Serrano-Andrés L, Siegbahn PEM, Stalring J, Thorsteinsson T, Veryazov V, Wierzbowska M, Widmark P-O (2003) MOLCAS Version 5.2, University of Lund, Lund, Sweden
52. Refson K (2000) Comput Phys Commun 126:310
53. Berendsen HJC, van der Spoel D, van Drunen R (1995) Comp Phys Comm 91:43–56
54. Jorgensen W, Maxwell DS, Tirado-Rives J (1996) J Am Chem Soc 118:11225
55. Jorgensen WL, Chandrasekhar J, Madura JD, Impey RW, Klein ML (1983) J Chem Phys 79:926
56. Chirlian LE, Francl MM (1987) J Comput Chem 8:894
57. Breneman M, Wiberg KB (1990) J Comput Chem 11:316
58. Zuev D, Bravaya KB, Crawford TD, Lindh R, Krilov AI (2011) J Chem Phys 134:034310
59. Gromov E, Burghardt I, Hynes I, Köppel H, Cederbaum LS (2007) J Photochem Photobiol A 190:241
60. Gromov E, Burghardt I, Köppel H, Cederbaum LS (2007) J Am Chem Soc 129:6798
61. Boggio-Pasqua M, Groenhof G (2012) J Phys Chem B 115:7021
62. Putschögl M, Zirak P, Penzkofer A (2008) Chem Phys 343:107
63. Rocha-Rinza T, Christiansenm O, Rajput H, Gopalan A, Rahbek DB, Andersen LH, Bochenkova AV, Granovsky AA, Bravaya KB, Nemukhin AV, Christiansen KL, Nielsen MB (2009) J Phys Chem A 113:9442
64. Nielsen IB, Boye-Peronne S, El Ghazaly MOA, Kristensen MB, Nielsen SB (2005) Anderson LH. Biophys J 89:2597
65. Zuev D, Bravaya KB, Crawford TD, Lindh R, Krylov AI (2011) J Chem Phys 134:034310
66. Gromov EV, Burghardt I, Hynes JT, Köppel H, Cederbaum LS (2007) J Photochem Photobiol A Chem 190:241
67. Muguruza González E, Guidoni L, Molteni C (2009) Phys Chem Chem Phys 11:4556
68. Sergi A, Crüting M, Ferrario M (2001) F Buda J Phys Chem B 105:4386
69. Naseem S, Laurent AD, Carroll EC, Vengris M, Kumauchi, Hoff MWD, Krylov AI, Larsen DS (2013) J Photochem Photobiol A Chem 270:43
70. Wang Y, Li H (2010) J Chem Phys 133:034108

Chapter 6
Development of a Massively Parallel QM/MM Approach Combined with a Theory of Solutions

Hideaki Takahashi and Nobuyuki Matubayasi

Abstract In this contributed article we review our method, referred to as QM/MM-ER, which combines the hybrid QM/MM simulation with the theory of energy representation. Our recent developments and applications related to the method are also introduced. First, we describe the parallel implementation of the Kohn-Sham DFT for the QM region that utilizes the real-space grids to represent the one-electron wave functions. Then, the efficiency of our code on a modern parallel machine is demonstrated for a large water cluster with an ice structure. Secondly, the theory of energy representation (ER) is formulated within the framework of the density functional theory of solutions and its application to the free energy analyses of the protein hydration is provided. Thirdly, we discuss the coupling of the QM/MM approach with the method of energy representation, where the formulation for free energy $\delta\mu$ due to the electron density fluctuation of a QM solute plays a key role. As a recent progress in QM/MM-ER we developed a rigorous free energy functional to compute free energy contribution $\delta\mu$. The outline of the method as well as its extension to the QM/MM simulation coupled with a second-order perturbation approach are described.

H. Takahashi (✉)
Department of Chemistry, Graduate School of Science,
Tohoku University, Sendai 980-8578, Miyagi, Japan
e-mail: hideaki@m.tohoku.ac.jp

N. Matubayasi (✉)
Division of Chemical Engineering, Graduate School of Engineering Science,
Osaka University, Toyonaka 560-8531, Osaka, Japan
e-mail: nobuyuki@cheng.es.osaka-u.ac.jp

J.-L. Rivail et al. (eds.), *Quantum Modeling of Complex Molecular Systems*,
Challenges and Advances in Computational Chemistry and Physics 21,
DOI 10.1007/978-3-319-21626-3_6

6.1 Introduction

Enzymes show their elaborate functions in promoting chemical reactions of solutes that are bound to specific sites embedded in the proteins [1, 2]. Interestingly, in most cases, the thermal fluctuation of the polymer chains of amino acids as well as water molecules surrounding the active sites plays a key role in organizing the chemical events. Thus, it is of fundamental importance in the fields of chemistry and biochemistry to study the role of the environment in a chemical reaction occurring in a confined region of a condensed system [3, 4]. Since the dominant path in a many-particle system is governed by the free energy, it is one of the major subjects in theoretical and computational chemistry to calculate the free energy change associated with the reaction path of interest.

The theoretical description of chemical reaction necessitates the method of quantum chemistry, however, its computational cost increases drastically with the number of electrons involved in the system. In addition, the free energy calculation by means of molecular simulation is also known as a demanding task because it requires a lot of configuration samplings for the intermediate points along the process. Thus, we have to overcome these difficulties simultaneously to compute free energy changes for chemical reactions in condensed phases [5].

The hybrid quantum mechanical/molecular mechanical (QM/MM) method is a promising approach [6–8] to reduce the cost in quantum chemical calculations without serious loss of accuracy. The QM/MM simulation has been extensively applied to various chemical events in solutions or biological systems and its efficiency and robustness are well established. As a notable feature in our development of a QM/MM code [9–11] we employed the real-space grid technique [12–14] to represent one-electron wave functions in Kohn-Sham density functional theory (KS-DFT) [15–17] for the QM subsystem to achieve high efficiencies in massively parallel computations. For the efficient calculation of the solvation free energy of a solute we resort to a theory of solutions, referred to as theory of energy representation (ER) [18–20], where the distribution function of the solute-solvent pair potential plays a role in constructing the free energy. By virtue of the approximate functional in the theory it is possible to compute free energy of solvation with a modest sampling in configuration space as compared to the method of molecular simulation such as free energy perturbation (FEP) or thermodynamic integration (TI) [21, 22]. By combining the QM/MM approach based on the real-space grid DFT with the theory of energy representation we developed a method termed QM/MM-ER [23, 24], where the energy distribution functions are to be generated through a set of QM/MM simulations. The point of the theoretical framework of QM/MM-ER lies in the decomposition of the total solvation free energy $\Delta\mu$ of a QM solute into two contributions $\Delta\bar{\mu}$ and $\delta\mu$ which are the free energies due, respectively, to the two-body and many-body interactions between QM and MM subsystems. The two-body contribution $\Delta\bar{\mu}$ can be calculated straightforwardly with the theory of energy representation and its efficiency has been well examined by performing a lot of classical simulations. On the other hand, we have to make some devices to evaluate the free energy $\delta\mu$ since the standard theory of solutions is based on the assumption that the solute-solvent interaction is pairwise additive [20, 25].

In a recent progress in the QM/MM-ER method we formulated a simple and rigorous functional [24] to evaluate the free energy $\delta\mu$ arising from the electron density fluctuation of a QM solute in solution. Importantly, this formulation provides a versatile tool which allows various extensions useful for several important applications. For instance, the effect of the electronic polarization of the solvent molecules on the solvation free energy can be treated within a unified framework. Further, the approach can also be applied to the free energy calculation for a QM/MM system where the electron density fluctuation of the QM subsystem is described with the second-order perturbation theory (PT2) [26].

In this issue we provide a review of the QM/MM-ER method and its recent progress as well. The organization of the article is as follows. In Sect. 6.2, we discuss the advantage of the real-space grid DFT in the parallel implementation by making comparison with the LCAO approach and demonstrate its efficiencies on modern parallel machines. In Sect. 6.3, we concisely review the theory of energy representation and its recent application to the analyses of the hydration free energies of proteins. The fourth section is devoted to describe the method to couple the QM/MM method with the theory of energy representation, where a particular emphasis is placed on the formulation to deal with the free energy contribution $\delta\mu$. The extension of the method to the QM/MM simulations coupled with perturbation approach is also illustrated.

6.2 Parallelization of Kohn-Sham DFT

Applications of the electronic structure calculation have so far been limited to relatively small systems due to its high numerical cost. Hence, the development of a method that enables us to treat large-scale systems extends drastically the scope of the theoretical approach. The state-of-the-art computer technology owes its high performance mainly to the massive integration of the CPU nodes connected with a high-speed network. Thus, the parallel implementation of the electronic structure calculation is one of the important issues in the quantum chemistry as well as in the solid state physics. In this section we briefly review the real-space grid approach which provides a simple framework suitable for massively parallel execution of DFT. The advantage of the method will be discussed by comparing it with the localized basis. We will also demonstrate the performance of the grid method on a modern parallel machine.

6.2.1 Density Functional Theory Using Real-Space Grids

Linear combination of atomic orbitals (LCAO) method [27] has been utilized as a common technique to represent one-electron wave functions that work as building blocks in density functional theories as well as in molecular orbital theory.

We consider here the computational cost and the parallel efficiency associated with the ab initio LCAO calculations. The most time-consuming part is the calculation that involves the two-electron integrals $<\chi_i \chi_j |1/r_{12}| \chi_k \chi_l>$ for all combinations of the indices i, j, k, and l attached to the basis functions $\{\chi_i\}$. Hence, a 4-fold loop with respect to the basis functions has to be processed in a code to evaluate the expectation values of the two-electron operators. A single basis function in a basis set is further given by a linear combination of primitive Gaussians with the same center in order to mimic the radial behavior of the corresponding Slater function. Therefore, the computational effort associated with the two-electron integrals increases drastically with the system size. It is important to note that most of the *off-diagonal* elements in the Hamiltonian matrix represented by the LCAO basis have non-zero values. As discussed below this causes a serious increase in the amount of data communications among the processors in the parallel computations. In the construction of the Fock matrix in the Hartree-Fock calculation the 4-fold loop for the two-electron integrals will be parallelized to yield an incomplete Fock matrix at each processor. Then, we reduce the matrix stored on each CPU to the master node to obtain complete Fock operator. Due to the non-zero elements in the off-diagonal part of the Fock matrix the whole of the matrix should be distributed to all processors to increment another SCF (self-consistent field) step. The global and mass communication among the processors are, thus, necessitated in the LCAO approach. This can be successfully applied to rather small molecules. In large systems, however, the parallel efficiency will be degraded seriously since the total communication costs is dominated by the off-diagonal part of the Hamiltonian matrix which increases drastically with the size of the matrix.

The plane wave basis, an eigenfunction of a kinetic energy operator, has been commonly used in the Kohn-Sham DFT (KS-DFT) calculations for periodic systems such as bulk crystals [28]. The matrix elements for the kinetic energy operator can be readily obtained in the reciprocal space; however, the exchange-correlation potential defined in the real space can be evaluated only after the total electron density is obtained in the real space. Hence, it is necessary to transform the one-electron wave functions in momentum space to those in the real space to complete the operation of the effective Hamiltonian of KS-DFT. To expedite the transformations of the wave functions between these spaces, the fast Fourier transformation (FFT) technique is utilized in most cases. The plane wave expansion works efficiently on a single CPU machine by virtue of a lot of sophisticated prevailing FFT algorithms. It is not, however, suitable in general for parallel computing especially on distributed memory architectures.

Here, we focus our attention to the parallelization of the electronic structure calculations using KS-DFT. The Kohn-Sham equation [16, 17], which gives the set of one-electron orbitals in the non-interacting system, is expressed as,

$$\left[-\frac{1}{2}\nabla^2 + \upsilon_{\mathrm{ext}}(\mathbf{r}) + \upsilon_{\mathrm{H}}(\mathbf{r}) + \upsilon_{\mathrm{xc}}(\mathbf{r})\right]\varphi_i(\mathbf{r}) = \varepsilon_i\varphi_i(\mathbf{r}), \tag{6.1}$$

where $\mathbf{r}$ is the position vector of the electron and the first and second terms in the bracket of the left hand side denote, respectively, the kinetic energy operator and the external potential such as Coulomb interaction between electrons and nuclei. The Hartree potential $\upsilon_{\mathrm{H}}(\mathbf{r})$ in Eq. (6.1) represents the classical Coulomb repulsion of the electron density $n(\mathbf{r})$. We note that $\upsilon_{\mathrm{H}}(\mathbf{r})$ as well as the external potential $\upsilon_{\mathrm{ext}}(\mathbf{r})$ are completely local in the real space. The exchange-correlation potential $\upsilon_{\mathrm{xc}}(\mathbf{r})$ describes the mean field originating from purely quantum mechanical interaction between electrons. Explicitly, the potential $\upsilon_{\mathrm{xc}}(\mathbf{r})$ is defined in terms of the exchange correlation functional $E_{\mathrm{xc}}[n]$, thus,

$$\upsilon_{\mathrm{xc}}[n](\mathbf{r}) = \frac{\delta E_{\mathrm{xc}}[n]}{\delta n(\mathbf{r})}. \tag{6.2}$$

Equation (6.2) clearly states that $\upsilon_{\mathrm{xc}}(\mathbf{r})$ is a local potential at position $\mathbf{r}$ for a given functional $E_{\mathrm{xc}}[n]$ of the electron density by its definition. In the practical approximation such as generalized gradient approximation (GGA) [17, 29], which gives a significant improvement on local density approximation (LDA), the argument of E_{xc} includes the absolute of gradient of $n(\mathbf{r})$ as an ingredient in addition to the electron density itself. Hence, the potential $\upsilon_{\mathrm{xc}}(\mathbf{r})$ is usually semi-local in practice. Operation of the kinetic energy operator also requires the evaluation of the Laplacian of the wave functions. Thus, the effective Hamiltonian in real-space representation has non zero elements only in the vicinity of diagonal region. Therefore, high parallel efficiency can be expected when the one-electron wave function is represented by a set of probability amplitudes defined on grid points which are uniformly distributed over a real space.

Chelikowsky et al. developed a scheme to solve the KS equation by means of the real-space grids [12, 13]. The point of their approach is to express the kinetic energy operator in Eq. (6.1) using the higher-order-finite-difference method. The Nth-order finite-difference representation at the grid point with index $(j,\ k,\ l)$ in three-dimensional space is generally given as

$$\begin{aligned}-\frac{1}{2}\nabla^2\varphi(j,k,l) = -\frac{1}{2h^2}\Bigg[&\sum_{n_1=-N}^{N} C_{n_1}\varphi(j+n_1,k,l) + \sum_{n_2=-N}^{N} C_{n_2}\varphi(j,k+n_2,l)\\ &+ \sum_{n_3=-N}^{N} C_{n_3}\varphi(j,k,l+n_3)\Bigg] + O\left(h^{2N+2}\right)\end{aligned} \tag{6.3}$$

where h is the grid width and C are the constant coefficients and directly determined from the Taylor series expansions of the wave functions. We provide in the following the procedure to determine the coefficients C in the case of $N = 2$ for instance. For the sake of simplicity we suppose a wave function $\varphi(x)$ with a one-dimensional variable x. The Taylor expansions for $\varphi(x)$ at the grid points $x + 2\,h$, $x + h$, $x - h$, and, $x - 2\,h$ are explicitly given as,

$$\varphi(x+2h) = \varphi(x) + \varphi^{(1)}(x) \cdot 2h + \frac{1}{2}\varphi^{(2)}(x) \cdot (2h)^2 + \frac{1}{6}\varphi^{(3)}(x) \cdot (2h)^3 + \cdots \tag{6.4a}$$

$$\varphi(x+h) = \varphi(x) + \varphi^{(1)}(x) \cdot h + \frac{1}{2}\varphi^{(2)}(x) \cdot h^2 + \frac{1}{6}\varphi^{(3)}(x) \cdot h^3 + \cdots \tag{6.4b}$$

$$\varphi(x-h) = \varphi(x) - \varphi^{(1)}(x) \cdot h + \frac{1}{2}\varphi^{(2)}(x) \cdot h^2 - \frac{1}{6}\varphi^{(3)}(x) \cdot h^3 + \cdots \tag{6.4c}$$

$$\varphi(x-2h) = \varphi(x) - \varphi^{(1)}(x) \cdot 2h + \frac{1}{2}\varphi^{(2)}(x) \cdot (2h)^2 - \frac{1}{6}\varphi^{(3)}(x) \cdot (2h)^3 + \cdots . \tag{6.4d}$$

Our purpose is to make $O(h^4)$ terms in Eqs. (6.4a)–(6.4d) to vanish in constructing $\varphi^{(2)}$. We note that the ratio of the coefficient of the h^4 term in Eq. (6.4a) to that in Eq. (6.4b) is 16. Hence, the $O(h^4)$ term vanishes when Eq. (6.4a) is mixed with Eq. (6.4b) with the weight of $-1/16$. Further, the terms of odd power of h vanish by summing Eqs. (6.4c) and (6.4d) with the same weights as the corresponding terms Eqs. (6.4b) and (6.4a). After simple algebraic manipulations we have an approximate expression for $\varphi^{(2)}$ with the error of $O(h^6)$ thus,

$$\begin{aligned}\varphi^{(2)}(x) = &-\frac{1}{12}\varphi(x-2h) + \frac{4}{3}\varphi(x-h) - \frac{5}{2}\varphi(x) + \frac{4}{3}\varphi(x+h) \\ &-\frac{1}{12}\varphi(x+2h) + O(h^6).\end{aligned} \tag{6.5}$$

It is worthy of note that the inclusion of only a few neighboring terms around $\varphi(x)$ contribute to a high accuracy in the evaluation of $\varphi^{(2)}(x)$. Hence, the increase in the accuracy does not cause serious non-locality of the operator. The coefficients for $N \geq 3$ can also be determined in a similar manner. For our practical applications we found that the choice of $N = 4$ is adequate to describe the kinetic energy operator.

We also discuss the parallel implementation of the Hartree potential $\upsilon_H(\mathbf{r})$ which is defined as

$$\upsilon_H(\mathbf{r}) = \int \frac{n(\mathbf{r}')}{|\mathbf{r}-\mathbf{r}'|} d\mathbf{r}'. \tag{6.6}$$

We here focus our attention to a non-periodic system. The direct evaluation of the integral in Eq. (6.6) with the real-space formalism requires a 2-fold loop with respect to the grid points, which results in a huge computational cost. Furthermore, in its parallelization, the whole data of the electron density $n(\mathbf{r})$, being stored in master node, should be distributed to all processors to construct a new potential $\upsilon_H(\mathbf{r})$ at each SCF step. A method which allows an efficient parallelization is to adopt the Poisson equation [14] given by

$$\nabla^2 \upsilon_{\mathrm{H}}(\mathbf{r}) = -4\pi n(\mathbf{r}). \tag{6.7}$$

With the finite-difference representation for the Laplacian in Eq. (6.7), we note that only the vicinity of the diagonal part of the matrix has non-zero values as discussed above. An outline for the numerical solution of Eq. (6.7) is illustrated as follows. Suppose that **A** denotes the Laplacian matrix in the real space, and $\boldsymbol{x}$ and $\boldsymbol{n}$ are the vectors representing the Hartree potential and the electron density, respectively, then the Poisson equation of Eq. (6.7) reads

$$\mathbf{Ax} = -4\pi\mathbf{n}. \tag{6.8}$$

It is easy to see that the solution of Eq. (6.8) for a given electron density $n(\mathbf{r})$ is equivalent to minimizing the value of

$$\langle \mathbf{x}|A|\mathbf{x}\rangle + 4\pi\langle \mathbf{x}|\mathbf{n}\rangle. \tag{6.9}$$

The minimization of Eq. (6.9) can be performed by the steepest descent (SD) or conjugate gradient (CG) procedure under some boundary conditions for **x**. We note that the amount of the data communication required in the multiplication of **A** to $\boldsymbol{x}$ is the same as that needed in the operation of the kinetic energy operator to only a single one-electron wave function. Thus, the Poisson equation provides an advantageous route to the parallelization of evaluating $\upsilon_{\mathrm{H}}(\mathbf{r})$ for finite systems.

With the real-space grids, however, it is difficult to describe the rapid behavior of the wave function near an atomic core region. A solution to this problem is to employ the pseudopotential which ensures the smoothness of the wave functions for valence electrons [28]. In general the pseudopotential is dependent on the angular momentum of the wavefucntion and hence, is in non-local form. Fortunately, the non-locality of the operator in the real space extends only over a confined region around an atomic core. Therefore, the use of the pseudopotential causes only a slight increase in the communication cost. Thus, it is shown that the real-space grid approach is advantageous to the parallel implementation especially for the non-periodic finite systems.

Here, we discuss about the key technique to achieve higher accuracy in performing KS-DFT calculations with the real-space grids. It is well known that the electronic energy of an atom in a real-space cell oscillates for the variation of the relative position of the atom with respect to the grid. This unpleasant situation is sometimes referred to as 'egg-box problem'. To alleviate the erroneous behavior originating from the finite grid size it is essential to introduce a multi-grid method which ensures higher resolutions in describing the wave functions specifically around the atomic cores. The multi-grid approach is efficient, however, it gives rise to a substantial increase in the computational cost for the integral that involves the non-local part of the pseudopotential. Ono and Hirose developed a timesaving double-grid approach [30] which allows an accurate integration with modest increase in the computational overhead. The point of their approach is to evaluate the values of the pseudo wave functions on the dense grids by interpolating those on the coarse grids. This approximation can be validated by the smooth nature of the pseudo-wave functions. On the contrary, the values of the pseudopotentials on the dense grids are being determined explicitly. With this approach the contribution of the wave function on the double grid can be cast into weight factors defined on the *course* grids. It should be emphasized that the weight factors remain constant during the SCF calculations unless the positions of the nuclei are unchanged. Thus, once the weight factors, which carry the information on the dense grids, are evaluated before entering the SCF procedure, there is no need to update the factors.

6.2.2 Performance on a Modern Parallel Machines

In general a modern parallel machine consists of a lot of CPUs connected with high-speed network facilities. Further, each CPU node itself is mostly constructed from multi cores working on a shared memory architecture (multi-core processor). The data communications among the nodes are usually conducted by using MPI interface, while the thread parallel within each node is processed by means of OpenMP (OMP) directives. Thus, the hybrid parallel with the combination of MPI and OMP provides a versatile scheme well suited to the prevailing parallel architectures. We implemented the hybrid parallelization on our code for the KS-DFT based on the real-space grid reinforced with the double-grid technique.

As a benchmark test of the parallel code developed by H. Takahashi et al. we performed the KS-DFT calculations for a water cluster with an ice structure. The number of water molecules contained in the cluster was 120 and the O–O distance between the neighboring water molecules was set at 2.76 Å. The ice structure used in the present calculation is illustrated in Fig. 6.1.

The cluster was embedded in a rectangular QM cell of the size: $L_x = 18.6$, $L_y = 21.3$, and $L_z = 23.9$ Å, where each ice surface was oriented parallel to that of the QM cell. The kinetic energy operator in Eq. (6.1) was represented by a 4th-order finite difference expression and the exchange-correlation potential υ_{xc} was evaluated using the BLYP functional [29, 31]. The interaction between valence electrons and nuclei is calculated by the pseudopotentials in the Kleinman-Bylander separable form [32]. The boundary conditions for the solution of the Poisson equation of Eq. (6.8) was constructed as the Coulomb potential at the boundary formed by the electron density inside the cell. To expedite the calculation of $\upsilon_H(\mathbf{r})$ the boundary condition at each SCF step is approximately determined as the Coulomb potential due to the ESP charges of the atoms formed by the electron density. The number of grid points were 112, 128, and 144 along the x, y, and z axes, respectively. The grid spacing h for each direction was set at 0.166 Å. Further, the double grid technique by Ono and Hirose [30] was adopted to the wave functions around the atomic cores. The width of the double grid was set at 0.2 h. In our preliminary calculations on smaller systems of ice we found that the real-space grid approach with these settings is comparable in accuracy to the LCAO calculation using aug-cc-pVDZ or aug-cc-pVTZ basis set [33].

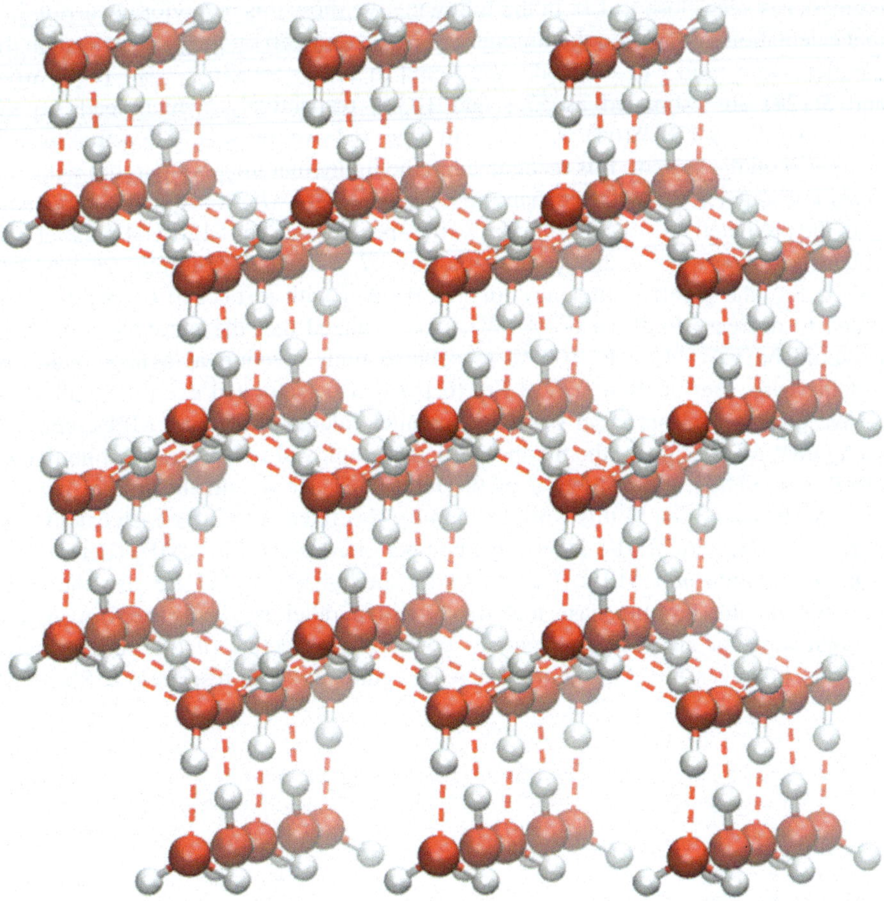

Fig. 6.1 *Ball* and *stick* representation of the ice structure which consists of 120 water molecules. The neighboring O–O distance is 2.76 Å. The geometrical parameters of the constituent water molecules are OH = 0.976 Å and HOH = 109.471°

The parallel computer available in the present calculation is a part of Cray XC30 and it consists of 32 CPU nodes with the same architecture (Intel Xeon E5: 2.30 GHz), where each node is equipped with 28 cores. By employing 64, 128, 256, and 512 cores of this machine we first examined the parallel efficiency of flat MPI. The speedup $S(p)$ in the case of MPI using p cores is given as

$$S(p) = \frac{1}{\frac{r}{p} + (1 - r)} \tag{6.10}$$

where r denotes the parallelization ratio of the code. It is expected from Eq. (6.10) that the value $S(p)$ is saturated soon with respect to the increasing number of

p unless r is very close to 1.0. In the following we show this behavior occurring in a real calculation. Measuring the timings at $p = 128$ and 256 on the present system the parallelization ratio r of our code was found to be $r = 99.8167...$ %. Then, $S(64)$ and $S(128)$ are estimated as 57.4 and 103.8, respectively, which leads to the speedup ratio $S(128)/S(64) = 1.809$. In Fig. 6.2 we present the actual results examined on the system. It is recognized in the figure that the MPI parallel with 128 cores is about 1.8 times faster than the 64 parallel, showing reasonable agreement with the predicted ratio 1.809. Similarly, the speedup ratio of 512 to 64 parallel was obtained as 4.607, while the actual value was 4.7. Since the ideal ratio for $S(512)/S(64)$ is 8.0, the parallel efficiency for 512 parallel with respect to 64 parallel was found to decrease to about 60 %. We, thus, realized that the parallelization ratio r should be very close to 100 % to achieve high speedup at a large value of p. Otherwise, the function $S(p)$ of Eq. (6.10) will be saturated even at small p.

We also carried out a set of hybrid parallel calculations using MPI combined with OMP. We designate the number of nodes employed in MPI as n, while that of OMP as m. Then, the number J of the total cores we use in the hybrid parallel amounts to $J = n \times m$. In this work the number J is fixed at 512 for each calculation of $n = 64$, 128, 256, or 512. Hence, the number of cores used in OMP is determined from the relation $m = 512/n$. The results for the hybrid parallel are also shown in Fig. 6.2. In the limiting case that the hybrid parallel is ideally performed, the elapsed time for an SCF step remains constant irrespective of the number n. In practice, however, the elapsed time decreases gradually with respect to n. Since the

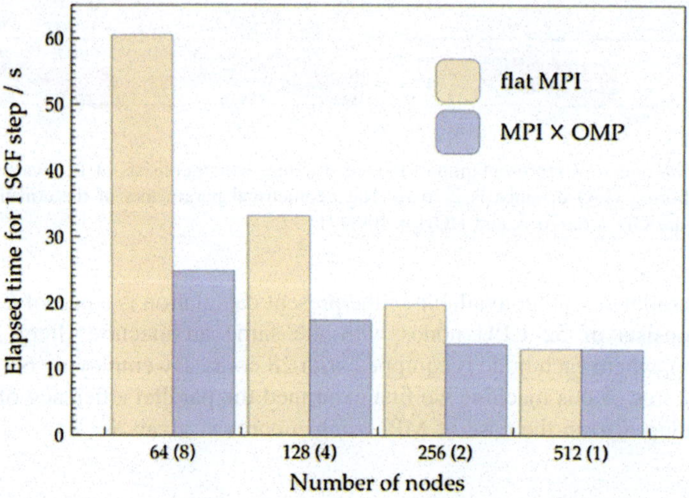

Fig. 6.2 The elapsed time necessitated for 1 SCF step in the Kohn-Sham DFT calculation for the variation of the number of nodes employed in the parallel machine. The parallel performance was assessed both for flat MPI and for hybrid parallel (MPI × OMP) computations. The integers in the parentheses in the *horizontal axis* indicate the number of cores used in the OMP calculations. The system is the cluster of 120 water molecules depicted in Fig. 6.1

thread parallel is applied only to the time-consuming loops, it is disadvantageous to assign much computational load to the OMP. On the other hand, putting large weight on the MPI leads to the increase in the communication overhead relative to the numerical task. Therefore, the load balance between MPI and OMP is crucial for the success of the hybrid parallel. In the present system we found that $(n, m) = (256, 2)$ or $(512, 1)$ is the appropriate choice to achieve high efficiency. Of course, the performance is dependent on the details of the computer architecture employed as well as the system of interest, and hence, the choice of (n, m) should be carefully examined according to the situation.

We conclude that the real-space grid approach is a superior method to achieve high parallel performance within the framework of Kohn-Sham DFT. Actually, it was demonstrated that using 128 cores in flat MPI affords us more than 100 times speedup in a realistic system. We should, however, make some remarks on the disadvantage of the method. In modern DFT the Hartree-Fock exchange plays a crucial role as an ingredient to density functionals for the accurate evaluation of molecular properties [34, 35]. A notable deficiency in the grid method lies in the high numerical effort associated with the calculation of the Hartree-Fock exchange energy. The real-space representation suffers from the serious non-locality of the orbital-based exchange operator since the wave functions are delocalized over the space. It is, hence, desirable to develop efficient methods for parallel implementation of the Hartree-Fock exchange.

6.3 Approach to the Free Energy of Solvation

In the context of statistical thermodynamics, the effect of solvent on the stability and reactivity of solute is quantified by the solvation free energy. This is the free-energy change for turning on the intermolecular interaction of the solute with the solvent and involves only the contribution from the potential energy. The free energy of solvation requires much computational demand with explicit solvent, however, when such standard techniques as free-energy perturbation and thermodynamic integration are employed [21, 22]. These methods are conducted in practice by introducing a number of intermediate states connecting the two states of the (pure) solvent without solute and the target solution with the solute. It should be noted that the initial and final states are of physical interest, while the intermediate states are often not. An efficient alternative is then to carry out the simulations only at the "end-point" states and use an approximate functional for the free energy [36–48]. Our strategy of free-energy computation belongs to this line. We approach the solvation free energy by combing the molecular simulation with a distribution-function theory of solutions. In the present section, we introduce the method of energy representation to compute the solvation free energy and describe an illustrative application to protein hydration.

6.3.1 Method of Energy Representation

As noted above, the method of energy representation [18–20, 49] is an "end-point" method for free-energy calculation. It is a density-functional method for solutions; while the density-functional theory of solution is typically formulated in terms of distribution functions defined over coordinate space [25, 50–54], the solution theory in the energy representation is developed with distribution functions of solute-solvent pair energy. In the energy-representation method, the solvation free energy is computed through an approximate functional consisting of distribution functions in two systems. One of the systems is the solution system of interest, and contains the solute and solvent molecules at full coupling of their interaction. The other is called reference-solvent system. In this system, the configurations of the solute and solvent are separately prepared, and the distribution functions are calculated by placing the solute in the solvent system as a test particle. In the energy-representation method, the molecular simulation is conduced only in the solution and reference-solvent systems, leading to the reduction of computational load. In the following, we show the explicit form of the free-energy functional currently employed [19, 49].

The solvation free energy $\Delta\mu$ is the free-energy change corresponding to the gradual insertion of the solute molecule. When the intra-molecular energy of the solute is $\Psi(\psi)$ and the total solvent-solvent energy is $U(\mathbf{X})$, $\Delta\mu$ is expressed as

$$\exp(-\beta\Delta\mu) = \frac{\int d\psi d\mathbf{X} \exp(-\beta\{\Psi(\psi) + \sum_i v(\psi, \mathbf{x}_i) + U(\mathbf{X})\})}{\int d\psi d\mathbf{X} \exp(-\beta\{\Psi(\psi) + U(\mathbf{X})\})} \tag{6.11}$$

where ψ denotes the solute configuration, $\mathbf{X}$ represents the solvent configuration collectively, $\mathbf{x}_i$ is the configuration of the i-th solvent molecule, v is the pair interaction potential between the solute and solvent, and β is the inverse of k_BT with the Boltzmann constant k_B and the temperature T. A restriction of attention to a certain set of solute intra-molecular state can be made simply by the corresponding alteration of the domain of integration over ψ. The energy representation is introduced by adopting the value of the solute-solvent interaction v of interest as the coordinate ε for the solute-solvent distribution. The instantaneous distribution $\hat{\rho}$ is defined as

$$\hat{\rho}(\varepsilon) = \sum_i \delta(v(\psi, \mathbf{x}_i) - \varepsilon) \tag{6.12}$$

where the sum is taken over the solvent molecules.

In the currently used version of the method of energy representation [19, 49], the solvation free energy $\Delta\mu$ is approximately expressed in terms of distribution functions constructed from $\hat{\rho}$ in the solution and reference-solvent systems. In our treatments, the solution system refers to the system in which the solute molecule interacts with the solvent under the solute-solvent interaction v of interest at full coupling. In the solution, the average distribution ρ of the v value is relevant in the approximate construction of $\Delta\mu$ and is given by

$$\rho(\varepsilon) = \langle \hat{\rho}(\varepsilon) \rangle \tag{6.13}$$

where $\langle \cdots \rangle$ represents the ensemble average in the solution system of interest. On the other hand, the reference solvent system denotes the system in which the solute does not interact with the solvent and the solvent configuration is generated only by the solvent-solvent interaction. At an instantaneous configuration of the reference solvent system, $\hat{\rho}$ is constructed by placing the solute molecule in the system as a test particle; at test-particle insertion, the solvent configuration generated from a simulation without solute is used as is and the solute is placed in the solvent system without disturbing the solvent configuration. The solute-solvent overlap is allowed and accounts for the excluded-volume effect in solvation. The average distribution ρ_0 and the correlation matrix χ_0 then appear in the approximate functional for $\Delta\mu$ and are expressed, respectively, as

$$\rho_0(\varepsilon) = \langle \hat{\rho}(\varepsilon) \rangle_0 \tag{6.14}$$

and

$$\chi_0(\varepsilon, \eta) = \langle \hat{\rho}(\varepsilon)\hat{\rho}(\eta) \rangle_0 - \langle \hat{\rho}(\varepsilon) \rangle_0 \langle \hat{\rho}(\eta) \rangle_0 \tag{6.15}$$

where $\langle \cdots \rangle_0$ represents the ensemble average in the reference-solvent system. In the sampling corresponding to $\langle \cdots \rangle_0$, the solute and solvent degrees of freedom are uncoupled from each other in the probability distribution.

An approximate functional for the solvation free energy $\Delta\mu$ is derived in Ref. [19]. The functional is constructed by adopting the PY (Percus-Yevick)-type approximation in the unfavorable region of solute-solvent interaction and the HNC (hypernetted chain)-type approximation in the favorable region, and $\Delta\mu$ is given by a set of definitions and equations listed as [19, 49]

$$\omega(\varepsilon) = -k_B T \, log\left(\frac{\rho(\varepsilon)}{\rho_0(\varepsilon)}\right) - \varepsilon \tag{6.16}$$

$$\sigma_0(\varepsilon) = -k_B T \int d\eta \left(\frac{\delta(\varepsilon - \eta)}{\rho_0(\varepsilon)} - (\chi_0)^{-1}(\varepsilon, \eta)\right)(\rho(\eta) - \rho_0(\eta)) \tag{6.17}$$

$$\begin{aligned} \Delta\mu = \int d\varepsilon \, \varepsilon\rho(\varepsilon) - k_B T \int d\varepsilon \Big[&(\rho(\varepsilon) - \rho_0(\varepsilon)) - \rho(\varepsilon) log\left(\frac{\rho(\varepsilon)}{\rho_0(\varepsilon)}\right) \\ &- \left(\frac{\rho(\varepsilon)}{\rho_0(\varepsilon)}\right) - \{\alpha(\varepsilon)F(\varepsilon) + (1 - \alpha(\varepsilon))F_0(\varepsilon)\}(\rho(\varepsilon) - \rho_0(\varepsilon)) \Big] \end{aligned} \tag{6.18}$$

$$F(\varepsilon) = \begin{cases} \beta\omega(\varepsilon) + 1 + \frac{\beta\omega(\varepsilon)}{exp(-\beta\omega(\varepsilon))-1} & (\text{when } \omega(\varepsilon) \leq 0) \\ \frac{1}{2}\beta\omega(\varepsilon) & (\text{when } \omega(\varepsilon) \geq 0) \end{cases} \quad (6.19)$$

$$F_0(\varepsilon) = \begin{cases} -log(1 - \beta\sigma_0(\varepsilon)) + 1 + \frac{log(1-\beta\sigma_0(\varepsilon))}{\beta\sigma_0(\varepsilon)} & (\text{when } \sigma_0(\varepsilon) \leq 0) \\ \frac{1}{2}\beta\sigma_0(\varepsilon) & (\text{when } \sigma_0(\varepsilon) \geq 0) \end{cases} \quad (6.20)$$

$$\alpha(\varepsilon) = \begin{cases} 1 & (\text{when } \rho(\varepsilon) \geq \rho_0(\varepsilon)) \\ 1 - \left(\frac{\rho(\varepsilon)-\rho_0(\varepsilon)}{\rho(\varepsilon)+\rho_0(\varepsilon)}\right)^2 & (\text{when } \rho(\varepsilon) \leq \rho_0(\varepsilon)) \end{cases} \quad (6.21)$$

Note that the first term of Eq. (6.18) is the average sum of the solute-solvent interaction energy. The rest of Eq. (6.18) is constructed from the distribution functions. See Appendix B of Ref. [20] and Appendix B of Ref. [49] for the scheme to determine $\sigma_0(\varepsilon)$ through Eq. (6.17). According to the benchmark tests for amino-acid analog solutes in solvent water, the error caused by the use of the approximate functional was observed not to be larger than the error due to the use of force field [55, 56].

Equation (6.13) is the average distribution (histogram) of the solute–solvent pair energy value ε in the solution system. Equation (6.14) is similarly the average distribution in the reference solvent and corresponds to the density of states for solute–solvent pair interaction. Equation (6.15) is a correlation matrix and describes the solvent–solvent correlation in the reference solvent. $\omega(\varepsilon)$ of Eq. (6.16) is an energy-represented version of the solvent-mediated part of solute–solvent potential of mean force. It vanishes when the solvent–solvent correlation is absent. $\sigma_0(\varepsilon)$ of Eq. (6.17) is also the solvent-mediated part of the response function of the solute–solvent distribution to the solute–solvent interaction in the reference solvent. Equations (6.19) and (6.20) introduce a combined PY-type and HNC-type approximation. They are the expressions in the solution and reference-solvent systems, respectively, and are mixed with a weighting function of Eq. (6.21). The PY approximation is formulated as the first-order perturbation of the solvent-mediated part of the solute-solvent pair distribution function expanded in terms of the change in the solvent local density caused by the solute insertion. The HNC approximation is, on the other hand, the first-order perturbation of the solvent-mediated part of the solute-solvent potential of mean force. It is recognized in treatments of simple liquids [25] that the PY and HNC approximations work well in the repulsive and attractive domains of the solute-solvent interaction, respectively. Equations (6.19) and (6.20) are energy-represented expressions of the conventional wisdom of the applicability of the PY and HNC approximations.

In the energy representation, the interaction energy between the solute and solvent is adopted for the one-dimensional coordinate of the distribution functions, and a functional for the solvation free energy is constructed from energy distribution functions in the solution and reference-solvent systems of interest. The introduction of the energy coordinate for distribution functions is a kind of coarse-graining procedure for reducing the information content of solute-solvent configuration. Since the set of configurations (structures) with the equal solute-solvent interaction energies are grouped into a unit in the energy representation, any approximate functional built in terms of energy distribution functions cannot violate the statistical-mechanical principle that the configurations with the same solute-solvent interaction energies contribute to the solvation free energy with the equal weights.

As seen in Eq. (6.12), each of the solute and solvent molecules is taken as a single unit in the energy representation. The molecule is treated as a whole, while the coordinate for the distribution functions is one-dimensional. No explicit reference is made to the detail of the molecular structure by focusing on the interaction energy. The following advantages then emerge in the method of energy representation.

Firstly, the method is straightforwardly applicable to molecules with intramolecular flexibility. The implementation is indifferent whether the molecule is rigid or flexible. The information of structural fluctuation of the molecule is adsorbed when the energy coordinate is introduced by Eq. (6.12). For large molecules constituting micellar, membrane, and protein systems, it is not allowed to neglect the molecular flexibility. In the method of energy representation, an additional and/or separate scheme is not necessary to be formulated for large, flexible species.

Secondly, the treatment of inhomogeneous system and clusters is straightforward. So far, the formulation does not assume the system homogeneity or the thermodynamic limit. The application to inhomogeneous and/or finite systems is then possible without modification. The binding of a molecule to such nanoscale structures as micelle, membrane, and protein can be viewed as a solvation in an inhomogeneous and partially finite, mixed solvent [57, 58]. The method of energy representation can thus be a useful approach to intermolecular correlation and association important in biological and interface sciences.

Thirdly, an accurate treatment is possible for supercritical fluid. In supercritical fluid, the solvent density and temperature can be varied over wide range and the solvent effect may act as a key to control a chemical process. It is well known that supercritical fluid can be described accurately when the whole molecule is treated as a single unit [25]. A multidimensional representation is necessary, however, in the usual coordinate space. By introducing the energy as the coordinate for distribution functions, the whole molecule can be taken as a single unit with keeping the description one-dimensional. The approximate functional given by Eqs. (6.13)–(6.21) incorporates the intermolecular correlation at the two-body level. The solvation free energy obtained is then exact to second order in the solvent density. Since the method is exact in the low-density regime, a formulation of a good approximation in the high-density regime leads to an accurate description over a wide range of solvent density.

Finally, the combination with the quantum-mechanical/molecular-mechanical (QM/MM) methodology can be performed. In QM/MM calculation, the many-body effect is introduced for the solute-solvent interaction and is beyond the applicability of conventional theories of solutions. In the method of energy representation, the fluctuation of the electronic state in response to the environment is viewed as a fluctuation of intramolecular degrees of freedom of the QM solute. The evaluation becomes feasible for the free energy for the many-body effect of the electronic fluctuation. In addition, Eq. (6.12) makes no reference to the functional form of the potential function. It refers only to the value of the potential energy, and there is no need for modification or deterioration of the electronic-state calculation. Thus, the treatment is possible for an arbitrary distribution of charges. The contraction to a set of point charges is not necessary, and the effect of the diffuse (cloud-like) nature of electronic distribution can be determined. The detail of the combination with the QM/MM methodology is given in Sect. 6.4.

6.3.2 Application to Protein Hydration

The structure of a protein is determined by the balance between its intramolecular energy (structural energy) and the free energy of solvation. Toward the structure prediction of protein, therefore, it is necessary to model the solvation free energy at high accuracy. MD simulation with explicit solvent meets this necessity. The free-energy calculation of hydration of such large molecule as protein is highly prohibitive, however, when standard and numerically exact methods such as free-energy perturbation and thermodynamic integration are to be conducted in all-atom treatment. Alternatively, the method of energy representation can be suitable for computing the free energy of protein hydration since it is an "end-point" method as described in the preceding subsection and reduces the computational load considerably. In this subsection, we provide an illustrative application of the energy-representation method to structure determination of protein complex [59].

Two protein complexes were examined: a complex consisting of bovine trypsin and CMTI-1 squash inhibitor (PDB ID of the complex: 1PPE [60]), and RNase SA in complex with barstar (PDB ID of the complex: 1AY7 [61]). They consist of 252 and 185 residues, respectively. We prepare a set of "decoy" structures as described in the following, and select the "right" structure through free-energy calculations with the energy-representation method.

The sets of complex structure models (decoys) were generated as follows. First, a rigid-body protein–protein docking program ZDOCK 3 was employed [62, 63]. By grid searching the rotational space at 6° and using the scoring function, the optimal translational position was determined for each orientation of one monomer relative to another. The 54,000 decoy structures thus generated were also re-ranked with ZRANK [64] equipped with more optimized scoring function. The decoys examined with all-atom simulation were then the top-ranked ones generated by ZDOCK and ZRANK, the one whose interface root-mean square distance ($RMSD_i$) is the smallest among the 54,000 decoys, and those with $RMSD_i$ closest to 2, 3, 4, 6, 8, and 10 Å; $RMSD_i$ was defined in the present calculation with respect to the Cα atoms in the interface residues in which at least one atom is located within 10 Å of the other protein in the crystal structure of the complex.

In addition to the above 9 structures, 4 more structures were subject to the analysis. Two of them were generated by Rosetta dock [65]. 10,000 decoys were constructed using the program "docking_protocol" in the Rosetta 3 package [65], which returns scores for the entire complex as well as the complex interface only. The top ranked decoys in the total score and in the interface score were picked up. The other two were the x-ray structure and the equilibrium structure in solution. To determine the latter, the crystal structure was solvated so that the distance between the outmost protein atom and the closest simulation box face in the initial setup were 10 Å. A 50-ns MD simulation was then performed, the last 10-ns trajectory was used to generate the average structure, and the snapshot closest to the average structure was employed.

Before starting the MD simulation, the side-chain optimization was conducted for each of the decoy structures generated by ZDOCK and Rosetta dock using the fixed-backbone design program in the Rosetta 3 package [65]. Each decoy structure was then subject to a 2,000-step energy minimization with the conjugate gradient method, and was solvated by water in a cubic box which is 2.5 times as large as the solute size along all the box axes. The system was brought to thermodynamic equilibrium at 300 K and 1 atm using Langevin thermostat and barostat. MD was carried out at a time step of 2 fs, and the electrostatic interaction was handled by the particle mesh Ewald method. After the energy minimization with a cut-off length of 12 Å, all of the protein atom coordinates were fixed and only water molecules were allowed to move. The water molecules around the fixed protein were sampled for 1 ns after a 1-ns equilibration run. A pure-water simulation was also performed over 100 ps with the same number of water molecules contained in the solute-water system. The solvent system was sampled every 100 fs, and the solute (decoy) insertion was conducted 1,000 times for each solvent configuration. With the above procedure, the solvation free energy was computed with 95 % error of several kcal/mol, which is small enough to distinguish the lowest free-energy decoy from the others.

The stability of each decoy structure is determined by the sum of the intra-complex energy E_c and the solvation free energy $\Delta\mu$. Figure 6.3 shows E_c $\Delta\mu$ and ($E_c + \Delta\mu$) against RMSD$_i$ for the 1PPE complex. The complex structures with low RMSD$_i$ have smaller E_c that the other decoys, and with addition of the solvation term, the ($E_c + \Delta\mu$) value for the equilibrium structure in solution becomes distinctively the smallest. The "right" structure for 1PPE can thus be singled out from Fig. 6.3. It should be noted that the crystal structure is a close but not the "right" one in solution. The protein structure in solution deviates from that in crystal typically by a few Å.

Figure 6.4 shows E_c $\Delta\mu$ and ($E_c + \Delta\mu$) against RMSD$_i$ for 1AY7. As in the 1PPE case, the structures with low RMSD$_i$ have small E_c and the smallest ($E_c + \Delta\mu$) is returned by the equilibrium structure in solution. Note that the structures of RMSD$_i$ = 8 and 10 Å have relatively low ($E_c + \Delta\mu$). They have high E_c with favorable effect of solvation. This is related to the fact that their interfaces within the complex are relatively small and that the complexes are well exposed to solvent water. It is seen furthermore that the structures around the equilibrium one have low

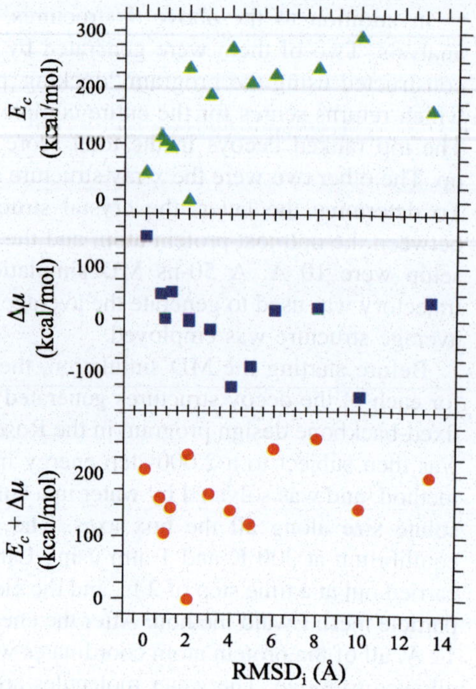

Fig. 6.3 Intra-complex energy E_c, solvation free energy $\Delta\mu$, and their sum ($E_c + \Delta\mu$) against RMSD$_i$ for 1PPE. Each value is shown as the difference from the value for the equilibrium structure in solution

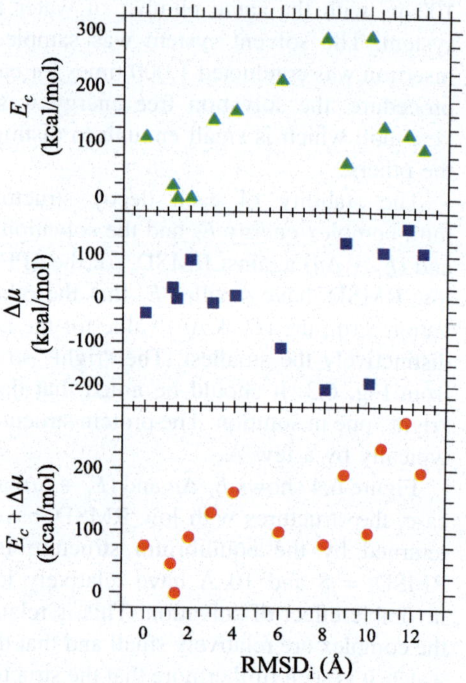

Fig. 6.4 Intra-complex energy E_c, solvation free energy $\Delta\mu$, and their sum ($E_c + \Delta\mu$) against RMSD$_i$ for 1AY7. Each value is shown as the difference from the value for the equilibrium structure in solution

E_c and high $\Delta\mu$. They achieve favorable interactions within the complex at expenses of relatively unfavorable interactions with solvent water.

In summary, the stable structures of protein complexes can be estimated first by the intra-complex energy and then singled out by adding the solvation effect. Both of the intra-complex and solvation effects are important, and the solvation free energy is necessary to identify the "right" structure in solution. To faithfully model the effect of solvent toward the structure prediction of protein, all-atom treatment is necessary and is becoming feasible through combination of large-scale molecular simulation and statistical mechanical theory of solutions.

6.4 QM/MM Method Combined with the Theory of Energy Representation

In 2004 we developed a method referred to as QM/MM-ER [23] for the purpose of computing solvation free energy of a QM solute in MM solvent by combining the hybrid QM/MM with the theory of energy representation. As described in Sect. 6.3 the standard theory of solutions [25] is based on the assumption that the solute-solvent interaction is pairwise additive. However, the QM/MM method involves the many-body interaction that originates from the electron density polarization of the QM subsystem. Hence, we have to make some device to manage the many-body effect within the framework of the energy representation.

The point of our method is to decompose the total solvation free energy into the contribution $\Delta\bar{\mu}$ due to the two-body interaction and the residual many-body contribution $\delta\mu$. For the free energy $\Delta\bar{\mu}$ the method of energy representation can be applied straightforwardly. In a recent progress in QM/MM-ER [24] we formulated a simple and rigorous functional to evaluate the free energy $\delta\mu$ by introducing a new energy coordinate that represents the energy shift of the whole system due to the polarization of the solute. The QM/MM-ER approach has so far been applied to various solution systems and its efficiency and accuracy have been well established [66–75]. The formulation to treat $\delta\mu$ provides a versatile theoretical framework which allows us subsequent various extensions [76, 77].

In the following we provide a formulation of QM/MM-ER with an emphasis placed on the method to calculate $\delta\mu$ and its extensions. The application of the method to the calculation of solvation free energy of H_2O molecule in water will also be presented.

6.4.1 *Treatment of the Many-Body Interaction*

The free energy change associated with a reaction in solution can be expressed in terms of the free energy change in the gas phase and the solvation free energies for the reactant and the product by virtue of the thermodynamic cycle. To illustrate the method of QM/MM-ER it is appropriate to start with defining the solvation free energy within the QM/MM approach [23].

Suppose that a solute molecule is described using some quantum chemical method and solvent molecules are represented by a classical force field. The total energy E_{tot} of a QM/MM system consists of three contributions as

$$E_{\mathrm{tot}} = E_{\mathrm{QM}} + E_{\mathrm{QM/MM}} + E_{\mathrm{MM}} \tag{6.22}$$

where E_{QM} and E_{MM} are the energies of the QM and MM subsystems. The cross term $E_{\mathrm{QM/MM}}$ in Eq. (6.22) describes the interaction energy between the two subsystems. It is assumed that the wave function Ψ of the solute in solution is the ground state eigenfunction for the Schrödinger equation

$$\left(\mathbf{H}_0 + \mathbf{V}_{\mathrm{pc}}[\mathbf{X}]\right)|\Psi\rangle = E|\Psi\rangle \tag{6.23}$$

where $\mathbf{H}_0$ is the electronic Hamiltonian including nuclear repulsion energy for the isolated solute and $\mathbf{V}_{\mathrm{pc}}$ is the Coulomb potential created by the point charges placed on the MM solvent molecules. $\mathbf{X}$ in Eq. (6.23) collectively denotes the set of the coordinates $\{\mathbf{x}_i\}$ for the constituent solvent molecules. In Eq. (6.23) we emphasize the dependence of $\mathbf{V}_{\mathrm{pc}}$ on the instantaneous solvent configuration $\mathbf{X}$. Accordingly, the wave function Ψ and E are also dependent on $\mathbf{X}$. The energy E_{QM} in Eq. (6.22) can be expressed as

$$E_{\mathrm{QM}} = \langle\Psi|\mathbf{H}_0|\Psi\rangle. \tag{6.24}$$

Here, we introduce the distortion energy E_{dist} of the QM solute in solution,

$$E_{\mathrm{dist}} = E_{\mathrm{QM}} - E_0 \tag{6.25}$$

where E_0 is the ground state energy of the solute at isolation. We note E_{dist} is always positive from the variational principle. The interaction energy $E_{\mathrm{QM/MM}}$ in Eq. (6.22) can be given by the sum of the electrostatic and van der Waals interactions, thus,

$$E_{\mathrm{QM/MM}} = E_{\mathrm{QM/MM}}^{\mathrm{ES}} + E_{\mathrm{QM/MM}}^{\mathrm{vdW}}. \tag{6.26}$$

$E_{\mathrm{QM/MM}}^{\mathrm{ES}}$ in Eq. (6.26) is the electrostatic interaction between the solute and the solvent, and is given as

$$\begin{aligned} E_{\mathrm{QM/MM}}^{\mathrm{ES}} &= \langle\Psi|\mathbf{V}_{\mathrm{pc}}[\mathbf{X}]|\Psi\rangle + \sum_{k\in\mathrm{QM}} z_k \cdot V_{\mathrm{pc}}[\mathbf{X}](\mathbf{R}_k) \\ &= \int d\mathbf{r}\, n[\mathbf{X}](\mathbf{r}) V_{\mathrm{pc}}[\mathbf{X}](\mathbf{r}) + \sum_{k\in\mathrm{QM}} z_k \cdot V_{\mathrm{pc}}[\mathbf{X}](\mathbf{R}_k) \end{aligned} \tag{6.27}$$

where z_k and $\mathbf{R}_k$ are, respectively, the nuclear charge and the position vector of the kth atom in the QM region, and notation $\mathbf{r}$ stands for the position vector of the electron. $n(\mathbf{r})$ in Eq. (6.27) is the electron density of the solute and is dependent on $\mathbf{X}$. The van der Waals interaction $E_{\mathrm{QM/MM}}^{\mathrm{vdW}}$ is conventionally evaluated by the Lennard-Jones (LJ) potential. The third term E_{MM} in Eq. (6.22) is in general described with the sum of the LJ potentials and the Coulomb interaction between fractional charges on the interaction sites.

With these preparations it is possible to express the solvation free energy $\Delta\mu$ of the QM solute in MM solvent, thus,

$$\exp(-\beta\Delta\mu) = \frac{\int d\mathbf{X}\exp\left(-\beta\left(E_{\mathrm{dist}}[\mathbf{X}] + E_{\mathrm{QM/MM}}[n[\mathbf{X}],\mathbf{X}] + E_{\mathrm{MM}}[\mathbf{X}]\right)\right)}{\int d\mathbf{X}\exp(-\beta E_{\mathrm{MM}}[\mathbf{X}])} \tag{6.28}$$

where β is the reciprocal of the product of Boltzmann constant k_{B} and the absolute temperature T. The interaction $E_{\mathrm{QM/MM}}$ in Eq. (6.28) can be written in completely pairwise form between solute and solvent. We note, however, $n(\mathbf{r})$ is dependent on whole of $\mathbf{X}$ and, hence, $E_{\mathrm{QM/MM}}$ is not pairwise additive. Furthermore, $E_{\mathrm{dist}}[\mathbf{X}]$ cannot be decomposed into sum of pair potentials. Thus, it is not straightforward to apply the theory of energy representation which assumes the pairwise interaction between solute and solvent.

To extract the free energy contribution $\Delta\bar{\mu}$ due to the two-body interaction between solute and solvent we reformulate the free energy $\Delta\mu$ as

$$\begin{aligned}\exp(-\beta\Delta\mu) &= \exp(-\beta\Delta\bar{\mu}) \times \exp(-\beta\delta\mu)\\ &= \frac{\int d\mathbf{X}\exp\left[-\beta\left(E_{\mathrm{QM/MM}}(\bar{n},\mathbf{X}) + E_{\mathrm{MM}}(\mathbf{X})\right)\right]}{\int d\mathbf{X}\exp[-\beta(E_{\mathrm{MM}}(\mathbf{X}))]}\\ &\quad\times \frac{\int d\mathbf{X}\exp\left[-\beta\left(E_{\mathrm{dist}}[\mathbf{X}] + E_{\mathrm{QM/MM}}(n[\mathbf{X}],\mathbf{X}) + E_{\mathrm{MM}}(\mathbf{X})\right)\right]}{\int d\mathbf{X}\exp\left[-\beta\left(E_{\mathrm{QM/MM}}(\bar{n},\mathbf{X}) + E_{\mathrm{MM}}(\mathbf{X})\right)\right]}.\end{aligned} \tag{6.29}$$

In Eq. (6.29) $\bar{n}(\mathbf{r})$ is an electron density fixed at an arbitrary distribution. Then, $\Delta\bar{\mu}$ describes the solvation free energy of a QM solute with the electron density $\bar{n}(\mathbf{r})$. Since $\bar{n}(\mathbf{r})$ is no longer dependent on $\mathbf{X}$, $\Delta\bar{\mu}$ is regarded as the free energy due to the two-body interactions. Thus, $\Delta\bar{\mu}$ can be straightforwardly evaluated using the approximate free energy functional provided in Eq. (6.18). On the other hand, $\delta\mu$ defined in Eq. (6.29) represents the free energy due to the fluctuation of the electron density around $\bar{n}(\mathbf{r})$ in response to the solvent motion. The free energy $\delta\mu$ is responsible for the many-body interaction. Therefore, the standard functional in the theory of energy representation cannot be applied to the computation of $\delta\mu$.

In a recent progress in QM/MM-ER we developed a simple and exact method to compute the free energy $\delta\mu$ within the framework of the theory of energy representation. The key of the method is to introduce a new energy coordinate η for the energy distribution functions which serve as fundamental variables to construct the free energy $\delta\mu$. Explicitly, η is defined as energy shift of the whole system due to the electronic polarization of the QM solute, thus,

$$\begin{aligned}\eta &= E_{\mathrm{dist}}[\mathbf{X}] + E_{\mathrm{QM/MM}}(n[\mathbf{X}], \mathbf{X}) - E_{\mathrm{QM/MM}}(\bar{n}, \mathbf{X}) \\ &= E_{\mathrm{dist}}[\mathbf{X}] + \sum_{i \in \mathrm{solvent}} \left(v(n[\mathbf{X}], \mathbf{x}_i) - v(\bar{n}, \mathbf{x}_i)\right).\end{aligned} \tag{6.30}$$

The energy distribution functions $P(\eta)$ and $P_0(\eta)$ are constructed, respectively, in the solution and the reference systems, thus,

$$P(\eta) = \frac{\int d\mathbf{X}\delta(\eta - H)\exp\left(-\beta\left(E_{\mathrm{dist}}[\mathbf{X}] + E_{\mathrm{QM/MM}}[n[\mathbf{X}], \mathbf{X}] + E_{\mathrm{MM}}[\mathbf{X}]\right)\right)}{\int d\mathbf{X}\exp\left(-\beta\left(E_{\mathrm{dist}}[\mathbf{X}] + E_{\mathrm{QM/MM}}[n[\mathbf{X}], \mathbf{X}] + E_{\mathrm{MM}}[\mathbf{X}]\right)\right)} \tag{6.31}$$

$$P_0(\eta) = \frac{\int d\mathbf{X}\delta(\eta - H)\exp\left(-\beta\left(E_{\mathrm{QM/MM}}[\bar{n}, \mathbf{X}] + E_{\mathrm{MM}}[\mathbf{X}]\right)\right)}{\int d\mathbf{X}\exp\left(-\beta\left(E_{\mathrm{QM/MM}}[\bar{n}, \mathbf{X}] + E_{\mathrm{MM}}[\mathbf{X}]\right)\right)} \tag{6.32}$$

where the variable H denotes the polarization energy defined by Eq. (6.30). In the solution system solvent molecules fully couple with the QM solute with the fluctuating electron density. In the reference system, on the other hand, the ensemble of the solvent is generated under the condition that solute electron density is fixed at $\bar{n}(\mathbf{r})$. By taking the logarithm of the ratio of these distribution functions we readily obtain the simple formulation for $\delta\mu$,

$$\delta\mu = k_B T \ln\left(\frac{P(\eta)}{P_0(\eta)}\right) + \eta. \tag{6.33}$$

It should be noted that Eq. (6.33) is identical in its form to the overlapping distribution method [22]. For later references, we denote the right hand side of Eq. (6.33) as $R(\eta)$. Since $\delta\mu$ is independent of the energy coordinate $R(\eta)$ must be constant with respect to the variation of η in principle. Hence, the constancy of $R(\eta)$ is a good measure of the statistical convergence. For the numerical robustness we multiply Eq. (6.33) by a weight function $W(\eta)$ and integrate with respect to η, which reads,

$$\begin{aligned}\delta\mu &= \int d\eta\, W(\eta)\left[k_B T \ln\left(\frac{P(\eta)}{P_0(\eta)}\right) + \eta\right] \\ &= \int d\eta\, W(\eta) R(\eta).\end{aligned} \tag{6.34}$$

The weight function $W(\eta)$ is arbitrary as long as it is normalized. It is apparent from Eq. (6.34) that $W(\eta)$ should have larger values in the region where $P(\eta)$ and $P_0(\eta)$ have sufficient overlap for adequate integration. Such function was proposed by Sakuraba and Matubayasi [78]. We emphasize that no approximation is used in the construction of Eq. (6.34). Thus, we formulated a simple but exact functional for the free energy due to the electron density fluctuation of a solute in solution in terms of the energy distribution functions.

In closing this subsection we make some remarks on the choice of the electron density $\bar{n}(\mathbf{r})$. From a physical view point it is natural to take the average distribution $\tilde{n}(\mathbf{r})$ of the electron density in solution as the reference electron density. Explicitly, $\tilde{n}(\mathbf{r})$ is yielded from the statistical average in solution system, thus,

$$\tilde{n}(\mathbf{r}) = \frac{\int d\mathbf{X}\, n[\mathbf{X}](\mathbf{r}) \exp\left(-\beta\left(E_{\text{dist}}[\mathbf{X}] + E_{\text{QM/MM}}[n[\mathbf{X}], \mathbf{X}] + E_{\text{MM}}[\mathbf{X}]\right)\right)}{\int d\mathbf{X} \exp\left(-\beta\left(E_{\text{dist}}[\mathbf{X}] + E_{\text{QM/MM}}[n[\mathbf{X}], \mathbf{X}] + E_{\text{MM}}[\mathbf{X}]\right)\right)}. \quad (6.35)$$

$\tilde{n}(\mathbf{r})$ is obviously an appropriate density for the computation of the free energy $\Delta\bar{\mu}$ and it also ensures the larger overlap between $P(\eta)$ and $P_0(\eta)$ which leads to the fast convergence in the free energy $\delta\mu$. For the numerical convenience it is advantageous to take the electron density n_0 at isolation because the computational effort to construct $\tilde{n}(\mathbf{r})$ can be avoided. As demonstrated in the following, the choice of the reference electron density does not affect the results of the solvation free energies seriously.

6.4.2 *Application of the QM/MM-ER Method: A Water Molecule*

As a benchmark test of the QM/MM-ER method described above we performed a set of simulations to compute solvation free energy of a QM water in MM water solvent. In these simulations we employed the electron density n_0 and the average distribution $\tilde{n}(\mathbf{r})$ as reference electron densities in Eq. (6.29) to examine the numerical robustness of our method.

The outline of the computational setups are as follows. For more details we refer the readers to our previous paper [24]. All the QM/MM simulations were performed with the code developed by Takahashi et al. [9–11], where the QM subsystem is described with the Kohn-Sham density functional theory utilizing the real-space grid method. The major methodological part of our grid approach was common to that provided in Sect. 6.2.1 except that the Hartree potential was constructed using FFT. For these calculations we utilized non-parallelized version of our program.

The structure of the QM solute was optimized by conducting the Gaussian 03 package with BLYP [29, 31]/aug-cc-pVDZ [33] levels of theory. The optimized geometrical parameters are OH = 0.975 Å and HOH = 104.2°. The position and the geometry of the QM solute was fixed during the simulations. The LJ parameter specified in SPC/E model [79] was assigned to the oxygen atom of the QM water to evaluate the van der Waals interaction $E_{\mathrm{QM/MM}}^{\mathrm{vdW}}$ in the QM/MM interaction $E_{\mathrm{QM/MM}}$.

The one-electron wave functions of the QM solute were contained within a cubic QM cell spanned by equally spaced 64 grids along each axis. The spacing h of the grid was set at 0.152 Å, which resulted in the QM cell size of $L_{\mathrm{QM}} = 9.71$ Å. We adopted the time-saving double-grid technique by Ono and Hirose around the atomic cores, for which the spacing was set at $h/5$. The exchange-correlation energy of the system was evaluated by the BLYP functional based on the GGA approximation.

The MM solvent was represented with 499 water molecules of SPC/E model that are contained within a periodic simulation box of the size $L = 24.6$ Å. The thermodynamic condition of the solvent was $\rho = 1.0$ g/cm^3 and 300 K. The constant-*NVT* ensemble was generated by means of the molecular dynamics simulations. The temperature control was performed by rescaling the velocities. The Nose-Hoover thermostat, that is a more rigorous approach, was not employed in our simulations since the properties of the thermostat will hardly affect the computational results. The internal structures of the solvent molecules were fixed during the simulation and their rotational motion were driven by using quaternion. The Newtonian equations of motion were solved numerically by the leap-frog algorithm [21] with the time step of 1 fs. The long range interactions between fractional charges on MM molecules were incorporated using the Ewald method [22]. On the other hand, the Coulombic interaction between QM and MM subsystems were treated with the bare form of $1/r$. Thus, the dipole-dipole interaction between solute and solvent was truncated at a long range. However, the dependence of the solvation free energy on the system size was found to be negligible in our previous calculations for QM H_2O molecule [23, 24]. For ionic molecules we safely compensated the free energy contribution due to the long range interaction using the Born's equation.

For the computation of the free energy $\Delta\bar{\mu}$ in Eq. (6.29) with the theory of energy representation, 200-ps QM/MM simulation was carried out to construct the energy distribution functions $\rho(\varepsilon)$ in the solution system, while 400-ps simulation was devoted for the distributions $\rho_0(\varepsilon)$ and $\chi_0(\eta,\zeta)$ in the pure solvent system. In the construction of these distribution functions the solvent molecules of which mass centers were within a sphere Ω of radius 11 Å were considered, where the center of the sphere was placed at the mass center of the solute. In the computation of $\delta\mu$ in Eq. (6.29), our simulations were conducted over 300 ps to yield the energy distribution functions defined in Eqs. (6.31) and (6.32), where all the solvent molecules are involved with the minimum image convention.

We first present the results for the QM/MM-ER simulations where the average electron distribution $\tilde{n}(\mathbf{r})$ was taken as the reference density $\bar{n}(\mathbf{r})$ in Eq. (6.29). Specifically, for the construction of $\tilde{n}(\mathbf{r})$ we performed a conventional QM/MM simulation for 100 ps after 5-ps equilibration. The energy distribution functions for the two-body interaction between the QM solute of the electron density $\tilde{n}(\mathbf{r})$ and MM solvent molecules are shown in Fig. 6.5. $\rho(\varepsilon)$ and $\rho_0(\varepsilon)$ are, respectively, for the

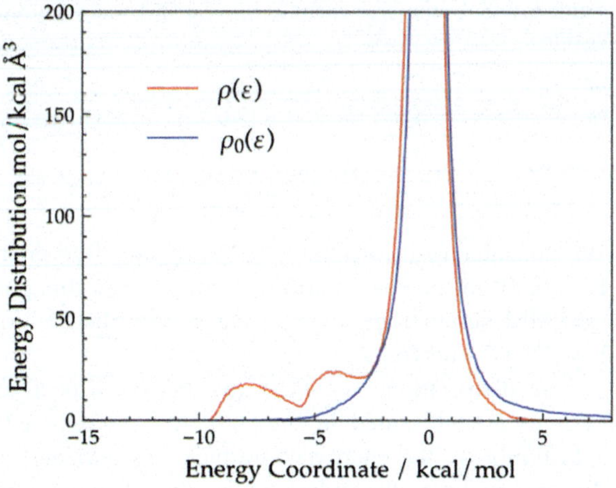

Fig. 6.5 Energy distribution functions $\rho(\varepsilon)$ and $\rho_0(\varepsilon)$ for the solution and the pure solvent systems, respectively, for the QM water solute with electron density fixed at $\tilde{n}(\mathbf{r})$. The distributions $\rho(\varepsilon)$ and $\rho_0(\varepsilon)$ are defined by Eqs. (6.13) and (6.14), respectively, in the text

solution and the pure solvent systems. The explicit definitions for $\rho(\varepsilon)$ and $\rho_0(\varepsilon)$ are presented in Eqs. (6.13) and (6.14) in Sect. 6.3. In the construction of these distribution functions we note that the computational effort is much less than the ordinary QM/MM simulations since there is no need to perform SCF procedure to optimize the electron density at each MD step. The peaks around the energy coordinate $\varepsilon = -8$ and -5 kcal/mol are both responsible for the hydrogen bonds (HB) between solute and solvent. Since the electrostatic field formed by the QM water is not identical in principle to that of the MM molecule, the HB energy curve for the QM solute which acts as a proton donor differs from that for the QM solute as a proton acceptor. In the present case the peaks around −8 and −5 kcal/mol are due to the QM solutes acting as a donor and an acceptor, respectively. In contrast to the distribution function in solution system the peaks ascribed to HBs completely vanish in the pure solvent system because the solvent configurations are being sampled in the absence of the solute molecule. The distributions $\rho(\varepsilon)$, $\rho_0(\varepsilon)$, and the correlation matrix $\chi_0(\varepsilon,\eta)$ defined by Eq. (6.15) in the pure solvent system were used as inputs to the free energy functional in Eq. (6.18). The free energy $\Delta\bar{\mu}$ due to the two-body interaction was, thus, computed as −9.7 kcal/mol. We note, however, that this free energy does not include the distortion energy $\bar{E}$ of the QM solute associated with the electron density polarization from n_0 to $\tilde{n}(\mathbf{r})$. The distortion energy can be simply computed from the statistical average of E_{dist} defined by Eq. (6.25), thus, $\bar{E} = \langle E_{dist} \rangle$. $\bar{E}$ was actually obtained as 4.4 kcal/mol through the same simulation to construct $\tilde{n}(\mathbf{r})$ in Eq. (6.35). Then, the solvation free energy of the solute with the *fixed* electron density $\tilde{n}(\mathbf{r})$ was estimated as $\bar{E} + \Delta\bar{\mu} = -5.3$ kcal/mol. Though the value is slightly larger than the experimental value of −6.3 kcal/mol [23, 80], it is adequate enough for the quantitative discussions in most cases. We summarize the results in Table 6.1 for readers' references.

The major source of the underestimate of the solvation free energy might be attributed to the neglect of the electron density fluctuation of the solute in solution.

Table 6.1 Solvation free energies $\Delta\mu$ of a QM water molecule and their components in units of kcal/mol

$\bar{n}$	$\bar{E}$	$\Delta\bar{\mu}$	$\delta\mu - \bar{E}$	$\Delta\mu$
$\tilde{n}$	4.4	−9.7	−1.6	−6.9
n_0	0.0	−4.0	−3.3	−7.3

We consider to incorporate the remaining free energy contribution $\delta\mu$ defined in Eq. (6.29) into the solvation free energy. For this purpose we performed additional QM/MM simulations to yield the energy distribution functions $P(\eta)$ and $P_0(\eta)$ in Eqs. (6.31) and (6.32).

The distributions for $P(\eta)$ and $P_0(\eta)$, respectively, in solution and reference systems are presented in Fig. 6.6. It is recognized that they have peaks around 3 kcal/mol on the energy coordinate. As indicated in Eq. (6.30) the energy coordinate η includes the polarization shift of the electron density from the fixed distribution $\bar{n}(\mathbf{r})$ $(= \tilde{n}(\mathbf{r}))$ as well as the distortion energy $E_{\mathrm{dist}}[\mathbf{X}]$, hence, it is natural that the distributions have peaks on the positive energy coordinate. The reasonable coincidence of $P(\eta)$ with $P_0(\eta)$ directly suggests that the electrostatic field produced by the average electron density $\tilde{n}(\mathbf{r})$ can faithfully reproduce the dynamics of the solvent under the influence of the solute with electron density $n(\mathbf{r})$. Also presented in the figure is the function of $R(\eta)$. We observe that $R(\eta)$ is reasonably constant at least in the region where both $P(\eta)$ and $P_0(\eta)$ have notable populations. The constancy of $R(\eta)$ implies the sufficient statistical convergence in the free energy $\delta\mu$.

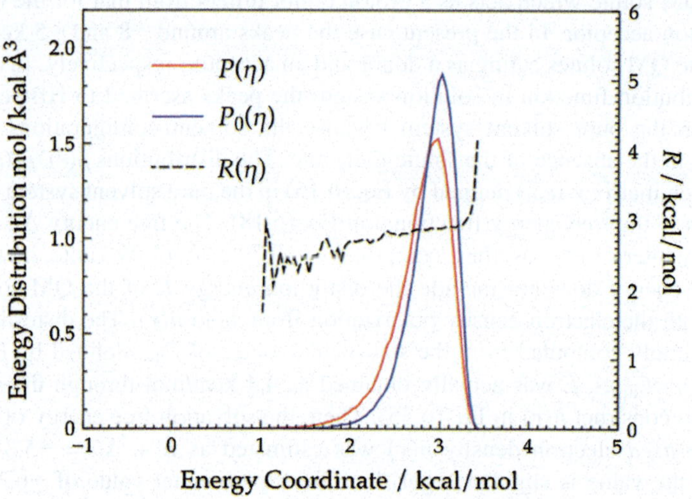

Fig. 6.6 The energy distribution functions $P(\eta)$ and $P_0(\eta)$ for the solution and the reference systems, respectively, for the QM water solute with electron density fixed at $\tilde{n}(\mathbf{r})$. The distributions $P(\eta)$ and $P_0(\eta)$ are defined by Eqs. (6.31) and (6.32), respectively, in the text. $R(\eta)$ is defined in Eq. (6.34)

By adopting $P(\eta)$ as the weight function $W(\eta)$ in Eq. (6.34) ($W(\eta) = P(\eta)$) we obtain the free energy $\delta\mu = 2.9$ kcal/mol. Since $\delta\mu$ includes the average distortion energy $\bar{E} = 4.4$ kcal/mol, the net free energy due to the electron density fluctuation can be evaluated as $\delta\mu - \bar{E} = -1.6$ kcal/mol. We obtained the total solvation free energy $\Delta\mu$ of the QM solute as -6.9 kcal/mol which excellently agrees with the experimental value of −6.3 kcal/mol. Our approach revealed that the electron density fluctuation gives more than 20 % contribution to the total solvation free energy of a water molecule. It is expected that the free energy $\delta\mu$ will make larger contributions in anionic molecules or transition states in particular since these species have larger polarizabilities in general. We also applied the weight function $W(\eta)$ proposed by Sakuraba and Matubayasi [78] for the integration of Eq. (6.34), which led to the free energy $\delta\mu = 2.9$ kcal/mol. We, thus, found that the free energy $\delta\mu$ is independent on the choice of the weight function suggesting the numerical robustness of Eq. (6.34).

Next, we adopted $n_0(\mathbf{r})$ as a reference electron density in Eq. (6.29) to examine the effect of the choice of the reference state on the solvation free energy. The distribution functions for the two-body interaction in the solution and the pure solvent systems are presented in Fig. 6.7. The distribution for the solution system has also two distinct peaks besides the one around $\sim$0 kcal/mol. These are, of course, responsible for the hydrogen bondings between the solute and solvent, however, the peak positions are slightly shifted toward the direction of larger energy coordinate. As a result the two-body free energy contribution $\Delta\bar{\mu}$ was obtained as −4.0 kcal/mol where we note the distortion energy $\bar{E}$ is 0.0 kcal/mol since electron density n_0 is taken as the reference density. This free energy can be directly compared with the value of −5.3 kcal/mol ($=\bar{E} + \Delta\bar{\mu}$) for the solute with electron density $\tilde{n}(\mathbf{r})$. Thus, the free energy contribution due to the *static* polarization of the solute was evaluated as −1.3 kcal/mol.

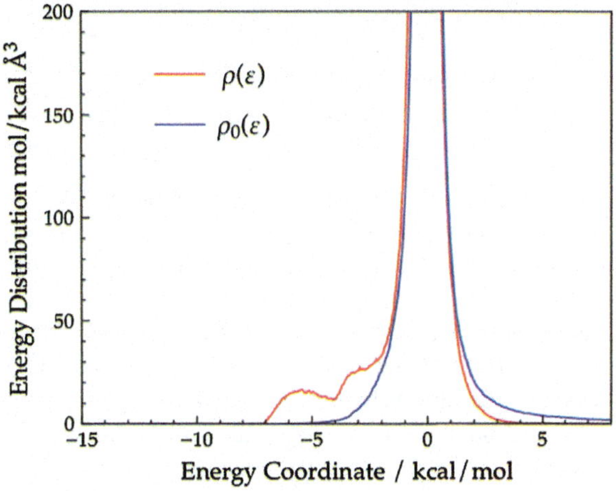

Fig. 6.7 The distribution functions synonymous with Fig. 6.5 except that the electron density of the QM water is fixed at $n_0(\mathbf{r})$

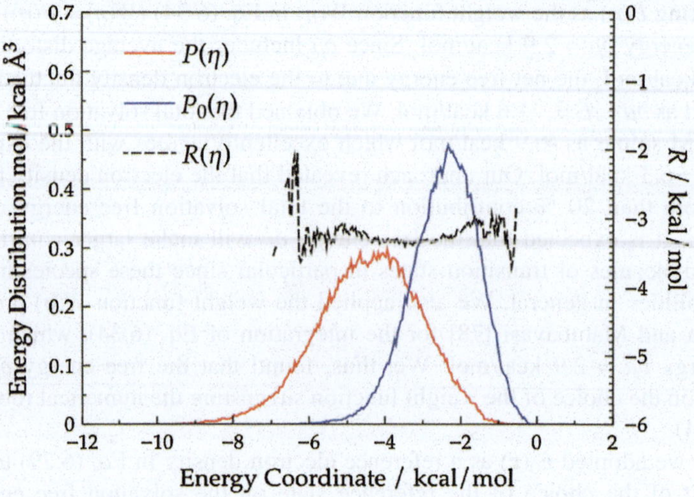

Fig. 6.8 The distribution functions synonymous with Fig. 6.6 except that the electron density of the QM water is fixed at $n_0(\mathbf{r})$

The distribution functions $P(\eta)$ and $P_0(\eta)$ for the solute with density n_0 are shown in Fig. 6.8. In contrast to the distributions for the solute with the density $\tilde{n}$ $P(\eta)$ in the solution system seriously splits from $P_0(\eta)$ in the reference system for the solute with n_0 density. This is, of course, attributed to the fact that the electrostatic field due to the solute with the fluctuating electron density n is substantially different from that due to the solute with n_0. The free energy functional of Eq. (6.34) with the weighting function $W(\eta)$ by Sakuraba et al. gives the free energy $\delta\mu = -3.3$ kcal/mol, which results in the total solvation free energy $\Delta\mu = -7.3$ kcal/mol. It was found that the method using n_0 as a reference electron density gives almost the same solvation free energy, demonstrating the robustness of the present approach.

6.4.3 *Perturbation Approach to the QM/MM-ER Method*

We have, so far, discussed the treatment of the free energy $\delta\mu$ due to the electron density fluctuation in solution. As a result a simple and exact functional has been formulated in terms of the energy distribution functions which describes $\delta\mu$. By virtue of the versatile form of the functional it can also be applied to the perturbation approach for the QM/MM-ER method. The major purpose of this subsection is to introduce our recent development which combines QM/MM-ER with a second-order perturbation theory (PT2) [76] and demonstrate its efficiency. With this method the free energy calculation can be substantially expedited since the computation of the polarization energy, the most time-consuming part in QM/MM simulations, is approximately evaluated with PT2.

The perturbation approach to the polarization of a QM molecule was first motivated by Francl [81] who applied PT2 to the Hartree-Fock method to construct the polarization energy of a QM molecule due to a single external charge. Later, Orozco et al. extended the Francl's work to the QM/MM simulations for condensed systems [82, 83]. The point of our approach in combining QM/MM-ER with PT2 is that the 2nd-order term in the PT2 energy instead of Eq. (6.30) is taken as the energy coordinate η to construct the energy distribution functions. As will be shown below, the perturbation expansion of the QM/MM interaction provides an excellent framework well suited to the application to QM/MM-ER. The notational conventions are almost common to those in Sect. 6.4.1. It seems appropriate to start with the review of 1-body perturbation theory in the context of QM/MM interaction.

It is quite natural to take the electric field $\mathbf{V}_{\rm pc}$ in Eq. (6.23) as a perturbation to the electronic Hamiltonian in the perturbation expansion. By the Rayleigh-Schrödinger perturbation theory the electronic energy $E_{\rm QM} + E^{\rm ES}_{\rm QM/MM}$ can be expressed as

$$E_{\rm QM} + E^{\rm ES}_{\rm QM/MM} = E^{(0)} + E^{(1)} + E^{(2)} + \cdots \tag{6.36}$$

where $E^{(n)}$ represents the nth-order term with respect to the perturbation. Specifically the unperturbed energy $E^{(0)}$ is identical to the electronic energy E_0 of the QM solute at isolation. The first order term $E^{(1)}$ corresponds to the electrostatic interaction between the unperturbed solute and the solvent, which can be expressed in terms of the 1-electron wave functions $\{\phi_i^{(0)}\}$ of the unperturbed system, thus,

$$E^{(1)} = \sum_i^{occ} \left\langle \phi_i^{(0)} \middle| \mathbf{V}_{\rm pc} \middle| \phi_i^{(0)} \right\rangle + \sum_{k \in {\rm QM}}^{\rm nuclei} z_k \cdot V_{\rm pc}(\mathbf{R}_k) \tag{6.37}$$

where the index i is for the occupied orbitals and the first term of right hand side of Eq. (6.37) is the Coulomb interaction of the field $\mathbf{V}_{\rm pc}$ and the electron density n_0 of the solute. The second term represents the interaction between $\mathbf{V}_{\rm pc}$ and the nuclear charge z_k of the kth atom. The 2nd-order term $E^{(2)}$ in the expansion of Eq. (6.36) is given by

$$E^{(2)} = \sum_i^{\rm occ} \sum_a^{\rm vir} \frac{1}{\varepsilon_i^{(0)} - \varepsilon_a^{(0)}} \left| \left\langle \phi_i^{(0)} \middle| \mathbf{V}_{\rm pc} \middle| \phi_a^{(0)} \right\rangle \right|^2 \tag{6.38}$$

where the index a runs through the virtual orbitals, and $\{\varepsilon_i^{(0)}\}$ express the set of eigenvalues for the corresponding orbitals. We note that the sum of the first order term $E^{(1)}$ and $E^{\rm vdW}_{\rm QM/MM}$ in Eq. (6.26) corresponds to the energy $E_{\rm QM/MM}(\bar{n},\mathbf{X})$ in Eq. (6.29) when the electron density n_0 at isolation is taken as $\bar{n}$. Further, the polarization energy η defined in Eq. (6.30) is replaced by the 2nd-order term $E^{(2)}$ provided that the perturbation expansion can be truncated up to the second term.

Of course, the accuracy can be systematically improved by incorporating the higher order terms into the coordinate η. Anyway, with the definition $\eta = E^{(2)}$ it is made possible to compute free energy contribution $\delta\mu$ with the method completely parallel to that presented in Sect. 6.4.2. Explicitly, the distribution functions $P(\eta)$ and $P_0(\eta)$ defined, respectively, in Eqs. (6.31) and (6.32) are rewritten in terms of the perturbation energies,

$$P_{\mathrm{PT2}}(\eta) = \frac{\int d\mathbf{X}\delta(\eta - E^{(2)}[\mathbf{X}])\exp\left(-\beta\left(E^{(1)}[\mathbf{X}] + E^{(2)}[\mathbf{X}] + E^{\mathrm{vdW}}_{\mathrm{QM/MM}}[\mathbf{X}] + E_{\mathrm{MM}}[\mathbf{X}]\right)\right)}{\int d\mathbf{X}\exp\left(-\beta\left(E^{(1)}[\mathbf{X}] + E^{(2)}[\mathbf{X}] + E^{\mathrm{vdW}}_{\mathrm{QM/MM}}[\mathbf{X}] + E_{\mathrm{MM}}[\mathbf{X}]\right)\right)} \tag{6.39}$$

$$P_{\mathrm{PT2},0}(\eta) = \frac{\int d\mathbf{X}\delta(\eta - E^{(2)}[\mathbf{X}])\exp\left(-\beta\left(E^{(1)}[\mathbf{X}] + E^{\mathrm{vdW}}_{\mathrm{QM/MM}}[\mathbf{X}] + E_{\mathrm{MM}}[\mathbf{X}]\right)\right)}{\int d\mathbf{X}\exp\left(-\beta\left(E^{(1)}[\mathbf{X}] + E^{\mathrm{vdW}}_{\mathrm{QM/MM}}[\mathbf{X}] + E_{\mathrm{MM}}[\mathbf{X}]\right)\right)}. \tag{6.40}$$

Since the ensemble of the configuration $\mathbf{X}$ generated in the reference system is equivalent to the solution system for $\Delta\bar{\mu}$, the distribution $P_{\mathrm{PT2},0}(\eta)$ of Eq. (6.40) can be constructed simultaneously through the simulation for the calculation of $\Delta\bar{\mu}$ with modest additional computational cost. The use of PT2 approach is advantageous since the SCF procedure, most time-consuming part of the QM/MM simulation, can be avoided. Once the zeroth-order wave functions $\{\phi_i^{(0)}\}$ have been obtained beforehand, there is no need to update the wave functions during the simulations provided that the geometry of the solute is fixed.

It is also helpful to describe the method to calculate the forces $\mathbf{F}^{(2)}$ responsible for the energy $E^{(2)}$ in Eq. (6.38). The straightforward differentiation of Eq. (6.38) with respect to the coordinate $\mathbf{x}_l$ of the lth site of the MM subsystem leads to

$$\mathbf{F}_l^{(2)} = -\sum_i^{\mathrm{occ.}}\sum_a^{\mathrm{vir.}}\frac{1}{\varepsilon_i^{(0)} - \varepsilon_a^{(0)}}\left\langle\phi_i^{(0)}\left|\frac{\partial\mathbf{V}_{\mathrm{pc}}[\mathbf{X}]}{\partial\mathbf{x}_l}\right|\phi_a^{(0)}\right\rangle\left\langle\phi_a^{(0)}\left|\mathbf{V}_{\mathrm{pc}}\right|\phi_i^{(0)}\right\rangle. \tag{6.41}$$

The direct evaluation of Eq. (6.41) necessitates the integration for each orbital pair of $\phi_i^{(0)}$ and $\phi_a^{(0)}$ for individual coordinate $\mathbf{x}_l$. To reduce the computational cost which involves the integration with respect to the electron coordinate $\mathbf{r}$, Eq. (6.41) is reformulated as

$$\begin{aligned}\mathbf{F}_l^{(2)} &= -\int d\mathbf{r}\,\frac{\partial}{\partial\mathbf{x}_l}\mathbf{V}_{\mathrm{pc}}[\mathbf{X}](\mathbf{r})\left[\sum_i^{\mathrm{occ.}}\sum_a^{\mathrm{vir.}}\frac{1}{\varepsilon_i^{(0)} - \varepsilon_a^{(0)}}\left\langle\phi_i^{(0)}\left|\mathbf{V}_{\mathrm{pc}}\right|\phi_a^{(0)}\right\rangle\phi_i^{(0)*}(\mathbf{r})\phi_a^{(0)}(\mathbf{r})\right]\\ &= -\int d\mathbf{r}\,\frac{\partial}{\partial\mathbf{x}_l}\mathbf{V}_{\mathrm{pc}}[\mathbf{X}](\mathbf{r})\Delta n(\mathbf{r})\,.\end{aligned} \tag{6.42}$$

In Eq. (6.42) the quantity in the square bracket represents the polarization density $\Delta n(\mathbf{r})$ of the QM solute and is common to every coordinate $\mathbf{x}_l$. Since $\Delta n(\mathbf{r})$ only needs to be calculated once at each MD step, the computational effort can be substantially decreased when Eq. (6.42) is utilized for the force calculation.

As a benchmark test for the method described above we performed simulations for a set of molecules which consists of three hydrocarbons (C_2H_6, C_2H_4, and C_2H_2), two alcohols (C_2H_5OH and C_6H_5OH), an ether ($(CH_3)_2O$), a nitrile (CH_3CN), NH_3, H_2O. To make comparisons we also carried out conventional QM/MM-ER simulations for the same set of molecules. The zeroth-order wave functions were constructed using Kohn-Sham DFT with a GGA functional.

The geometries for the molecules in the benchmark set were optimized under the polarizable continuum model (PCM) environment to mimic the influence of water solvent. The exchange-correlation energy was evaluated using B3LYP functional [31, 34] and aug-cc-pVTZ basis set [33] was used. These optimizations were performed conducting Gaussian 09 package. The geometries of the solutes were kept fixed during the simulations. The zeroth-order wave function for each solute was constructed with the Kohn-Sham DFT utilizing the real-space grids. In the evaluation of Eq. (6.38) we truncated the sum at the eigenvalue of 0.2 au for the virtual orbitals. The other computational details for the QM subsystem are common to those presented in Sect. 6.4.2.

The MM subsystem was prepared in the following way. First, 500 water molecules represented with SPC/E model were contained within a cubic simulation box with periodic boundary conditions. The size L of the simulation box was, then, adjusted to reproduce the density 1.0 cm^3 of water in ambient condition, which led to $L = 24.6$ Å. To embed a solute molecule into the solvent we removed solvent molecules of which total volume amounts approximately to that of the solute. The molecular volumes of the solutes were those estimated in the standard PCM calculations [84] in Gaussian 09 with B3LYP/aug-cc-pVDZ level of theory. As a result the number of water molecules in the box was reduced to 496 $\sim$ 499 according to the size of the solute molecule. The van der Waals interactions between solutes and solvent molecules were calculated using the OPLS-AA force field [85].

For the free energy contribution $\delta\mu$, we performed 200-ps simulations in the solution and reference systems to construct the energy distribution functions $P_{\mathrm{PT2}}(\eta)$ and $P_{\mathrm{PT2},0}(\eta)$ in Eqs. (6.39) and (6.40), respectively. For the construction of these distributions all the solvent molecules in the simulation box were taken into consideration with the minimum image convention. For the free energy contribution $\Delta\bar{\mu}$ due to two-body interactions, 200-ps simulation was carried out to yield the energy distribution function $\rho(\varepsilon)$ in the solution system, while 100-ps simulation was devoted for the pure solvent system, in which the solute was randomly inserted into the solvent at every 100 step (100 fs) to construct $\rho_0(\varepsilon)$ and correlation matrix $\chi_0(\varepsilon, \eta)$. 10^3 times insertions were made at every 100 fs by randomly generating the x-, y-, and z-coordinates of the mass center of the solute molecule at fixed orientation. Consequently, 10^6 solvent configurations were sampled out of a 100-ps

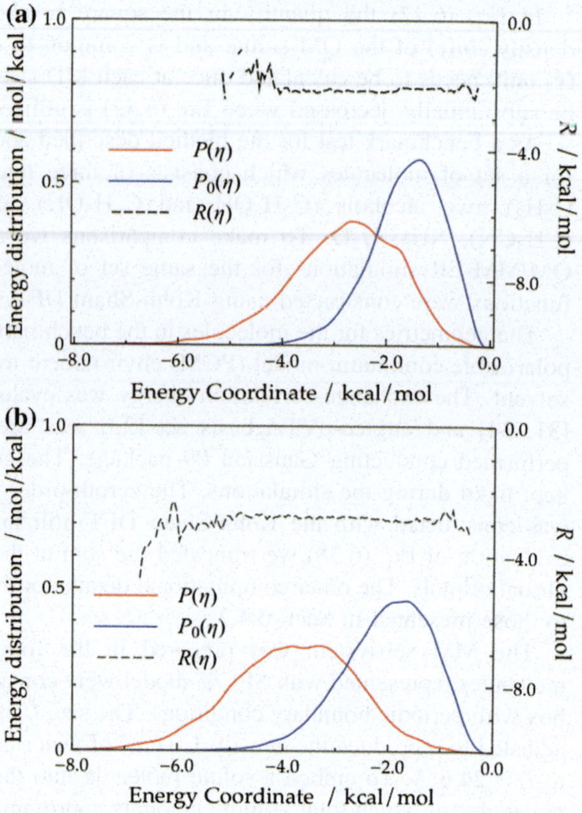

Fig. 6.9 Distribution functions of the polarization energy η of Ethanol in water solution. $P(\eta)$ and $P_0(\eta)$ are for the solution and the reference systems, respectively. **a** Polarization energy η is evaluated with (**a**) PT2 approach and with (**b**) KS-SCF

simulation. For the constructions of the distribution functions $\rho(\varepsilon)$, $\rho_0(\varepsilon)$, and $\chi_0(\varepsilon, \eta)$, we considered the solvent molecules of which mass centers were located within a sphere Ω of radius 11 Å, where the center of sphere Ω was placed at the mass center of the solute.

We provide in Fig. 6.9a the energy distribution functions $P_{\text{PT2}}(\eta)$ and $P_{0,\text{PT2}}(\eta)$ for an Ethanol molecule in water. For the construction of these functions the PT2 approach described above was employed. To make comparisons we also performed a set of QM/MM simulations that utilized the KS-SCF procedure at each MD step. These distribution functions $P(\eta)$ and $P_0(\eta)$ are shown in Fig. 6.9b. We found that overall behaviors of the PT2 distributions reasonably agree with those of the KS-SCF suggesting the efficiency of the method. It is recognized, however, that the distributions given by SCF procedures are slightly broader than the corresponding distributions obtained with PT2 method. The discrepancies can be attributed to the truncation at the second order in the perturbation expansion and also to the finite sum in the virtual orbitals in Eqs. (6.38) and (6.42). Using the PT2 distribution functions $P_{\text{PT2}}(\eta)$ and $P_{0,\text{PT2}}(\eta)$ the free energy $\delta\mu$ due to the electron density fluctuation of Ethanol in water was evaluated as −2.1 kcal/mol showing reasonable agreement with the value −2.8 kcal/mol given by the distributions $P(\eta)$ and $P_0(\eta)$.

The slight underestimation of the free energy $\delta\mu$ in PT2 approach will be a consequence of the truncation of the higher order terms. The two-body contribution $\Delta\bar{\mu}$ for Ethanol was computed as −1.7 kcal/mol using Eq. (6.18). The sum of the free energies $\Delta\bar{\mu}$ and $\delta\mu$ gives the total solvation free energies $\Delta\mu_{\mathrm{PT2}} = -3.8$ kcal/mol and $\Delta\mu_{\mathrm{SCF}} = -4.5$ kcal/mol. These values show excellent agreements with the experimental value of −5.01 kcal/mol [86].

In Fig. 6.10a we show the results of the PT2 approach compared with those of the KS-SCF calculations for the benchmark set that consists of 9 molecules. First of all, it is recognized in the figure that overall agreement between PT2 and KS-SCF is good. It fails seriously, however, to reproduce the hydration free energy $\Delta\mu$ of Acetonitrile CH_3CN. Explicitly, $\Delta\mu_{\mathrm{PT2}}$ was obtained as −11.2 kcal/mol, while the experimental value is −3.86 kcal/mol [86]. Thus, the PT2 method apparently overestimates the affinity of Acetonitrile to water solvent. In contrast, the standard QM/MM-ER method that uses KS-SCF soundly produces the hydration free energy of −5.1 kcal/mol. The source of the error in the PT2 calculation would be the strong electric field $\mathbf{V}_{\mathrm{pc}}$ formed by the solvent molecules. The average dipole moment of Acetonitrile was computed as 5.53 D in the present QM/MM simulation, while those of the other molecules except for ions were found to be less than 2.59 D (H_2O

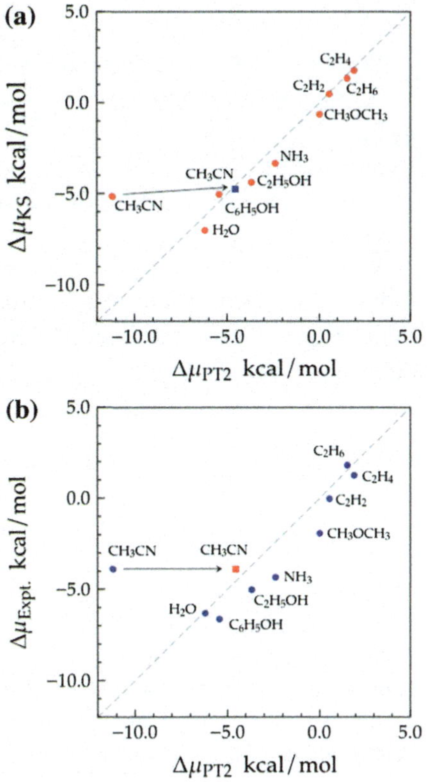

Fig. 6.10 Comparison of the solvation free energies for 9 neutral molecules **a** between the PT2 method ($\Delta\mu_{\mathrm{PT2}}$) and KS-SCF ($\Delta\mu_{\mathrm{KS}}$), **b** between the PT2 method ($\Delta\mu_{\mathrm{PT2}}$) and experiment ($\Delta\mu_{\mathrm{Expt.}}$) [86]. The arrows in the figures corresponds to the variation of the zeroth-order Hamiltonian from $\mathbf{H}_0$ to $\mathbf{H}'_0 = \mathbf{H}_0 + \tilde{\mathbf{V}}_{\mathrm{pc}}$

in solution). Thus, the large dipole moment of CH_3CN gives rise to a formation of a strong electric field of the surrounding polar solvent, which resulted in a poor description with the perturbation approach. It should also be noted that the PT2 calculations in the benchmark set mostly underestimate the absolute values of $\Delta\mu$ as compared to KS-SCF. This is, off course, due to the truncation in the perturbation expansion. The PT2 results are also compared with experimental values in Fig. 6.10b. In contrast to the comparison between PT2 and KS-SCF the agreement between PT2 and experiment is not so fine. This can be attributed to the errors inherent in the force field as well as in the exchange-correlation functional used in DFT calculations. Anyway, it was demonstrated that the PT2 approach is adequate to calculate solvation free energy that deserves for quantitative discussions at least for solutes with modest dipole moments.

A time-consuming, but promising method to solve the problem revealed for Acetonitrile is to take a biased Hamiltonian $\mathbf{H}'_0 = \mathbf{H}_0 + \bar{\mathbf{V}}_{pc}$ instead of $\mathbf{H}_0$ as a zeroth-order Hamiltonian for the PT2 approach. Here, an appropriate choice for $\bar{\mathbf{V}}_{pc}$ is, for instance, the statistical average $\tilde{\mathbf{V}}_{pc}$ of the electrostatic field, thus,

$$\tilde{\mathbf{V}}_{pc}(\mathbf{r}) = \frac{\int d\mathbf{X}\, \mathbf{V}_{pc}[\mathbf{X}](\mathbf{r}) \exp\left[-\beta\left(E_{dist}[\mathbf{X}] + E_{QM/MM}(n[\mathbf{X}], \mathbf{X}) + E_{MM}(\mathbf{X})\right)\right]}{\int d\mathbf{X} \exp\left[-\beta\left(E_{dist}[\mathbf{X}] + E_{QM/MM}(n[\mathbf{X}], \mathbf{X}) + E_{MM}(\mathbf{X})\right)\right]}. \tag{6.43}$$

We note that $\tilde{\mathbf{V}}_{pc}$ is no longer dependent on the configuration $\mathbf{X}$. Correspondingly, the perturbation $\mathbf{V}'_{pc}[\mathbf{X}](\mathbf{r})$ for the biased Hamiltonian is $\mathbf{V}_{pc}[\mathbf{X}]$ subtracted by $\tilde{\mathbf{V}}_{pc}$, that is,

$$\mathbf{V}'_{pc}[\mathbf{X}](\mathbf{r}) = \mathbf{V}_{pc}[\mathbf{X}](\mathbf{r}) - \tilde{\mathbf{V}}_{pc}(\mathbf{r}). \tag{6.44}$$

With respect to the perturbation defined by Eq. (6.44) it is also possible to formulate the perturbation expansions completely parallel to Eqs. (6.36)–(6.38). Since $\tilde{\mathbf{V}}_{pc}$ is the average potential for $\mathbf{V}_{pc}[\mathbf{X}]$ it is naturally expected that the difference potential $\mathbf{V}_{pc}[\mathbf{X}](\mathbf{r}) - \tilde{\mathbf{V}}_{pc}(\mathbf{r})$ is modest enough as a perturbation. For Acetonitrile we constructed the average potential $\tilde{\mathbf{V}}_{pc}$ performing an additional QM/MM simulations for 500 ps. A much less statistics would be adequate, however, since $\tilde{\mathbf{V}}_{pc}$ merely defines an intermediate states for solvation. For the perturbation $\mathbf{V}'_{pc}[\mathbf{X}](\mathbf{r}) = \mathbf{V}_{pc}[\mathbf{X}](\mathbf{r}) - \tilde{\mathbf{V}}_{pc}(\mathbf{r})$ the distribution functions $P_{PT2}(\eta)$ and $P_{0,PT2}(\eta)$ defined, respectively, in Eqs. (6.39) and (6.40) were constructed. As a consequence of the use of $\mathbf{V}'_{pc}[\mathbf{X}](\mathbf{r})$ as a perturbation, the distribution $P_{PT2}(\eta)$ perfectly overlaps with $P_{0,PT2}(\eta)$ in contrast to the original approach. Using these distributions the free energy $\delta\mu$ was evaluated as −0.7 kcal/mol. The two-body contribution $\Delta\bar{\mu}$ was computed as −3.8 kcal/mol for the solute with electron density $\tilde{n}$ that was determined as the ground state eigenfunction for the zeroth-order Hamiltonian $\mathbf{H}'_0$. Then, the total solvation free energy $\Delta\mu$ was given as

−4.5 kcal/mol that is very close to the experimental value (−3.86 kcal/mol). The variation of the free energies (KS-SCF vs PT2) when the perturbation is changed from $\mathbf{V}_{pc}[\mathbf{X}]$ to $\mathbf{V}'_{pc}[\mathbf{X}](\mathbf{r})$ is shown by an arrow in Fig. 6.10a. We note that zeroth-order Hamiltonian for KS-SCF is also changed in the construction of the new point (blue square dot in the figure). It was, thus, demonstrated that the introduction of the biased Hamiltonian significantly improves the applicability of the method.

We close this subsection with a discussion for the parallelization of the PT2 approach. The data communications with respect to the wave functions can be avoided in PT2 since the set of the unperturbed wave functions $\{\phi_i^{(0)}\}$ is not updated throughout the simulation provided that the molecular geometry of the solute is fixed. The QM/MM simulation based on PT2 is, therefore, quite amenable to parallelization both on the shared and distributed memory architectures. For the shared-memory machines the double loops associated with the calculations of $E^{(2)}$ and $\mathbf{F}_l^{(2)}$ defined, respectively, by Eqs. (6.38) and (6.42) are the major targets for the thread parallel which can be readily implemented using standard OMP directives. In the MPI parallelization, we first divide the zeroth-order wave functions either in the orbital space or in the real space, and, distribute the portions to computational nodes that constitute a parallel machine. Once the wave functions are assigned to the nodes before simulation, one only needs to reduce at each MD step the scalar values evaluated individually on the nodes to the master node to construct the complete values for $E^{(2)}$ and $\mathbf{F}_l^{(2)}$. It is expected that high parallel efficiency can be achieved with modest additional effort for programming.

We examined the efficiencies of the OMP parallel executed for a water molecule and a benzene dimer. The computer node used was the Intel Xeon (E5-2680, 2.7 GHz) with 16 cores. As a specific treatment to expedite the force calculation, we reduced the number of grids from 64 to 32 along each direction by taking average of the polarization density Δn in Eq. (6.42). In our measurement on a single core for a QM water molecule in 499 MM solvent water molecules we found ~73 % of the total computational time was devoted to the construction of the electrostatic potential $\mathbf{V}_{pc}$, and the residual time was used mostly for the evaluation of the PT2

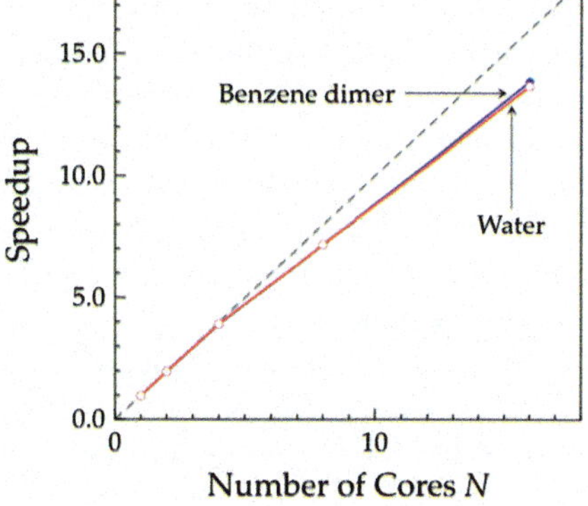

Fig. 6.11 Speedups measured for a Benzene dimer and a water molecule with respect to the increasing number N of cores in a multi-core processor (N = 1, 2, 4, 8, and 16)

forces. We show in Fig. 6.11 the plot of the speedup with respect to the increasing number N of cores up to $N = 16$. The figure shows that the speedup increases almost linearly with increasing N irrespective of the number of electrons in the QM systems. At $N = 16$, the parallel efficiencies for Benzene dimer and a water molecule are 87 and 86 %, respectively. Thus, it was demonstrated that QM/MM-PT2 is amenable to parallel implementation and shows an excellent performance on a shared memory architecture. The fine tuning of the OMP code to avoid the 'cache miss' which occurs in the memory access will further improve the efficiency.

6.4.4 Application to the Free Energy Analyses for the Hydration of Benzene

The method of QM/MM-PT2 coupled with the theory of solutions so far discussed provides us with an excellent theoretical framework to analyze solvation free energies. In a recent work, we applied the method described in Sect. 6.4.3 to study the role of the π electrons in a hydration process. Explicitly, the free energy $\delta\mu$ associated with the hydration of Benzene was decomposed into the contributions $\delta\mu_\pi$ and $\delta\mu_\sigma$ due, respectively, to the π and σ electrons. In the following we illustrate the outline of the work with an emphasis placed on the methodological aspect.

The interaction between an aromatic ring and a σ bond attracts a broad interest in the context of the XH/π interaction (X = C, N, or O) [87, 88]. It is well recognized that XH/π interactions are ubiquitous in protein systems and plays a decisive role in organizing their elaborate functionalities and structures in solutions. Since Benzene constitutes one of fundamental building blocks in organic molecules, the interaction of a σ bond with a Benzene ring can be considered as a representative XH/π interaction. In a pioneering work by Suzuki et al. it was proved experimentally for C_6H_6–H_2O dimer that the OH bond of the water molecule pointing toward the center of the aromatic ring forms a weak *hydrogen bond* with the aromatic ring. The detailed energetics for the interaction was studied by Tsuzuki et al. who performed a set of high level ab initio calculations [89]. They concluded that the major source of the attractive interaction in the dimer is the induction effect of the electronically polarized OH bond on the aromatic ring, though its absolute energy ($\sim$3 kcal/mol) is much smaller than the conventional hydrogen bondings. Thus, it is apparent that a weak hydrogen bond is formed between an aromatic ring and a polarized σ bond in a static isolated dimer. However, the role of the XH/π interaction in condensed environment is not well examined nor understood on the quantitative basis. This issue has a direct relevance with a well-known fact that benzene has a negative solvation free energy ($\Delta\mu = -0.87$ kcal/mol) though benzene is known as a typical non-polar molecule.

Here, we report the results of the free energy analyses based on QM/MM-PT2 coupled with the theory of energy representation, which provides an insight into the role of the OH/π interaction in the solvation process. To perform the analyses we first need to introduce the concepts of the static and dynamic polarizations of a solute in solution. The static polarization is defined as the electronic distortion of a solute under the influence of the average electric field $\tilde{\mathbf{V}}_{\mathrm{pc}}(\mathbf{r})$ given by Eq. (6.43). Remind that $\tilde{\mathbf{V}}_{\mathrm{pc}}(\mathbf{r})$ is independent of the solvent configuration $\mathbf{X}$. The polarizable continuum model (PCM) [84] or reference interaction site model (RISM) combined with SCF [90, 91] is the method that realizes the static polarization of a solute applying the mean-field approximations for the solvent on the basis of the fundamental electromagnetism or the theory of solutions, respectively. Then, the dynamic polarization is due to the residual potential $\mathbf{V}'_{\mathrm{pc}}[\mathbf{X}](\mathbf{r}) = \mathbf{V}_{\mathrm{pc}}[\mathbf{X}](\mathbf{r}) - \tilde{\mathbf{V}}_{\mathrm{pc}}(\mathbf{r})$ in Eq. (6.43). Thus, the dynamic polarization arises from the electron density fluctuation of a solute in response to the potential $\mathbf{V}'_{\mathrm{pc}}[\mathbf{X}](\mathbf{r})$ dependent on the instantaneous solvent configuration.

Our major concern in the present analyses lies in the free energy contribution $\delta\mu$ due to the dynamic polarization. The polarization energy pertaining to the dynamic polarization can be approximated with the perturbation method as

$$E'^{(2)} = \sum_{i}^{\mathrm{occ}} \sum_{a}^{\mathrm{vir}} \frac{1}{\varepsilon_i'^{(0)} - \varepsilon_a'^{(0)}} \left| \left\langle \phi_i'^{(0)} \middle| \mathbf{V'}_{\mathrm{pc}}[\mathbf{X}] \middle| \phi_a'^{(0)} \right\rangle \right|^2 \tag{6.45}$$

when the perturbation expansion is truncated at the second order (PT2). Note that Eq. (6.45) is completely parallel in form to Eq. (6.38). The prime notation is attached to the superscript of each term in Eq. (6.45) to specify that the biased Hamiltonian is used in the PT2 expansion. Then, it is possible to decompose naturally the energy $E'^{(2)}$ due to the dynamic polarization into the contributions from σ and π electrons, which reads,

$$\begin{aligned} E'^{(2)} &= E_{\pi}'^{(2)} + E_{\sigma}'^{(2)} \\ &= \left(\sum_{i \in \pi}^{occ.} + \sum_{i \in \sigma}^{occ.} \right) \sum_{a}^{vir.} \frac{1}{\varepsilon_i'^{(0)} - \varepsilon_a'^{(0)}} \left| \left\langle \varphi_i'^{(0)} \middle| \mathbf{V}'_{\mathrm{pc}}[\mathbf{X}] \middle| \varphi_a'^{(0)} \right\rangle \right|^2 . \end{aligned} \tag{6.46}$$

It is apparent that the term $E_{\pi}'^{(2)}$ in Eq. (6.46) represents the energy shift which arises from the polarization of the π electrons of a solute. The same is true for the polarization of σ electrons. Using the spatial distribution functions $\rho(\mathbf{X})$ for the configuration $\mathbf{X}$ of the solvent it is possible to formulate the Kirkwood's charging equation for the individual free energies $\delta\mu_\pi$ and $\delta\mu_\sigma$, thus,

$$\begin{aligned} \delta\mu &= \delta\mu_\pi + \delta\mu_\sigma \\ &= \int_0^1 d\lambda \int d\mathbf{X} \rho(\mathbf{X};\lambda) \frac{d}{d\lambda} E_{\pi,\lambda}'^{(2)}[\mathbf{X}] + \int d\lambda \int d\mathbf{X} \rho(\mathbf{X};\lambda) \frac{d}{d\lambda} E_{\sigma,\lambda}'^{(2)}[\mathbf{X}] \end{aligned} \tag{6.47}$$

The coupling parameter λ in Eq. (6.47) is introduced to describe the degree of the electron density fluctuation of a solute which interacts with the environment.

At $\lambda = 0$ the solute electron density is fixed at an electron density specified by the bias potential $\tilde{\mathbf{V}}_{\mathrm{pc}}(\mathbf{r})$, while the solute electron density at $\lambda = 1$ fully couples with the surrounding electric field $\mathbf{V}'_{\mathrm{pc}}[\mathbf{X}](\mathbf{r})$. The λ-dependence of the distribution function $\rho(\mathbf{X})$ as well as the energies $E_\pi'^{(2)}$ and $E_\sigma'^{(2)}$ are explicitly denoted in the equation. We note that Eq. (6.47) expressed in terms of the distribution function $\rho(\mathbf{X})$ is very simple in form. It is, however, infeasible to treat it numerically since $\mathbf{X}$ consists of huge number of variables which represent the coordinates of the constituent molecules. Fortunately, it is possible to reformulate Eq. (6.47) in terms of the distribution functions $P(\eta)$ for the polarization energy η of only one dimension. Explicitly, η is defined as $\eta = E_\pi'^{(2)}$ for π electrons. The free energy $\delta\mu_\pi$ for instance can be expressed in the form parallel to the Kirkwood's charging equation, thus,

$$\delta\mu_\pi = \int_0^1 d\lambda \int d\eta P_\pi(\eta;\lambda)\frac{d}{d\lambda}E_{\pi,\lambda}'^{(2)}(\eta). \tag{6.48}$$

For the derivation of Eq. (6.48) we refer the readers to the Appendix of Ref [77]. The free energy $\delta\mu_\sigma$ for σ electrons can, of course, be given in the same form as Eq. (6.48). It should be noted that the fluctuation of the π electrons in response to the external field correlates with that of the σ electrons to a certain degree. Therefore, it is necessitated to introduce the indirect part of the solute-solvent potential $\omega_\pi(\eta,\lambda)$ of the mean force which describes the effects of this correlation. Due to the potential $\omega_\pi(\eta,\lambda)$ the free energy $\delta\mu_\pi$ cannot be evaluated with Eq. (6.34). Explicitly, $\omega_\pi(\eta,\lambda)$ is defined as

$$\omega_\pi(\eta;\lambda) = -k_B T \log\left(\frac{P_\pi(\eta;\lambda)}{P_\pi(\eta;0)}\right) - \eta. \tag{6.49}$$

$P_\pi(\eta;\lambda)$ in Eq. (6.49) is the distribution function of η and dependent on the coupling parameter. Using the local potential $\omega_\pi(\eta,\lambda)$ as well as the distribution functions the free energy $\delta\mu_\pi$ is expressed as

$$\begin{aligned}\delta\mu_\pi = -k_B T \int d\eta \Big[\big((P_\pi(\eta) - P_{0,\pi}(\eta)\big) + \beta\omega_\pi(\eta)P_\pi(\eta) \\ -\beta\int_0^1 d\lambda\omega_\pi(\eta;\lambda)\big(P_\pi(\eta) - P_{0,\pi}(\eta)\big)\Big)\Big].\end{aligned} \tag{6.50}$$

$P_\pi(\eta)$ and $P_{0,\pi}(\eta)$ in Eq. (6.50) are the distribution functions of the energy coordinate η (= $E_\pi'^{(2)}$) in the solution ($\lambda = 1$) and the reference ($\lambda = 0$) systems, respectively. As expected Eq. (6.50) has the same form as Eq. (6.18). In the construction of the free energy using an approximate functional, however, it is possible to adopt an reduced form of the functional since there is no need to consider the effect of the exclusion volume for the free energy. Explicitly, the integration of

$\omega_\pi(\eta;\lambda)$ with respect to λ in Eq. (6.50) is expressed in terms of the potential $\omega_\pi(\eta;\lambda)$ only at $\lambda = 1$, thus,

$$\beta \int_0^1 \omega_\pi(\eta;\lambda)d\lambda = \begin{cases} \beta\omega_\pi(\eta) + 1 + \frac{\beta\omega_\pi(\eta)}{\exp(-\beta\omega_\pi(\eta))-1} & (\omega_\pi(\eta) \leq 0) \\ \frac{1}{2}\beta\omega_\pi(\eta) & (\omega_\pi(\eta) \geq 0). \end{cases} \tag{6.51}$$

In the construction of Eq. (6.51) the PY approximation is applied to the unfavorable region ($\omega_\pi(\eta) \leq 0$), while HNC is used for the favorable region ($\omega_\pi(\eta) \geq 0$) as usual. The equations parallel to Eqs. (6.48)–(6.51) can also be adopted to the calculation of the free energy $\delta\mu_\sigma$ associated with the dynamic polarization of σ electrons. As described above the potential $\omega_\pi(\eta;\lambda)$ carries the information for the correlation between π and σ electrons, and hence, the free energy decomposition into the contributions to σ and π electrons can only be performed in a formal fashion. It should also be stressed, however, that it is the potential $\omega_\pi(\eta;\lambda)$ which enables one to decompose the free energy $\delta\mu$ in a formal but natural manner. In the following we present the results of the decomposition analyses applied to Benzene in aqueous solution.

The σ–π decomposition was performed with our real-space grid DFT program. The computational details for the QM/MM-PT2 simulations as well as the method of energy representation are almost common to those described in Sect. 6.4.3. Specifically, we employed GROMOS 43A1 force field [92] to evaluate the van der Waals interaction for Benzene. We refer the readers to Ref. [77] for the other specifics of the computational setups.

First, we assessed the role of the static polarization of Benzene. As shown in Table 6.2, the solvation free energy $\Delta\bar{\mu}$ was obtained as 0.84 kcal/mol for Benzene with the fixed electron density which corresponds to the biased Hamiltonian $\mathbf{H}_0' = \mathbf{H}_0 + \tilde{\mathbf{V}}_{\mathrm{pc}}(\mathbf{r})$. Thus, the calculation which considers only the effect of the static polarization fails to reproduce the experimental value of −0.87 kcal/mol [86]. The free energy $\delta\mu$ due to the dynamic polarization was evaluated as −1.3 kcal/mol using Eq. (6.34). It was, thus, revealed that inclusion of free energy $\delta\mu$ realize the negative hydration free energy of −0.47 kcal/mol which reasonably agrees with the experimental value. The present calculation clearly suggests the importance of the dynamic polarization in the hydration of Benzene.

We, further, decompose the free energy $\delta\mu$ into the contributions $\delta\mu_\pi$ and $\delta\mu_\sigma$ defined in Eq. (6.50). The energy distribution functions used to construct these free energies are presented in Fig. 6.12. It is worthy of note that the distribution $P_\pi(\eta)$ for π electrons in the solution system has a long tail in the region of low energy coordinate, which shows a clear contrast to the distribution in the reference system.

Table 6.2 Hydration free energy of Benzene and its components in units of kcal/mol

	$\Delta\bar{\mu}$	$\delta\mu$	$\delta\mu_\pi$	$\Delta\mu_\sigma$	$\Delta\mu$	$\Delta\mu^a_{\mathrm{expt.}}$
Benzene	0.84	−1.30	−0.94	−0.35	−0.47	−0.87

[a]Experimental value provided in Ref. [86]

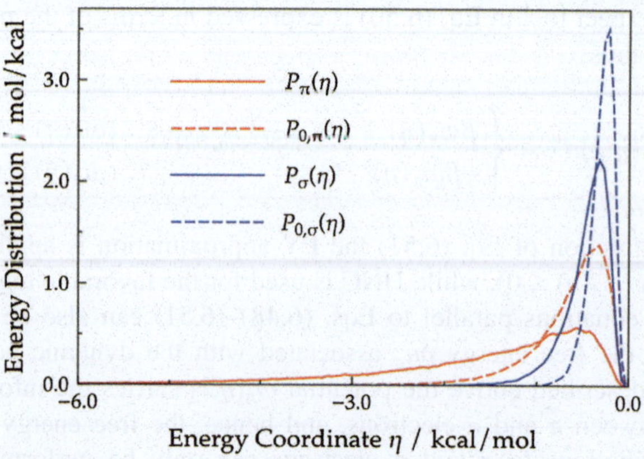

Fig. 6.12 Distribution functions of the polarization energy η of benzene in water solution. $P_\pi(\eta)$ and $P_{0,\pi}(\eta)$ are the distributions of the energy $E_\pi^{\prime(2)}$ defined by Eq. (6.46) in the text for π electrons in the solution and the reference systems, while $P_\sigma(\eta)$ and $P_{0,\sigma}(\eta)$ are those for σ electrons

This demonstrates the fact that the instantaneous solvent configurations yielded in the solution system are substantially different from those in the reference system. Using these distribution functions the free energies $\delta\mu_\pi$ and $\delta\mu_\sigma$ were evaluated as −0.94 and −0.35 kcal/mol, respectively. The sum of the individual free energies $\delta\mu_\pi$ and $\delta\mu_\sigma$ amounts to −1.29 kcal/mol and it excellently agrees with the exact value $\delta\mu$ of −1.3 kcal/mol. This suggests the robustness of the method of the free energy decomposition as well as the approximate functional given by Eq. (6.51). The free energy contribution of the π electrons due to dynamic polarization was found to be almost three times as large as that of σ electrons although the number of σ orbitals are four times as large as the π orbitals. The role of the fluctuation of the π electrons in the hydration of benzene was, thus, clarified on the quantitative basis.

6.5 Summary and the Outlook

We summarize this chapter by reviewing the major results. In Sect. 6.2.1 we provided the outline of the real-space grid Kohn-Sham DFT (RS-DFT) which aims at massively parallel implementation. Since the Hamiltonian matrix of the KS-DFT is dominated by the diagonal part in the real-space representation the parallelization can be achieved only by local communications between neighboring nodes in contrast to the LCAO approach. In Sect. 6.2.2 we assessed a parallel performance of our RS-DFT code on a modern parallel machine equipped with 512 cores for a water cluster with an ice structure. To avoid the global communications associated with FFT we instead employ the Poisson equation to construct the Hartree potential. Then, we achieved a high parallelization ratio of ∼99.8 % measured on this system. It was also revealed, however, that the speedup will be saturated immediately with respect to the increasing number of cores used. Hence, the parallelization ration should be very close to 100 % to obtain desired speedup when one employs thousands of cores.

In Sect. 6.3.1 we introduced a novel theory of solutions, referred to as the theory of energy representation, where the distribution functions of the solute-solvent interaction potential serves as fundamental variables in the density functional theory of solutions. It makes a contrast to the conventional theory of solutions that employs the spatial distribution functions of the solvent around a solute to construct the solvation free energy of the solute. Since a whole molecule is being treated as a single entity in the theory of energy representation there is no need to rely on the artificial concept of the interaction sites. Importantly, by virtue of this notable feature, the theory can be naturally combined with quantum mechanical approaches because the spatial diffuseness of the electron density, one of the inherent properties of the quantum mechanical objects, can be taken into consideration without reducing it to a set of point charges on atoms. It was demonstrated in Sect. 6.3.2 that the method can be successfully applied to the analyses of the hydration free energy of proteins.

We illustrate in Sect. 6.4.1 the method to combine the hybrid QM/MM approach with the theory of energy representation. The standard theory of energy representation is based on the assumption that the solute-solvent interaction is pairwise as well as the conventional theory, hence, some device should be developed to combine QM/MM with the theory of solutions. A promising approach to solve the problem is to extract the two-body contribution $\Delta\bar{\mu}$ from the total solvation free energy $\Delta\mu$. Then, the remaining many-body contribution $\delta\mu$ is treated separately with the theory of energy representation. For the evaluation of $\Delta\bar{\mu}$ the approximate free energy functional based on the energy representation can be applied straightforwardly. In a recent development, we formulate a simple but rigorous functional to compute $\delta\mu$ due to the electron density fluctuation of a QM solute in solution. Thus, the basic framework of the QM/MM-ER method was established. The application of QM/MM-ER to the hydration of a QM water molecule to an MM water solvent was presented in Sect. 6.4.2, which demonstrated the accuracy and the numerical robustness of the method.

The equation to compute $\delta\mu$ has a versatile form which allows various methodological extensions. Making use of the formulation we combine the QM/MM-ER approach with a perturbation approach (PT2) for the purpose of accelerating the free energy calculations. This method also provides an excellent framework for analyzing the free energy contribution responsible for the affinity of Benzene to water solvent. Our analyses successfully revealed the role of the fluctuation of π electrons in the hydration of Benzene on the basis of quantitative discussion.

As demonstrated in Sect. 6.2 the Kohn-Sham DFT utilizing the real-space grid technique enables ones to perform QM/MM simulations involving large scale QM molecules. Furthermore, the computational effort associated with the statistical average to construct free energy can be substantially reduced by virtue of the energy representation method. Thus, QM/MM-ER based on the real-space grid DFT provides us with a practically powerful tool to investigate various chemical events occurring in solutions and in biological systems. Actually, calculation of the solvation free energy of a whole ATP (adenosine triphosphate) molecule is now proceeding for the purpose of analyzing the free energy of hydrolysis of ATP in water using massively parallel machines. According to our benchmark test calculations, in the near future some protein molecule as a whole can be treated as a QM solute in conjunction with the next generation massively parallel machine.

Acknowledgments Manuscript for the contribution to the book: *Quantum Modeling of Complex Molecular Systems* (Springer).

References

1. Warshel A, Levitt M (1976) J Mol Biol 103:227
2. Warshel A (1991) Computer modeling of chemical reactions in enzymes and solutions. Wiley, New York
3. Cramer CJ, Truhlar DG (1994) Structure and reactivity in aqueous solution. ACS symposium series, vol 568, ACS, Washington, DC
4. Tapia O, Bertrán J (1996) Solvent effects and chemical reactivity. Kluwer Academic, Dordrecht
5. Berne BJ, Ciccotti G, Coker DF (1998) Classical and quantum dynamics in condensed phase simulations. World Scientific, Singapore
6. Combined QM/MM calculations in chemistry and biochemistry. J Mol Struct THEOCHEM 632 (2003) (Special Issue, edited by Ruiz- López MF)
7. Canuto S (2008) Solvation effects on molecules and biomolecules. Challenges and advances in computational chemistry and physics, vol. 6. Springer, London
8. Canuto S (2010) Combining quantum mechanics and molecular mechanics: some recent progresses in QM/MM methods. Advances in quantum chemistry, vol. 59. Elsevier, Oxford in UK
9. Takahashi H, Hori T, Wakabayashi T, Nitta T (2000) Chem Lett 29:222
10. Takahashi H, Hori T, Wakabayashi T, Nitta T (2001) J Phys Chem A 105:4351
11. Takahashi H, Hori T, Hashimoto H, Nitta T (2001) J Comp Chem 22:1252
12. Chelikowsky JR, Troullier N, Saad Y (1994) Phys Rev Lett 72:1240
13. Chelikowsky JR, Troullier N, Wu K, Saad Y (1994) Phys Rev B 50:11355
14. Hirose K, Ono T, Fujimoto Y, Tsukamoto S (2005) First-principles calculations in real-space formalism. Imerial College Press, Singapore
15. Hohenberg P, Kohn W (1964) Phys Rev 136:B864
16. Kohn W, Sham LJ (1965) Phys Rev 140:A1133
17. Parr RG, Yang W (1989) Density-functional theory of atoms and molecules. Oxford University Press, New York
18. Matubayasi N, Nakahara M (2000) J Chem Phys 113:6070
19. Matubayasi N, Nakahara M (2002) J Chem Phys 117:3605 (erratum, J Chem Phys 118:2446 (2003))
20. Matubayasi N, Nakahara M (2003) J Chem Phys 119:9686
21. Allen MP, Tildesley DJ (1987) Computer simulation of liquids. Oxford University Press, Oxford
22. Frenkel D, Smit B (2002) Understanding molecular simulation, from algorithms to applications, 2nd edn. Academic Press, London
23. Takahashi H, Matubayasi N, Nakahara M, Nitta T (2004) J Chem Phys 121:3989

24. Takahashi H, Omi A, Morita A, Matubayasi N (2012) J Chem Phys 136:214503
25. Hansen JP, McDonald IR (1986) Theory of simple liquids, 2nd edn. Academic Press, London
26. Suzuoka D, Takahashi H, Morita A (2014) J Chem Phys 140:134111
27. Szabo A, Ostlund NS (1996) Modern quantum chemistry, introduction to advanced electronic structure theory. Dover, New York
28. Martin RM (2008) Electronic structure: basic theory and practical methods. Cambridge University Press, Cambridge
29. Becke AD (1988) Phys Rev A 38:3098
30. Ono T, Hirose K (1999) Phys Rev Lett 82:5016
31. Lee C, Yang W, Parr RG (1988) Phys Rev B 37:785
32. Kleinman L, Bylander DM (1982) Phys Rev Lett 48:1425
33. Dunning TH (1989) J Chem Phys 90:1007
34. Becke AD (1993) J Chem Phys 98:5648
35. Burke K (2012) J Chem Phys 136:15091
36. Levy RM, Belhadj M, Kitchen DB (1991) J Chem Phys 95:3627
37. Luzhkov V, Warshel A (1992) J Comput Chem 13:199
38. Åqvist J, Medina C, Samuelsson JE (1994) Protein Eng 7:385
39. Carlson HA, Jorgensen WL (1995) J Phys Chem 99:10667
40. Kast SM (2001) Phys Chem Chem Phys 3:5087
41. Vener MV, Leontyev IV, Dyakov YA, Basilevsky MV, Newton MD (2002) J Phys Chem B 106:13078
42. Galván IF, Sanchez ML, Martin ME, Olivares del Valle FJ, Aguilar MA (2003) J Chem Phys 118:255
43. Freedman H, Truong TN (2004) J Chem Phys 121:2187
44. Higashi M, Hayashi S, Kato S (2007) J Chem Phys 126:144503
45. Chuev GN, Fedorov MV, Crain J (2007) Chem Phys Lett 448:198
46. Yamamoto T (2008) J Chem Phys 129:244104
47. Frolov AI, Ratkova EL, Palmer DS, Fedorov MV (2011) J Phys Chem B 115:6011
48. Lin B, Pettitt BM (2011) J Comput Chem 32:878
49. Sakuraba S, Matubayasi N (2014) J Comput Chem 35:1592
50. Henderson D (1992) Fundamentals of inhomogeneous fluids. Dekker, New York
51. Chandler D, McCoy JD, Singer SJ (1986) J Chem Phys 85:5971
52. Ramirez R, Gebauer R, Mareschal M, Borgis D (2002) Phys Rev E 66:031206
53. Liu Y, Zhao S, Wu J (2013) J Chem Theory Comp 9:1896
54. Sergiievskyi VP, Jeanmairet G, Levesque M, Borgis D (2014) J Phys Chem Lett 5:1935
55. Karino Y, Fedorov MV, Matubayasi N (2010) Chem Phys Lett 496:351
56. Karino Y, Matubayasi N (2013) Phys Chem Chem Phys 15:4377
57. Matubayasi N, Liang KK, Nakahara M (2006) J Chem Phys 124:154908
58. Matubayasi M, Shinoda W, Nakahara M (2008) J Chem Phys 128:195107
59. Takemura K, Guo H, Sakuraba S, Matubayasi N, Kitao A (2012) J Chem Phys 137:215105
60. Bode W, Greyling HJ, Huber R, Otlewski J, Wilusz T (1989) FEBS Lett 242:285
61. Sevcik J, Urbanikova L, Dauter Z, Wilson KS (1998) Acta Crystallogr Sect D Biol Crystallogr 54:954
62. Chen R, Weng ZP (2002) Proteins 47:281
63. Pierce BG, Hourai Y, Weng ZP (2011) PLoS ONE 6:e24657
64. Pierce B, Weng ZP (2007) Proteins 67:1078
65. Leaver-Fay A, Tyka M, Lewis SM, Lange OF, Thompson J, Jacak R, Kaufman K, Renfrew PD, Smith CA, Sheffler W, Davis IW, Cooper S, Treuille A, Mandell DJ, Richter F, Ban YEA, Fleishman SJ, Corn JE, Kim DE, Lyskov S, Berrondo M, Mentzer S, Popovic Z, Havranek JJ, Karanicolas J, Das R, Meiler J, Kortemme T, Gray JJ, Kuhlman B, Baker D, Bradley P (2011) Methods Enzymol 487:545
66. Takahashi H, Satou W, Hori T, Nitta T (2005) J Chem Phys 122:044504
67. Takahashi H, Kawashima Y, Nitta T, Matubayasi N (2005) J Chem Phys 123:124504
68. Hori T, Takahashi H, Nakano M, Nitta T, Yang W (2005) Chem Phys Lett 419:240

69. Hori T, Takahashi H, Furukawa S, Nakano M, Yang W (2007) J Chem Phys B 111:581
70. Takahashi H, Ohno H, Yamauchi T, Kishi R, Furukawa S, Nakano M, Matubayasi N (2008) J Chem Phys 128:064507
71. Takahashi H, Ohno H, Kishi R, Nakano M, Matubayasi N (2008) J Chem Phys 129:205103
72. Takahashi H, Miki F, Ohno H, Kishi R, Ohta S, Furukawa S, Nakano M (2009) J Math Chem 46:781
73. Takahashi H, Maruyama K, Karino Y, Morita A, Nakano M, Jungwirth P, Matubayasi N (2011) J Phys Chem B 115:4745
74. Takahashi H, Iwata Y, Kishi R, Nakano M (2011) Int J Quantum Chem 111:1748
75. Matubayasi N, Takahashi H (2012) J Chem Phys 136:044505
76. Suzuoka D, Takahashi H, Morita A (2014) J Chem Phys 140:134111
77. Takahashi H, Suzuoka D, Morita A (2015) J Chem Theory Comput 11:1181
78. Sakuraba S, Matubayasi N (2011) J Chem Phys 135:114108
79. Berendsen HJC, Grigera JR, Straatsma TP (1987) J Phys Chem 91:6269
80. IAPWS Industrial Formulation 1997 for the thermodynamic properties of water and steam (1997). Erlangen, Germany
81. Francl MM (1985) J Phys Chem 89:428
82. Luque FJ, Orozco M (1998) J Comput Chem 19:866
83. Cubero E, Luke FJ, Orozco M, Gao J (2003) J Phys Chem B 107:1664
84. Tomasi J, Mennucci B, Cammi R (2005) Chem Rev 105:2999
85. Jorgensen WL, Maxwell DS, Tirado-Rives J (1996) J Am Chem Soc 118:11225
86. Gallicchio E, Zhang LY, Levy RM (2002) J Comput Chem 23:517
87. Meyer EA, Castellano RK, Diederich F (2003) Angew Chem Int Ed 39:1210
88. Ma JC, Dougherty DA (1997) Chem Rev 97:1303
89. Tsuzuki S, Honda K, Uchimaru T, Mikami M, Tanabe K (1984) J Am Chem Soc 106:4102
90. Ten-no S, Hirata F, Kato S (1994) J Chem Phys 100:7443
91. Sato H, Hirata F (1999) J Phys Chem B 103:6596
92. van Gunsteren WF, Billeter SR, Eising AA, Hünenberger PH, Krüger P, Mark AE, Scott WRP, Tironi IG (1996) Biomolecular simulation: the GROMOS96 manual and user guide. vdf Hochschulverslag, ETH Zürich Switzerland

Chapter 7
Structure and Electronic Properties of Liquids and Complex Molecular Systems in Solution: Coupling Many-Body Energy Decomposition Schemes to Born-Oppenheimer Molecular Dynamics

Benedito J.C. Cabral, K. Coutinho and S. Canuto

Abstract A discussion on the structure, dynamics and electronic properties of liquids and complex molecular systems in solution is presented. Special emphasis is placed on the sequential coupling of electronic structure calculations to Monte Carlo and Molecular dynamics sampling procedures. A promising approach to investigate the electronic absorption spectra of liquids and molecular solutions relying on a many-body energy decomposition scheme is presented and some applications to hydrogen bonding liquids are discussed. The possibility to parametrize classical force fields by using information generated by first principles molecular dynamics is investigated and preliminary results for the structure of chlorophyll-c_2 in liquid methanol relying on this approach are reported.

7.1 Introduction

The study of electronic properties is of fundamental importance for understanding chemical reactivity in condensed phases. For many systems of interest chemical reactions take place in solution, where the chemical kinetics and energy transfer

B.J.C. Cabral (✉)
Grupo de Fisica Matematica da, Universidade de Lisboa,
Av. Prof. Gama Pinto 2, 1649-003 Lisbon, Portugal
e-mail: bjcabral@fc.ul.pt

K. Coutinho · S. Canuto
Instituto de Fisica da, Universidade de São Paulo, São Paulo, Brazil
e-mail: kaline@if.usp.br

S. Canuto
e-mail: canuto@if.usp.br

J.-L. Rivail et al. (eds.), *Quantum Modeling of Complex Molecular Systems*,
Challenges and Advances in Computational Chemistry and Physics 21,
DOI 10.1007/978-3-319-21626-3_7

processes are assisted by the environment. The importance of environmental effects fostered the development of theoretical methods to take into account these effects on the structure, dynamics and electronic properties of complex molecular systems. Simplified representations of the solvent by a continuum medium of dielectric constant ε led to the formulation of self-consistent reaction-field (SCRF) methods [1–6]. The SCRF approaches were very important for the analysis of solvent effects on the electronic properties of complex molecular systems. They can be seen as a first formulation of general multi-scale methods, which are characterised by the definition of a quantum mechanical (QM) sub-system embedded in a classical environment [7, 8] described by a dielectric constant ε (QM/ε). The importance of thermal effects motivated the development of different methodologies where electronic structure calculations are coupled to statistical mechanics sampling procedures such as Monte Carlo and Molecular Dynamics. These approaches are part of quantum mechanics/molecular mechanics (QM/MM) methodologies [7–9]. Further developments of QM/MM methods opened the possibility to investigate environment effects in large systems of biological interest [10, 11]. An alternative formulation of QM/MM methods rely on the a posteriori calculation of the electronic properties for a quantum system in a classical environment by using configurations generated by statistical mechanics sampling [12–14].

Monte Carlo and Molecular Dynamics statistical mechanics sampling can be carried out by using energy and forces derived from classical force fields or from first principles quantum mechanical calculations. The application of classical molecular dynamics to complex biological systems is now a standard procedure that is only limited by the reliability of the classical force fields. By adopting a first principles approach, this technique may overcome, in many cases, the limitations of classical force fields to represent the interactions in condensed phase. The applications of first principles molecular dynamics to the study of very large systems are, however, limited by computational costs. Recent advances in the analysis of the electronic structure theory, specifically of incremental procedures [15–32], opened the way to the application of high level electronic structure methods to complex systems in condensed phases [18, 20, 24, 27, 29]. Incremental methods usually rely on the partition of the system in smaller fragments and applications to systems with thousand of atoms were reported [21]. The partition procedures are anchored on the decomposition of the total energy in many-body energy (MBE) contributions, which describe the interactions between the fragments of the system [21, 24]. The possibility of partitioning or definition of fragments relying on localization procedures by using the electronic density [33] or natural transition/localized orbitals [28] has been also exploited. Incremental methods can be applied to generate the dynamics as well as to the analysis of the electronic structure for configurations generated by classical molecular dynamics. An interesting development concerns the application of MBE schemes to the ab initio calculation of the electronic absorption spectra [18, 20, 24, 28, 34–36].

In this chapter we will review some recent applications of MBE decomposition schemes to the calculation of the electronic absorption spectra of hydrogen bonding liquids. The possibility to parametrize classical force fields by using information generated by first principles molecular dynamics is discussed. Preliminary results relying on this approach for the structure of chlorophyll-c_2 in liquid methanol are also presented.

7.2 Electronic Properties of Liquids and Molecular Solutions

7.2.1 *Sequential QM/MM Calculation of the Electronic Properties*

A posteriori calculations of the electronic properties of complex molecular systems in condensed phases by using configurations generated by statistical mechanics sampling defines a QM/MM procedure with some specific features. There is no need to address the well known issue concerning the definition of the interface between the QM and the MM partitions [9, 37, 38], which is particularly difficult when the QM and MM system are covalently linked [37, 38]. The a posteriori method for the calculation of the electronic properties is not necessarily the same method used in the sampling procedure. Thus, high level electronic structure calculation methods can be used by adopting a sequential QM/MM methodology. A promising approach associated with QM/MM procedures is the coupling of many-body energy decomposition schemes to configurations generated by statistical mechanical sampling [24]. This approach opens the way to high level ab initio calculations of electronic properties in condensed phase [26, 34, 35].

7.2.2 *MBE Schemes for the Calculation of Electronic Excitation Energies: The Frenkel Exciton Hamiltonian Formalism*

Ab initio electronic structure calculations in condensed phases can be carried out by adopting fragmentation schemes [22–24, 29]. In general, these schemes rely on the decomposition of the total ground state energy $E[0]$ in many-body contributions accordingly to the following equation

$$E[0] = \sum_{i=1}^{N_f} E[i] + \sum_{i=1}^{N_f} \sum_{j<i}^{N_{f-1}} \Delta E[ij] + \sum_{i=1}^{N_f} \sum_{j<i}^{N_{f-1}} \sum_{k<j}^{N_{f-2}} \Delta E[ijk] + \ldots - C[0] \quad (7.1)$$

where N_f is the number of fragments (or partitions) in which the system is decomposed, and the different contributions are given by

$$\Delta E[ij] = E[ij] - E[i] - E[j] \tag{7.2}$$

$$\Delta E[ijk] = E[ijk] - E[ij] - E[ik] - E[jk] - E[i] - E[j] - E[k] \tag{7.3}$$

$C[0]$ is the self-interaction energy of the charge background in which the expansion is carried out. Implicit in the introduction of a charge background in MBE decomposition schemes is also the partition of the system into quantum (QM) and classical (MM) parts. Therefore, a MBE decomposition scheme with electrostatic embedding is closely related to a QM/MM approach where polarization effects on the QM system due to the charge background (the MM partition) are taken into account.

As discussed below, the introduction of polarization effects even through a simplified representation of the environment significantly improves the convergence of the expansion 1. Some aspects concerning many-body energy decomposition schemes are worth noticing. Although the definition of a fragment may be difficult when covalent bond breaking and formation is involved, for some systems it is relatively simple. This is the case of molecular aggregates or clusters for which we can assume that each fragment corresponds to a single or multiple monomeric units. The convergence of the expansion for the total energy (Eq. 7.1) is in general quite slow. This is mainly due, for each term of the expansion, to the absence of polarization effects from the environment. Moreover, when one-electron basis-sets are used higher-body terms in the expansion are contaminated with basis-set superposition errors (BSSE) [23, 39]. The convergence of the MBE decomposition scheme can be improved by including an approximate representation of the environment for each term in the expansion 1. BSSE can be systematically reduced by using larger basis-set or explicitly correlated basis-sets [40].

The importance of polarization effects for the convergence of the MBE expansion for two water hexamers is illustrated in Fig. 7.1. The energy was calculated at the CCSD(T)/cc-pVTZ level with two different charge embeddings that were defined by using the TIP3P model for water [41, 42] and the PCM [5] charges for a water molecule. The PCM charges were determined for a single water molecule in a continuum medium with the dielectric constant of liquid water. Merz-Singh-Kollman charges [43] calculated at the MP2/aug-cc-pVDZ level were adopted. At this level, the water dipole moment is 2.22 D. At the CCSD(T)/cc-pVTZ level, and by defining the computer time for N = 6 as 100, the times for N = 1, 2, 3 and 4 are, 0.07, 2.6, 26, and 90, respectively. However, it should be observed that these ratios are very dependent on the scaling with N of the specific theoretical level.

The results for $|\Delta E(\mathrm{N}) - E(\mathrm{FULL})|$ shows the importance of carrying out the MBE expansion by taking into account polarization effects from the environment (data for this difference with no embedding are represented by NOEMB). The results also indicate that by using a 2-Body approximation (N = 2), $|\Delta E(\mathrm{N}) - E(\mathrm{FULL})|$ is

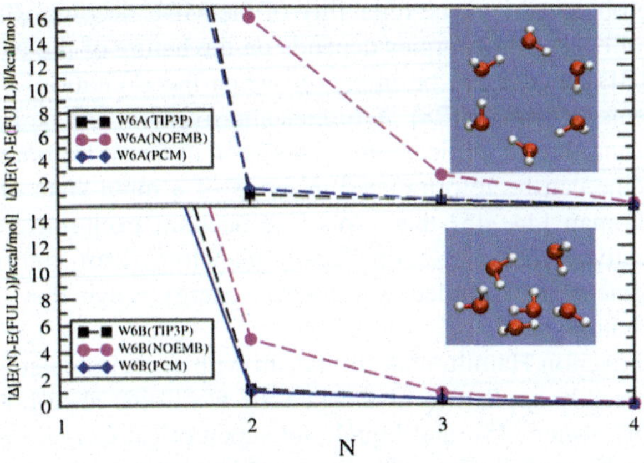

Fig. 7.1 Convergence of the MBE decomposition scheme for two water hexamers. Energies were calculated at the CCSD(T)/cc-pVTZ level. The embedding charges correspond to the TIP3P model [41, 42] and PCM approach [5]. Calculations with no embedding are represented as NOEMB

close to 1 kcal/mol. Interestingly, the convergence of the series is not very dependent on the choice of the embedding charges, at least when the results relying on TIP3P and PCM are compared. However, an improved representation of the environment in which the MBE expansion is carried out remains an open issue. Specifically, the construction of a polarizable and self-consistent environment should be exploited.

Excitation energies can be calculated as the difference between the total energies for the ground and excited states.

$$\Delta E = (E[*] - C[*]) - (E[0] - C[0]) \tag{7.4}$$

where $E[*]$ and $E[0]$ are the excited and ground state energies, respectively. $C[*]$ is the charge background self-interaction energy contribution to excited state.

By applying MBE schemes, $E[*]$ is expanded as

$$E[*] = \sum_{i=1}^{N_f} E[i^*] + \sum_{i=1}^{N_f} \sum_{j<i}^{N_{f-1}} E[(ij)^*] + \sum_{i=1}^{N_f} \sum_{j<i}^{N_{f-1}} \sum_{k<j}^{N_{f-2}} E[(ijk)^*] + \cdots \tag{7.5}$$

In the previous expression, only one specific fragment for each term of the sums is excited. For the one-body contribution when the fragment i is excited, all the other fragments for $j \neq i$ are in the ground state. For the two-body contributions, when the (ij) fragment is excited all the other (kl) pairs for $i \neq k$ and $l \neq j$ are in the ground state.

MBE decomposition schemes were recently applied to calculate the electronic absorption spectra of clusters [24, 25], liquids [34–36], and molecular solutions [26].

It should be observed that the reliability of the MBE decomposition scheme for the prediciton of excitation energies depends on the nature of the excited state. The accuracy of MBE scheme can be improved when the excitations are localized on specific fragments or orbitals [28]. A further improvement can be accomplished by taking into account the excitonic coupling between different fragments [24, 34–36]. The MBE scheme can be reformulated in terms of a multi-state effective Frenkel Exciton Hamiltonian [44, 45] that takes into account excitonic coupling. In the multi-state effective Frenkel Exciton Hamiltonian formalism, multiple excitations from each fragment are included, yielding an energy range that depends on the number of excitations.

The Frenkel exciton Hamiltonian formalism with energies calculated with MBE schemes was recently applied to investigate the electronic absorption spectrum of liquid HCN [34], water [35], and liquid and supercritical CO_2 [36].

The effective Frenkel Exciton Hamiltonian [44, 45] can be written as:

$$H = \sum_{i=1}^{N_f}\sum_{k\in i}^{N_s} E_i^k |i,k> <k,i| + \sum_{i\neq j}^{N_f}\sum_{k\in i}^{N_s}\sum_{l\in j}^{N_s} J_{ij}^{kl}|i,k> <j,l| \tag{7.6}$$

where N_f is the number of fragments or monomers in the system and N_s is the number of excited states for each fragment. The excited state $|k>$ in the fragment i with energy E_i^k is represented by $|i,k>$. J_{ij}^{kl} is the coupling between the kth and lth excited states on the ith and jth fragments, respectively. The dipole [d-d] excitonic coupling between states k and l, on fragments i and j is approximated by the interaction between their transition dipoles

$$J_{ij}^{kl}\frac{1}{\mathbf{R}_{ij}^3}\left[\mathbf{d}_i^k \cdot \mathbf{d}_j^l - 3(\mathbf{d}_i^k \cdot \hat{\mathbf{R}}_{ij})(\mathbf{d}_j^k \cdot \hat{\mathbf{R}}_{ij})\right] \tag{7.7}$$

where $\mathbf{R}_{ij}$ is a vector from fragment i to fragment j, $\mathbf{d}_i^k = <\mathbf{i},\mathbf{k}|\hat{\mathbf{d}}|\mathbf{i},\mathbf{0}>$, $\hat{\mathbf{d}}$ is the dipole moment operator, and $|i,0>$ represents the ground state wavefunction of fragment i.

The off-diagonal terms of H are set to zero for excitations taking place on the same fragment. Thus, the matrix elements $H_{\alpha\beta}$ are given by:

$$H_{\alpha\beta} = \begin{cases} E_i^k & \text{if } i = j \wedge k = l \\ J_{ij}^{kl} & \text{if } i \neq j \\ 0 & \text{if } i = j \end{cases} \tag{7.8}$$

where the k, l, i and j have the same meaning as above, and the α and β indexes are given by $\alpha = (i-1) \times N_s + k$ and $\beta = (j-1) \times N_s + l$. Diagonalization of the Hamiltonian matrix yields a set of $N_f \times N_s$ eigenvalues, and from the eigenvectors $c_{\alpha\beta}$ of the matrix $H_{\alpha\beta}$, the transition dipole of state β can be written as $d_\beta = \sum_\alpha d_\alpha c_{\alpha\beta}$, where d_α's are the transition dipole moments of the individual states of each fragment. From the transition dipole moments the oscillator strengths from the ground to the excited state β are calculated through the expression $f_\beta = \frac{2}{3} E_\beta d_\beta^2$, where E_β is the excitation energy of thc state β.

The one-body [1B] approximation for the energies of the ground state (E_0[1B]) and excited state k of the fragment i ($E_i^k[1B]$) are given, respectively, by $E_0[1B] = \sum_{i=1}^{N_f} E_0[i] - C[0]$ and $E_i^k[1B] = E_i^k + \sum_{j \neq i}^{N_f - 1} E_0[j] - C[j^*]$. $C[0]$ and $C[j^*]$ are constants related to the contribution of the electrostatic environment to the total energies of the ground and excited states, respectively [24]. The diagonal elements of the Frenkel exciton Hamiltonian are then approximated by $E_i^k[1B]$. After diagonalization and representing the eigenvalues of H by $\tilde{E}_i^k[1B]$ the excitation energies from the ground to the excited state k are calculated as $\omega_i^k[1B] = \tilde{E}_i^k[1B] - E_0[1B]$.

7.3 Applications

7.3.1 Electronic Properties of Hydrogen Bonding Liquids

Hydrogen bond (HB) networks play a relevant role on the dynamics of chemical and biochemical reactions in solution. The structure of HB network is determined by many-body energy cooperativity associated with polarization effects. Electronic density fluctuations induced by thermal effects or photoexcitation, assist and drive energy transfer processes in solution [46]. An interesting aspect related to the formation of HB networks is the role that they may have played in the origin of self-organized structures associated with life [47]. Therefore, the structure, dynamics and more recently, the electronic properties of these complex HB networks have been the subject of several fundamental investigations. In this section, we review recent applications of MBE decomposition schemes to the calculation of the electronic absorption spectra of liquid water and HCN.

7.3.1.1 Ab Initio Calculation of Electronic Absorption of Liquid Water: Coupling MBE Decomposition Schemes to First Principles Molecular Dynamics

The HB network of water [48, 49] exhibits a high ability to reorganize itself in different environments or thermodynamic conditions. The liability of the HB network is illustrated by the multiples phases of ice [50] and the reorganization of water molecules around charged [51–53] and hydrophobic species [54, 55]. Interestingly, recent investigations indicate that nanoconfined water exhibits electronic properties quite different from normal water [56, 57]. These more recent works support the relationship between the structure of the HB network and the electronic properties of water.

Electronic absorption in water has been the subject of several experimental [58–60] and theoretical investigations [61–71]. The electronic absorption of liquid water at ambient conditions is characterized by a ~0.8 eV blue shift of the first absorption maximum relative to the observed position for the gas-phase water monomer [58, 59]. Moreover, a gradual red shift (up to ~0.6 eV) in the absorption band edge is observed by increasing the temperature from 25 to 400 °C [60].

Several works on the electronic absorption of liquid water were reported [24, 61–63, 65]. Here, we will couple MBE decomposition schemes for the calculation of excitation energies [18, 23, 34, 35] to configuration generated by Born-Oppenheimer molecular dynamics (BOMD).

BOMD were generated with the Becke (B) [72] and Lee-Yang-Parr (LYP) [73] functional for exchange and correlation, respectively (BOMD/BLYP, and with a modified BLYP functional (BLYP-D3) that includes an empirical correction (D3) [74–76] for the dispersion interactions (BOMD/BLYP-D3). The dynamics was carried out with the hybrid Gaussian and plane-wave method GPW [77] as implemented in the QUICKSTEP module [78] of the CP2K program [78]. Goedecker, Teter, and Hutter (GTH) norm-conserving pseudopotentials [79] were used for representing the core electrons and only valence electrons were explicitly included in the quantum mechanical density functional theory (DFT) calculation of the forces to generate the dynamics. Further technical details on the BOMD runs can be found in Ref. [35].

In Fig. 7.2 we present the results for the radial distribution functions (RDFs) of water predicted by BOMD/BLYP and BOMD/BLYP-D3. Comparison with recent experimental data [80] shows that the inclusion of empirical corrections for the dispersion interactions leads to a much better agreement with experiment. In keeping with previous first principles molecular dynamics investigations the BLYP

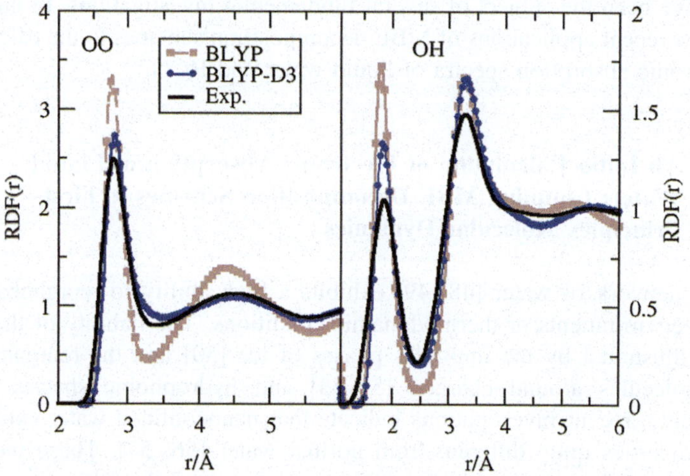

Fig. 7.2 Partial radial distribution functions [RDF(r)] for liquid water from different models. *Left panel* O-O; *right panel* O-H. Experimental curve from Soper [80]

exchange-correlation functional leads to a more structured water in comparison with experiment. The importance of dispersion interactions in liquid water was discussed by several studies [81–86].

For the a posteriori calculation of excitation energies using configurations generated by molecular dynamics, a quantum system, including explicitly a given number n of molecules is defined. This supermolecular structure is embedded in the electrostatic environment of the remaining molecules, that are represented by point charges. It is important to assess how the results depend on the size of the quantum system (n) and on the choice of the embeddeding charge background. Actually, previous analysis indicated a very weak dependence of the electronic absorption spectrum on the choice of the embedding charges [24, 71] that is presently represented by TIP3P [41, 42] charges.

Excitation energies were calculated with the equation of motion coupled cluster theory with single and double excitations (EOM-CCSD) [87] and the Dunning's correlation-consistent d-aug-cc-pVDZ and aug-cc-pVTZ basis-sets [88].

As shown in Fig. 7.3 the results are not significantly dependent on the size of the quantum system n. Another fundamental aspect concerns the dependence of the electronic absorption spectrum on the electrostatic background. This is illustrated in Fig. 7.4 where we report results with (EMB) and withouth charge embedding (NOEMB) of the quantum system.

Polarization effects significantly contribute to blue shift the maximum position of the first absorption band of liquid water. Therefore, electrostatic interactions are at the origin of the experimentally observed ~0.8 eV blue shift of the first absorption maximum of liquid water relative to the gas-phase water molecule [58, 59]. It should be noticed that hydrogen bonding in liquid water leads to a small stretching of the OH bond relative to the gas-phase value. In addition, this structural change leads to a small red shift of the first maximum relative to the isolated molecule [35]. Therefore, the overall shift is determined by a competition between

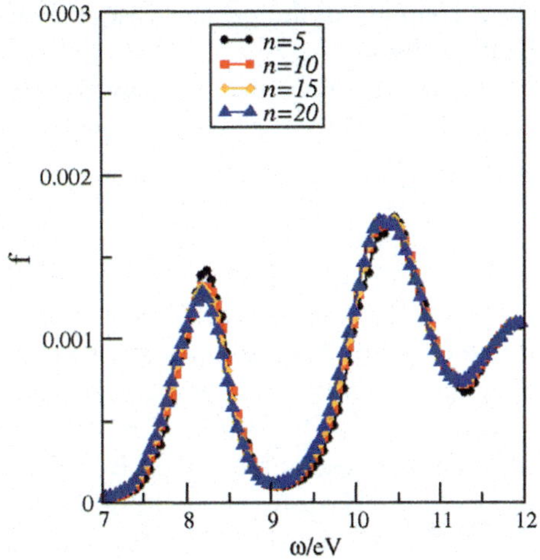

Fig. 7.3 Dependence of the excitation spectra of water on the size of the quantum system n for calculations with the BLYP-D3 functional and the d-aug-cc-pVDZ basis-set

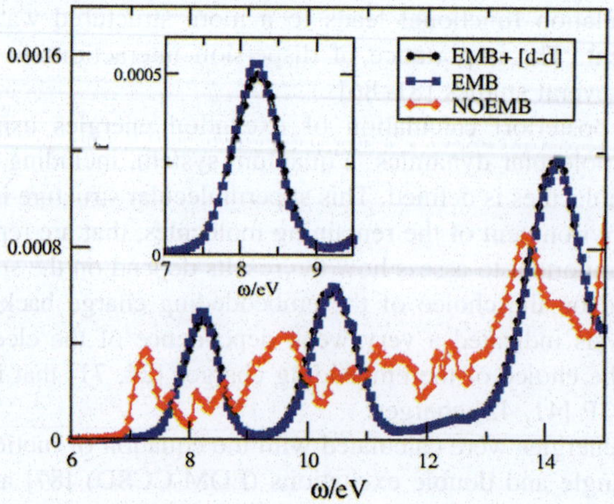

Fig. 7.4 Dependence of the excitation spectra of water on the charge embedding for a system with n = 15 explicit water molecules. EOM-CCSD/d-aug-cc-pVDZ calculations with a number of excited states N_s = 20 for each fragment (or water monomer) The inset panel illustrates the role played by [d-d] excitonic couplings

OH stretching that contributes to a small red shift and electrostatic interactions that contribute to a blue shift. The role played by coordination and Coulombic effects on the optical spectrum of water has been investigated by Hermann et al. [68]. However, it should be stressed the importance of taking into account small structural changes of the water molecules due to hydrogen bonding for a quantitative prediction of the electronic properties of water in bulk phases as well as in the water liquid-gas interface, where a non-uniform distribution of hydrogen bonds and monomeric dipole moments was observed [89].

The role played by dipole-dipole [d-d] excitonic coupling (see Eq. 7.7) is also illustrated in the inset panel of Fig. 7.4. The results indicate that the main effect of [d-d] coupling in the present case is a small blue shift of the first absorption band peak position.

Figure 7.5 shows the result for the electronic absorption of liquid water relying on EOM-CCSD/aug-cc-pVTZ calculations. The 1-Body calculations and a quantum system with n = 20 explicit water molecules were used. A total number of N_s = 30 excited states were considered for each water monomer. The vertical bars are experimental data for liquid water from Heller et al. [58] (Exp1) and Hayashi et al. [59] (Exp2). Quite similar results for BOMD/BLYP and BOMD/BLYP-D3 are observed. The maxima positions are at 8.3, 10.5, and 13.9 eV with the d-aug-cc-pVTZ basis-set. Comparison with the first set of experimental data (Exp1) [58] shows a very good agreement, particularly for the first peak position that is only blue shifted by ∼0.1 eV relative to Exp1 (8.2 eV). For the second and third maxima, our results relying on BOMD/dapvdz are in better agreement with those reported by Hayashi et al. [59]. In this case, our BOMD values (BLYP and BLYP-D3) are red shifted by 0.1 and 0.4 eV in comparison with Exp2 [59].

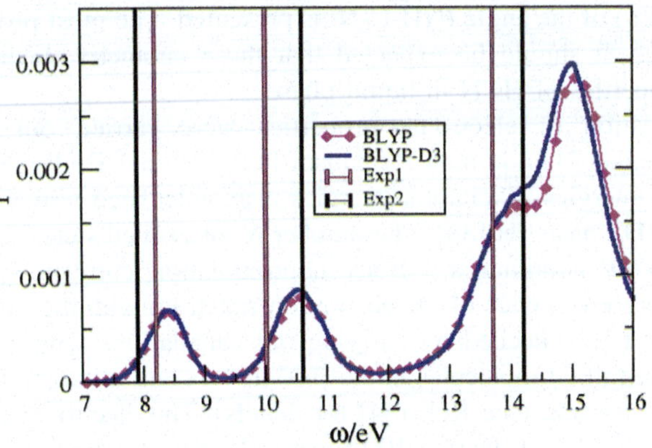

Fig. 7.5 EOM-CCSD/aug-cc-pVTZ results for the absorption spectrum of liquid water. Calculations with $n = 20$. *Vertical lines* are the experimental peak positions from Heller [58] (Exp1) and Hayashi [59] (Exp2)

7.3.1.2 Electronic Absorption in Liquid HCN

HCN is a strong dipolar species with a gas-phase dipole moment of 3.0 D [90] that may form HB networks with a rich topological structure consisting of polymerized chains, ramified, and cyclic aggregates [91, 92]. The presence of polar domains in strongly dipolar fluids has also been investigated by several works [93–95] and the structure and energetic properties of the HCN monomer and clusters [96–102] have been reported.

Results for the structure of liquid HCN from BOMD have been previously reported [34]. Here, we will focus on some specific structural changes of the HCN monomer in the liquid phase and also on the coupling of MBE schemes to calculate its electronic absorption spectrum. Deviations from the linearity of the HCN monomer are observed in condensed phase. This is illustrated in Fig. 7.6 where the

Fig. 7.6 Distribution N(θ) of the angle θ (H-C-N) in liquid HCN illustrating deviations from the linearity relative to the gas phase monomer geometry

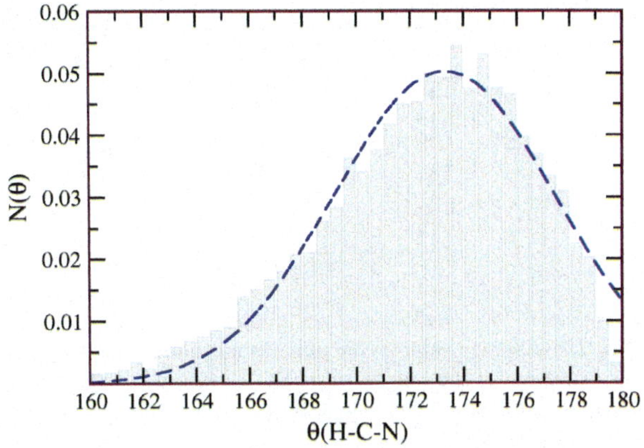

distribution N(θ) of the angle θ (H-C-N) is presented. The most probable value of θ is 173 degrees. It should be expected that these geometry deviations affect the electronic properties of HCN in liquid phase.

The calculation of electronic properties were carried out by adopting a multi-state Frenkel Hamiltonian formalism [44] with a 1-Body approximation for the excitation energies. Excitation energies were calculated with EOM-CCSD and the aug-cc-pVDZ basis set [88]. The number N_s of excited states was nine for each fragment that by adopting a 1-Body approximation corresponds to one HCN molecule. The convergence of the absorption spectrum with the number n of HCN molecules has been checked by carrying out calculations with several values of n from 2 to the full 64 molecules in the BOMD box and different definitions of the supermolecular system (see Ref. [34] for details). The electronic absorption spectrum was calculated by selecting 100 supermolecular structures from BOMD. In a first approach (named as no embedded single box), the system was defined by the n explicit HCN molecules and the electrostatic background of the remaining 64-n molecules were represented by point charges corresponding to a classical force field model [103]. In the second approach (named as embedded supercell) the BOMD box was replicated around the central one thus defining a supercell with 1728 (64×3^3) molecules. It should be noticed that by adopting the second approach each HCN molecule experiences nearly the same electrostatic environment, in contrast with the first approach, where the fragments near the edges of the box experience an assymetric environment.

Figure 7.7 illustrates the behaviour of the average value of the maxima position for the two first excitations of liquid HCN with the number n of HCN molecules

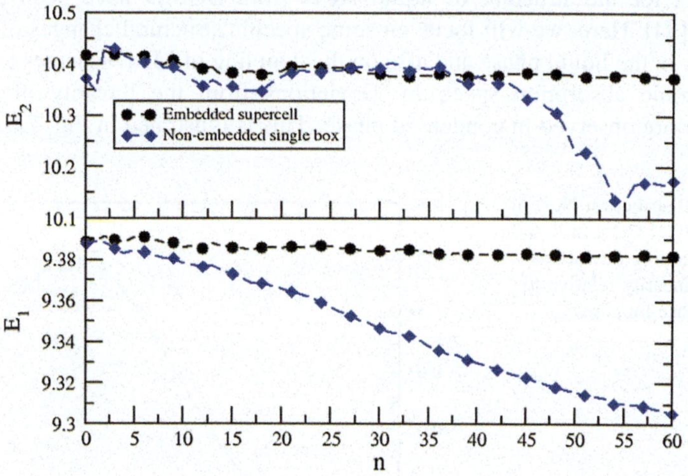

Fig. 7.7 Average value of the maxima position for the two first excitation energies of liquid HCN as a function of the number n of HCN monomers explictly included in the quantum system by using two different embeddings. *Black circles* embeddeed supercell; *blue diamonds* no embedded single box (see text)

explictly included in the quantum system. The results strongly indicate that by adopting the embedded supercell approach no significant dependence with n is observed. If the no embedded single box approach is adopted, a correct estimate for the average value of the excitations can be made by using a relatively small number of HCN monomers. However, the results deviate quickly from the correct value with increasing n, particularly for the first excitation energy. Consequently, the importance of checking the dependence of the results with n as well as the adoption of the embedded supercell approach should be stressed.

Figure 7.8 shows the spectrum of liquid HCN calculated by using the 1-body approximation and EOM-CCSD energies. The embedded supercell approach was used with $n = 64$ explicit HCN molecules in the quantum system. The vertical lines are the values for the gas phase monomer.

The spectrum of liquid HCN is characterized by the presence of two allowed bands centered at 9.4 and 10.4 eV. A weakly allowed band on the red edge of the first allowed one (see inset panel) is also observed. Further details on the nature of the excitations defining each band can be found in Ref. [34]. Here, we will focus on some specific aspects that stress the importance of the presently adopted approach. By coupling a MBE decomposition scheme to BOMD configurations the electronic spectra can be calculated at the EOM-CCSD/d-aug-cc-pVDZ level for systems with a significant number of fragments (n = 64 HCN monomers). The results based on the embedded supercell supermolecular systems show no significant dependence on the number n of HCN monomers included in the quantum system and supports the adequacy of the present approach when the number of monomers in the BOMD is relatively small.

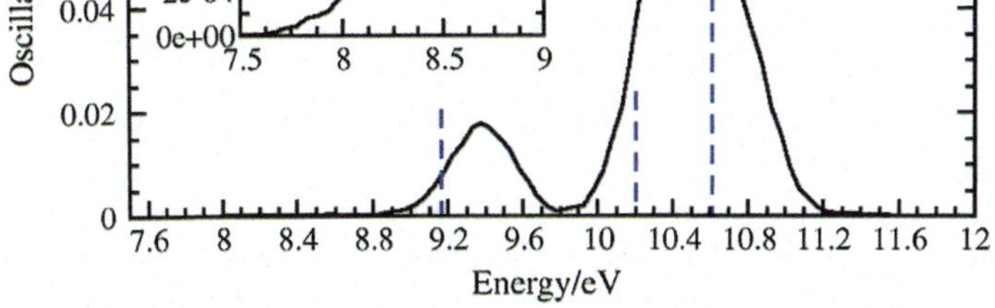

Fig. 7.8 Electronic absorption spectrum of liquid HCN. Results relying on EOM-CCSD/d-aug-cc-pVDZ and the 1-Body approximation. Inset panel: weakly allowed low energy excitation band of liquid HCN. *Vertical lines* are excitation energies for the gas phase HCN molecule

Comparison between the maxima positions for the two first absorption bands of liquid HCN at 9.4 and 10.4 eV relative to the gas phase values show that they are shifted to the blue by 0.2 and 0.4 eV, respectively. This is a small shift in comparison with liquid water, where a ~0.8 eV blue shift is observed for the maximum position of the first band relative to the gas phase value [58, 59]. The small shift for liquid HCN is possibly explained by the geometry changes from the gas to the liquid phase and also to the nature of the first excited states of HCN. As our results for the structure indicate (see Fig. 7.6), some bending of the HCN molecule is observed in liquid HCN as well as a small increase of the CN bond distances. Both changes are also observed for the geometry of the HCN in the first excited adiabatic state. Moreover, very weak interactions between transition dipole moments corresponding to the first excitations in liquid HCN preserve the local nature of the gas phase HCN excitation. Our results also indicate that these localized excitations are not very dependent on the interactions with the liquid environment.

7.3.2 *Structure and Electronic Properties of Chlorophyll-C_2 in Liquid Methanol*

Electronic properties of chlorophylls and related systems are of fundamental interest to understand the molecular mechanisms of energy and charge transfer in complex antenna and photosynthetic reaction centers [104]. Several studies were dedicated to investigate the electronic absorption spectra of photosynthetic chromophores (see König and Neugebauer [105] for a recent review). Most of the available experimental information on the electronic properties of chlorophylls is determined in solution. This feature fostered theoretical studies on the electronic properties of photosynthetic chromophores in solution, [106–114] or in interaction with hydrogen bonding species, [113, 115–122] or with the protein environment [123, 124].

Recently, some works pointed out the importance of new photosynthetic species that show an absorption spectrum at lower energies, specifically in the 700–750 nm region or beyond [125]. This is related to the role that may be played by new chromophores in artificial photosynthesis, by expanding the spectral absorption region, as well as in photoprotection, by increasing photoabsorption at lower energies. A close related issue concerns the role played by hydrogen bonding on the structural mechanisms and photoprotective function of water-soluble chlorophyll-binding proteins [126] and synthetic analogues of natural chlorophylls aggregates in aqueous media [127].

In contrast with other chlorophylls, chlorophyll-c_2 (chlc_2) is characterized by the absence of the isoprenoid tail and the similarity with porphyrins (no reduction of the D ring). Although chlc_2 is not soluble in water it is soluble in methanol and the absorption spectrum of chlc_2 in liquid methanol is known [128]. The structure of chlc_2 is illustrated in Fig. 7.9.

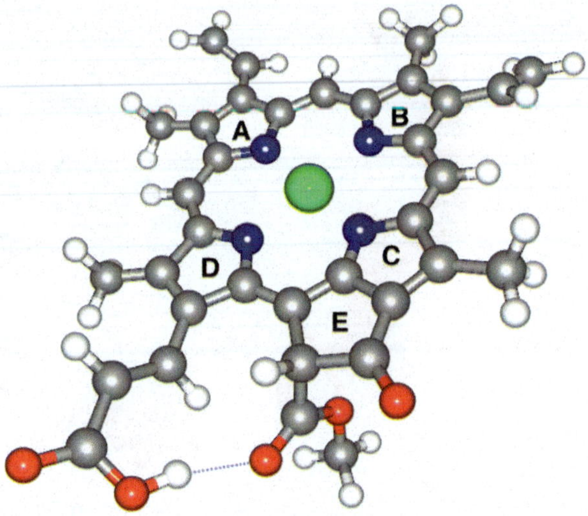

Fig. 7.9 Structure of chlorophyll-c_2

Theoretical investigations on the electronic absorption [112] and ionization spectra of chlc$_2$ [113] in methanol were reported. These studies were based on the sequential Monte Carlo(MC)/Quantum mechanics approach [13] and the calculations were carried out with the gas-phase optimized structure of chlc$_2$ (that was kept rigid during the MC sampling) and a classical force field to represent the interactions in the chlc$_2$ methanol-system [112]. More recently, we have carried out a Born-Oppenheimer Molecular Dynamics (BOMD) [129] of chlc$_2$ in liquid methanol [114]. BOMD was carried out with the Perdew, Burke, and Ernzerhof (PBE) exchange-correlation functional [130]. Time dependent density functional theory (TDDFT) calculations of the excitation energies were carried out with the B3LYP functional.

It was verified that the structure of chlc$_2$ is characterized by a displacement of the Mg atom from the macrocycle plane. Moreover, the orientation of the methanol molecules relative to the macrocycle plane is different when we compare the side with a protruding Mg atom and the other one, where the methanol molecules exhibit a solvatophobic behaviour with the methyl groups oriented towards the macrocycle plane. These features are illustrated in Fig. 7.10.

Deviation of Mg from the macrocycle plane has been experimentally observed in magnesium phthalocyanine crystals [131] and an aqua(phthalocyanine)magnesium compound [132]. X-ray single crystals studies indicate that Mg can be significantly displaced from the macrocycle plane and that the deviation increases with decreasing temperature [131]. For the aqua(phthalocyanine)magnesium, deviations of ~0.45 Å of Mg from the N_4 plane were observed [132]. Another related system is ethyl chlorophyllide-a dyhydrate for which the crystal X-ray structure was reported [133]. In this case the displacement of the Mg atom from the N atoms plane is 0.39 Å [133]. Our prediction for chlc$_2$ in liquid methanol (0.39 Å) coincides with this value.

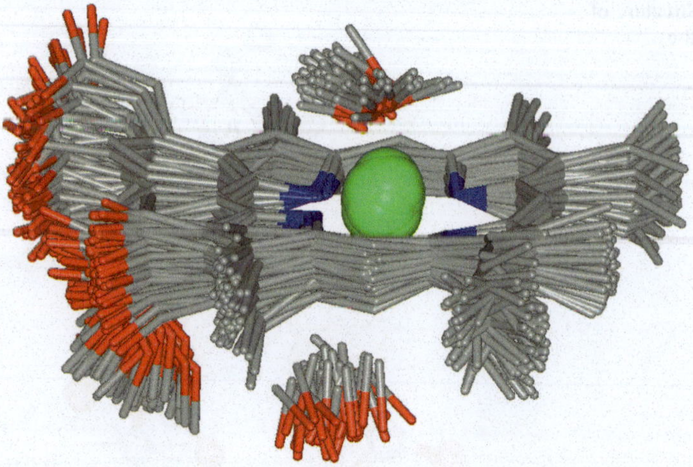

Fig. 7.10 Superposition of configurations from BOMD illustrating the displacement of Mg from the macrocycle plane and the different orientations of the methanol molecules on each side of the macrocycle

The structure of chlc$_2$ in liquid methanol, specifically the coordination of the chromophore metallic center to the methanol molecules can be discussed by representing the radial distribution function [RDF(r)] related to Mg-O interactions. A relevant issue concerns the coordination number of the metalic center. BOMD results for the structure indicate that the coordination number for Mg of chlc$_2$ in liquid methanol is five [114] a feature that seems to be related with the displacement of the metallic center from the macrocycle plane. This is in contrast with the results from force field calculations that indicate a coordination number of six (four nitrogen atoms plus two oxygen atoms). Several works discussed the coordination of the central Mg atom of porphyrins and chlorophylls in different solvents and environments [112, 118, 134–138].

A question that naturally arises concerns the limitations of the available classical force field for describing the interactions between chlc$_2$ and methanol [112]. Taking into consideration the good agreement between the structure predicted by BOMD and some experimental information for the structure and coordination number of the Mg atom, it could be of interest to design a parametrization procedure for the construction of a new classical force field. This procedure could rely on the data for the structure generated by BOMD.

As discussed above, the MC simulation for generating the structures of chlorophyll-c$_2$ in liquid methanol used a rigid geometry for the solute. With the Mg atom localized at the center of the ring the coordination number obtained after the simulation was six methanol molecules. This raises the question of the adequacy of the optimized geometry and the classical force field used in the MC simulation. However, by using the geometry of the solute where the Mg atom is lifted out-of-plane by 0.39 Å, as obtained as the most probable value in the BOMD, and repeating the MC simulation with the same force field and still a rigid solute geometry leads to the coordination of five. The good agreement between the structure predicted by BOMD and the reparametrized force field is illustrated for the Mg-O RDF in Fig. 7.11.

In addition, only mild changes in the calculated absorption spectrum are seen. This suggests that, in this case, adapting the solute to the solvated situation obtained with ab initio dynamics essentially corrects the limitations of the classical force field. However, a more complex situation may arise. In the case of free base phthalocyanine the average and the distribution results obtained from a BOMD for the bond distances, bond angles and torsion angles were used to reparametrize the GROMOS53a6 [139] force field. Preliminary results for the electronic absorption spectrum [140, 141] well reproduced the data from the BOMD for free base phthalocyanine [142].

The electronic absorption spectrum of chlorophyll-c_2 in liquid methanol is presented in Fig. 7.12, where results with only the charge embedding (CE) are compared with results with CE + 20 explicit methanol molecules. By using only the charge embedding the maxima position of the Soret band is at 457 nm in good agreement with the experimental data of Jeffrey and Shibata (451 nm) [128]. The result based on the CE + 20 methanol molecule for the maximum position of the Soret band is 464 nm. Although this result is red shifted by 13 nm relative to the data reported by Jeffrey and Shibata [128], it is in good agreement with the experimental value for chlc_2 in 2-methyl tetrahydrofuran at 77 K (466 nm) [143]. Overall, a good agreement is observed between the present data relying on a combination of BOMD with TDDFT and experimental information thus supporting the adequacy of the present theoretical approach to investigate the electronic properties of complex supramolecular species in solution.

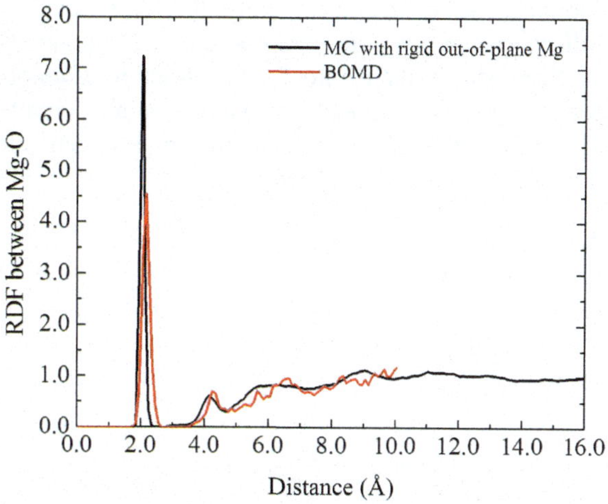

Fig. 7.11 Mg-O radial distribution functions describing the correlation between the oxygen atom of methanol and the Mg atom of chlorophyll-c_2. *Black line* reparametrized force field; *red line* BOMD

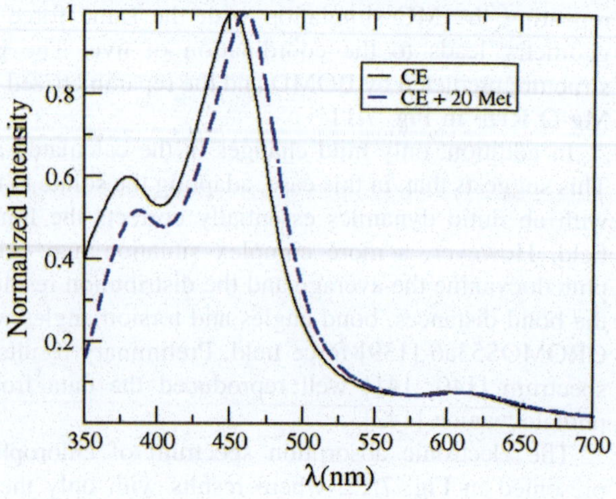

Fig. 7.12 Electronic absorption spectrum of chlorophyll-c_2 in liquid methanol. Calculations with embedding TIP3P charges (CE, *black line*) and CE + 20 explicit methanol molecules (*dashed-blue line*)

7.4 Conclusions and Perspectives

The possibility to carry out high level ab initio studies of electronic properties in liquid phase by coupling many-body energy (MBE) decomposition schemes to configurations generated by statistical mechanical sampling procedures was discussed and some recent applications to hydrogen bond liquids was reviewed. A particularly powerful approach is defined by the combination of MBE schemes with Born-Oppenheimer Molecular Dynamics. This combination overcomes the well known limitations of classical force fields to model the complex interactions in condensed phase and, at the same time, make possible the application of high level ab initio methods to systems with hundreds of atoms. However, MBE schemes should be improved. Specifically, the representation of the embedding by using a self-consistent procedure with polarisable charges should be exploited. In addition, as illustrated by the results for the structure of chlorophyll-c_2 in liquid methanol, a promising route to model very large systems could be based on the parametrization of classical force fields by using information generated by first principles molecular dymamics for smaller sub-systems.

Ackonwledgement The authors gratefully acknowledge support by FCT (Portugal), CNPq, CAPES, FAPESP, INCT-FCx, NAP-FCx, and BioMol (Brazil).

References

1. Tapia O, Goscinsky O (1975) Mol Phys 29:1653
2. Rivail JL, Rinaldi D (1976) Chem Phys 18:233
3. Cramer CJ, Truhlar DG (1999) Chem Rev 99:2161
4. Cramer CJ, Truhlar DG (2008) Acc Chem Res 41:760

5. Tomasi J, Mennucci B, Cammi R (2005) Chem Rev 105:2999
6. Tomasi J (2009) Theor Chem Acc 123:157
7. Warshel A, Levitt M (1976) J Mol Biol 103:227
8. Field MJ, Bash PA, Karplus M (1990) J Comput Chem 11:700
9. Friesner RA, Guallar R (2005) Ann Rev Phys Chem 56:389
10. Senn HM, Thiel W (2009) Angew Chem Int Ed 48:1198
11. Monari A, Rivail JL, Assfeld X (2013) Acc Chem Res 46:596
12. Coutinho K, Canuto S (1997) Adv Quantum Chem 28:89
13. Coutinho K, Canuto S, Zerner MC (2000) J Chem Phys 112:9874
14. Coutinho K, Canuto S (2000) J Chem Phys 113:9132
15. Kitaura K, Ikeo E, Asada T, Nakano T, Uebayasi M (1999) Chem Phys Lett 313:701
16. Fedorov DG, Kitaura K (2005) J Chem Phys 122:54108
17. Zhao Y, Truhlar D (2005) J Chem Theory Comput 1:415
18. Chiba M, Fedorov DG, Kitaura K (2007) J Chem Phys 127:104108
19. Dahlke EE, Truhlar DG (2008) J Chem Theory Comput 3:46
20. Chiba M, Koido T (2010) J Chem Phys 133:044113
21. Fedorov D, Kitaura K (2007) J Phys Chem A 111:6904
22. Fedorov DG, Kitaura K (2009) J Chem Phys 171106:171106
23. Mata RA, Stoll H (2008) Chem Phys Lett 465:136
24. Mata RA, Stoll H, Cabral BJC (2009) J Chem Theory Comput 5:1829
25. Ferreira C, Martiniano HFMC, Cabral BJC, Aquilanti V (2011) Int J Quantum Chem 111:1824
26. Mata RA (2010) Mol Phys 108:381
27. Mata RA, Cabral BJC (2010) Adv Quantum Chem 59:5041
28. Mata RA, Stoll H (2011) J Chem Phys 134:034122
29. Fedorov DG, Kitaura K (2012) J Phys Chem A 116:704
30. Gordon MS, Fedorov DG, Pruitt SR, Slipchenko LV (2012) Chem Rev 112:632
31. Ma Y, Ma H (2013) J Phys Chem A 117:3655
32. Song GL, Li ZH, Fan KN (2013) J Chem Theory Comput 9:1992
33. Yang W (1991) Phys Rev Lett 66:1438
34. Martiniano HFMC, Cabral BJC (2013) Chem Phys Lett 555:119
35. Martiniano HFMC, Galamba N, Cabral BJC (2014) J Chem Phys 140:164511
36. Cabral BJC, Rivelino R, Coutinho K, Canuto S (2015) J Chem Phys 142:024504
37. Assfeld X, Rivail J-L (1996) Chem Phys Lett 263:100
38. Ferre N, Assfeld X, Rivail J-L (2002) J Comput Chem 23:610
39. Kamiya M, Hirata S, Valiev M (2008) J Chem Phys 128:074103
40. Petterson KA, Allen TB, Werner HJ (2008) J Chem Phys 128:084102
41. Jorgensen WL (1986) J Phys Chem 90:1276
42. Jorgensen WL, Maxwell DS, Tirado-Rives J (1996) J Am Chem Soc 118:11225
43. Singh UC, Kollman PA (1984) J Comp Chem 5:129
44. Cheng YC, Fleming GR (2009) Annu Rev Phys Chem 60:241
45. Renger T, May V, Ku O (2001) Phys Rep 343:137
46. Bakker HJ, Planken PCM, Lagendijk A (1990) Nature 347:745
47. Brédy R, Bernard JM, Chen L, Montagne G, Li B, Martin S (2009) J Chem Phys 130:114305
48. Finney JL (2004) Philos Trans R Soc Lond B Biol Sci 359:1145
49. Mishima O, Stanley HE (1998) Nature 396:329
50. Salzmann CG, Radaelli PG, Hallbrucker A, Mayer E, Finney JL (2006) Science 311:1758
51. Soper AK, Weckström K (2006) Biophys Chem 124:180
52. Mancinelli R, Botti A, Bruni F, Ricci MA, Soper AK (2007) J Phys Chem B 111:13570
53. Galamba N (2012) J Phys Chem B 116:5242
54. Mateus MPS, Galamba N, Cabral BJC (2012) J Chem Phys 136:014507
55. Galamba N (2013) J Phys Chem B 117:2153
56. Reiter GF, Kolesnikov AI, Paddison SJ, Platzman PM, Moravsky AP, Adams MA, Mayers J (2012) Phys Rev B 85:045403

57. Reiter GF, Deb A, Sakurai Y (2013) Phys Rev Lett 111:036803
58. Heller JM, Hamm RN, Birkhoff RD, Painter LR (1974) J Chem Phys 60:3483
59. Hayashi H, Watanabe N, Udagawa Y, Kao CC (1998) J Chem Phys 108:823
60. Marin TW, Takahashi K, Bartels DM (2006) J Chem Phys 125:104314
61. Hahn P, Schmidt W, Seino K, Preuss M, Bechstedt F, Bernholc J (2005) Phys Rev Lett 94:037404
62. Aschi M, D'Abramo M, Di Teodoro C, Di Nola A, Amadei A (2005) ChemPhysChem 6:53
63. Garbuio V, Cascella M, Reining L, Sole R, Pulci O (2006) Phys Rev Lett 97:137402
64. Cabral do Couto P, Cabral BJC (2007) J Chem Phys 126:014509
65. D'Abramo M, Di Nola A, Aschi M, Amadei A (2008) J Chem Phys 128:021103
66. Brancato G, Rega N, Barone V (2008) Phys Rev Lett 100:107401
67. Lu D, Gygi F, Galli G (2008) Phys Rev Lett 100:147601
68. Hermann A, Schmidt WG, Schwerdtfeger P (2008) Phys Rev Lett 100:207403
69. Cabral do Couto P, Chipman DM (2010) J Chem Phys 132:244307
70. Cabral do Couto P, Chipman DM (2012) J Chem Phys 137:184301
71. Mata RA, Cabral BJC, Millot C, Coutinho K, Canuto S (2009) J Chem Phys 130:014505
72. Becke AD (1993) J Chem Phys 98:564
73. Lee C, Yang W, Parr RG (1988) Phys Rev B 37:785
74. Grimme S (2006) J Comput Chem 27:1787
75. Grimme S, Antony J, Ehrlich S, Krieg H (2010) J Chem Phys 132:154104
76. Risthaus T, Grimme S (2013) J Chem Theory Comput 9:1580
77. Lippert G, Hutter J, Parrinello M (1997) Mol Phys 92:477
78. VandeVondele J, Krack M, Mohemed F, Parrinello M, Chassaing T, Hutter J (2005) Comput Phys Commun 167:103
79. Goedecker S, Teter M, Hutter J (1996) Phys Rev B 54:1703
80. Soper AK (2013) ISRN Physical Chemistry 2013. Article ID 279463 http://dx.doi.org/10.1155/2013/279463
81. Santra B, Michaelides A, Scheer M (2009) J Chem Phys 131:124509
82. Lin I-C, Seitsonen AP, Coutinho-Neto MD, Tavernelli I, Rothlisberger U (2009) J Phys Chem B 113:1127
83. Jonchiere R, Seitsonen AP, Ferlat G, Saitta AM, Vuilleumier R (2011) J. Chem. Phys. 135:154503
84. Yoo S, Xantheas SS (2011) J Chem Phys 134:121105
85. Lin I-C, Seitsonen AP, Tavernelli I, Rothlisberger U (2012) J Chem Theory Comput 8:3902
86. Ma Z, Zhang Y, Tuckerman ME (2012) J Chem Phys 137:044506
87. Stanton JF, Bartlett RJ (1994) J Chem Phys 100:4695
88. Woon DE, Dunning TH Jr (1993) J Chem Phys 98:1358
89. Galamba N, Cabral BJC (2008) J Am Chem Soc 130:17955
90. Esposti C, Cazzoli G, Damiani D, Favero P, Strumia F (1998) Infrared Phys 28:21
91. Matthews CN (1991) Orig Life Evol Biosph 21:421
92. Mamajanov I, Herzfeld J (2009) J Chem Phys 130:134503
93. Shelton DP (2005) J Chem Phys 123:084502
94. Shelton DP (2008) J Chem Phys 129:134501
95. Cabral BJC (2000) J Chem Phys 112:4351
96. King BF, Farrar TC, Weinhold F (1995) J Chem Phys 103:348
97. Stone AJ, Buckingham AD, Fowler PW (1997) J Chem Phys 107:1030
98. Cabaleiro-Lago EM, Ros MA (1998) J Chem Phys 109:8398
99. Rivelino R, Chaudhuri P, Canuto S (2003) J Chem Phys 118:10593
100. Sanchez M, Provasi PF, Aucar GA, Alkorta I, Elguero J (2005) J Phys Chem B 109:18189
101. Provasi PF, Aucar GA, Sanchez M, Alkorta I, Elguero J, Sauer SPA (2005) J Phys Chem A 109:6555
102. Adrian-Scotto M, Vasilescu D (2007) J Mol Struct (Theochem) 803:45
103. Kotdawala RR, Ozgur Y, Kazantzis N, Thompson RW (2007) Mol Sim 33:843

104. Blankenship R (2014) Molecular Mechanisms of Photosynthesis. Wiley-Blackwell, New York
105. König C, Neugebauer J (2012) ChemPhysChem 13:386
106. Seely G, Jensen R (1965) Spectrochim Acta 21:1835
107. Agostiano A, Cosma P, Trotta M, Monsu L, Micali N (2002) J Phys Chem B 106:12820
108. Ellervee A, Linnanto J, Freiberg A (2004) Chem Phys Lett 394:80
109. Linnanto J, Korppi-Tommola J (2006) Phys Chem Chem Phys 8:663
110. Renger T, Grundkötter B, Madjet MEA, Müh F (2008) Proc Natl Acad Sci USA 105:13235
111. Karki K, Roccatano D (2011) J Chem Theory Comput 7:1131
112. Jaramillo P, Coutinho K, Cabral BJC, Canuto S (2011) Chem Phys Lett 516:250
113. Jaramillo P, Coutinho K, Cabral BJC, Canuto S (2012) Chem Phys Lett 546:67
114. Cabral BJC, Coutinho K, Canuto S (2014) J Chem Phys 138:225102
115. Bernas A, Grand D, Hautecloque S, Myasoedova T (1984) Chem Phys Lett 104:105
116. Pandey A, Datta SN (2005) J Phys Chem B 109:9066
117. Ben Fredj A, Ben Lakhdar Z, Ruiz-López MF (2008) Chem Commun 14:718
118. Ben Fredj A, Ben Lakhdar Z, Ruiz-López MF (2009) Chem Phys Lett 472:243
119. Ben Fredj A, Ruiz-López MF (2010) J Phys Chem B 114:681
120. Uyeda G, Williams JC, Roman M, Mattioli TA, Allen JP (2010) Biochemistry 49:1146
121. Yamijala SRKCS, Periyasamy G, Pati SK (2011) J Phys Chem A 115:12298
122. Rutkowska-Zbik D, Korona T (2012) J Chem Theory Comput 8:2972
123. Curutchet C, Muñoz-Losa A, Monti S, Kongsted J, Scholes GD, Mennucci B (2009) J Chem Theory Comput 5:1838
124. Curutchet C, Kongsted J, Muñoz-Losa A, Hossein-Nejad H, Scholes GD, Mennucci B (2011) J Am Chem Soc 133:3078
125. Chen M, Blankenship RE (2011) Trends Plant Sci 16:427
126. Horigome D, Satoh H, Itoh N, Mitsunaga K, Oonishi I, Nakagawa A, Uchida A (2007) J Biol Chem 282:6525
127. Miyatake T, Tamiaki H (2010) Coord Chem Rev 254:2593
128. Jeffrey S (1969) Biochim Biophys Acta Gen Subj 177:456
129. Marx D, Hutter J (2009) Ab Initio molecular dynamics: basic theory and advanced methods. Cambridge University Press, Cambridge
130. Perdew JP, Burke K, Ernzerhof M (1996) Phys Rev Lett 77:3865
131. Janczak J, Kubiak R (2001) Polyhedron 20:2901
132. Janczak J, Idemori YM (2002) Acta Crystallogr Sect C: Cryst Struct Commun 58:m549
133. Chow HC, Serlin R, Strouse CE (1975) J Am Chem Soc 97:7230
134. Timkovich R, Tulinsky A (1969) J Am Chem Soc 91:4430
135. Umetsu M (1999) Biochim Biophys Acta 1410:19
136. Oba T, Tamiaki H (2002) Photosynth Res 74:1
137. Oba T, Tamiaki H (2005) Bioorg Med Chem 13:5733
138. Heimdal J, Jensen KP, Devarajan A, Ryde U (2007) J Biol Inorg Chem 12:49
139. Oostenbrink C, Villa A, Mark AE, van Gunsteren WF (2004) J Comput Chem 25:1656
140. Cruzeiro VWD (2014) MSc Dissertation, Instituto de Fsica, Universidade de So Paulo
141. Cruzeiro VWD, Cabral BJC, Coutinho K, Canuto S (2014) Chem Phys Lett 595:97–102 (in preparation)
142. Cabral BJC, Cruzeiro VWD, Coutinho K, Canuto S (2014) Chem Phys Lett 596–596:97
143. Niedzwiedzki DM, Jiang J, Lo CS, Blankenship RE (2014) Photosynth Res 120:1–2 (in press)

Chapter 8
Free Energy Gradient Method and Its Recent Related Developments: Free Energy Optimization and Vibrational Frequency Analysis in Solution

Yukichi Kitamura, Norio Takenaka, Yoshiyuki Koyano and Masataka Nagaoka

Abstract To obtain stable states (SS) and transition states (TS) of chemical reaction system in condensed state at a finite temperature, the free energy gradient (FEG) method was proposed in 1998 as an optimization method on a multidimensional free energy surface (FES). This is analogous to the method for the Born Oppenheimer potential energy surface (PES) considered by ab initio molecular orbital (MO) calculation, and utilizes the force and Hessian on the FES with respect to the coordinates of a solute molecule, which can be adiabatically calculated by molecular dynamics (MD) method. In this chapter, we reviewed the FEG methodology that is the method for estimating molecular properties based on the free energy (FE) landscape in condensed state and also discussed a future perspective for the improvement and the extension of the theoretical methods. We believe that a family of the FEG methodologies should become more efficient as one promising strategic setting and will play important roles to survey condensed state chemistry on the basis of recent supercomputing technology.

List of Acronyms

ASEC	Aaveraged Solvent Electrostatic Potential
ASEP	Average Solvent Electrostatic Configuration
BFGS	Broyden-Fletcher-Goldfarb-Shanno
CASSCF	Complete Active Space Self-Consistent Field
CIS	Configuration Interaction Singles

Y. Kitamura · N. Takenaka · Y. Koyano · M. Nagaoka (✉)
Graduate School of Information Science, Nagoya University, Chikusa-Ku, Nagoya 464-8601, Japan
e-mail: mnagaoka@is.nagoya-u.ac.jp

M. Nagaoka
Core Research for Evolutional Science and Technology, Japan Science and Technology Agency, Honmachi, Kawaguchi 332-0012, Japan

N. Takenaka · M. Nagaoka
ESICB, Kyoto University, Kyodai Katsura, Nishikyo-ku, Kyoto 615-8520, Japan

J.-L. Rivail et al. (eds.), *Quantum Modeling of Complex Molecular Systems*, Challenges and Advances in Computational Chemistry and Physics 21,
DOI 10.1007/978-3-319-21626-3_8

CM	Classical Mechanical
COSMO	COnductor-like Screening MOdel
CPCM	Conductor-like Polarizable Continuum model
DCM	Dielectric Continuum Model
DFT	Density Functional Theory
DOF	Degrees Of Freedom
DOS	Density Of States
ER	Energy Representation
est	Electrostatic
EVB	Empirical Valence Bond
FE	Free Energy
FEG	Free Energy Gradient
FEH	Free Energy Hessian
FEP	Free Energy Perturbation
FES	Free Energy Surface
FMO	Fragment Molecular Orbital
FTIR	Fourier Transform InfraRed
HF	Hartree-Fock
ICN	Integrated Coordination Number
IFIE	Inter-Fragment Interaction Energy
INM	Instantaneous Normal Mode
IR	InfraRed
LJ	Lennard-Jones
MC	Monte Carlo
MD	Molecular Dynamics
MFA	Mean Field Approximation
MO	Molecular Orbital
MP2	Second-order Moller-Plesset perturbation theory
mw	Mass-weighted
NAM	Number-Adaptive Multiscale
NEB	Nudged Elastic Band
NF	Neutral-Form
PE	Potential Energy
PES	Potential Energy Surface
PPII	Polyproline II
PS	Product-State
QM	Quantum Mechanical
QM/MM	Quantum Mechanical/Molecular Mechanical Method
RDF	Radial Distribution Function
Red moon	Rare event-driving mechanism of our necessity
RS	Reactant-State
SCF	Self-Consistent Field
SD	StanDard
SEI	Solid Electrolyte Interphase

SS	Stable State
TS	Transition State
vdW	Van der Waals
VFA	Vibrational Frequency Analysis
ZW	ZWitterion

8.1 Introduction

Today, it is indispensable and common in modern chemistry to deal with molecules as the quantum systems that consist of a couple of classical mechanical (CM) nuclei and quantum mechanical (QM) electrons, for understanding chemical phenomena deeply. Such QM approaches can provide us the microscopic information such as the structural information (e.g. stable state (SS) and transition state (TS)) and chemical properties (e.g. electric or magnetic external fields and internal perturbations such as a nuclear or electron spin) of chemical reaction systems. However, from the point of view of computational efforts, it remains difficult to directly apply the QM approaches to large reaction systems such as the solution (or biological) ones that we are interested in. Thus, to treat these whole reaction systems in solution and biological environment, it is very useful in many cases to employ a multiscale model such as the quantum mechanical/molecular mechanical (QM/MM) methods, which are often combined with molecular dynamics (MD) or Monte Carlo (MC).

Nevertheless, the potential energy surfaces (PESs) for the condensed system are too high dimensional, and then it is almost impossible to search even SS of solute molecule with respect to all degrees of freedom for the whole system. Therefore, the statistical treatments have been utilized to represent the effective properties per a single molecule under the experimental conditions. In particular, the free energy of the solute system is one of important thermodynamic quantities and is related to a number of chemical properties, i.e., a reaction rate constant, equilibrium constant and so on. So far, the free energy surface (FES) has been often estimated approximately along only a certain reaction coordinate since FES makes it possible to consider effectively a vast number of solvent molecules. Given such concepts, the Free Energy Gradient (FEG) methodology has been proposed as SS and TS optimization method in condensed systems [1–3]. In this methodology, we consider the gradient on the multidimensional FES with respect to the coordinates of a reactive part (usually solute molecule) to solve this problem. The contributions of the degrees of freedom of a remainder part (usually solvent molecules) are included in the terms of the ensemble-averaged ones under a certain thermodynamic environment. If we know the first and second derivatives on the FES, i.e., the force and Hessian, the "reaction ergodography" [4, 5] would become possible in solution [1–3, 6–11], and could be appropriately utilized to both the optimization problems on the FES and researches of spectroscopic characteristics, i.e., the searching of equilibrium structures of solute molecules in solution and vibrational (IR and Raman) spectra in solution.

In this chapter, we review the essential aspects of the FEG methodology and some applications to clarify the solvent effects on the molecular properties (e.g. structure and spectroscopy) in condensed systems. In addition, we briefly review the theoretical methods related to the FEG methodology to improve and extend them for the solution chemical reaction system.

8.2 Methodology

In this section, starting with concise description of the standard QM/MM treatment, we have defined the derivatives on FES as statistical averages using a number of QM/MM-MD trajectories, i.e., free energy gradient (FEG) and free energy hessian (FEH). Then, we introduce and discuss the original and basic treatment to find any stationary states on the FES, i.e. equilibrium molecular structures where the first derivatives with respect to them should be zero. We also touch on a simple optimization scheme using the quasi-Newton method and a reaction path search connecting between SS and TS. Further, we introduce the dual approach to vibrational frequency analysis (VFA) using two kinds of "FEH" matrices. Finally, we have also proposed a protocol to theoretically explore the chemical reactions in solution.

8.2.1 The Quantum Mechanical/Molecular Mechanical (QM/MM) Method

The hybrid Quantum Mechanical/Molecular Mechanical (QM/MM) approach is one of multiscale models for complex chemical reactions in solution and in proteins. In the QM/MM method, it is common that only the reactive parts in the whole solution reaction system are treated quantum-mechanically, while the other parts are molecular-mechanically [12–16]. The effective Hamiltonian $\hat{H}$ of the whole system consists of three terms:

$$\hat{H} = \hat{H}_{\mathrm{QM}} + \hat{H}_{\mathrm{MM}} + \hat{H}_{\mathrm{QM/MM}}, \tag{8.1}$$

where the first two terms $\hat{H}_{\mathrm{QM}}$ and $\hat{H}_{\mathrm{MM}}$ stand for the standard Hamiltonian of the QM and MM systems, respectively, while the last QM/MM term $\hat{H}_{\mathrm{QM/MM}}$ represents the interaction between the QM and MM system, and is expressed as a sum of electrostatic (est) and van der Waals (vdW) terms:

$$\hat{H}_{\mathrm{QM/MM}} = \hat{H}^{\mathrm{est}}_{\mathrm{QM/MM}} + \hat{H}^{\mathrm{vdW}}_{\mathrm{QM/MM}}. \tag{8.2}$$

The first term $\hat{H}^{\mathrm{est}}_{\mathrm{QM/MM}}$ is given by

$$\hat{H}^{\text{est}}_{\text{QM/MM}} = \sum_M q_M V_{\text{QM}}(\mathbf{R}_M), \tag{8.3}$$

with the potential

$$V_{\text{QM}}(\mathbf{R}_M) = -\sum_x \frac{1}{R'_{sM}} + \sum_A \frac{Z_A}{R_{AM}}, \tag{8.4}$$

which is represented as the sum of coulomb interactions among the electrons and cores of QM atoms [17]. In these expressions, q_M is the atomic point electric charge on the M-th MM atom in solvent molecules, whose position is denoted by $\mathbf{R}_M$, R'_{sM} is the distance between the s-th QM electron of the solute molecule and the M-th MM atom, Z_A is the core charge of the A-th QM atom, and R_{AM} is the distance between the A-th QM atom and the M-th MM one. The vdW term $\hat{H}^{\text{vdW}}_{\text{QM/MM}}$, which is included to introduce effectively quantum dispersion and repulsion, is usually expressed using a number of 6–12 Lennard-Jones (LJ) functions:

$$\hat{H}^{\text{vdW}}_{\text{QM/MM}} = \sum_A \sum_M \varepsilon_{AM} \left[\left(\frac{r_{AM}}{R_{AM}} \right)^{12} - 2 \left(\frac{r_{AM}}{R_{AM}} \right)^{6} \right], \tag{8.5}$$

where ε_{AM} and r_{AM} are a couple of LJ parameters for the A-th QM atom interacting with the M-th MM atom. The potential energy V of the total system is then obtained by

$$V = \langle \Psi | \hat{H}_{\text{QM}} + \hat{H}_{\text{QM/MM}} | \Psi \rangle + V_{\text{MM}} = V_{\text{SB}} + V_{\text{MM}}. \tag{8.6}$$

Once an effective Hamiltonian for the system has been defined, the wave function Ψ and energy can be evaluated by minimizing the V_{SB} in Eq. (8.6) with respect to the molecular orbital coefficients of the ground state wave function using a self-consistent (SCF) procedure to solve the effective Schrödinger equation.

Until now, our laboratory has been developing a minimal QM/MM implementation combining AMBER with Gaussian [18, 19] or PAICS [20] for the purpose of performing ab initio QM/MM-MD simulation.

8.2.2 *Analytical Force and Hessian on Free Energy Surface (FES)*

The force (vector) on the FES, i.e., the potential surface of mean force, can be calculated by time-averaging the instantaneous forces acting on each atom of a solute molecule over the equilibrium distribution for all solvent molecules. Furthermore, we can also evaluate the Hessian matrix on the FES, i.e., the second derivative matrix with respect to the solute coordinates on the FES.

The forces on the FES are defined as follows:

$$\mathbf{F}^{\mathrm{FE}}(\mathbf{q}) = -\frac{\partial G(\mathbf{q})}{\partial \mathbf{q}} = -\left\langle \frac{\partial V_{\mathrm{SB}}(\mathbf{q})}{\partial \mathbf{q}} \right\rangle_{\mathrm{B},\mathbf{q}}, \tag{8.7}$$

where $\mathbf{F}^{\mathrm{FE}}$ ($\mathbf{q}$) is a force vector on the FES, $\mathbf{q}$ denotes a vector of the solute mass-weighted (mw) coordinates, $G(\mathbf{q})$ is a free energy (FE) function of the solute at a structure $\mathbf{q}$, V_{SB} is the sum of both the potential energy of the solute and the interaction energy between the solute and solvent molecules. The brackets $<\cdots>_{\mathrm{B},\mathbf{q}}$ denote the time average along an equilibrium molecular dynamics (MD) trajectory constraining the solute molecular structure $\mathbf{q}$, which is equal to the ensemble average by the equilibrium distribution, i.e., the multiple integral with respect to the degrees of freedom (DOF) of the whole solution minus the solute internal DOF. In the MD simulation, it is usual that the forces acting against each atom of the solute molecule by all solvent molecules are always calculated every time step. We can, therefore, evaluate the force vector on the FES (Eq. (8.7)) without any additional effort in the MD method combined with the QM/MM methodology (i.e., the QM/MM-MD method).

Similarly, the Hessian matrix on the FES ("FE-Hessian") is obtained by

$$\mathbf{H}^{\mathrm{FE}}(\mathbf{q}) = \frac{\partial^2 G(\mathbf{q})}{\partial \mathbf{q} \partial \mathbf{q}} \tag{8.8}$$

$$= \left\langle \frac{\partial^2 V_{\mathrm{SB}}(\mathbf{q})}{\partial \mathbf{q}\partial \mathbf{q}} \right\rangle_{\mathrm{B},\mathbf{q}} - \beta \left[\left\langle \frac{\partial V_{\mathrm{SB}}(\mathbf{q})}{\partial \mathbf{q}} \frac{\partial V_{\mathrm{SB}}^{\mathrm{T}}(\mathbf{q})}{\partial \mathbf{q}} \right\rangle_{\mathrm{B},\mathbf{q}} - \left\langle \frac{\partial V_{\mathrm{SB}}(\mathbf{q})}{\partial \mathbf{q}} \right\rangle \left\langle \frac{\partial V_{\mathrm{SB}}(\mathbf{q})}{\partial \mathbf{q}} \right\rangle^{\mathrm{T}}_{\mathrm{B},\mathbf{q}} \right], \tag{8.9}$$

where β means $1/k_{\mathrm{B}}T$, and the "superscript" T denotes the transposition. While the second and third terms in the RHS of Eq. (8.9) can be calculated out of those quantities usually evaluated in MD simulations, the first term evaluation needs an additional procedure where a number of the "instantaneous" second derivatives on PES at the solute structure $\mathbf{q}$ should be evaluated including the "instantaneous" solvent influence [1–3]. The approximation to neglect the second and third terms of the force fluctuation in Eq. (8.9), which was originally introduced in the averaged solvent electrostatic potential (ASEP)/MD [21, 22] treatment and also assumed in the average solvent electrostatic configuration (ASEC) [23, 24], was often adopted in the free energetic studies. Then, the approximate FE-Hessian is obtained in the QM/MM formalism (Eqs. (8.1)–(8.6)),

$$\mathbf{H}^{\mathrm{FE}} = \left\langle \frac{\partial^2 V_{\mathrm{SB}}(\mathbf{q})}{\partial \mathbf{q}\partial \mathbf{q}} \right\rangle_{\mathrm{B},\mathbf{q}}, \tag{8.10}$$

thereby, the FE-Hessian matrix $\mathbf{H}^{\mathrm{FE}}$ is obtained as an equilibrium ensemble average of instantaneous Hessian matrices at a given solute conformation in solution, i.e., $\mathbf{q}$.

Thus, each matrix element of $\mathbf{H}^{\mathrm{FE}}$, which is a $3N \times 3N$ matrix containing all the second derivatives of potential energy V with respect to the solute mw Cartesian coordinates $\mathbf{q}$, can be expressed as follows,

$$\frac{\partial^2 V_{\mathrm{SB}}(\mathbf{q})}{\partial \mathbf{q} \partial \mathbf{q}} = \left(\frac{\partial^2 \langle \Psi | \hat{H}_{\mathrm{QM}} + \hat{H}^{\mathrm{est}}_{\mathrm{QM/MM}} | \Psi \rangle}{\partial q_{\alpha A} \partial q_{\beta B}} + \frac{\partial^2 \hat{H}^{\mathrm{vdW}}_{\mathrm{QM/MM}}}{\partial q_{\alpha A} \partial q_{\beta B}} \right) \tag{8.11a}$$

$$= \begin{pmatrix} \frac{\partial^2 \langle \Psi | \hat{H}_{\mathrm{QM}} + \hat{H}^{\mathrm{est}}_{\mathrm{QM/MM}} | \Psi \rangle}{\partial q_{x1} \partial q_{x1}} + \frac{\partial^2 \hat{H}^{\mathrm{vdW}}_{\mathrm{QM/MM}}}{\partial q_{x1} \partial q_{x1}} & \cdots & \frac{\partial^2 \langle \Psi | \hat{H}_{\mathrm{QM}} + \hat{H}^{\mathrm{est}}_{\mathrm{QM/MM}} | \Psi \rangle}{\partial q_{x1} \partial q_{zN}} + \frac{\partial^2 \hat{H}^{\mathrm{vdW}}_{\mathrm{QM/MM}}}{\partial q_{x1} \partial q_{zN}} \\ \vdots & \ddots & \vdots \\ \frac{\partial^2 \langle \Psi | \hat{H}_{\mathrm{QM}} + \hat{H}^{\mathrm{est}}_{\mathrm{QM/MM}} | \Psi \rangle}{\partial q_{zN} \partial q_{x1}} + \frac{\partial^2 \hat{H}^{\mathrm{vdW}}_{\mathrm{QM/MM}}}{\partial q_{zN} \partial q_{x1}} & \cdots & \frac{\partial^2 \langle \Psi | \hat{H}_{\mathrm{QM}} + \hat{H}^{\mathrm{est}}_{\mathrm{QM/MM}} | \Psi \rangle}{\partial q_{zN} \partial q_{zN}} + \frac{\partial^2 \hat{H}^{\mathrm{vdW}}_{\mathrm{QM/MM}}}{\partial q_{zN} \partial q_{zN}} \end{pmatrix}, \tag{8.11b}$$

with

$$\mathbf{q} = \{q_{\alpha A}\} = \{\sqrt{m_1}x_1, \sqrt{m_1}y_1, \sqrt{m_1}z_1, \cdots, \sqrt{m_A}\alpha_A, \cdots, \sqrt{m_N}x_N, \sqrt{m_N}y_N, \sqrt{m_N}z_N\}, \tag{8.11c}$$

where subscripts A and B are the indices of the QM atoms (which have each integral number from 1 to N), and α and β denote one of x, y and z. m_A is the mass of the A-th QM atom.

Further, it is understood that each matrix element consists of the components originating in the pure QM, the est and the vdW contribution. The est components are conveniently computed by the quantum chemical calculation package. For instance, in GAUSSIAN program [25], several approximate methods of electronic state calculations are available, e.g., the Hartree-Fock (HF), second-order Møller-Plesset perturbation theory (MP2), configuration interaction field (CIS), complete active space self-consistent field (CASSCF) method, and the density functional theory (DFT) methods. On the other hand, since the vdW components are expressed as such analytical functions of the mw Cartesian coordinate variables involved in the same atom ($A = B$) as follows,

$$\frac{\partial^2 \hat{H}^{\mathrm{vdW}}_{\mathrm{QM/MM}}}{\partial q_{\alpha A} \partial q_{\beta A}} = \begin{cases} \sum_M \varepsilon_{AM} \left[24 \left(7 \left(\frac{r_{AM}}{R_{AM}} \right)^{12} - 4 \left(\frac{r_{AM}}{R_{AM}} \right)^{6} \right) \frac{(q_{\alpha A} - q_{\alpha M})^2}{R^4_{AM}} \right. \\ \qquad \left. -12 \left(\left(\frac{r_{AM}}{R_{AM}} \right)^{12} - \left(\frac{r_{AM}}{R_{AM}} \right)^{6} \right) \frac{1}{R^2_{AM}} \right] & (\alpha = \beta), \\ \sum_M \varepsilon_{AM} \left[24 \left(7 \left(\frac{r_{AM}}{R_{AM}} \right)^{12} - 4 \left(\frac{r_{AM}}{R_{AM}} \right)^{6} \right) \frac{(q_{\alpha A} - q_{\alpha M})(q_{\beta A} - q_{\beta M})}{R^4_{AM}} \right] & (\alpha \neq \beta), \end{cases} \tag{8.12}$$

where M denotes an integral number discriminating the MM atoms, and ε_{AM} and r_{AM} are a couple of LJ parameters for the A-th QM atom interacting with the M-th MM atom, and R_{AM} is the distance between the A-th QM atom and the M-th MM one.

8.2.2.1 Structural Optimization on FES

By utilizing forces on FES, we can identify the SS and TS structures in solution with full optimization with respect to all coordinates of the solute molecules. For example, if we adopt the quasi-Newton method with the following Broyden-Fletcher-Goldfarb-Shanno (BFGS) procedure [26–29] for structural optimization scheme in the FEG method, the $(i+1)$-th reactant structure $\mathbf{q}_{i+1}$ is taken as,

$$\mathbf{q}_{i+1} = \mathbf{q}_i + \Delta\mathbf{q}_i = \mathbf{q}_i - \alpha_i \mathbf{H}_i^{-1}\mathbf{F}^{\mathrm{FE}}(\mathbf{q}_i), \tag{8.13}$$

where α_i is the step length and the matrix H_i^{-1} is the approximate representation of the inverse Hessian matrix, which is defined as,

$$\mathbf{H}_i^{-1} = \left[\mathbf{I} - \frac{\mathbf{s}_i\mathbf{y}_i^{\mathrm{T}}}{\mathbf{y}_i^{\mathrm{T}}\mathbf{s}_i}\right]\mathbf{H}_{i-1}^{-1}\left[\mathbf{I} - \frac{\mathbf{y}_i\mathbf{s}_i^{\mathrm{T}}}{\mathbf{y}_i^{\mathrm{T}}\mathbf{s}_i}\right] + \frac{\mathbf{s}_i\mathbf{s}_i^{\mathrm{T}}}{\mathbf{y}_i^{\mathrm{T}}\mathbf{s}_i}, \tag{8.14}$$

where,

$$\mathbf{s}_i = \mathbf{q}_i - \mathbf{q}_{i-1}, \tag{8.15}$$

and

$$\mathbf{y}_i = \mathbf{F}^{\mathrm{FE}}(\mathbf{q}_i) - \mathbf{F}^{\mathrm{FE}}(\mathbf{q}_{i-1}). \tag{8.16}$$

Then, the optimization procedure is executed in the following processes,

(Step 1) Start with the initial structure $\mathbf{q}_k$, with $k = 1$.

(Step 2) For $\mathbf{q}_k$, calculate the forces on the FES $\mathbf{F}^{\mathrm{FE}}(\mathbf{q}_k)$ and the free energy change ΔA_k by using free energy perturbation (FEP) method [30], which is described as follows:

$$\Delta A_k = A_k - A_{k-1} = -k_{\mathrm{B}}T\ln\langle\exp[-\beta\{V_{\mathrm{SB}}(\mathbf{q}_k) - V_{\mathrm{SB}}(\mathbf{q}_{k-1})\}]\rangle_{k-1}, \tag{8.17}$$

(Step 3) Find the stationary structure, displacing $\mathbf{q}_k$ by the amount $\Delta\mathbf{q}_k$ using the forces,

$$\Delta\mathbf{q}_k = \alpha_k \mathbf{H}_k^{-1}\mathbf{F}^{\mathrm{FE}}(\mathbf{q}_k), \tag{8.18}$$

where α_k is a constant. If the force $\mathbf{F}^{\mathrm{FE}}(\mathbf{q}_k)$ is satisfied with the following zero-gradient condition,

$$\mathbf{F}^{\mathrm{FE}}(\mathbf{q}_k) = -\left\langle \frac{\partial V_{\mathrm{SB}}(\mathbf{q}_k)}{\partial \mathbf{q}_k} \right\rangle_{\mathrm{B},\mathbf{q}_k} = 0 \tag{8.19}$$

Then, finish the optimization procedure with the optimized structure $\mathbf{q}_{\mathrm{eq}}^{\mathrm{FE}} = \mathbf{q}_k$. If not,

(Step 4) Set $\mathbf{q}_{k+1} = \mathbf{q}_k + \Delta\mathbf{q}_k$, $k = k + 1$ and return to step 2.

8.2.2.2 TS Structure and Reaction Path Search on FES

By combining the FEG method with the nudged elastic band (NEB) method (FEG-NEB) (FEG-NEB) [11], we can also obtain the minimum energy pathway which connects the reactant-state (RS) and product-state (PS) passing through the transition-state (TS) for an elementary chemical reaction. The NEB method is a widely used interpolation algorithm and optimizes the "chain-of-conformations" $\{\mathbf{q}_j | j = 1, \ldots, N_{\mathrm{node}}\}$ which is a sequence of structural conformations (nodes) generated from two terminal conformations corresponding to RS and PS. To keep the nodes uniformly spaced along the path, the NEB method utilizes the force projections along the path direction and the off-path direction by a spring which is put in between each pair of neighboring nodes. Thus, in the FEG-NEB method, $\mathbf{F}^{\mathrm{FE}}(\mathbf{q}_j)$ are calculated in the following way.

$$\mathbf{F}^{\mathrm{FE}}(\mathbf{q}_j) = \mathbf{F}^{FE,//}(\mathbf{q}_j) + \mathbf{F}^{\mathrm{FE},\perp}(\mathbf{q}_j), \tag{8.20}$$

$$\begin{cases} \mathbf{F}^{FE,//}(\mathbf{q}_j) = k\left(|\mathbf{q}_{j+1} - \mathbf{q}_j| - |\mathbf{q}_j - \mathbf{q}_{j-1}|\right) \cdot (\boldsymbol{\tau}_j \boldsymbol{\tau}_j) \\ \mathbf{F}^{\mathrm{FE},\perp}(\mathbf{q}_j) = -\frac{\partial A(\mathbf{q}_j)}{\partial \mathbf{q}_j}(1 - \boldsymbol{\tau}_j \boldsymbol{\tau}_j) \end{cases}, \tag{8.21}$$

and

$$\boldsymbol{\tau}_j = \frac{\mathbf{q}_{j+1} - \mathbf{q}_j}{|\mathbf{q}_{j+1} - \mathbf{q}_j|}, \tag{8.22}$$

where $\mathbf{F}^{\mathrm{FE},//}(\mathbf{q}_j)$ denotes the parallel component of the force vector on the FES for the j-th conformation at present optimization step, while $\mathbf{F}^{\mathrm{FE},\perp}(\mathbf{q}_j)$ is the perpendicular one. k and $\boldsymbol{\tau}_j$ represent the spring constant and the projection vector to the direction of the reaction coordinate, respectively. For example, according to the same optimization procedure previously mentioned (Eqs. (2.7)–(2.13)), the optimization cycles are repeated self-consistently until all conformations $\{\mathbf{q}_j\}$ are optimized so as to be satisfied independently with the zero gradient conditions for all nodes,

$$\left\{ \mathbf{F}^{\mathrm{FE}}\left(\mathbf{q}_j\right) = 0 \quad \middle| j = 0, \ldots, N_{\mathrm{node}} \right\}. \tag{8.23}$$

8.2.2.3 Dual Approach to Vibrational Frequency Analysis (VFA)

To investigate vibrational properties of solute molecules in solution, we have proposed a new theoretical method as a direct extension of the FEG one, i.e., the "dual" approach to the vibrational frequency analysis (VFA) [31]. By employing the dual VFA approach, we can simultaneously obtain the effective vibrational normal modes and the vibrational spectra in solution, which uses complementarily two kinds of Hessian matrices obtained by the equilibrium QM/MM-MD trajectories, that is, a unique Hessian on the FES (i.e., the FE-Hessian) and a sequence of instantaneous ones (i.e., the instantaneous normal mode Hessians: INM-Hessians). Figure 8.1 shows a schematic chart of the dual VFA approach. First, we execute the QM/MM-MD simulation and collect many solvent conformations around the solute molecule being fixed at **q**, sequentially numbered. Second, we obtain an FE-Hessian as the average of instantaneous Hessian matrices at those conformations and then, by diagonalizing the FE-Hessian (cf. Eq. (8.11a)), we can obtain a set of FE normal coordinates $\{Q_i\}$ and FE vibrational frequencies $\{\omega_i\}$ of the solute molecule in solution.

The effective vibrational frequencies $\{\omega_i^{\mathrm{FE}}\}$ are estimated by diagonalizing the FE-Hessian matrix $\mathbf{H}^{\mathrm{FE}}$ (i.e., Eq. (8.10)),

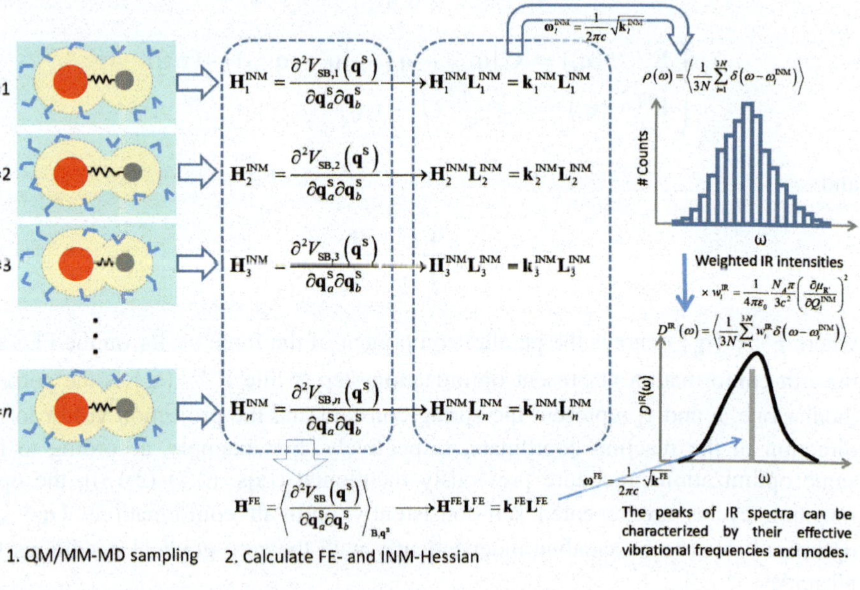

Fig. 8.1 Schematic flow chart of the dual approach

$$\mathbf{U}^{-1}\mathbf{H}^{\mathrm{FE}}\mathbf{U} = \mathbf{k}^{\mathrm{FE}}\mathbf{I} = \left[k_1^{\mathrm{FE}}\, k_2^{\mathrm{FE}} \cdots k_{3N}^{\mathrm{FE}}\right] \cdot \mathbf{I}, \tag{8.24}$$

where $\mathbf{U}$ and $\mathbf{I}$ are 3 N-dimentional unitary and identity matrices for an N-atomic solute molecule, respectively. $\mathbf{k}^{\mathrm{FE}}$ is a diagonal matrix consisting of the eigenvalues $\{k_i\}$ of $\mathbf{H}^{\mathrm{FE}}$ and is related with a matrix of effective vibrational frequencies

$$\boldsymbol{\omega}^{\mathrm{FE}} = \left[\omega_i^{\mathrm{FE}}\right] = \left[\frac{1}{2\pi c}\sqrt{k_i^{\mathrm{FE}}}\right], \tag{8.25a}$$

with the speed of light c. The matrix of the corresponding vibrational normal coordinates is expressed as follows,

$$\mathbf{Q} \equiv [Q_i] = \left[\sum_{A=1}^{N}\sum_{\alpha}^{x,y,z} u_{i,A\alpha}\Delta q_{A\alpha}\right] = \mathbf{U}\Delta\mathbf{q}, \tag{8.25b}$$

where $\Delta\mathbf{q}$ is the mw Cartesian displacement vector. Further, a set of 3 N vibrational vectors $\mathbf{L}^{\mathrm{FE}}$ can be also obtained by

$$\mathbf{H}^{\mathrm{FE}}\mathbf{L}^{\mathrm{FE}} = \mathbf{k}^{\mathrm{FE}}\mathbf{L}^{\mathrm{FE}}; \quad \mathbf{L}^{\mathrm{FE}} \equiv \left[\mathbf{L}_1^{\mathrm{FE}}\,\mathbf{L}_2^{\mathrm{FE}} \cdots \mathbf{L}_{3N}^{\mathrm{FE}}\right]. \tag{8.25c}$$

On the other hand, one can calculate the approximate vibrational spectra via the instantaneous normal mode (INM) analysis [32–35]. The l-th INMs are defined by diagonalizing an instantaneous Hessian ("INM-Hessian") matrix at the l-th solvent conformation $\mathbf{R}^l = \{\mathbf{R}_M\}^l$ in the sequence of conformations. The corresponding l-th eigenvalues $\left\{k_{i,l}^{\mathrm{INM}}\right\}$ $(i = 1, \cdots, 3N)$ are obtained by

$$\mathbf{k}_l^{\mathrm{INM}}\mathbf{I} = \left[k_{i,l}^{\mathrm{INM}}\right] \cdot \mathbf{I} = \mathbf{U}_{\mathbf{R}^l}^{-1}\mathbf{H}^{\mathrm{INM}}\mathbf{U}_{\mathbf{R}^l}, \tag{8.26}$$

with

$$\mathbf{H}_l^{\mathrm{INM}} = \left.\frac{\partial^2 V_{\mathrm{SB}}(\mathbf{q})}{\partial\mathbf{q}\partial\mathbf{q}}\right|_{\mathbf{R}^l}, \tag{8.27}$$

and

$$\boldsymbol{\omega}_l^{\mathrm{INM}} = \left[\omega_{i,l}^{\mathrm{INM}}\right] = \left[\frac{1}{2\pi c}\sqrt{k_{i,l}^{\mathrm{INM}}}\right]; \quad \mathbf{H}_l^{\mathrm{INM}}\mathbf{L}_l^{\mathrm{INM}} = \mathbf{k}_l^{\mathrm{INM}}\mathbf{L}_l^{\mathrm{INM}}. \tag{8.28}$$

To characterize the ensemble-averaged vibrational frequency distribution, the vibrational INM density of states (DOS) are defined as follows;

$$\rho(\omega) = \left\langle \frac{1}{3N - N_{\rm m}} \sum_{i=1}^{3N-N_{\rm m}} \delta\left(\omega - \omega_i^{\rm INM}\right) \right\rangle_{\rm B,q} \tag{8.29a}$$

$$= \frac{1}{L(3N - N_{\rm m})} \sum_{l=1}^{L} \sum_{i=1}^{3N-N_{\rm m}} \delta\left(\omega - \omega_{i,l}^{\rm INM}\right). \tag{8.29b}$$

Here, $N_{\rm m}$ shows the number of normal modes with imaginary frequencies and L is the number of solvent conformations considered. In addition, to obtain infrared (IR) spectra, we define the vibrational INM DOS weighted by the dipole moment derivatives $\partial\mu_l/\partial Q_{i,l}^{\rm INM}$,

$$D^{\rm IR}(\omega) = \left\langle \frac{1}{3N - N_{\rm m}} \sum_{i=1}^{3N-N_{\rm m}} \delta\left(\omega - \omega_i^{\rm INM}\right) w_i^{\rm IR} \right\rangle_{\rm B,q}, \tag{8.30a}$$

$$= \frac{1}{L(3N - N_{\rm m})} \sum_{l=1}^{L} \sum_{i=1}^{3N-N_{\rm m}} \delta\left(\omega - \omega_{i,l}^{\rm INM}\right) w_{i,l}^{\rm IR}, \tag{8.30b}$$

with

$$w_{i,l}^{\rm IR} = \frac{1}{4\pi\varepsilon_0} \frac{N_{\rm A}\pi}{3c^2} \left(\frac{\partial\mu_l}{\partial Q_{i,l}^{\rm INM}} \right)^2, \tag{8.31}$$

where ε_0, $N_{\rm A}$ and μ_l are vacuum permittivity, Avogadro's number and the net dipole moment of the solute molecule at an instantaneous solvent conformation $\mathbf{R}^l$. Moreover, we can estimate the ensemble-averaged IR intensities $I_{\rm aq}$ in solution to extract the contribution of the absorption coefficient for IR spectra from the total spectral line as follows,

$$I_{\rm aq}(\omega) = \frac{D^{\rm IR}(\omega)}{\rho(\omega)} = \frac{\left\langle \frac{1}{3N-N_{\rm m}} \sum_{l=1}^{3N-N_{\rm m}} \delta\left(\omega - \omega_i^{\rm INM}\right) w_i^{\rm IR} \right\rangle_{\rm B,q}}{\left\langle \frac{1}{3N-N_{\rm m}} \sum_{i=1}^{3N-N_{\rm m}} \delta\left(\omega - \omega_i^{\rm INM}\right) \right\rangle_{\rm B,q}}. \tag{8.32}$$

8.2.3 Research Protocol for Exploring FES in Solution

For isolated molecules, when their molecular properties are investigated by quantum chemical (QM) calculations, we optimize initially its molecular structure and

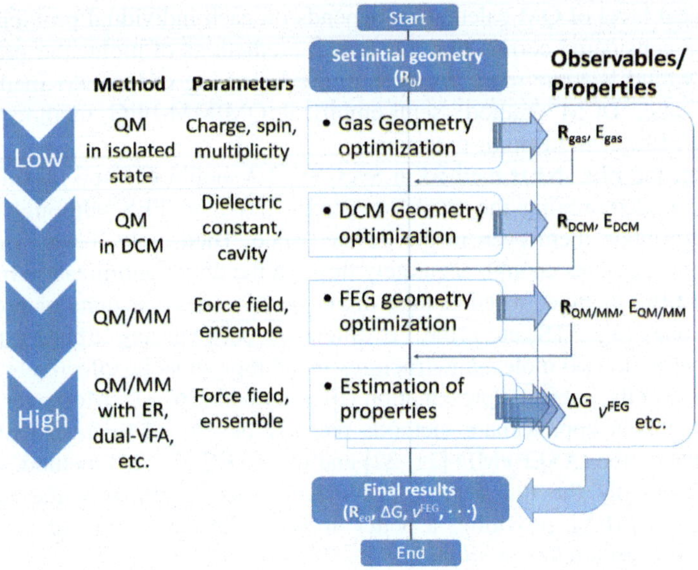

Fig. 8.2 Schematic flow chart of the calculation protocol for exploring the molecular properties of solute molecule in solution

then execute additional calculation to estimate the specific properties that we want to know, e.g., the electron density analysis, the vibrational frequency analysis and so on. Moreover, it is usual for us to apply the higher-level theoretical method step by step, if further improvement is required in describing their electronic states in chemical accuracy.

Also, in principle, the procedure should be the same for those molecules in solution. Figure 8.2 shows the schematic flow chart of the calculation protocol for exploring the molecular properties of solute molecules in solution. First, we would execute the geometry optimization of the solute molecule *in the isolated state*, i.e., gas geometry optimization. Then, the DCM (dielectric continuum model) geometry optimization is to be executed by assuming a dielectric constant for the solvent, starting from the final structure obtained in advance by the gas geometry optimization as the initial guess geometry, i.e., a prior treatment in solution with a lower theoretical level. Finally, the FEG geometry optimization is executed by adopting the structure obtained by the DCM geometry optimization as the initial guess geometry. Hence, we can obtain efficiently more accurate stable molecular structures in solution. In particular, this step-by-step treatment is so important and economical because the FEG method is necessarily time-consuming and then expensive due to the necessity to collect many solvent conformations in estimating the force on FES. As we need more QM calculation, such treatment would become more serious and important.

After those stable molecular structures are optimized by the FEG method, we estimate their molecular properties of our interest. Truly, the selection of a theoretical level of QM calculation depends on each individual problem, but it is in general considered correct that theoretical predictions of molecular properties in solution would become more accurate in the following order: QM method in the isolated state; DCM method; semi-empirical QM/MM-FEG method; ab initio QM/MM-FEG method (high precision).

Further, the FEG-NEB method in Sect. 8.2.2.2 should become useful for many systems for determining the transition states (TSs) on FES, although it is also difficult to obtain them even in the isolated state. Then, only if we would obtain such stationary states enough accurately through the direct optimization by ab initio QM/MM-FEG method, whether they might be SSs (i.e., reactants or products) or unstable ones (e.g., TS, etc.), the FE difference between any arbitrary two states corresponding to two molecular structures in solution must be effectively evaluated by employing the energy representation (ER) method [36, 37]. Moreover, there are several excellent approximate methods recently developed to investigate solvent effects, that is, the (ASEP)/MD [21, 22] and the ASEC [23, 24] method, which can reduce drastically the number of QM calculations by utilizing the mean field approximate (MFA), providing a better approximate description of the quantum characteristics of solutes in solution.

Hence, in near future, the FEG methodology will become a systematic theoretical strategy as a basic tool to understand chemical reactive molecules in solution and their FESs would be able to be automatically explored computationally by combining the FEG method with other effective methods.

8.3 Applications

In this section, we have applied the FEG method to several chemical reaction systems for investigating the solvent effects on the structural and vibrational properties of solute molecules in aqueous solution.

8.3.1 Structural Optimization of Equilibrium Structures and Reaction Path

8.3.1.1 Zwitterionic Glycine in Aqueous Solution

First, we show how the FEG method realizes the FE changes through the process of structural optimization in solution at a finite temperature. As an example, we take the glycine zwitterion (ZW) in aqueous solution [6] which is the most simple but interesting amino acid. To describe the glycine ZW potential energy surface in aqueous solution, we employed the empirical valence bond (EVB) method which was prepared so as to reproduce a set of energies and forces calculated at the

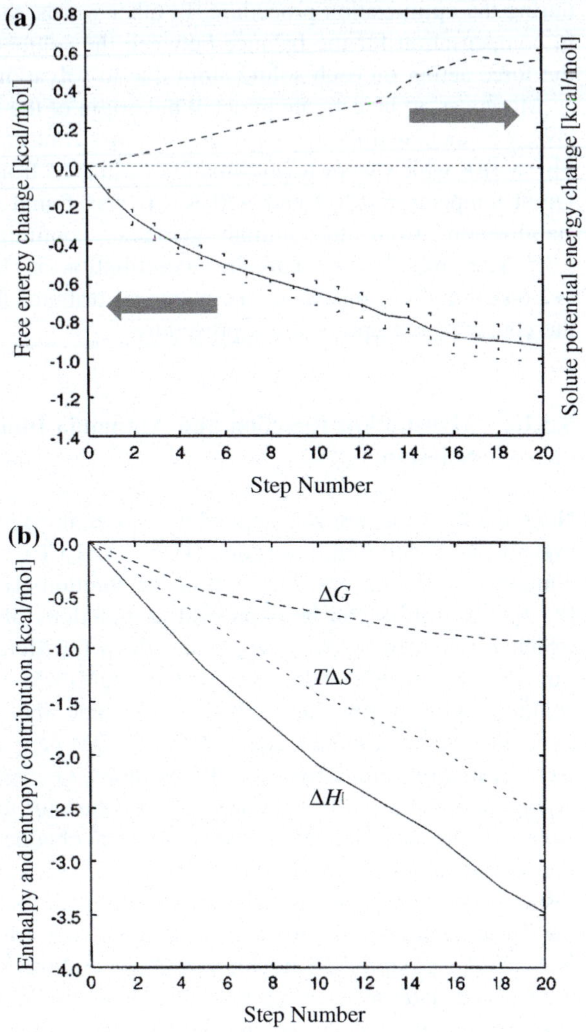

Fig. 8.3 **a** Free energy (*solid curve* with *error bars*) and solute potential energy (*dashed curve*) changes of the stabilization of glycine ZW in aqueous solution using the FEG. **b** Changes in the enthalpy and entropy contributions of the stabilization of glycine ZW in solution using the FEG method (Reproduced with permission from Okuyama-Yoshida et al. [6]. Copyright © 2000 American Institute of Physics)

HF/6-31 + G* level of theory. Then, the geometry optimized in the gas phase at the HF/6-31 + G* level of theory was adopted as the initial one.

In Fig. 8.3, the changes of (a) the FE, the potential energy (PE) and (b) the enthalpy and entropy contributions of the stabilization of glycine ZW in aqueous solution are shown as functions of the optimization step number. As the optimization step number increases, the FE decreases and becomes almost flat for step numbers greater than 18. At the final step 20, the FE change of stabilization from the initial geometry results in −0.9 kcal mol^{-1}. On the other hand, the PE increases as the step number increases (see the dashed curve in Fig. 8.3a). This means that the destabilization of the solute PE is offset by the larger hydration energy change during the optimization procedure. In other words, the stable structure is optimized in compensation for the balance between the solute potential energy gradient and the force acting on each solute atom due to solvation [9].

Moreover, to investigate in detail the origin of the FE stabilization of the glycine ZW, we calculated the enthalpy and entropy contributions to the stabilization (see Fig. 8.3b), which were approximately estimated from the FE changes at two different temperature (300 and 320 K). It was found that the enthalpy changes of stabilization from the initial geometry optimized in the gas phase is -3.5 kcal mol^{-1}. The enthalpy contribution is larger than the entropy one (+2.6 kcal mol^{-1}), thus the enthalpy one is attributed to the stabilization in the FE in the course of the optimization procedure.

8.3.1.2 Menshutkin Reaction and Ammonia Ionization in Aqueous Solution

Next, the FEG method was applied to explore the TSs for two asymmetric reaction systems, i.e., Menshutkin reaction ($H_3N + CH_3Cl \rightarrow H_3NCH_3^+ + Cl^-$) [7] and the ammonia ionization reaction in aqueous solution ($H_3N + H_2O \rightarrow H_4N^+ + OH^-$) [8, 9]. FE profiles can be expressed as functions of certain reaction coordinates, which are defined by $R_{C\text{–}Cl}$–$R_{N\text{–}C}$ and $R_{N\text{–}H}$ for Menshutkin and ammonia ionization reaction, respectively. We used the QM/MM-MD method to describe both solution systems for the purpose to include the solvent molecular structures explicitly. In the original studies [7–9], the QM parts of the whole solution systems were treated by semi-empirical MO methods, i.e., AM1 and PM3 level of theory.

Figure 8.4 shows the optimized TS structure for the Menshutkin reaction in the isolated state and that in aqueous solution obtained by the FEG method. The optimized values of $R_{N\text{–}C}$ and $R_{C\text{–}Cl}$ for TS in solution are 1.97 and 2.07 Å, respectively. In comparison with those lengths in the isolated state, it is found that the TS in solution shifts toward the reactant side. Moreover, the distance between NH_3 and Cl group of TS in solution ($R_{N\text{–}Cl} = 4.04$ Å) becomes larger than that in the isolated state ($R_{N\text{–}Cl} = 3.90$ Å). Such structural deformation should enhance the charge separation of the solute complex, and consequently brings about the FE stability of TS. In fact, from the radial distribution functions (RDFs) with respect to

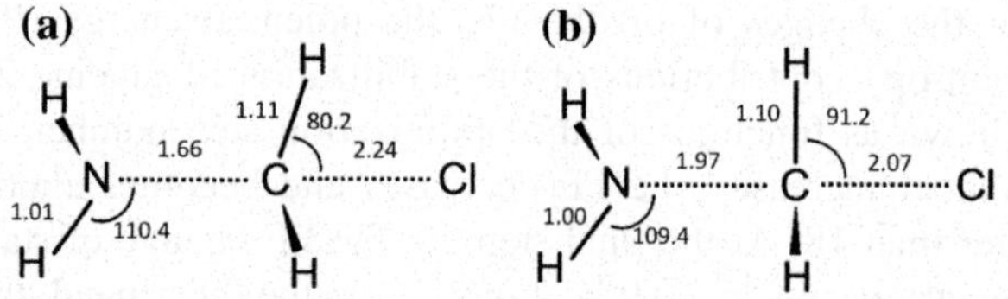

Fig. 8.4 Optimized TS geometries for the Menshutkin reaction (**a**) in the isolated state and (**b**) in aqueous solution obtained by the FEG method. Bond length is in Å and bond angle is in degree (Reproduced with permission from Hirao et al. [7]. Copyright © 2001 Elsevier B.V.)

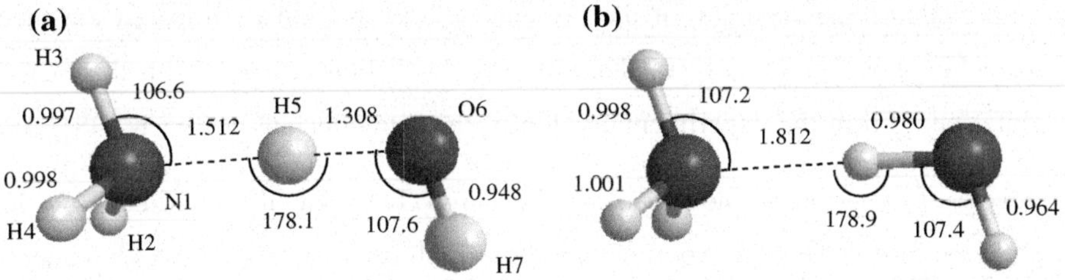

Fig. 8.5 Optimized geometries of the $HN_3\cdots H_2O$ molecule pair in aqueous solution at (**a**) the TS and (**b**) the SS by the FEG method. Bond length is in Å and bond angle is in degree (Reproduced with permission from Nagaoka et al. [9]. Copyright © 2006 American chemical society)

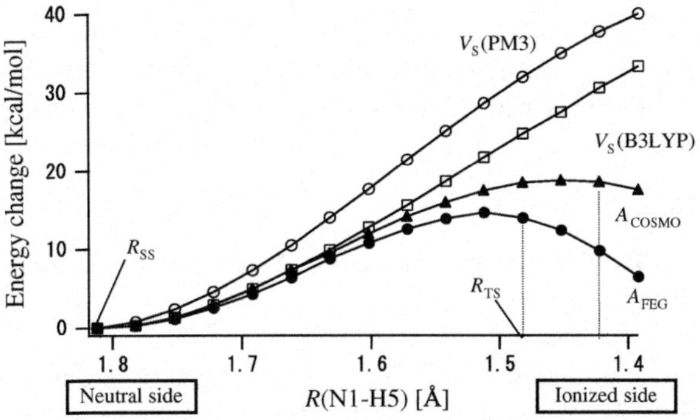

Fig. 8.6 Energy profiles in aqueous solution by the FEG method (A_{FEG}, *closed circles*), and the COSMO method (A_{COSMO}, *closed triangles*), and in the isolated state at PM3 (V_S(PM3), *opened circles*) level of theory along the reaction coordinate R(N1-H5) according to the FEG method (Reproduced with permission from Nagaoka et al. [9]. Copyright © 2006 American chemical society)

the Cl–Ow and Cl–Hw distances, it was observed that the strength of the interaction between Cl and water molecules increases by the structural optimization (for details, see Ref. [7]).

By optimizing all the structural parameters except the reaction coordinate R_{N-H} [9], the TS structure of the $H_3N\cdots H_2O$ molecule pair in aqueous solution (Fig. 8.5a) was obtained as the structure at the maximum of the resultant FE profile in solution (Fig. 8.6). It should be noted that the TS structure that does *not* exist in the isolated state was determined theoretically, *for the first time*, in aqueous solution. The R(N1-O6) of TS is 2.820 Å which is longer than that of SS (2.792 Å, see Fig. 8.5b), and it was also observed that the dipole moment of the SS becomes larger according to the ionization progress, i.e., those of the SS and TS are 4.004 D and 7.032 D, respectively. It is shown clearly that the $H_3N\cdots H_2O$ molecule pair's polarization makes the bare potential energy itself destabilized (V_S(PM3) in Fig. 8.6) but the pair in aqueous solution is *free-energetically* stabilized by ambient water molecules.

To evaluate the accuracy of the present method, we have compared the theoretical FE of activation $\Delta A_f^\ddagger$ of the ammonia ionization reaction with the experimental one. From the experimental value of the rate constant at 295 K [38], $\Delta A_f^\ddagger$ can be estimated to be 9.57 kcal mol^{-1}. It is true that the activation energy difference of 5.13 kcal mol^{-1} might be large between the theoretical value (14.7 kcal mol^{-1}) and the experimental one, which is referred to as $\Delta\Delta A_f^\ddagger$. This can be mainly explained by insufficiency of QM level of theory. The PE of the bare $H_3N\cdots H_2O$ pair at R(N1-H5) = 1.512 Å (Fig. 8.5a) by the B3LYP/6-31(d) level of theory, then, became a value, 24.23 kcal mol^{-1}, which is smaller than the PM3 value, 28.65 kcal mol^{-1}. Because the PE difference between PM3 and B3LYP/6-31G(d) (i.e., 4.42 kcal mol^{-1}) is almost the same amount as 5.14 kcal mol^{-1}, which means that $\Delta\Delta A_f^\ddagger$ might be originating mainly in the quality difference of electronic state calculations for the reactants themselves. Also, the TS dipole moment 5.834 D at B3LYP/6-31G(d) is larger than 5.027 D at PM3, and it is also considered that the dielectric stabilization is brought about by the former.

8.3.1.3 Intramolecular Proton Transfer of Glycine in Aqueous Solution

Through the above applications of QM/MM-FEG method combined with the EVB and the semi-empirical MO method to the chemical species in aqueous solution, it was clearly understood that the structural optimization of some stationary states (SS and TS) on the FES is inevitable to obtain accurate information with respect to a chemical reaction process in solution. However, the conventional QM/MM-FEG method has still three issues unresolved for its wider practical use:

(i) It is necessary to develop some convenient and reasonable FEG strategy reinforced directly with ab initio MO method, for the purpose to further improve the accuracy and applicability of the FEG method. This is because, from the viewpoint of chemical accuracy, it is well known that the semi-empirical MO methods would sometimes bring about inaccurate description for the electronic states of solute molecules.
(ii) In most of real applications, it is difficult to choose a suitable set of well-defined variables for reaction coordinates, except special cases, and sometimes it is troublesome to define a reasonable mapping with more than two variables.
(iii) The conventional VFA in the FEG method can characterize only the averaged behavior of solutes in solution, providing a set of average values of vibrational frequencies with their representative normal-mode vectors.

To resolve the first issue (i), we have recently developed an AMBER–GAUSSIAN interface (AG-IF) [18] with the FEG method orientation. Moreover, for the second one (ii), we have proposed ab initio QM/MM-FEG method combined with the NEB

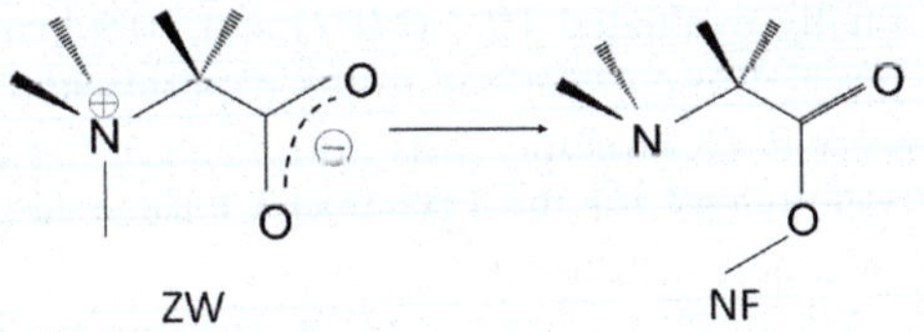

Fig. 8.7 Glycine isomers in aqueous solution

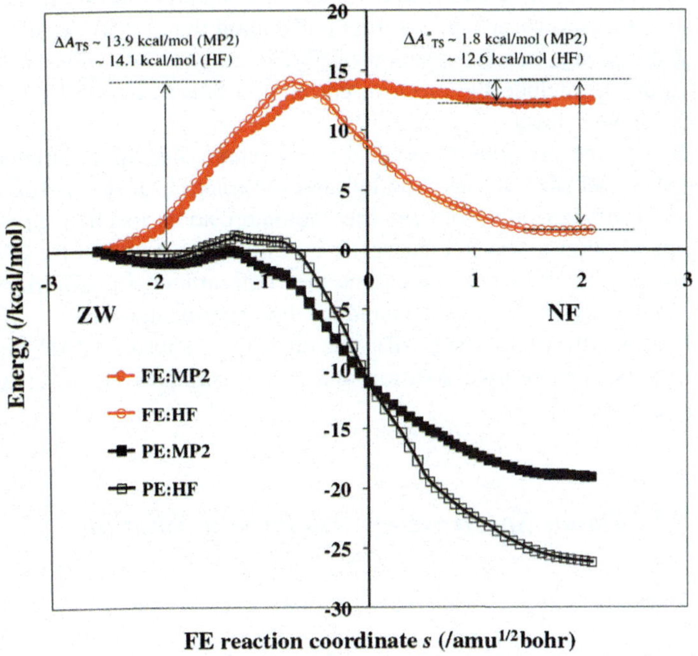

Fig. 8.8 Free Energy (FE) and potential energy (PE) profiles for the intramolecular proton transfer process of glycine molecule in aqueous solution at the HF/6-31+G(d) and the MP2/6-31+G(d) level along the reaction coordinate *s* obtained by the FEG method at the HF/6-31+G(d) level (Reproduced with permission from Takenaka et al. [11]. Copyright © 2011 Springer-Verlag)

method (ab initio QM/MM-FEG-NEB method) [11]. For demonstration, this method has been applied to the intramolecular proton transfer reaction from ZW to neutral-form (NF) of glycine in aqueous solution (Fig. 8.7).

Figure 8.8 shows the FE profiles and PE ones, at the HF/6-31+G(d) and the MP2/6-31+G(d) level of theory, along the FE reaction coordinate obtained by the FEG method at the HF/6-31+G(d) level. The FEs of activation from the forward reaction from ZW to NF, ΔA_{TS}s, were estimated to be 13.9 and 14.1 kcal mol^{-1} at the MP2 and the HF level, respectively. To compare them with the experimental value (14.4 kcal mol^{-1} [39]), we have also estimated the FE contribution from the degrees of freedom for the intramolecular vibration of glycine in solution. As a result, ΔA_{TS}s were finally evaluated 10.7 (MP2) and 10.9 kcal mol^{-1} (HF), which are slightly underestimated in comparison to the experimental one. This should be partly because the present TS structure was optimized at the HF level, employing the same set of LJ parameters for the QM/MM LJ force field throughout all the structures along the reaction coordinate.

On the other hand, the FE of activation of the inverse reaction, ΔA^*_{TS}, was 1.8 kcal mol^{-1} (MP2) and is in good agreement with the estimation by the Car-Parrinello MD method ($\sim$1.5 kcal mol^{-1} [40]). On the contrary, the FE overestimation at the HF level could be understood mainly due to the incorrectness of the PE itself as shown in Fig. 8.8. It is understood that, for discussion of the FE stability of the proton transfer reaction of glycine in aqueous solution, it is essentially necessary to include the electron correlation effect at least by the MO calculation at the MP2 level.

According to the previous theoretical study employing the polarizable continuum solvent model [41], it was reported that the rotation of the carboxylic group would play a fundamental role on the tautomerization mechanism of glycine molecule. In addition to the free energetic investigations between the ZW and NF process, it might be necessary to examine the configurational equilibrium between those NF conformers in fully understanding the experimental one.

Finally, as for the third issue (iii), the dual VFA approach [31] proposed in Sect. 8.2.2.3, is one possible solution and some examples are discussed in the following sections.

8.3.2 *Vibrational Spectroscopy Analysis in Solution*

8.3.2.1 Water

For an application to the vibrational spectroscopy analysis, we took an H_2O molecule in liquid water [42]. Initially, the structure of the H_2O molecule in water was optimized by the standard FEG method for the H_2O geometry to satisfy the zero-FEG condition (cf. Eq. (8.19)) using the FE-Hessian matrix $\mathbf{H}^{FE}$ (cf. Eq. 8.10). Then, to estimate INM-Hessian matrices for the vibrational frequency analysis (VFA) at the optimized structure $\mathbf{q}^{FE}_{eq}$ on FES, we executed ab initio QM/MM-MD simulation to apply the dual VFA (cf. Sect. 8.2.2.3) approach to the present H_2O system.

A set of three effective vibrational normal mode frequencies of the OH stretching and the HOH bending motion has been obtained by diagonalizing $\mathbf{H}^{FE}$, which are scaled by the recommended factor of 0.9418 [43]. In Fig. 8.9, they are shown as three grey bars with the heights in proportion to their infrared (IR) intensities (Eq. (8.32)). The vibrational frequencies of the symmetric and asymmetric stretching modes were 3287 and 3387 cm^{-1}, showing large red-shifts, −354 and −393 cm^{-1}, respectively, due to the environmental change from in the isolated state to in liquid phase. On the contrary, the bending modes was 1676 cm^{-1}, showing a

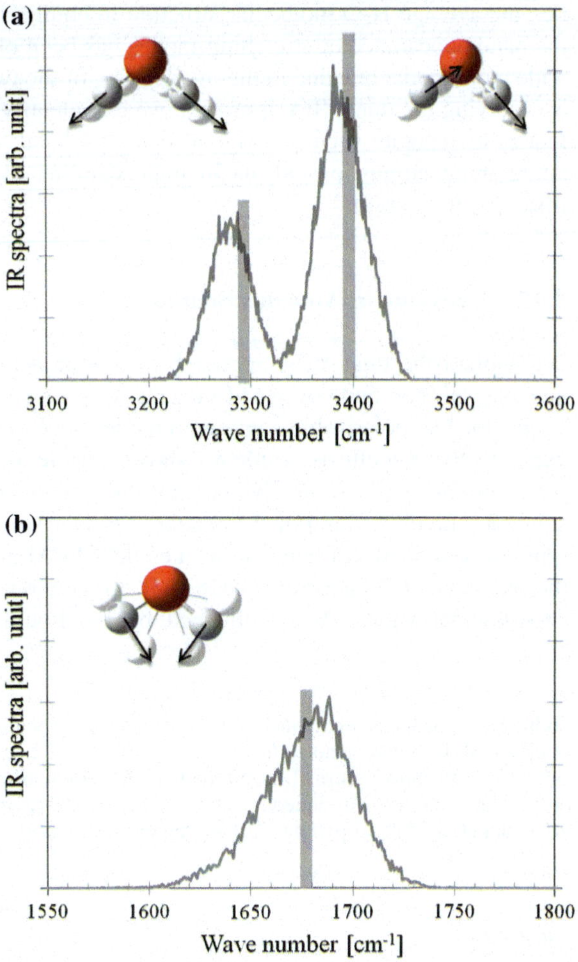

Fig. 8.9 Calculated IR spectra obtained by diagonalizing the FE-Hessian matrix (the *grey bars*) and by the density of states (DOS) of IR intensities prepared by diagonalizing the INM-Hessian matrices along an equilibrium QM/MM-MD trajectory (the *grey jagged curves*). **a** The OH stretching and **b** the HOH bending frequencies (Reprinted table with permission from Georg et al. [42]. Copyright © 2014 Springer-Verlag)

slight blue-shift, +77 cm^{-1}. These shifts stem from the electrostatic interactions of a water molecule with ambient water molecules [44] and the tendency is consistent with previous studies of liquid H_2O [44–47].

In addition, we have also evaluated the INM-DOS of IR spectra in liquid water with respect to the "instantaneous" vibrational normal mode frequencies (cf. Eq. (8.30a)) along an equilibrium QM/MM-MD trajectory. In the high frequency region (>1600 cm^{-1}), the IR spectra of liquid water consist of two main bands, corresponding to (a) the stretching motions of the covalent OH bonds and (b) the bending motion with respect to the HOH angle (Fig. 8.9). The bandwidths of these three bands were $\sim 50\ cm^{-1}$ (full width at half-maximum (FWHM)), reflecting only the electron density fluctuations which depends on "instantaneous" solvent conformations around the "solute" H_2O molecule involved. This is because the molecular structure of the H_2O molecule was kept as that one optimized on FES, i.e., the average H_2O molecular structure in equilibrium liquid water. It is true that the band appearance of stretching modes has been observed in experiments to be a wide monomodal one due to the overlapping of the two stretching modes. However, by applying the dual VFA approach, we can identify these fundamental vibrational modes to separate such an original complex band to the individual bands, facilitating more atomistic analysis in understanding the molecular vibrational spectroscopy in solution.

8.3.2.2 Glycine in Aqueous Solution

We applied the dual VFA approach to a neutral form (NF) glycine molecule in aqueous solution and compared the calculation results with those estimated by the conductor-like polarizable continuum model (CPCM) method in order to extract the explicit solvation effects. Table 8.1 shows a triple of typical vibrational frequencies (ω_{gas}, ω_{CPCM}, ω_{FE}) of a glycine molecule in the isolated state and in aqueous solution with their vibrational frequency shifts, ($\Delta\omega_{CPCM}$, $\Delta\omega_{FE}$), evaluated by two types of solute-solvent interactions, i.e., the CPCM and QM/MM method, scaled by the recommended factor of 0.9418 [43]. In addition, they were compared with the experimental values $\Delta\omega_{exp}$ obtained by the Fourier transform infrared (FT-IR)

Table 8.1 Calculated vibrational frequencies (Normal vibrational frequencies (in cm^{-1}) scaled by 0.9418 [43]. It should be noted that these by the FE- and the CPCM-Hessian are calculated at the optimized structures by the FEG and the CPCM method, respectively) ω (cm^{-1}) and their shifts from in gas phase to in aqueous solution $\Delta\omega$ (cm^{-1}) (Reprinted table with permission from Kitamura et al. [31]. Copyright © 2014 American chemical society)

Mode	Gas	In water			Frequency shift	
	ω_{gas}	ω_{CPCM}	ω_{FE}	$\Delta\omega_{CPCM}$ [a]	$\Delta\omega_{FE}$ [b]	$\Delta\omega_{exp}$
Stretching						
$\nu_{as}(NH_2)$	3476	3351	3416	−125	−60	−13[c]
$\nu_s(NH_2)$	3381	3262	3322	−119	−59	
$\nu(OH)$	3570	3047	3077	−523	−493	−320[c]
$\nu_{as}(CH_2)$	3011	2990	3028	−21	+17	
$\nu_s(CH_2)$	2958	2936	2972	−22	+14	
$\nu(CO)$	1708	1657	1691	−51	−17	−39[c]
Bending						
$\delta(COH)$	1341	1232	1338	−9	+97	
$\delta(COH) + \gamma_s(NH_2)$	886	897	990	+11	+104	
$\gamma(COH)$	606	563	997	−43	+391	+416[d]
$\gamma_{as}(NH_2)$	225	221	505	−4	+280	

[a] $\Delta\omega_{CPCM} = \omega_{CPCM} - \omega_{gas}$
[b] $\Delta\omega_{FE} = \omega_{FE} - \omega_{gas}$
[c] FTIR data for Gly-(H_2O) and Gly-$(H_2O)_2$ cluster in Ar matrix [48]
[d] FTIR data for Gly-(1-methyluracil) cluster in Ar matrix [49]

spectroscopy measurement of glycine-$(H_2O)_n$ (n = 1,2) [48] and glycine-(1-methyluracil) H-bond complexes [49].

Both $\Delta\omega_{CPCM}$ and $\Delta\omega_{FE}$ for the stretching modes (>1650 cm^{-1}) were red-shifted with the exception of those $\Delta\omega_{FE}$ of CH_2 group, while $\Delta\omega_{FE}$ agreed well with $\Delta\omega_{exp}$ in comparison with $\Delta\omega_{CPCM}$ (see Table 8.1). On the other hand, for the four bending modes (≤1650 cm^{-1}), all ω_{FE}s showed rather large blue-shifts in solution. In particular, ω_{FE} of γ(COH) showed the largest blue-shift (+391 cm^{-1}) in the four, being clearly consistent with $\Delta\omega_{exp}$ (+416 cm^{-1}). In contrast to $\Delta\omega_{FE}$, $\Delta\omega_{CPCM}$ for the bending modes were very small in any cases, showing that the inclusion of explicit effect of HBs is inevitable to reproduce vibrational frequency shifts for the bending modes.

In order to clarify the microscopic origin of such vibrational frequency shifts $\Delta\omega_{FE}$, we have investigated two essential effects, i.e., the structural relaxation of glycine molecule itself and the explicit solvation structure around it (for details, see Ref [31]). According to these results, it is considered that the mechanism of frequency shifts of the stretching-modes would be attributed to the softening (or hardening) of vibrational frequency modes due to the bond length elongation (or shortening). On the contrary, the large blue-shifts of ω_{FE} of the bending-modes in aqueous solution can be understood to be brought about by the microscopic solute-solvent interactions. The mechanism must be explicitly explained by the steric hindrance by the HB water molecules, since the bending motions generally occur oscillatorily orthogonal to the HB direction and the stronger HB formation should make their motions more difficult to occur.

Moreover, Fig. 8.10 shows the vibrational INM DOS $D^{IR}(\omega)$ (Eq. (8.30a)) weighed with w_i^{IR} of each vibrational mode in the high-frequency range (2800–3600 cm^{-1}) to clarify the influences of the explicit solvent water molecules on the

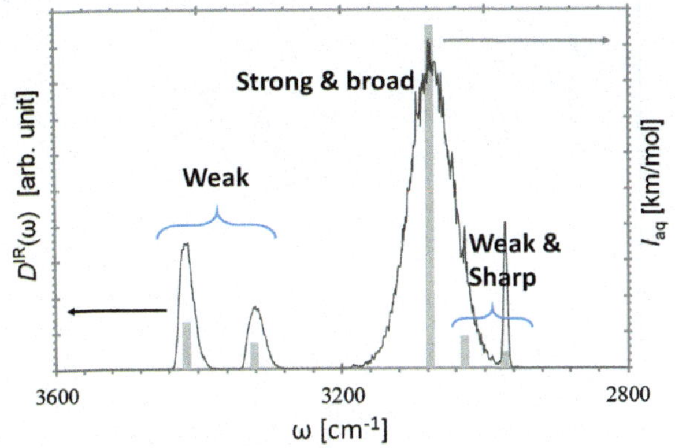

Fig. 8.10 The approximate IR spectra: the vibrational INM DOS $D^{IR}(\omega)$ weighed with w_i^{IR} in high-frequency range. Those average IR intensities I_{aq} at the vibrational frequencies estimated by the VFA approach are also indicated by the *gray bar* (Reproduced with permission from Kitamura et al. [31]. Copyright © 2014 American chemical society)

vibrational spectra. The relatively broad peak in the range of 3000–3160 cm^{-1} should correspond to the stretching motion of the –OH group (3077 cm^{-1}, see Table 8.1), where its full width at half-maximum (FWHM) is ~ 80 cm^{-1}. On the contrary, in the range of 3280–3440 cm^{-1}, there were two less broad peaks with each FWHM (~ 30 cm^{-1}) less than a half the value of –OH group. Actually, it was found that such two peaks, i.e., 3322 and 3416 cm^{-1}, correspond to the symmetric and the asymmetric modes of $-NH_2$ group, respectively. In contrast to the above three modes, the two modes of $>CH_2$ group, i.e., 3028 and 2972 cm^{-1}, were found to become very sharp since they were hardly influenced electrostatically even in aqueous environment.

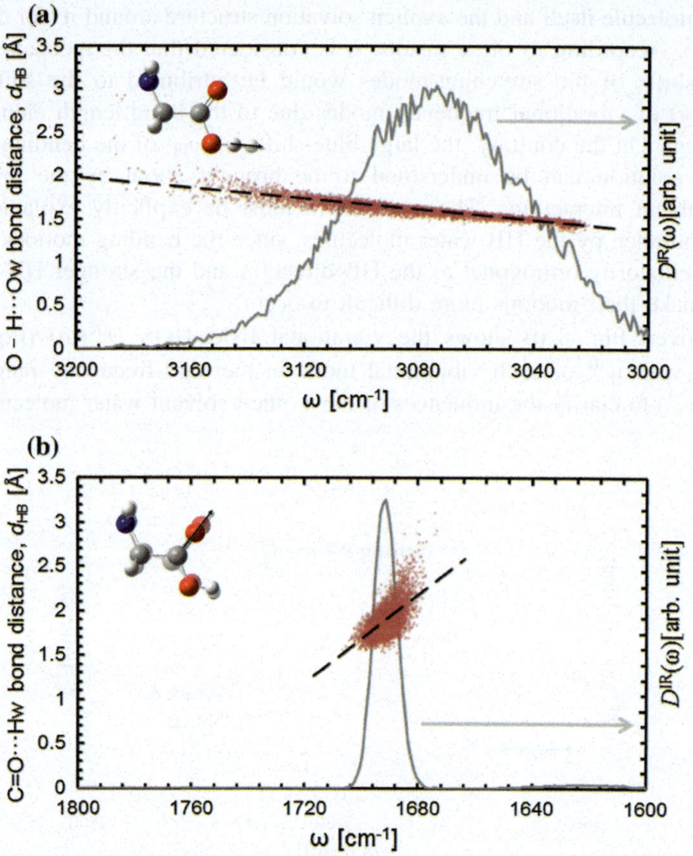

Fig. 8.11 IR spectra of O–H and C=O stretching modes. Scatter plots of the lengths of nearest-neighboring H or O atom of water molecules versus the O–H and C=O stretching frequencies. Linear fits (*dashed lines*) are (**a**) $d_{O-H\cdots O_w} = 0.0031\ \nu(OH) - 7.9532$; $R = 0.901$ and (**b**) $d_{C=O\cdots H_w} = -\ 0.0249\ \nu(CO) + 43.936$; $R = -\ 0.623$ (Reproduced with permission from Kitamura et al. [31]. Copyright © 2014 American chemical society)

We have investigated the microscopic origin of the broadening of spectra in solution. For the purpose, we have focused on the IR spectra of the C=O and O–H stretching modes. Figure 8.11 shows the scatter plots of the intermolecular HB distances d_{HB} with the nearest-neighboring O (or H) atoms in the water molecule (the left ordinate) and $D^{\mathrm{IR}}(\omega)$ (the right ordinate), where the abscissa is taken as the vibrational frequency. In this case, we have also shown the linearly-fitted curves, i.e., $d_{\mathrm{O-H\cdots O_w}} = 0.0031 \cdot \nu(\mathrm{OH}) - 7.9532$ and $d_{\mathrm{C=O\cdots H_w}} = -0.0249 \cdot \nu(\mathrm{CO}) + 43.936$. As a result, ν(OH) was positively correlated with the O–H···$\mathrm{O_w}$ distance $d_{\mathrm{O-H\cdots O_w}}$, with its correlation coefficient $R = 0.901$, indicating the strong correlation. This positive correlation indicates that ν(OH) is red-shifted as the O–H···$\mathrm{O_w}$ distance becomes smaller, which is ascribed to the weakening of O–H bond due to the decrease of the electron density of O–H bond induced by the negative point-charge on the O atom of a closer water molecule. On the other hand, the slope of the fitted curve for ν(CO) became negative, showing relatively weak correlation ($R = -0.623$). Contrary to an above tendency of the O–H group, ν(CO) is blue-shifted as the intermolecular HB distance (the C=O···$\mathrm{H_w}$ distance $d_{\mathrm{C=O\cdots H_w}}$) becomes smaller and the electron densities in the C=O σ and π bonding orbitals increase slightly due to the polarization but these changes are smaller than the change in the O–H bonding orbital.

To interpret and explain those vibrational frequency changes occurring upon HB formation, we have investigated the electron density of glycine molecule in aqueous solution via the natural bond orbital (NBO) analysis and found that the contribution of hyperconjugative interaction energy increases by the polarization of C=O group. As a result, it was considered that ν(CO) is blue-shifted due to the stabilization by the hyperconjugative interaction, while ν(OH) is red-shifted by the destabilization of the corresponding orbitals.

In this way, by using the dual approach, the vibrational spectra in solution can be, in principle, separated to identify the individual contributions of each vibrational normal modes. Moreover, by combining it with the FEG method and other additional analyses, e.g., solvation structures and the electron density analysis, the dual approach can provide us fundamental and essential information to understand the general features of experimental vibrational spectra in various molecular systems in solution. In conclusion, the present approaches are quite useful to interpret the microscopic origin of the experimental vibrational spectra.

8.4 Recent Related Developments

Accuracy of the QM/MM-MD method depends on not only the description of the QM region but also the interaction between the QM and MM ones. Because the charge distribution in the MM region is described by a number of point charges, the standard (SD) QM/MM electrostatic interaction, which often contributes the main environmental effect on the QM molecules, pose serious theoretical and technical problems in describing the boundary between the QM and MM regions, relating to the short range behaviors due to the absence of the QM characteristics, that is, the Pauli repulsion. This problem is known as the electron spill-out problem [50, 51]. In addition, it is also important to select the parameters of the empirical functions such as the Lennard-Jones interactions [10].

Under the circumstances, a number of theoretical methods have been already developed to improve the QM/MM-MD method, e.g., the modification of the semi-empirical QM Hamiltonians [7, 52–54], the optimization of the QM/MM empirical parameters [10] and the replacement of the empirical repulsion potential functions [55]. However, these methods need the numerical values of some reasonable reference quantities to optimize the parameters for some specific molecular systems. Moreover, it is usually hard to obtain the reference experimental or computational ones in solution. It is, therefore, reasonable and plausible as a second best strategy that the closer MM solvent molecules around the QM solute should be included into the QM region to avoid the serious problems in the boundary between QM and MM regions. This is because the most serious problem is originating in the quantum-mechanical behaviors. On the basis of such strategy, we have developed the number-adaptive multiscale (NAM) QM/MM-MD [56, 57] and the QM/MM-MD method combined with the fragment molecular orbital (FMO) one, i.e., FMO-QM/MM-MD method [20].

8.4.1 *Number Adaptive Multiscale (NAM) Method*

The adaptive multiscale QM/MM-MD methods [58–61] are straightforward approaches to include the closer MM solvent molecules around the QM solute into the QM region. For these methods, it is common to introduce the transition (T) region between the QM and MM ones to smooth the forces exerted on the solvent molecules when they move from the QM region to the MM one or vice versa. Recently, we have proposed a new treatment of the adaptive multiscale QM/MM-MD method, that is, the number-adaptive multiscale (NAM) QM/MM-MD method [56, 57]. This method defines both the QM and T regions by the "number" of solvent molecules around the solute, although most of previous methods adopted conventionally by the "distance" from the solute, i.e., the "distance-based" definition. Thus, the NAM QM/MM-MD method is suitable for the parallel computing since the number of QM calculations is constant at each time step during the MD simulation [56]. For more information about the theoretical details and the algorithm, see Refs. [56, 57].

We have applied the NAM method to the pure water system [56] and a glycine ZW ZW molecule in aqueous solution [57], and calculated the radial distribution functions (RDFs) between the QM solute molecule and the MM solvent ones for verifying the applicability of the NAM QM/MM-MD method. Figure 8.12 displays the RDFs between the oxygen atoms of the solute water molecule and other MM solvent oxygen atoms by (a) the SD methods and (b) the NAM ones at the HF, BLYP and B3LYP methods with the 6-31+G(d,p) basis set. We also display in

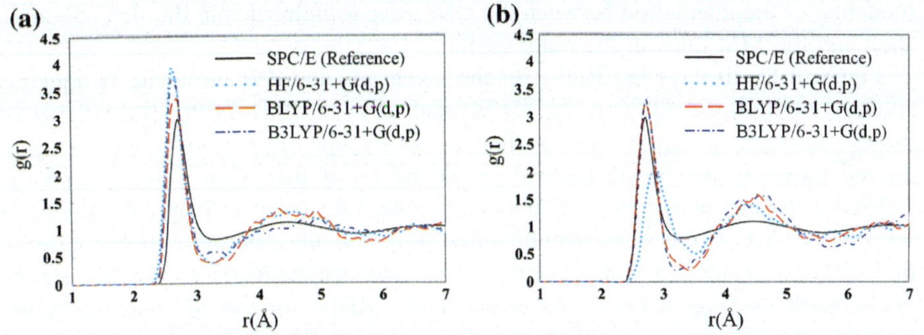

Fig. 8.12 Radial distribution functions $g(\mathrm{r})$ of the QM water molecule relative to all other MM solvent ones in pure water by (**a**) the SD QM/MM-MD method and (**b**) the NAM ones at the HF, BLYP and B3LYP methods with the 6-31+G(d,p) basis set. As a reference, $g(\mathrm{r})$ by the full MM-MD method with the SPC/E water model is also shown (Reproduced with permission from Takenaka et al. [56]. Copyright © 2012 Elsevier B.V.)

Fig. 8.12 the RDF by the full MM-MD method with the SPC/E water model as a reference since it was calibrated to correspond very well to the experimental results [62]. In the RDFs by the SD method (Fig. 8.12a), the first peak is found to exceed the reference (black curve) due to the inaccurate intermolecular interactions between one QM water molecule and close MM water ones in the solute. On the other hand, those by the NAM method with the BLYP and B3LYP methods (Fig. 8.12b) are found to be closer to the reference (black curve), while that with the HF method is lowered due to the neglect of the electronic correlation effect in the QM calculation and the second peaks cannot be accurately reproduced by the NAM method. In this study, we have included four water molecules in the QM region as the first solvation shell, and this discrepancy is probably due to the inaccurate

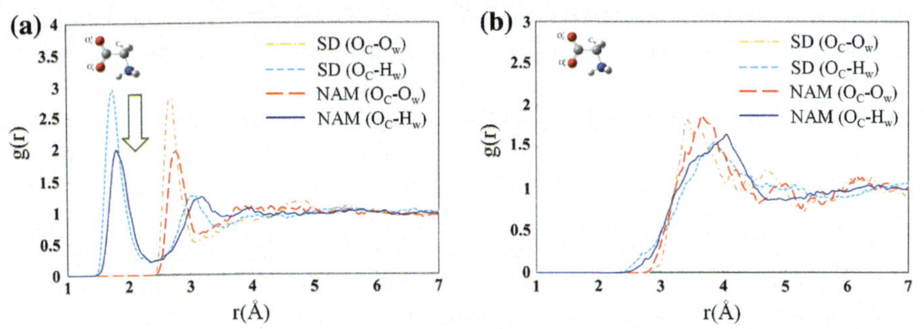

Fig. 8.13 Radial distribution functions $g(\mathrm{r})$ for glycine ZW in aqueous solution of (**a**) oxygen atoms (O_C) in the $-COO^-$ group and (**b**) a carbon atom (C_α) in the $>CH_2$ group with respect to the oxygen (Ow) and hydrogen (Hw) atoms of the solvent water molecules (Reproduced with permission from Takenaka et al. [57]. Copyright © 2012 American institute of physics)

modeling of the interaction between the QM solvent molecules in the first solvation shell and the MM ones in the outer region.

Figure 8.13 displays the RDFs for the solute glycine ZW molecule in aqueous solution by both the SD and the NAM method at the B3LYP/6-31G(d) level of theory. As seen in Fig. 8.13a, the first peaks by the SD method were found to exceed those by the NAM method in the RDFs of the $-COO^-$ group and the estimated hydration number was obtained to be 4.8, which is in good agreement with 4.4 ± 0.2 of the experimental ones [63]. On the contrary, in the case of the $>CH_2$ group, there was no clear difference between the RDFs by the SD and the NAM methods (Fig. 8.13b). The errors in the RDFs for the SD and the NAM method were different according to the chemical properties of the functional groups involved, stemming from their trapping extent of electrons occurring due to the neglect of the Pauli repulsion at the short distance. In particular, it should be noticed that the $>CH_2$ group shows no clear overestimation since the $>CH_2$ group does not form the hydrogen-bond with the solvent water molecule for the hydrophobic nature.

In conclusion, by treating quantum mechanically the closer solvent water molecules around the solute molecule via the present NAM method, the QM description of the solute-solvent interactions would be essentially able to be improved systematically and reasonably without additional adjustments of the empirical parameters.

8.4.2 *Fragment MO-QM/MM-MD Method*

For dealing with large molecular systems, fragment-based QM methods have been well-established to open up and widen the active field of QM researches. One among them is the fragment molecular orbital (FMO) method by Kitaura and co-workers [64, 65]. As its application with MD simulation, a full ab initio FMO-MD method was developed with an application by Komeiji et al. [66, 67]. However, even with the FMO method, such direct QM-MD calculation still demands great computational cost. Under the circumstances, we have presented a QM/MM treatment based on ab initio FMO method, i.e., the ab initio FMO-QM/MM method, and implemented an AMBER-PAICS interface which combines a famous MD program AMBER with an FMO program PAICS [68, 69] to execute direct FMO-QM/MM-MD simulations. For more information about the theory and algorithm [20].

As an example of a solution system, we treated one alanine dipeptide (Fig. 8.14) in aqueous solution and investigated the stability of an alanine dipeptide in the polyproline II (PP_{II}) conformation by using two systems with two different sizes of the QM region. For the system with a larger QM region (the large QM system), all molecules within 5 Å from the alanine dipeptide were included into the QM region and the definition of the QM region remains unchanged during the MD simulation.

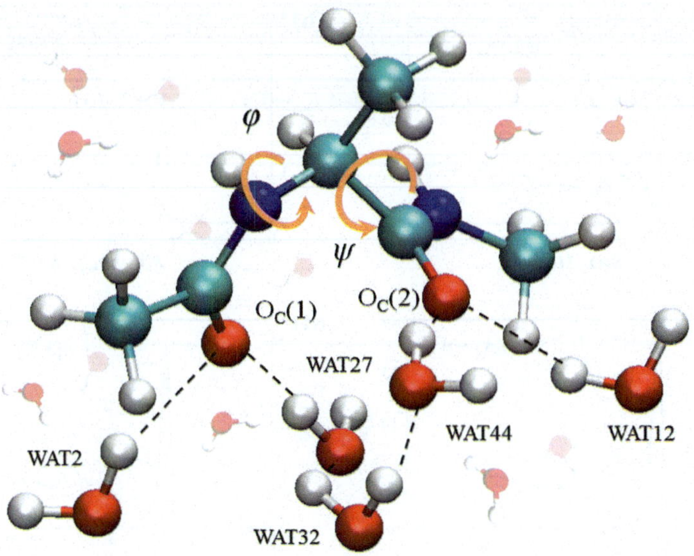

Fig. 8.14 Typical snapshot structure of the alanine dipeptide in aqueous solution. Two peptide C=O groups of the alanine dipeptide are forming directly hydrogen bonds with four H_2O molecules (WAT2, WAT12, WAT27, and WAT44). The local backbone conformation of the alanine dipeptide was defined by two dihedral angles, φ and ψ (Reproduced with permission from Okamoto et al. [20]. Copyright © 2013 The chemical society of Japan)

For the system with a smaller QM region (the small QM system), only the alanine dipeptide was treated as the QM molecule.

From the FMO-QM/MM MD simulation for the large QM system, we calculated the inter-fragment interactions (IFIEs) between the alanine dipeptide and the surrounding QM H_2O molecules. For 5 H_2O molecules, i.e., WAT2, WAT12, WAT27, WAT32, and WAT44 (Fig. 8.14), the averaged IFIEs (ca. -10 kcal mol^{-1}) are stronger than those of the other H_2O molecules. From a typical snapshot of alanine dipeptide and these 5 H_2O molecules (Fig. 8.14), those 4 H_2O molecules interact directly with two peptide C=O group of the alanine dipeptide, while a *triple-water bridge* is observed between two peptide C=O group via WAT32 molecule, i.e., C=O $\cdots(H_2O)_3\cdots$O=C.

In Fig. 8.15, shown are the RDFs and the integrated coordination numbers (ICNs) with respect to the lengths $R(O_C\text{–}H_W)$ and $R(O_C\text{–}O_W)$, where O_C stands for the oxygen atom in two peptide C=O groups of the alanine dipeptide and the oxygen and hydrogen atoms of solvent H_2O molecule are abbreviated to O_W and H_W, respectively. For the O_C–H_W RDF (Fig. 8.15a), the ICN up to the first minimum ($R(O_C\text{–}H_W)$ = 2.5 Å) is 4.4. The first peak of the O_C–H_W RDF reflects the formation of the direct hydrogen bonds with 4 H_2O molecules. For the O_C–H_W RDF (Fig. 8.15b), the first minimum are located at ($R(O_C\text{–}H_W)$ = 3.4 Å) and the integration of the first peak gives an ICN of 5.4. From these results, it is found that

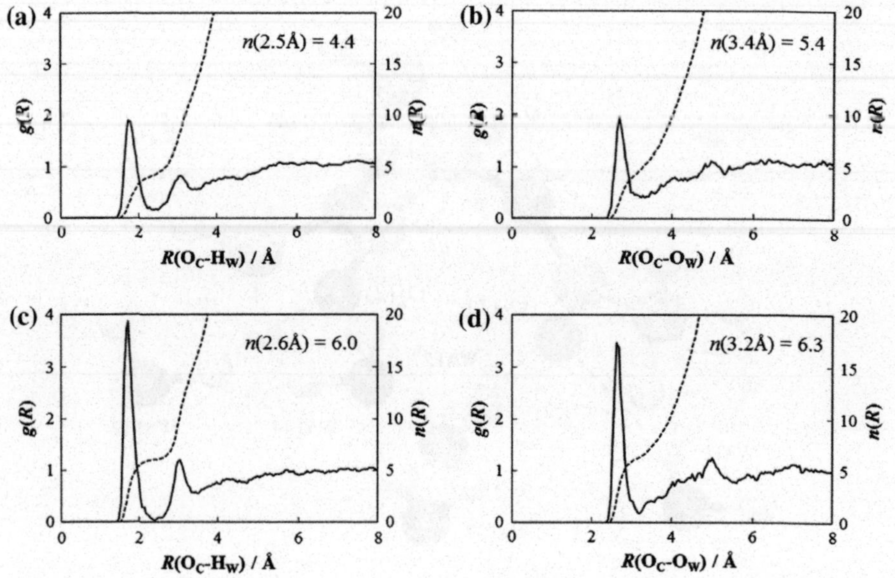

Fig. 8.15 RDF (*solid curve*) and ICN (*broken line*) between solute alanine dipeptide and solvent H_2O molecule with respect to Oc–Hw and Oc–Ow for (**a, b**) the large QM system and (**c, d**) the small QM system (Reproduced with permission from Okamoto et al. [20]. Copyright © 2013 The chemical society of Japan)

these 5 H_2O molecules contribute significantly to the stabilization of the alanine dipeptide in aqueous solution.

On the other hand, for the smaller QM system, three kinds of *double-water bridges* were observed, i.e., C=O $\cdots(H_2O)_2\cdots$O=C, C=O $\cdots(H_2O)_2$ H–N, and N–H $\cdots(H_2O)_2\cdots$O=C. For the O_C–H_W RDF (Fig. 8.15c), the first peak and minimum are observed at $R(O_C\text{–}H_W) = 1.7$ and 2.4 Å, respectively, and integration of the first peak yields an ICN of 6.0. For the O_C–O_W RDF (Fig. 8.15d), the ICN integrating over the first peak ($R(O_C\text{–}H_W) = 2.6$ Å) to the minimum ($R(O_C\text{–}H_W) = 3.2$ Å) is 6.3. The well-defined first peak and minimum for O_C–H_W and O_C–O_W RDFs correspond to the formation of *double-water bridges* around the alanine dipeptide. For the small QM system, the O_C–H_W and O_C–O_W RDFs appear more structured than those for the large QM system, with a higher first peak and a deeper first minimum.

Therefore, it is concluded that the intermolecular description of the QM/MM boundary depends strongly on the QM region size in the QM/MM model.

8.4.3 Future Aspects: Fragment MO-QM/MM Methodology with NAM and Hybrid MC/MD Reaction Method

Within a QM/MM framework, the molecular system is divided into QM and MM regions. For example, in a solution system containing both QM and MM solvent molecules, the forces exerted on the atoms near the QM/MM boundary are inaccurate because of the transferability problems between the QM and MM force models. The adaptive multiscale QM/MM methods such as the NAM method are very useful to obtain a proper boundary representation between the QM and MM regions. In the NAM method, we introduce the partitioning of the molecular system by using the "number" of solvent molecules around the solute and the QM/MM calculations for individual partitions can be entirely independent each other in the parallel-computing. However, the applicability of the present NAM method are limited, since the QM/MM calculations for partitions with large-sized QM regions require high computational cost. Under the circumstances, we are recently developing a new FMO-QM/MM treatment combined with the NAM method in order to reduce the computational times.

Moreover, we have recently proposed and developed the hybrid Monte Carlo (MC)/Molecular Dynamics (MD) reaction method [70] to realize the practical atomistic simulation method for a massive complex chemical reacting systems. The concept of the hybrid MC/MD reaction method is shown in Fig. 8.16. First, we execute MD simulation in the region R until one (or some) pair of atoms meets the necessary conditions and select (instantaneous) configuration state $\boldsymbol{r}$. Next, one of pairs is virtually reacted to generate a (instantaneous) configuration state $\boldsymbol{s}$, relaxing

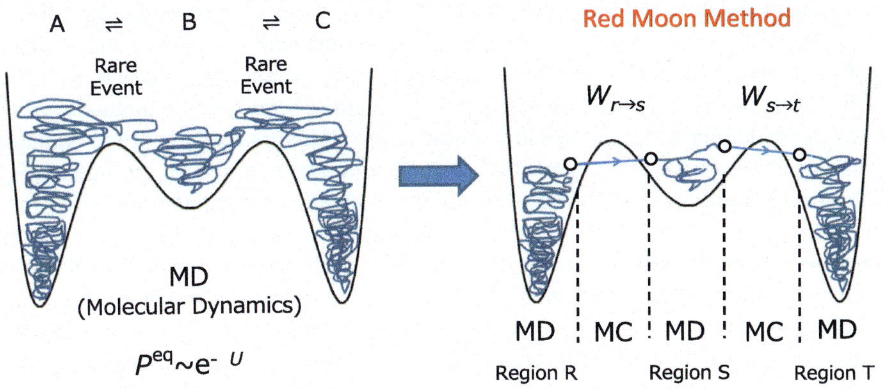

Fig. 8.16 Schematic representation of the hybrid reaction method (*Red moon* method). In *left* figure, the configurational distribution in equilibrium P^{eq} is proportional to the exponential factor $\exp[-\beta U]$ where U is the global potential function. On the other hand, in *right* figure, those are by regional ones. The connecting points of the two kind of moves (*open circles*) are selected according to the criterion of chemical reaction. $W_{r\to s}$ and $W_{s\to t}$ are the transition probabilities from the state r to s and from s to t, respectively (Reproduced with permission from Nagaoka et al. [70]. Copyright © 2013 Elsevier B.V.)

the whole system through short MD simulation. In this step, after computing the potential energy difference ΔU_{rs} ($= U_s - U_r$) between the reactant and product, the present reaction event is accepted (or rejected) according to the transition probability under the Metropolis scheme. In this method, a Monte Carlo (MC) method is applied to drive chemical reaction processes which are rare events, while the proper molecular motions (molecular translation, rotation and vibration) are directly treated by using molecular dynamics (MD) method. By repeating such a combined cycle (MC/MD cycle) consisting of a couple of MC and MD treatment, the present method can enhance the change of states that is brought about by chemical reactions, to achieve a "long-time" simulation of large-scale chemically reactive systems. This method is so efficient that 2-chlorobutane racemization process [70], the solid electrolyte interphase (SEI) film formation in lithium-ion battery [71] and the membrane nanostructure in interfacial polycondensation [72] have been studied successfully. The hybrid MC/MD reaction method is called the Red Moon method from the acronyms of "Rare event-driving mechanism of our necessity", and our intention of the naming is described in Ref. [70].

8.5 Summary and Conclusions

In this chapter, we have reviewed in general the free energy gradient methodology. A family of the free energy gradient methods can provide us some essential properties of molecules in condensed state, such as molecular structures in stationary states and their vibrational spectroscopic information including the atomistic solvent effects under the thermodynamic condition, which must be helpful to understand the nature of chemical reactions in solution. This methodology should be systematized as the research protocol for automatically exploring the reaction pathway of an elementary chemical reaction on the free energy surface in solution. This should be accomplished by combining it with several specific methods which have been demonstrated by specific applications and were explained with suitable references. Thus, the development of theoretical methods based on the free energy landscape would bring about a systematic development of computational chemistry in condensed phase. This strategy is analogous to that developed in quantum chemistry to explore the Born-Oppenheimer potential energy surfaceIp of a reacting species in the isolated state. Moreover, we have proposed a novel method, i.e., hybrid MC/MD reaction method, to directly simulate such complex chemical reaction systems that an "effectively" enormous number of elementary ones are participating in. By employing the hybrid MC/MD reaction method, the stereochemical control of aggregation-state chemical reactions and the structural control of super-nano scale aggregates would be realized within a practical computational time. We are now trying to deal with and study more extensive chemical problems from the viewpoint of theoretical chemistry by combining the protocol for exploring elementary chemical reaction pathways on a free energy surface.

References

1. Nagaoka M, Okuyama-Yoshida N, Yamabe T (1998) J Phys Chem A 102:8202
2. Okuyama-Yoshida N, Nagaoka M, Yamabe T (1998) Int J Quantum Chem 70:95
3. Okuyama-Yoshida N, Nagaoka M, Yamabe T (1998) J Phys Chem A 102:282
4. Fukui K (1981) Acc Chem Res 14:363
5. Fukui K (1982) Angew Chem Int Ed 21:801
6. Okuyama-Yoshida N, Kataoka K, Nagaoka M, Yamabe T (2000) J Chem Phys 113:3519
7. Hirao H, Nagae Y, Nagaoka M (2001) Chem Phys Lett 348:350
8. Nagae Y, Oishi Y, Naruse N, Nagaoka M (2003) J Chem Phys 119:7972
9. Nagaoka M, Nagae Y, Koyano Y, Oishi Y (2006) J Phys Chem A 110:4555
10. Koyano Y, Takenaka N, Nakagawa Y, Nagaoka M (2010) Bull Chem Soc Jpn 83:486
11. Takenaka N, Kitamura Y, Koyano Y, Asada T, Nagaoka M (2011) Theor Chem Acc 130:215
12. Warshel A, Levitt M (1976) J Mol Biol 103:227
13. Singh UC, Kollman PA (1986) J Comp Chem 7:718
14. Field MJ, Bash PA, Karplus MA (1990) J Comput Chem 11:700
15. Thery V, Rinaldi D, Rivail JL, Maigret B, Ferenczy GG (1994) J Comput Chem 15:269
16. Monard G, Loos M, Thery V, Baka K, Rivail JL (1996) Int J Quantum Chem 58:153
17. Luque FJ, Reuter N, Cartier A, Ruiz-López MF (2000) J Phys Chem A 104:10923
18. Okamoto T, Yamada K, Koyano Y, Asada T, Koga N, Nagaoka M (2011) J Comput Chem 32:932
19. Yamada K, Koyano Y, Okamoto T, Asada T, Koga N, Nagaoka M (2011) J Comput Chem 32:3092
20. Okamoto T, Ishikawa T, Koyano Y, Yamamoto N, Kuwata K, Nagaoka M (2013) Bull Chem Soc Jpn 86:210
21. Martin ME, Sánchez ML, Olivares del Valle FJ, Aguilar MA (2002) J Chem Phys 116:1613
22. Galván IF, Sánchez ML, Martín ME, Olivares del Valle FJ, Aguilar MA (2003) J Chem Phys 118:255
23. Coutinho K, Georg HC, Fonseca TL, Ludwig V, Canuto S (2007) Chem Phys Lett 437:148
24. Georg HC, Canuto S (2012) J Phys Chem B 116:11247
25. Frisch MJ et al (2009) Gaussian 09, Revision A.02; Gaussian Inc, Wallingford
26. Broyden GC (1969) Comp J 12:94
27. Fletcher R (1970) Comp J 13:317
28. Goldfarb D (1970) Math Comp 24:23
29. Shanno FD (1970) Math Comp 24:647
30. Zwanzig RW (1954) J Chem Phys 22:1420
31. Kitamura Y, Takenaka N, Koyano Y, Nagaoka M (2014) J Chem Theor Comp 10:3369
32. Buchner M, Ladanyi BM, Stratt RM (1992) J Chem Phys 97:8522
33. Cho M, Fleming GR, Saito S, Ohmine I, Stratt RM (1994) J Chem Phys 100:6672
34. Kindt JT, Schmuttenmaer CA (1997) J Chem Phys 106:4389
35. Melzer A, Schella A, Schablinski J, Block D, Piel A (2013) Phys Rev E 87:033107
36. Matubayasi N, Nakahara M (2000) J Chem Phys 113:6070
37. Matubayasi N, Nakahara M (2002) J Chem Phys 117:3605
38. Friess SL, Lewis ES, Weissberger A (1963) Investigation of rate and mechanism of reactions. Technique of organic chemistry. Interscience vol 8-II
39. Slifkin MA, Ali SM (1984) J Mol Liq 28:215
40. Leung K, Rempe SB (2005) J Chem Phys 122:184506
41. Tuñón I, Silla E, Okamoto T, Ruiz-López MF (2000) Chem Phys Lett 321:433
42. Georg HC, Fernandes TS, Canuto S, Takenaka N, Kitamura Y, Nagaoka M (2014) A combination of the sequential QM/MM and the free energy gradient methodologies with applications. In: Leszczynski J, Shukla MK (eds) Practical aspects of computational chemistry III. Springer, Heidelberg, pp 231–247
43. Merrick JP, Moran D, Radom L (2007) J Phys Chem A 111:11683

44. Hermansson K (1993) J Chem Phys 99:861
45. Benedict WS, Gailar N, Plyler EK (1956) J Chem Phys 24:1139
46. Ford TA, Falk M (1968) Can J Chem 46:3579
47. Bertic JE, Lan Z (1996) Appl Spectrosc 50:1047
48. Ramaekers R, Pajak J, Lambie B, Maes G (2004) J Chem Phys 120:4182
49. Boeckx B, Maes G (2012) J Phys Chem B 116:11890
50. Tu Y, Laaksonen A (2000) J Chem Phys 113:11264
51. Laio A, VandeVondele J, Rothlisberger U (2002) J Chem Phys 116:6941
52. Takenaka N, Koyano Y, Nakagawa Y, Nagaoka M (2009) J Comput Chem 31:1287
53. Takenaka N, Koyano Y, Nagaoka M (2010) Chem Phys Lett 18:119
54. Koyano Y, Takenaka N, Nakagawa Y, Nagaoka M (2010) J Comput Chem 31:2628
55. Chalmet S, Ruiz-López MF (2000) Chem Phys Lett 329:154
56. Takenaka N, Kitamura Y, Koyano Y, Nagaoka M (2012) Chem Phys Lett 524:56
57. Takenaka N, Kitamura Y, Koyano Y, Nagaoka M (2012) J Chem Phys 137:024501
58. Kerdcharoen T, Liedl KR, Rode BM (1996) Chem Phys 211:313
59. Kerdcharoen T, Morokuma K (2002) Chem Phys Lett 355:257
60. Heyden A, Lin H, Truhlar DG (2007) J Phys Chem B 111:2231
61. Bulo RE, Ensing B, Sikkema J, Visscher L (2009) J Chem Theory Comput 5:2212
62. Berendsen HJC, Grigera JR, Straatsma TP (1987) J Phys Chem 91:6269
63. Kameda Y, Ebara H, Usuki T, Uemura O, Misawa M (1996) Bull Chem Soc Jpn 69:1495
64. Kitaura K, Ikeo E, Asada T, Nakano T, Uebayasi M (1999) Chem Phys Lett 313:701
65. Fedorov DG, Kitaura K (2009) The fragment molecular orbital method: practical applications to large molecular systems. CRC Press, Boca Raton. doi:10.1201/9781420078497
66. Komeiji Y, Nakano T, Fukuzawa K, Ueno Y, Inadomi Y, Nemoto T, Uebayasi M, Fedorov DG, Kitaura K (2003) Chem Phys Lett 372:342
67. Komeiji Y, Inadomi Y, Nakano T (2004) Comput Biol Chem 28:155
68. Ishikawa T, Ishikura T, Kuwata K (2009) J Comput Chem 30:2594
69. Ishikawa T (2010) Parallelized ab initio calculation system based on FMO (PAICS). http://www.paics.net/. Accessed 19 May 2010
70. Nagaoka M, Suzuki Y, Okamoto T, Takenaka N (2013) Chem Phys Lett 538:80
71. Takenaka N, Suzuki Y, Sakai H, Nagaoka M (2014) J Phys Chem C 118:10874
72. Suzuki Y, Koyano Y, Nagaoka M (2015) J Phys Chem B 119:6776

Chapter 9
Towards an Accurate Model for Halogens in Aqueous Solutions

M.I. Bernal-Uruchurtu, A. Alcaraz Torres, F.A. Batista Romero and R. Hernández-Lamoneda

Abstract The overwhelming progress and constant evolution of computational and theoretical methods in chemistry have provided us a more detailed molecular description of some complex molecular properties. Our group has intended to do so for an old problem: the solvatochromic properties of halogens in aqueous systems. There are beautiful experiments that show how sensitive Br_2 and Cl_2 are to the structure of the environment around them. In this chapter, we present the tests and calculations performed with different theoretical methods to identify their reliability as pieces of a multi-scale study aimed to address open questions related with this phenomenon. We used different approaches to explicitly take into account the solvent effect and tested several theoretical methods on the solvatochromic effect of small clusters. The combination of a semiempirical Born-Oppenheimer molecular dynamics study (SEBOMD) of Br_2 in liquid water solution using PM3-PIF and then, the evaluation of the effect the closest water molecules have on the shifts are presented. This is a first step towards a robust multi-scale protocol ad hoc designed for these systems.

9.1 Introduction

In the last decade, detailed experiments on the spectroscopic behavior of halogens in aqueous environments demanded the concurrence of theoretical methods to provide a rationale of the results in terms of the molecular properties of its components. These experiments [1–4] increased the amount of information confirming the remarkable sensitivity of chlorine, bromine and iodine as sensors of their molecular environment.

M.I. Bernal-Uruchurtu (✉) · A. Alcaraz Torres · F.A. Batista Romero · R. Hernández-Lamoneda
Centro de Investigaciones Químicas, Universidad Autónoma del Estado de Morelos, 62209 Cuernavaca, Mexico
e-mail: mabel@uaem.mx

J.-L. Rivail et al. (eds.), *Quantum Modeling of Complex Molecular Systems*, Challenges and Advances in Computational Chemistry and Physics 21,
DOI 10.1007/978-3-319-21626-3_9

Indeed, at the end of the 19th century the first observations of the large shift of the electronic spectra of halogens in solution were published [5, 6]. A favored hypothesis, aimed to explain the color of solutions containing halogens in different solvents, claimed that the colors observed were only related to properties of the solvent. In 1950 Mulliken contested this idea with the proposal of solute-solvent complexes [7]. His suggestion of charge transfer complexes was later proved wrong; nonetheless it prompted the search of a better understanding of their nature, structure and stability. Mulliken's suggestion of charge transfer complexes was later reinforced by the structural studies of Hassel's group on the structures of adducts formed between oxygenated solvents such as 1,4 dioxane and bromine [8]. The stability of these donor-acceptor complexes was thought to have the same origin as charge-transfer structures, i.e., transfer of negative electric charge from a donor molecule to an acceptor molecule. Few years later, the very first semi-empirical quantum calculations on the Cl_2–H_2O system found a strong charge polarization of the dihalogen and little charge transfer from one molecule to the other [9].

During the 80s and 90s, the advance of experimental techniques well suited to the study of small clusters in the gas phase and in noble gas matrices was fundamental for the proper characterization of the interaction between dihalogens molecules and water. The molecular beam spectroscopy studies done, in particular by A. Legon's group, allowed for a definitive characterization of the forces responsible of the stability of Lewis base donation to dihalogens and inter-halogen molecules [10–14]. The alignment of the lone pair on the oxygen atom of the donor molecule with the axis of the covalent bond in the dihalogen acceptor, led to the identification of a new type of intermolecular interaction: *'halogen-bond'*. Groups studying molecular interactions in crystals soon adopted this term [15]. The IUPAC definition of this force recognizes the parallelism between hydrogen- (HB) and halogen-bond (XB). Its general properties are:

- The physical forces involved in the halogen bond must include electrostatic and inductive forces in addition to London dispersion forces.
- The atoms Y and X are covalently bound to one another, and B···X–Y is polarized so that the X atom becomes more electropositive (i.e. the partial positive charge δ^+ increases).
- The lengths of the X–Y bond and, to a lesser extent, the bonds involved in B deviate from their equilibrium values.
- The stronger the halogen bond, the more nearly linear is the B···X–Y arrangement and the shorter the B···X distance.
- The interaction energy per halogen bond is greater than at least a few times kT, where T is the temperature of the observation, in order to ensure its stability.

The spectroscopic behavior of halogens in aqueous solution is further complicated by two aspects, first the relatively low solubility of halogens in liquid water and second, the rich chemistry of dihalogens in these media. Bromine has the highest solubility in liquid water, ≈0.2 M, ($Cl_2 \approx 0.05$ M and $I_2 \approx 0.001$ M) but, as the others, it decomposes in solution. The hydrolysis products of X_2 in water depend on pH and lead to the formation of the oxyhalogen acids, HOX and its

$$X_2 + H_2O \leftrightarrows HXO + H^+ + X^-$$
$$X_2 + X^- \leftrightarrows X_3^-$$

Scheme 9.1 Hydrolysis reactions of dihalogens

anions (See Scheme 9.1). This problem is particularly true for chlorine, whose K_{eq} is at least five orders of magnitude larger than those of bromine and iodine. Furthermore, a concurring equilibrium forms the X_3^- ion with a large extinction coefficient and a well-known signature in UV-vis spectra.

The gas phase spectra of dihalogens are dominated by two valence excited states close in energy that result in an overlapping spectra. The solvent chosen can modify the position of the maximum in the spectra. The solvatochromic phenomena observed at the end of the 19th century has not yet been fully understood. The work of Ham in 1954 confirmed that the formation of a donor-acceptor complex between the solvent and the dihalogen was responsible for the observed blue shift in the spectra [16]. The experimental studies of Janda's and Apkarian's groups recorded the shift in the absorption spectra of chlorine and bromine in liquid water solution, ice solution and clathrates. The shifts observed in aqueous liquid and ice solutions of halogens is similar; 550, 1750 and 2820 cm^{-1} for Cl_2, Br_2 and I_2 respectively, whereas the shifts recorded for the bromine and iodine clathrate are at least 50 % smaller [17]. For some years our group has pursued an explanation of the differences observed in different media.

To understand how all the halogen and water molecular properties act together resulting in the observed condensed phase phenomena, we have used several computational chemistry methodologies. As our work has found, choosing a methodology that pinpoints the relative relevance each feature has on the macroscopic behavior, is a difficult task. In this work we describe the results obtained using different methods to model the aqueous environment in condensed phases and the problems encountered with their use for large systems or precise spectroscopic predictions. In some sense, this exploration has rendered the pieces that might be useful to build a high-performance multi-scale model for the study of halogens in aqueous systems.

The interactions present in halogen-bonded systems are in general weak, thus the need to use quantum mechanical methods that include electron correlation for an accurate description. In particular, these interactions are sensitive, not only to the basis set used but also to the level of electron correlation used, the counterpoise correction and the inclusion of spin-orbit effects for bromine and iodine. In contrast with the widely accepted idea that hydrogen bonded systems are reasonably well described with Moller-Plesset perturbation (MP2) methods, halogen bond interactions are overestimated at this level. Therefore, for an accurate description of these systems a coupled cluster method is advised, preferably with single and double excitations (CCSD) or with perturbative triple excitations (CCSD(T)). This problem pertains also to density functional theory methods (DFT) where some methods lead to better results than others and the choice of the best functional is not straightforward. DFT methods have the advantage over wave function methods that time dependent DFT (TD-DFT) offers access to excited states in a considerably less costly manner than CI or CC methods.

The need to validate the model for halogens in aqueous media using spectroscopic predictions makes necessary an accurate description of the electronic structure of the halogen and its environment or at least, the effect the latter has on the solute. The dynamic and thermal effects of the aqueous system also need to be captured in a realistic way. For these reasons, designing an accurate model for these systems is a true challenge. Furthermore, a well-suited model for this system requires attributes such as:

- *Accuracy*: The structure, stability and spectroscopic properties must be well reproduced. This needs a proper description of the HB in aqueous system, as well as, the interactions formed between a water molecule and a halogen one (XB), and a weaker one, hydrogen-halogen interaction (HX) [18]. For excitation energies and shifts the tolerance has to be larger than for other properties but a requirement might be that the method discern with no error the blue shifting configurations from the red shifting ones and, of course, the predicted average shifts must coincide, at least in the order of magnitude, with the experimental ones.
- *Flexibility*. The model selected for the study must be able to reproduce most of the structural changes induced by specific molecular interactions. HB and XB result in significant structural changes that favor the stability of the system through non-additive effects. Furthermore, XB and also HB result in a modification of some covalent bonds as a pre-requisite for reactivity [19].
- *Polarizability*. The large polarizability of halogen molecules is one of the factors leading to the interaction with water, and the polarizability of water itself is a property that has been found as crucial for the complex formation. Instead of a phenomenological treatment of molecular polarizability as many refined classical models have, it would be convenient to count with a more realistic charge distribution and mobility that would also provide the anisotropic element of the interactions present in the system.
- *Low-cost*. The reliability of a computational study of halogens in condensed phases requires a proper sampling of the configurational space. This only is achieved if a very large number of configurations or a long trajectory is obtained. Thus, very costly methods or full ab initio or on-the-fly DFT simulations are not a reasonable option.

9.2 Brief Review of Models Aimed to Halogen-Bonded Systems

An indicator of the acceptance halogen-bonding as a relevant variable for the better understanding of many interesting phenomena, from: molecular recognition, drug design, crystal engineering to basic physical chemistry of halogens, is the rise in the literature of approximate models during the past few years. As expected, several classical models aimed to study phenomena in condensed phases have been developed. Among them it is possible to distinguish those using general parameters for halogen atoms like *MMFF* (Merck Molecular Force Field) from those that focus on the peculiarities of halogen bonding like its polarizability and anisotropy. For example, in 2011 Ibrahim proposed a form in which the sigma-hole[1] notion [20] could be added to standard MM (molecular mechanics) models [21]. The electropositive region in a family of halobenzenes was modeled using an extra positive point charge (EP) placed at an optimized distance of the halogen atom. This approach led to a good reproduction of halogen bonds between halobenzenes and a series of Lewis bases. However, it's probable that EPs alone are not sufficient to reproduce the anisotropy of the interaction for a halogen bonds formed with dihalogens that exhibit a large quadrupole moment like bromine.

Another widely used model, the optimized potentials for liquid simulations (OPLS), was unable to reproduce XB due to the negative charge placed in halogens for the generation of the dipole moment of molecules bearing them. The addition of a site with a positive charge to the halogen atoms corrected their intermolecular behavior and the updated model might be used in combination with other water models [22]. The studies of Jorgensen's group on halobenzenes hydration revealed that XB, albeit not having a large impact on the total solute-solvent interaction energy, modified the orientation of water molecules close to the halogen-ending side of the aromatic derivative and favored specific interactions with active sites in a model protein.

An explicit sigma-hole (ESH) was added to chlorine, bromine and iodine in the Amber force field [23]. In this model, ESH was also designed as a massless point charge. The fine balance between the magnitude of ESH charge and its position with respect to the van der Waals boundary of the halogen atom were the main issues related to the improvement of the gas-phase calculations and protein–ligand geometry optimizations. As in many cases, an increased accuracy might require a revision of the universal parameters by a careful parameterization targeted to the specific molecules/problems. Du et al. developed a different set of parameters, also compatible with the Amber force field but considering the anisotropic effects present in halogen bonds [24]. A sphere of negative charge and an ellipsoid of positive charge represent the charge distribution characteristic of halogens. The parameters were fitted to reproduce the ab initio electrostatic potential obtained for several individual halomolecules. The performance of the model consisted in verifying that the essential features of the ab initio potential energy surface of one halogen-bonded pair was correct.

The *SIBFA* model [25, 26] (sum of interactions between fragments ab initio computed) obtains the energy of a system as the contribution of electrostatic multipolar, short-range repulsion, polarization, charge transfer and dispersion.

---

[1]**sigma-hole**: A region of electropositive potential due to the depletion of electron density occurring at the ends of the covalent sigma X-Y bond. It can favorably interact with a lone pair of electrons on a heteroatom from the donor.

The multipolar components coming from ab initio calculations were obtained using Stone's analysis and the explicit inclusion of them provides the anisotropic property to the polarizable model required for a good description of a sigma-hole driven interaction between halobenzenes and water and Mg(II) [27].

Schofield and Jordan [28] developed a polarizable force field specifically fitted to describe the interaction between bromine and water. In the COS (charges on springs) model, polarization is achieved by placing a single charge-on-a-spring at the bond midpoint of the Br_2 molecule and fitting the force constant of the spring and its value to reproduce the ab initio polarizability of this molecule. The bromine-water interaction terms were fitted to reproduce the electrostatic, polarization, dispersion and exchange repulsion contributions to the total interaction energy of Br_2-water pairs. The quality of the fitting allows them to have average errors for binding energy reproduction of only 0.076 kcal mol^{-1}. This is, indeed, a very good fit however, its accuracy is difficult to evaluate here due to the fact that the accuracy of the resolution-of-the-identity second-order Moller-Plesset perturbation (RIMP2) for calculating the interaction energies of the complex is not reported. Therefore, if a first approximation is to consider it as accurate as a MP2 calculation, it will overestimate the interaction energy of this complex by ≈5 %. The performance of *COS* model was validated with the reproduction of the ab initio obtained PES of Br_2 in the polyhedral cages that form the crystal structure of clathrates and the calculation of the binding energy from the minima found for each cage. The polarizable model, overestimates it by 12–20 % when compared with their best reference: a local coupled cluster singles doubles (triples) LCCSD(T)/ AVDZ, AVTZ, AVQZ. (aug-cc-pV(D,T,Q)Z) Another finding reported by them is that *MMFF* underestimates the interaction between Br_2 and the water cages by a large amount. This is a very good example of the advantages of having an ad hoc model fitted to reproduce good quality ab initio data instead of a general method. Their model is used to compare the stability difference between three polymorphic structures for bromine clathrate. Molecular dynamics (MD) simulation considered rigid molecules in all cases and provided a reasonable good stability order considering that only the enthalpic contribution to the stability was evaluated from their work.

On the other hand, it is possible to find that approximate quantum based methods and models have appeared in the past 2 years. Kubař et al. presented a self consistent charge density functional based tight-binding (*SCC-DFTB)* method for halogens [29]. These methods derived from DFT have showed interesting results as an alternative to semiempirical methods to describe complex molecular systems. More recently the group developing them obtained parameters for halogens. Their performance for a set of halogenated organic molecules might be further improved and there are not yet results for non-bonded interactions as the halogen-bond.

The computational cost/performance ratio is one of the most attractive features of semiempirical methods and for that reason they have been recognized as a limited but still useful way to approach the study of complex systems. However, the accuracy limitations on the description of non-bonded interactions have been the main deterrent for their use. In the past years, several new parameterizations have been made and empirical corrections added for dispersion and hydrogen bonding, which have improved the quality of the results for some systems, but several non-physical artifacts remain in the description of geometries far from the equilibrium ones. Several years ago, the origin of those artifacts in PM3 was pointed to the Gaussian correction functions (GCF) [30] introduced to create a minimum in the H-bond region of the water dimer and correct the excessive repulsion MNDO-type methods had. Indeed, the replacement of the GCF for a simpler form, a parametrizable interaction function (Eq. 9.1) (*PIF*) [31], fitted to reproduce high-quality ab initio data opened a procedure for obtaining ad hoc semiempirical models [32–34].

$$PIF = \sum_{A,B}^{\text{inter}} g(A,B) = \sum_{A,B}^{\text{inter}} \alpha_{AB} \boldsymbol{e}^{-(\beta_{AB} R_{AB})} + \frac{\chi_{AB}}{\boldsymbol{R}_{AB}^{6}} + \frac{\delta_{AB}}{\boldsymbol{R}_{AB}^{8}} + \frac{\varepsilon_{AB}}{\boldsymbol{R}_{AB}^{10}} \tag{9.1}$$

Equation 9.1 Analytical form of the parameterizable interaction function.

Models like PM3-PIF are very well suited for the description of water clusters, liquid water and the solvation of a wide range of molecules [35]. The model built this way can be used as a quantum semiempirical force field where the molecular flexibility and polarizability are natural features at the core of the model. The lacking components of molecular interaction, mainly dispersion, are included in an effective way through the parameterization of the PIF function to MP2 potential energy surfaces of the system of interest. We have recently published the parameters for the Br–O and Br–H interaction [36]. The tests of their performance showed, once again, a considerable improvement over the geometry and interaction energies of the specific interactions of this system, XB and XH, with respect to PM3. Additional to the interaction of a Br_2–H_2O complex, a thorough set of tests to validate its use for the study of bromine in aqueous systems was performed. In a forthcoming section we will present additional details of its performance and some new applications.

This overview of the available models capable of treating halogen-bonded systems shows three important things: first, considering the physical ingredients of the interaction, i.e., anisotropic charge distribution and polarization, results in better performing models; second, serious validation of an approximate model, classic or quantum, requires an analysis of its performance in molecular dynamics, and third, general models for halogen bonding tend to estimate results with lower accuracy than those developed for a specific system or problem.

9.3 The Stepwise-Approach to a Multi-scale Model

Ab initio molecular dynamics and/or Car-Parrinello-type simulations seem a promising option to model the physical chemistry of halogens in aqueous systems due to their robustness, and the fact that they might be useful to simulate spectroscopic properties and reactivity. However, the size of the systems needed to study, either its reactivity or its spectroscopy, is beyond the current capabilities of those methods. Besides, in the literature there are some examples where this option does not always provide physical results [37–39]. To have an idea of the complexities of the system we are interested in, let's consider for example, the formation of bromine clathrates; starting from a saturated solution of bromine in liquid water upon clathrate formation there is an increase of the Br_2/H_2O ratio by 40. Thus, modeling a system big enough to look into this interesting process is not amenable to all standard numerical simulation protocols. The pieces of information gathered about these systems, with well-known quantum and classical methodologies, are useful to build up a hierarchical scheme of the molecular properties needed to model this complex system. This will be helpful to evaluate at least two important factors: the convenience of including or not, a particular molecular feature in the reproduction of a particular property and, the accuracy of a protocol for studying these complex systems.

In the next sections we present an account of the main findings of each step towards a better understanding of the physical chemistry of halogens in aqueous media. The order chosen considers the different levels of complexity in the description of the aqueous environment as well as some of the available methods to look into the effect this environment has on the spectroscopic properties of bromine.

9.3.1 Discrete and Small Number of Solvent Molecules

The effect a single water molecule has on the spectroscopic properties of bromine, was studied by Hernández-Lamoneda et al. with ab initio methods [40]. From the experimental work done by Legon's group it was clear that a non-negligible force was responsible of the complex formation. The interaction energies calculated using CCSD(T) methods and large basis sets showed that the interaction responsible for the water-bromine complex formation is stronger than most "weak interactions": 15.23 kJ mol^{-1}, this amounts to approx. 60 % of the interaction energy in the water dimer. However, when zero point energy (ZPE) is considered the energy difference between these interactions becomes smaller [41].

The prediction of the well depths is dependent on the size of the basis set selected and the complete basis set limit extrapolated value is 15.02 kJ mol^{-1}. Curiously enough, the overestimation observed with MP2 methods and double-ζ basis sets results in a reasonable agreement with more refined methods and larger basis sets, making of MP2 a reasonable choice for the study of larger systems.

On the other hand, DFT methodologies have shown that for the same system, the interaction energies and quite probably, other properties are not only dependent on the size of the basis set but also on the functional chosen. Two independent studies showed that B3LYP and BHHLYP with 6-311+G(d,p) and 6-31++G(d,p) basis sets respectively, overestimated the interaction energy, 16.99 and 24.48 kJ mol^{-1} correspondingly [42, 43]. Thus, the election of this convenient method, in terms of computational cost, implies a severe compromise on interaction energies.

In fact, two different works looked into the stable structures of clusters formed with dihalogens and water clusters, one using DFT [42] and the other ab initio methods [18]. The results differ in the structures identified as the global minimum for each number of water molecules. MP2/AVDZ methods identify as minima structures where water molecules form themselves the most stable structure for that water stoichiometry and the dihalogen forms a strong XB with one of the water molecules. Additional stability arises from with the formation of the weak XH interaction. That DFT study reports as stable structures some that are close in energy to the ab initio minima but differ in the arrangement of the water molecules. It also found some structures that are not ab initio minima. A recurring difference between similar DFT and ab initio structures is that the former methods do not stabilize the XH interaction found in [18].

DFT offers, through TD-DFT, advantages on the ease of calculation of the transition energies to the excited states compared with restricted methods such as RMP2 or RCCSD(T). Since the study of halogens in condensed phase solutions is aimed to understand the solvatochromic effect, here we decided to perform test calculations with the ubiquitous B3LYP functional and a larger basis set than for MP2 calculations, AVTZ instead of AVDZ. An exhaustive exploration of different levels within the DFT framework is beyond the scope of this work.

In Fig. 9.1, it is possible to find the optimized geometries for six selected clusters, the ab initio structures were obtained at the MP2/AVDZ level [18] and the DFT ones at the B3LYP/AVTZ level. Only two of the three minima found for the 1:1 complex were considered in this work (the missing one differs only in the orientation of the water molecule). As expected the XB interaction leads to a more stable structure than the one stabilized by a XH. The decomposition of the interaction energy using the symmetry adapted perturbation theory (SAPT) for this two structures reveals that they are quite different; the electrostatic component of the XB structure contributes with 60 % of the interaction energy, the induction with 10 % and dispersion with 30 %; whereas for the XH bonded complex it was found that the main contribution to its stability was dispersion with 55 % followed by 31 % of electrostatic and 14 % of induction contribution. This difference between XB and XH interaction is useful to understand why this interaction is not that important in DFT optimized structures. The shifts calculated for small clusters are presented in Table 9.1.

The TD-DFT calculation of the transition energies corresponds to the average of the lowest lying singlet excited states, whereas RMP2 and CCSD(T) correspond to the triplet states. The comparison is feasible since singlet and triplet states arise from the same electronic configuration. Also notice that since the triplet state corresponds to the ground state for that multiplicity it is possible to use open-shell ground state methods. In general, a good agreement is observed between ab initio and DFT but the shifts predicted by TD-DFT are 10 % smaller than the ab initio ones. However, for the structures strongly dependent on XH interactions the errors are quite large, 65 and 30 % for 1b and 3b respectively. Both methods coincide in predicting considerably larger shifts for these small clusters than the ones observed for bromine in solution.

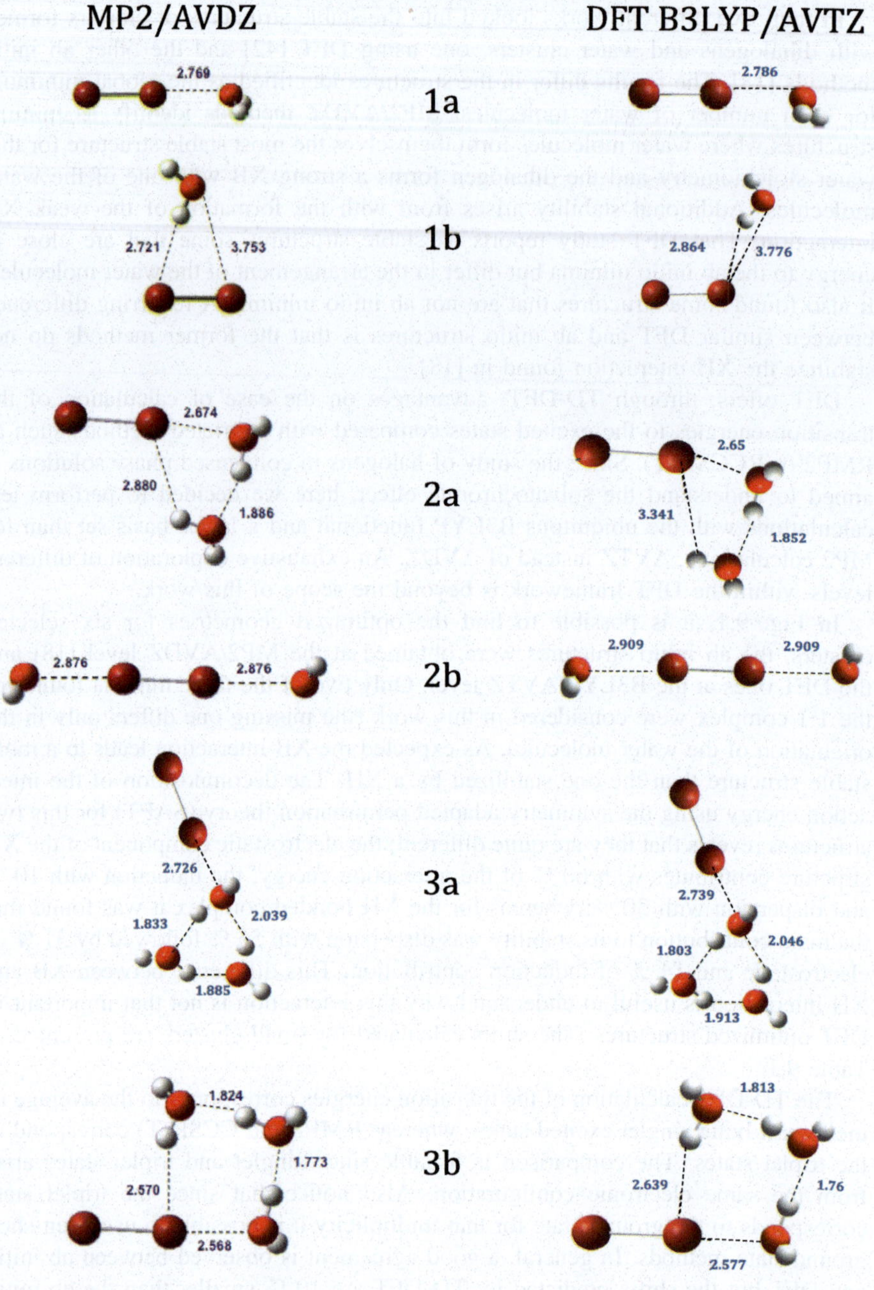

Fig. 9.1 Optimized structures of selected stable clusters of Br_2-$(H_2O)_n$ with n = 1,3. The values of the intermolecular distances are in Å

Table 9.1 $\Delta\omega$ in cm^{-1} calculated for optimized geometries of Br_2-$(H_2O)_n$ with n = 1–3 shown in Fig. 9.1

	RMP2[a]	RCCSD(T)[b]	TD-DFT[c]
Br_2–H_2O (a)	2052	1983	1788
Br_2–H_2O (b)	640	571	201
Br_2–$(H_2O)_2$ (a)	2892	2724	2567
Br_2–$(H_2O)_2$ (b)	3626	3471	3018
Br_2–$(H_2O)_3$ (a)	2372	2294	2074
Br_2–$(H_2O)_3$ (b)	6739	4465	3109

[a]Restricted MP2/AVDZ
[b]MP2/AVDZ//RCCSD(T)
[c]TD-DFT B3LYP/AVTZ

9.3.2 Continuum Model of the Aqueous Environment

During the last 40 years it has been possible to witness an important evolution on the way the environment around a solute molecule is described. The reaction field approach, the effect a continuous dielectric medium has on the charge distribution of a molecule that polarizes back the dielectric and generates a reaction potential, is a standard scheme to consider the solvent effects on many molecular properties. Most modern continuum models obtain through a self-consistent cycle the wave function of the molecule affected by the reaction potential thus the self-consistent reaction field acronym (SCRF). Solvatochromic effects have been more or less successfully explained using: from Onsager's to more refined models like Nancy SCRF [44], Tomasi's polarizable continuum model (PCM) [45], Cramer and Truhlar's SMx models [46].

Early tests on the field effects characterized by dielectric constant values ranging from 0 to 100 had on Cl_2 shifts resulted in a small blue shift (30–80 cm^{-1}) for equilibrium transition i.e., the Cl–Cl distance was optimized considering the field around the molecule. These calculations were done using the Nancy SCRF model [47] implemented in Gaussian03. Similar behavior was found with PCM on Gaussian09 [48, 49] for the non-equilibrium vertical transitions of bromine but with larger values.

The dielectric constant differences between liquid water, ice and clathrates are not known but the effect of adding a reaction field with the dielectric constant of liquid water allowed us to evaluate the effect it might have over an already XB or XH interacting complex. As several studies have shown, the explicit inclusion of water molecules allows having the short range effect of hydration while the continuum contributes to the electrostatic and polarization long range effects [50–53]. In Table 9.2 we present the shifts calculated for the non-equilibrium vertical transitions of the small clusters described previously using TD-DFT at the B3LYP/AVTZ level on structures optimized in a vacuum and in the cavity. As it can be seen the size of the shifts increases as a consequence of the continuum field over the clusters although there is not a clear trend in the amount increased. However, this observation

Table 9.2 $\Delta\omega$ in cm^{-1} of small Br_2-$(H_2O)_n$ with n = 1–3 in vacuum and in a PCM solvent cavity. Structures are shown in Fig. 9.1. The reference in gas phase corresponds to the B3LYP/AVTZ optimized geometry for bromine

	TD-DFT	
	$\Delta\omega$ in vacuum	$\Delta\omega$ in RF
Br_2	0	434
Br_2–H_2O (a)	1788	2629
Br_2–H_2O (b)	201	549
Br_2–$(H_2O)_2$ (a)	2567	3203
Br_2–$(H_2O)_2$ (b)	3018	3832
Br_2–$(H_2O)_3$ (a)	2074	2873
Br_2–$(H_2O)_3$ (b)	3109	3402

could be used in cluster models of the liquid or clathrates to take into account the effect of the surrounding environment.

9.3.3 A Simple Electrostatic Model of the Environment

As the size of the system increases the kind of problems that require to be considered for modeling these systems is different. As the previous sections mentioned, for small clusters the ability to accurately describe the geometry or the shifts of a particular method has is of crucial importance. In the case of clathrates an open question was the effect the possible different orientations of the halogen in the cages had on the measured shift. Polyhedral cages with 20, 24, 26, or 28 water molecules form the clathrate structure. Combinations of different proportions of these building blocks create the crystal cell of the clathrate. The size of the system makes unpractical the study of the potential energy surface (PES) of the dihalogen inside those cages with standard ab initio methodology. For that reason we decided to use a minimalist model of the cages. The simplest consideration possible of the effect the structure of the environment around Br_2 has on its spectroscopic behavior was to use a crude representation of water molecules. For this, a dipole could replace each one of the molecules forming a clathrate cage[2] (5^{12}, $5^{12}6^2$, $5^{12}6^3$, $5^{12}6^4$) however, there is no way of computing individual dipole moments without an approximation of the charge distribution of each molecule on the cage. A simpler approach might then be the placement of point charges at the O and H positions in the cage. These partial charges might be obtained from the best available electronic density of each cage, using one of the many available charge model [36].

The effect this electrostatic model of a cage has on the transition energies was evaluated using the equations of motion (EOM)–CCSD method [54] as implemented in Molpro [55]. Constraining the vibrational mode of Br_2 and sampling all the possible orientations of the diatomic in each cage, it is possible to show that the anisotropy of the interaction, combined with the particular symmetry of each cage,

[2]Nomenclature 5^n6^m indicates a water cage composed of *n* pentagonal and *m* hexagonal faces.

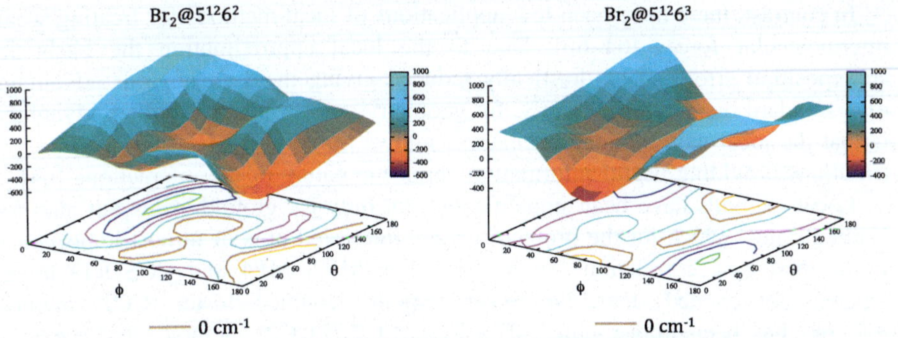

Fig. 9.2 EOM:CCSD calculated shifts (in cm^{-1}) of Br_2 inside point charge simulated $5^{12}6^2$ and $5^{12}6^3$ clathrate cages. θ and ϕ (in degrees) correspond to the orientation of the dihalogen with respect to the main symmetry axis of the cage

results in an irregular pattern of red- and blue-shifting regions as shown on Fig. 9.2. The latter correspond well with the regions where the electrostatic field created by the point charges stabilizes the ground state and vice versa.

This crude model predicts blue shifts in the right order of magnitude of the experimental ones. However it has clear shortcomings due to the lack of other relevant components of the interaction between the cage and Br_2 such as exchange and dispersion.

9.3.4 Local Correlation Methods

The main motivation behind local correlation methods is to be able to treat large molecular systems within full ab initio formalism. In this respect they represent an alternative to hybrid approaches, such as quantum mechanics/molecular mechanics (QM/MM), where one has to deal with the non-trivial problem of the coupling between the boundary dividing the regions treated with different levels of theory. The key ingredient of the local correlation methods [56–58] is the localization of both, occupied and virtual spaces, such that by restricting the electronic excitations within localized domains it is possible to achieve linear scaling with the size of the system [59]. In fact, it has been shown that when using local methods, the bottleneck of the calculation becomes the Hartree-Fock procedure eventually. In order to overcome this limitation the use of density fitting approximations for the two-electron integrals becomes mandatory [60, 61].

So far, the most successful applications of local correlation methods have been the calculation of basic structural information: geometries, frequencies, dipole moments, NMR chemical shifts and reaction energies for systems composed of over a hundred atoms and few thousand basis functions. In those cases it has been shown that 98–99 % of the correlation energy can be recovered [62]. It has also been used within QM/MM approaches to model enzymatic reactions [63].

In contrast, there have been few applications of local methods for treating weak intermolecular forces [64–66]. Due to the local approximation the basis set superposition error can be drastically reduced giving them an obvious advantage over standard methods particularly, for geometry optimizations. Another advantage is that the local nature of the excitations allows a partitioning of the energy contributions providing useful information about the nature of the interactions [64].

Local methods have been implemented not only for ground states but also for excited states, which for the problem we are interested in is of utmost importance. In the first case correlation can be treated at MP2, MP4 and CCSD(T) levels whereas for excited states the linear response coupled-cluster (CC2) method [67, 68] has been implemented. The local CC2 (LCC2) method has been successfully tested against canonical CC2 for a variety of molecules having deviations of less than 0.05 eV in the transition energies [67, 68]. Furthermore, it can deal with complicated cases such as Rydberg states and charge transfer excitations, which are a challenge for TD-DFT methodology.

In Table 9.3 the calculated shifts using LCC2 for small clusters are compared with the values coming from the highly accurate RCCSD(T).

In general, the agreement between LCC2 and RCCSD(T) is quite good except for two clusters: 1b and 3b. As it is possible to see in Fig. 9.1 for both geometries the XH interaction plays a key role for their stability. To understand the large differences for these two cases, one has to take into account the fact that transition energies are quite sensitive to the XB and XH distances, thus a vertical transition starting on a slightly displaced position in the ground state PES might result in large changes due to the steep slope of the excited state PES in that particular region. The LCC2 optimized geometries are free of BSSE error in contrast with RCCSD(T) and TD-DFT at the B3LYP/AVTZ (see Table 9.1). For structure 1b the Br–H is 0.1 Å longer in the local case and for 3b it is 0.05 Å. However, to confirm that the large difference found for 1b is due to the difference in geometry, we use the local optimized geometry in a standard RCCSD(T) calculation of the shift. The result, 11 cm^{-1} is much closer to the LCC2 value.

Table 9.3 $\Delta\omega$ in cm^{-1} of small Br_2-$(H_2O)_n$ with n = 1–3 using local and standard ab initio methods. Structures are shown in Fig. 9.1. The reference in gas phase corresponds to the optimized geometry for bromine with each method

	RMP2	RCCSD(T)	LCC2
Br_2–H_2O (a)	2052	1983	1980
Br_2–H_2O (b)	640	571	−70
Br_2–$(H_2O)_2$ (a)	2892	2724	2463
Br_2–$(H_2O)_2$ (b)	3626	3471	3209
Br_2–$(H_2O)_3$ (a)	2372	2294	2252
Br_2–$(H_2O)_3$ (b)	6739	4465	2108

9.3.5 Semiempirical Born-Oppenheimer Molecular Dynamics of Br_2 in Liquid Water

The PM3-PIF models that have been developed in the past decade were aimed to enable more realistic studies of solutes in aqueous solutions. The fact that at the center of the model there is a correct structural and energetic description of the intermolecular forces governing the physical and chemical behavior of a condensed phase makes of these PM3-PIF models an excellent choice for a large exploration of the dynamic and structural properties of solutions. A clear obstacle for an ample use of these models is the need to obtain specific atomic pair parameters for each system and, as recently found, they are not necessarily transferable [34]. However, the advantage of having an inexpensive model including implicit polarizability and molecular flexibility with a satisfactory description of interactions makes worth the parameterization challenge.

As we recently reported, fitting the PIF parameters for Br–O and Br–H was not exempt of its own difficulties. In particular, the use of the regular PIF function (Eq. 9.1) resulted in a poor description of the PES of bromine in the cages. For that reason an additional term was added to the function resulting in an improved fitting and description of the 1:1 complex PES as well as the Br_2@cages PES. The PIF form used for Br–O and Br–H pairs only is:

$$PIF = \sum_{A,B}^{inter} g_{PIF}(A,B) = \sum_{A,B}^{inter} \alpha_{AB} e^{-(\beta_{AB} R_{AB})} + \frac{\chi_{AB}}{R_{AB}^6} + \frac{\delta_{AB}}{R_{AB}^8} + \frac{\varepsilon_{AB}}{R_{AB}^{10}} + \frac{\eta_{AB}}{R_{AB}^{12}} \quad (9.2)$$

Equation 9.2 PIF function used only for the Br–O and Br–H pairs.

PM3-PIF was implemented in DivCon, a semiempirical linear scaling code and then coupled with Tinker [69], a molecular dynamics code. With the DivCon/Tinker set of programs we calculated molecular dynamics trajectories for a system formed with 210 water molecules and one Br_2 molecule for a total of 150 ps. Several interesting questions about bromine in liquid water solution might be addressed with this study. For example: Is the hydration of a non polar but polarizable solute, hydrophobic or hydrophilic?, To what extent the spectroscopic behavior of bromine in liquid water is the result of the closest neighbors interactions? In this work we will address the latter since it pertains to the discussion about the performance of different methods to evaluate the transition energies to the excited states and we will leave a more detailed description of bromine hydration to a forthcoming report.

With the trajectory obtained for the system described above in a NVE ensemble at 298 K, using periodic boundary conditions and Ewald correction and considering the density of this dilute solution close to that of pure water, we obtained the radial distribution function (rdf). The first peak of the $g_{Br\text{–}O}$ and $g_{Br\text{–}H}$ distributions start around 2.4 Å and have the first minimum close to 5 Å, both peaks are centered around the same distance, ≈3 Å. The closest molecules to bromine are those in the

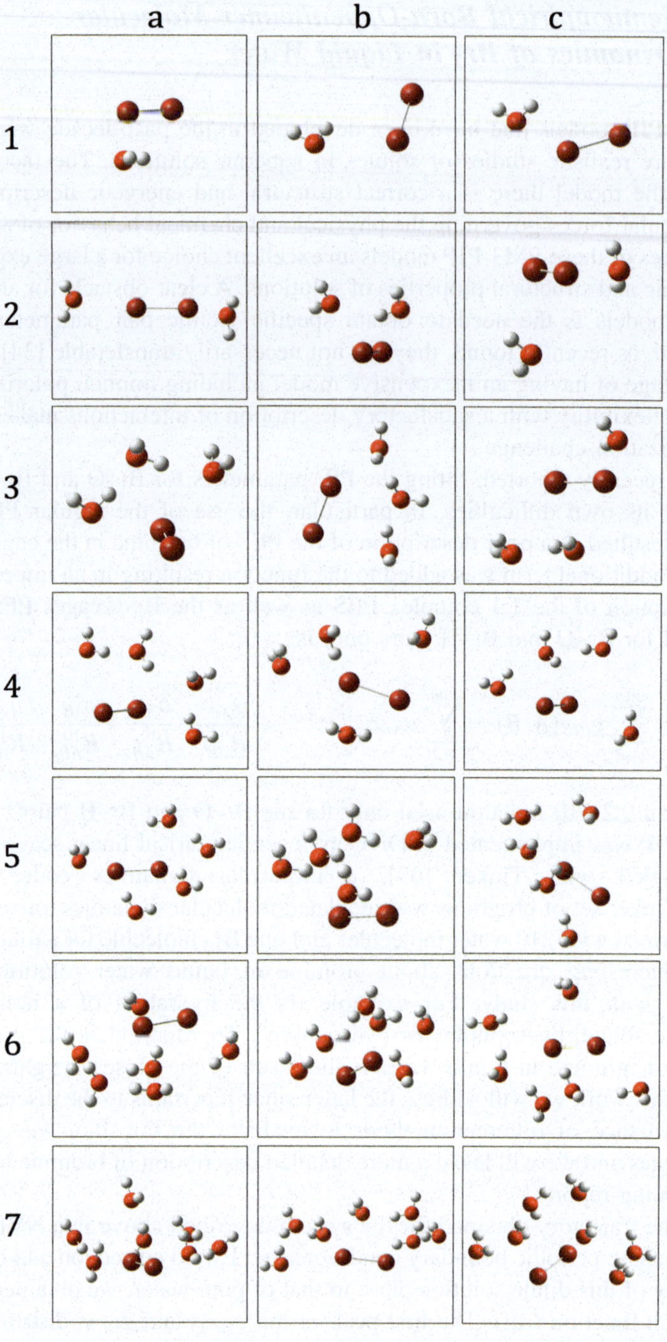

Fig. 9.3 Instantaneous structures of bromine molecule and its closest water molecule neighbors using PM3:PIF

inner part of this peak, between the beginning of the peak and its maxima, their number ranged from 1 to 7. Three different instantaneous configurations with 1–7 water molecules were extracted and their structures are shown in Fig. 9.3. Using as reference the average Br–Br distance during the MD trajectory, 2.45 Å, their shifts were calculated and in Table 9.4 we present the results using the three different methods above described: EOM-CCSD, LCC2 and TD-DFT at the B3LYP/AVTZ.

The LCC2 shifts are in good agreement with EOM-CCSD, most differences are within a 350 cm^{-1} ($\sim$1 kcal mol^{-1}) error. To further confirm the reliability of LCC2 a couple of structures where the shift difference was large were calculated with the RCCSD(T). For both cases these values lay between the other two calculations but closer to the more accurate EOM-CCSD. The results obtained with TD-DFT show a very large red shift. This result was quite unexpected, given the good performance TD-DFT calculations had shown for the equilibrium structures shifts. This kind of errors might be due to transitions of electrons between two separated regions of space or between orbitals of different spatial extend. We would like to emphasize that these results do not prescribe the use of TD-DFT for this kind of systems but that a more careful evaluation of the origin of these large errors is absolutely required.

Table 9.4 $\Delta\omega$ in cm^{-1} for the closest water molecules to Br_2 in 21 instantaneous MD configurations shown in Fig. 9.2. The reference in gas phase corresponds to the optimized geometry for bromine with each method

$(H_2O)_n$	Structure	EOM-CCSD	LT-DF-LCC2	TD-DFT
1	a	22	−3	−4366
	b	37	1	−4358
	c	505	387	−3904
2	a	−370	−702	−4896
	b	31	3	−4353
	c	700	757	−3910
3	a	−746	−1104	−5967
	b	292	210	−4358
	c	−308	−431	−4980
4	a	207	156	−4214
	b	566	488	−3858
	c	990	904	−3451
5	a	−225	−479	−4761
	b	319	78	−4351
	c	58	−265	−4500
6	a	−7	−340	−4360
	b	−467	−707	−5093
	c	−628	−952	−5515
7	a	565	–	−4110
	b	304	−80	−4264
	c	794	496	−3641

9.4 Conclusions

The various tests we presented here on the performance of standard ab initio, local and DFT methods to predict the shifts of bromine—water clusters and different solvent models, manifest clearly the challenge this system embodies. However, the advantages and limitations each combination of methods and level of calculation presented for drawing a physically correct representation of the phenomena is of great interest for deciding on the best way to build a multi-scale protocol to study halogens in aqueous media.

It is clear now that the solvatochromic effect is the result of a combination of specific interactions like, XB, XH and HB and bulk effects. The continuum model of the solvent showed that the latter are not a negligible contribution to the total shift, thus the calculations from clusters coming from numerical simulation studies like SEBOMD, should take this into account.

The relevance of the correct treatment of the dispersion driven interaction XH is now evident. The fact that a common DFT calculation, like the one tested here, predicts a longer H···Br distance than ab initio methods leads to structural changes in the optimized structures. However, the large differences observed for the structures coming from the liquid simulations, strongly suggest that TD-DFT at the level presented, fails to represent the excited state interactions.

The comparison of the local method LCC2 with refined standard ab initio results indicates that the former is a good choice for the study of excited states in large systems. LCC2 tends to predict red shifts slightly larger than RCCSD(T) by an average of less than 350 cm^{-1} ($\sim$1 kcal mol^{-1}). This is only a problem for small shifts, which for any methodology is a challenge.

Future work is aimed to include the thermal and dynamic effects on the shifts of bromine in aqueous systems like clathrates.

Acknowledgments Several colleagues have enriched the insight gained up to now in this subject; we would like to thank A. Apkarian, N. Halberstadt, K. C. Janda, O. Roncero and M. F. Ruiz-López for many valuable comments and discussions. This work has been funded by CONACYT 79975 and 128065. AAT and FABR gratefully acknowledges scholarships from CONACYT.

References

1. Kerenskaya G, Goldschleger IU, Apkarian VA, Janda KC (2006) Spectroscopic signatures of halogens in clathrate hydrate cages. 1. Bromine. J Phys Chem A 110(51):13792–13798. doi:10.1021/jp064523q
2. Kerenskaya G, Goldschleger IU, Apkarian VA, Fleischer E, Janda KC (2007) Spectroscopic signatures of halogens in clathrate hydrate cages. 2. Iodine. J Phys Chem A 111(43):10969–10976. doi:10.1021/jp0747306
3. Janda KC, Kerenskaya G, Goldschleger IU, Apkarian VA, Fleischer E (2008) UV-visible and resonance raman spectroscopy of halogen molecules in clathrate-hydrates. In: Proceedings of

the 6th international conference on gas hydrates (ICGH 2008), Vancouver, British Columbia, Canada, 6–10 July 2008

4. Goldschleger IU, Kerenskaya G, Senekerimyan V, Janda KC, Apkarian VA (2008) Dynamical interrogation of the hydration cage of bromine in single crystal clathrate hydrates versus water. Phys Chem Chem Phys 10(48):7226–7232. doi:10.1039/b811529j
5. Beckmann E (1889) Change of color of iodine solutions. Z Phyzik Chem 5:76
6. Gautier C, Charpy M (1890) Bromine color in solution. Compt Rend 110(189):645
7. Mulliken RS (1950) Structures of complexes formed by halogen molecules with aromatic and with oxygenated solvents. J Am Chem Soc 72(1):600–608. doi:10.1021/ja01157a151
8. Hassel O (1970) Structural aspects of interatomic charge-transfer bonding. Nobel Lecture, Oslo
9. Fredin L, Nelander B (1973) An infrared and CNDO study of a water-chlorine complex. J Mol Struct 16(2):217–224
10. Dahl T, Roeggen I (1996) An analysis of electron donor-acceptor complexes:H2O-F2, H2O-Cl2, and H2O-ClF. J Am Chem Soc 118(17):4152–4158. doi:10.1021/ja9537890
11. Legon AC (1998) pi-Electron "Donor-acceptor" complexes B…ClF and the existence of the "chlorine bond". Chem Eur J 4(10):1890–1897
12. Legon AC (1999) Prereactive complexes of dihalogens XY with lewis bases B in the gas phase: a systematic case for the halogen analogue B···XY of the hydrogen bond B···HX. Angew Chem Int Ed 38(18):2686–2714
13. Davey JB, Legon AC, Thumwood JMA (2001) Interaction of water and dichlorine in the gas phase: an investigation of H_2O-Cl_2 by rotational spectroscopy and ab initio calculations. J Chem Phys 114(14):6190–6202. doi:10.1063/1.1354178
14. Legon AC, Thumwood JMA, Waclawik ER (2002) The interaction of water and dibromine in the gas phase: an investigation of the complex H_2O-Br_2 by rotational spectroscopy and ab initio calculations. Chem Eur J 8(4):940–950
15. Legon AC (2010) The halogen bond: an interim perspective. Phys Chem Chem Phys 12(28):7736–7747. doi:10.1039/c002129f
16. Ham J (1954) The spectra of iodine solutions. III. The3Π1-1Σ+ transition of the iodine molecule in solution. J Am Chem Soc 76(15):3886–3887. doi:10.1021/ja01644a004
17. Bernal-Uruchurtu MI, Kerenskaya G, Janda KC (2009) Structure, spectroscopy and dynamics of halogen molecules interacting with water. Int Rev Phys Chem 28(2):223–265. doi:10.1080/01442350903017302
18. Bernal-Uruchurtu MI, Hernández-Lamoneda R, Janda KC (2009) On the unusual properties of halogen bonds: a detailed ab initio study of X-2-(H_2O)(1-5) clusters (X = Cl and Br). J Phys Chem A 113(19):5496–5505
19. Santoyo-Flores JJ, Cedillo A, Bernal-Uruchurtu MI (2012) Br_2 dissociation in water clusters: the catalytic role of water. Theor Chem Acc 132(1). doi:10.1007/s00214-012-1313-9
20. Clark T, Hennemann M, Murray J, Politzer P (2007) Halogen bonding: the s-hole. J Mol Model 13(2):291–296
21. Ibrahim MA (2011) Molecular mechanical study of halogen bonding in drug discovery. J Comput Chem 32(12):2564–2574. doi:10.1002/jcc.21836
22. Jorgensen WL, Schyman P (2012) Treatment of halogen bonding in the OPLS-AA force field; application to potent anti-HIV agents. J Chem Theory Comp 8(10):3895–3801. doi:10.1021/ct300180w
23. Kolář M, Hobza P (2012) On extension of the current biomolecular empirical force field for the description of halogen bonds. J Chem Theory Comp 8(4):1325–1333. doi:10.1021/ct2008389
24. Du L, Gao J, Bi F, Wang L, Liu C (2013) A polarizable ellipsoidal force field for halogen bonds. J Comput Chem 34(23):2032–2040. doi:10.1002/jcc.23362
25. Gresh N (2006) Development, validation, and applications of anisotropic polarizable molecular mechanics to study ligand and drug-receptor interactions. Curr Pharm Des 12(17):2121–2158. doi:10.2174/138161206777585256

26. Piquemal JP, Chevreau H, Gresh N (2007) Toward a separate reproduction of the contributions to the Hartree-Fock and DFT intermolecular interaction energies by polarizable molecular mechanics with the SIBFA potential. J Chem Theory Comp 3(3): 824–837. doi:10.1021/ct7000182
27. El Hage K, Piquemal JP, Hobaika Z, Maroun RG, Gresh N (2013) Could an anisotropic molecular mechanics/dynamics potential account for sigma hole effects in the complexes of halogenated compounds? J Comput Chem 34(13):1125–1135. doi:10.1002/jcc.23242
28. Schofield DP, Jordan KD (2009) Molecular dynamics simulations of bromine clathrate hydrates. J Phys Chem A 113(26):7431–7438. doi:10.1021/jp900237j
29. Kubař T, Bodrog Z, Gaus M, Köhler C, Aradi B, Frauenheim T, Elstner M (2013) Parametrization of the SCC-DFTB method for halogens. J Chem Theory Comp 9(7): 2939–2949. doi:10.1021/ct4001922
30. Csonka GI, Ángyán JG (1997) The origin of the problems with the PM3 core repulsion function. J Mol Struct (Theochem) 393(1–3):31–38. doi:10.1016/S0166-1280(96)04872-5
31. Bernal-Uruchurtu MI, Martins-Costa MTC, Millot C, Ruiz-López MF (2000) Improving description of hydrogen bonds at the semiempirical level: water-water interactions as test case. J Comput Chem 21(7):572–581
32. Harb W, Bernal-Uruchurtu MI, Ruiz-López MF (2004) An improved semiempirical method for hydrated systems. Theor Chem Acc 112(4):204–216. doi:10.1007/s00214-004-0576-1
33. Arillo-Flores OI, Ruiz-López MF, Bernal-Uruchurtu MI (2007) Can semi-empirical models describe HCl dissociation in water? Theor Chem Acc 118(2):425–435. doi:10.1007/s00214-007-0280-z
34. Marion A, Monard G, Ruiz-Lopez MF, Ingrosso F (2014) Water interactions with hydrophobic groups: assessment and recalibration of semiempirical molecular orbital methods. J Chem Phys 141(3):034106. doi:10.1063/1.4886655
35. Monard G, Bernal-Uruchurtu MI, van der Vaart A, Merz KM Jr, Ruiz-Lopez MF (2005) Simulation of liquid water using semiempirical hamiltonians and the divide and conquer approach. J Phys Chem A 109(15):3425–3432. doi:10.1021/jp0459099
36. Bernal-Uruchurtu MI, Janda KC, Hernandez-Lamoneda R (2015) Motion of Br_2 molecules in clathrate cages. A computational study of the dynamic effects on its spectroscopic behavior. J Phys Chem A 119(3):452–459. doi:10.1021/jp5082092
37. Hammond JR, Govind N, Kowalski K, Austschbach J, Xantheas SS (2009) Accurate dipole polarizabilities for water clusters n = 2–12 at the coupled-cluster level of theory and benchmarking of various density functionals. J Chem Phys 131:214103. doi:10.1063/1.3263604
38. Sprik M, Hutter J, Parrinello M (1996) Ab initio molecular dynamics simulation of liquid water: comparison of three gradient-corrected density functionals. J Chem Phys 105:1142. doi:10.1063/1.471957
39. VandeVondele J, Mohamed F, Krack M, Hutter J, Sprik M, Parrinello M (2005) The influence of temperature and density functional models in ab initio molecular dynamics simulation of liquid water. J Chem Phys 122:014515. doi:10.1063/1.1828433
40. Hernandez-Lamoneda R, Rosas VH, Bernal-Uruchurtu MI, Halberstadt N, Janda KC (2008) Two-dimensional H_2O-Cl_2 and H_2O-Br_2 potential surfaces: an ab initio study of ground and valence excited electronic states. J Phys Chem A 112(1):89–96. doi:10.1021/jp077074i
41. Franklin-Mergarejo R, Rubayo-Soneira J, Halberstadt N, Ayed T, Bernal-Uruchurtu MI, Hernandez-Lamoneda R, Janda KC (2011) Large shift and small broadening of Br_2 valence band upon dimer formation with H_2O: an ab initio study. J Phys Chem A 115(23):5983–5991. doi:10.1021/jp110389z
42. Pathak AK, Mukherjee T, Maity DK (2008) Microhydration of X_2 gas (X = Cl, Br, and I): a theoretical study on X_2;nH_2O Clusters (n = 1–8). J Phys Chem A 112(4):744–751. doi:10.1021/jp076594a
43. Alkorta I, Rozas I, Elguero J (1998) Charge-transfer complexes between dihalogen compounds and electron donors. J Phys Chem A 102(46):9278–9285. doi:10.1021/jp982251o

44. Ruiz-López MF (2008) The multipole moment expansion solvent continuum model: a brief review. In: Canuto S (ed) Solvation effects on molecules and biomolecules, vol 6. Challenges and advances in computational chemistry and physics. Springer, Netherlands, pp 23–38. doi:10.1007/978-1-4020-8270-2_2
45. Tomasi J (1994) Application of continuum solvation models based on a quantum mechanical hamiltonian. Struct React Aqueous Solution 568:10–23. doi:10.1021/bk-1994-0568.ch002
46. Cramer CJ, Truhlar DG (2008) A universal approach to solvation modeling. Acc Chem Res 41 (6):760–768. doi:10.1021/ar800019z
47. Rinaldi D, Bouchy A, Rivail JL, Dillet V (2004) A self-consistent reaction field model of solvation using distributed multipoles. I. Energy and energy derivatives. J Chem Phys 120(5):2343–2350. doi:10.1063/1.1635355
48. Frisch MJ, Trucks GW, Schlegel HB, Scuseria GE, Robb MA, et al (2009) Gaussian 09 Revision C.01. Gaussian, Ing, Wallingford, CT
49. Scalmani G, Frisch MJ, Mennucci B, Tomasi J, Cammi R, Barone V (2006) Geometries and properties of excited states in the gas phase and in solution: theory and application of a time-dependent density functional theory polarizable continuum model. J Chem Phys 124(9):094107. doi:10.1063/1.2173258
50. Bernal Uruchurtu MI, Ruiz-López MF (2007) Eigen and Zundel ions in aqueous environments. A theoretical study using semi-empirical force fields. In: Hernández-Lamoneda R (ed) Beyond standard quantum chemistry: applications from gas to condensded phases. Transworld Research Network, Kerala, pp 65–85
51. Serrano-Andres L, Fulscher MP, Karlstrom G (1997) Solvent effects on electronic spectra studied by multiconfigurational perturbation theory. Int J Quantum Chem 65(2):167–181
52. Han WG, Liu T, Himo F, Toutchkine A, Bashford D, Hahn KM, Noodleman L (2003) A theoretical study of the UV/visible absorption and emission solvatochromic properties of solvent-sensitive dyes. ChemPhysChem 4(10):1084–1094. doi:10.1002/cphc.200300801
53. Kelly CP, Cramer CJ, Truhlar DG (2006) Adding explicit solvent molecules to continuum solvent calculations for the calculation of aqueous acid dissociation constants. J Phys Chem A 110(7):2493–2499. doi:10.1021/j055336f
54. Korona T, Werner H-J (2003) Local treatment of electron excitations in the EOM-CCSD method. J Chem Phys 118(7):3006–3019. doi:10.1063/1.1537718
55. Werner H-J, Knowles PJ, Amos RD, Bernhardsson A, Berning A, et al (2002) Molpro, 3 edn
56. Pulay P (1983) Localizability of dynamic electron correlation. Chem Phys Lett 100(2): 151–154
57. Saebo S, Pulay P (1993) Local treatment of electron correlation. Ann Rev Phys Chem 44: 213–236
58. Werner HJ, Pflüger H-J (2006) Ann Rep Comp Chem 2:53–80
59. Hetzer G, Werner HJ (1999) Low-order scaling local electron correlation methods. I. Linear scaling local MP2. J Chem Phys 111:5691–5705. doi:10.1063/1.479957
60. Werner HJ, Manby FR, Knowles PJ (2003) Fast linear scaling second-order Møller-Plesset perturbation theory (MP2) using local and density fitting approximations. J Chem Phys 118:8149. doi:10.1063/1.1564816
61. Polly R, Werner HJ, Manby FR, Knowles PJ (2006) Fast Hartree-Fock theory using local density fitting approximations. Mol Phys 102:2311–2321. doi:10.1080/0026897042000274801
62. Werner H-J, Pflüger K (2006) Chapter 4 On the selection of domains and orbital pairs in local correlation treatments. In: David CS (ed) Annual reports in computational chemistry, vol 2. Elsevier, Amsterdam. pp 53–80. doi:10.1016/S1574-1400(06)02004-4
63. Dieterich JM, Werner HJ, Mata RA, Metz S, Thiel W (2010) Reductive half-reaction of aldehyde oxidoreductase toward acetaldehyde: ab initio and free energy quantum mechanical/molecular mechanical calculations. J Chem Phys 132:035101. doi:10.1063/1.3280164
64. Schütz M, Rauhut G, Werner HJ (1998) Local treatment of electron correlation in molecular clusters: structures and stabilities of $(H_2O)n$, n = 2-4. J Phys Chem A 102:5997–6003

65. Runeberg N, Schutz M, Werner HJ (1999) The aurophilic attraction as interpreted by local correlation methods. J Chem Phys 110(15):7210–7215. doi:10.1063/1.478665
66. Magnko L, Schweizer M, Rauhut G, Schutz M, Stoll H, Werner HJ (2002) A comparison of metallophilic attraction in (X-M-PH_3)(2) (M = Cu, Ag, Au; X = H, Cl). Phys Chem Chem Phys 4(6):1006–1013. doi:10.1039/b110624d
67. Kats D, Korona T, Schutz M (2006) Local CC2 electronic excitation energies for large molecules with density fitting. J Chem Phys 125(10):104106. doi:10.1063/1.2339021
68. Kats D, Schutz M (2009) A multistate local coupled cluster CC2 response method based on the Laplace transform. J Chem Phys 131(12):124117. doi:10.1063/1.3237134
69. Ponder J (2005) Tinker. 5.1 edn., Washington

Chapter 10
Theoretical Studies of the Solvation of Abundant Toxic Mercury Species in Aqueous Media

J.I. Amaro-Estrada and A. Ramírez-Solís

Abstract We present the main results of a comprehensive research program related to the identification of the solvation patterns of abundant toxic Hg-containing species in aqueous media. Two different solvation models have been used. A systematic study of stepwise hydration using cluster models at the Density Functional Theory level with the B3PW91 exchange-correlation functional. We address solvation free energies, optimized geometries, vibrational frequencies and Hg-coordination patterns for the Hg(II)XY–$(H_2O)_n$ (X,Y = Cl,OH; n ≤ 24) complexes. One to three direct Hg-water interactions appear along the hydration process. A stable pentacoordinated Hg trigonal bipyramid structures arises from $n = 15$. The first solvation shell is fully formed with 22 and 24 water molecules for HgClOH and $Hg(OH)_2$ species respectively. The thermal stability and the persistence of the trigonal bipyramid coordination around Hg of fully solvated structures Hg(II)XY-$(H_2O)_n$ (X,Y = Cl,OH; n = 24) has been verified using Born-Oppenheimer molecular dynamics simulations at the B3PW91/6-311G** level.

Abbreviations

DFT	Density Functional Theory
BO-MD	Born Oppenheimer Molecular Dynamics
MC	Monte Carlo
MP2	2nd order Möller Plesset Perturbation Theory
CCSD(T)	Coupled Cluster singles, doubles and perturbative triple excitations
RECP	Relativistic Effective Core Potentials
AVDZ	aug-cc-pVDZ basis set
AVTZ	aug-cc-pVTZ basis set
AVQZ	aug-cc-pVTZ basis set
BSSE	Basis Set Superposition Error

J.I. Amaro-Estrada · A. Ramírez-Solís (✉)
Departamento de Física, Facultad de Ciencias,
Universidad Autónoma Del Estado de Morelos,
Av. Universidad 1001. Cuernavaca, 62209 Morelos, Mexico
e-mail: alex@uaem.mx

J.-L. Rivail et al. (eds.), *Quantum Modeling of Complex Molecular Systems*,
Challenges and Advances in Computational Chemistry and Physics 21,
DOI 10.1007/978-3-319-21626-3_10

GGA	Generalized Gradient Approximation
NPA	Natural Population Analysis
NBO	Natural Bond Orbital
CBS	Complete Basis Set limit
CP	Counterpoise correction
PCM	Polarizable Continuum Model

10.1 Introduction

As a consequence of the growing industrial activity in the last century, the concentration of mercury levels in ecosystems worldwide has increased in a significant way. Surface soils, plants and marine environments [1] have accumulated high levels of this toxic metal even in remote regions such as the Arctic [2]. From the ecological point of view it is crucial to find relevant information on the fate, transport and pathways of mercury species in these biological environments. It is difficult to establish the exact amount of mercury worldwide, however, approximately 90 % of Hg in the atmosphere has been found as elemental Hg(0), although the ratio of free monatomic vapor varies continually [3, 4] due to oxidation processes to Hg^{2+} forms that occur through various physical and chemical mechanisms. Many adverse effects of Hg accumulation in humans are well known, particularly on the nervous and immune systems.

One of the most important entrance pathways of mercury complexes into the food chain is through photosynthetic organisms and bacteria, which are then ingested by higher predators (marine mammals, fish, polar bears, seabirds, etc.) leading to a process of mercury accumulation and biomagnification. Previous studies about the mercury bioaccumulation [5] and methylation [6] processes have shown that $HgCl_2$, HgClOH and $Hg(OH)_2$ are the most abundant Hg-containing inorganic complexes in aqueous environments.

The bioavailability of deposited mercury is followed by a crucial biochemical step in which one or several Hg-containing species cross the cell membrane by passive or active diffusion. The intrinsic selective properties of the cell membrane suggest that passage for Hg-containing species is dependent on the specific type of ligands bonded to the Hg atom. In this way, Gutknecht [7] and Myers et al. [8] compared the permeability rates of several mercury complexes with a variety of inorganic and organic ligands ($HgCl_2$, $HgCl_3^-$, $HgCl_4^{2-}$, $Hg(OH)_2$ and HgClOH) through laboratory-synthesized lipid bilayer membranes. These studies showed that lipid membranes are highly permeable to $HgCl_2$, $Hg(OH)_2$ and HgClOH showing a permeability twenty times larger than the permeability to water and more than a million times more permeable than to Na^+, K^+ and Cl^- [7].

The molecular processes involved in the transmembrane passage of Hg-containing molecules as well as the relative stability of these Hg-containing species are two fundamental issues related to the question of Hg cellular uptake

[9–11]. We address here some questions related to the first issue from the molecular perspective dealing with the most common Hg-containing neutral species in aqueous environments.

As a first step, we have delved into some issues related to the bioavailability of $HgCl_2$. In particular, using quantum chemical methods [12, 13] we have studied how is this species solvated, how many water molecules belong to its first solvation shell and what is the interaction energy with its water environment. Another crucial question related to the bioavailability is whether $HgCl_2$ can be considered as a water-dressed molecule or not during the trans-membrane transport process making it available to the cell interior. In this regard we recently addressed the solvation of $HgCl_2$ through Density Functional Theory (DFT) studies using stepwise cluster solvation including up to 24 water molecules [12], and through Monte Carlo simulations of the aqueous solution [13]. We found that the hydrogen bond network is crucial to allow orbital-driven interactions between Hg(II) and the water molecules. Born–Oppenheimer molecular dynamics simulations of the $HgCl_2$–$(H_2O)_{24}$ cluster showed that an $HgCl_2$–$(H_2O)_3$ *effective solute* appears with a trigonal bipyramid configuration. The remaining 21 water molecules form a first solvation shell around it, in the shape of a water-clathrate with external water molecules rarely exchanging positions with those belonging to the *effective solute*. Hg directly interacts with 3 water molecules with donor–acceptor type Hg–O bonds, while all other interactions are of hydrogen bond type.

As an example of how detailed information at the molecular level can be of help to understand the main processes at work in these complex metal-membrane interactions, we recall here that a refined study on the solvation mechanism of As $(OH)_3$ has allowed us to better understand how this neutral toxic species can enter the cell via transmembranal aquaglyceroporines [14]. It is important to note that the interaction of arsenious acid with the extracellular portion of aquaglyceroporines might be facilitated due to its singular amphipathic solvation pattern. Therefore, accurate information at the molecular level concerning the solvation pattern of Hg-containing molecules in aqueous environments can be used to provide new insights into the type of trans-membrane cellular uptake of these toxic species.

In this respect we report the main structural and energetic features of the stepwise hydration of $Hg(OH)_2$ [15] and HgClOH [16, 17] complexes. Once we reach the optimized structure of the first aqueous solvation shell, we include the temperature effects on the system through DFT Born-Oppenheimer molecular dynamics simulations. The study of the intra and inter-molecular interactions using a cluster description of the aqueous solvation of Hg-containing molecules is the first step to reach a more realistic description through Monte Carlo (MC) simulations of the condensed phase. Recently, this MC approach was used by some of us to accurately describe aqueous solvation of $As(OH)_3$ [14] and $HgCl_2$ [13] in the liquid phase. Based on an adequate description of the microsolvated species in the gas phase, the next stage in our research is the development of refined interatomic interaction potentials needed to carry out classical Monte Carlo simulations of both Hg complexes in aqueous solution as we have done for the $HgCl_2$ molecule [13]. The next section presents the methods and computational details, in section III we show the results and the section IV we present the main conclusions and some perspectives.

10.2 Methods and Computational Details

10.2.1 *Cluster Stepwise Solvation of $HgCl_2$, HgClOH and $Hg(OH)_2$*

Since we are interested on the accurate description of solvation free energies, geometrical parameters and frequencies of small clusters Hg(II)XY-$(H_2O)_n$ with X, Y = Cl, OH (up to n = 5), we carried out calculations at the 2nd order Möller-Plesset perturbation theory (MP2) level. The chlorine [18] and mercury [19] atoms were treated with the Stuttgart-Dresden relativistic effective core potentials (RECP) in combination with their adapted valence basis sets augmented by a set of polarization and diffuse functions: *s* exponent 0.007, *p* exponent 0.0013, *d* exponent 0.0025 for Hg, and *p* exponent 0.005477, *d* exponents 0.22, 0.0643, f exponent 0.706 for Cl. The oxygen and hydrogen atoms have been described with the aug-cc-pVDZ (AVDZ), aug-cc-pVTZ (AVTZ) and the very large aug-cc-pVQZ (AVQZ) [20] basis sets. The counterpoise (CP) correction was applied to all binding energies to account for the basis set superposition error (BSSE). With the AVQZ basis sets these calculations lead to a molecular basis of 893 orbitals for the tetrahydrated [HgClOH–$(H_2O)_4$] complex. Clearly, the study of larger clusters (n > 5) at the MP2/aug-cc-pVnZ level is not feasible from the computational point of view and a more affordable DFT-based theoretical approach is needed. Therefore we performed a careful calibration of the exchange-correlation functional using the MP2/AVTZ results for $HgCl_2$ with one to three water molecules as references. The hybrid B3PW91 functional turned out to be the best among ten other Generalized Gradient Approximation (GGA), hybrid and meta-GGA exchange-functionals.

In order to complete the first solvation shell around the HgXY (X,Y = Cl,OH) molecules, the geometry optimizations were performed using a stepwise solvation scheme at the DFT level with the hybrid B3PW91 functional, and the nature of the stationary points (all minima) was verified with analytical frequency calculations for all *n*. Again, the mercury [19] and chlorine [18] atoms were treated with the Stuttgart-Köln relativistic effective core potentials in combination with their adapted valence basis sets. Since many more water molecules were added to form the first solvation shell around all the mercury species, the oxygen and hydrogen atoms have been described with the 6-311G(d,p) basis set. At this point it is important to note that we also made calculations using larger basis sets including diffuse functions (6-311 ++G(d,p)); we found that the Gibbs free energies thus obtained are considerably affected by the basis set superposition error. When the CP correction was applied to these energies we obtained values close to those obtained with the 6-311G(d,p) basis sets, so we decided to use the latter without the CP correction for our systematic studies. The electronic density at the B3PW91 level of the optimized structures was analyzed using the Natural Population Analysis (NPA) and the Natural Bond Orbital (NBO) schemes [21]. All the MP2, DFT and NPA calculations were performed with the Gaussian03 program [22].

10.2.2 DFT Born-Oppenheimer Molecular Dynamics

To assess the thermal stability of Hg(II)XY–$(H_2O)_n$ (X,Y = Cl,OH) fully solvated structures ($n = 24$). Born-Oppenheimer molecular dynamics (BO-MD) simulations at the B3PW91/6-311G** level were carried out with the Geraldyn2.1 code based on the method implemented by Raynaud et al. [23], which uses the velocity-Verlet integration scheme [24]. For the integration of the equations of motion a time step of 0.5 fs. was chosen. The stability of $Hg(OH)_2$-$(H_2O)_{24}$ structure was tested at T = 300 K while for HgClOH–$(H_2O)_{24}$ structure the simulation was development at T = 700 K. The systems are equilibrated after 2–3 ps and this has been verified for all cases. Statistical results are obtained after equilibration has been achieved. Note that this temperature (700 K) is higher than the critical point of water but it involves a rather small thermal energy of ca. 1.5 kcal/mol. We chose this 700 K temperature in order to test the stability of hydrated systems as we have done before for $HgCl_2$. Initially we performed both simulations at 700 K, however the $Hg(OH)_2$–$(H_2O)_{24}$ structure was not stable and water molecules evaporated very early in this high temperature simulation. These simulations were performed in the canonical ensemble and a Nosé–Hoover chain of thermostats [25, 26] was used to control the temperatures during the 8–10 ps of the simulations. The total MD simulations took around 100 CPU days on 32 processors@2.8 GHz running the Linux versions of Geraldyn2.1-G03.

10.3 Results and Discussion

10.3.1 Stepwise Solvation with Few Water Molecules

10.3.1.1 HgClOH

As the first step we studied the structural and energetic properties of the isolated molecule. The main optimized geometrical parameters are a quasi-linear O–Hg–Cl angle of ca. 176° and bond distances of ca. 1.94 Å for Hg–O, 2.30 Å for Hg–Cl and 0.96 Å for H–O. Table 10.1 presents the optimized MP2 parameters with the various basis sets and Fig. 10.1 schematically depicts the optimal geometry. Note that the Hg-O distance goes from 1.98 Å with the AVDZ to ca. 1.91 Å with the AVQZ basis sets, which shows the importance of the (mainly oxygen) basis set quality for this intramolecular geometric parameter. Although the dipole moment and NPA atomic charges were already reported in ref [13], we shall briefly

Table 10.1 Optimized geometrical parameters of HgClOH at the MP2 level

Basis	Hg-O	Hg–Cl	H–O	O–Hg–Cl	Hg–O–Hg
AVDZ	1.982	2.306	0.971	176.6	108.7
AVTZ	1.943	2.303	0.964	176.5	109.2
AVQZ	1.907	2.300	0.961	176.1	111.0

Distances in Å, angles in degrees

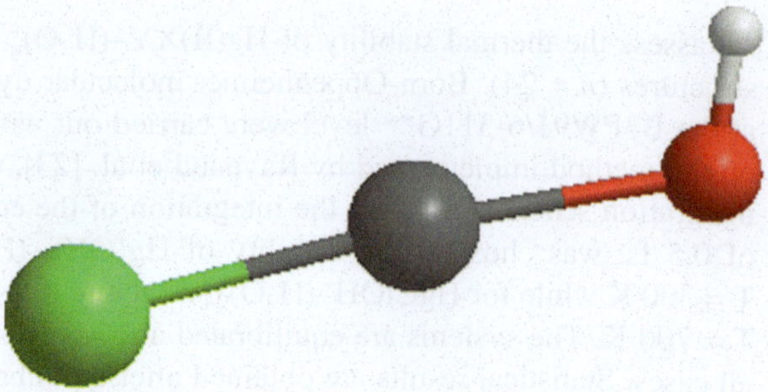

Fig. 10.1 Optimized MP2 geometry of isolated HgClOH molecule

comment these. The atomic charges are –0.49 for Cl, +0.96 for Hg, +0.50 for H and –0.97 for O. As pointed out previously [13], HgClOH has a rather large dipole moment of 2.13 D, even larger than that (1.85 D) of the water molecule. We recall here that the dipole moment of $HgCl_2$ is almost zero due to its nearly linear geometry [12]. For the HgClOH case, the exchange of the OH unit for the chlorine atom significantly modifies the dipole moment. Therefore, in an aqueous environment, it could be expected that HgClOH interacts more strongly than $HgCl_2$ with the dipole moments of the solvating water molecules.

The interaction of HgClOH with water was studied via stepwise solvation by adding a number (n) of water molecules to the system, with $n = 1,2,3,4$. Special care was taken in order to insure that all the Cl–Hg–O–O_w (O_w stands for water oxygens) dihedral angles were properly sampled with each additional water molecule, thus we report here only absolute minima in the corresponding MP2 potential energy surfaces of the microsolvated species. The stability of the optimized structures with varying number of water molecules was verified by the absence of imaginary frequencies at the MP2 level with the various basis sets used.

When a single water molecule is added around HgClOH (see Fig. 10.2, top-left) the optimized geometry looks, at first sight, like the one computed by Shepler et al. [27] and by Castro et al. [12] for $HgCl_2$. An interaction is also found between Hg and the water oxygen (2.68 Å), but the Cl–Hg–O_w angle is significantly larger due to the asymmetry of the present solute and the first (weak) hydrogen bond interaction, at a distance of 2.60 Å, appears with the OH moiety of HgClOH. The Cl–Hg–OH angle increases by 2.2° compared to HgClOH in the gas phase. The Hg–OH bond length increases 0.02 Å while the Hg–Cl bond remains unchanged. As for the $HgCl_2$–H_2O case, the nature of the Hg–O_w interaction is electrostatic [12], with an oxygen lone pair pointing towards mercury.

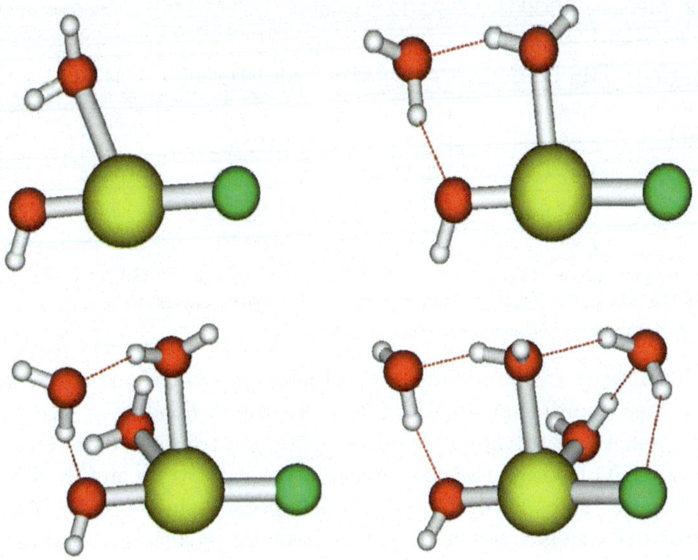

Fig. 10.2 Optimized geometry of HgClOH solvated by one (*top-left*), two (*top-right*), three (*bottom-left*) and four (*bottom-right*) water molecules

A second water molecule was then added to the previous structure using many initial configurations, some of them in such a way that two Hg–OH_2 interactions are formed at the beginning. This was done using very many dihedral angles of the second water molecule with respect to the first one; however the optimizations do not maintain this structure. The optimized structure (Fig. 10.2, top-right) arising from all the possible initial configurations resembles again the one computed by Shepler et al. [27] and Castro et al. [12] for $HgCl_2$; however, there is an important qualitative difference since the new H_2O molecule is here stabilized by two hydrogen bonds, one between its hydrogen and the oxygen of HgClOH, and another with the first water molecule.

Addition of a third water molecule leads to the growth of the hydrogen bond network (Fig. 10.2, bottom-left). As could be expected, these structures do not look like the ones calculated for $HgCl_2$ [12, 19] due to the strong preference of water to establish hydrogen bonds with the OH moiety of HgClOH. Only at this stage two direct Hg–O_w interactions appear, at 2.59 Å respectively, leading to a tetra-coordinated mercury atom. A fourth water molecule was then added to the previous structure using many initial configurations. This was done using very many dihedral angles of the fourth water molecule with respect to the other three an to the intramolecular Cl–Hg–O axis. It is only when the fourth water molecule is added that a hydrogen-halogen bond is established with the solute via the chlorine atom. Figure 10.2 (bottom-right) shows the optimized structure, where it can be seen that each water molecule is stabilized by one to three hydrogen bonds. At this stage two direct different Hg–O_w interactions are found (at 2.48 and 2.64 Å).

Table 10.2 MP2/aug-RECP(Hg,Cl)-AVTZ optimized geometrical parameters of gaseous and solvated HgClOH–$(H_2O)_n$ systems

n	Hg–Cl	Hg–O(H)	Cl–Hg–O(H)	Cl–Hg–O_w	Hg–O_w	Hg–O_w
0	2.30	1.94	176.5		–	–
1	2.30	1.96	178.7	110.47	2.60	–
2	2.31	1.97	176.5	95.77	2.47	3.70
3	2.31	2.00	174.5	95.49	2.59	2.59
4	2.34	1.99	174.1	90.29	2.48	2.64

The water oxygen atoms (O_w) are those directly linked to Hg. Distances in Å and angles in degrees. Cl–Hg–O_w is the angle concerning the water oxygen closest to Hg

Note the asymmetry of the hydrogen bond network, where one of the water molecules (the one coordinated to Cl) is also hydrogen bonded to two other water molecules. When four solvating water molecules are considered the Hg–Cl bond is stretched (2.34 Å) from its isolated molecule value (2.30 Å) and the Cl–Hg–O(H) angle slightly decreases (174.1°) from the gas phase value (176.5°). Clearly, only minor structural changes are induced in the solute by the surrounding microsolvation environment. Table 10.2 summarizes the previous results with the RECP (Hg,Cl)-AVTZ basis sets at the MP2 level.

We have obtained the vibrational spectra for the isolated molecule as well as for all the HgClOH–$(H_2O)_n$ optimized complexes. Results can be reviewed at the MP2/aug-RECP(Hg,Cl)-AVTZ level as supplementary material in ref. [16] along with the corresponding infrared intensities. Note the presence of additional normal modes when a new water molecule is added to cluster, one of them appearing around 1600 wavenumbers and two more between 3000 and 4000 cm^{-1}, corresponding to the bending, the symmetric and the antisymmetric stretching normal modes of the water molecule. Table 10.3 shows the evolution of the main modes as functions of n (n = 0,1,2,3,4). In particular, when four water molecules surround HgClOH, the frequencies of the Hg–Cl and Hg–O(H) stretching intramolecular modes decrease by 30 and 89 cm^{-1}, respectively. On the other hand the Cl–Hg–O bending mode is barely modified by the microsolvation.

The strength of the water-solute interaction cannot be assessed from the incremental binding energies since these also include the water-water interaction

Table 10.3 MP2/aug-RECP(Hg,Cl)-AVTZ harmonic frequencies of selected vibrational (S for stretching, B for bending) modes of gaseous and solvated HgClOH–$(H_2O)_n$ systems

N	Hg–Cl S	Hg-O(H) S	Cl–Hg–O(H) B	Cl–Hg–O_w B	Hg–O_w S	Hg–O_w S
0	383	636	140			
1	379	610	138	76	152	
2	371	591	132	66	183	
3	363	564	136	75	211	171
4[a]	353	547	121	66	190	106

The water oxygen atoms (O_w) are those directly linked to Hg. All energies in wavenumbers. Cl–Hg–O_w is the angle concerning the water oxygen closest to Hg energies, which arise from the growing network of hydrogen bonds between them. A more reliable measure of the strength of the solute-water interaction are the solute-O_W stretching frequencies. Note that with three solvating water molecules the strongest water-solute interaction occurs, since the largest Hg–O_w stretching frequency of 211 cm^{-1} is found for $n = 3$. This is also consistent with the fact that for n = 3 we find two water molecules that are equidistant from the solute (at 2.59A), while for n = 2 and 4 one of the water molecules lies significantly further away from Hg.

Since Shepler et al. [27] showed that the complete basis set (CBS) results obtained with the MP2 approach lead to nearly the same (within 0.5 kcal/mol) incremental binding energies for the microsolvation of $HgCl_2$ and $HgBr_2$ as those obtained with the considerably more expensive MP2/CBS + [CCSD(T)/AVDZ − MP2/AVDZ] hybrid method. Table 10.4 presents the evolution of the incremental binding energies for the HgClOH–$(H_2O)_n$ complexes as functions of the basis set quality. As can be seen, both the binding energy ($n = 1$) and the MP2 incremental binding energies (for larger n), increase with basis set quality if one considers the Counterpoise (CP) -uncorrected MP2 energies. However, as shown before for the microsolvation of $HgCl_2$ [27], the basis set superposition errors are quite large and they increase with basis set quality for all n, so that the CP corrections are crucial to provide reliable energetic values. We stress that using the CP-corrected energies; a much smaller dispersion of the incremental binding energies is found for each n as a function of basis set quality. Also note that the trend is reversed with respect to the CP-uncorrected values since the incremental binding energies slightly decrease with basis set quality for a given value of n. For the present case the counterpoise corrected MP2/CBS limit values are 5.20, 11.49, 6.63 and 10.00 kcal/mol. At this point we recall that the dipole moment of HgClOH is much larger than that of $HgCl_2$ and, in spite of this, the first two values are quite close to those (6.98, 11.45 kcal/mol) obtained for the microsolvation of $HgCl_2$ with one and two water molecules; also note that the third incremental binding energy is slightly smaller than for the $HgCl_2$ case [27]. This reveals that the dominant intermolecular interactions at work for the solvation of both species are of the same type. For the incremental binding energies this is obvious since these arise mainly from the build-up of the hydrogen bond network around the solute, but we stress

Table 10.4 Counterpoise corrected MP2 incremental binding energies (kcal/mol) of the minimum-energy HgClOH–$(H_2O)_n$ structures with different basis sets, aug-RECP(Hg,Cl) valence basis for Hg and Cl used throughout

Basis set	*n = 1*	*N = 2*	*n = 3*	*N = 4*
AVDZ	(9.38)	(13.85)	(10.05)	(12.80)
	6.76	11.89	7.49	10.60
AVTZ	(11.19)	(14.77)	(11.63)	(15.25)
	6.45	11.95	6.69	10.01
AVQZ	(14.48)	(16.70)	(15.65)	(16.71)
	5.19	11.49	6.64	10.00
CBS limit	5.20	11.49	6.63	10.00

Counterpoise uncorrected values in parentheses. Complete basis set (CBS) limit values are also shown that this is also true even for the interaction of the mercury species with the first water molecule.

10.3.2 *Stepwise Solvation up to the First Water Shell*

10.3.2.1 $Hg(OH)_2$

As the number of solvating water molecules increases the Hg-O(H) distances grow up to 2.09 Å and the O–Hg–O angle shows a large change from quasi-linearity at 166° for $n = 12$. Table 10.5 shows the main optimized geometrical parameters for the $Hg(OH)_2$ -$(H_2O)_n$ clusters, with particular emphasis on the variation of the distances and the angle between Hg and lateral O(H) moieties of the solute, as well as the distances of the water oxygen atoms (O_w) directly coordinated to Hg. When a single water molecule is added to $Hg(OH)_2$, the optimized structure is qualitatively different to the one computed by Castro et al. for the $HgCl_2$ molecule [12] since no direct Hg–O_w interaction is found; however, it is similar to one obtained for HgClOH [16] (Fig. 10.2, top-left) at the MP2 level. An attractive interaction of the water hydrogen atom with the electronegative part of the solute (one of the oxygen atoms) arises, producing a slight stretching in the Hg–O bond (2.01 Å).

The first direct interaction between Hg and a water oxygen (2.59 Å) was found for $n = 2$, explaining the fact that the solvation free energy when $n = 1$ (−0.3 kcal/mol, see Table 10.6) is smaller than the value of −2.7 kcal/mol we

Table 10.5 Main optimized geometrical parameters of gaseous and solvated $Hg(OH)_2$ systems. Water oxygen atoms (O_w) are those directly linked to Hg

n	Hg–O (H)	Hg–O (H)	O(H)–Hg–O(H) angle	Hg–O_w d1	Hg–O_w d2	Hg–O_w d3
0	1.98	1.98	175.8	–	–	–
1	2.01	1.98	177.7	2.82	–	–
2	2.02	1.99	179.1	2.59	–	–
3	2.02	2.02	173.8	2.52	–	–
4	2.02	2.02	174.0	2.63	2.97	–
5	2.04	2.03	178.2	2.68	2.70	–
6	2.04	2.02	178.7	2.65	2.75	–
8	2.04	2.04	172.7	2.59	3.03	3.19
12	2.06	2.05	166.4	2.66	2.72	3.00
15	2.07	2.03	177.8	2.58	2.71	3.22
16	2.09	2.09	175.7	2.63	2.66	2.68
24	2.04	2.09	177.3	2.75	2.79	2.80
24[a]	2.03	2.04	171.7	2.97	3.02	3.06

Distances in Å and angle in degrees

[a] Average structure obtained with the BO-DFT-MD simulation at 300 K

Table 10.6 Gibbs microsolvation free energies and incremental microsolvation free energies (kcal/mol), number of direct Hg–O(water) interactions and hydrogen bonds versus the number of water molecules in each $Hg(OH)_2$–$(H_2O)_n$ cluster

Water molecules	Direct Hg–O_w interactions	Hydrogen bonds	ΔG^o_{solv}	Incremental energy
1	0	1	−0.30	0.30
2	1	2	−4.42	−4.12
3	1	4	−6.53	−2.11
4	2	5	−11.04	−4.51
5	2	7	−13.63	−2.59
6	2	8	−12.43	+1.20
7	2	10	−21.41	−8.98
8	2	12	−24.71	−3.3
12	2	19	−32.82	−8.11
15	2	24	−35.45	−2.63
16	3	27	−38.68	−3.23
24	3	41	−60.93	−22.25

$\Delta G^o_{solv} = [G^o(\text{system with } n \text{ water molecules}) - nG^o(H_2O) - G^o(Hg(OH)_2)]$

reported for the $HgCl_2$–H_2O case [12]. The nature of the water-mercury interaction was investigated by computing the Wiberg bond indexes. Interestingly, the Wiberg index value (0.1463) for $n = 2$ case is larger compared with the 0.0927 value when $n = 1$ in line with a stronger orbital interaction. For $n = 2$ the O–Hg-O angle reaches its maximum value (179.1°) but a decrease of 5.3° appears for $n = 3$ due to the symmetric hydrogen bonds that appear on both sides of the solute.

Although as expected, the microsolvation free energies increase monotonically, the incremental microsolvation energies show an interesting sawtooth pattern as the number of water molecule increases. We have found that this pattern is related to two factors: the Hg–O_w interactions and the particular way in which the hydrogen bonding between water molecules is established for each n. Particularly, the absolute value of the incremental energy for the trimer (2.11 kcal/mol for $n = 3$) is lower compared with the value obtained for the dimer (4.12 kcal/mol for $n = 2$), despite of having two new hydrogen bonds in the network. This suggests a strong contribution from the direct mercury-water interaction to the solvation energy for $n = 2$. This phenomenon also appears when n changes from 4 to 5 and when n changes from 7 to 8. A remarkable case is when n increases from 5 to 6, where the change even shows a sign reversal. A careful examination of the global minima for these structures reveals the reason of this apparent anomaly in the incremental microsolvation energy trend. We found that two structural features explain this fact: the water-water distances are larger in the hydrogen bond network for $n = 6$ and, most importantly, a rather large reorientation of one of the solute hydrogen atoms (i.e., a significantly modified O–Hg–O–H dihedral angle with respect to the solute geometry when $n = 5$) contributes to a decrease in the incremental free microsolvation energy when n goes from 5 to 6.

A second Hg-O(water) interaction is observed with the addition of a fourth water molecule (see Fig. 10.3, top), although this new direct interaction is more evident for $n = 5$ as shown by the Wiberg indexes of 0.1381 and 0.1313. The variation of the geometrical parameters in the solute seems to play an important role on the evolution of the water hydrogen bond network when the size of the cluster is increased. When n changes from 4 to 5, a slight increase of 4.2° in the O-Hg-O angle allows the entrance of an additional water molecule into the cluster (see Fig. 10.3, right). Each water molecule interacts with other two neighbouring water molecules, leading to an important increase of the ΔG^{o}_{solv} value (−13.63 kcal/mol) as a consequence of the cumulative interactions between them in the solvating network.

Figure 10.3 (bottom) presents the optimized geometries of the solute solvated by six and eight water molecules. In the former case, two Hg–O(water) direct interactions (2.65, 2.75 Å) and eight hydrogen bonds appear in the network while, in the latter case, an additional Hg–O(water) weak interaction is observed at 3.19 Å. As a consequence, the O–Hg–O angle decreases (172.7°) and the free microsolvation energy is now -24.71 kcal/mol. This clearly highlights the energetic importance of the direct Hg-water interactions over the formation of hydrogen bonds in the solvating network. From $n = 16$ onwards the third direct interaction Hg–O_w appears, which is again supported by the Wiberg indices of 0.1179, 0.1204 and 0.1097 obtained with the NBO scheme. This is in line with orbital–driven interactions.

We have found that from $n = 16$ up to the complete the first solvation shell ($n = 24$), the coordination geometry of mercury is a trigonal bipyramid, as shown in Fig. 10.4. Note that for $n = 24$ the direct coordination Hg–O_w optimized distances are

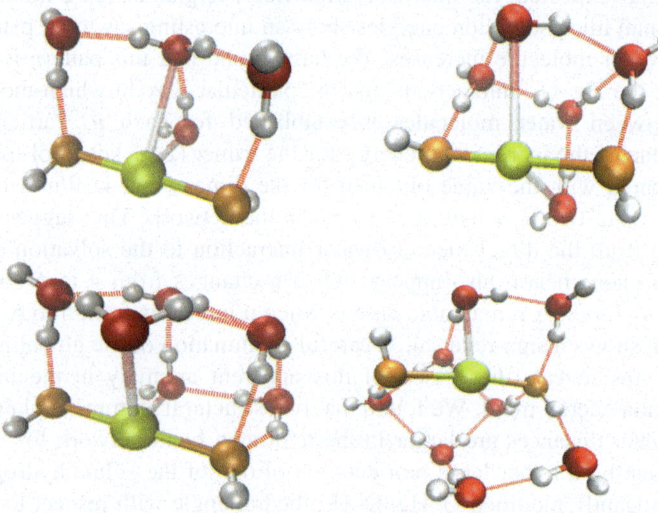

Fig. 10.3 Optimized geometries of $Hg(OH)_2$ solvated by four (*top-left*), five (*top-right*), six (*bottom-left*) and eight (*bottom-right*) water molecules

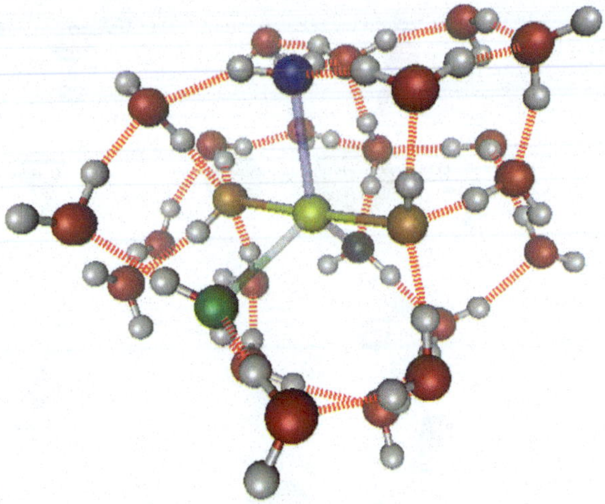

Fig. 10.4 Optimized geometry of $Hg(OH)_2$ solvated by 24 water molecules at 0 K. We highlight the three water oxygens in the equatorial plane of the trigonal bipyramid (*blue*, *green* and *black* atoms) and the solute oxygen atoms are shown in gold

2.75, 2.79 and 2.80 Å. Unlike the $HgCl_2–(H_2O)_{24}$ case, where both the *equatorial* and the *apical* coordination patterns were found as stable structures [12], the fully solvated $Hg(OH)_2$ presents only the *apical* structure, where the solute oxygen atoms are in the apical positions of the triangular bipyramid pattern (O–Hg–O angle of 177.3°). We explored different hydration configurations for $n = 24$ in order to find a solute-equatorial structure but, in all cases, this search lead to axial-solute stable structures, all of these very close to our reported geometry (which has the lowest energy). The microsolvation free energy for $n = 24$ is found to be -60.93 kcal.mol^{-1}.

10.3.2.2 HgClOH

For $n = 1$ and $n = 2$ (Fig. 10.5), we find that the B3PW91 optimized structures are consistent with those reported at the MP2 level (Fig. 10.2). The intra and inter-molecular distances at the DFT level are slightly larger in all cases, while the Cl–Hg–O(H) angle slightly oscillates depending on the n value. The first direct mercury-water interaction (2.70 Å) appears from the beginning of the solvation process with $n = 1$.

For $n = 3$ and $n = 4$ we found two slightly different minima when compared to those computed at the MP2 level (see Fig. 10.5). While the MP2 scheme predicts two direct water-Hg orbital interactions for $n = 3$, the B3PW91 scheme favors the formation of two new hydrogen bonds between the water molecules. For $n = 4$ the difference with the MP2 optimized structure is qualitatively less important, since both the DFT and the ab initio methods yield two direct water-Hg orbital interactions, the latter also predicting the first hydrogen-halogen bridge. Therefore, for small n values, with this hybrid DFT method we find a faster evolution of the hydrogen bond network, mainly due to more significant contributions of the water-water interactions with respect to the water-mercury and water–halogen interactions.

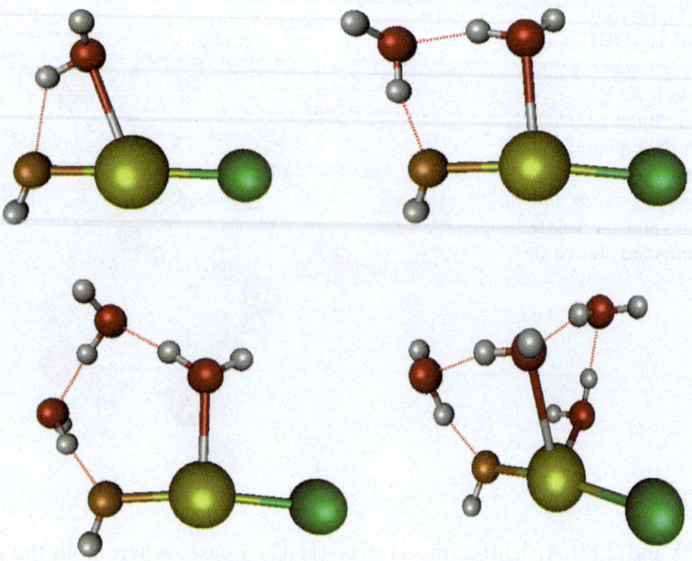

Fig. 10.5 B3PW91 geometries of HgClOH solvated by one (*top-left*), two (*top-right*), three (*bottom-left*) and four (*bottom-right*) water molecules

It is crucial to emphasize that the incremental energies (5.25 and 10.90 kcal/mol) computed using the B3PW91 scheme and the AVTZ basis sets are in excellent agreement with the counterpoise-corrected MP2 complete basis set values reported previously for n = 1 and 2 (5.20 and 11.49 kcal/mol). On the other hand, for $n = 3$ the presence of two new hydrogen bonds between the H_2O molecules at DFT level gives rise to a slightly larger incremental binding energy value (8.16 kcal/mol) compared with the incremental CC-MP2 value (6.69 kcal). This trend is reversed when the fourth water molecule is added to the cluster. In that case, the structure computed at DFT level has a lower incremental binding energy (6.53 kcal/mol) than the CC-MP2 optimized value (10.01 kcal/mol) for the MP2. Here note that, while MP2 calculations considerably overestimate the interaction energies due to the basis set superposition error, the much faster B3PW91/AVTZ calculations directly lead to reasonably accurate incremental interaction energies even without the counterpoise correction. Note, however, that B3PW91/6-311G** approach leads to overestimated values. So keeping in mind that the B3PW91 method provides an approximate description of the aqueous solvation for this metallic species, it is clear that the study of much larger clusters cannot be achieved at the MP2/AVTZ level. Therefore, we continue to address the study of larger clusters with the B3PW91/6-311G** approach.

A comparison of the evolution of the total Gibbs free water binding energies for the HgClOH–$(H_2O)_n$ and $Hg(OH)_2$–$(H_2O)_n$ clusters is presented in Fig. 10.6. For the HgClOH–$(H_2O)_n$ case some interesting aspects can be observed. For instance, the absolute value of the incremental free energy for $n = 2$ (-1.94 kcal/mol) is nearly

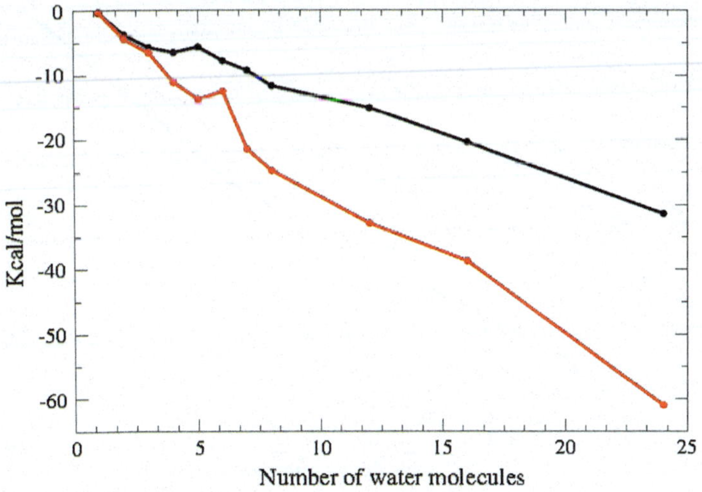

Fig. 10.6 Gibbs free water binding energies versus number of water molecules in HgClOH–$(H_2O)_n$ (*black curve*) and $Hg(OH)_2$–$(H_2O)_n$ (*red curve*) clusters

five times larger compared with the value for $n = 1$ (-0.40 kcal/mol), this suggests that the Hg–O(water) interaction plays a more important role in the trimer ($n = 2$) than in the dimer ($n = 1$) case, in spite of the fact that the thermal correction to the Gibbs free energy for the trimer is much larger (16.9 vs. 2.9 kcal/mol) than for the dimer. This difference seemed large but a careful systematic examination of the thermal corrections for systems with varying number of water molecules reveals that, for every added water molecule, the thermal correction to the Gibbs free energy grows ca. 15 kcal/mol. This fact is also consistent with the shorter Hg–O_w distance for the trimer (2.5 Å) than for the dimer (2.7 Å). Another interesting feature is that the water interaction energy slightly decreases when going from n = 4 to n = 5; this apparently anomalous behavior was also found for the $Hg(OH)_2$–$(H_2O)_n$ solvation when going from n = 5 to n = 6 and was discussed in the previous section.

For HgClOH–$(H_2O)_n$ with n > 20, note that the near-constant changes of 2–3 kcal/mol per additional water are consistent with the fact that the first solvation shell has been fully formed, i.e., that each additional water molecule simply builds new hydrogen bonds with water molecules that constitute the first solvation shell and have little effect on the internal "core" geometry of the cluster.

Overall these curves reveal that $Hg(OH)_2$ interacts more strongly with water environment than HgClOH.

At this point it is interesting to analyze the role played by cuadrupole moments on the total water binding energies. In order to do so, we plot in Fig. 10.7 the sum of the squares of all the components of the cuadrupole moments *vs*. number of water molecules in the HgClOH–$(H_2O)_n$ and $Hg(OH)_2$–$(H_2O)_n$ clusters. It is found that there is no clear correlation between the magnitude of the cuadrupole moments with

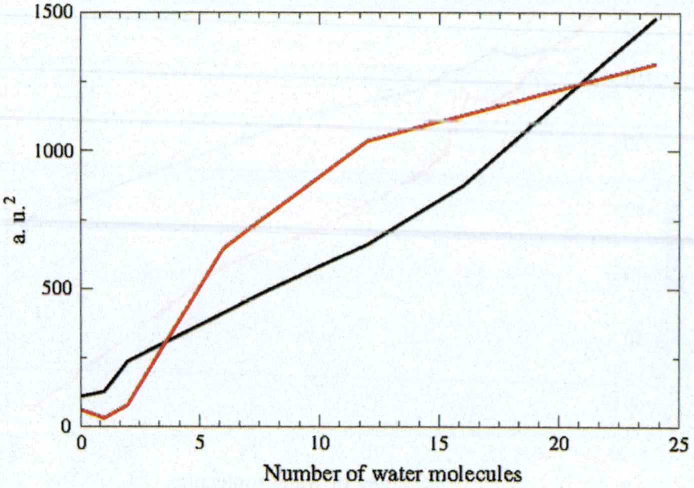

Fig. 10.7 Sum of the squares of the components of the cuadrupole moments vs. number of water molecules in HgClOH–$(H_2O)_n$ (*black curve*) and $Hg(OH)_2$–$(H_2O)_n$ (*red curve*) clusters

the Gibbs free binding energies as the number of water molecules increases. If one is interested in the bulk description for the solvated systems, it is clear that nothing can be said with such a small number of explicit water molecules since many more of them are needed to built the second, third and subsequent solvation shells.

Figure 10.8 shows the optimized geometries of the solute solvated by twelve and sixteen water molecules. Note the presence of three Hg–O_w (water) direct interactions (2.41, 2.53 and 2.62 Å) leading to a close packed trigonal bipyramid pattern around Hg when *n* reaches 16 and this local solvation pattern remains for larger values of *n*.

We found that 22 water molecules are needed to build the first solvation shell. When this point is reached, the addition of one extra water molecule has to be done by making two new hydrogen bonds with the previously existing water network, so that the 23rd water molecule is not directly interacting with the HgClOH solute but

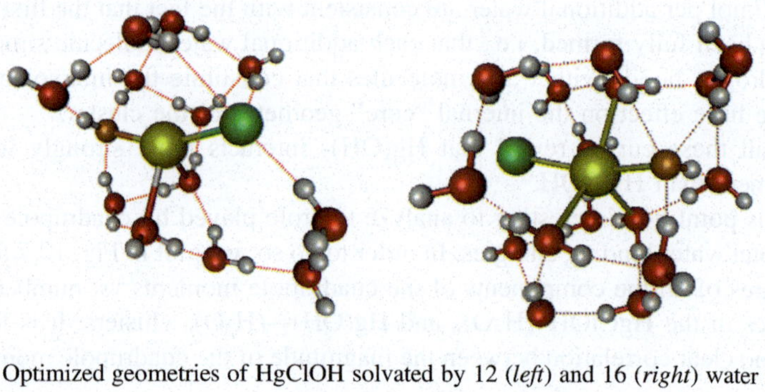

Fig. 10.8 Optimized geometries of HgClOH solvated by 12 (*left*) and 16 (*right*) water molecules

merely serving as a water-water bridge. On the other hand, the optimized structure with 21 water molecules leaves a region around HgClOH without the effect of aqueous solvation, in other words, leading to an incomplete surface coverage of the solute by the solvent.

The largest solvated complex we considered comprises 24 water molecules (Fig. 10.9). Note the *apical* coordination pattern where the Cl, Hg and O atoms of the solute recover a nearly-linear configuration; a similar coordination has been reported for $Hg(OH)_2$–$(H_2O)_{24}$ [15] and $HgCl_2$-$(H_2O)_{24}$ [12]. We recall that Castro et al. [12], besides finding the apical structure, also found a stable equatorial trigonal bipyramid pattern for the $HgCl_2$ case. Therefore, we explored different configurations in order to find an equivalent solute-equatorial structure. However, only stable apical solvated structures were found having very similar (within 3 kcal/mol) Gibbs free energies; note that we present here only the one corresponding to the lowest Gibbs free energy (ca. −32 kcal/mol).

In order to address the possible effects of the polar aqueous medium on the geometry of the optimized structures including explicit water molecules we performed calculations using the Polarizable Continuum Model (PCM) of Tomasi et al. [25] as implemented in Gaussian09; we used the same 6-311G(d,p) basis sets as above. As is well known, in the PCM approach the solvent, in this case water, is represented by a continuous medium which is characterized by its dielectric constant ($\varepsilon = 78.355$). Thus, we re-optimized the structures determined as minima *in vacuo* for selected values of *n*. It can be seen that the PCM optimized intermolecular distances are overestimated. For the bare the solute intramolecular Hg–Cl and Hg–O(H) PCM distances are 0.1 and 0.06 Å longer. However, when explicit water molecules are considered, the overestimation of the Hg–O_w distances is surprisingly large, especially for the smaller clusters. To emphasize this, note that

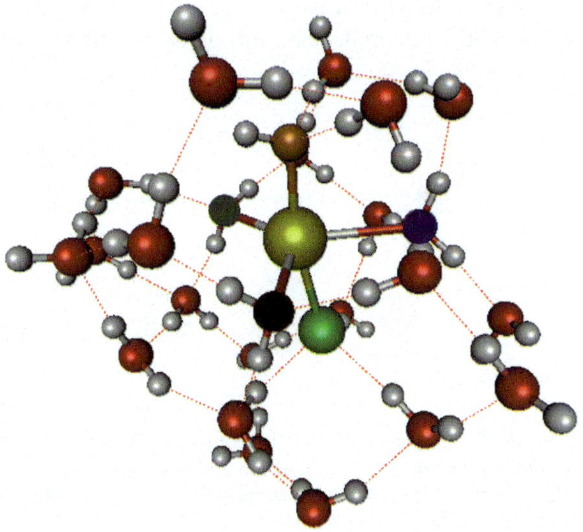

Fig. 10.9 Optimized geometry of HgClOH solvated by 24 water molecules at 0 K. We highlight the three water oxygens in the equatorial plane of the trigonal bipyramid (*blue*, *dark green* and *black* atoms) and the solute oxygen atom is shown in gold

for n = 1, 2, 3, 6, and 12 the differences with the shortest optimized B3PW91 Hg–O_w distances are 1.19, 1.03, 1.02, 1.36 and 0.45 Å, respectively. It is clear that the PCM scheme significantly favors the separation of the explicit water molecules from the solute, leading to a rather loosely bound solvating water network around HgClOH in a clathrate manner. As an example of this, Table S4 in the Electronic Supplementary information in ref. [17] shows the vibrational frequencies obtained for the optimized B3PW91/6-311G** HgClOH–H_2O complex *in vacuo* and with the PCM scheme with water as continuous solvent. Figure 10.10 presents the B3PW91-PCM optimized geometry for the HgClOH–$(H_2O)_{1,3}$ cases, where the closest Hg–O_w distances are 3.89 and 3.49 Å vs. 2.69 and 2.47 Å obtained *in vacuo*. For the HgClOH–(H_2O) PCM case we stress the total absence of orbital interaction between Hg and the water oxygen, clearly showing that the O_w lone pair solvation by the PCM scheme is energetically more favorable than the interaction with Hg.

Let us now focus on the energetic aspect of the PCM results of the Gibbs free energies at 298 K of the solute, of a single water molecule and of the optimized structures for selected values of n, both *in vacuo* and in the polarizable medium. For n = 0, the difference of the Gibbs free energy *in vacuo* and in the polarizable medium is -38.8 kcal/mol, which provides an approximation to the solvation energy of this species in water. Here we have employed the usual definition of the incremental water binding energies [27], i.e., ΔG^o = [G^o(HgClOH with n water molecules)-$nG^o(H_2O)$–G^o(HgClOH)], One might be tempted to obtain the incremental water binding energies for $n > 0$ considering the polarizable medium. However, it can be easily shown that incremental water binding energies obtained with the PCM approach are fundamentally wrong for a simple reason. First of all, note that the continuous solvent model leads to a nearly constant decrease of the Gibbs free energies of ca. 12 kcal/mol for all *n,* roughly corresponding to the solvation energy of the bare solute. This fact is explained by the near-constant increase in the size of the surface which defines the solute cavity since, with each additional water molecule considered as solute, three new spheres (one with the

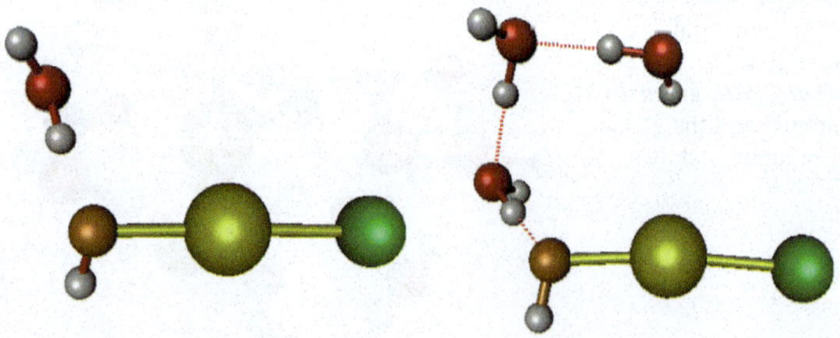

Fig. 10.10 Optimized B3PW91 geometries of the HgClOH–H_2O (*left*) and HgClOH–$(H_2O)_3$ (*right*) complexes with the PCM scheme with water as a continuous solvent. Note the absence of Hg–O_w interaction in the first case and the much longer Hg–Ow distance in the latter as compared with the structure in Fig. 10.5

oxygen atom radius and two with hydrogen radii) are added to the cavity generation algorithm of the PCM code.

Secondly, the sum of the Gibbs free energy of the fragments, HgClOH plus n times that of an isolated water in the PCM scheme is obviously more negative than that of the HgClOH–$(H_2O)_n$ complex. Note that this difference becomes even larger for larger n, thus suggesting that these complexes never get to form. This is so because the polarizable medium solvation of the isolated water molecules (or even of the frozen-geometry-$(H_2O)_n$ water clusters derived from the optimized HgClOH–$(H_2O)_n$ structures), with the PCM approach always leads to lower energies due to the larger surface area of the separated fragments. This is the fundamental reason why previous studies on explicit aqueous microsolvation of mercury mono and dihalides [27] as well as of thiosulfuric acid and its tautomeric anions [28] did not address the energetic differences with the polarizable medium model. We emphasize once again that incremental water binding energies cannot be interpreted as incremental solvation energies since, with increasing n values, they include the water-water attractive interactions which are not directly related to the stabilization of the solute by the presence of the neighboring aqueous environment.

Before going to the dynamical results we point out that we tried to address the effects of dispersion-corrected density functionals on structures and energies. Disappointingly we found that no quantum chemistry program available to us (Gaussian09, Molpro-2010, NWChem or CRYSTAL) is capable of using dispersion-corrected density functionals for mercury containing molecules. However, at this point we recall that, while the M06 functional was not only designed to account for dispersion corrections it partially accounts for these effects; the comparison of the M06 and the ab initio MP2 results described earlier in this study indicate that there are small differences with the B3PW91 ones, suggesting that dispersion effects are small on the presently optimized HgClOH–$(H_2O)_n$ complexes.

10.3.3 DFT Born-Oppenheimer Molecular Dynamics

10.3.3.1 $Hg(OH)_2$

Finally, as part of a complete analysis of the first solvation shell, we carried out a Born–Oppenheimer molecular dynamics simulation at the B3PW91 level and 300 K starting from the $Hg(OH)_2$–$(H_2O)_{24}$ optimized structure (see Fig. 10.4). Two important dynamical results must be highlighted. The first one is that the solvation shell remains stable throughout the 8 ps of the simulation despite the non-negligible thermal effects at room temperature. Secondly, while a clathrate-like structure is obtained at the end of the simulation (see Fig. 10.11), the trigonal bypiramid pattern remains with the three equatorial waters oscillating around their equilibrium positions coordinated to Hg.

Due to the thermal effects the average equatorial Hg–O_w distances are somewhat larger than the static ones (2.97, 3.02 and 3.06 Å). In Table 10.5 it is also possible

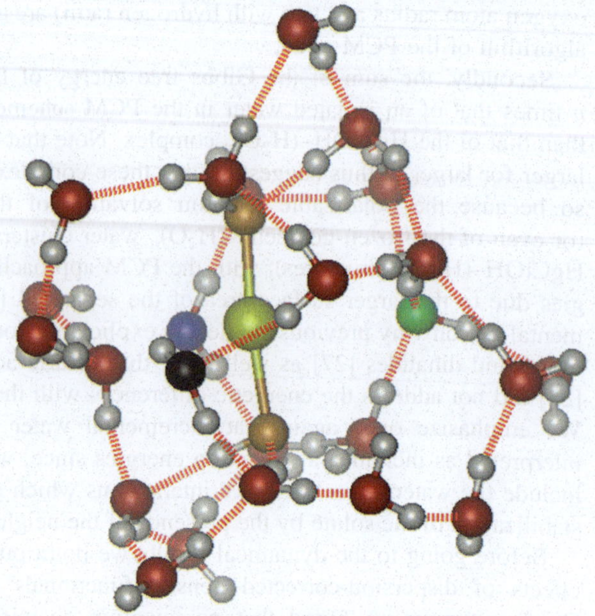

Fig. 10.11 Geometry of Hg $(OH)_2$ solvated by 24 water molecules obtained at the end of the BO-DFT-MD simulation at 300 K. We highlight the three water oxygens in the equatorial plane of the trigonal bipyramid (*blue*, *green* and *black* atoms) and the solute oxygen atoms are shown in gold

to observe that the Hg–O average distances (2.03 and 2.04 Å) are similar to those of the static DFT optimized structure and that the quasi-linearity of the solute is only slightly modified (171.1°). Figure 10.12 shows the temporal evolution of the Hg–O_w distances where large oscillations around the average values can be seen.

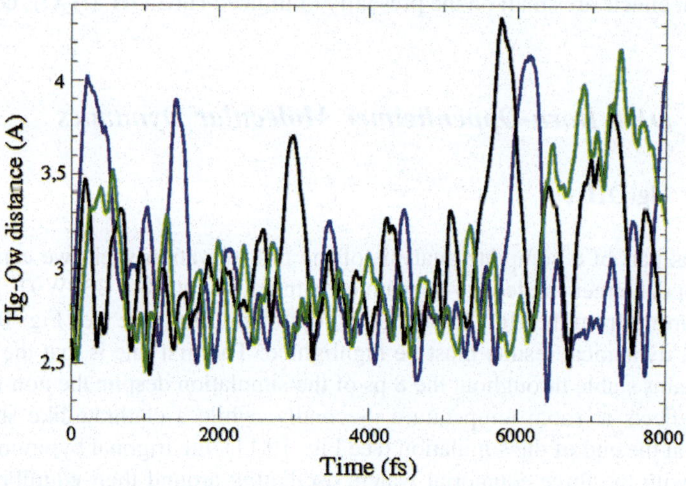

Fig. 10.12 Hg–O_w distance evolution for the three equatorial water molecules in the trigonal bipyramid solvation pattern. The *blue*, *green* and *black curves* correspond to the atoms highlighted in Figs. 10.5 and 10.10

We note that after 5 ps the Hg–O_w oscillations become larger, suggesting that water exchange could happen if longer simulation times were computationally accessible. The rather large amplitude (2 Å) of the oscillations also reveals that the Hg-water interactions are relatively weak but, at the same time, strong enough to keep the trigonal bipyramid motif throughout the simulation.

Figure 10.13 shows the temporal evolution and the average value of the O–Hg–O angle of the solute throughout the simulation. It can be seen that large amplitude oscillations (>17°) occur every 1.0–1.5 ps, while small amplitude oscillations (<10°) have a much shorter period of around 0.1–0.2 ps. However, the average O–Hg–O angle is 171.1°, in line with the average trigonal bipyramid coordination of Hg throughout the simulation

These results allow us to say that the static cluster model is valid to provide a first account of the aqueous solvation with the first water shell around the solute. This also gives us confidence about the estimation of the Gibbs free microsolvation energy obtained from the static optimizations.

10.3.3.2 HgClOH

To address the role of thermal effects on the fully solvated structure, a Born–Oppenheimer MD simulation at the B3PW91 level part of that high temperature (700 K) was carried out. The simulation started from the HgClOH–$(H_2O)_{24}$ optimized structure with random velocities consistent with the fixed temperature. A particularly important feature of the dynamical results is the stability of the cluster-derived first solvation shell. Throughout the MD simulation we observed the persistence of a trigonal bipyramid coordination pattern around Hg, as shown by

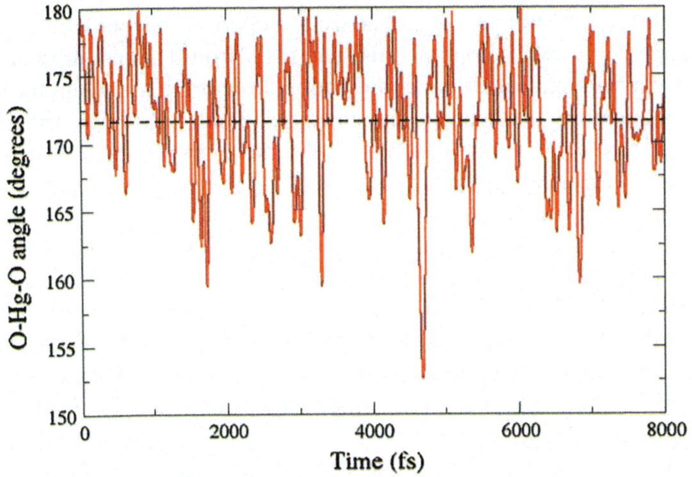

Fig. 10.13 O–Hg–O angle evolution and the average value (*dashed line*) at T = 300 K

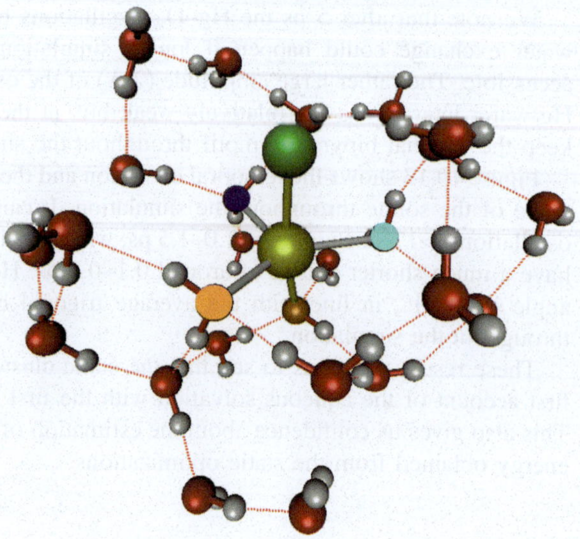

Fig. 10.14 Representative geometry of HgClOH solvated by 24 water molecules with BO-DFT-MD simulation at 700 K. We highlight the three oxygens that directly interact with the mercury (*blue*, *orange*, turquoise) during the simulation. The solute oxygen atom is shown in gold

the representative geometry depicted in Fig. 10.14 Note however that up to seven water molecules actively participate in setting up of this pattern throughout the simulation. This BO-MD simulation allows us to examine the evolution of the main geometrical parameters, in particular, the behavior of intermolecular Hg–O(water) distances.

Figure 10.15 (Top) shows the Hg–O_w distance evolution for the seven water oxygens that directly interact with mercury during the BO-MD simulation as well as the solute Cl–Hg–O angle. Note that the first water exchanges, occurring very early ($t < 0.8$ ps) in the simulation, take out two of the equatorial water molecules (black and green curves) and these molecules never come back to interact directly with Hg again. On the other hand, exactly the opposite happens for a water molecule (orange curve) whose optimal distance is far from Hg and, after around 1 ps, comes into the trigonal equatorial coordination around Hg to remain there until the end of the simulation. Intermediate situations are found, where two water molecules (pink and light blue curves) originally not coordinated to Hg, intermittently build up the trigonal coordination environment around Hg. Only one (purple curve) of the seven closest water molecules briefly participates in the formation the trigonal coordination environment around Hg; it quickly approaches Hg while another molecule (light blue curve) recedes from Hg due to steric impediments, to finally move away (all this in less than 0.4 ps). However, we emphasize that after 5 ps the Hg–O_w oscillations become smaller and the number of water exchanges decreases due to thermalization. Therefore, using the last 5 ps of the simulation we can roughly estimate that one water exchange takes place every 10 ps after thermalization has been achieved. This 10^{11} s^{-1} rate can be compared with the water exchange rate in the first solvation shell of metal ions in aqueous solution. For instance, in the case of the bare Cs^+ ion the mean lifetime is 200 ps [29]. As could

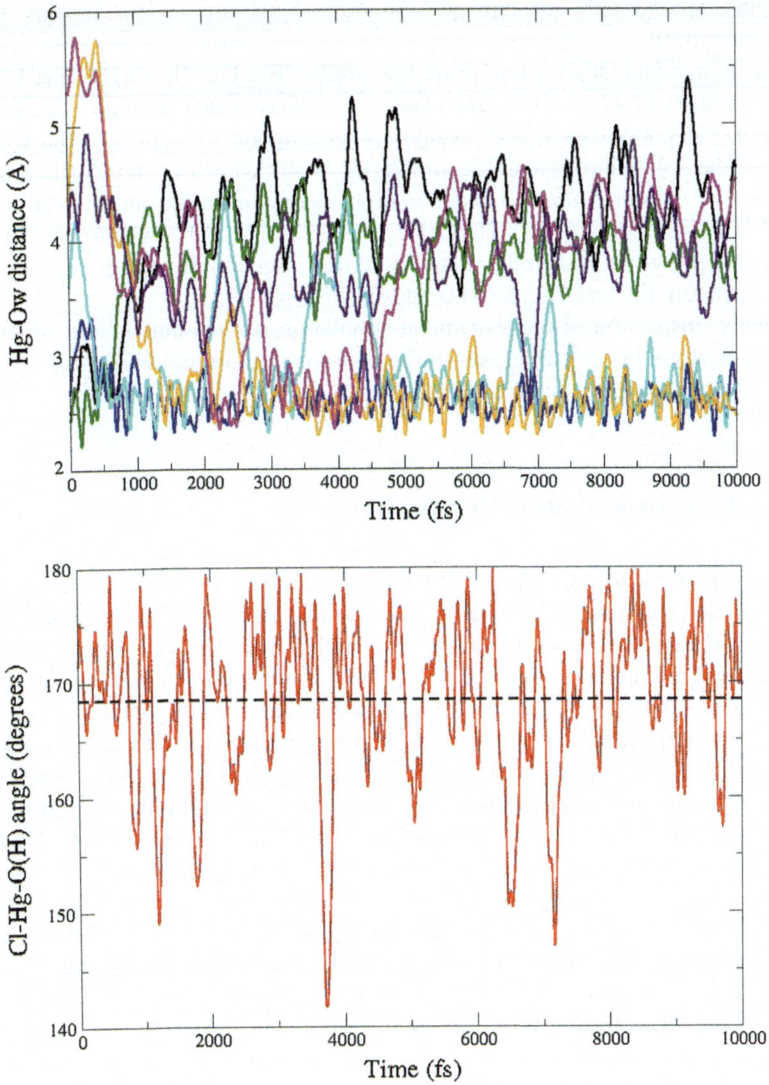

Fig. 10.15 Hg–O_w distance evolution for the oxygen atoms directly interacting with Hg during the BO-MD simulation. Lower panel, solute Cl–Hg–O(H) angle evolution and the average value (*dashed line*) at T = 700 K

be expected, the much shorter lifetime we estimate here for molecules in the first solvation shell is in agreement with the smaller interaction energy of water with the neutral HgClOH molecule as compared to that of the ion-water system. However, we stress that: a) much longer simulation times would be needed to extract statistical data to determine a reliable average water-exchange period and, b) these mean lifetimes apply to gas-phase clusters and all the many body effects arising from the second and subsequent hydration shells are absent in this type of simulations [30].

The dynamical description provides similar Hg–Cl, Hg-O(H) and Cl–Hg-O average values (2.47, 2.10 Å and 168.7°) compared with the static (2.46, 2.10 Å and 170.0°) B3PW91/6-311G** values. The temporal evolution of the Cl–Hg–O angle of the solute throughout the simulation is shown in Fig. 10.15 (bottom). The presence of chlorine atom produces a slight decrease on the angle average value (168.7°) compared with the $Hg(OH)_2$ case (171.1°) [15], where two hydroxyl groups are present. On the other hand, note that large amplitude oscillations (>18°) occur between 1.5 and 2.0, 2.7–4.0 and 6.2–7.5 ps.

Finally, inspection of the dynamical simulation reveals that 17 out of 24 water molecules never get to participate in the coordination around Hg, simply breaking and making inter-water hydrogen bonds.

10.4 Conclusions and Perspectives

Previous theoretical studies [12–14] have showed the relevance to obtain adequate information at the molecular level to describe the aqueous solvation mechanism of abundant Hg-containing neutral molecules. In this work we present the main results of a detailed research program about the study of the solvation patterns of $Hg(OH)_2$ [15] and HgClOH [16, 17], two of the most common Hg-containing species in marine environments. As a first step in the approach to the aqueous solvation problem of HgClOH, we analyse the results of a systematic MP2 study of the structures, vibrational frequencies and the incremental binding energies of the HgClOH–$(H_2O)_n$ (n = 1-4) microsolvated species. A series of augmented correlation-consistent basis sets was used in order to obtain complete basis set limit binding energies including the CP correction to account for the basis set superposition error. The basis set superposition errors are quite large and they increase with basis set quality for all n, so that the CP corrections are crucial to provide reliable energetic values. The MP2/CBS-CP incremental binding energies are 5.2, 11.5, 6.6 and 10.0 kcal/mol for $n = 1$–4. Although two direct Hg–O_w orbital-driven interactions with water oxygens appear for $n = 3$ and $n = 4$, the intramolecular geometry of HgClOH is barely modified by the microsolvation environment. The vibrational analysis reveals that the strongest water-solute interaction appears for $n = 3$.

On a second stage we have analyzed the evolution of geometrical parameters and the microsolvation free energies for the optimized Hg(II)XY–$(H_2O)_n$ (X,Y = Cl, OH) complexes with $n \leq 24$ at B3PW91/6-311G** level. After an exhaustive search, we obtained the absolute minima in the potential energy surfaces of Hg $(OH)_2$-$(H_2O)_n$ and HgClOH–$(H_2O)_n$ complexes with $n < 6$. For larger n the optimized structures might be local or global minima, since the number of possibilities of hydrogen bonds in the surrounding network dramatically increases with cluster size. Nevertheless, the optimized structures are within a few kcal/mol of the absolute minima in each case, the difference being due to small changes in the orientation or coordination of the intra-water hydrogen bond network.

For $Hg(OH)_2$ system, the Hg-O solute distances remain practically unchanged, the O–Hg–O angle of the solute depends on cluster size. Three direct Hg–O_w orbital-driven interactions appear from $n = 16$ up to $n = 24$, when the first solvation shell is complete. We obtained the Wiberg indices to establish the presence of mercury-water oxygen direct interactions for each n value, which have also allowed us to ascertain their orbital-driven nature. A strong dependence of the solvation free energy with respect to the number of hydrogen bonds in the water network is observed. The final microsolvation energy of the "apical" optimized structure is estimated to be around −61 kcal/mol, approximately 26 kcal/mol larger than the value we reported previously for a $HgCl_2$–$(H_2O)_{24}$ structure [12]. This rather large difference between the $Hg(OH)_2$–$(H_2O)_{24}$ and $HgCl_2$–$(H_2O)_{24}$ systems arises due to the presence of the hydroxyl groups in the former which allow the formation of four solute-water hydrogen bonds (two with each OH moiety), whereas the halogen-hydrogen bonds are less numerous and weaker in the latter case. Thus, the first aqueous solvation shell presents a stronger interaction with the $Hg(OH)_2$ solute. This suggests that the transmembrane passage of $Hg(OH)_2$ into the cell via simple diffusion is less favorable compared to the case when the metal is coordinated with two Cl groups. Through natural population analysis, we found that the singly-ionic character of mercury remains practically unaltered along the stepwise solvation process, thus pointing to a Hg(I) state instead of the usually presumed Hg (II) oxidation state for $Hg(OH)_2$.

We found that 22 water molecules are necessary to fully form the first solvation shell around HgClOH molecule. For small n DFT energetic and structural optimized parameters are in close agreement with counterpoise-corrected MP2 calculations. The intramolecular Hg-O and Hg–Cl solute distances practically remain unchanged along the solvation process; however the Cl–Hg-O angle of the solute shows a larger dependence with the n value. Three short direct Hg–O(water) orbital-driven equatorial interactions appear from $n = 16$ onwards. The pentacoordinated HgClOH trigonal bipyramid coordination around Hg is on the same way those found in the microhydration patterns for $Hg(OH)_2$ and $HgCl_2$ [12]. On average, two new hydrogen bonds appear when a new water molecule is added to the network, even for the larger clusters.

When considering the effects of a continuous polarizable solvent through the PCM scheme, we found that the optimized geometries lead to very large Hg–O_w distances due to the larger stabilization produced by the continuum solvation of the explicit water molecules. For the larger clusters the PCM optimized structures show a loosely bound water network around HgClOH, in a clathrate-like manner.

The role of dispersion-corrected density functionals could not the assessed since none of the quantum chemistry codes we have access to can deal with mercury-containing systems. However, comparison of the B3PW91 geometries and water binding energies with the MP2 and M06 results for small n suggest that the dispersion corrections to them should actually be small.

Finally, we explored in detail the thermal stability of the optimized Hg(II)XY-$(H_2O)_n$ (X,Y = Cl, OH) structures by means of a DFT Born-Oppenheimer molecular dynamics simulations. We found that when temperature effects are taken into account the solvating water network is slightly altered and clathrate-like structures appear where the Hg–O_w distances are slightly larger than the static optimized values. The trigonal bipyramid structures remains stable throughout the simulations and rather large-amplitude oscillations of the Hg–O_w distances in the equatorial plane are observed. This suggests that possible water exchange could occur in the first solvation shell if longer simulations times were computationally accessible.

We emphasize that these static and dynamical descriptions of the first solvation shell of Hg-containing species using a stepwise hydration scheme yield a restricted view when compared with the full solvation that occurs in the liquid phase, however they represent a crucial step in order to obtain a better description of the aqueous solvation through Monte Carlo simulations of the condensed phase.

Acknowledgments The authors thank support from the ECOS-ANUIES/CONACYT Mexican-French cooperation program through project M10-P01. ARS also thanks support from CONACYT basic science project No. 130931. AEJI thanks a graduate scholarship from CONACYT.

References

1. Gray JE, Hines ME (2006) Mercury: Distribution, transport, and geochemical and microbial transformations from natural and anthropogenic sources. Appl Geochem 21:1819
2. AMAP Arctic Pollution (2011) Arctic monitoring and assessment programme (AMAP), Oslo
3. Fitzgerald WF, Mason RP (1997) Mercury and its effects on environment and biology, vol 34, 1st ed. Marcel Dekker, Inc., New York, pp 53–111
4. Kim KH, Hanson PJ, Barnett MO, Lindberg SE (1997) Mercury and its effects on environment and biology, 1st edn. Marcel Dekker, New York, pp 185–212
5. Morel FMM, Krapiel AML, Amyot M (1998) The chemical cycle and bioaccumulation of Mercury. Annu Rev Ecol Syst 29:543
6. Larose C, Dommergue A, De Angelis M, Cossa D, Averty B, Marusczak N, Soumis N, Schneider D, Ferrari C (2010) Springtime changes in snow chemistry lead to new insights into mercury methylation in the artic. Geochim Cosmochim Acta 74:6263
7. Gutknecht JJ (1981) Inorganic mercury transport through lipid bilayer membranes. Membr Biol 61:61
8. Myers GJ, Davidson PW, Cox C, Shamlaye CF, Palumbo D, Cernichiari E, Sloane-Reeves J, Wilding GE, Kost J, Huang LS, Clarkson TW (2003) Prenatal methylmercury exposure from ocean fish consumption in the Seychelles child development study. Lancet 361:1686
9. Barkay T, Gillman M, Turner RR (1997) Effects of dissolved organic carbon and salinity of bioavailability of mercury Appl. Environ Microbiol 63:4267
10. Golding GR, Kelly CA, Sparling R, Loewen PC, Barkay T (2007) Evaluation of mercury toxicity as a predictor of mercury bioavailability Environ. Sci Technol 41:5685–5692
11. Barkay T, Turner RR, Rasmussen LD, Kelly CA, Rudd JWM (1998) Lux-facilitated detection of mercury in natural waters. In: La Rossa R (ed) Bioluminescent Protocols. Humana Press, Inc., Totowa, pp 231–246

12. Castro L, Dommergue A, Renard A, Ferrari C, Ramírez-Solís A, Maron L (2011) Theoretical study of the solvation of $HgCl_2$, HgClOH, $Hg(OH)_2$ and $HgCl_3$: a density functional theory cluster approach. Phys Chem Chem Phys 13:16772
13. Hernández-Cobos J, Ramírez-Solís A, Maron L, Ortega-Blake I (2012) Theoretical study of the aqueous solvation of HgCl2: Monte Carlo simulations using second-order Moller-Plesset-derived flexible polarizable interaction potentials. J. Chem. Phys. 136:014502
14. Hernández-Cobos J, Vargas MC, Ramírez-Solís A, Ortega-Blake I (2010) Aqueous solvation of $As(OH)_3$: a Monte Carlo study with flexible polarizable classical interaction potentials. J Chem Phys 133:114501
15. Amaro-Estrada JI, Maron L, Ramírez-Solís A (2013) Aqueous solvation of $Hg(OH)_2$ Energetic and dynamical density functional theory studies of the $Hg(OH)_2$–$(H2O)_n$ ($n = 1 - 24$) structures. J Phys Chem A 117:9069
16. Amaro-Estrada JI, Ramírez-Solís A (2013) Aqueous microsolvation of HgClOH. A systematic MP2 study of the HgClOH-$(H_2O)_n$ species with $n = 1 - 4$. Comput Theor Chem 1006:45
17. Amaro-Estrada JI, Maron L, Ramírez-Solís A (2014) Aqueous solvation of HgClOH. Stepwise DFT solvation and Born-Oppenheimer molecular dynamics studies of the HgClOH-$(H_2O)_{24}$ complex. Phys Chem Chem Phys 16:8455
18. Bergner A, Dolg M, Kuechle W, Stoll H, Preuss H (1993) *Ab initio* energy-adjusted pseudopotentials for elements of groups 13–17. Mol Phys 80:1431
19. Kuechle W, Dolg M, Stoll H, Press H (1991) *Ab initio* pseudopotentials for Hg through Rn. Mol Phys 74:1245
20. Dunning TH (1998) Gaussian basis sets for use in correlated molecular calculations. The atoms boron through neon and hydrogen. J Chem Phys 90:1007
21. Reed AE, Curtiss LA, Weinhold F (1998) Intermolecular Interactions from a Natural Bond Orbital. Donor-Accept Viewp Chem Rev 88:899
22. Frisch MJ, Trucks GW, Schlegel HB, Scuseria GE, Robb MA, Cheeseman J, Zakrzewski VG, Montgomery JA, Stratman RE, Burant JC et al (2003) Gaussian 03. Gaussian Inc, Pittsburgh
23. Raynaud C, Maron L, Daudey JP, Jolibois F (2004) Reconsidering Car-Parrinello molecular dynamics using direct propagation of molecular orbitals developed upon Gaussian type atomic orbitals. Phys Chem Chem Phys 6:4226
24. Verlet L (1967) Computer "experiments" on classical fluids. I. Thermodynamical properties of Lennard-Jones molecules. Phys Rev 159:98
25. Nosé S (1984) A unified formulation of the constant temperature molecular-dynamics methods. J Chem Phys 81:511
26. Hoover WG (1985) Canonical dynamics: equilibrium phase-space distributions. Phys Rev A 31:1695
27. Shepler BC, Wright AD, Balanov NVB, Peterson KA (2007) Aqueous microsolvation of mercury halide species. J Phys Chem 111:11342
28. Steudel R, Steudel Y (2009) J Phys Chem A 113:9920
29. Helm L, Merbach AE (1999) Water exchange on metal ions: experiments and simulations. Coord Chem Rev 187:151
30. Ramírez-Solís A, Maron L (2012) Int J Quant Chem 112:3484

Chapter 11
Advances in QM/MM Molecular Dynamics Simulations of Chemical Processes at Aqueous Interfaces

Marilia T.C. Martins-Costa and Manuel F. Ruiz-López

Abstract We review recent studies carried out in our group on the modeling of aqueous interfaces using Molecular Dynamics simulations with a combined Quantum Mechanics and Molecular Mechanics force-field (QM/MM). We first present the methodology and we comment on some ongoing developments. Since in the QM/MM approach the adsorbed molecule is described quantum mechanically, this computational scheme has allowed us to get insights on interface solvation effects on molecular properties. In particular, we have shown that polarization phenomena at the air–water interface may produce larger effects than polarization in bulk water. This finding contrasts with the usual assumption that polarity at liquid interfaces is close to the arithmetic average of the polarity of the two bulk phases, and that solvation effects at the air–water interface should be similar to the effects in a low polar solvent such as butyl ether. A summary of previous results is presented with some selected examples that are briefly discussed, and which include systems of atmospheric interest at the air–water interface, as well as systems of biological relevance at a water-organic interface. Then, we report some new results for a series of small volatile organic compounds at the air–water interface, namely methyl chloride, acetonitrile and methanol. These molecules share a similar structure but display quite different behaviors at the interface; the discussion focuses on the orientational dynamics and the solvation effects on reactivity indices. Finally, some conclusions and future directions in this exciting field are presented.

Acronyms

B3LYP	Becke's 3-parameter-Lee-Yang-Parr exchange-correlation functional defined in reference [64]
HF	Hartree-Fock method
HOMO	Highest occupied molecular orbital
LUMO	Lowest unoccupied molecular orbital
MM	Molecular mechanics

M.T.C. Martins-Costa · M.F. Ruiz-López (✉)
SRSMC, University of Lorraine and CNRS, BP 70239,
54506 Vandoeuvre-lès-Nancy, France
e-mail: manuel.ruiz@univ-lorraine.fr

J.-L. Rivail et al. (eds.), *Quantum Modeling of Complex Molecular Systems*,
Challenges and Advances in Computational Chemistry and Physics 21,
DOI 10.1007/978-3-319-21626-3_11

MD	Molecular dynamics
NVT	Canonical ensemble (constant number of particles, constant volume, constant temperature)
QM	Quantum mechanics
ROS	Reactive oxygen species
SCF	Self-consistent field method
SFG	Sum frequency generation
VOC	Volatile organic compound

11.1 Introduction

The study of solvation phenomena is fundamental to rationalize the mechanisms and energetics of chemical processes in condensed phases, which may be very different in comparison with equivalent gas phase reactions [1]. The huge amount of experimental and theoretical work that has been accumulated in the last decades has allowed for significant improvements in our understanding of the physics behind solvation effects in water or in traditional organic solvents. However, a number of important challenges remain to be addressed as, for instance, how solvation influences the reactivity in non-conventional media such as supercritical fluids, ionic liquids, etc. Besides, the role of solvation on the chemical properties of molecules interacting with a liquid interface is still poorly known. This is unfortunate because liquid interfaces, and in particular aqueous interfaces (air–water interface, water-hydrophobic interfaces), are widespread in nature, from the atmosphere to the Earth's surface, and from inorganic sources to cells and living organisms.

Since the beginning of Quantum Chemistry, computational methods have strongly contributed to the understanding of molecular properties and chemical reactions, as well as to elucidate solvation phenomena. The theoretical study of solvent effects with quantum chemical methods started in the 1970s decade. One of the simplest models was probably the so-called *supermolecule approach*, introduced by Pullman and Pullman [2]. In this model, the solvent molecules in the first solvation shell are explicitly considered in the quantum chemical calculation of the molecule of interest. Nevertheless, the major contribution to the theoretical study of solvated molecules came from the use of dielectric solvent models, first implemented in semiempirical quantum theory by Rivail and Rinaldi [3, 4]. The dielectric solvent models, also known as self-consistent reaction field models or polarizable continuum models, were strongly ameliorated and completed in the 1980s by several groups in Nancy [5–7], Pisa [8–10] and Minnesota [11, 12], to cite the most well-known examples; several reviews on these and on other related models are available [13–15] and for a comparison between them, see [16]. The first ab initio calculations of transition structures in solution [17, 18] opened the way to computational studies of reaction mechanisms in the liquid state, with an increasing number of applications reported each year.

However, dielectric solvation models do suffer from a too simple description of the solute-solvent interactions, and also from a static vision of the solvated system, which is supposed to be at thermodynamic equilibrium. A step forward was done after the publication in 1990 of a paper by Field et al. [19] on the QM/MM model, which was an extension of the seminal work reported by Warshel and Levitt in 1976 [20] (the acronym QM/MM stands for Quantum Mechanics/Molecular Mechanics). The QM/MM model was initially developed to carry out Molecular Dynamics (MD) simulations of large molecules such as proteins using a semi-empirical description of the active site. But extension to *first-principles* studies of ions and molecules in solution [21–23] was reported soon after Karplus' work and algorithms to simulate chemical reaction trajectories and analyze non-equilibrium effects in solution based on a *rare event* sampling approach were developed [24, 25].

Only until very recently, has the QM/MM MD approach been applied to analyze solvation effects on chemical properties of molecules adsorbed at liquid surfaces and interfaces [26–31]. The main aim of this chapter is to summarize the current state of knowledge in the field. The rest of the chapter is organized as follows. We briefly present the QM/MM MD methodology, summarizing the most important equations of this approach and pointing out some ongoing developments. The Results and Discussion section is divided in two subsections. In the first one, we review our previous studies on interfaces, outline the main conclusions that were reached and emphasize their potential impact. We then report new results for methane derivatives (MeX with X = Cl, CN, OH) interacting with the water surface. These molecules are simple volatile organic compounds of atmospheric relevance and they have been chosen here specifically to illustrate the role of interface orientational effects.

11.2 The QM/MM Model for Liquid Interfaces

The QM/MM method has been described in detail in many places [32–36]. In addition, this book contains several chapters with specific applications of the method (see in particular the chapter by B. Mennucci). Hence, we will just summarize here the main lines of our model for interface simulations.

In QM/MM models of solutions, solute and solvent molecules are described explicitly. However, while the solute and (if required) a few solvent molecules are described using Quantum Mechanics techniques, their environment is described classically using a force-field from Molecular Mechanics. Since (by construction) there are no covalent bonds between the QM and MM subsystems, the total Hamiltonian is simply written:

$$H = H_{QM} + H_{MM} + H_{QM/MM} \tag{11.1}$$

Here, H_{QM}, H_{MM} and $H_{QM/MM}$ represent, respectively, the Hamiltonians for the QM and MM subsystems and for the interaction between them, which contains an electrostatic and a non-electrostatic part:

$$H_{QM/MM} = H_{QM/MM}^{elec} + H_{QM/MM}^{non-elec} \tag{11.2}$$

The latter is usually described through a Lennard-Jones potential including the atomic interactions between the QM nucleus n and the MM center s:

$$H_{QM/MM}^{non-elec} = \sum_{n}\sum_{s} 4\varepsilon_{ns}\left[\left(\frac{\sigma_{ns}}{R_{ns}}\right)^{12} - \left(\frac{\sigma_{ns}}{R_{ns}}\right)^{6}\right] \tag{11.3}$$

The repulsion term (represented by the 12-term) may also be computed using pseudo-potentials that explicitly depend on the electron coordinates [37], which is useful to avoid an overestimation of the electronic polarization in some cases (in particular when very large basis sets are employed). The non-electrostatic QM/MM parameters are usually taken from the MM force-field, but if necessary they can be refined using mean-field procedures [38].

The QM/MM Hamiltonian can be used to carry out Molecular Dynamics simulations of a complex system. In the case of liquid interfaces, the simulation box contains the solute and solvent molecules and one must apply appropriate periodic boundary conditions. Typically, for air–water interface simulations, we use a cubic box with periodic boundary conditions in the X and Y directions, whereas for liquid/liquid interfaces, we use a rectangle cuboid interface with periodic boundary conditions in the three directions. An example of simulation box for a liquid-liquid interface is illustrated in Fig. 11.1. The solute's wave function is computed "on the fly" at each time step of the simulation using the terms in the whole Hamiltonian that explicitly depend on the solute's electronic coordinates (the Born-Oppenheimer approximation is assumed in this model). To accelerate the convergence of the wavefunction calculation, the initial guess in the SCF iterative procedure is taken from the previous step in the simulation, or better, using an extrapolated density matrix from the last three or four steps [39]. The forces acting on QM nuclei and on MM centers are evaluated analytically, and the classical equations of motion are solved to obtain a set of new atomic positions and velocities.

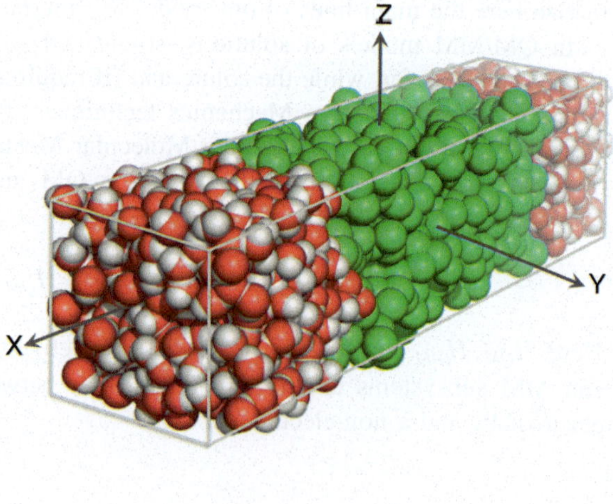

Fig. 11.1 Example of simulation box for a liquid–liquid interface. Periodic boundary conditions are applied along the three axes. Two liquid–liquid interfaces are obtained

11.2.1 Polarization at Liquid Interfaces

Some experiments have suggested that the polarization of a molecule at a liquid/liquid interface is close to the arithmetic average of the polarization in the two bulk phases [40, 41]. Other experiments, however, have shown that polarization at interfaces is highly dependent on molecular orientation [42] and theoretical calculations have recently demonstrated that solvation effects at the air–water interface on electronic properties can be larger than solvation effects in bulk water [43]. From these and other related studies, it has become clear that in order to achieve a realistic description of chemical compounds adsorbed at aqueous interfaces, it is crucial to deal with instantaneous polarization effects.

In the QM/MM approach, the polarization of the QM subsystem is introduced in a natural way through the modification of its Hamiltonian to account for electrostatic embedding. In contrast, explicit polarization of the MM subsystem is barely considered and most QM/MM simulations of condensed phases employ effective potentials optimized for bulk solvation properties. These MM force-fields are not necessarily well adapted for liquid interfaces and the use of polarizable MM force-fields will need to be envisaged in the future.

When the MM force-field includes polarization terms, the equations given above has to be slightly modified. Detailed expressions for the case of induced dipole moments were reported before [36]. Basically, the QM/MM term must include an additional contribution:

$$H_{QM/MM} = H^{elec}_{QM/MM} + H^{pol}_{QM/MM} + H^{non-elec}_{QM/MM} \tag{11.4}$$

defined as:

$$H^{pol}_{QM/MM} = -\sum_{s} \overrightarrow{\mu}^{ind}_{s} \cdot \overrightarrow{\mathrm{E}}_{QM}(\overrightarrow{R}_{s}) \tag{11.5}$$

where $\overrightarrow{\mu}^{ind}_{s}$ represents the induce dipole moment in MM center s (see below) and $\overrightarrow{\mathrm{E}}_{QM}(\overrightarrow{R}_{s})$ the electric field created by the QM system on center s. The MM Hamiltonian must also include a polarization term, which must account for the energy needed to create the induced dipoles, usually derived within the context of linear response theory. It is possible to show that in that case, the sum of the QM/MM and MM polarization terms lead to:

$$H^{pol}_{QM/MM} + H^{pol}_{MM} = -\frac{1}{2}\sum_{s} \overrightarrow{\mu}^{ind}_{s} \cdot \left(\overrightarrow{\mathrm{E}}_{MM}(\overrightarrow{R}_{s}) + \overrightarrow{\mathrm{E}}_{QM}(\overrightarrow{R}_{s})\right) \tag{11.6}$$

where $\overrightarrow{\mathrm{E}}_{MM}(\overrightarrow{R}_{s})$ is the electric field on center s due to the other MM centers (charges, permanent dipole, induced dipoles). The induced dipole moments are obtained from the polarizabilities and electric field contributions as:

$$\overrightarrow{\mu}_s^{ind} = \alpha_s \left(\overrightarrow{E}_{MM}(\overrightarrow{R}_s) + \overrightarrow{E}_{QM}(\overrightarrow{R}_s) - \sum_{t \neq s} \mathbf{T}_{st} \cdot \overrightarrow{\mu}_t^{ind} \right) \tag{11.7}$$

where the tensor $\mathbf{T}_{st}$ is defined by

$$\mathbf{T}_{st} = \frac{1}{R_{st}^3} \left(1 - \frac{3\overrightarrow{R}_{st}\overrightarrow{R}_{st}}{R_{st}^2} \right) \tag{11.8}$$

with:

$$\overrightarrow{R}_{st} = \overrightarrow{R}_s - \overrightarrow{R}_t \tag{11.9}$$

11.2.2 Free Energy Calculations

The calculation of free energy is necessary to describe the kinetics of chemical reactions or the energetics of interface adsorption and accommodation processes, for instance. In our approach, the free energy profiles are usually computed using the umbrella sampling technique [44]. The use of the metadynamics technique [45] can be envisaged too and the implementation in our code is underway. This type of calculations is very time consuming because many trajectories have to be calculated along the reaction coordinate by varying the energy minimum position of the bias potential by small steps. As a consequence, a rigorous QM/MM calculation of the free energy profile can hardly be computed at very high QM levels. We have proposed an approximated *dual-level* method [46] in which high-level QM/MM results can be estimated by adding a perturbation correction to the potential of mean force obtained at a lower QM/MM level (or even at MM level). Let us assume that the required free energy profile has been obtained using a simple method A supposed to provide a reasonably good approximation for the system sampling. The significance of the term "reasonably good approximation" will become clearer below. To estimate the free energy correction in changing the Hamiltonian from level A to the higher-level B, we use free energy perturbation theory according to which:

$$W^B(\xi) - W^A(\xi) = -k_B T \ln \left\langle e^{-\frac{U^B(\xi) - U^A(\xi)}{k_B T}} \right\rangle \tag{11.10}$$

Here k_B is the Boltzmann constant, T is the temperature and $U^{A,B}$ represents the potential energy created by A or B. The ensemble average is calculated over a set of system configurations for a particular value of the reaction coordinate ξ. This is the major advantage of the method because one does not need to carry out high-level calculations all along the reaction coordinate but only on some selected points.

The potential energy differences in the exponential may be too large because they correspond to different methods and it is possible to compute them to some appropriate reference (for instance for infinite separated systems). In addition, the previous equation can be reformulated to make more apparent the physical contributions to the free energy. If we define the average quantity:

$$\left\langle U^{BA}(\xi)\right\rangle = \left\langle U^{B}(\xi) - U^{A}(\xi)\right\rangle \tag{11.11}$$

we can write:

$$W^{B}(\xi) - W^{A}(\xi) = \left\langle U^{BA}(\xi)\right\rangle - k_B T\ \ln\left\langle e^{-\frac{\delta\Delta U^{BA}(\xi)}{k_B T}}\right\rangle \tag{11.12}$$

where the first term represents the correction to the potential energy, and the second one is connected with entropic and thermal corrections since it accounts for the fluctuations of the potential energy difference with respect to its average value:

$$\delta\Delta U^{BA}(\xi) = U^{B}(\xi) - U^{A}(\xi) - \left\langle U^{BA}(\xi)\right\rangle \tag{11.13}$$

11.2.3 Multi-scale and Adaptative QM/MM Schemes

The QM/MM method, and the polarizable continuum method as well, are usually considered as prototypical examples of the so-called multi-scale approaches. They combine two different description levels for the chemical system; in both cases, a quantum part interacts with a classical part. Indeed, the QM/MM method can easily be extended to multi-scale schemes that include more than two description levels. Examples of three level schemes are the QM/MM/Continuum [47] and QM/QM'/MM approaches [48, 49]. In the later case, the system is divided into two QM parts, which may be described with the same or different methods, and a classical MM part. Dielectric continuum models for liquid interfaces are already available [43, 50, 51] and a QM/MM/Continuum partition could be imagined in this case too, for instance to describe a solute-solvent cluster interacting with a polarizable dielectric medium. Here, however, we will focus on the QM/QM'/MM partition. There is not a general scheme for this kind of approach and different algorithms can be employed to describe the interaction between subsystems. The main issue is the calculation of the interaction between two quantum subsystems that are described at QM (possibly different) theoretical levels.

Let us consider two QM subsystems A and B (molecules or ensemble of molecules) in a model environment described by MM centers, as represented in Fig. 11.2. The QM/MM interactions will not be considered explicitly in this case since they parallel the equations described above. At sufficiently long distances, roughly when the electronic clouds of A and B do not significantly overlap, it is

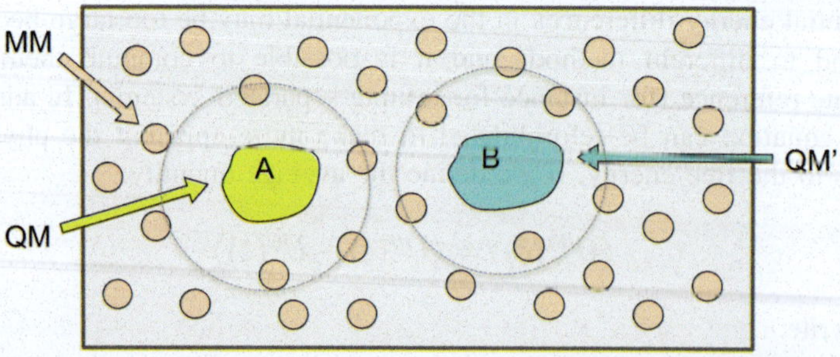

Fig. 11.2 Scheme for a molecular system described with a QM/QM'/MM partition

possible to describe the A–B interaction using perturbation theory. As in the QM/MM method, in the QM/QM' scheme we separate the A–B intermolecular interactions into electrostatic and non-electrostatic terms, even though in this case, both A and B are described quantum mechanically:

$$E_{QM/QM'}^{AB} = E_{QM/QM'}^{AB,\,elec} + E_{QM/QM'}^{AB,\,non-elec} \tag{11.14}$$

The non-electrostatic term is approximated by a Lennard-Jones potential. The electrostatic interaction is calculated using the wavefunctions of the molecules, ψ_A and ψ_B, through perturbation theory. Since the systems are (individually) described quantum mechanically, they are polarizable and their wavefunctions must be obtained iteratively. With this aim, the calculation of ψ_A is carried out using a modified Hamiltonian $H_{QM}^A + V^B$ that includes the electrostatic potential created by B, and the result is used to compute ψ_B in the same way. The process is continued until convergence of the total energy (initially, gas phase wavefunctions can be used). Then, the electrostatic-polarization energy of the A–B system can be written:

$$E_{QM/QM}^{AB,\,elec} = \sum_{a \in A, b \in B} \frac{Z_a Z_b}{R_{ab}} + \sum_{a \in A} \int \frac{Z_a \rho_B}{R_{ai}} + \sum_{b \in B} \int \frac{Z_b \rho_A}{R_{bj}} + \int \frac{\rho_A \rho_B}{r_{ij}} + E_{QM}^{A,pol} + E_{QM}^{B,pol} \tag{11.15}$$

where the last terms account for the polarization energy of A and B; they are computed as the difference of the expectation values of the molecular Hamiltonians for A and B using the polarized and unpolarized wavefunctions. In many cases, the electrostatic potentials $V^{A(B)}$ can be approximated using a classical charge distribution, for instance using a multipolar expansion or even net atomic charges $q_{a(b)}$ that fit the molecular electrostatic potential. It may be noted than, in principle, the total interaction energy in this scheme could be obtained using a density functional together with the total density $\rho_A + \rho_B$. In addition, frozen densities for one or for the two QM systems could optionally be chosen, and the approach then resembles that developed by Wesolowsky et al. [52].

The above equations are valid for large intermolecular distances. At short distance, especially when the molecules strongly interact or react to form other chemical species, a single QM calculation for the whole A–B system becomes necessary. The transition from a QM/QM' partition to a single QM one can be carried out through the use of the so-called adaptative algorithms, which have been developed by several groups, basically within the QM/MM framework. An excellent review is presented in another chapter of this book and we will not describe these techniques here (see chapters by Jiang et al. and by Pezeshki and Lin). Let us mention, nevertheless, that such adaptative approaches can be customized to deal with QM/QM'/MM partitions in order to account for diffusion of the QM and QM' parts. The fundamental idea is to define an intermediate or buffer region where the energy of the whole system changes smoothly from the long-range to the short-range expressions. There are however a number of technical problems that need to be fixed before the adaptative methodologies can be used in general applications and further developments are still required in this domain.

11.3 Results and Discussion

In this section, we first present a summary of previous applications of our QM/MM approach to several problems that include molecules and radicals of atmospheric interest interacting with the air–water interface, and molecules of biological relevance at water-hydrophobic interfaces. Then, we will present some new results for a series of volatile organic compounds (MeCl, MeCN and MeOH) at the air–water interface; for these systems, in particular, we will discuss the orientational dynamics and the effect of the water medium on their electronic properties.

11.3.1 Summary of Previous Studies

Most of our previous studies have been devoted to reactive oxygen species (ROS) at the air–water interface because such species are ubiquitous and play a crucial role in atmospheric chemistry, in environmental processes, water treatment technologies and biochemical reactions. The complex chemistries associated to these species, and their interconnection across different reaction media, have recently been reviewed [53]. The stability of ozone, molecular oxygen, hydrogen peroxide, hydroxyl and hydroperoxyl radicals, and other related compounds at the air–water interface had been established through classical molecular dynamics simulations [54–56]. Those studies, in particular, suggested that many of the compounds could accumulate at the surface of cloud water droplets, influencing in this way the overall chemistry of the troposphere. Recently, combined QM/MM MD simulations have confirmed the marked affinity of ROS species such as HO_2 [27] and ozone [30] for the air–water interface.

Many volatile organic compounds (VOCs) display also aqueous interface affinity. For instance, using the *dual-level* QM/MM method to estimate the free energy profile for the uptake of formaldehyde by a water droplet (high//low QM levels were B3LYP/6-611+G*//HF/STO-3G), we predicted a free energy minimum at the interface of about −1.5 kcal/mol with respect to the free energy in bulk water, corresponding to a significant interface accumulation. Since both ROS and VOC species are stabilized at the air–water interface, they can react with each other and the effective reaction rates are expected to be significantly higher than in gas phase.

QM/MM MD simulations have also allowed us to acquire critical knowledge on how interface solvation modifies the molecular properties and the reactivity of the adsorbed solutes. The most striking finding is probably the fact that, in contrast to common assumptions, the molecular properties at the interface are not necessarily intermediate between the properties in gas phase and bulk water. For instance, the HOMO and LUMO frontier orbital energies of several compounds have been shown [27] to undergo larger modifications at the interface compared to bulk, which suggests that their reactivity could also be more affected by the interface than by the bulk. These results can be rationalized in terms of the polarization field using simple dielectric models [43]. On one hand, the interface polarization leads to the appearance of a background electrostatic potential. On the other hand, it induces an inhomogeneous electric field, whose magnitude is generally smaller than the corresponding field in bulk water but which can present larger gradients. In addition, the interface asymmetry favors specific orientations of the solute and restrains its rotational dynamics.

The simplest view of a polarized interface is illustrated in Fig. 11.3. A proton donor such as HO_2 is oriented in such a way that the proton points towards the water surface. The surface is then polarized and creates a negative electrostatic potential above it, destabilizing the solute molecular orbitals [27]. The opposite situation applies for proton acceptors such as H_2CO [28] or other aldehydes [31]. Obviously, this simple scheme becomes more complicated in the general case. In particular, the energy shift of the orbitals is not constant because the electrostatic potential is accompanied by an electric field, an electric field gradient, etc. Nevertheless, this feature can be at the origin of significant modifications of the

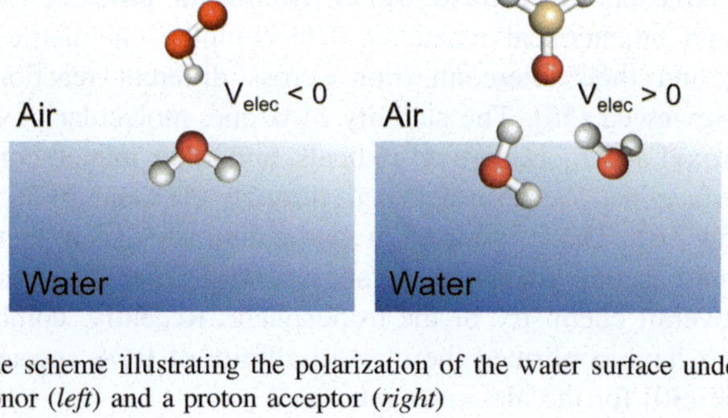

Fig. 11.3 Simple scheme illustrating the polarization of the water surface under the interaction with a proton donor (*left*) and a proton acceptor (*right*)

intrinsic chemical reactivity at the interface, which may enhance or inhibit a process to a larger extent than solvation effects in bulk water [28].

Free energy calculations for the uptake of HO_2 and of its conjugated basis, the superoxide anion O_2^-, made possible an estimation of the pK_a and redox potentials at the air–water interface [27]. The QM/MM calculations for the interface pK_a of HO_2 yield $\sim 6.3 \pm 0.5$ (experimental value is 4.8 in bulk water), whereas estimation of the redox potential of the O_2/O_2^- couple at the interface yields −0.65 eV (experimental value is −0.33 eV in bulk water). Obviously, the precise definition of these quantities at the interface is not straightforward since it implicates a system characterized by large fluctuations and non-equilibrium phenomena (see below); indeed, some usual chemical concepts in bulk solution may need to be revisited when handling with liquid interfaces.

Photochemical processes can also be modified by solvation at the air–water interface. In a recent work [30], it has been shown that ozone-water interactions at the surface of cloud water droplets favors the light absorption and the photolysis of ozone in the UV and visible regions. This effect, combined with a more efficient reaction of the photolytic product, namely the oxygen atom, with the surrounding water molecules, leads to an increase of the OH radical formation rate by about three-four orders of magnitude compared to the same process in the gas phase. Since the OH radical is one of the most important atmospheric species (it is usually referred to as the detergent of the troposphere), this finding might have significant implications for our understanding of the overall oxidation capacity of the atmosphere. The interface promoted photolysis of benzaldehyde into the radicals C_6H_5 and HCO has also been described through QM/MM calculations [31]; in this case, the photolysis rate constant is predicted to increase by one order of magnitude with respect to gas phase, and due to the accumulation of benzaldehyde at the water surface, the photolysis rate should increase by as much as six orders of magnitude. Experimental work is now required to confirm these theoretical predictions and the impact they might have on the global tropospheric chemistry, which is difficult to quantify because many other factors need to be incorporated into the calculations (altitude effects, UV penetration in clouds, size and distribution of water droplets and aqueous aerosols, etc.).

The study of aqueous interfaces is also relevant for the understanding of many biological mechanisms. In particular, water-hydrophobic interfaces represent a suitable model to investigate how cell membranes work. Passive diffusion of amino acids has attracted attention in recent years to analyze the permeability properties of the lipid bilayer but the mechanism of transfer is still unclear. QM/MM MD simulations for the phase transfer of glycine and valine across a water/CCl_4 interface [26, 29] has led to some unexpected conclusions. It was found that the neutral and zwitterionic tautomers of glycine and valine are roughly isoergonic in the organic layer close to the interface. This finding suggests in turn that the water-to-CCl_4 uptake follows a two-step mechanism: adsorption−neutralization of the zwitterion close to the interface, with some water molecules surrounding the aminoacid in the organic layer (see Fig. 11.4), followed by the uptake of the nonionized tautomer into the organic phase.

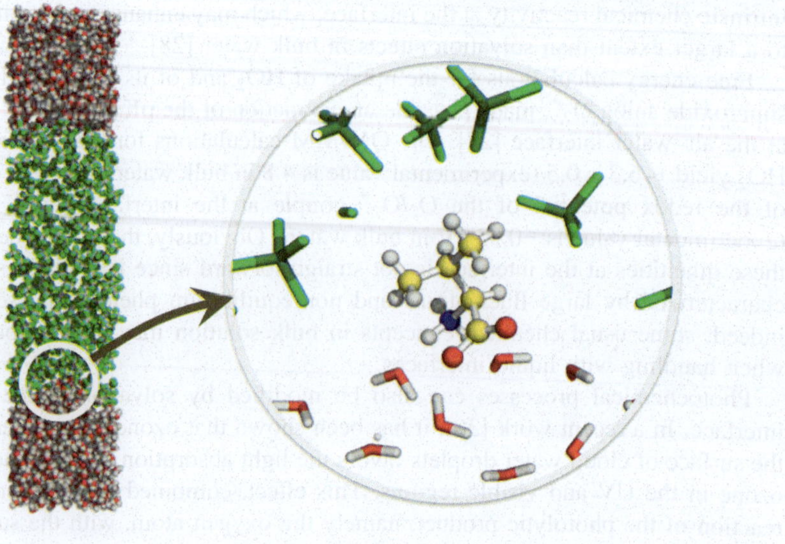

Fig. 11.4 Valine at the water-CCl_4 interface. The image shows the formation of a water default when zwitterionic valine enters the organic layer from the water one prior to neutralization

11.3.2 Study of Some Volatile Organic Compounds (Me-X, X = Cl, CN, OH) of Atmospheric Relevance at the Air–Water Interface

Methyl chloride (MeCl), acetonitrile (MeCN) and methanol (MeOH) are present in the troposphere and participate to oxidation chemistry. Methyl chloride is emitted by both biogenic (biomass burning, coastal salt marshes, tropical plants, etc.) and anthropogenic sources (industry, fuel combustion, etc.). It is quite abundant and represents a main source of organic chlorine in the atmosphere with a long tropospheric lifetime (>1 year), which renders possible its transport into the stratosphere and its participation (through photolysis) to ozone depletion processes. Acetonitrile is supposed to be mainly emitted from biomass burning (indeed, like hydrogen cyanide as well, it can be used as an indicator for biomass burning emissions) although a detailed knowledge of sources and sinks for this molecule is still missing. It is relatively stable and its lifetime in the atmosphere has been estimated between 0.45 and 1.5 years (therefore, as methyl chloride, acetonitrile can reach the stratosphere). Methanol is the second most abundant organic gas (after methane) in the atmosphere but its atmospheric budget is poorly understood. The main source seems to be plant growth, while other sources include plant decay, biomass burning, atmospheric oxidation of volatile organic compounds, and anthropogenic activities. Its lifetime is about 10 days (oxidation by OH being the major sink) and it represents a significant source of formaldehyde.

Despite a similar molecular structure, these three organic compounds present significant differences in terms of polarity and chemical reactivity and therefore the study of their interactions with the air–water interface, and the possible atmospheric implications, is interesting. Indeed, methyl chloride and methanol at the liquid water-vapor interface have been the subject of previous theoretical and experimental investigations [57–60], which focused on the preferred orientations and the thermodynamics of the adsorption process. In the present work, we have carried out QM/MM MD simulations for methyl chloride, acetonitrile, and methanol trying to get further insights into the solvation effects of the interface on the electronic properties of the systems, as well as on the orientational dynamics.

The main characteristics of the simulations are as follows. We assume the NVT ensemble (T = 298 K, Nosé–Hoover thermostat [61, 62]) using a box containing the solute and 499 water molecules (TIP3P model [63]). The box size is (in Å) 24.689 × 24.689 × 130. Periodic boundary conditions are used along the X and Y directions. The solutes are described at the B3LYP level [64] using the 6-311+G(d) basis set [65]. The time step is 0.25 fs. After equilibration, the simulations were carried out for 300 ps. The simulations were done using the Gaussian 09 program [66] for the QM calculations, Tinker 4.2 [67] for the MD simulations and the interface developed by us [68].

(a) Analysis of the solute dynamics

Figure 11.5 shows the position of the solute's center of mass with respect to the average interface along the 300 ps MD simulation for the three solutes. As shown, acetonitrile and methanol remain at the interface all along the 300 ps simulation (they enter from time to time into the aqueous layer but return to the interface after a few picosenconds). This is not the case for methyl chloride, which desorbs from the interface after about 220 ps. Classical MD simulations with a polarizable classical

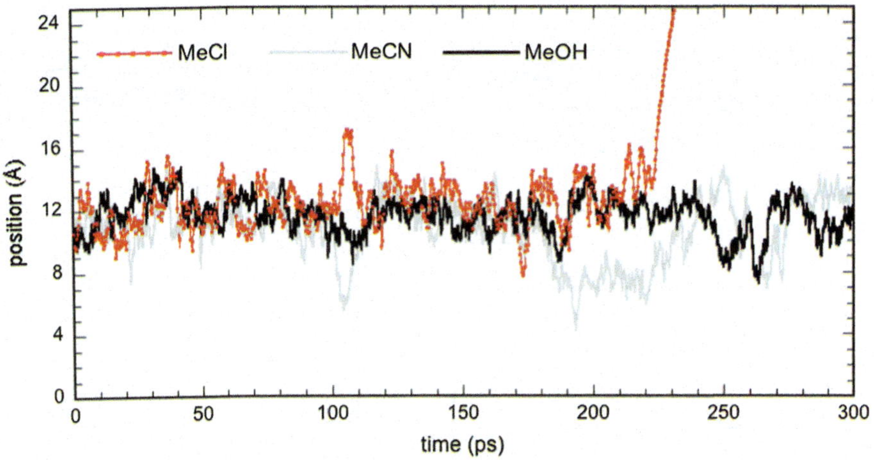

Fig. 11.5 Position of the center of mass of the organic molecules along the QM/MM simulations at the air–water interface. The average interface is at 12 Å

force-field [60] reported a residence lifetime for methyl chloride at the liquid water-vapor interface of 140 ps. Since our QM/MM simulations are computationally costly, it was not possible to obtain a statistically meaningful value for residence lifetimes, but the calculated trajectory for methyl chloride is consistent with the predictions of much longer classical simulations. The results clearly suggest that methyl chloride, in comparison with the other two molecules, interacts weakly with the interface, and this conclusion is further confirmed by the analysis of the radial distributions functions in Fig. 11.6. Both acetonitrile and methanol form stable hydrogen bonds with water molecules. The peaks in the radial distribution functions of methanol are quite intense and show that methanol behaves as both a proton acceptor and a proton donor though, as expected, the MeOH–O_{water} hydrogen-bond is the strongest one. The radial distribution functions for methyl chloride exhibit a quite small feature at about 2.4 Å for the MeCl–HOH interaction; methyl protons are able to form weak hydrogen-bonds with water molecules [58] but no significant features have been found for this interaction in our simulations at 298 K. Note that for the analysis of the radial distribution functions of MeCl, as well as for the analysis that will be carried out below, we have only considered the part of the trajectory in which the system is clearly at the interface.

As said above, an important feature of solvation at the water surface is the possibility to get some preferred orientations for the adsorbed molecules. To investigate this issue, the angular probability distributions have been calculated and they are plotted in Fig. 11.7. The distributions for methyl chloride and acetonitrile present small orientational preferences suggesting that both molecules experience high mobility. Orientations in which methyl chloride is parallel to the interface ($\theta \sim 90°$) and those with the C–Cl bond pointing towards the air layer ($0 < \theta < 90°$) seem to be slightly preferred, in agreement with previous studies (see below). In the

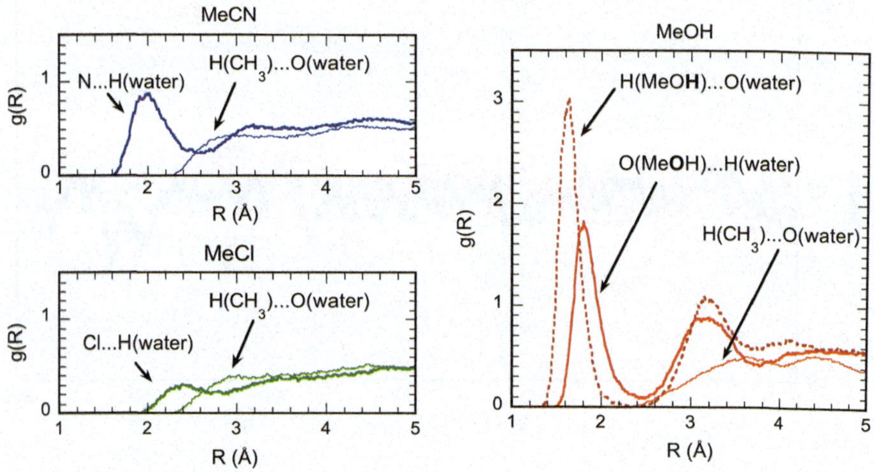

Fig. 11.6 Radial distribution functions for hydrogen-bonds with water molecules

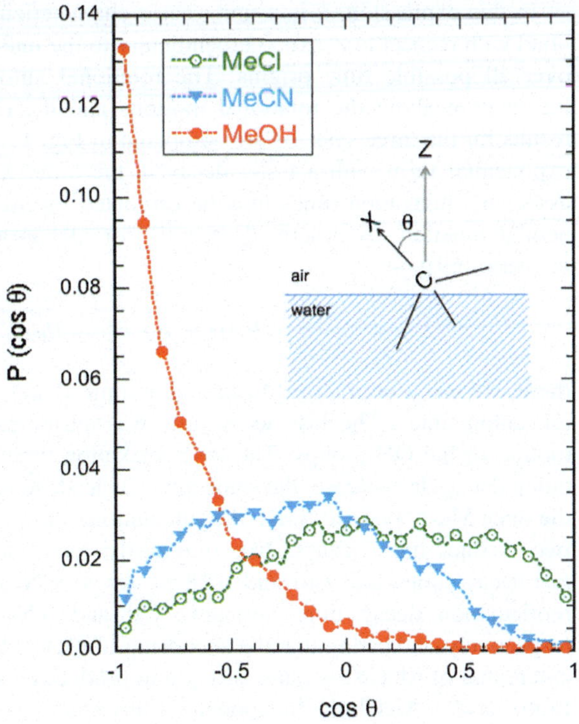

Fig. 11.7 Angular probability distributions for MeX (X = Cl, CN, OH) molecules at the air–water interface

case of acetonitrile, in contrast, the C–CN bond is rather pointing towards the water layer. The distribution for methanol is very different in regard to the two other molecules. The C–OH bond is clearly pointing towards the water layer, with a strong probability maximum at $\theta \sim 180°$.

These findings are consistent with vibrational sum frequency generation experiments [57], which were able to detect methanol at the water surface and suggested that it adsorbs in an ordered layer. In the same study, however, methyl chloride could not be detected at the surface. Our results also agree with a recent theoretical study on methyl chloride using a classical polarizable force-field, which showed that despite some preferential orientation with the methyl groups directed towards the water surface, the distribution is rather broad and relatively flat. Other simulations have led to comparable conclusions for MeCl orientation [58, 59].

To further characterize the rotational dynamics of the molecules, we have used a second-order rotational autocorrelation function, which is usually employed in pump-probe experiments and is defined by the second-order Legendre polynomial:

$$P_2(t) = \left\langle \frac{1}{2}(3\cos^2(\vec{u}_o \cdot \vec{u}_t) - 1 \right\rangle \quad (11.16)$$

In this expression, $\vec{u}$ is a unit vector characterizing the orientation of the C–X bond with respect to Z-axis (perpendicular to the interface) and the average is taken over all possible time origins. The rotational autocorrelation functions provide information about the rotational motions and the characteristic time scales. The results for the three systems are compared in Fig. 11.8. They exhibit a more or less exponential form with a faster decay in the order MeCl > MeCN > MeOH. The associated relaxation times may be estimated by fitting these curves by a biexponential function, as described recently for OH groups of water at the air/water interface [69]:

$$P_2(t) = Ae^{-t/\tau_1} + Be^{-t/\tau_2} \tag{11.17}$$

(here, we use a normalized form and assume B = 1 − A). The authors found two relaxation times. The fast decay time was attributed to the short-time librational motion of the OH groups. The long relaxation time was attributed to the reorientation decay. In our case, the same analysis leads to quite interesting differences for the three Me-X systems. First, the calculations show that MeCN and MeOH display two distinct time decays, like previously found for the water molecules. The fast-decay times are 0.40 and 0.28 ps for MeCN and MeOH, respectively. The reorientation decay times amount 3.70 and 6.50 ps for MeCN and MeOH, respectively, indicating that the anchoring of MeOH to the interface, and to a lesser extent that of MeCN, is quite strong and produces a significant orientational effect. In the case of MeCl, the biexponential fit leads to two small relaxation times (0.26 and 0.93 ps) and suggests a different behavior for this system compared to the other two. Indeed, the rotational autocorrelation function of MeCl is satisfactorily fitted by a single exponential decay, the corresponding decay time being 0.36 ps, which

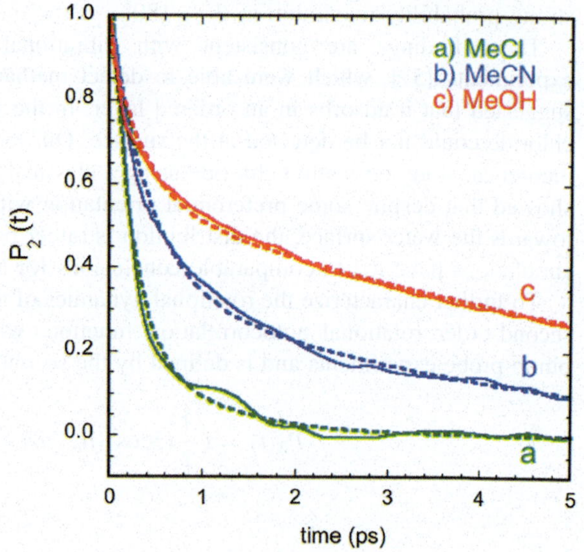

Fig. 11.8 Second-order rotational autocorrelation functions and biexponential fits

Table 11.1 Time-averages of the electronic properties of the studied volatile organic compounds at the air–water interface compared to gas phase values for the optimized structures. The reactivity indices μ, η and ω correspond, respectively to the chemical potential $((\varepsilon_{HOMO} + \varepsilon_{LUMO})/2)$, the hardness $(\varepsilon_{LUMO} - \varepsilon_{HOMO})$ and the electrophilicity (μ^2/η)

	Dipole (D)	HOMO (au)	LUMO (au)	μ (eV)	η (eV)	ω (eV)
MeCl gas	2.21	−0.3036	−0.0103	−4.27	7.98	2.28
interface	2.54	−0.3026	−0.0027	−4.15	8.16	2.11
MeCN gas	4.05	−0.3406	−0.0116	−4.79	8.95	2.56
interface	5.26	−0.3488	−0.0024	−4.78	9.42	2.42
MeOH gas	1.89	−0.2843	−0.0025	−3.90	7.67	1.99
interface	2.77	−0.2938	0.0149	−3.79	8.40	1.73

confirms that this molecule lacks long-time correlation and exhibits disordered adsorption at the air–water interface.

(a) Analysis of electronic properties

Some time-averaged electronic properties are summarized in Table 11.1 where they are compared to the gas phase values for the optimized molecular structure. Figure 11.9 displays the time fluctuations of the induced dipole moments for the three molecules along the simulations at the interface.

One striking result is that the calculated induced dipole moments do not show a trivial correlation with the gas phase dipole moments (as could be anticipated with a dielectric model for molecules that exhibit a similar structure). The induced dipole moment is quite large for acetonitrile (average 1.2 D), which also displays the largest dipole moment in gas phase. However, the induced dipole for methanol is

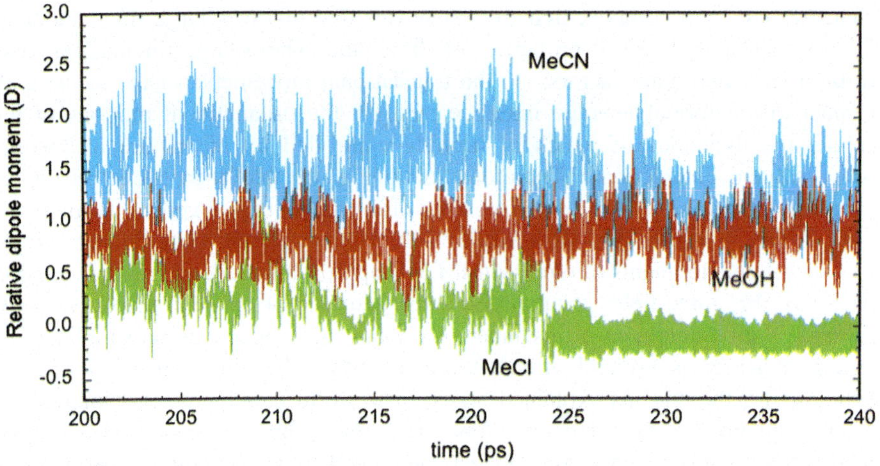

Fig. 11.9 Time fluctuations of the induced dipole moments obtained as the difference between the instantaneous dipole moment along the QM/MM MD simulation and the gas phase dipole moment calculated for the optimized structures

only slightly smaller (average 0.9 D) despite the fact that in gas phase, its dipole moment is less than half that of acetonitrile. Likewise, the induced dipole for methyl chloride (average 0.3 D) is much smaller than that for methanol, although the two molecules have a similar dipole moment in the gas phase; in fact, at the interface, their relative polarity is inversed.

The comparison of the dipole moments shows that the most polar species are not necessarily those undergoing the largest polarization effect of the aqueous surface. Similar remarks can be done if one looks at the reactivity indices. Note for instance that the largest solvation effect corresponds in general to methanol while the smallest one corresponds to methyl chloride, even if there are some exceptions (e.g. the lowest ω is found for acetonitrile). These trends, together with the structural properties discussed above, highlight the role of hydrogen-bonding with water molecules and the resulting orientational features characterizing the interaction with the aqueous surface. Such properties cannot be predicted through electrostatic considerations as derived from a simple dielectric model. Indeed, this observation emphasizes the necessity to deal with explicit solvent molecules, such as in the QM/MM model proposed here for the theoretical study of aqueous interfaces.

11.4 Conclusions and Future Directions

The study of chemical properties and chemical reactions of molecules adsorbed at aqueous interfaces is still in its infancy. The first results obtained with the QM/MM MD approach show that interface solvation phenomena cannot be understood (even qualitatively) from the usual concepts of bulk solutions. Water interfaces (and more generally liquid interfaces) have to be considered as a medium on their own, characterized by a complex dynamics behavior that involves large fluctuations of the physical and chemical properties of the solute. These large fluctuations arise from at least two main factors: (1) the translational movements of the solute with respect to the formal average interface, and (2) the orientational motions of the solute, the dynamics of which is strongly controlled by the nature of specific solute-solvent interactions. Therefore, obtaining reliable theoretical results of chemical properties at liquid interfaces is a difficult task that in general requires the use of explicit solvation models and the achievement of statistical simulations.

Current developments of our QM/MM MD approach include the implementation of polarizable force-fields for the MM subsystem and of efficient adaptative algorithms. Explicit treatment of instantaneous polarization effects in the simulation of liquid interfaces is probably much more important than in simulations of bulk solutions for the reasons explained above. It is worth mentioning that many classical simulations with polarizable force-fields have already been carried out; in many cases the results are close (at least qualitatively) to those with non-polarizable potentials, but in some instances their use has been shown to be compulsory [70].

The study of chemical reactions is another exciting field of application of the QM/MM MD approach. Many experiments have provided evidence that certain chemical reactions may be enhanced by solvation at aqueous interfaces (see for instance [71–86]) but these observations are still poorly understood. Many issues need to be carefully investigated in this regard. One intriguing question, in particular, concerns the role of dynamic effects on the apparent kinetic constants. Transition State Theory assumes chemical equilibrium between the reactants and the transition state, and formally this hypothesis is not fulfilled in aqueous media due to the slow component of water relaxation. At the water surface, in addition, the translation and orientation coordinates of the chemical system can intervene on the definition of the adiabatic reaction path and their coupling with the internal coordinates of the reactants deserves further attention.

The experimental techniques to study the adsorption thermodynamics and reactions at the water surface have been summarized in a recent review [87]. Recently, increasing interest has focused on vibrational sum-frequency generation and related heterodyne-detected vibrational or electronic sum frequency generation (SFG) methods [88–91]. These techniques are very promising and substantial developments have been done in recent years [92–98]. Application of the QM/MM MD methodology to study SFG phenomena in connection with experiments is very exciting and this represents certainly one the most biggest challenges for the foreseeable future.

References

1. Reichardt C (2003) Solvents and solvent effects in organic chemistry, 3rd edn. Wiley-VCH Verlag GmbH & Co. KGaA, Weinheim
2. Pullman A, Pullman B (1974) Q Rev Biophys 7:505–566
3. Rinaldi D, Rivail JL (1973) Theor Chim Acta 32:57
4. Rivail JL, Rinaldi D (1976) Chem Phys 18:233
5. Rinaldi D, Ruiz-López MF, Rivail JL (1983) J Chem Phys 78:834
6. Rivail JL, Terryn B, Rinaldi D, Ruiz-Lopez MF (1985) J Mol Struct THEOCHEM 120:387
7. Dillet V, Rinaldi D, Rivail JL (1994) J Phys Chem 98:5034–5039
8. Miertus S, Scrocco E, Tomasi J (1981) Chem Phys 55:117–129
9. Tomasi J, Persico M (1994) Chem Rev 94:2027–2094
10. Tomasi J, Mennucci B, Cancès E (1999) J Mol Struct THEOCHEM 464:211–226
11. Cramer CJ, Truhlar DG (1992) J Comput-Aided Mol Des 6:629–666
12. Chambers CC, Hawkins GD, Cramer CJ, Truhlar DG (1996) J Phys Chem 100:16385–16398
13. Cramer CJ, Truhlar DG (1999) Chem Rev 99:2161–2200
14. Tomasi J, Mennucci B, Cammi R (2005) Chem Rev 105:2999–3093
15. Ruiz-López MF (2008) In: Canuto S (ed) Solvation effects on molecules and biomolecules: Computational methods and applications, Springer, Berlin p 23–38
16. Curutchet C, Cramer CJ, Truhlar DG, Ruiz-López MF, Rinaldi D, Orozco M, Luque FJ (2003) J Comput Chem 24:284
17. Ruiz-López MF, Assfeld X, García JI, Mayoral JA, Salvatella L (1993) J Am Chem Soc 115:8780
18. Pappalardo RR, Sánchez-Marcos E, Ruiz-López MF, Rinaldi D, Rivail JL (1993) J Am Chem Soc 115:3722–3730
19. Field MJ, Bash PA, Karplus M (1990) J Comput Chem 11:700–733
20. Warshel A, Levitt M (1976) J Mol Biol 103:227–249

21. Tuñón I, Costa MTC, Millot C, Ruiz-López MF (1995) J Mol Mod 1:196–201
22. Tuñón I, Martins-Costa MTC, Millot C, Ruiz-Lopez MF (1995) Chem Phys Lett 241:450–456
23. Stanton RV, Hartsough DS, Merz KM (1993) J Phys Chem 97:11868–11870
24. Strnad M, Martins-Costa MTC, Millot C, Tuñón I, Ruiz-López MF, Rivail JL (1997) J Chem Phys 106:3643
25. Tuñón I, Martins-Costa MTC, Millot C, Ruiz-López MF (1997) J Chem Phys 106:3633
26. Martins-Costa MTC, Ruiz-Lopez MF (2011) Phys Chem Chem Phys 13:11579–11582
27. Martins-Costa MTC, Anglada JM, Francisco JS, Ruiz-Lopez M (2012) Angew Chem Int Ed 51:5413–5417
28. Martins-Costa MTC, Anglada JM, Francisco JS, Ruiz-Lopez MF (2012) J Am Chem Soc 134:11821–11827
29. Martins-Costa MTC, Ruiz-Lopez MF (2013) J Phys Chem B 117:12469–12474
30. Anglada JM, Martins-Costa M, Ruiz-Lopez MF, Francisco JS (2014) Proc Natl Acad Sci USA 111:11618–11623
31. Martins-Costa MTC, Garcia-Prieto FF, Ruiz-Lopez MF (2015) Org Biomol Chem 13:1673–1679
32. van der Kamp MW, Mulholland AJ (2013) Biochemistry 52:2708–2728
33. Hu H, Yang WT (2008) Annu Rev Phys Chem 59:573–601
34. Lin H, Truhlar DG (2007) Theoret Chem Acc 117:185–199
35. Senn HM, Thiel W (2007) In: Reiher M (ed) Atomistic approaches in modern biology: from quantum chemistry to molecular simulations. Springer, Berlin pp 173–290
36. Ruiz-López MF, Rivail JL (1998) In: Schleyer PvR (ed) Encyclopedia of computational chemistry. Wiley, New York, p 437
37. Chalmet S, Ruiz-Lopez MF (2000) Chem Phys Lett 329:154
38. Martin ME, Aguilar MA, Chalmet S, Ruiz-Lopez MF (2002) Chem Phys 284:607
39. Chalmet S, Ruiz-López MF (1999) J Chem Phys 111:1117–1125
40. Wang HF, Borguet E, Eisenthal KB (1998) J Phys Chem B 102:4927–4932
41. Wang HF, Borguet E, Eisenthal KB (1997) J Phys Chem A 101:713–718
42. Sen S, Yamaguchi S, Tahara T (2009) Angew Chem Int Ed 48:6439–6442
43. Martins-Costa MC, Ruiz-Lopez M (2015) Theoret Chem Acc 134:1–7
44. Torrie GM, Valleau JP (1977) J Comput Phys 23:187
45. Laio A, Parrinello M (2002) Proc Natl Acad Sci USA 99:12562–12566
46. Retegan M, Martins-Costa M, Ruiz-López MF (2010) J Chem Phys 133:064103
47. Chalmet S, Rinaldi D, Ruiz-López MF (2001) Int J Quantum Chem 84:559
48. Gogonea V, Westerhoff LM, Merz KM Jr (2000) J Chem Phys 113:5604–5613
49. Cui Q, Guo H, Karplus M (2002) J Chem Phys 117:5617–5631
50. Mozgawa K, Mennucci B, Frediani L (2014) J Phys Chem C 118:4715–4725
51. Kelly CP, Cramer CJ, Truhlar DG (2004) J Phys Chem B 108:12882–12897
52. Wesolowski T, Muller RP, Warshel A (1996) J Phys Chem 100:15444–15449
53. Anglada JM, Martins-Costa M, Francisco JS, Ruiz-Lopez MF (2015) Acc Chem Res 48:575–583
54. Roeselova M, Jungwirth P, Tobias DJ, Gerber RB (2003) J Phys Chem B 107:12690–12699
55. Mundy CJ, Kuo IFW, Tuckerman ME, Lee HS, Tobias DJ (2009) Chem Phys Lett 481:2–8
56. Vácha R, Slavíček P, Mucha M, Finlayson-Pitts BJ, Jungwirth P (2003) J Phys Chem 108:11573
57. Harper K, Minofar B, Sierra-Hernandez MR, Casillas-Ituarte NN, Roeselova M, Allen HC (2009) J Phys Chem A 113:2015–2024
58. Pasalic H, Roeselova M, Lischka H (2011) J Phys Chem B 115:1807–1816
59. Habartova A, Valsaraj KT, Roeselova M (2013) J Phys Chem A 117:9205–9215
60. Habartova A, Obisesan A, Minofar B, Roeselova M (2014) Theoret Chem Acc 133
61. Nosé S (1984) J Chem Phys 81:511
62. Hoover WG (1985) Phys Rev A 31:1695

63. Jorgensen WL, Chandrashekar J, Madura JD, Impey WR, Klein ML (1983) J Chem Phys 79:926–935
64. Becke AD (1993) J Chem Phys 98:5648–5652
65. Krishnan R, Binkley JS, Seeger R, Pople JA (1980) J Chem Phys 72:650–654
66. Frisch MJ, Trucks GW, Schlegel HB, Scuseria GE, Robb MA, Cheeseman JR, Scalmani G, Barone V, Mennucci B, Petersson GA, Nakatsuji H, Caricato M, Li X, Hratchian HP, Izmaylov AF, Bloino J, Zheng G, Sonnenberg JL, Hada M, Ehara M, Toyota K, Fukuda R, Hasegawa J, Ishida M, Nakajima T, Honda Y, Kitao O, Nakai H, Vreven T, Montgomery JA Jr, Peralta JE, Ogliaro F, Bearpark MJ, Heyd J, Brothers EN, Kudin KN, Staroverov VN, Kobayashi R, Normand J, Raghavachari K, Rendell AP, Burant JC, Iyengar SS, Tomasi J, Cossi M, Rega N, Millam NJ, Klene M, Knox JE, Cross JB, Bakken V, Adamo C, Jaramillo J, Gomperts R, Stratmann RE, Yazyev O, Austin AJ, Cammi R, Pomelli C, Ochterski JW, Martin RL, Morokuma K, Zakrzewski VG, Voth GA, Salvador P, Dannenberg JJ, Dapprich S, Daniels AD, Farkas Ö, Foresman JB, Ortiz JV, Cioslowski J, Fox DJ (2009) Gaussian 09. Gaussian Inc, Wallingford
67. Ponder JW (2004) TINKER: software tools for molecular design. Washington University School of Medicine, Saint Louis
68. Martins-Costa MTC (2014) A Gaussian 09/Tinker 4.2 interface for hybrid QM/MM applications. University of Lorraine—CNRS
69. Vila Verde A, Bolhuis PG, Campen RK (2012) J Phys Chem B 116:9467–9481
70. Jungwirth P, Tobias DJ (2006) Chem Rev 106:1259–1281
71. Enami S, Sakamoto Y, Colussi AJ (2014) Proc Natl Acad Sci USA 111:623–628
72. Enami S, Mishra H, Hoffmann MR, Colussi AJ (2012) J Phys Chem A 116:6027–6032
73. Enami S, Stewart LA, Hoffmann MR, Colussi AJ (2010) J Phys Chem Lett 1:3488–3493
74. Mishra H, Enami S, Nielsen RJ, Stewart LA, Hoffmann MR, Goddard WA III, Colussi AJ (2012) Proc Natl Acad Sci USA 109:18679–18683
75. Enami S, Hoffmann MR, Colussi AJ (2012) J Phys Chem Lett 3:3102–3108
76. Griffith EC, Vaida V (2013) J Am Chem Soc 135:710–716
77. Henderson EA, Donaldson DJ (2012) J Phys Chem A 116:423–429
78. Jubb AM, Hua W, Allen HC (2012) Annu Rev Phys Chem 63:107–130
79. Nissenson P, Knox CJH, Finlayson-Pitts BJ, Phillips LF, Dabdub D (2006) Phys Chem Chem Phys 8:4700–4710
80. Donaldson DJ, Valsaraj KT (2010) Environ Sci Technol 44:865–873
81. Donaldson DJ, Vaida V (2006) Chem Rev 106:1445–1461
82. Strekowski RS, Remorov R, George C (2003) J Phys Chem A 107:2497–2504
83. Mmereki BT, Donaldson DJ (2003) J Phys Chem A 107:11038–11042
84. Mmereki BT, Donaldson DJ, Gilman JB, Eliason TL, Vaida V (2004) Atmos Environ 38:6091–6103
85. Kolb CE, Cox RA, Abbatt JPD, Ammann M, Davis EJ, Donaldson DJ, Garrett BC, George C, Griffiths PT, Hanson DR, Kulmala M, McFiggans G, Poschl U, Riipinen I, Rossi MJ, Rudich Y, Wagner PE, Winkler PM, Worsnop DR (2010) O' Dowd CD. Atmos Chem Phys 10:10561–10605
86. Raja S, Valsaraj KT (2005) J Air Waste Manage Assoc 55:1345–1355
87. Valsaraj KT (2012)
88. Superfine R, Huang JY, Shen YR (1990) Opt Lett 15:1276–1278
89. Yamaguchi S, Tahara T (2008) J Chem Phys 129:101102
90. Nihonyanagi S, Ishiyama T, Lee T-k, Yamaguchi S, Bonn M, Morita A, Tahara T (2011) J Am Chem Soc 133:16875–16880
91. Stiopkin IV, Jayathilake HD, Bordenyuk AN, Benderskii AV (2008) J Am Chem Soc 130:2271–2275
92. Backus EH, Eichler A, Kleyn AW, Bonn M (2005) Science 310:1790–1793
93. Fayer MD, Levinger NE (2010) Annu Rev Anal Chem 3:89–107

94. Bakker H, Skinner J (2009) Chem Rev 110:1498–1517
95. Bredenbeck J, Ghosh A, Nienhuys H-K, Bonn M (2009) Acc Chem Res 42:1332–1342
96. Singh PC, Nihonyanagi S, Yamaguchi S, Tahara T (2012) J Chem Phys 137:094706
97. Xiong W, Laaser JE, Mehlenbacher RD, Zanni MT (2011) Proc Natl Acad Sci USA 108:20902–20907
98. Zhang Z, Piatkowski L, Bakker HJ, Bonn M (2011) Nature Chem 3:888–893

Chapter 12
QM/MM Approaches for the Modeling of Photoinduced Processes in Biological Systems

Benedetta Mennucci

Abstract Multiscale QM/MM approaches are nowadays a well-established computational tool to study properties and processes of supramolecular systems. In this chapter, an overview of the extension of these methods to photoinduced processes in biological systems will be presented and discussed. The attention will be focused on the strategies which can be used to properly describe the static and dynamic effects that the environment exerts on the electronic states involved in the processes. Specific problems related to the modeling of stationary properties and correlations, as well as reactive events will be analyzed and the computational tools developed so far within the QM/MM framework to solve them will be described.

12.1 Introduction

In the last decades, the investigation of photoinduced phenomena in (bio)systems has enormously benefitted from the development of new spectroscopic techniques: the real time scale of many processes has been discovered and it has been possible to clarify their mechanism and their relationship with the molecular structure and the influence of the chemical environment. Laser pulses are used not only to activate the photochemical reactivity but also to analyze the excited molecular states and their dynamics (pump-probe experiments) and to achieve control on the reactive events. Such a significant progress has however also led to new questions arising from the shortening of the investigated time and length scales towards the quantum realm. In biological systems, in particular, these progresses have led to a completely new way of looking at biological processes in which quantum mechanics plays the major role to gain a biological advantage [1]. In this context, it is clear that only theoretical methods and computational tools that can accurately

B. Mennucci (✉)
Department of Chemistry, University of Pisa, Via G. Moruzzi 13, 56124 Pisa, Italy
e-mail: benedetta.mennucci@unipi.it
URL: http://www.dcci.unipi.it/molecolab/

J.-L. Rivail et al. (eds.), *Quantum Modeling of Complex Molecular Systems*, Challenges and Advances in Computational Chemistry and Physics 21,
DOI 10.1007/978-3-319-21626-3_12

describe the quantum nature of the phenomena but still including a large part of the supramolecular system can try to answer these questions.

The hybrid methods which combine quantum-mechanical (QM) and classical descriptions are surely one of the mostly well-suited strategies in this context. Two main families of hybrid methods can be defined according to the model used to describe the classical part of the system. Either continuum or atomistic formulations can be introduced where, in the first case, the classical subsystem is described as a dielectric medium while, in the second case, a Molecular Mechanics (MM) formulation is generally adopted. While QM/continuum methods have been largely and successfully applied to molecular solutes in liquid solutions [2–5], QM/MM formulations have been more often used in the field of structured (biological) environments [6–10] even if the study of chemical reaction dynamics in solution represents another important field of applications of the method [11, 12].

QM/MM methods have been strongly correlated to biosystems and biological processes since their birth: the 1976 seminal work by Warshel and Levitt was a study of an enzymatic reaction [13]. The reasons for this connection are clear: on the one hand biological systems are too large to be treated with only quantum-mechanical approaches (even if of semiempirical nature) and on the other hand their processes, even when they can be localized in a small subsystem, cannot be accurately described if we do not include the effects of, at least, the local environment.

In the last decades, QM/MM methods have enormously extended their applicability and improved their accuracy: nowadays they represent a well established computational strategy not only in biology but also in many fields of chemistry, physics, and materials science, just to quote the main ones. The primary role they have achieved has been unequivocally shown by the 2013 Nobel Prize in Chemistry assigned to Karplus, Levitt and Warshel "for the development of multiscale models for complex chemical systems". The official statement used by the Nobel Committee contains the two words which best describe the QM/MM philosophy and its main realm of application, namely "multiscale" and "complex". The multiscale nature in fact is an intrinsic characteristic of all approaches, which combining two (or more) different levels of descriptions, automatically assume that the system can be partitioned in shells characterized by different spatial scales. The "complexity" is instead the characteristic of all systems, which combine a large number of components with a rich network of interactions among them. For the characteristics that will be detailed in the following sections, the QM/MM approach is "ideal" to face the complexity by cleverly defining artificial boundaries which allow the application of a multiscale description where the "small" scale involves an accurate QM description and the "large" scale acts as a perturbation which can be treated with a simplified, e.g. classical, approach.

The photoinduced processes we focus on in this contribution are not only a clear example of this combination of complexity and multiscale character but they also introduce a further level of analysis through the additional external perturbation (the electromagnetic radiation) which induces a dynamic process characterized by a specific time. This further "temporal" scale which sums up to the original "spatial" scale may introduce a separation in the responses of the different components of the supramolecular system which the multiscale model has to properly take into account. A simple example is the photoinduced electronic excitation of a solute in a polar solvent. The differences in the characteristic response times of the various degrees of freedom of the solvent may lead to a solvation regime in which the slow components (i.e., those arising from translations and rotations of the solvent molecules) are no longer equilibrated with the solute upon excitation. The resulting "nonequilibrium" regime will possibly evolve into a new equilibrium in which the all the solvent degrees of freedom including the slow ones have relaxed. This simple example of temporal multiscale phenomenon, when acting within a biosystem brings in important applications. For example, by conjugating solvatochromic fluorescent probes to biologically relevant molecules, it is possible to obtain valuable information regarding the functions, activities and interactions of such species in the context of living systems with spatial and temporal resolution. Furthermore, the excitation process of an embedded chromophore can initiate important processes within the biosystem such as photo-induced proton/electron and energy transfers and photochemical reactions.

In this contribution the applicability of QM/MM methods to photo-induced processes in biosystems will be critically analyzed with particular attention to the limits and potentials that a classical description of the environment presents. The presentation will be divided in three parts, the first devoted to the main aspects of the QM/MM strategy, the second analyzing its applications in the simulation of stationary properties and correlations and the third reporting on the extension to reactive processes. Finally, some conclusive remarks will be added.

12.2 The Main Components of QM/MM Methods

The QM/MM formalism can accommodate almost any combination of QM and MM methods. Traditionally semi-empirical QM methods have been most popular and they are still largely used even if many current QM/MM applications use density-functional theory (DFT) as the QM method owing to its favorable computational-effort/accuracy ratio. In the last years, various highly correlated methods have also been coupled with an MM description of the environment. In all cases the description of the hybrid system is made possible replacing the Hamiltonian operator representing the QM part only ("H_0") with a new or effective Hamiltonian ("H_{eff}") including an additional interacting term, as a result a modified Schrödinger equation has to be solved for the QM subsystem, namely:

$$\hat{H}_{eff}|\Psi\rangle = \left(\hat{H}_0 + \hat{H}_{env}\right)|\Psi\rangle \tag{12.1}$$

where H_{env} contains both the QM-MM interactions and those within the MM components, namely:

$$\hat{H}_{env} = \hat{H}_{QM/MM} + \hat{H}_{MM} \tag{12.2}$$

The definition of the two terms present in Eq. (12.2) depends on the selected MM force field (FF) as well as the "embedding" scheme.

As far as the choice of the FF is concerned, all those available in the literature can, in principle, be coupled with a QM description. As it regards the real interacting operator ($H_{QM/MM}$) different strategies (or embedding schemes) can be introduced. The simplest and less accurate one is to neglect the explicit interaction between the QM and the MM subsystems in the QM problem and approximate it at MM level adding a further contribution to H_{MM}. This approach, also known as "mechanical embedding" is rarely used in the modeling of photoinduced processes as it does not allow to include the effects of the environment in the description of the electronic process of interest if not indirectly by solving the Schrödinger equation for a geometry obtained optimizing the full system at MM level.

The most common QM/MM strategy for simulating photoinduced processes is the one known as "electrostatic" embedding where the $H_{QM/MM}$ term contains the electrostatic interaction between the QM subsystem and the atomic charges of the FF characterizing the MM subsystem. From a QM point of view, this term corresponds to a one-electron operator which is generally collected together with the one representing the electron-nuclei interaction in the QM subsystem. Different formulations of partial atomic charges can be used but commonly they are computed by constraining the charges to reproduce calculated physical observables: the standard procedure is to perform a QM calculation of the molecule under investigation, evaluate its electrostatic potential on a grid, and derive the ESP charges with a (possibly restrained) fitting procedure so that they reproduce the QM potential [14].

The introduction of the QM/MM electrostatic operator in Eq. (12.1) allows for the QM part to be polarized by the classical subsystem but this latter cannot respond back to this polarization. To account for this mutual polarization effects induced dipoles can be added to the permanent charges [15–19]: within this context, each MM site is described in terms of an atomic charge and an atomic polarizability. As a result, the Hamiltonians reported in Eq. (12.2) become:

$$\begin{aligned} \hat{H}_{QM/MM} &= \hat{H}^{el}_{QM/MM} + \hat{H}^{pol}_{QM/MM} = \sum_m q_m \hat{V}(\mathbf{r}_m) - \sum_a \mu_a^{ind} \cdot \hat{E}_a^{solute}(\mathbf{r}_a) \\ \hat{H}_{MM} &= \hat{H}^{el}_{MM} + \hat{H}^{pol}_{MM} = \sum_m \sum_{n>m} \frac{q_m q_n}{r_{mn}} - \frac{1}{2} \sum_a \mu_a^{ind} \cdot \sum_m \frac{q_m (\mathbf{r}_a - \mathbf{r}_m)}{|\mathbf{r}_a - \mathbf{r}_m|^3} \end{aligned} \tag{12.3}$$

where $V(\mathrm{r}_m)$ and E_a^{solute} represent the electrostatic potential and the electric field operators due to the QM system calculated at the MM sites. On the other hand, in Eq. (12.3) H^{el}_{MM} describes the electrostatic self-energy of the MM charges, while H^{pol}_{MM} represents the polarization interaction between such charges and the induced dipoles. We recall that the former term enters in the effective Hamiltonian only as a constant energetic term, while the latter depends on the QM wavefunction through the induced dipoles. The dipoles induced on each MM polarizable site can be obtained assuming a linear approximation, neglecting any contribution of magnetic character, and using an isotropic polarizability for each selected point in the MM part of the system. The electric field that determines such dipoles contains a sum of contributions from the QM subsystem, from the MM point charges and from the induced dipole moments. This mutual polarization between the dipoles can be solved through a matrix inversion approach, by introducing a matrix equation of the type:

$$\mathbf{R}\boldsymbol{\mu} = \mathbf{E} \tag{12.4}$$

where the vector **E** collects the electric field from the QM subsystem and the MM permanent charge distribution. The form of matrix **R** is determined uniquely by the position of the polarizable MM sites and the values used for the point polarizabilities α_i, namely:

$$\mathbf{R} = \begin{bmatrix} \alpha_1^{-1} & 0 & \cdots & 0 \\ 0 & \alpha_2^{-1} & \cdots & 0 \\ \vdots & \vdots & \ddots & \vdots \\ 0 & 0 & \cdots & \alpha_N^{-1} \end{bmatrix} + \begin{bmatrix} 0 & T_{12} & \cdots & T_{1N} \\ T_{21} & 0 & \cdots & T_{2N} \\ \vdots & \vdots & \ddots & \vdots \\ T_{N1} & T_{N2} & \cdots & 0 \end{bmatrix} \tag{12.5}$$

where N is the number of polarizable sites. The dipole field tensor T_{pq} is defined as

$$T_{pq} = \frac{\mathbf{1}}{r_{pq}^3} - \frac{3}{r_{pq}^5} \begin{bmatrix} x^2 & xy & xz \\ yx & y^2 & yz \\ zx & zy & z^2 \end{bmatrix} \tag{12.6}$$

where r_{pq} is the distance between site p and q, and x, y and z are the Cartesian components of the vector connecting the two.

It is evident that the computational results of the method critically depend on the quality of the parameterization and in particular on the point polarizabilities usually associated with atoms. The atomic polarizability parameters are generally obtained by fitting to either experimental or QM molecular polarizabilities or QM electrostatic potentials. The methods can also be divided into two groups: additive and interactive models, depending on the level of interactions permitted between induced dipoles [20]. In the additive approach, polarizable sites are allowed to respond to an external electric field but not to permanent and induced multipoles on other sites within a molecule. In nonadditive, also called interactive, polarization models, instead, each of a molecule's polarizable sites is allowed to respond to an external electric field not only from other molecules but from other sites within the same molecule. Consequently, all interacting sites polarize themselves. Under certain conditions, two inducible dipoles at short distances can cause a "polarization catastrophe"; this effect can be imputed to the use of point polarizabilities instead of diffuse charge distributions. To avoid this problem, the 1–2 and 1–3 bonded polarization interactions can be turned off; alternatively, one can apply distance-dependent damping for interactions on short distances or use both procedures. Several schemes for damping interactions between inducible dipoles at short distances have been proposed by Thole by introducing distance-dependent screening functions to be used in the calculation of the dipole tensor T_{pq} of Eq. (12.6) [21, 22]. We also recall that the introduction of induced dipoles requires to re-parameterize the permanent charges: the same ESP approach can be used but now it is the sum of the potential generated by induced dipoles and charges on the grid points which must reproduce the calculated QM potential [23].

The "induced dipole" formulation is an example of the so-called "polarizable embedding" but it is not the only possible choice. There are in fact various alternatives schemes to simulate the polarization of the MM subsystem, such as the fluctuating charges [24, 25], the classical Drude oscillators [26], or the Electronic Response of the Surroundings (ESR) [27] which mixes a non-polarizable MM scheme with a polarizable continuum model characterized by a dielectric constant extrapolated at infinite frequency.

In all the embedding schemes presented so far the effects that the classical part of the system exerts on the QM part are of electrostatic (and induction) nature. Additional nonelectrostatic (van der Waals) terms are generally described by empirical potentials independent of the QM electronic degrees of freedom, thus not affecting the QM wavefunction. As in standard MM methods, the van der Waals interaction is typically described by a Lennard–Jones (LJ) potential, namely

$$H^{LJ}_{QM/MM} = 4\sum_{i \in QM}\sum_{j \in MM} \varepsilon_{ij}\left[\left(\frac{\sigma_{ij}}{r_{ij}}\right)^{12} - \left(\frac{\sigma_{ij}}{r_{ij}}\right)^{6}\right] \tag{12.7}$$

so that suitable parameters are needed also for the QM atoms.

Other specificities appear, instead, any time the boundary between the QM and MM subsystems passes through a covalent bond. This is quite common in biomolecular systems; for example it is the case when the process of interest involves protein residues but it can also happen when only cofactors are involved but they are chemically bonded to the embedding macromolecule. In all these cases, the cut bond of the QM atom must be capped. In the literature, there are many different "boundary" strategies but the most popular are those using (i) link-atom schemes, and (ii) localized-orbital schemes [28]. In the most common strategy (i), an additional atomic center (the link atom usually represented by an hydrogen atom), that is not part of the real system is introduced and covalently bound to the QM atom involved in the original bond to saturate its free valence [29]. In the strategy (ii), instead, a set of suitably oriented localized orbitals are placed at the boundary and generally kept frozen [30, 31]. To reduce at most the artifacts following from these boundary schemes, the bond being cut should be unpolar and not involved in conjugative interactions, a typical "good" candidate is an aliphatic, C-C bond.

For electrostatic embedding, overpolarization of the QM density by the MM charges close to the cut can become an important problem and should be prevented. Different approaches have been proposed especially within the link-atom scheme [32]. A possible solution is to delete the one-electron integrals associated with the link atoms or the MM point charges in the link region; alternatively, these point charges can be shifted and redistributed in the nearby atoms [33] or substituted with Gaussian charge distributions [34]. In the case of polarizable embedding, a smoothing function can be also used for QM/induced dipole interaction.

When a photoinduced process is simulated with QM/MM methods, a new issue has to be dealt with. This issue refers to possible delays in the response of the environment with respect to the fast times which characterize the changes induced in the QM part e.g. during an electronic excitation, and as a consequence of an electron or energy transfer. As commented in the introduction, the delays involve the inertial components of the environment response as the dynamic (i.e. electronic) ones are generally assumed to be sufficiently fast to immediately follow any change in the QM charge distribution. In electrostatic embedding schemes the inclusion of the inertial component of the response is taken into account in an automatic way by assuming that the spatial distribution of the MM charges does not change, neither their values, during the investigated process. As far as it concerns the dynamic component, a polarizable embedding is instead necessary as we have to allow the fast part of the environment response to change: this can be obtained, for example, using the induced dipoles which can adapt to the new charge distribution. In principle, also the van der Waals interactions should be affected during a photoinduced process as the electronic charge distribution of the QM part will strongly change an accordingly also their dispersion and repulsion interactions included in the LJ terms. As a matter of fact, these possible changes are generally neglected and the LJ terms are calculated once using the parameters corresponding to the "unperturbed" configuration of the system. Indeed, it is now clear that especially dispersion can play an important role in photoinduced processes but properly accounting for that would require translating the dispersion interactions between the QM and the MM part into a QM operator exactly as for the electrostatic+polarization effects and this is not yet available in most of the QM/MM formulations.

In the above description of QM/MM methods, we have implicitly adopted what is called as the "additive formulation". An alternative "subtractive formulation" is however possible. In their standard implementation, subtractive QM/MM schemes require three independent calculations: an MM calculation on the entire system (S) and two calculations on the inner subsystem (I), one at the selected QM level and one at the same MM level used for the entire system. The QM/MM energy is then obtained as:

$$E^{sub}_{QM/MM} = E_{MM}(S) + E_{QM}(I) - E_{MM}(I) \tag{12.8}$$

The main advantage of this formulation is its simplicity: no explicit QM–MM coupling terms are needed but the standard QM and MM procedures can be used without any modifications and the implementation is thus straightforward. Equation (12.8) can also properly handle the presence of link atoms. The main disadvantages of a subtractive scheme is that a complete set of MM parameters for the inner subsystem is necessary, and more severely, the coupling between the subsystems is handled entirely at the MM level (i.e. it correspond to the "mechanical embedding"). As a matter of fact, something equivalent to the electronic embedding can be recovered as done in one of the most popular subtractive methods, the ONIOM (our n-layered integrated molecular orbital and molecular mechanics) method by Morokuma and co-workers [35] More recently, ONIOM has also been further extended to a polarizable embedding based on induced dipoles [36].

12.3 Stationary Properties and Correlations: Specific Versus Bulk Effects

A large part of the QM/MM applications to photoinduced processes are focused on the calculations of "stationary" properties, namely transition energies and dipoles, absorption/emission and electronic circular dichroism spectra, etc.

All these properties are sensitive to both specific and bulk effects due to the environment. The former are generally short-range interactions involving the QM atoms more exposed to the environment and the so-called first "solvation shells", i.e. those molecules, fragments or chemical groups of the environment, which are within few angstroms of distance from the QM region. Bulk effects instead collect all the other effects and they can also be described as "mean field" effects as they do not explicitly depend on the specific configuration of the environment but they are the result of a sort of averaging.

QM/MM approaches in principle are well suited to properly account for both specific and bulk effects. There are however some issues that have to be dealt with in order to obtain a correct description.

As it regards specific effects, the use of an atomistic description of the environment through the MM approach should automatically lead to a proper description of possible interactions among the QM atoms and those in the first "solvation shells". However, this is valid only if the configuration of those shells is correct: in fact if the structure of the system is not well determined possible artificial interactions among the QM and the MM parts could appear and largely affect the accuracy of the description. Moreover, the effects of such interactions should be properly included in the QM Hamiltonian: if a mechanical embedding is used this is not possible but even if the most common electrostatic embedding is into play only the electrostatic component of the interaction can be properly accounted for. This is generally enough when charged and/or strong hydrogen-bonding groups are present in the environment. A more complete picture should in principle be achieved if a polarizable embedding is used: in that case in fact the environment can better "adapt" to the QM charge distribution and to its changes during the photo-induced process. Also in this case, however, dispersive interactions as well as charge-transfer effects between the QM and the MM part are not included. When these interactions are known to be important a possible strategy to adopt is to enlarge the boundary so to include in the QM subsystem also the parts of the environment which more strongly interact with the QM atoms involved in the process of interest. In all cases, however the most delicate aspect is the correct parameterization of the FF used. All the available FFs in fact have been optimized within a completely classical framework: there is therefore no assurance that they can be properly extended to described QM-classical interactions. Fortunately, the results obtained so far by combining the most popular FFs with QM approaches seem to show that indeed the electrostatic (+polarization) effects are described in a reasonably correct way, however this aspect still need to be further explored in the future to eventually develop FFs specifically optimized for being combined with QM descriptions. As a matter of fact, examples in this direction have been already presented and successfully applied also to biosystems. One of the most popular strategies is the so-called Effective Fragment Potentials (EFP) approach by Gordon and coworkers [37]. Separate calculations on model systems are used to derive a set of one-electron terms (the EFP) that account for the electrostatic, inductive, and repulsive interactions of a moiety. The EFP is then incorporated into the Hamiltonian of a QM calculation, where it describes the effects of the environment on the QM part. Originally designed to model the solvent environment in QM calculations, the scope of the EFP method was subsequently extended to treat covalent bonds across the boundary. Another example along this direction is the X-Pol method developed collaboratively by Gao and Truhlar [38]. In this method, the system is partitioned into individual subsystems, called fragments or blocks, which can be an individual molecule, a molecular fragment, or even a subset of the atomic orbitals on a single atom or on a group of atoms. The distinct feature of the X-Pol method is the incorporation of many-body polarization through performing a double SCF calculation; that is, self-consistent calculations of fragments in the electrostatic field of the other fragments usually represented by the Mulliken charges (often scaled by an empirical factor).

Moving now to bulk effects, the problem apparently seems of easier solution: to account for longer-range effects we just need to enlarge the MM subsystem! This is certainly true but another fundamental aspect has to be taken into account. To correctly simulate bulk effects in fact it is necessary to introduce averages. In a statistical language, the bulk can be defined as an artificial representation, which mimics the average over all possible configurations of the environment. To achieve such a representation we can either generate many statistical replicas of the system or propagate the system in time and then average on a large enough set of uncorrelated configurations. In both cases, the QM/MM strategy has to be coupled with computational simulations either using a stochastic or a deterministic method. In biological applications, both Monte Carlo (MC) and Molecular Dynamics (MD) formulations are often used within a purely classical formulation. This means that the QM/MM strategy have to be preceded by a MC or MD simulation of the system from which many different configurations (snapshots) are extracted and used for the following QM/MM calculations. In such a way, the fluctuations in the structure of the system are completely determined at classical level while the QM description is applied to many "frozen-geometry" systems. This procedure is extremely popular as easy to implement and computationally efficient: the simulation is performed at classical level and the QM calculation is restricted to a limited number of configurations. How large this number should be depends both on the specific process to be studied and the type of environment. In general, numbers of configurations of the order of few hundreds are enough to properly account for the structural fluctuations of standard biological environments (such as proteins or DNA).

By adopting this decoupled approach (classical simulation followed by QM/MM calculations) it is possible to reproduce UV-Vis absorption spectra of embedded chromophores including the inhomogeneous broadening of the spectral band associated with the environment reorganization energy corresponding to change in the solvation energy passing from the nonequilibrium to the equilibrium solvation regimes. If an MD simulation is used, various snapshots are extracted from the trajectory to represent different configurations of the system. By calculating transition energies and dipoles of the embedded chromophore at the QM/MM level for each of the configuration, fluctuations of the environment and of the resulting transition energies and dipoles are automatically included in the analysis. A convolution of the spectra calculated for the different local environments will mimic the inhomogeneous broadening.

Much more delicate is instead the simulation of correlation functions generally used to reproduce spectral densities through their Fourier transform: in those cases the number of QM/MM calculations to be performed increases a lot as thousands of correlated configurations have to be explored. Spectral densities are particularly important in the simulation of photoinduced processes of multichromophoric systems such as those present in the pigment-protein complexes (PPC) used by natural photosynthetic organisms to collect solar light. In these light-harvesting antennae, the interacting pigments, usually (bacterio)chlorophylls, absorb light and transfer the resulting excitation energy to other PPCs, called reaction centers (RC), where the energy is used to initiate charge separation processes [39]. The environment (namely protein and solvent, and eventually the membrane as most PPC are membrane bound proteins) plays a fundamental role in this process. First of all specific and bulk effects can differentially change the excitation energies (commonly denoted "site energies") of each chomophore and affect their excitonic interactions (by mostly screening the electronic couplings). Besides this tuning of excitation energies and screening of the couplings, the protein has to dissipate the excess energy of the excitons when they relax downward to the excited states of the RC. This dissipation is achieved primarily by protein vibrations that absorb the excess energy and distribute it over many degrees of freedom. The coupling of pigment excited states to the protein vibrations is characterized by the spectral density $J(\omega)$ and the energy associated with equilibration of the environment after excitation is quantified by the reorganization energy defined as:

$$\lambda = \int_0^{\infty} \frac{J(\omega)}{\omega} d\omega \tag{12.9}$$

Both the smooth continuous density of low-frequency vibrational modes, arising from protein and solvent, and the discrete vibrational modes of intra- or intermolecular vibrations can participate in the energy transfer [40]. Much interest has been focused on the influence of a slowly relaxing environment associated with the smooth part of the spectral density on coherent exciton dynamics. Studies have shown that such an environment can protect quantum coherence [41] and hence support coherent energy transfer for longer times than environments inducing purely stochastic (Markovian) exciton evolution. Non-Markovian exciton dynamics are also expected when the electronic system couples to selective intra- and inter-molecular vibrations. Indeed, theoretical studies [42, 43] have found that discrete non-equilibrium vibrations that are resonant with differences in energy of electronic eigenstates can also sustain coherent dynamics for long timescales consistent with those reported in experiments [44, 45].

The spectral density J(ω) can be obtained by the coupling of classical MD and QM/MM calculations: what has to be computed is the auto correlation function of the site energies fluctuations. This approach has been used for different light-harvesting systems using both electronic and polarizable embeddings [46–48]. Due to the large computational cost of this simulation which requires thousands of QM/MM calculations on different configurations extracted from the MD, semiempirical methods are generally used as QM method. This aspect when combined with that of using a classical FF to generate the intra- and intermolecular modes, seems to show important limitations: while the positions of the peaks of $J(\omega)$ are reproduced in reasonable agreement with experiments, the intensities corresponding to the most coupled vibrations are generally overestimated. This shows that the fluctuations predicted by the decoupled two-step approach are too large. These findings are not unexpected, as most of the classical FFs have been parameterized to describe average structures and not temperature-dependent fluctuations of intramolecular motions. Moreover, the same FF have not been explicitly optimized for describing the common light-harvesting pigments ((bacterio)chlorophylls, bilins and carotenoids). A much better description of the spectral densities would be obtained with QM/MM dynamics but the cost of such simulations is still extremely high even when using a semiempirical QM level (which on the other hand should be first benchmarked to be sure that can properly describe the coupling of the intramolecular distortions with the electronic properties of the chromophore).

12.4 Reactive Processes

Even if the two-step strategy running first a classical (MD or MC) simulation followed by a set of QM/MM calculations has shown to be successful in the simulation of many (photoinduced) processes in (bio)systems, in some cases a more refined formulations is necessary. For example, if a reactive event is into play, such as in a photochemical reaction or a photoinduced proton/electron transfer, the purely classical description is not applicable [49–52]. To properly model chemical processes either geometry optimizations or MD approaches can be used. Both techniques mainly involve an iterative scheme in which the energy, the forces (i.e., the first derivatives of the energy with respect to the geometric variables) are used to generate a new geometry (together with the calculation of velocities in the case of MD). This procedure is reiterated until the chosen convergence criteria are fulfilled (in the case of geometry optimizations) or the total simulation time reaches the defined value (in the case of MD). The formal extension of geometry optimizations or MDs to a QM/MM description is simple within the Born-Oppenheimer (BO) approximation: at each step of the iterative scheme the QM energy and forces come from a converged SCF calculation. However the computational cost can enormously increase. This for example implies that the time windows that can be explored with BO-MD have to be significantly reduced. If we think that now classical MDs can relatively easily simulate up to tens of microseconds, hybrid QM/MM simulations can rarely go beyond tens of nanoseconds (if a very approximated QM approach is used) or hundreds of picoseconds if accurate DFT or other QM methods are used. In the application of QM/MM for reactive processes, in addition to the time issue, another aspect can become difficult to manage. During a reactive event in fact, an a priori definition of the boundary between the QM and the MM parts is not always possible especially if we want to keep the "expensive" QM region as smallest as possible. In the recent years, adaptive QM/MM strategies have been developed in which the QM region is defined "on the fly" during the dynamics. It must be noted that the instantaneous switching of an atom in and out of the QM reactive region has raised concern in the past. This is because from one time step to the next there will inevitably be an abrupt change in the forces on that atom. In dynamics studies this discontinuity can influence the integration with respect to time and conceivably lead to nonphysical behavior. To circumvent this a more robust integration method can be used and/or ad hoc algorithms have been developed for smoothing the forces and the potential of the atoms switching regions in order to conserve both energy and momentum.

A further complexity in the simulation of photochemistry and more in general of excited state photoinduced dynamics is that they are intrinsically nonadiabatic processes, in which the coupling between the nuclear and electronic motion leads to nonradiative transitions between electronic states. A generally applicable approach for this purpose is the mixed quantum-classical dynamics in which the nuclear motion is described by classical trajectories obtained in the framework of molecular dynamics "on the fly" combined with Tully's surface hopping (TSH) procedure

[53]. In the TSH approach, a trajectory runs on a given potential energy surface (PES) and at the same time the state probabilities are computed, usually by numerical integration of the electronic Schrödinger equation. Hops to other surfaces take place stochastically, according to the computed probabilities. In order to simulate the spread of the QM initial wavepacket in the coordinate and momentum space, as well as the thermal distributions, one has to run many trajectories, with a suitable sampling of initial conditions. The forces and nonadiabatic couplings required to propagate nuclear trajectories can be obtained using different ab initio or semiempirical methods [54, 55], and, in the last years, different approaches for the nonadiabatic dynamics in the framework of the time-dependent density functional theory (TDDFT) have been developed [56–58] and successively coupled with a MM description of the environment [59–61].

Recently, an alternative approach to nonadiabatic dynamics, known as full multiple spawning (FMS) [62], has been extended to an hybrid QM/MM scheme [63]. The FMS method is designed to treat problems involving dynamics on multiple electronic states by employing the quantum mechanical nonadiabatic couplings that trigger the transfer of population between electronic states while maintaining some connection to classical dynamics in the description of the nuclear wave function. Specifically, the nuclear wave function is expanded in a basis set of complex Gaussian functions, called trajectory basis functions (TBFs). The phase space center of each of these TBFs evolves according to classical mechanics. PESs, governing the propagation of the nuclear basis function, and nonadiabatic coupling are calculated every time step "on-the-fly". During nonadiabatic events, the nuclear basis set expands as needed ("spawning"), and throughout the dynamics, the amplitudes of the nuclear basis functions are determined by solving the nuclear Schrödinger equation in the time-evolving basis set.

In all cases, one needs at least a schematic knowledge of the potential energy surfaces of the embedded system, i.e. some cuts along the most important internal coordinates and/or critical points, such as minima, transition states and minimum energy conical intersections. This is indeed a challenging problem, since it combines the still open issues of electronic calculations for excited states and reacting systems (first of all a balanced treatment of electron correlation at different geometries and for different states) with the complexity of condensed phase chemical environments. It is evident that here embedding schemes going beyond the mechanical one are compulsory. In principle, the polarizable scheme should also be preferred as it allows for a state-specific response of the environment. The enhanced realism introduced by mutual polarization effects between the QM and the MM parts, presents however a problem that has largely limited the applications to problems of real biological interest. Polarizable QM/MM models (such QM/continuum models) present in fact a theoretical ambiguity, which becomes important in multistate treatments, and originates from the mixing of classical and QM degrees of freedom: if the polarization of the medium is computed in a state-specific way, and the chromophore is in turn polarized by the reaction field of the medium, the computed electronic states are no longer orthogonal and, in quasi-degeneracy situations, cannot be easily identified. This may disturb a univocal definition of the PESs and of their crossings. In specific and limited situations, like in vertical transitions (either from a ground, in absorption, and an excited state in emission), the ambiguity can be solved by introducing a net separation of the environment response into inertial and dynamic components and assuming that only the inertial components can change: this leads to what we have described before as nonequilibrium solvation regime. A possible more general solution to the problem is to go beyond a PES-based approach such as in the Car–Parrinello MD (CP-MD) [64] in which the wave functions are treated as fictitious dynamic variables within a Lagrangian scheme and follow the motion of the nuclei "on the fly". QM/MM approaches based on CP-MD have been developed and successfully applied to biomolecular simulations but they are limited to the electrostatic embedding [65]. The Lagrangian scheme can instead be generalized so to account for additional dynamic variables which represent the evolution of the solvent polarization. In such a way both the inertial and the dynamic components of the environment response would be properly included in the calculations being the inertial component treated on equal footing with the nuclei of the QM subsystem while the dynamic component following their motion "on the fly".

12.5 Concluding Remarks

We have presented a necessarily shortened and simplified overview of hybrid QM/MM methods by focusing on the aspects which are more specific for their applications to photoinduced processes in biosystems. We have tried to show that the QM/MM strategy can be effectively used to describe both "specific" and "bulk" effects of the environment as well as to account for "dynamic" effects which are required to simulate reactive processes involving different electronic states. We have also shown that some important problems are still open and further developments of the model are necessary in order to obtain accurate simulations of photoinduced processes. In the near future, we may expect improvements in the efficiency and applicability of QM/MM dynamics as well an expansion in the applicability of polarizable MM schemes.

Moreover, new force fields optimally suited to be combined with QM calculations are expected as well alternative strategies to standard MM-FF such as those along the directions opened by fragmentation methods or QM/QM embedding schemes such the proposed wave-function theory in wave-function theory (WFT-in-WFT) [66], wave-function theory-in-DFT (WFT-in-DFT) [67, 68], and density-functional theory in density-functional theory (DFT-in-DFT) frozen-density embeddings (FDE) [69, 70]. Whatever direction the progresses in the field will follow, it is evident that the most important key is the dialog between theory and experiments. It is exactly thanks to this, that the elementary acts of photochemical and photophysical processes are now observable and interpretable on the spatial and temporal scale of a single molecule.

References

1. Lambert N, Chen Y-N, Cheng Y-C et al (2012) Quantum biology. Nat Phys 9:10–18
2. Cramer C, Truhlar D (1999) Implicit solvation models: equilibria, structure, spectra, and dynamics. Chem Rev 99:2161–2200
3. Tomasi J, Mennucci B, Cammi R (2005) Quantum mechanical continuum solvation models. Chem Rev 105:2999–3093
4. Rivail JL, Rinaldi D (1996) Liquid-state quantum chemistry: computational applications of the polarizable continuum models. In: Jerzy L (ed) Computational chemistry: reviews of current trends. Computational chemistry, review of current trends, pp 139–174. World Scientific Publishing, Singapore
5. Ruiz-López MF (2008) The multipole moment expansion solvent continuum model. In: Canuto S (ed) Solvation effects on molecules and biomolecules: computational methods and applications. Springer Series: Challenges and Advances in Computational Chemistry and Physics, pp 23–28. Springer
6. Gao J (1996) Hybrid quantum and molecular mechanical simulations: An alternative avenue to solvent effects in organic chemistry. Accounts Chem Res 29:298–305
7. Riccardi D, Schaefer P, Yang et al (2006) Development of effective Quantum Mechanical/Molecular Mechanical (QM/MM) methods for complex biological processes. J Phys Chem B 110:6458–6469
8. Senn HM, Thiel W (2009) QM/MM methods for biomolecular systems. Angew Chem Int Edit 48:1198–1229
9. Lin H, Truhlar DG (2007) QM/MM: what have we learned, where are we, and where do we go from here? Theoret Chem Acc 117:185–199
10. Mennucci B (2013) Modeling environment effects on spectroscopies through QM/classical models. Phys Chem Chem Phys 15:6583
11. Tuñon I, Martins-Costa MTC, Millot C, Ruiz-López MF (1997) Molecular dynamics simulations of elementary chemical processes in liquid water using density functional and molecular mechanics potentials. I. Proton transfer in strongly H-bonded complexes. J Chem Phys 106:3633–3642
12. Strnad M, Martins-Costa MTC, Millot C, Tuñon I, Ruiz-Lopez MF, Rivail JL (1997) Molecular dynamics simulations of elementary chemical processes in liquid water using density functional and molecular mechanics potentials. II. Charge separation processes. J Chem Phys 106:3643–3657
13. Warshel A, Levitt M (1976) Theoretical studies of enzymic reactions: dielectric, electrostatic and steric stabilization of the carbonium ion in the reaction of lysozyme. J Mol Biol 103:227–249
14. Bayly CI, Cieplak P, Cornell W, Kollman PA (1993) A well-behaved electrostatic potential based method using charge restraints for deriving atomic charges: the RESP model. J Phys Chem 97:10269–10280
15. Thompson MA, Schenter GK (1995) Excited states of the bacteriochlorophyll b dimer of Rhodopseudomonas viridis: A QM/MM study of the photosynthetic reaction center that includes MM polarization. J Phys Chem 99:6374–6386
16. Illingworth CJR, Gooding SR, Winn PJ et al (2006) Classical polarization in hybrid QM/MM methods. J Phys Chem A 110:6487–6497
17. Curutchet C, Munoz-Losa A, Monti S et al (2009) Electronic energy transfer in condensed phase studied by a polarizable QM/MM model. J Chem Theory Comput 5:1838–1848
18. Olsen J, Kongsted J (2011) Molecular properties through polarizable embedding. Adv Quantum Chem 61:107–143
19. Ruiz-Lopez MF, Rivail JL (1998) Combined quantum mechanics and molecular mechanics approaches to chemical and biochemical reactivity. In: Schleyer PVR (ed) Encyclopedia of computational chemistry, Wiley & sons, vol 1, pp 437–448

20. Cieplak P, Dupradeau FJ, Duan Y, Wang J (2009) Polarization effects in molecular mechanical force fields. J Phys Condens Matter 21:333102
21. Thole B (1981) Molecular polarizabilities calculated with a modified dipole interaction. Chem Phys 59:341–50
22. van Duijnen P, Swart M (1998) Molecular and atomic polarizabilities: thole's model revisited. J Phys Chem A 102:2399–2407
23. Wang J, Cieplak P, Li J et al (2011) Development of polarizable models for molecular mechanical calculations I: parameterization of atomic polarizability. J Phys Chem B 115:3091–3099
24. Bryce RA, Buesnel R, Hillier IH, Burton NA (1997) A solvation model using a hybrid quantum mechanical/molecular mechanical potential with fluctuating solvent charges. Chem Phys Lett 279:367–371
25. Lipparini F, Barone V (2011) Polarizable force fields and polarizable continuum model: a fluctuating charges/PCM approach. I: theory and implementation. J Chem Theory Comput 7:3711–3724
26. Boulanger E, Thiel W (2012) Solvent boundary potentials for hybrid QM/MM computations using classical drude oscillators: a fully polarizable model. J Chem Theor Comput 8:4527–4538
27. Jacquemin D, Perpète EA, Laurent AD et al (2009) Spectral properties of self-assembled squaraine-tetralactam: a theoretical assessment. Phys Chem Chem Phys 11:1258–1262
28. Reuter N, Dejaegere A, Maigret B, Karplus M (2000) Frontier bonds in QM/MM methods: a comparison of different approaches. J Phys Chem A 104:1720–1735
29. Field MJ, Bash PA, Karplus M (1990) A combined quantum mechanical and molecular mechanical potential for molecular dynamics simulations. J Comput Chem 11:700–733
30. Thery V, Rinaldi D, Rivail JL, Maigret B, Ferenczy GG (1994) Quantum mechanical computations on very large molecular systems: The local self-consistent field method. J Comput Chem 15:269–282
31. Monari A, Rivail J-L, Assfeld X (2013) Theoretical modeling of large molecular systems. Advances in the local self consistent field method for mixed quantum mechanics/molecular mechanics calculations. Acc Chem Res 46:596–603
32. Lin H, Truhlar DG (2005) Redistributed charge and dipole schemes for combined quantum mechanical and molecular mechanical calculations. J Phys Chem A 109:3991–4004
33. Sherwood P (2003) QUASI: a general purpose implementation of the QM/MM approach and its application to problems in catalysis. J Mol Struct Theochem 632:1–28
34. Amara P, Field MJ (2003) Evaluation of an ab initio quantum mechanical/molecular mechanical hybrid-potential link-atom method. Theor Chem Acc 109:43–52
35. Chung LW, Hirao H, Li X, Morokuma K (2011) The ONIOM method: its foundation and applications to metalloenzymes and photobiology. Wiley Interdisc Rev Comput Mol Sci 2:327–350
36. Caprasecca S, Jurinovich S, Viani L et al (2014) Geometry optimization in polarizable QM/MM models: the induced dipole formulation. J Chem Theory Comput 10:1588–1598
37. Gordon MS, Fedorov DG, Pruitt SR, Slipchenko LV (2012) Fragmentation methods: a route to accurate calculations on large systems. Chem Rev 112:632–672
38. Xie W, Song L, Truhlar DG, Gao J (2008) The variational explicit polarization potential and analytical first derivative of energy: towards a next generation force field. J Chem Phys 128:234108
39. Blankeship RE (2014) Molecular mechanisms of photosynthesis, 2nd edn. Wiley-Blackwell
40. Ishizaki A, Fleming GR (2009) Theoretical examination of quantum coherence in a photosynthetic system at physiological temperature. P Natl Acad Sci USA 106:17255–17260
41. Lee H, Cheng Y-C, Fleming GR (2007) Coherence dynamics in photosynthesis: protein protection of excitonic coherence. Science 316:1462–1465
42. Kolli A, O'Reilly EJ, Scholes GD, Olaya-Castro A (2012) The fundamental role of quantized vibrations in coherent light harvesting by cryptophyte algae. J Chem Phys 137:174109

43. Chin AW, Prior J, Rosenbach R et al (2013) The role of non-equilibrium vibrational structures in electronic coherence and recoherence in pigment–protein complexes. Nat Phys 9:113–118
44. Engel GS, Calhoun TR, Read EL et al (2007) Evidence for wavelike energy transfer through quantum coherence in photosynthetic systems. Nature 446:782–786
45. Collini E, Wong CY, Wilk KE et al (2010) Coherently wired light-harvesting in photosynthetic marine algae at ambient temperature. Nature 463:644–647
46. Rivera E, Montemayor D, Masia M, Coker DF (2013) Influence of site-dependent pigment-protein interactions on excitation energy transfer in photosynthetic light harvesting. J Phys Chem B 117:5510–5521
47. Viani L, Corbella M, Curutchet C, O'Really EJ, Olaya-Castro A, Mennucci B (2014) Molecular basis of the exciton–phonon interactions in the PE545 light-harvesting complex. Phys Chem Chem Phys 16:16302–16311
48. Aghtar M, Strumpfer J, Olbrich C et al (2014) Different types of vibrations interacting with electronic excitations in phycoerythrin 545 and Fenna–Matthews–Olson antenna systems. J Phys Chem Lett 5:3131–3137
49. Garavelli M (2006) Computational organic photochemistry: strategy, achievements and perspectives. Theoret Chem Acc 116:87–105
50. Hammes-Schiffer S, Stuchebrukhov AA (2010) Theory of coupled electron and proton transfer reactions. Chem Rev 110:6939–6960
51. Lasorne B, Worth GA, Robb MA (2011) Excited-state dynamics. Wiley Interdisc Rev Comput Mol Sci 1:460–475
52. Domcke W, Yarkony DR (2012) Role of conical intersections in molecular spectroscopy and photoinduced chemical dynamics. Annu Rev Phys Chem 63:325–352
53. Tully JC (1990) Molecular dynamics with electronic transitions. J Chem Phys 93:1061
54. Barbatti M (2011) Nonadiabatic dynamics with trajectory surface hopping method. Wiley Interdisc Rev Comput Mol Sci 1:620–633
55. Persico M, Granucci G (2014) An overview of nonadiabatic dynamics simulations methods, with focus on the direct approach versus the fitting of potential energy surfaces. Theor Chem Acc 133:1526
56. Curchod BFE, Rothlisberger U, Tavernelli I (2013) Trajectory-based nonadiabatic dynamics with time-dependent density functional theory. Chem Phys Chem 14:1314–1340
57. Werner U, Mitrić R, Suzuki T, Bonačić-Koutecký V (2008) Nonadiabatic dynamics within the time dependent density functional theory: ultrafast photodynamics in pyrazine. Chem Phys 349:319–324
58. Tapavicza E, Bellchambers GD, Vincent JC, Furche F (2013) Ab initio non-adiabatic molecular dynamics. Phys Chem Chem Phys 15:18336–18348
59. Carloni P, Rothlisberger U, Parrinello M (2002) The role and perspective of ab initio molecular dynamics in the study of biological systems. Acc Chem Res 35:455–464
60. Tavernelli I, Curchod BFE, Rothlisberger U (2011) Nonadiabatic molecular dynamics with solvent effects: a LR-TDDFT QM/MM study of ruthenium (II) tris (bipyridine) in water. Chem Phys 391:101–109
61. Wohlgemuth M, Bonačić-Koutecký V, Mitrić R (2011) Time-dependent density functional theory excited state nonadiabatic dynamics combined with quantum mechanical/molecular mechanical approach: photodynamics of indole in water. J Chem Phys 135:054105
62. Ben Nun M, Quenneville J, Martinez TJ (2000) Ab initio multiple spawning: photochemistry from first principles quantum molecular dynamics. J Phys Chem A 104:5161–5175
63. Punwong C, Owens J, Martinez TJ (2014) Direct QM/MM excited-state dynamics of retinal protonated schiff base in isolation and methanol solution. J Phys Chem B. doi:10.1021/jp5038798
64. Car R, Parrinello M (1985) Unified approach for molecular dynamics and density-functional theory. Phys Rev Lett 55:2471–2474
65. Laio A, VandeVondele J, Rothlisberger U (2002) A Hamiltonian electrostatic coupling scheme for hybrid Car-Parrinello molecular dynamics simulations. J Chem Phys 116:6941–6947. doi:10.1063/1.1462041

66. Höfener S, Visscher L (2012) Calculation of electronic excitations using wave-function in wave-function frozen-density embedding. J Chem Phys 137:204120
67. Goodpaster JD, Barnes TA, Manby FR, Miller TF (2014) Accurate and systematically improvable density functional theory embedding for correlated wavefunctions. J Chem Phys 140:18A507
68. Daday C, König C, Neugebauer J, Filippi C (2014) Wavefunction in density functional theory embedding for excited states: which wavefunctions, which densities? Chem Phys Chem 15:3205–3217
69. Casida ME, Wesolowski TA (2003) Generalization of the Kohn-Sham equations with constrained electron density formalism and its time-dependent response theory formulation. Int J Quantum Chem 96:577–588
70. Neugebauer J (2007) Couplings between electronic transitions in a subsystem formulation of time-dependent density functional theory. J Chem Phys 126:134116–134112

Chapter 13
The Non Empirical Local Self Consistent Field Method: Application to Quantum Mechanics/Molecular Mechanics (QM/MM) Modeling of Large Biomolecular Systems

Jean-Louis Rivail, Antonio Monari and Xavier Assfeld

Abstract The present chapter starts with an analysis of the problems encountered when applying a mixed Quantum Mechanics/Molecular Mechanics to a large molecular system, which cannot be approached at a full quantum level of computation and a review of the possible solutions. A Non Empirical Local Self Consistent Field methodology, allowing computing at any quantum chemical level a part of a very large molecule interacting with the rest of this molecule is described in some detail. This approach is illustrated by various applications to the spectroscopic properties of various bio-macromolecules. Finally, and as a test case we will focus on the QM/MM modelling of spectroscopic and photophysical properties of exogenous chromophores interacting with DNA. Hence, we will show how the combination of high-level QM/MM methods with Molecular Dynamics simulations allows us to gain unprecedented insights in the process of DNA Photosensitization that is of paramount importance to understand the induction of DNA photolesions and to unravel novel anticancer therapeutic strategies.

J.-L. Rivail (✉) · A. Monari · X. Assfeld
Theory-Modeling-Simulation, SRSMC, Université de Lorraine, Vandoeuvre-lès-Nancy, France
e-mail: Jean-Louis.Rivail@univ-lorraine.fr

X. Assfeld
e-mail: xavier.assfeld@univ-lorraine.fr

J.-L. Rivail · A. Monari · X. Assfeld
CNRS, Theory-Modeling-Simulation, SRSMC, Vandoeuvre-lès-Nancy, France

J.-L. Rivail et al. (eds.), *Quantum Modeling of Complex Molecular Systems*,
Challenges and Advances in Computational Chemistry and Physics 21,
DOI 10.1007/978-3-319-21626-3_13

13.1 Introduction

Macromolecules are very diverse in composition and function and they occur in many fields of chemistry and biochemistry. The usual applications of synthetic polymers mainly deal with their physical properties: molecular weight, conformations, van des Waals interactions… which don't require detailed knowledge of the electronic structure and can be approached by classical computational force fields which are at the basis of what one usually calls molecular mechanics. Nevertheless, one may be interested in the chemical reactivity of some region of the polymer, such as structural defects, and a quantum computation may be the only way to get the reliable chemical information. The other, very important class of macromolecules contains the innumerable biomacromolecules: polypeptides, enzymes, nucleic acids. The understanding of their role in life usually requires the knowledge of the electronic structure and often the reactivity of at least well defined parts of the large system they constitute. This knowledge can only be reached by means of quantum chemical computations.

The main problem with macromolecules, from a quantum chemist's point of view is their size: hundreds or more often thousands of atoms, and their treatment by means of standard quantum chemical codes is still unrealistic, even on modern supercomputers. The computational time increases as the third, or even fifth power of the size of the basis set used for the computation, according to the method used, and the usual basis sets require, in the average, 5 to *ca.* 20 functions per atom other than hydrogen. In addition, a full treatment of macromolecular systems often implies molecular dynamics or Monte-Carlo statistical simulations which may require the computation of millions of configurations. Therefore, special computational strategies have to be devised in order to get the information of interest in such very large systems. This field is at the present time the object of numerous methodological approaches which witness the importance of the question, and several reviews or general papers, focussed on various aspects of the problem, have been published recently [1–6].

For the time being, a large variety of methods are available to model large molecular systems.

The empirical force fields of molecular mechanics (MM) allow the computation of the energy variations of a system following the geometry changes: bond lengths, bond angles, torsional motions. They include the interaction energy between non-chemically bonded atoms: electrostatic and van der Waals interactions, hydrogen bonding. Most of the parameters entering the expression of the energy of a molecular conformation are derived from quantum chemical computations on model molecules and the apparent charges of the atoms are parameterized to reproduce the interaction energies, so that it is usually admitted that they take into account an average value of the induction energy. Classical force fields allow to simulate vibrational spectra, and to perform molecular dynamics on the whole system, including the presence of solvent molecules. There are quite a few well known force fields [7–10] which enter computational packages ready to use and they have very modest computational requirements even for very large systems. Some more elaborate force fields include a polarizability term which takes explicitly into account the contribution of the induction energy [11–13] but, in principle the inclusion of electronic induced moments on the elementary components (atoms) of a molecule requires an iterative process which makes the computations more time consuming.

When one is concerned with the electronic structure of a molecule, quantum computations become compulsory and the computational requirements become far more important than those of molecular mechanics. Nevertheless, some recent methodological developments made quantum computations affordable on moderately large molecular systems. Most of them are known under the general label of linear scaling. The aim of these methods is to develop an algorithm for which the computational time increases more or less linearly or at least at a power close to one with the number of atoms (low order scaling) [14, 15]. The basic idea is to replace the large square matrix having the size of the whole basis set to be diagonalized, by many small matrices corresponding to small connected fragments of the system. The so-called "Divide- and-Conquer" methods [16] are becoming quite efficient for such approaches. They allow computations on pretty large systems, in particular at the semi-empirical quantum level [17].

Macromolecular systems or solutes are also characterized by an inherent flexibility, showing the possibility to assume different conformations with almost equal weights. Dynamic effects become much more important than for isolated small molecules. Such systems therefore are usually treated by means of molecular dynamics, following the Born-Oppenheimer or the Car-Parrinello scheme. The Car-Parrinello approach [18] is well suited for the simulation of large systems although it requires rather large computational facilities and time. One may anticipate that such approaches become more common, in particular for molecular dynamics studies in competition with quantum computations based on the usual Born-Oppenheimer approximation. It is applied to some biomolecular systems [19] but, owing to the size of the systems of interest and to make them reasonably tractable on present day's computers, most of the applications don't involve a uniform quantum treatment. Instead, they take advantage of a hybrid quantum mechanical/molecular mechanical (QM/MM) computational scheme as developed below [20].

In this paper we present an overview of the various strategies developed to insert a quantum computation on the chemically active part of a large system into a molecular mechanical description, leading to the so-called QM/MM approaches, with a particular emphasis on the Local Self Consistent Field method developed in our group.

The common features of these methods will be presented as an introduction. They will be classified according to the nature of the quantum-classical interactions considered and by the way the connection between these two parts is treated.

13.2 Hybrid Methods for Electronic Properties of the Active Part of a Macromolecule

In the past, when the computational facilities were far below the present ones, some studies were limited to model systems in which the whole system was represented by a small molecule cut from the macromolecule with hydrogen atoms saturating the dangling bonds [21]. But it is well established that the rest of the system cannot be discarded as it interacts with the centre of interest by means of mechanical constraints, electrostatic and induction interactions. It became important to devise some computational schemes allowing taking into account the influence of the part of the system, which cannot be considered at the quantum level. The simple idea is to perform an embedding of the chemical interesting part into its surroundings treated by means of a less demanding method. The most common kind of such low level method refers to molecular mechanics giving rise to a family of treatments known under the generic name of quantum mechanical/molecular mechanical (QM/MM) methods [1, 6].

The interactions occurring in such embedding can be divided in two kinds of contributions: steric (geometric constrains) and electrostatic.

The steric contributions contain the non-bonded van der Waals interactions between MM and QM atoms and, when the MM and the QM parts are chemically bonded, the contributions to the energy changes of the deformations of these bonds (stretching, angular deformation and dihedral rotation). When the interaction of the QM subsystem with the MM one are limited to this kind of contributions, one speaks of *mechanical embedding*.

The electrostatic contributions are mainly the Coulomb interactions between the point charges defined on the MM atoms by the force-field and the electrons and nuclei in the QM region. The interactions between the fixed MM point charges and the electrons of the QM system are taken into account by including the QM/MM electrostatic interaction into the Fock operator. This constitutes the so-called *electrostatic embedding*.

Finally, if the electrostatic properties of the MM part are allowed to change under the influence of the changes of local electric field, one faces the *polarization embedding* which requires a more detailed expression of the perturbation term of the Fock operator.

13.2.1 Mechanical Embedding: The IMOMM-Type Approach

Macromolecules, in particular biomacromolecules such as proteins, have a secondary and tertiary structure which may impose a mechanical constraint on the subsystem to be studied. A typical example, in which these effects are taken into account, is given by the popular IMOMM method [22]. The calculation of the energy *E(S)* of the full system involves 3 steps:

- MM computation of the whole system *S*. Energy: $E_{MM}(S)$
- MM computation of the quantum subsystem *Q* at fixed geometry. Energy: $E_{MM}(Q)$
- High level computation of the quantum subsystem at the same geometry. Energy: $E_{QM}(S)$

The QM/MM energy of the system is then defined by:

$$E(S) = E_{MM}(S) - E_{MM}(Q) + E_{QM}(Q) \tag{13.1}$$

With this approach, the electrostatic interaction between the quantum and the molecular mechanical parts is not taken in account during the quantum computation. It is why the method belongs to a mechanical embedding.

Note that any other QM/MM method reduces to mechanical embedding if the electrostatic interactions between the QM and the MM parts of the system are neglected.

13.2.2 *Electrostatic Embedding, QM/MM Treatment of the System as a Whole*

The electrostatic influence of the atomic charges of the classical moiety on the quantum subsystem may be quite noticeable as it is known in the case of solvent effects treated by a QM/MM approach. It may therefore be important to develop methodologies in which the perturbation created by the classical part is incorporated into the Hamiltonian of the quantum one. Therefore, the wavefunction, energy and energy derivatives are obtained with the help of a general Hamiltonian which writes:

$$H = H_{MM} + H_{QM} + H_{QM-MM} \tag{13.2}$$

where the usual Fock or Kohn-Sham operators derive from this Hamiltonian.

When the classical part of the system is described with a usual force field in which the electrostatic interactions make use of adjusted point charges on the atoms, one speaks of *electrostatic embedding*. The latter contribution is described by the $H_{QM\text{-}MM}$ term of the Hamiltonian, in particular, taking into account the interaction of the spin-orbitals describing the QM electrons with the MM fixed point charges. Most of the QM/MM methods, even the modified IMOMM one, consider this embedding to be enough for a correct description of the system. One has to note that the MM point charges are derived in order to reproduce condensed phase properties, i.e. they are adjusted in such a way that one can consider that they contain an average of the induction interaction. This is the reason why the electrostatic embedding is generally considered without the polarization refinement. However, when the QM part undergoes a severe electronic rearrangement such as a chemical reaction or an electronic transition, the charge distribution of the MM fragment should be allowed to adapt itself to this new situation, at least in the vicinity of the QM part, and the latter is precisely a case in which the polarization embedding becomes crucial. To treat such effects one can describe the MM part by using the so called polarizable force fields, that require a more important computational cost, an alternative solution, particularly efficient in the case of electronic transitions, will be described later in this review.

In such framework, apart from the case of solutions in which the solute can be considered as not chemically bonded to the solvent, the use of QM/MM methods poses the problem of the treatment of the chemical bonds between the quantum and the classical subsystems. A large variety of strategies have been proposed to allow such dichotomy to be as chemically meaningful as possible and the detailed expression of the $H_{QM\text{-}MM}$ term may vary from one method to another.

13.2.2.1 The Link or Connection Atom

The simplest idea to separate a chemically significant group of atoms from the macromolecule is to cut covalent carbon-carbon single bonds and to saturate each dangling bonds by a hydrogen atom called *link atom*. The first attempts to make the molecular mechanics packages able to perform QM/MM computations were built upon this link atom process [23–25]. The term $H_{QM\text{-}MM}$ then takes the form:

$$H_{QM-MM} = -\sum\sum\frac{q_k}{r_{ki}} \qquad (13.3)$$

where q_k stands for the atomic charge of a classical atom k and r_{ki} for the distance between atom k and electron i. A classical van der Waals term corresponding to the non bonded non electrostatic interactions between the quantum and the classical atoms has to be added to the electronic energy obtained after the resolution of the Hartree-Fock or Kohn-Sham equations in order to simulate the interaction which takes place between the actual electronic clouds. After each such step, the bonds between the quantum and the classical subsystems have to be mechanically restored. Special care is required to parameterize these bonds as well as the electrostatic properties of the classical atoms bonded to the quantum part. This problem is carefully analysed in reference [24]. Energy, and energy derivatives are then readily computed and the method allows geometry optimizations and molecular dynamics.

The use of hydrogen link atoms is quite popular, especially when the separation between the subsystems is achieved by cutting carbon-carbon single bonds, such as the C_α–C_β bonds in proteins.

In an alternative slightly different approach [26] the electronic structure of the quantum subsystem is reduced to atomic partial charges which interact with the classical ones.

The presence of this link atom introduces some distortions in the electric field (overpolarization) and some unphysical degrees of freedom which have to be corrected [27], especially when energy derivatives are needed, for geometry optimizations or molecular dynamics.

To overcome the difficulty introduced by the hydrogen link atom which is located in between the actual carbon atoms of the bond to be cut, the use of an adjusted connection atom, located at the place of the first classical atom, has been proposed [28]. The properties of this connection atom are adjusted in such a way that the bond with a neighbouring carbon atom belonging to the quantum subsystem mimics an actual C–C single bond. Regarding the classical subsystem, this atom is assumed to behave like any other similar carbon atom. The first implementation of the method has been done within the framework of semi-empirical quantum methods. The same philosophy led to the use of a "dummy" (CH_3) capping group of the bond linking both systems [28]. Several other tricks have been introduced to overcome the inconveniences of the original link atom scheme [29, 30].

The alternative ab initio methodology takes advantages of the pseudopotential theory by adjusting the properties of a pseudo carbon or pseudo fluorine atom which replaces the actual carbon atom at the boundary position on the MM subsystem [31–33]. Another related approach deals with "Effective group potentials" [34].

13.2.2.2 Methods Involving the Bonds Linking the Subsystems

One of the basic assumptions of the NDDO semi-empirical methods is the orthogonality of the atomic orbitals. Therefore, if one describes the bond between the quantum and the classical part by a strictly localized bond orbital (SLBO), i.e. a localized orbital free of localization tails [35] hosting 2 electrons, the molecular orbitals of the quantum part are orthogonal to the localized one and are able to describe correctly the chemical group of interest. This property has been put forward by Warshel and Levitt in a basic paper [36].

In practice, the localized orbital is extracted from a model small molecule computed with the same semi-empirical method and incorporated into the QM/MM scheme. This is the basis of the Local Self-Consistent Field [37] which consists in solving the Hartree-Fock equations relative to the quantum subsystem in which the electronic interactions with the electrons of the localized bond as well as the classical point charges are included. The adaptation of the PM3 semi-empirical scheme [38] to the AMBER force-field [9] and the analytical expression of the energy derivatives and changes of coordinates led to the so-called classical-quantum force-field (CQFF) [39]. In this force-field, the first atom belonging to the classical part and linked to the quantum part by means of the

SLBO is sometimes called a quanto-classical atom. This modelling tool has demonstrated its efficiency in several applications such as the study of a peptide hydrolysis by an enzyme (Thermolysin) [40]

In this approach, the localized molecular orbital can be analysed as a combination of two hybrid atomic orbitals, one on the quantum atom, one on the classical one, each pointing towards the other atom. On the basis of this remark, Gao et al. [41] developed a generalized hybrid orbital (GHO) method in which 4 hybrid atomic orbitals are defined on the MM (quanto-classical) frontier atom. The orbital pointing towards the QM atom, called the active hybrid orbital, is combined with the atomic orbitals of the QM subsystem in the quantum computation. The three others, called auxiliary orbitals are directed along the bond of the atom with its classical neighbours. They contribute to the total density matrix by three diagonal terms adjusted to reproduce the atomic partial charge of the boundary atom. This algorithm has been coupled with the CHARMM force-field [10].

This scheme has been extended to ab initio [42] and self-consistent-charge density functional tight binding quantum methods [43].

The delicate part of these methods is the parameterization of the boundary MM atom to allow an integrated force field. In the GHO method, adjusted charges and dipoles are introduced along the classical bonds of the boundary atom [44].

13.2.2.3 The Non-empirical Local Self Consistent Field (LSCF)

The basic ideas of the Local Self Consistent Field using a strictly localized bond orbital to describe the bonds separating the quantum to the classical subsystems have been extended to the ab initio or density functional levels of computation [45], but in this case a difficulty appears due to the fact that overlap between atomic orbitals is no longer neglected. Therefore, the molecular orbitals of the quantum subsystem have to be kept orthogonal to each SLBO. This can be achieved by an orthogonalization of the basis set to the SLBOs, but owing to the fact that some functions of the set enter the SLBOs, a linear dependency appears between the orthogonalized functions. This inconvenience can be overcame by means of a canonical orthogonalization [46] which yields a set of orthogonal, linearly independent basis functions which can be used to develop the molecular orbitals of the quantum subsystem.

The solution of the problem consists in optimizing a monodeterminantal wavefunction in the orbital approximation with a peculiarity that some orbitals are externally imposed and that they should remain constant during the optimization procedure, i.e. frozen. It is noteworthy that these frozen orbitals can be monatomic, diatomic or polyatomic [47]. They can also be occupied or empty [48, 49] but in QM/MM computations they are doubly occupied SLBOs. The coefficients defining these frozen orbitals are the only data needed to start the computation.

The detailed mathematical procedure corresponding to this method is developed in reference [45]. Let us just mention the various steps of the algorithm.

We assume that the quantum subsystem is linked to the classical one by L single bonds to which L strictly localized bond orbitals (SLBO) developed in the basis set used for the quantum computation are attached. Without any loss of generality we can assume that these functions are orthogonal

- Step 1: orthogonalization of the N functions of the basis set to the L SLBOs. The initial basis set is transformed by a square $N \times N$ matrix $\boldsymbol{M}$. One gets a set of N non linearly independent functions since there are L combinations built with the same functions.
- Step 2: Perform a canonical orthogonalization [46] by a $(N \times (N - L))$ matrix $\boldsymbol{X}$ so that the rectangular matrix $\boldsymbol{B} = \boldsymbol{M} \cdot \boldsymbol{X}$ directly transforms the initial basis set into the set of basis functions which are linearly independent, orthogonal and orthogonal to the SLBOs.
- Step 3: Define the Fock (or Kohn–Sham) matrix $\boldsymbol{F}$ with the trial density matrix $\boldsymbol{P}^{\boldsymbol{Q}}$ corresponding to the quantum subsystem and the constant density matrix relative to the SLBOs $\boldsymbol{P}^{\boldsymbol{L}}$.
- Step 4: Compute $\boldsymbol{F}' = \boldsymbol{B}^{+} \cdot \boldsymbol{F} \cdot \boldsymbol{B}$ where $\boldsymbol{B}^{+}$ is the transposed of $\boldsymbol{B}$
- Step 5: Diagonalize $\boldsymbol{F}'$: $\Lambda = \boldsymbol{C}^{+} \cdot \boldsymbol{F}' \cdot \boldsymbol{C}$. Λ is the diagonal (eigenvalues) matrix.
- Step 6: Exit test. If not satisfied go back to Step 3.

The Fock matrix is the sum of the usual matrix defined by the basis set functions and the terms representing the interaction of the electrons with the point charges in the classical part of the system.

In order to develop a user friendly code, the transformations of the SLBO when its orientation changes have been implemented as well as its derivatives to allow a full optimization of geometry [50]. In the first implementation of this approach it appeared that during this optimization, the bond linking the quantum to the classical system was systematically too short and the curvature of the potential energy surface (PES) around the minimum was different from that which is obtained by a full QM calculation, whatever be the method used to localize the orbital. It is clear that this is a consequence of the fact that the nucleus of the quanto-classical atom of charge Z is replaced by a charge +1 since this atom contributes for one electron to the SLBO. Therefore, the interaction between nuclei is underestimated and, in addition, the variation of the overlap between the basis functions with respect to the bond length is not taken into account. This defect has been first corrected by introducing a 5 parameters empirical interaction potential for the frontier bond [50] of the form:

$$E_{X-Y} = \left(A + Br + Cr^2\right)e^{Dr} + \frac{E}{r} \tag{13.4}$$

where r is the distance between the two atoms forming the bond. The parameters have been adjusted for any couple of C, O, N atoms, either at the quanto-classical or quantum position and for various hybridization states of the carbon atom.

To set up a non-empirical method and then to avoid the use of an empirical potential, the analysis of the factors affecting the energy variations of the system when the length of the frontier bond is varied showed that the discrepancy comes from the fact that the quanto-classical atom is treated as a pseudo one electron atom. The consideration of the inner shell electrons by means of frozen core orbitals [51, 52] allowed us to solve this problem. In the case of a carbon quanto-classical atom the solution was to switch the nuclear charge to +3 and add two electrons 1s in the QM system. The core electrons are no longer frozen and the description of the frontier bond is still correct. This modified LSCF method is called LSCF + 3 by contrast to the initial approach called LSCF + 1. The scheme has been tested [51] considering the energy variations of the ethane molecule when the C–C bond length is varied in a representation in which one CH_3 group is quantum and one is classical. Results have shown that the LSCF + 3 scheme reproduces well the position of the minimum as well as the curvature obtained with the full quantum results, the error on the minimum distance being of the order of 0.1 Å. Of course, the agreement ceases to be good at long interatomic distances because the SLBO extracted from a model molecule at the equilibrium bond length is only valid in the vicinity of the interatomic distance at which it has been obtained. This scheme has been successfully applied to the C–N peptide bond where the C atom is at the MM frontier. It has also been extended to the C–N peptide bonds in which the quanto-classical atom is the nitrogen atom [5, 53]. In this case, it appears that in addition to the electron involved in the frontier bond and the 1s electrons of the inner shell, one must consider two extra valence electrons in the orbital which is conjugated with the C = O π orbital. This procedure has been tested on a simple capped diglycine system [5]. The full QM results, obtained at DFT level with the B3LYP exchange-correlation functional and the 6-311G basis set, are compared with the case where one of the peptides is treated at MM level (CHARMM force field). In particular we compare the situation where the carbon (LSCF + 3), or the nitrogen (LSCF + 5) atom are treated as the quanto-classical atom, respectively. The agreement on the equilibrium bond length and on the force constant rests within the error bars of the force-field. One therefore developed a LSCF + 5 method, free of extra parameters, which opens the way to a symmetric description of the amino acid residues without any arbitrariness. In a QM/MM framework this non empirical LSCF method opens the way to directly cut through a peptide bond, currently taking into account the electron delocalization occurring at the amide bond. It has to be noted that link atom approaches, which use hydrogen atoms to saturate the dangling bonds are of course not able to reproduce this feature.

This reasoning is expected to work for heavier quanto classical atoms by considering more core electrons. It has already been applied to silicon with a +11 nuclear charge and a pseudo-potential describing the 10 core electrons [54].

The LSCF method developed here uses strictly localized molecular orbitals which are obtained by using any standard localization procedure [55–59] followed by the removal of the tails of orthogonalization and delocalization, and renormalizing. A study of the results obtained on small reference molecules by using these localization methods has been done [60], and the comparison with the use of extremely localized molecular orbitals (ELMO) [61–63] allows us to assess the reliability of the methodology. The study shows that the results obtained are rather comparable, but the use of the ELMO technique gives the best results compared with full QM computations.

Various possible improvements have been considered. For instance mixing bonding and antibonding bond orbitals [36] led to an Optimized LSCF method [64] in which each SLBO is combined with its corresponding Strictly Localized Anti Bonding Orbital.

13.2.3 Polarization Embedding

Most of the standard force-fields are now perfectly validated for modelling the properties of systems in the ground state since their parameters, in particular those describing the electrostatic interactions are derived from quantum mechanical computations on model molecules so that they are considered as taking into account an averaged value of induced polarization. However, several force-fields add induced polarization effects to the usual electrostatic interactions [12, 13, 65–70] in view of having a more realistic model of interatomic interactions. The methods presented above are perfectly adapted to working with such MM tools, although the computational time becomes obviously larger so that they are not used in usual ground state studies. There is nevertheless some cases in which electronic polarization changes cannot be ignored such as the study of vertical electronic transitions in chromophores or solutes [11, 68, 69] because the induced electronic moments of the surroundings are able to follow the fast changes of the solute's properties, contrarily to electrostatic interactions which can be considered as frozen during the electronic transition (Franck-Condon principle) because they are bound to the motion of the molecules. The only quantity required to take this effect into account is the change of interaction energy which can be evaluated simply by a non-polar, polarizable continuum characterized by a dielectric constant extrapolated at infinite frequency, or by the square of the refractive index. The quantum subsystem is assumed to be placed in a cavity surrounded by this continuum. This is the principle of the so-called "Electronic Response of the Surroundings" (ERS) [71] which has been applied successfully to the modelling of absorption spectra and circular dichroism of biological macromolecules [71–76] showing that the role of the induced polarization changes is far from being negligible. It is also noteworthy to cite that the addition of the polarization continuum in the computation of ground state does not introduce significant changes. Indeed, as shown by Very et al. [74], if one considers the QM atomic charges, the variation induced by ERS is only of 0.1 %, while the effect of changing the classical force field may induces variations of about 5 %.

13.3 Applications of the Non Empirical Local Self Consistent Field to Bio macromolecular systems: Unravelling DNA Photosensitization

QM/MM and the non-empirical Local Self Consistent Field methodology described above has been applied to various systems, in particular biological ones. The inclusion of ERS has also allowed to efficiently tackle excited state properties and hence modelling photophysics and photochemistry of complex systems. As an example we may cite the modelling of physico-chemical properties of proteinic compounds [72], in particular plastocyanins [73] together with the influence of specific mutations on the active site [77], the spectroscopy of human serum albumin [5, 78] and the interactions of various exogenous molecules with DNA [74, 76–83]. As an example of application, we develop below the modelling of the photophysical and photochemical properties of drugs interacting with DNA. In particular we are interested to the so called DNA photosensitization, in which different chromophores (the sensitizers) interact with DNA and may, upon absorption of visible or UVA light, induce the formation of DNA lesions, such as cyclobutane dimers [84, 85]. The mechanisms by which the sensitization proceeds are quite complex, usually involve the participation of the sensitizer triplet manifold, and may include excited state electron transfer (Type I photosensitization); production of reactive oxygen species, such as $^1O_2$ followed by subsequent DNA oxidation (Type II photosensitization); or energy transfer to the DNA nucleobases (Triplet Photosensitization).

The lesions produced upon photosensitization may be so widespread as to induce the cellular apoptosis. As a consequence photosensitization is nowadays used in clinical anticancer therapy following the protocol of the so-called photodynamic therapy. Indeed, photodynamic therapy, thanks to the combined use of light and sensitizer has an incomparable higher spatial selectivity compared to conventional chemotherapeutic drugs, and hence secondary effects are strongly limited. The complex processes taking place during photosensitization require a very carefully tailored modelling able to provide on the one hand a balanced description of all the excited states involved, and on the other hand to properly take into account the effects of the strong inhomogeneous environment constituted by DNA. Indeed, the DNA environment is far from being innocent and in such cases, such as the very well known "light-switch effect", can completely switch on or off the sensitizer's luminescence. [86] Furthermore dynamic and structural effects, in particular connected with the formation of non-covalent DNA/sensitizer aggregates need to be properly described. In this situation QM/MM methods and their coupling with classical molecular dynamics (MD) represent the best strategy to tackle such a problem and are able to provide a proper description of the photosensitization with an atomistic and electronic resolution. As representative examples in this book chapter we report the study of two well known organic DNA sensitizers, harmane cation [87, 88] and benzophenone [89, 90].

13.3.1 Spectroscopical Properties of Harmane Interacting with DNA

Harmane (Fig. 13.1) is a member of the β-carboline family and is known to interact with DNA acting as a photosensitizer; in particular it may induce the cleavage of the glycosidic bond. Thanks to its interaction with DNA it has remarkable antibiotic and antiparasitary properties that make it attractive for the pharmaceutical industry. [91, 92] In aqueous solution neutral harmane (Har) exists in equilibrium with its cation (Har +) and both species contribute to the absorption spectrum. In particular the cation gives a maximum, due to the absorption to the lowest S1 state at about 370 nm, while the neutral has a blue shifted band at 340 nm. Upon DNA addition and because of the strongly negative charged backbone, the equilibrium is displaced toward the cationic form, which is the one interacting with the macromolecule.

Etienne et al. [87] have reported a careful QM/MM study of the spectral properties of the harmane cation in water and interacting with DNA. In particular as far as the aqueous solution is concerned, absorption spectrum can be obtained with two different protocols:

- In the "static" approach the ground state equilibrium geometry is optimized, with solvent treated implicitly with a continuum method, for instance polarizable continuum method (PCM). Excited states are then obtained as vertical transition form the equilibrium geometry
- In the "dynamic" approach the chromophore is explicitly embedded in a solvation box and the ground state conformational space is explored by means of classical MD. Representative snapshots are then extracted and excited states are

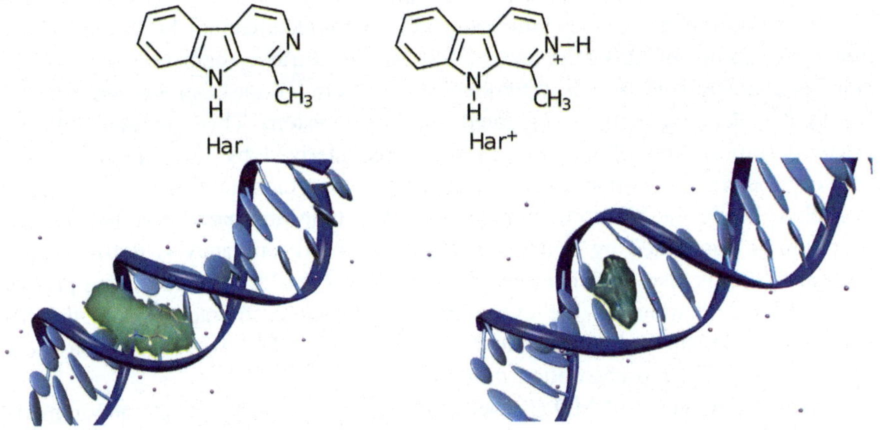

Fig. 13.1 2D Chemical Structure of Har and Har+. Representative structure of the sensitizers interacting with DNA, extracted from molecular dynamics are also given. Two stable interaction modes (intercalation and minor groove-binding) have been identified

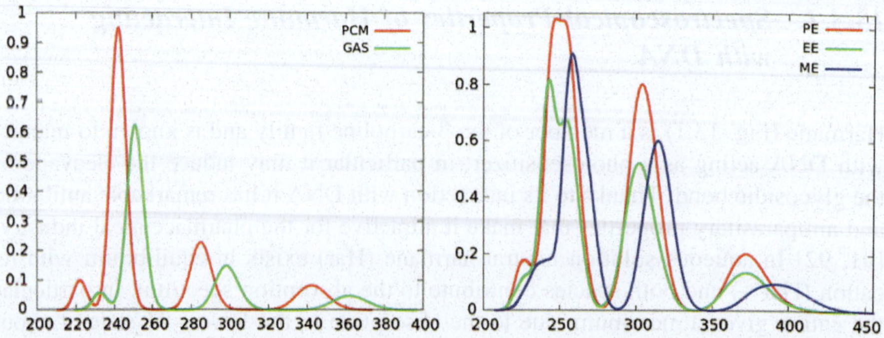

Fig. 13.2 Static (*left*) and Dynamic (*right*) TD-DFT spectra for harmane in water solution [86]. Wavelengths in nm, intensities in arbitrary units. PE: polarization embedding, EE: electrostatic embedding, ME: mechanical embedding

calculated at QM/MM level of theory on each of them. The final spectrum will be the convolution of vertical transitions of individual snapshots.

The spectra calculated for Har+ in water solution at time dependent density functional theory (TD-DFT) with the static and the dynamic approaches are reported in Fig. 13.2. Surprisingly, the static approach fails totally in reproducing the experimental results, and gives an absorption band far too much blue-shifted. Paradoxically, even the gas-phase calculation, i.e. without the PCM approach, seems to give a better agreement with experiment. Furthermore, as reported in [87] this fact should not be ascribed to a failure of the TD-DFT since different functionals all draw the same picture. On the contrary, the dynamical approach gives a very good agreement with the experiment provided that the electrostatic and polarizable embedding are taken into account.

It appears obvious that dynamic and vibronic effects cannot be discarded, and that the coupling of QM/MM and MD is able to retrieve them properly. Indeed, Har+ is characterized by a low frequency ($\sim$73 cm^{-1}) out-of-plane vibration that breaks the planarity of the conjugated three-ring system. This situation induces a larger destabilization of the ground state characterized by bonding interactions between the rings [87] than of the excited states in which antibonding orbitals are occupied, hence the red-shift compared to the static approach. The fact that the spectrum is dominated by dynamic effects and the discrepancy is not due to the solvent model is also evidenced by the results of Fig. 13.3. Indeed if we perform pure QM calculations on the chromophore geometries extracted from the MD simulation with the solvent implicitly described by PCM we once again nicely reproduce the 370 nm absorption maximum.

Furthermore the use of MD allowed to evidence the two stable interaction modes between Har+ and DNA, namely intercalation and minor-groove binding (Fig. 13.1 lower panel). Although, a systematic calculation of the binding free energy was not performed the stability of the two modes appears comparable. The dynamic approach also nicely reproduce the spectroscopical features, that once again do not

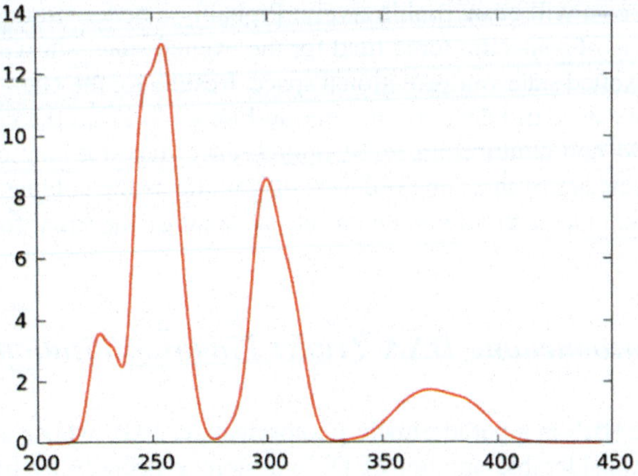

Fig. 13.3 Har+ TD-DFT spectrum obtained as a convolution of vertical transitions from the MD trajectory. The solvent is treated implicitly with PCM. Wavelengths in nm intensities in arbitrary units

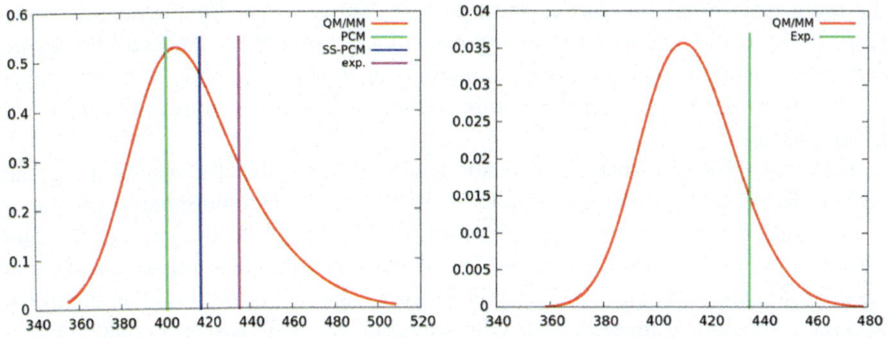

Fig. 13.4 Dynamic QM/MM TD-DFT calculation of the Har+ fluorescence spectra in water (*left*) and in DNA (*right*). For water solution the comparison with experimental results and with equilibrium and non-equilibrium PCM is given. Wavelengths in nm, intensities in arbitrary units

allow to discriminate between the two modes, suggesting a coexistence of the two aggregates.

The same considerations sketched for the absorption spectrum also hold for emission and namely fluorescence [89]. In the static approach the spectrum can easily be obtained by optimizing the geometry of the excited state and taking into account solvent relaxation using equilibrium PCM. Conversely, in the dynamic QM/MM approach the situation is much more complex. From a methodological point of view the emission spectrum at QM/MM level can be obtained by optimizing the excited state geometries for each of the snapshots, however relaxing the solvent would give rise to an optimization with too many degrees of freedom, so that the computational cost will grow prohibitively. Probably a better strategy would be the parameterization of a specific force field for the excited state, allowing to efficiently sampling the excited state conformational space. Results for the Har+ fluorescence in water and in presence of DNA are reported in Fig. 13.4. From the comparison with equilibrium and non-equilibrium PCM fluorescence maxima, one can see that the dynamical effects are of the same order of magnitude as the solvent relaxation, and hence both phenomena should be taken into account on an equal footing [89].

13.3.2 Benzophenone DNA Triplet Photosensitization

Benzophenone (BP) is a paradigmatic photosensitizer [93], whose action is known to produce a considerable amount of DNA lesions especially by triplet photosensitization. Indeed BP, like many aryl-ketones is known to have an easy intersystem crossing with efficient population of the triplet manifold [94, 95]. The BP-centred triplet state may subsequent transfer its energy to a nearby thymine triggering a cascade of events leading to cyclobutane thymine dimers. [93] Despite its paradigmatic role as sensitizer no structure of the BP/DNA was known or crystallized, the use of MD simulation has allowed [88] to identify two stable interaction modes (Fig. 13.5). If the first mode, minor groove binding is rather standard, the second one represented an entirely novel non-covalent interaction mode that was called double-insertion. In this mode an entire base-pair is ejected from the Watson and Crick pairing.

QM and QM/MM analysis also allowed to identify that the ejected bases are stabilized by strong hydrogen bonds with the backbone phosphate, so that to compensate the rather large helical distortion. In addition to the simple structure, once again a proper QM/MM modelling of the spectroscopical properties shows a specific signature on the UVA/visible range of the electronic circular dichroism spectra of the double-inserted benzophenone (Fig. 13.6). This aspect could allow an easy and straightforward experimental discrimination between the two modes.

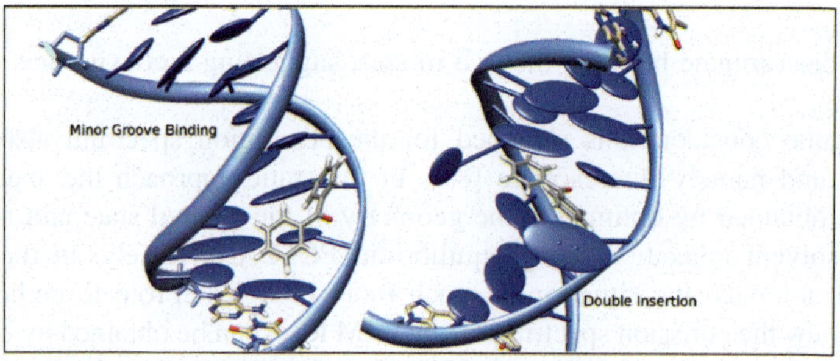

Fig. 13.5 Two representative snapshots of minor groove-binding (*left*) and double-insertion (*right*) obtained with classical MD [88]

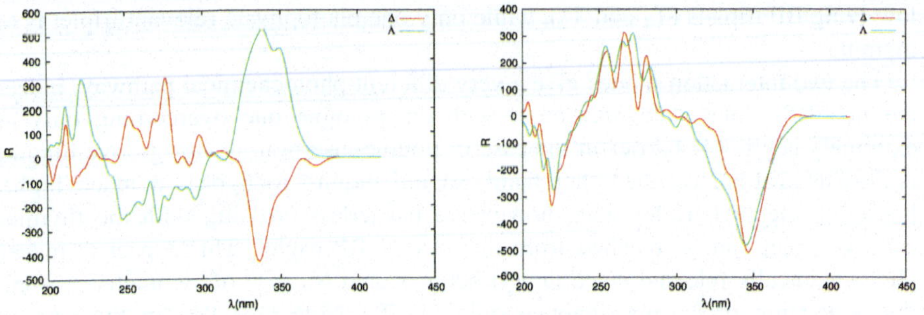

Fig. 13.6 QM/MM and TD-DFT electronic circular dichroism for the minor groove (*left*) and double-insertion (*right*). Wavelengths in nm intensities in arbitrary units

In addition to the spectroscopical and structural properties it is important to analyse the photophysical and photochemical pathways leading to triplet photosensitization and the differences between the two modes. Indeed, because of the close proximity with the DNA nucleobases the dominant energy transfer mechanism will be of the Dexter type [90]. To analyse this situation Dumont et al. [90] have followed the potential energy profile of the different triplet states on a approximate coordinate ξ connecting the equilibrium geometry of the triplet centred on BP (ξ = 0) with the one of the triplet centred on the thymine (ξ = 1). Results are presented in Fig. 13.7, the change of the wavefunction nature being monitored by natural transition orbital (NTO) [96]. One may recognize the presence of two

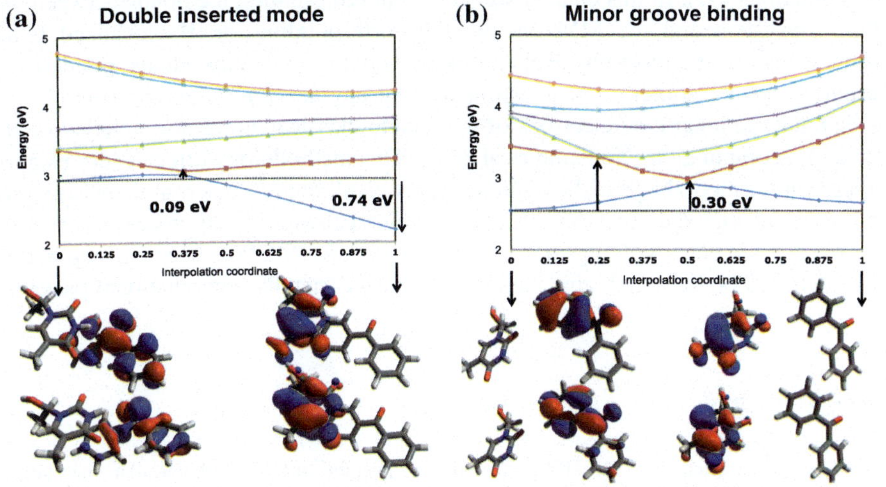

Fig. 13.7 Potential energy surfaces along the simplified coordinate ξ describing the triplet transfer from BP to thymine. Scheme **a** represents the double inserted mode, and **b** the minor-groove binding. The nature of the lowest triplet is evidenced through occupied and virtual NTO reported for ξ = 0 and 1

low-lying BP triplets (T_1 and T_2), while only one photophysic relevant triplet is on thymine.

The two interaction modes give a very different photochemical pathway, indeed the first BP triplet state may cross with the thymine one overcoming only an extremely negligible barrier in the case of double-insertion. Minor groove-binding photosensitization, on the other hand, should require to bypass a much higher barrier in the first triplet state, however a barrierless crossing with the thymine centred triplet can be reached from the second BP triplet. This aspect is indeed photochemically relevant since in aryl-ketones, and because of symmetry reasons, the population of T_1 may happen only via T_2. Note also that in the case of double-insertion T_2 crosses the thymine centred triplet already very close to $\xi = 0$, hence a second extremely efficient channel to the population of the thymine triplet may be hypothesized. Overall, this study has for the first time unravelled at atomistic and electronic levels the photosensitization processes and has confirmed the great efficiency of BP in inducing the population of the DNA triplets.

13.4 Conclusion

The Local Self Consistent Field (LSCF) methodology is based upon the very chemically intuitive concept of localized bond orbital. It opens the way of various studies in which part of an electronic system deserves a special treatment. For instance, it has been used to study the core-ionized and the core-excited states of macromolecules [48, 49]. It is also well adapted to constrained density functional computations [97]. In this chapter we show that representing the bonds between the QM subsystem and the MM one by the corresponding strictly localized bond orbitals insures the most physical continuity between both subsystems. In addition, it allows a full geometry optimization without any empirical parametrization provided that the inner core electrons of the quanto-classical atoms are included in the quantum computation (Non empirical SCF). This methodology offers a very elegant approach for modelling biochemical reactions involving bio-macromolecules as it is the case for enzymes or DNA. It also permits a theoretical approach of spectroscopic phenomena involving such systems [73–83, 87–90, 96]. The next step in this research will be to include this method into a user friendly computational package.

References

1. Monard G, Merz KM Jr (1999) Combined quantum mechanical/molecular mechanical methodologies applied to biomolecular systems. Acc Chem Res 32:904–911
2. Lin H, Truhlar DG (2007) QM/MM: what we have learned, where are we, and where do we go from here? Theor Chem Acc 117:185–199
3. Senn HM, Thiel W (2009) QM/MM Methods for biomolecular systems. Angew Chem Int Ed 48:1198–1229

4. Gordon MS, Fedorov DG, Pruitt SR, Slipchenko LV (2012) Fragmentation methods: a route to accurate calculations on large molecular systems. Chem Rev 112:632–672
5. Monari A, Rivail JL, Assfeld X (2013) Theoretical modeling of large molecular systems. Advances in the local self consistent field method for mixed quantum mechanics/molecular mechanics calculations. Acc Chem Res 46:596–603
6. Akimov AV, Prezhdo OV (2015) Large scale computation in chemistry: a bird's eye view of a vibrant field. Chem Rev. doi:10.1021/cr500524c
7. van Gunsteren WF, Berendsen HJC (1987) Groningen molecular simulation (GROMOS) library manual biomos, Groningen
8. Allinger NL, Yuh YH, Lii JH (1989) Molecular mechanics. The MM3 force field for hydrocarbons. J Am Chem Soc 111:8551–8566
9. Pearlman DA, Case DA, Caldwell JW, Ross WS, Cheatham TE III, DeBolt S, Ferguson D, Seibel G, Kollman P (1995) AMBER, a package of computer programs for applying molecular mechanics, normal mode analysis, molecular dynamics and free energy calculations to simulate structural and energetic properties of molecules. Comput Phys Commun 91:1–41
10. MacKerell AD Jr, Bashford D, Bellott M, Dunbrack RL Jr, Evanseck JD, Field MJ, Fischer S, Gao J, Guo H, Ha S, Joseph-McCarthy D, Kuchnir L, Kuczera K, Lau FTK, Mattos C, Michnick S, Ngo T, Nguyen DT, Prodhom B, Reiher WE III, Roux B, Schlenkrich M, Smith JC, Stote R, Straub J, Watanabe W, Wiórkiewicz-Kuczera J, Yin D, Karplus M (1998) All-atom empirical potential for molecular modeling and dynamics studies of proteins. J Phys Chem B 102:3586–3616
11. Dang LX, Chang T-M (1997) Molecular dynamics study of water clusters, liquid and liquid-vapor interface of water with many-body potentials. J Chem Phys 106:8149–8159
12. Dupuis M, Aida M, Kawashima Y, Hirao K (2002) A polarizable mixed Hamiltonian model of electronic structure for micro-solvated excited states. I Energy and gradients formulation and application to formaldehyde ($^1A_2$). J Chem Phys 117:1242–1255
13. Gresh N, Andrés Cisneros G, Darden TA, Piquemal J-P (2007) Anisotropic, polarizable molecular mechanics studies of inter- and intramolecular interactions and ligand-macromolecule-complexes. A bottom-up strategy. J Chem Theory Comput 3:1960–1986
14. Dixon SL, Merz KM Jr (1996) Semiempirical molecular-orbital calculations with linear system size scaling. J Chem Phys 104:6643–6649
15. Daniels AD, Millam JM, Scuseria GE (1997) Semiempirical methods with conjugate gradient density matrix search to replace diagonalization for molecular systems containing thousand of atoms. J. Chem. Phys. 107:425–431
16. Yang W, Lee TS (1995) A Density matrix divide-and-conquer approach for electronic structure calculations of large molecules. J Chem Phys 103:5674–5678
17. Monard G, Bernal-Uruchurtu MI, van der Vaart A, Merz KM Jr, Ruiz-López MF (2005) Simulation of liquid water Using semiempirical hamiltonians and the divide and conquer approach. J Phys Chem A 109:3425–3432
18. Car R, Parrinello M (1985) Unified approach for molecular dynamics and density-functional theory. Phys Rev Lett 55:2471–2474
19. Carloni P, Rothlisberger U, Parrinello M (2002) The role and perspective of ab initio molecular dynamics in the study of biological systems. Acc Chem Res 35:455–464
20. Garrec J, Patel C, Rothlisberger U, Dumont E (2012) Insight into intrastrand cross-link lesions of DNA from QM/MM molecular dynamics simulations. J Am Chem Soc 134:2111–2119
21. Mestres J, Duran M, Bertran J (1996) Characterization of the transition state for the hydride transfer in a model of the flavoprotein reductase class of enzymes. Biorgan Chem 24:69–80
22. Maseras F, Morokuma K (1995) IMOMM, a new integrated ab initio + molecular mechanics geometry optimization scheme of equilibrium structure and transition states. J Comput Chem 16:1170–1179
23. Singh UC, Kollman PA (1986) A combined quantum ab initio mechanical and molecular mechanical method for carrying out simulations on complex molecular systems: Applications to the $CH_3Cl + Cl^-$ exchange reaction and gas phase protonation of polyethers. J Comput Chem 7:718–730

24. Field MJ, Bash PA, Karplus M (1990) A combined quantum mechanical and molecular mechanical potential for molecular dynamics simulations. J Comput Chem 11:700–733
25. Eurenius KP, Chatfield DC, Brooks BR (1996) Enzyme mechanisms with hybrid quantum and molecular mechanical potentials. I. Theoretical considerations. Int J Quantum Chem 60:1189–1200
26. Kaminski GA, Jorgensen WL (1998) A quantum mechanical and molecular mechanical method based on CM1A charges: applications to solvent effects on organic equilibria and reactions. J Phys Chem B 102:1787–1796
27. Ferré N, Olivucci M (2003) The amide bond: pitfalls and drawback of the link atom scheme. J Mol Struct (Theochem) 632:71–82
28. Antes I, Thiel W (1999) Adjusted connection atoms for combined quantum mechanical and molecular mechanical methods. J Phys Chem A 103:9290–9295
29. Swart M (2003) AddRemove: a new link model for use in QM/MM studies. Int J Quantum Chem 91:177–183
30. Das D, Eurenius KP, Billings EM, Sherwood P, Chatfield DC, Hodoscek M, Brook BR (2002) Optimization of quantum mechanical molecular mechanical partitioning schemes: Gaussian delocalization of molecular mechanical charges and the double link atom method. J Chem Phys 117:10534–10547
31. Di Labio GA, Hurley MM, Christiansen PA (2002) Simple one-electron quantum capping potentials for use in hybrid QM/MM studies of biological molecules. J Chem Phys 116:9578–9584
32. Zhang Y, Lee T-S, Yang W (1999) A pseudobond approach to combining quantum mechanical and molecular mechanical methods. J Chem Phys 110:46–54
33. Xiao C, Zhang Y (2007) Design-atom approach for quantum mechanical/molecular mechanical covalent boundary: a design-carbon atom with five valence electrons. J Chem Phys 127:124102
34. Poteau R, Ortega Y, Alary F, Ramirez Solis A, Barthelat J-C, Daudey JP (2001) Effective group potentials. 1. Method J Phys Chem A 105:198–205
35. Ferenczy GG, Rivail J-L, Surján PR, Náray-Szabó G (1992) NDDO fragment self-consistent-field approximation for large electronic systems. J Comput Chem 13:830–837
36. Warshel A, Levitt M (1976) Theoretical studies of enzymic reactions: Dielectric, electrostatic and steric stabilization of the carbonium ion in the reaction of lysozyme. J Mol Biol 103:227–249
37. Théry V, Rinaldi D, Rivail J-L, Maigret B, Ferenczy GG (1994) Quantum mechanical computations on very large molecular systems: the local self-consistent field method. J Comput Chem 15:269–282
38. Stewart JJP (1989) Optimization of parameters for semiempirical methods I. Method J Comput Chem 10:209–220
39. Monard G, Loos M, Théry V, Baka K, Rivail JL (1996) Hybrid classical quantum force field for modeling very large molecules. Int J Quantum Chem 58:153–159
40. Antonczak S, Monard G, Ruiz-López MF, Rivail J-L (1998) Modeling of peptide hydrolysis by thermolysin. A semiempirical and QM/MM Study. J Am Chem Soc 120:8825–8833
41. Gao G, Amara P, Alhambra C, Field MJ (1998) A generalized hybrid orbital (GHO) method for the treatment of boundary atoms in combined QM/MM calculations. J Phys Chem A 102:4714–4721
42. Pu J, Gao J, Truhlar DG (2004) Generalized hybrid orbital (GHO) method for combining ab initio Hartree-Fock wave functions with molecular mechanics. J Phys Chem A 108:632–650
43. Pu J, Gao J, Truhlar DG (2004) Combining self-consistent-charge density-functional tight-binding (SCC-DFTB) with molecular mechanics by the generalized hybrid orbital (GHO) method. J Phys Chem A 108:5454–5463
44. Pu J, Truhlar DG (2005) Redristributed charge and dipole schemes for combined quantum mechanical and molecular mechanical calculations. J Phys Chem A 109:3991–4004
45. Assfeld X, Rivail J-L (1996) Quantum chemical computations on parts of large molecules: the ab initio local self consistent field method. Chem Phys Lett 263:100–106

46. Szabo A, Ostlund NS (1989) Modern quantum chemistry. Mc Graw Hill, New York, p 144
47. Moreau Y, Loos P-F, Assfeld X (2004) Solvent effects on the asymmetric Diels-Alder reaction between cyclopentadiene and (-)-menthyl acrylate revisited with the three-layer hybrid local self-consistent field/molecular mechanics/self-consistent reaction field method. Theor Chem Acc 112:228–239
48. Ferré N, Assfeld X (2002) Application of the local self-consistent-field method to core-ionized and core-excited molecules, polymers, and proteins: True orthogonality between ground and excited states. J Chem Phys 117:4119–4125
49. Loos PF, Assfeld X (2007) Core-ionized and core-excited states of macromolecules. Int J Quantum Chem 107:2243–2252
50. Ferré N, Assfeld X, Rivail J-L (2002) Specific force field determination for the hybrid ab initio QM/MM LSCF method. J Comput Chem 23:610–624
51. Fornili A, Loos P-F, Sironi M, Assfeld X (2006) Frozen core orbitals as an alternative to specific frontier bond potential in hybrid quantum mechanics/molecular mechanics methods. Chem Phys Lett 427:236–240
52. Loos P-F, Fornili A, Sironi M, Assfeld X (2007) Removing extra frontier parameters in QM/MM methods: a tentative with the local Self-consistent field approach. Comput Lett 4:473–486
53. Loos P-F, Assfeld X (2007) On the frontier bond location in the QM/MM description of peptides and proteins. AIP Conf Proc 963:308–315
54. Ferré N, Assfeld X (2003) A new three-layer hybrid method (LSCF/MM/Madelung) devoted to the study of chemical reactivity in zeolites. Preliminary results. J Mol Struct (Theochem) 632:83–90
55. Boys SF (1960) Construction of some molecular orbitals to be approximately invariant for changes from one molecule to another. Rev Mod Phys 32:296–299
56. Foster JM, Boys SF (1960) Canonical configurational interaction procedure. Rev Mod Phys 32:300–302
57. Magnasco V, Perico A (1967) Uniform localization of atomic and molecular orbitals. I. J Chem Phys 47:971–981
58. Weinstein H, Pauncz R (1968) Molecular orbital set determined by a localization procedure. Symp Faraday Soc 2:23–31
59. Pipek J, Mezey PG (1989) A fast intrinsic localization procedure applicable for ab initio and semiempirical linear combination of atomic orbital wave functions. J Chem Phys 90:4916–4926
60. Fornili A, Moreau Y, Sironi M, Assfeld X (2006) On the suitability of strictly localized orbitals for hybrid QM/MM calculations. J Comput Chem 27:515–523
61. Sironi M, Famulari A (2000) An orthogonal approach to determine extremely localised molecular orbitals. Theor Chem Acc 103:417–422
62. Fornili A, Sironi M, Raimondi M (2003) Determination of extremely localized molecular orbitals and their application to quantum mechanics/molecular mechanics methods and to the study of intramolecular hydrogen bonding. J Mol Struct (Theochem) 632:157–172
63. Genoni A, Ghitti M, Pieraccini S, Sironi M (2005) A novel extremely localized molecular orbitals based technique for the one-electron density matrix computation. Chem Phys Lett 415:256–260
64. Loos PF, Assfeld X (2007) Self-consistent strictly localized orbitals. J Chem Theory Comput 3:1047–1053
65. Celebi N, Ángyán JG, Dehez F, Millot C, Chipot C (2000) Distributed polarizabilities derived from induction energies: a finite perturbation approach. J. Chem. Phys. 112:2709–2717
66. Dehez F, Soetens JC, Chipot C, Ángyán JG, Millot C (2000) Determination of distributed polarizabilities from a statistical analysis of induction energies. J Phys Chem A 104:1293–1303
67. Dehez F, Chipot C, Millot C, Ángyán JG (2001) Fast and accurate determination of induction energies: reduction of topologically distributed polarizability models. Chem Phys Lett 338:180–188

68. Thompson MA (1996) QM/MMpol: a consistent model for solute/solvent polarization. Application to the aqueous solvation and spectroscopy of formaldehyde, acetaldehyde and acetone. J Phys Chem 100:14492–14507
69. Dupuis M, Kawashima Y, Hirao K (2002) A polarizable mixed Hamiltonian model of electronic structure for micro-solvated excited states. II Application to the blue shift of the H_2CO $1(\pi^* \leftarrow n)$ excitation in water. J Chem Phys 117:1256–1268
70. Xie W, Song L, Truhlar DG (2008) Gao J The variational explicit polarization potential and analytical first derivatives of energy: toward a new generation force field. J Chem Phys 128:234108
71. Jacquemin D, Perpète EA, Laurent AD, Assfeld X, Adamo C (2009) Spectral properties of self-assembled squaraine-tetralactam: a theoretical assessment. Phys Chem Chem Phys 11:1258–1262
72. Laurent AD, Assfeld X (2010) Effect of the enhanced cyan fluorescent proteinic framework on the UV/visible absorption spectra of some chromophores. Interdiscip Sci Comput Life Sci 2:38–47
73. Monari A, Very T, Rivail J-L, Assfeld X (2012) A QM/MM study on the spinach plastocyanin: Redox properties and Absorption spectra. Comp. Theoret. Chem. 990:119–125
74. Very T, Despax S, Hébraud P, Monari A, Assfeld X (2012) Spectral properties of polypyridyl Ruthenium complex intercalated in DNA: theoretical insights on the surrounding effects for $[Ru(dppz)(bpy)_2]^{2+}$ Phys. Chem Chem Phys 14:12496–12504
75. Chantzis A, Very T, Monari A, Assfeld X (2012) Improved treatment of surrounding effects: UV/vis absorption properties of a solvated Ru(II) complex. J Chem Theory Comp 8:1536–1541
76. Gattuso H, Assfeld X, Monari A (2015) Modeling DNA electronic circular dichroism by QM/MM methods and Frenkel Hamiltonian. Theor Chem Acc 134:36
77. Monari A, Very T, Rivail J-L, Assfeld X (2012) Effects of mutations on the absorption spectrum of copper proteins: A QM/MM study. Theor Chem Acc 131:1221
78. Etienne T, Assfeld X, Monari A (2014) QM/MM calculation of absorption spectra of complex systems: the case of human serum albumin Comp. Theoret Chem 1040–1041:360–366
79. Chantzis A, Very T, Despax S, Issenhut J-T, Boeglin A, Hébraud P, Pfeffer M, Monari A, Assfeld X (2014) UV-vis absorption spectrum of a novel Ru(II) complex intercalated in DNA: [Ru(2,2′-bipy)(dppz)(2,2′-ArPy)]. J Mol Model 20:2082
80. Chantzis A, Very T, Daniel C, Monari A, Assfeld X (2013) Theoretical evidence of photo-induced large transfer from DNA to intercalated ruthenium (II) organometallic complex Chem. Phys Lett 578:133–137
81. Very T, Ambrosek D, Otsuka M, Gourlauen C, Assfeld X, Monari A, Daniel C (2014) Photophysical properties of ruthenium (II) polypyridyl DNA intercalators: effects of the molecular surroundings investigated by theory. Chem Eur J 40:12901–12909
82. Huix Rottlant M, Dumont E, Ferré N, Monari A (2015) Photophysics of acetophenone interacting with DNA: why the road to photosensitization is open. Photochem Photobiol 91:323–330
83. Dumont E, Monari A (2015) Interaction of palamtine with DNA: an environmentally controlled drug. J Phys Chem B 119:410–419
84. Epe B (2012) DNA damage spectra induced by photosensitization Photochem. Photobiol Sci 11:98–106
85. Cadet J, Wagner R (2013) DNA base damage by reactive oxygen species, oxidizing agents, and UV radiation Cold Spring Harb. Perspect Biol 5:a12599
86. Friedman AE, Chambron JC, Sauvage J-P, Turro NJ, Barton JK (1990) Molecular light switch for DNA: $Ru(bipy)_2(dppz)$. J Am Chem Soc 112:4960
87. Etienne T, Very T, Perpète EA, Monari A, Assfeld X (2013) A QM/MM study of harmane in water solution and interacting with DNA: the crucial role of dynamic effects. J Phys Chem B 117:4973–4980

88. Etienne T, Gattuso H, Monari A, Assfeld X (2014) QM/MM modelling of harmane cation fluorescence spectrum in water solution and interacting with DNA Comp. Theor Chem 1040–1041:367–372
89. Dumont E, Monari A (2013) Benzophenone and DNA: evidence for a double insertion mode and its spectral signature. J Phys Chem Lett 4:4119–4124
90. Dumont E, Wibowo M, Roca-Sanjuan D, Garavelli M, Assfeld X, Monari A (2015) esolving the benzophenone DNA-photosensitization mechanism at QM/MM level. J Phys Chem Lett 6:576–580
91. Gonzalez MM, Arnjberg J, Denofrio MP, Erra-Balsells R, Ogilby PR, Cabreizo FM (2009) One- and two-photon excitation of of b-carbline in aqueous solution: ph-dependent spectroscopy, photochemistry and photophysics. J Phys Chem A 113:6648–6656
92. Paul BK, Guchait N (2011) Exploring the strength, mode, dynamics and kinetics of binding interaction of a cationic biological photosensitizers with DNA: implication on dissociation of the drug-DNA complex with detergent sequestration. J Phys Chem B 115:11938–11949
93. Cuquerella MC, Lhiaubet-Vallet V, Cadet J, Miranda MA (2012) Benzophenone photosensitized DNA damage. Acc Chem Res 45:1558–1570
94. Huix-Rottlant M, Siri D, Ferré N (2013) Theoretical study of the photochemical generation of triplet acetophenone Phys. Chem Chem Phys 15:19293–19300
95. Sergentu D-C, Maurice R, Havenith RWA, Broer R, Roca-Sanjuan D (2014) Computational determination of the dominant triplet population mechanism in photoexcited benzophenone. Phys Chem Chem Phys 16:25393–25403
96. Etienne T, Assfeld X, Monari A (2014) Toward a quantitative assessment of electronic transition's charge-transfer character. J Chem Theory Comp 10:3896–3905
97. Kalduk B, Kowalczyk T, Van Voorhis T (2012) Constrained density functional theory. Chem Rev 112:321–370

Chapter 14
Computational Study of the Initial Step in the Mechanism of Dehaloperoxidase A: Determination of the Protonation Scheme at the Active Site and the Movement of the His55 Residue

Fiorentina Bottinelli, Patricia Saenz-Méndez and Oscar N. Ventura

Abstract Dehalopeoxidase A (DHP A) is a detoxifying enzyme found in the marine worm Amphitrite ornata. This enzyme converts halophenols found in the environment where the worm lives, into quinones by dehalogenation. The enzyme has globin structure and function, but works also as a peroxidase in the presence of H_2O_2 which binds to the iron present in the heme group. The initial step in the enzymatic reaction path is the transformation of the heme Fe(III) ion into a ferryl (Fe = O) moiety. A distal histidine, His55, is crucial for this process. His55 can occupy two positions, either in the distal pocket of the active center ("closed"), or exposed to the solvent ("open"). NMR experiments show that His55 moves between those positions in the resting state of the enzyme. For this process to occur it is necessary that a gate, composed of a triad Asn37-Lys36-Lys51 and two carboxylates on the heme group, suffer a conformational change before and after the passage of the histidine. We examined computationally this process at the B3LYP/6-31G(d,p) level, within a PCM simulated aqueous environment. This analysis leads us to propose a correction of the experimental structure of the enzyme determined by X-ray crystallography and offers an explanation for different conformations of the twin carboxylates at the heme group observed in the crystals. This new proposal agrees with the experimentally determined electron density distributions and explains the role of the His55 as a functional hook for the peroxide in the aqueous media.

Abbreviations

DHP Dehaloperoxidase
RMSD Root mean square deviation

F. Bottinelli · P. Saenz-Méndez · O.N. Ventura (✉)
Computational Chemistry and Biology Group (CCBG), Detema,
Facultad de Química, UdelaR, Cc1157, 11800 Montevideo, Uruguay
e-mail: oscar.n.ventura@gmail.com

J.-L. Rivail et al. (eds.), *Quantum Modeling of Complex Molecular Systems*,
Challenges and Advances in Computational Chemistry and Physics 21,
DOI 10.1007/978-3-319-21626-3_14

14.1 Introduction

The terebellid polychaete *Amphitrite ornata* inhabits polluted environments containing toxic haloaromatics produced by other organisms such as *Notomastus lobatus*. It has been shown that its ability to survive is due to the presence of a detoxyfying enzyme, hemoglobin Dehaloperoxidase (DHP) present in two forms encoded by different genes, dhpA and dhpB [1, 2].

Peroxidases (EC number 1.11.1.x) are a large family of enzymes reducing hydrogen peroxide or organic hydroperoxides to water or alcohols respectively. Most peroxidases are structurally similar to cytochrome c peroxidase (CCP), as for example horseradish peroxidase (HRP) and lignin peroxidase (LiP). Active sites contain usually a heme factor or redox-active cysteine or selenocysteine residues. The general reaction in these enzymes can be written as

$$\mathrm{ROOR'} + \text{electron donor}\,(2\,\mathrm{e}-) + 2\mathrm{H}+ \;\rightarrow\; \mathrm{ROH} + \mathrm{R'OH} \qquad (14.1)$$

DHP A however belongs to the family of the globins and its folding is practically the same as that of myoglobin [1, 3]. DHP was known for many years as the coelomic hemoglobin of *A. ornata* [4]. It was not until 1996 that the enzyme was rediscovered as a peroxidase [5]. Due to its relative abundance in *A. ornata* and structural similarity to the globins, it is now thought that the protein can act both as an oxygen carrier and a peroxidase. At first it was thought that DHP A is composed of two identical subunits, but it was reported recently that it is indeed a monomer in solution [6].

DHP A transforms mono-, di- and tri-substituted halogenated phenols into dehalogenated quinones in the presence of H_2O_2 [5], according to the reaction

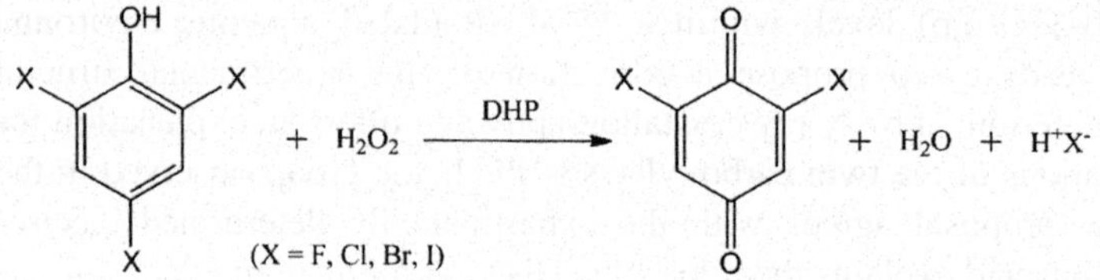

The crystallographic structure of DHP A was determined La Count et al. [3] and de Serrano et al. [7] (see Fig. 14.1). The heme factor with the central Fe atom can be appreciated in the figure, as well as several residues which are important for the reaction.

As said before, it was initially thought that the enzyme worked as a dimer, but this is no longer the case. The monomers observed in the crystallographic structure do interact at an interface where Lys and other residues link through hydrogen bonding. These interface regions are located far away from the catalytic region. The structure of the monomers and a detail of the active site are better observed in Fig. 14.1.

As in other peroxidases, iron in the heme group is present as Fe(III). Two histidine residues, called proximal (His89) and distal (His55) are present in the

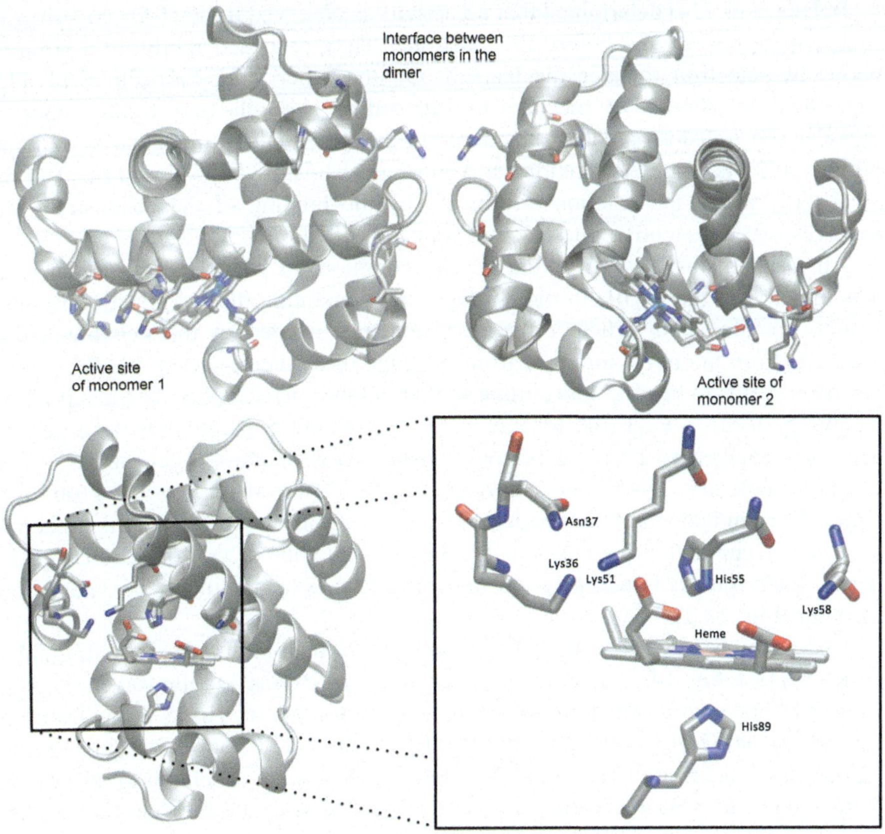

Fig. 14.1 Crystallographic structure of the dimer, showing the interface between monomers, and the active center in each monomer (*above*). Structure of one of the monomers in the crystallographically determined dimer, and detail of the residues participating in the active site (*below*)

active site. His89 is coordinated to Fe(III) in the heme group, which sixth coordination position is filled with a water molecule, later substituted by a peroxide for the reaction to proceed. His55 has the ability to move between "open" and "closed" positions, as will be described later.

The location and structure of the substrate binding site is not completely clear as yet. It was thought originally that the distal pocket, i.e. the region where His55 is located, could be the binding site. However, it was later found that this pocket is actually a site for inhibition. If substrate is placed at this site, His55 is forced into the solvent and precluded to participate in the catalysis. An alternative site was proposed to be located on the surface of the protein, at the interface of both subunits, near Trp120. However, this suggestion does not explain that the enzyme can work as a monomer. More recently, Zhao et al. found that there seems to be indeed an internal binding site [1, 3, 8–13].

Belyea et al. [14] determined that no activity is observed if the H_2O_2 co-substrate is added to the enzyme before the substrate. They proposed therefore a simultaneous two-electron transfer mechanism. On the other side, Franzen et al. [15] suggested an alternative mechanism proceeding through two tightly coupled one-electron transfer oxidation reactions. In this mechanism the substrate remains bound until it is completely oxidized. A third mechanism was proposed by Osborne et al. [16] where the reaction is initiated by the binding of the cosubstrate and involves two consecutive one-electron transfer oxidations.

All mechanisms have an initial step in common, in which hydrogen peroxide acts directly on the Fe(III) in the heme group, generating the ferryl cation radical. Doubts remain about whether the extra electron needed in the dehalogenation proceeds from the ferryl group or from the edge of the heme group, but the role of peroxide seems to be clear and similar to that in other peroxidases. The distal His55 residue performs a vital role for this part of the reaction. It functions as a general acid-base catalyst to promote proton transfer between the oxygen atoms in the peroxide molecule, O–O bond cleavage, and formation of the Fe = O group and water. The fundamental importance of His55 is clear from the fact that the reaction does not occur if this residue is substituted by another or conformationally hindered. Therefore, it is necessary to understand its behavior in the context of the reaction (Fig. 14.2).

His55 can be observed at two different conformations in the crystallographic structures (see Fig. 14.3). On the one hand, it can be exposed to the solvent in what is called "open" conformation. When in this conformation, His55 is far from the iron in the heme group (more than 9 Å according to LaCount et al. [3]). On the second hand, the ring of His55 can be rotated around the $C_\alpha C_\beta$ bond to occupy the distal cavity, now nearer to the iron cation, in what is called the "closed" conformation. NMR studies in solution [12] showed that in the absence of the substrate, His55 constantly moves between the open and closed conformations.

Pivoting around the $C_\alpha C_\beta$ bond might occur either toward the inside of the active site, nearer to the hemo group, or to the opposite site. We will call these directions below and above the plane of the His ring. Rotation above the plane is restricted by the presence of a tyrosine residue (Tyr38) effectively preventing this alternative. On the other side, rotation below the plane is hindered by a complex gate structure, consisting of a triad Asn37-Lys36-Lys51, two carboxylates linked to the heme group and Lys58. This structure presents some flexibility and at least two conformations can be observed in the crystallographic structures. On the one hand, there is a structure which we will call O (for ordered) in which the triad exhibits closer distances between the residues (a more compact structure) and the carboxylate groups are almost coplanar (see Fig. 14.3a). On the other hand, there is a D (for disordered) structure, where the triad is less compact, with larger distances between the groups, and the planes of the carboxylate groups are parallel (see Fig. 14.3b).

None of the crystal structures available [2, 3, 7, 17] present enough resolution to assign reasonably the position of protons. Additionally, some theoretical calculations have been performed [18, 19] but always assuming the position of the protons

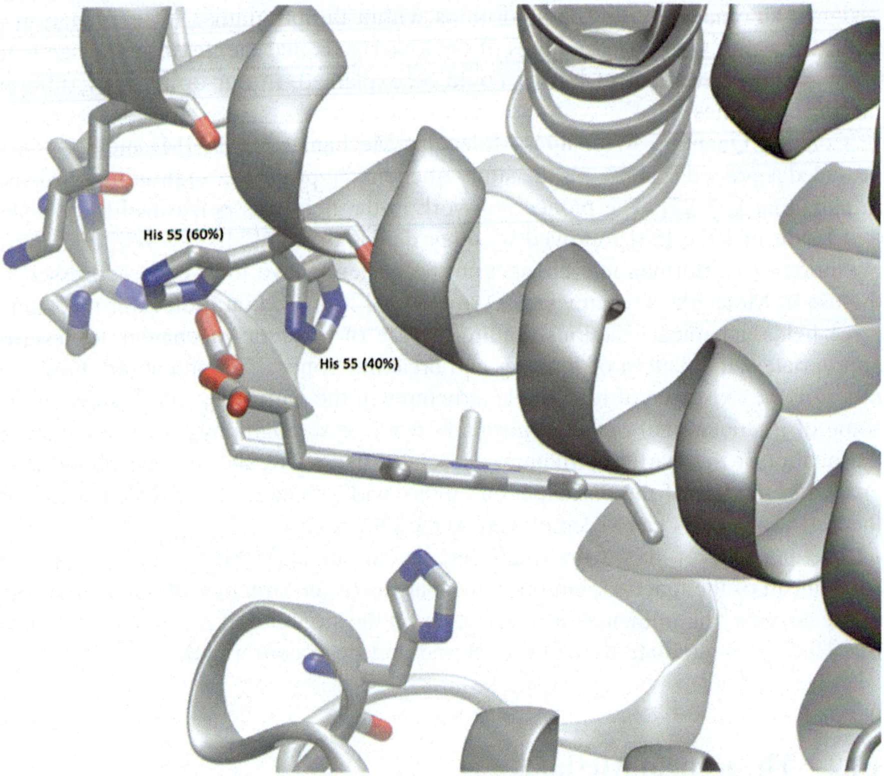

Fig. 14.2 Open (observed in 60 % of the structures) and closed conformation of His55 (observed in 40 % of the structures)

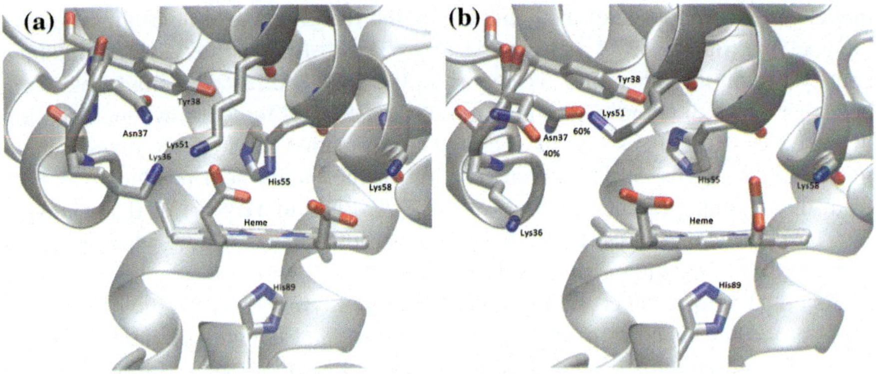

Fig. 14.3 Ordered (a) and disordered (b) conformations of the triad residues and the heme carboxylates

assigned automatically by the algorithms within the programs. An explanation of the existence of O and D structures of the lower gate, and the flexibility it needs for allowing the movement of His55, could be explained, in our opinion by different protonation states.

Coupled Quantum Mechanics/Molecular Mechanics (QM/MM) methods have provided a general scheme for the study of chemical processes in proteins, enzymes in particular [20–22]. The pioneering work in the field was performed by Warshel and Levitt in 1976 [23] followed by other developments [24, 25]. While temporal evolution of conformational changes in the proteins have long been addressed by the use of Molecular Dynamics (MD) methods [26] based on molecular mechanics force fields, chemical reactions require the use of quantum mechanics to describe appropriately the pattern of forming and breaking bonds. The procedure, however, relies on the existence of reasonable structures of the native enzyme (an eventually some of its mutations) obtained normally from crystallography. An assessment of the quality of the starting structure, especially the active site, can be obtained by analyzing models of the participating amino acids, since in this case more precise theoretical methods can be employed in the QM region.

The purpose of this work therefore, is not the QM/MM study of the full mechanism of the reaction, but only the analysis of the structure of the resting state of the enzyme, the influence of protonation on the structure of the gate and the way that His55 can alternate between its closed and open conformations.

14.2 Theoretical Methods

Our interest in this study is to determine the protonation state and general structure of the triad-carboxylates gate and the way in which His55 interacts with the residues while it pivots between the open and closed conformations. Therefore, we chose a model system composed of the heme group, the His55 and the important residues in the gate, the triad Asn37-Lys36-Lys51 on one side of the carboxylates of the heme and Lys58, on the other side. To represent the His89 we used an imidazole which occupies the sixth coordination position of the Fe(III). Moreover, we included later in those cases where His55 is in its closed conformation the Tyr38 to investigate the interaction with His55. The initial position of the amino acid residues was taken from the crystallographic structure of DHP A, PDB id 1EW6 [3]. It is shown in Fig. 14.4. We chose that structure because it was obtained at pH 7.0 which is the one with maximum enzyme activity. [15]

As noted before, the crystallographic structure does not include the protons. The usual procedure to perform molecular mechanics and dynamics calculations is to allow the protonation of the charged residues based on some simple rules for saturation. The present case is certainly difficult for such simple models. For instance, if both carboxylates were deprotonated, as is usually the assumption in non-acidic media, electronic repulsion would be high between them and the structure should reflect this fact.

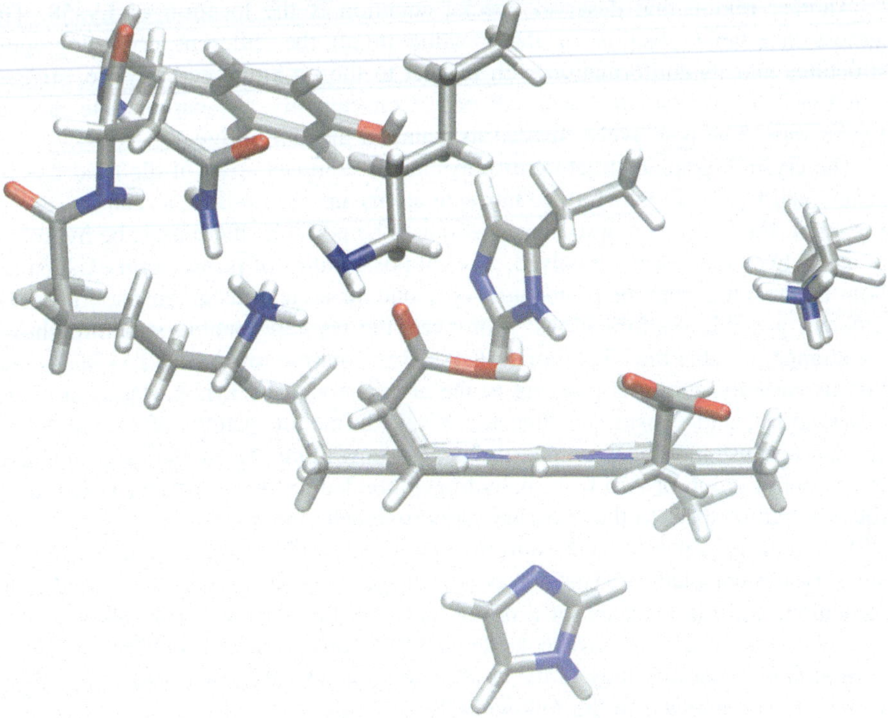

Fig. 14.4 Model employed to perform the different calculations shown in this work

Instead, the orientation of the carboxylates heads in some of the crystallographic structures is almost coplanar, with two oxygen atoms opposed head to head. Additionally, the distance between both oxygen atoms in the crystallographic structure is not very large—about 4 Å— making plausible the assumption that they share a proton in between. If the proton would not be present, the negatively charged residues would tend to repel and rotate, to ease the repulsion. Then the planes of the carboxylates would be almost perpendicular and not parallel. The hypothesis of the existence of an intermediate proton was tested by optimization of four structures, characterized by the position of His55 and the presence or absence of a proton between the carboxylate heads. The naming of the initial structures is shown in Table 14.1.

Table 14.1 Models used to study the protonation state of the heme carboxylates

Model	His55	Intermediate proton	Tyr38
I	Open	Present	Absent
II	Open	Absent	Absent
III	Closed	Present	Present
IV	Closed	Absent	Present

Another region that deserves special attention is the location of Lys58. The almost invariable position of that residue in all the different crystallographic structures and its conformation with respect to the proximal carboxylate, suggest that Lys58 is participating in a salt pair. Consequently, the carboxylate next to Lys58 should be negatively charged to maintain the salt bridge.

The crystallographic structure displays also a compact array of the triad Lys36, Asn37 and Lys51, in which three nitrogen atoms are face to face, as can be seen in Fig. 14.4. For a structure as compact as this to be possible, there must be hydrogen bonds between the atoms involved. Several possibilities of protonation exist, since both Lys residues may be protonated, only one, or none, thus given rise to several different possible conformations. Moreover, the crystallographic structure shows the strange fact that the NH2 group of Asn37 is pointing toward the Lys heads and that its carbonyl group is pointing to the aromatic ring of Tyr38. Based on these observations, and considering that the X-ray diffraction pattern of O and N are similar, because their electron density are, it seems more reasonable to assume that the carbonyl group of Asn37 12 is hydrogen bonded to the protonated Lys, a 180° rotation with respect to the experimentally assigned structure.

Different protonation and conformational possibilities were then tested. To simplify the calculations, His55 was considered to be always in the closed conformation, and the presence of a proton between the carboxylate heads was also assumed. Lys58 was always considered to be protonated. Therefore, we used several models for the study of the triad: model I as in Table 14.1, and some others which are summarized in the following Table 14.2.

Finally, we studied the protonation of His55 and its opening mechanism. Histidines may be neutral, protonated (either at Nδ or at Nε), or charged (diprotonated). Charged His has no free nitrogen to interact with H_2O_2 in the first step of the reaction. Thus we discarded this species and focused our study on models with a neutral His and different schemes of nitrogen protonation. As we mentioned already in the case of N and O in Asn, the X-ray diffraction of C and N are similar and therefore it is difficult to distinguish the positions of both atoms. This means that histidines could be rotated with respect to the assignment reported on the basis of the 13 crystallographic structure without modifying unreasonably the observed electron density distributions. In order to study in depth the conformation of His55,

Table 14.2 Models used to study the conformation and protonation state of Lys36, Asn37 and Lys51 triad

Model	Lys36	Asn37[a]	Lys51
III	Protonated	Crystallographic	Neutral
V	Protonated	Rotated	Neutral
VI	Neutral	Crystallographic	Protonated
VII	Neutral	Rotated	Protonated
VIII	Protonated	Crystallographic	Protonated
IX	Protonated	Rotated	Protonated

[a]"Crystallographic" and "rotated" labels identify respectively the structure as assigned on the basis of the crystallographic determination and the 180° rotated structure

Table 14.3 Models used to study the structure of His55

Model	Conformation[a]	N protonated
III	Crystallographic	δ
X	Rotated	δ
XI	Crystallographic	ε
XII	Rotated	ε

[a]"Crystallographic" and "rotated" labels identify respectively the structure as assigned on the basis of the crystallographic determination and the 180° rotated structure

four models were optimized, as described in Table 14..3. In all those models we assumed the presence of the proton between the carboxylate heads, and chose Lys58 to be charged. Furthermore, we considered in all cases that Lys36 is also charged, Asn37 rotated and Lys51 neutral. Finally we studied the opening mechanism of His55 by a potential energy surface (PES) scan, varying the dihedral H-Cα-Cβ-Cγ in 10° until reaching the dihedral angle of the open conformation. This study was performed for models III, X and XII.

Full geometry optimization of the models with the conformations and protonation states given in Tables 14.1, 14.2, and 14.3 was performed at the density functional level, using the B3LYP [27] method. The effective core potential type basis set Lanl2dz was used for the Fe atom and the 6-31G(d,p) Pople basis set was used for the other atoms. The coordinates of the alpha carbon of the amino acids were held fixed in all the calculations and so was the porphyrin ring in the heme group. All calculations were carried out with the enzyme immersed into an aqueous environment. The 14 continuum PCM method of solvation of Tomasi and coworkers [28] was employed for these calculations.

Root-mean-square deviation (RMSD) data were calculated considering only the C, N and O atoms that were not fixed during the optimizations. All calculations were performed using the G09 suit of programs [29].

14.3 Results and Discussion

14.3.1 Variation of the Protonation State of Carboxylates

The first analysis performed was on the protonation of carboxylates while His55 is in the open conformation, models I and II. As shown in Fig. 14.5, the carboxylate heads remain facing each other and in an almost coplanar conformation with a reduced O–O distance when a proton is shared between them. At the same time, the charged lysines (Lys36 and Lys58) keep closer to each other at a distance of about 2.7 Å. On the contrary, if the proton is absent, those lysines approach the carboxylates slightly, and the distance between the carboxylate heads increases considerably (6.2 Å) and adopting now a non-coplanar conformation.

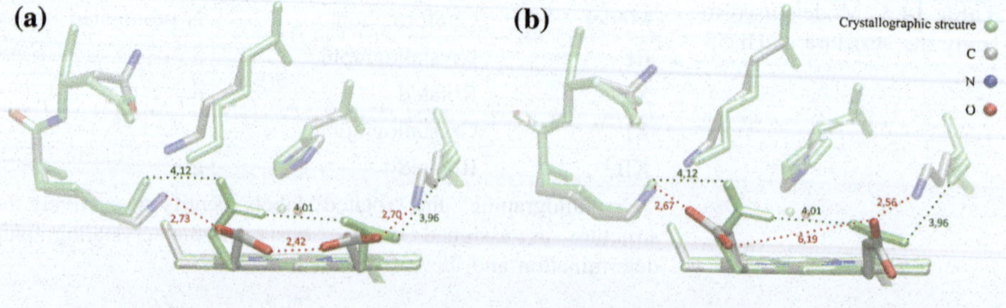

Fig. 14.5 Optimized geometries of Model I (**a**) and Model II (**b**) showing the conformation adopted by the carboxylate heads. Both conformations are superimposed on the crystallographic structure exhibiting the almost coplanar carboxylates. Hydrogen atoms are not shown

The results obtained with those models with closed His55 are similar, models III and IV (see Fig. 14.6). On the model with the intermediate proton the oxygen atoms remain faced at a distance of 2.6 Å but the distance between the carboxylate and Lys36 increases slightly. Additionally, in the model where both carboxylates are negative charged, the distances between them increases until a distance of 5.7 Å, and equally to the open conformation the distance between them and the charged lysines decreases until a distance of approximately 2.7 Å.

The number of atoms in the four models is different, because of the presence of the additional proton in I and III. Therefore we cannot compare directly their energies. However, RMSD values with respect to the crystallographic structure, which were calculated for all models, can be compared. For those cases in which both carboxylates are charged (Model II and Model IV, RMSD 1.0 in both cases), the RMSD value is slightly lower than for those cases in which the carboxylates are sharing a proton (Model I and Model III, RMSD 1.1 in both cases). This 10 % difference as well as the much better fitting of the carboxylate heads to the crystallographic structure, are reasonable clues pointing to the necessity of the presence of the proton in between the oxygen atoms at the pH at which the enzyme was crystallized. This proton allows the gate formed by both carboxylates to be locked in position, with the oxygen atoms facing each other.

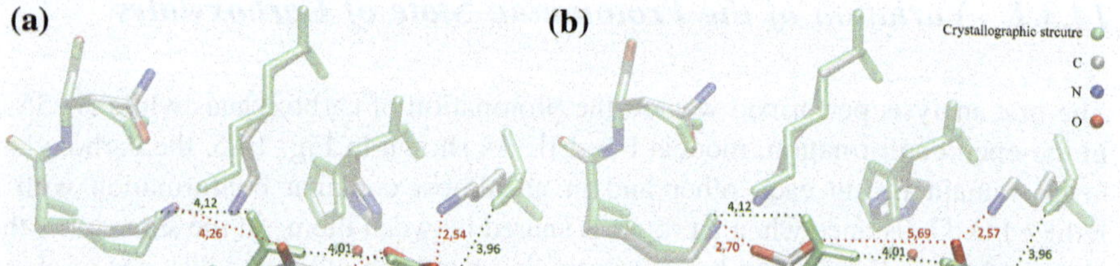

Fig. 14.6 Optimized geometries of Model III (**a**) and Model IV (**b**) showing the conformation adopted by the carboxylate heads. Both conformations are superimposed on the crystallographic structure exhibiting the almost coplanar carboxylates. Hydrogen atoms are not shown

14.3.2 Triad Lys36, Asn37 and Lys51

Two conformations of Asn37 were studied for each of the possible protonation states of Lys36 and Lys51. For the three possible states of protonation, the starting geometry of Asn37 was obtained from the crystal structure, models III, VI and VIII. All optimizations resulted in Asn37 being rotated from the initial model, with the carbonyl oriented in the direction of Lys36 and Lys51. Consequently, the NH_2 group of Asn37 ends up pointing toward the electron cloud of the aromatic ring of Tyr38. The optimized structures of the models are shown in Fig. 14.7 showing how Asn37 is rotated with respect to the crystallographic structure.

According to the distances of the optimized geometries (see Table 14.4), the distances of Lys51 to the other two amino acids increase considerably in the models where both lysine residues are charged, while the distance between Asn37 and Lys36 is relatively constant. Because of the repulsion between both cation heads, Lys51 is then exposed to the solvent, thereby loosening the compact structure observed in the crystal. Even though RMSD values are acceptable for both structures, we can dismiss this protonation model because it does not reproduce appropriately the crystallographic structure.

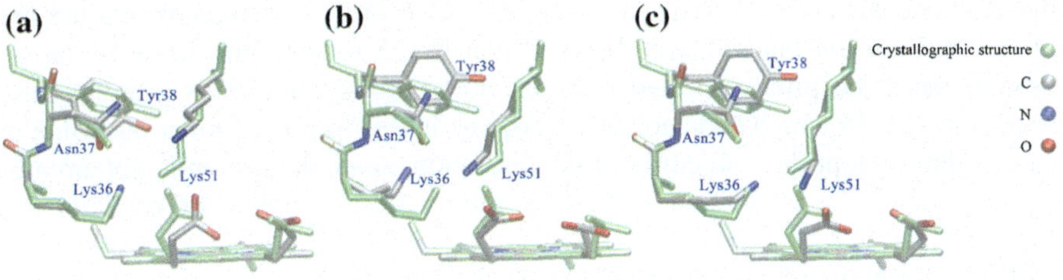

Fig. 14.7 Optimized geometries of models IX (**a**), VII (**b**) and V (**c**) as compared to crystallographic structure. Hydrogen atoms are not shown

Table 14.4 Distances between the amino acids Lys36, Asn37, and Lys51, RMSD values and relative energy

Model	Distance (Å)			RMSD	ΔE(kcal/mol)
	Lys36-Asn37	Lys51-Asn37	Lys36-Lys51		
Crystallography	2.86	3.45	3.51		
III	2.87	4.28	2.70	1.0	0.0
V	2.78	4.54	2.66	1.1	1.8
VI	6.38	4.59	5.12	1.3	18.3
VII	3.97	2.84	5,94	1.6	8.1
VIII	2.73	4.79	4.68	1.0	
IX	2.72	6.66	7.59	1.2	

Once established that one of the lysine residues is indeed neutral, we focused our attention on analyzing the models in which Lys51 is charged and Lys36 neutral, models VI and VII. In both of them, the triangle formed by the triad appears totally distorted on inspection. RMSD values lead us to the same conclusion, these values being the highest among all the models. Models III to VII have the same number of electrons and consequently their energy values can be compared as shown also in Table 14.4, reaching the same conclusion. Therefore, RMSD and energy values as well as the distances between the amino acids are consistent with the conclusion that Lys36 must be charged and Lys51 must be neutral in the structure of the enzyme at this pH.

14.3.3 Protonation, Conformation and Movement of His55

All models of the containing His55 in its closed conformation are showing Fig. 14.8. It is important to notice that all these models have a similarly small value of RMSD (see Table 14.5), so that this parameter is not useful to tell the models apart.

When His55 is protonated at Nδ in the conformation of the crystal structure (Model III), only one hydrogen bond is formed, between the Nε and the oxygen of the H_2O coordinated to Fe(III)with a distance of 1.74 Å. When analyzing the PES of the opening mechanism, we observed that His55 rotates until its Nε(without a proton) faces the proton located between both carboxylates. This proton is then trapped by His55 which ends up with charge +1 in the open conformation, while the carboxylates adopt the non coplanar configuration, since the proton is not anymore binding them. The activation energy for this process is about 49 kcal/mol (See Fig. 14.8).

His55 forms two hydrogen bonds in what we called Model X, one between the proton in Nδ and the oxygen of the hydroxyl group of Tyr38 (2.09 Å) and the other between the non-protonated Nε and a H of the H_2O coordinated to Fe(III) (1.72 Å). The stabilization that both hydrogen bonds provides in this structure is reflected in the energy values, this structure being at least 8 kcal/mol more stable than the other three analyzed when His55 is in its closed conformation. Analogously to the case of model III, we observed that His55 rotates while moving from the closed to the open conformation. Nevertheless, in this case His55 does not face the proton located between both carboxylate groups. This is due to the fact that the carboxylate groups bound by the proton open a bit their structure to allow the passage of the His55 without this residue trapping the proton. Therefore, the histidine remains neutral with a proton in Nδ in its open conformation. The energy threshold is in this case about 51.5 kcal/mol (Fig. 14.9).

Finally, both models with the protonated Nε are initially less stable than Model X. In Model XI there is no hydrogen bond in the structure and the Cε-H bond is pointing to the O of H_2O while in Model XII Nε-H forms a hydrogen bond with the O of H_2O (2.00 Å). The energy threshold for the opening mechanism of

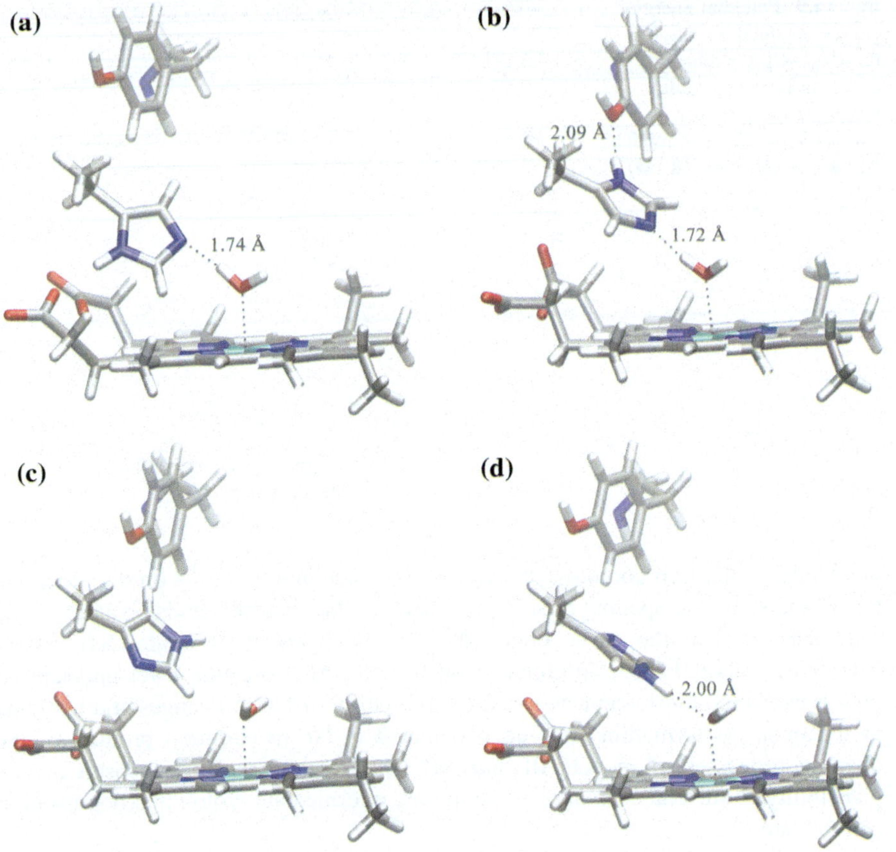

Fig. 14.8 Optimized models III (**a**), X (**b**), XI (**c**) and XII (**d**)

Table 14.5 RMSD values and relative energy (in kcal/mol) of the models containing His55 in the closed conformation

Model	RMSD (Å)	ΔE (kcal/mol)
III	1.0	11.5
X	1.0	0.0
XI	0.9	10.7
XII	0.9	8.6

model XII is 33.4 kcal/mol, see Fig. 14.9. In this case the final structure is the open conformation protonated in Nε, rotated with respect of the structure determined on the basis of the crystallography analysis.

According to the data obtained, a reasonable proposal for the structure and energy barriers would be the following. The closed form of His55 is as in model X, (b) in Fig. 14.8, where the His55 is rotated with respect to the proposed crystallographic structure and is protonated at Nδ.Two hydrogen bond interactions bind

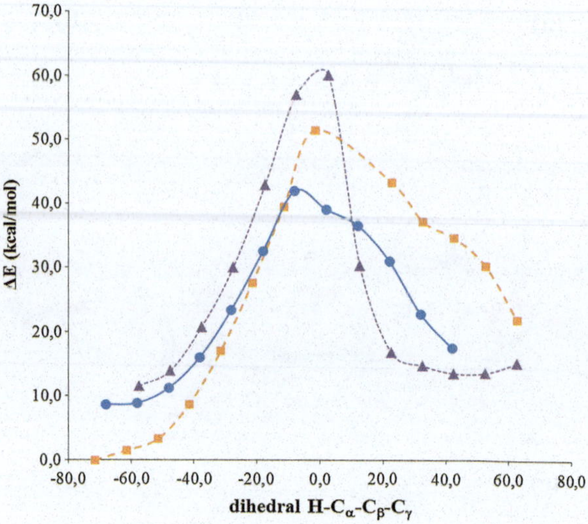

Fig. 14.9 Potential energy curves obtained by varying the H-Cα-Cβ-Cγ dihedral angle (in degrees) while reoptimizing the other geometric variables in models III (▲), X (■) and XII (●)

this His55 to the iron coordinated water on one side and to Tyr38 on the other. As this His55 starts the opening process, it suffers a rotation which ends up at about the same energy than the curve obtained from the opening of model XII, (d) in Fig. 14.8, where His55 is protonated on the Nε. At this point a yet undisclosed proton exchange must occur, where the protonation of His55 changes and histidine in the open conformation ends up protonated at Nε (a common product of the opening processes of models III and XII. Our research is now focused on the plausibility of such an exchange involving the intermediate proton that bounds both carboxylates.

These conclusions are not final, because they may be affected by the mechanical constraints exerted by the surroundings may play a leading role in such conformational changes (mechanical embedding). More elaborated studies, including the whole of the protein in actual QM/M studies, are necessary to confirm or reject the latter proposal.

14.4 Conclusions

We examined different in silico models of the DHP A active site with a PCM simulated environment. Different protonation and conformational possibilities were tested. Based on the results obtained, we can conclude that some corrections are needed regarding the conformation of Asn37 in the triad and His55, as derived from the diffraction data. We have concluded that the Asn37 conformation must be rotated 180º from the position assigned on the basis of the crystallographic studies. We concluded also that a single proton is shared between both carboxylates in the heme group, thus assuring that they share an almost coplanar structure. Moreover, the structure of the His55 in the closed conformation has been shown to be protonated at Nδ and rotated with respect to the structure determined on the basis of the crystallographic analysis. Finally, we are able to propose a consistent model of the opening mechanism of the His55 starting from model X in Fig. 14.8. The His55 molecule, initially in the conformation of model X, moves until a destabilization similar to the one found for the movement of His55 in the conformation of model XII is reached. At this stage we propose –but have not demonstrated yet—that a proton exchange must occur, involving the proton that holds the carboxylates together. His55 then loses its Nδ proton and is protonated at Nε, this being the final protonation scheme of His55 in the open conformation. We do not consider the 33 kcal/mol activation barrier obtained for the movement of His55 to be a reliable value, since better methods and basis sets, as well as inclusion of a larger model of the active site is necessary. However, within the limitations of the present calculations, models X and XII appear to be reasonable candidates for QM/MM studies of the enzymatic reaction path.

Acknowledgments We thank the Comisión Sectorial de Investigación Científica (CSIC, Universidad de la República, Uruguay, Proyecto de Grupo CSIC655), Programa de Desarrollo de las Ciencias Básicas (PEDECIBA) and Agencia Nacional de Investigación e Innovación (ANII) for financial support.

References

1. Lebioda L, LaCount MW, Zhang E et al (1999) An enzymatic globin from a marine worm. Nature 401:445–446
2. Han K, Woodin SA, Lincoln DE et al (2001) Amphitrite ornata, a marine worm, contains two dehaloperoxidase genes. Mar Biotechnol 3:287–292
3. LaCount MW, Zhang E, Chen YP et al (2000) The crystal structure and amino acid sequence of dehaloperoxidase from Amphitrite ornata indicate common ancestry with globins. J Biol Chem 275:18712–18716
4. Weber RE, Mangum C, Steinman H et al (1977) Hemoglobins of two terebellid polychaetes: Enoplobranchus sanguineus and Amphitrite ornata. Comp Biochem Physiol A Physiol 56:179–187
5. Chen YP, Woodin SA, Lincoln DE et al (1996) An unusual dehalogenating peroxidase from the marine terebellid polychaete Amphitrite ornata. J Biol Chem 271:4609–4612
6. Thompson MK, Franzen S, Davis MV et al (2011) Dehaloperoxidase-hemoglobin from amphitrite ornata is primarily a monomer in solution. J Phys Chem B 115:4266–4272
7. De Serrano V, Chen Z, Davis MF et al (2007) X-ray crystal structural analysis of the binding site in the ferric and oxyferrous forms of the recombinant heme dehaloperoxidase cloned from Amphitrite ornata. Acta Crystallogr D Biol Crystallogr 63:1094–1101
8. Nienhaus K, Deng P, Belyea J et al (2006) Spectroscopic study of substrate binding to the carbonmonoxy form of dehaloperoxidase from amphitrite ornata. J Phys Chem B 110:13264–13276
9. Thompson MK, Davis MF, de Serrano V et al (2010) Internal binding of halogenated phenols in dehaloperoxidase-hemoglobin inhibits peroxidase function. Biophys J 99:1586–1595
10. Davis MF, Bobay BG, Franzen S (2010) Determination of separate inhibitor and substrate binding sites in the dehaloperoxidase-hemoglobin from Amphitrite ornata. Biochemistry 49:1199–1206

11. Davis MF, Gracz H, Vendeix FAP et al (2009) Different modes of binding of mono-, di-, and trihalogenated phenols to the hemoglobin dehaloperoxidase from amphitrite ornata. Biochemistry 48:2164–2172
12. Nicoletti FP, Thompson MK, Howes BD et al (2010) New insights into the role of distal histidine flexibility in ligand stabilization of dehaloperoxidase-hemoglobin from amphitrite ornata. Biochemistry 49:1903–1912
13. Nienhaus K, Nickel E, Davis MF et al (2008) Determinants of substrate internalization in the distal pocket of dehaloperoxidase hemoglobin of Amphitrite ornata. Biochemistry 47: 12985–12994
14. Belyea J, Gilvey LB, Davis MF et al (2005) Enzyme function of the globin dehaloperoxidase from Amphitrite ornata is activated by substrate binding. Biochemistry 44:15637–15644
15. Franzen S, Gilvey LB, Belyea JL (2007) The pH dependence of the activity of dehaloperoxidase from Amphitrite ornata. Biochimica et Biophysica Acta—Proteins Proteomics 1774:121–130
16. Osborne RL, Coggins MK, Raner GM et al (2009) The mechanism of oxidative halophenol dehalogenation by Amphitrite ornata dehaloperoxidase is initiated by H_2O_2 binding and involves two consecutive one-electron steps: Role of ferryl intermediates. Biochemistry 48:4231–4238
17. Chen Z, de Serrano V, Betts L et al (2009) Distal histidine conformational flexibility in dehaloperoxidase from Amphitrite ornata. Acta Crystallogr D Biol Crystallogr 65:34–40
18. Zhao J, de Serrano V, Dumarieh R et al (2012) The role of the distal histidine in H_2O_2 activation and heme protection in both peroxidase and globin functions. J Phys Chem B 116:12065–12077
19. D'Antonio EL, D'Antonio J, de Serrano V et al (2011) Functional consequences of the creation of an Asp-His-Fe triad in a 3/3 globin. Biochemistry 50:9664–9680
20. Van der Kamp MW, Mulholland AJ (2013) Combined quantum mechanics/molecular mechanics (QM/MM methods in computational enzymology. Biochemistry 52:2708–2728
21. Brunk E, Rothlisberger U (2015) Mixed quantum mechanical/molecular mechanical molecular dynamics simulations of biological systems in ground and electronically excited states. Chem Rev. doi:10.1021/cr500628b
22. Van der Kamp MW, Shaw KE, Woods CJ et al (2008) Biomolecular simulation and modelling: status, progress and prospects. J R Soc Interface 5:S173–S190
23. Warshel A, Levitt M (1976) Theoretical studies of enzymic reactions—dielectric, electrostatic and steric stabilization of carbonium-ion in reaction of lysozyme. J Mol Biol 103:227–249
24. Lonsdale R, Ranaghan KE, Mulholland AJ (2010) Computational enzymology. Chem Commun 46:2354–2372
25. Senn HM, Thiel W (2009) QM/MM Methods for Biomolecular Systems. Angew Chem Int Ed 48:1198–1229
26. Karplus M, McCammon A (2002) Molecular dynamics simulations of biomolecules. Nat Struct Mol Biol 9:646–652
27. Becke AD (1993) A new mixing of Hartree-Fock and local density-functional theories. J Chem Phys 98:1372–1377
28. Tomasi J, Mennucci B, Cammi R (2005) Quantum mechanical continuum solvation models. Chem Rev 105:2999–3094
29. Frisch MJT, Trucks GW, Schlegel HB et al (2009) Gaussian 09, Revision D.01. Gaussian Inc, Wallingford

Chapter 15
Exploring Chemical Reactivity in Enzyme Catalyzed Processes Using QM/MM Methods: An Application to Dihydrofolate Reductase

J. Javier Ruiz-Pernía, Vicent Moliner and Iñaki Tuñón

Abstract Enzymes are the catalysts used by living organisms to accelerate chemical processes under physiological conditions. In this chapter, we illustrate the current view about the origin of their extraordinary rate enhancement based on molecular simulations and, in particular, on methods based on the combination of Quantum Mechanics and Molecular Mechanics potentials which provide a solution to treat the chemical reactivity of these large and complex molecular systems. Computational studies on Dihydrofolate Reductase have been selected as a conductor wire to present the evolution and difficulties to model chemical reactivity in enzymes. The results discussed here show that experimental observations can be currently understood within the framework of Transition State Theory provided that the adequate simulations are carried out. Protein dynamics, quantum tunnelling effects and conformational diversity are essential ingredients to explain the complex behaviour of these amazing molecular machineries.

Acronyms

2D-PMF	Two dimensional potential of mean force
AM1	Austin model 1
AM1-SRP	Austin model 1—specific reaction parameters
BsDHFR	Geobacillus stearothermophilus Dihydrofolate Reductase
DAD	Donor acceptor distance
DFT	Density futctional Theory
DHF	7,8-dihydrofolate
DHFR	Dihydrofolate Reductase
EA-VTST	Ensemble-averaged variational transition state theory

J. Javier Ruiz-Pernía · V. Moliner (✉)
Departament de Química Física i Analítica, Universitat Jaume I, 12071 Castelló, Spain
e-mail: moliner@uji.es

I. Tuñón (✉)
Departament de Química Física, Universitat de València, 46100 Burjasot, Spain
e-mail: ignacio.tunon@uv.es

J.-L. Rivail et al. (eds.), *Quantum Modeling of Complex Molecular Systems*,
Challenges and Advances in Computational Chemistry and Physics 21,
DOI 10.1007/978-3-319-21626-3_15

EcDHFR	Escherichia Coli Dihydrofolate Reductase
E_{elec}	Electronic energy
EVB	Empirical valence bond
FEP	Free energy perturbation
GHO	Generalized hybrid orbital
IRC	Intrinsic reaction coordinate
KIE	Kinetic isotope effect
LDH	Lactate Dihydrogenase
LSCF	Local self-consistent field
MD	Molecular dynamics
MM	Molecular mechanics
NADPH	Nicotinamide adenine dinucleotide phosphate
PES	Potential energy surface
PM3	Parameterized model 3
PMF	Potential of mean force
PS	Product state
QM	Quantum mechanics
QM-FEP	Quantum mechanics free energy perturbation
QM/MM	Quantum mechanical/molecular mechanics
RS	Reactant state
THF	5,6,7,8-tetrahydrofolate
TS	Transition state
TST	Transition state theory
V	Potential energy
V_{nn}	Nuclear repulsion energy
WHAM	Weighted histogram analysis method

15.1 Introduction

Enzymes are the most efficient catalysts, capable of working at mild conditions of temperature and pressure, but also showing other important advantages with respect to non-natural catalysts such as their chemo-, regio- and stereoselectivity. Moreover, the use of biocatalysts for one-pot transformations synthetic methods allows decreasing the number of energy consuming steps such as separation and purification of intermediates, with the obvious consequent economical savings. These features justify the fact that natural biochemical processes in living organisms are a major source of inspiration for design of new biocatalysts with potential applications in industry. The design of new enzymes provides a rigorous test of our understanding of how naturally occurring enzymes work [1–4]. On the other side, since most of reactions taking place in living organisms utilize enzymes to accelerate chemical processes, making them compatible with life, a detailed knowledge of the origin of catalysis in these complex macromolecular machines can be used for the synthesis of inhibitors with applications in biomedicine. Combination of experimental tools such as protein engineering techniques, with Computational Chemistry methods has revealed as a promising option to study the chemical reactivity of natural enzyme catalysed processes and their applications.

In this chapter, we will illustrate the state of the art of the computational methods devoted to the study of chemical reactivity in enzyme catalysed processes. It will be shown how the use of computational sophisticated tools are required to describe such complex systems. The adequate computational protocol to explore enzyme reactivity will be dictated by the fact that enzymes are huge macromolecules, with noticeable flexibility required to acquire the different steps of the full catalytic process from binding of reactants to the product release, and also by the electronic and quantum complexity of the reactions catalysed in their actives sites.

Transition-state stabilization in a pre-organized active site seems to be the major source of catalysis although, as previously mentioned, some degree of protein flexibility is needed to reach the maximum catalytic efficiency [5–8]. The seminal Pauling's postulate [9], that explained the origin of enzymes catalytic activity based on the complementarity between the enzyme's active site and the transition structure, has been rationalized in terms of the electrostatic stabilization that the protein provides for the transient structures that appear during the transformation of reactants into products and, in particular, for the transition state (TS) [8]. Nevertheless, the protein has to adopt different conformations from the substrates binding step to the products release, passing through the chemical step. Obviously, protein flexibility is necessary to understand how a cavity created during millions of years to interact and stabilize the TS of a chemical reaction, is also capable to accommodate, for instance, reactant state structures. Thus, enzymes must be stable to retain their three-dimensional structure but flexible enough to change among the different conformations relevant at each step of the catalytic cycle [10, 11]. Moreover, the presence of different conformations in any of the states along the full catalytic process is essential to explain experimental results such as the temperature dependence of kinetic isotope effects (KIEs) [12, 13] or the significant different rate constants observed in single-molecule spectroscopy [14]. The presence of a wide distribution of rates in individual enzyme molecules has also been predicted by computer simulations on different enzymatic systems [15–19]. Modelling enzyme reactivity must then consider the inclusion of TS stabilizing interactions derived from electrostatic and non-electrostatic effects with an adequate theoretical framework, as it will be shown in the present chapter. However, the lack of a detailed understanding of the link between sequence, structure, flexibility, and function still prevents the complete understanding of enzyme catalytic processes and the strategies to design new biocatalysts [20].

In this chapter, Dihydrofolate reductase (DHFR) has been selected as a conductor wire to present the evolution and difficulties to model chemical reactions in enzymes, from the early calculations based at semiempirical level, carried out in gas phase, to the more sophisticated simulations based on hybrid Quantum Mechanical/Molecular Mechanics (QM/MM) schemes [21, 22]. DHFR offers an excellent opportunity for the analysis of computational strategies applied to the study of enzyme catalysis and, in fact, it has been the benchmark of many experimental and theoretical studies devoted to reveal the origin of enzyme catalysis.

DHFR (E.C. 1.5.1.3) catalyses the NADPH-dependent reduction of 7,8-dihydrofolate (DHF) to 5,6,7,8-tetrahydrofolate (THF) in both bacterial and vertebrate cells. This enzyme is necessary for maintaining intracellular pools of THF and its derivatives, which are essential cofactors in the biosynthesis of several amino acids. This property has made DHFR a clinical target for antitumor and anti-infective therapy, and numerous inhibitors have been described [23, 24]. Nevertheless, the details of its molecular reaction mechanism are still unclear and can only be hypothesised at present [25, 26]. Information derived from complex computational models will be crucial to evolve in this sense. We will show how an adequate method within a realistic model, taking into account the pKa of the titratable aminoacids, the flexibility of the protein, the size of the system or the level of theory used to describe the QM region, must be used to obtain reliable conclusions. It will be demonstrated the need of performing statistical simulations to sample the full conformational space of all states involved in the reaction, that allow getting free energies and averaged properties directly comparable with experimental data. Finally, keeping in mind that the chemical step of the DHFR catalysed reaction involves the transfer of a light particle, inclusion of quantum tunnelling effects will be required to properly estimate the phenomenological free energy barrier.

15.2 Potential Energy Surfaces: The Electronic Complexity of Enzyme Catalysed Reactions

In order to study a chemical reaction, accurate values of the electronic energy of the system, as a function of its nuclear coordinates, have to be obtained. This function, known as potential energy surface, PES, will provide the relationship between the energy and the geometry of the system under study.

Then, by selecting the proper distinguished reaction coordinates, a chemical reaction can be described as the evolution of the function, $V(R)$, from one minimum to another, both corresponding to stable geometries of the system such as reactants, products or possible intermediates. The point on the PES corresponding to the maximum energy of the minimum energy path connecting two minima will be characterized as a saddle point of order one, or TS structure (see Fig. 15.1).

In order to solve the PES, Quantum Chemistry, the discipline in which quantum mechanics (QM) principles are used to rationalize and predict chemical behaviour and, in particular, to study chemical reaction processes, must be used. Nonetheless, significant obstacles remain preventing the routine use of QM to provide quantitative understanding of complicated chemical systems. The problems raised because describing processes that involve bond-breaking and bond-forming, charge transfer, and/or electronic excitation, require a QM treatment but, due to the very demanding computational cost, their application is still limited to not too large

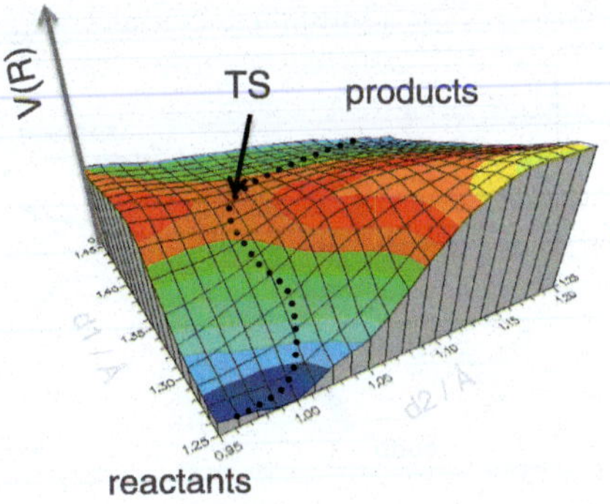

$$V(R) = E_{elec}(R) + V_{nn}(R)$$

Fig. 15.1 Schematic representation of a potential energy surface (PES) of a single step chemical reaction. V is the molecular potential energy, E_{elec} the electronic energy and V_{nn} the nuclear repulsion energy

systems. On the other side, models based on classical Molecular Mechanical (MM) force fields, widely used in molecular dynamics (MD) simulations of large systems, are not adequate to describe a chemical reaction. Algorithms that combine QM and MM (QM/MM) [21, 22] provide the way to sort out the problem; the solute molecule or the reaction center relevant to the formation and breaking processes of chemical bonds is described quantum mechanically, while the solvent and protein environments are treated by a MM force field. It must be pointed out that other strategies based on Empirical Valence Bond methods [27, 28] or reactive force fields [29], have been also applied to solve the problem.

15.2.1 Gas Phase Calculations

Historically, first insights into enzymatic reaction mechanisms were obtained through gas phase calculations, using standard programs of quantum chemistry to obtain PESs. Thus, for example, in the early nineties, the reaction catalyzed by DHFR (see Fig. 15.2) was studied in our group by means of semiempirical Hamiltonians (AM1 and PM3) in gas phase [30, 31].

The predicted structure of the TS localized for the hydride transfer step in a model of the DHFR, supported that the endo conformation of the donor and acceptor rings. This seemed a general feature on hydride transfer steps rationalized in terms of the frontier orbital overlap of the donor and acceptor centers in the saddle point [32–38]. Apart from this, and information regarding the distance between the donor and acceptor carbon atoms at the TS, no much more information could be derived from exploration of the PESs in gas phase with a reduced model. In order to understand the influence and behaviour that the environment has on the reaction, the inclusion of a realistic protein model was required.

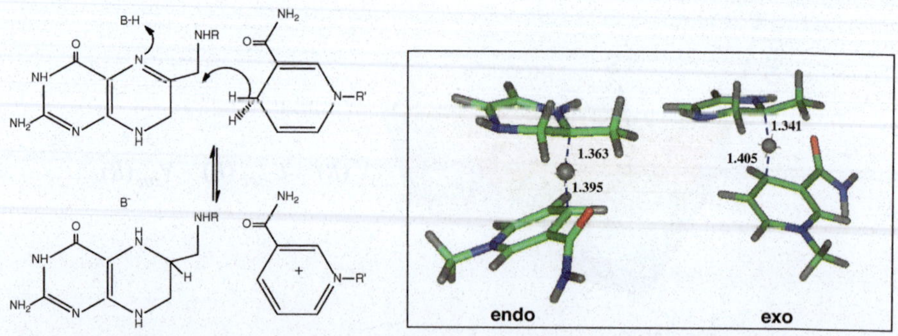

Fig. 15.2 Schematic representation of the chemical step reaction catalyzed by DHFR (*left panel*) and TS structures for the hydride transfer step obtained in a gas phase model at AM1 semiempirical level (*right panel*). Distances in Å

15.2.2 Inclusion of Environment Effects: The Quantum Chemical Cluster Approach

The properties of stationary structures of enzymatic processes can be different from those obtained in gas phase calculations because, obviously, the interaction with the environment are not considered in the latter. Then, a more realistic picture of enzyme catalyzed reactions can be obtained including a small part of the active centre into the calculations. The problem is that in this *cluster* or *supermolecule* models, the optimised structures do not necessary fit into the enzyme active site and computing artefacts can be obtained. A common strategy is to anchor some key atoms of the enzyme to their crystallographic positions and then optimize the rest of the coordinates of the model [39]. Nevertheless, this is an approximate solution that presents several deficiencies. First, the result will be dependent on the initial X-ray structure that, in many cases, is far from the real structure of the protein-substrate complex at the TS. Second, long-range effects on the nuclear and electronic polarisation of the chemical system are not included in the calculations. Third, and probably the most dramatic deficiency, the enzyme flexibility is not properly taken into account. And finally, the computational cost of these calculations rapidly increases, as more atoms of the environment are explicitly included.

Nevertheless, *cluster models* are still broadly used with models with up to 200 atoms, placed in a dielectric cavity and treated at the highest possible QM level, which so far has mainly meant hybrid density functional theory (DFT) [40–42]. The most common application of these models is on systems where the chemical reaction implies complex electronic states as the ones appeared in enzymes containing transition metals where more than one electronic state is involved and high level QM methods are required to describe the reaction. Nevertheless, due to the computational cost, such studies have embraced topics devoted to the modelling of the first coordination sphere of the active site to perform an exploration of the molecular mechanism solving problems of stereoselectivity [42], up to the development of biomimetic, or bioinspired, catalytic systems [43, 44]. But, in these cases, as discussed above, the effect of the rest of the protein and/or solvent environment is not included in the calculations and, moreover, one of the requisites of a proper modelling of an enzyme, the flexibility of the system, is not taken into account.

15.2.3 Hybrid QM/MM Potentials Based Models

There was a major breakthrough in the methods trying to mimic the role of the enzyme in calculations by combining QM Hamiltonians and MM force fields; namely the QM/MM methods [21, 22]. In these methods, a small portion of the system is described by QM, the region where the most important chemical changes (i.e. bond breaking and forming processes) take place, while the rest of the system is described by means of MM potentials. In this way a very large number of atoms can be explicitly considered in the calculations. The division of the full system in two subsystems may require cutting a covalent bond. Then, the problem of fulfilling the valence of the quantum atom placed in the QM/MM boundary has to be treated by means of different techniques such as adding hydrogen link atoms [45], frozen orbitals on each QM boundary atom (*Local self-consistent field, LSCF*) [46–51], or the use of hybrid atomic orbitals as basis functions on the boundary atoms of the MM fragment (Generalized Hybrid Orbital, GHO) [52]. All in all, the combination of two levels of theory, by means of the appropriated coupling terms, allows obtaining the wave function of the quantum subsystem, and thus any related property under the influence of the environment. It is important to note that there are also subtractive schemes, such as the originally implemented in ONIOM methods [53, 54], in which the wave function of the QM subsystem is not polarized under the influence of the MM subsystem. These methods have been successfully applied to systems, such as organocatalysts, where the steric effects of the ligands can be determinant [55–57].

The existence of a QM/MM potential energy function, which combines reliability and computational efficiency, is not the only requirement to deal with chemical reactions. To describe such processes we should be able to locate and characterize a set of stationary structures (RS, TS, PS and possible intermediates) that defines a particular reaction mechanism. This is complicated by the large dimensionality of the PES when the environment is explicitly included. In a first approximation to follow a chemical process, a distinguished geometrical coordinate can be chosen and energy minimisations would be carried out for different values of this coordinate. Thus, the minima on a potential energy profile corresponded with reactants, possible intermediates or products, while the maximum would represent an approximate TS of the reaction. Obviously, this procedure may not be always convenient as far as the true reaction coordinate can be different from the selected one. Most of the programs using hybrid calculations used to contain only algorithms to locate energy minima. The direct location of a saddle point of index one is much more difficult and was not included in commercial package of programs until more recently. Application of QM/MM methods to real chemical problems demands tools for location and characterisation of saddle points, intrinsic reaction coordinates, hessians and vibrational frequencies for very large flexible systems. Nevertheless, the high dimensionality of the surface prevents the calculation of the full Hessian matrix and thus the proper location and characterization of saddle points of index one.

In the late nineties, we carried out the first QM/MM study where a TS structure was located and characterized as a first order saddle point for a fully flexible molecular system on the reaction catalyzed by Lactate Dihydrogenase (LDH) [58]. In order to locate the TS a new algorithm, GRACE [15], was employed with an explicit hessian for a sub-set of atoms (QM subsystem plus some residues in the active-site region) by central finite-differencing of the gradient vector. The single negative eigenvalue of the resultant hessian corresponded to the force constant associated to the transition vector. The remaining atoms of the MM enzyme environment were continually relaxed. TS optimisation in the core was continued until no element of the gradient vector of the entire system (core plus environment) was larger than a certain threshold value. The intrinsic reaction coordinate (IRC) was computed in both directions from a saddle point to confirm that it was the expected reaction TS. The stationary structures obtained in this way could be characterized as a true stationary structures on the PES (all the first derivatives are equal to zero) and having the correct number of imaginary frequencies in the Hessian matrix of the core (i.e. zero for minima and only one for transition state). Shortly after this study, a similar computational scheme was employed to study the catalytic mechanisms of the hydride transfer step in DHFR [59]. In this case, 53 atoms were treated by the AM1 semiempirical Hamiltonian and the rest of protein and solvent water molecules (ca. 3000 atoms) were treated by MM force fields. In contrast with previous calculations in gas phase, the protein environment selectively molds the substrate in a conformation close to the exo transition structure (see Fig. 15.2). At the Michaelis complex, the enzyme compresses the substrate and the cofactor into a conformation close to the transition structure, thus facilitating the hydride transfer. The population analysis obtained by means of the *in vacuum* and the hybrid QM/MM methods renders similar net atomic charges for the donor and acceptor fragments, suggesting that a hydrogen more than a hydride ion was transferred. Since the environment effect was included in the simulations, information could be obtained on the role of key residues such as Asp27 (see Fig. 15.3) in stabilizing the cationic pteridine ring. The ordered structure observed for the X-ray crystallographic water molecules supported the hypothesis of an indirect transfer of a proton from Asp27 residue to the N5 atom of the substrate, analyzed in more detail in a subsequent study [60]. Moreover, a good agreement between experimental and QM/MM primary KIEs was found.

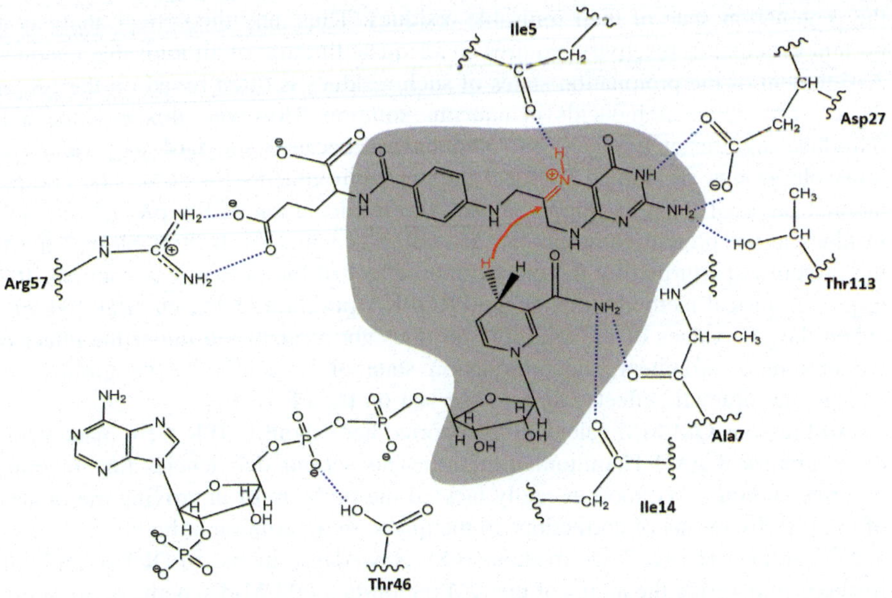

Fig. 15.3 Schematic representation of the hydride transfer step from nicotinamide ring of NADPH to the pteridine ring of DHF-H+in the active site of E. coli DHFR. A possible QM-MM partitioning employed in QM/MM calculations on this system is shown in grey

15.2.4 Limitations of Analysis Based on PESs

The use of different QM/MM molecular models to study the DHFR catalyzed reaction allowed getting some other preliminary conclusions. For reactions involving a large number of degrees of freedom it is not correct to refer to one unique TS structure. The TS should be considered as an average of nearly degenerated structures. Although other authors as Truhlar and co-workers had already pointed out this conclusion for molecular systems in condensed media [61], these results were one of the first QM/MM studies that demonstrated how the transition state for the enzymatic reaction must be represented as an average of the properties of these many, nearly degenerate TS structures. This insight emerged only as a consequence of the flexible model of the active site employed in the study. The observation of slightly different conformations obtained from different QM/MM models was in agreement with our previous QM/MM studies on LDH [15], where different minimum energy paths with noticeable different potential energy barriers were located starting from different initial configurations obtained by MD simulations.

Another common problem that can rise in theoretical studies of enzyme reaction mechanisms is that the most expensive computational simulation may give meaningless results if erroneous charges are assigned to the titratable aminoacids of the protein. Properties of enzymes are pH-dependent because their efficacy depends on the protonation state of their ionisable residues. Thus, any theoretical analysis of protein structure or reactivity requires an adequate titration of all ionisable residues. Assignment of the protonation states of such residues is often based on the pK_a of the corresponding aminoacids in aqueous solution. However, this practice may introduce significant artefacts into simulations because standard pK_a values of ionisable groups are shifted by local protein environments [62, 63]. The computational protocol must be improved with the recalculation of the pK_a of titratable residues in the protein environment, according to which each titratable residue in the protein is perturbed by the electrostatic effect of the protein environment [62, 63], or empirical methods such as the PROPKA program of Jensen et al. [64–66]. When the pK_a values of the ionisable residues are recalculated under the effect of the protein environment, the protonation state of several ionisable groups can change and dramatic effects can be observed on the PESs [67].

All previous QM/MM calculations reported here about DHFR were made using a semiempirical AM1 Hamiltonian that presents serious deficiencies for obtaining accurate energies. We more recently tackled the problem of improving the quality of the PES by means of corrections to the quantum description at higher level such as DFT or post-Hartree Fock methods [68]. Nowadays, the use of DFT or ab initio methods to describe the atoms of the QM region in a QM/MM scheme is becoming more and more popular due to the efficiency of the algorithms and the increase of the computational power. Another possible strategy is based on the reparametrization of semiempirical Hamiltonians to accurately reproduce the energetics of a particular reaction. Indeed, AM1-Specific Reaction Parameters (AM1-SRP) for the hydride transfer catalysed by DHFR have been developed by Major and co-workers paving the way to accurate and cheap simulations on this system [69].

15.3 Free Energy Surfaces: Statistical Simulations

As pointed out above, a fundamental difficulty of modelling large molecular systems derived from the fact that exploration of a single projection of the PES is not enough to get magnitudes directly comparable to experiment. The PES contains a myriad of stationary structures mainly due to the great number of possible conformations accessible to the enzyme and the solvent molecules, as already indicated above. Furthermore, enzymes are usually very flexible molecular systems and in fact, sidechain and backbone movements are, in many cases, responsible of some of the steps of the full catalytic process. Thus, a statistical ensemble of minima and transition structures must be explored to properly define the reactant state (RS), TS and product state (PS) and entropic and thermal effects should be included in the analysis. Free energies associated to the transformation between different states can be extracted using different techniques applied to molecular simulations [70].

Transition State Theory (TST), in its conventional [71] or generalized formulation, [72] has provided an adequate theoretical framework to study enzyme catalyzed reactions in the past decades [73]. According to TST, the rate of a chemical reaction can be obtained from the equilibrium forward flux across a hypersurface that divides the reactant region from the product region [74]. This dividing surface is defined by a particular value of the reaction coordinate ($\xi = \xi^{\ddagger}$), which is assumed to be separable from the remaining degrees of freedom of the system. Thus, according to TST, the rate constant of a given reaction is given by:

$$k_r = \kappa \cdot \gamma(\xi) \frac{k_B T}{h} K^0 e^{-\frac{\Delta G^{\ddagger}(\xi)}{RT}} \tag{15.1}$$

where k_B and h are the Boltzmann's and Planck's constants, T the absolute temperature and K° accounts for the standard state. $\Delta G^{\ddagger}$ is the free energy difference between the TS and the RS. γ is the transmission coefficient that accounts for the non-separability of the reaction coordinate from the remaining degrees of freedom. This parameter being lower than unity indicates the presence of trajectories recrossing the dividing surface defined by the selected reaction coordinate. Tunneling contributions, κ, can be important in enzymatic reactions involving the transfer of light particles, such as proton, hydrogen or hydride transfers (H transfers) [17]. According to Eq. (15.1), an increase in the rate constant for the catalyzed reaction relative to the uncatalyzed reaction can be due to a decrease in the free energy barrier or to an increase in the prefactor. Changes in the activation free energy seem to be the main contribution to catalysis [6, 12, 75], while prefactors are similar in the catalyzed and uncatalyzed processes [17, 76, 77]. However, it must be stressed that there is no an unambiguous way to separate the effects of the enzyme on the activation free energy from those on the transmission coefficient because both depend on the choice of the reaction coordinate [17].

A common approach to compute free energies for systems with a small number of degrees of freedom is based on normal analysis methods applied to the located stationary points on the explored PESs. The knowledge of local regions around these points can provide thermodynamic information about the system. Nevertheless, as stated above, enzymatic systems contains a large number of degrees of freedom and the associated PES can be very rugged, preventing the use of the harmonic approximation. A single structure belonging to a particular state does not offer information of other possible conformations describing the same state. In the worst of the possible scenarios, if the located structures were far from the lowest energy conformation of the same state, the results would be completely wrong. Then, statistical simulations based on Monte Carlo or Molecular Dynamics methods must be used to obtain detailed information from which averaged and thermodynamical properties can be properly derived [70]. In particular, free energies associated to the transformation from the RS to the TS (the activation free energy) and to the PS (the reaction free energy) can be extracted using different techniques applied to molecular simulations.

15.3.1 *Free Energy Perturbation (FEP)*

Quantum mechanics free energy perturbation (QM-FEP) method, developed by Kollman and coworkers [78, 79], is a popular technique to get reaction free energy profiles for enzymatic reactions. In this approach the reaction path is obtained for a gas phase model of the active site. Coordinates and charges of the chemical system are afterwards used into purely classical simulations where only the changes in the environment are sampled. FEP is then used to get the reaction and activation free energies as the sum of the gas phase reaction energy and environment free energy contributions. The main advantage of this approach is that high-level quantum methods can be used for the gas phase calculations. The main drawbacks are: (i) the lack of the chemical system flexibility contribution to the free energy and (ii) during the simulation of the reaction path, the environment is not incorporated in the calculation.

A similar but more coupled scheme is obtained if the reaction path is determined in the active site by means of QM/MM iterative optimisations along a reaction path [80]. In our approach the reaction path is obtained as an IRC traced from a TS structure located in the active site and considering the influence of the full enzymatic environment. The configurations for which the free energy difference on enzyme catalyzed reactions can be estimated correspond to those structures obtained along the QM/MM IRC calculation and are thus characterized by a single coordinate:

$$s = \left[\sum_{i \in QM} m_i \left((x_i - x_{0i})^2 + (y_i - y_{0i})^2 - (z_i - z_{0i})^2 \right) \right]^{1/2} \tag{15.2}$$

where x_{0i}, y_{0i}, z_{0i} are the Cartesian coordinates of the QM atoms in the transition state structure while x_i, y_i and z_i are the coordinates of a structure belonging to the IRC, and m_i are the masses of the atoms. Within this treatment the free energy relative to the reactant can be expressed as a function of the s coordinate as:

$$\begin{aligned} \Delta G_{FEP}(s^{\mathrm{R}} \to s^j) &= \Delta E_{QM}^{0,R\to j} + \Delta G_{QM/MM}^{R\to j} \\ &= \left(E_{QM}^0(s^{\mathrm{j}}) - E_{QM}^0(s^R) \right) - k_B T \sum_{i=R}^{i=j-1} \ln \left\langle \exp \frac{1}{k_B T} \left(\mathrm{E}_{\mathrm{QM/MM}}(s^{\mathrm{i+1}}) - \mathrm{E}_{\mathrm{QM/MM}}(s^{\mathrm{i}}) \right) \right\rangle_{MM,i} \end{aligned} \tag{15.3}$$

where E_{QM}^0 is the gas-phase energy of the QM subsystem, k_B is the Boltzmann's constant and T the temperature. The QM/MM interaction term of the free energy difference between two different values of s is obtained by averaging the QM/MM interaction energy (including the polarization energy) over all the MM coordinates of the system obtained for a particular value of the s coordinate. Then, a term to include the contribution of the vibration of all the QM coordinates (except s) can be added to the free energy profile, assuming a quantum harmonic treatment for the vibrational modes of the QM subsystem in the field created by the rest of the enzyme. In this way the chemical system flexibility is also considered. The quasiclassical free energy difference can be obtained adding the contribution of the QM vibrational degrees of freedom as:

$$G^{QC}(s^j) - G^{QC}(s^R) = \Delta E_{QM}^{0,R\to j} + \Delta G_{QM/MM}^{R\to j} + \Delta G_{vib,QM}^{j} \tag{15.4}$$

where

$$\Delta G_{vib,QM}^{j} = \sum_{x=1}^{3N-7}\left[\frac{1}{2}h\nu_x^j + k_B T\ln\left(1 - e^{-\frac{h\nu_x^j}{k_B T}}\right)\right] - \sum_{x=1}^{3N-6}\left[\frac{1}{2}h\nu_x^R + k_B T\ln\left(1 - e^{-\frac{h\nu_x^R}{k_B T}}\right)\right] \tag{15.5}$$

The frequencies of the QM subsystem can be obtained projecting out the contribution of the reaction coordinate in the Hessian for all the structures but the reactant. The activation free energy is then obtained as:

$$\Delta G^{\ddagger} = \max_j\left(G^{QC}(s^j) - G^{QC}(s^R)\right) \tag{15.6}$$

15.3.2 Potential of Mean Force (PMF)

Another popular technique to obtain the free energies profiles is based on the calculation of the Potential of Mean Force, PMF, associated to a particular distinguished reaction coordinate [70]. A proper selection of this coordinate is critical to obtain meaningful free energy profiles. Selection of this reaction coordinate should be based on the exploration of the PES including the environment, or even better, on IRCs traced down to the corresponding products and reactants valleys from transition structures located and characterized in the enzyme active site.

The main advantage of the PMF technique is that contributions of all the degrees of freedom of the system, except the reaction coordinate, are naturally considered and then it provides a magnitude more easily comparable to the activation free energy. The PMF is in fact related to the normalized probability of finding the system at a particular value of the chosen coordinate by Eq. (15.7):

$$W(\xi) = C - kT\ln\int \rho(\mathbf{r}^N)\delta(\xi(\mathbf{r}^N) - \xi)d\mathbf{r}^N \tag{15.7}$$

The activation free energy can be then expressed as [81]:

$$\Delta G^{\ddagger}(\xi) = W(\xi^{\ddagger}) - \left[W(\xi^{R}) + G_{\xi}(\xi^{R})\right] \tag{15.8}$$

where the superscripts indicate the value of the reaction coordinate at the RS and TS and $G_{\xi}(\xi^{R})$ is the free energy associated with setting the reaction coordinate to a specific value at the reactant. Normally this last term makes a small contribution [82] and the activation free energy is directly estimated from the PMF change between the maximum of the profile and the reactant's minimum, correcting for the quantized nature of the vibrations of the QM subsystem:

$$\Delta G^{\ddagger}(\xi) \approx W(\xi^{\ddagger}) - W(\xi^{R}) + \Delta G_{vib,QM} = \Delta W^{\ddagger}(\xi) + \Delta G_{vib,QM} \tag{15.9}$$

The selection of the reaction coordinate is trivial when the mechanism can be driven by a single valence geometric coordinate or a simple combination of them (as the anti-symmetric combination of two interatomic distances). However this may not be the case for an enzyme-catalyzed reaction, were several distances are usually involved in the process. In some cases, the method can be improved by computing two dimensional PMF (2D-PMF) using two coordinates. The 2D-PMF is related to the probability of finding the system at particular values of two coordinates:

$$W(\xi_1, \xi_2) = C' - kT \ln \int \rho(\mathbf{r}^N)\delta(\xi_1(\mathbf{r}^N) - \xi_1)\delta(\xi_2(\mathbf{r}^N) - \xi_2)d\mathbf{r}^N \tag{15.10}$$

In order to obtain the full probability distribution, independent MD simulations (windows) are run around particular values of the selected coordinates applying a biasing potential using assuming an harmonic dependence, known as the umbrella sampling approach [83, 84]. Then the values of the variables sampled during the different simulations are pieced together to construct a full distribution function employing the weighted histogram analysis method (WHAM) [84].

In the case of the hydride transfer catalysed by *Escherichia coli* DHFR (EcDHFR), one-dimensional PMFs were computed using the antisymmetric combination of distances describing the hydride transfer, $\xi = d(C_{donor}H) - d(C_{acceptor}H)$, as the reaction coordinate. The umbrella sampling approach was used, with the system restrained to remain close to the desired value of the reaction coordinate by means of the addition of a harmonic potential. As stated above, the probability distributions obtained from MD simulations within each individual window were combined by means of the WHAM method. Twenty picoseconds of relaxation and 40 ps of production MD, with a time step of 0.5 fs, in the canonical ensemble [NVT (number of molecules, volume, temperature), with a reference temperature of 300 K] and the Langevin–Verlet integrator [85] were used in the simulations.

Figure 15.4 shows the classical mechanical AM1-SRP/MM PMF started from five different structures of the TS, selected from snapshots of a long QM/MM MD

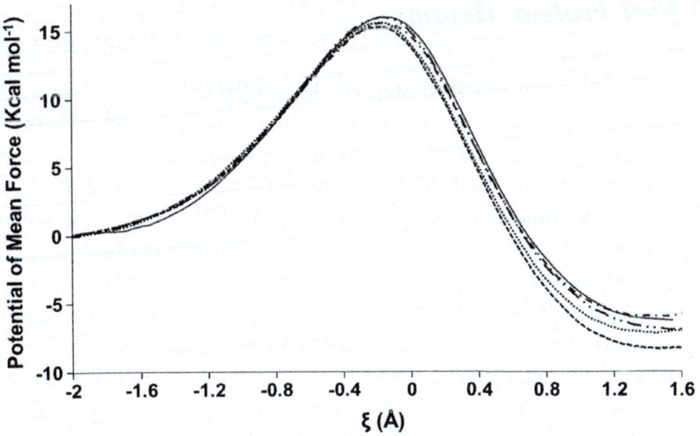

Fig. 15.4 AM1-SRP/MM potential of mean force (PMF) obtained from five different structures of the TS. Reaction coordinate, $\xi = d(C_{donor}H) - d(C_{acceptor}H)$

simulation with the reaction coordinate restrained to the value obtained for the transition state (TS) of the first PMF. The observed deviations are within the expected uncertainty of the method. From these PMFs, the classical mechanical activation free-energy barrier, $\Delta W^{\ddagger}$, is 15.8 kcal·mol^{-1}, from which a $\Delta G^{\ddagger}(\xi)$ value of 14.6 kcal·mol^{-1} can be derived (see Eq. 15.9). This value was in excellent agreement with the free energy barrier derived from the experimental rate constant (14.3 kcal·mol^{-1}) [86].

15.4 Dynamic Complexity

As mentioned above, enzymes are not rigid structures that simply accommodate the substrates needed for a chemical reaction. Enzyme dynamics is complex and its motions span a wide range of time-scales, from fast bond stretching motions that require femtoseconds to large conformational changes happening in milliseconds or even larger time-scales [10]. During the enzymatic turnover protein motions can be involved at different stages. For example, substrate binding and product release may require the motion of protein loops that control the accessibility of the active site. The enzymatic turnover can be then seen as a path on a multidimensional free energy surface involving conformational and chemical coordinates [87]. Because of the large number of degrees of freedom these surfaces are rough and within a particular subset of configurations, identified as a valley by the value of a global conformational coordinate in a particular free energy landscape, it is possible to identify a manifold of sub-valleys characterized by more local coordinates, establishing in this way hierarchy of complexity [88].

15.4.1 Fast Protein Motions

Focusing in the chemical reaction step of the catalytic cycle, a particular reactant conformation can be transformed into a product conformation moving along the chemical coordinate. This coordinate is in principle a collective variable that involves the participation of all the atoms of the system. Such kind of coordinates are employed for example in the context of Empirical Valence Bond calculations of enzymatic catalysis where the reaction free energy profile is obtained as a function of the energy gap between the diabatic curves that define the reactants and products states [27, 28]. In other QM/MM schemes the reaction coordinate is usually selected as a combination of a few geometrical parameters, for example the distances associated to the chemical bonds that formed and/or broken during the chemical reaction, as made in Fig. 15.4. When the free energy profile is obtained along the reaction coordinate, which is the quantity needed to determine the rate constant in TST (Eq. 15.1), all the remaining degrees of freedom of the system are assumed to be in equilibrium at each value of the coordinate. It is precisely the validity of this equilibrium assumption that is under debate in the field of enzymatic catalysis. According to some authors fast protein motions, usually termed as 'promoting vibrations', taking place in the active site could be coupled to the reaction coordinate making necessary an explicit treatment of them in the calculation of the reaction coordinate [89, 90]. These 'promoting motions' could favour the approach of donor and acceptor atoms in H-transfer reactions, assisting then the chemical processes. Pressure effects on chemical reactions and temperature dependence of Kinetic Isotope Effects (KIEs) could be the signature of these promoting motions [89], although it should be stressed that these effects could be also analysed within a theoretical framework based in TST [91]. It should be noted that, in principle, protein motions are also incorporated in TST, either in the definition of the reaction coordinate or as thermal fluctuations.

The role of protein motions during the chemical step can be monitored using 2D free energy surfaces as a function of a chemical coordinate and an environmental coordinate. We investigated the hydride transfer reaction catalysed by EcDHFR using this approach [92]. As shown above, the chemical coordinate (ξ) was simply the antisymmetric combination of the distances of the transferred hydride to the donor and the acceptor atoms, while the environmental coordinate (s_e) was the antisymmetric combination of the electrostatic potential created by the protein and solvent atoms on the donor and acceptor atoms. This coordinate captures those protein motions affecting the reaction energetic and thus those motions that can be relevant for the chemical processes [93]. Figure 15.5 shows the free energy surface for the hydride transfer in DHFR as a function of the two coordinates. The minimum free energy path illustrates how the reaction proceeds with changes both in the chemical system and in the environment. Basically the reaction takes places with a small reorganization of the environment followed by the transfer of the hydride from the donor to the acceptor and a final environmental relaxation to the product valley. The motion along the environmental coordinate can be characterized with an

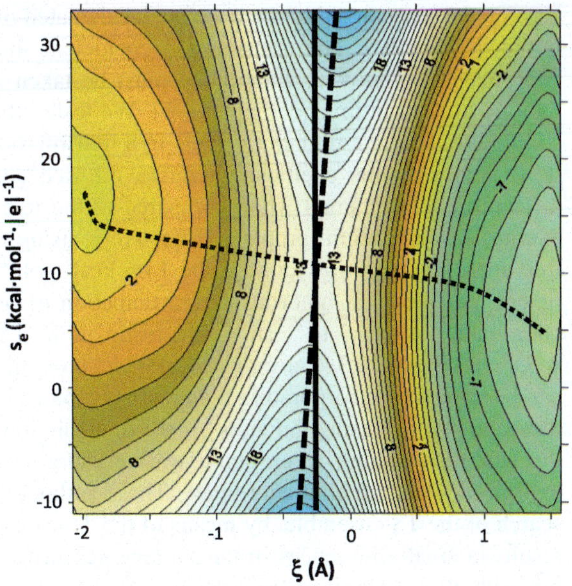

Fig. 15.5 2D Free energy surface for the hydride transfer reaction catalyzed by EcDHFR as a function of a chemical (ξ) and solvent (s_e) coordinate (see text for details). The *dotted line* represents the minimum free energy path and the *straight lines* correspond to the TS previously obtained just as a function of the chemical coordinate (*continuous line*) or as a function of both coordinates (*dashed*)

average frequency of 240 cm^{-1}, that corresponds to the reorganization of hydrogen bond interactions in the active site.

An interesting feature observed in this equal-foot treatment of substrate and protein motions is the similarity observed between the dividing surface (or TS) obtained when only the chemical coordinate is employed and that obtained in the two-dimensional treatment (when protein motions are explicitly considered). This similarity is a consequence of the fact that in reactive trajectories slower environmental motions precede the H-transfer and thus the TS obtained under the equilibrium treatment for the chemical coordinate is a reasonable approximation to that obtained with a more complete definition of the reaction coordinate. It should be taken into account that in TST the equilibrium assumption is needed to describe only the reactants and transition states (allowing the determination of their relative probabilities or free energy difference) and not along the whole reaction path.

The quality of the reaction coordinate chosen to explore the reaction process can be tested by means of the computation of the transmission coefficient (see Eq. 15.1) that accounts from the fraction of trajectories that recross a particular dividing surface back to the reactants. This transmission coefficient can be evaluated according to the positive flux formulation [94], assuming that the trajectory is initiated at the barrier top with forward momentum along the reaction coordinate:

$$\gamma(t) = \frac{\langle j_+ \theta[z(+t)] \rangle - \langle j_+ \theta[z(-t)] \rangle}{\langle j_+ \rangle} \tag{15.11}$$

where ξ is the reaction coordinate, j_+ represents the initially positive flux at $t = 0$ and $\theta(\xi)$ is a step function equal to one in the product side of the reaction coordinate and zero on the reactant side. The average is calculated over many activated trajectories initiated at the TS. The transmission coefficient should approach unity when the coordinate is improved, although it must be taken into account that other effects may be present in this coefficient [95]. We tested that in the case of EcDHFR the simple chemical coordinate ξ results in a transmission coefficient of 0.57 at 300 K [86]. Although this coefficient reduces the rate constant by a factor of $\sim$2 its influence is really small when compared to the role played by the activation free energy, an equilibrium property. Effectively, using the ξ coordinate the activation free energy barrier was found to be 14.6 kcal·mol^{-1}, while the transmission coefficient indicates that ignoring the participation of protein motions in the reaction coordinate leads to an underestimation of the barrier of only $\Delta G = -RT \cdot \ln\gamma \sim 0.3$ kcal·mol^{-1}. Major and co-workers showed that the coordinate can be improved incorporating the donor-acceptor distance and the hybridization of these atoms, increasing the value of the transmission coefficient up to 0.76 [96]. In this case the resulting TS ensemble was very close to that obtained using simple the antisymmetric stretching coordinate. These authors claimed that using a nonbiased search of the TS ensemble, by means of the Transition Path Sampling scheme [97], results in small differences in the average geometrical description of the ensemble but with almost identical free energy.

During the last decade we have studied a number of enzymatic reactions, not only H-transfer processes, and compared the transmission coefficient of the catalyzed reaction with that of the counterpart process in aqueous solution [77, 98, 99]. A useful approach for the calculation of the transmission coefficient is provided by Grote-Hynes theory, where this coefficient can be obtained from a generalized Langevin equation where a time-dependent friction acts on the advance of the system along the reaction coordinate [100]:

$$\omega_r^2 - \omega_{eq}^2 + \omega_r \int_0^\infty \zeta_{TS}(t) e^{-\omega_r t} dt = 0 \tag{15.12}$$

$$\gamma_{GH} = \frac{\omega_r}{\omega_{eq}} \tag{15.13}$$

where ω_{eq} and ω_r are the frequencies for the motions over the barrier top obtained under the equilibrium approach and the actual reactive value, respectively and $\zeta_{TS}(t)$ is the friction kernel obtained at the TS. Analysis of this friction kernel offers the opportunity for a systematic improvement of the reaction coordinate characterizing those motions that are more strongly coupled to the reaction coordinate.

Our results based in the use of Grote-Hynes theory indicate that the transmission coefficients are systematically larger for the enzymatic reactions than for the counterpart processes in aqueous solution. Although the ratio between both quantities is small, discarding then any consideration about an important contribution of dynamical effects to catalysis, this systematic observation points to a general characteristic of enzymatic processes: if the enzymatic active site is prepared to stabilize the TS, then less environmental motions would be required to evolve the system from reactants to the TS, resulting in smaller coupling to the reaction coordinate and then in a larger value (closer to unity) of the transmission coefficient. This picture agrees with the preorganization concept introduced by Warshel [101]. The same conclusion was reached after the comparison of the 2D free energy surfaces obtained for the hydride transfer step in EcDHFR (Fig. 15.5) and in aqueous solution [92]. In the latter case the displacement along the environmental coordinate (s) was significantly larger than for the enzymatic reaction, confirming that much larger environmental motions are required to accommodate the TS charge distribution in solution than in the enzyme. A preorganized active site minimizes the participation of environmental motions in the process, something reflected in the larger value of the transmission coefficient corresponding to a distinguished reaction coordinate defined exclusively by the coordinates of the chemical system.

As a general conclusion of this section, the dynamical complexity of the active site in an enzymatic reaction can be satisfactorily considered within the framework of TST, either including protein vibrational motions in the definition of the reaction coordinate or treating them as equilibrium thermal fluctuations and incorporating a recrossing transmission coefficient into the formalism. None of the reactions analysed until now has shown conclusive evidences of a severe failure of the equilibrium approximation. Such a behaviour could be expected for chemical reactions with a very low free energy barrier (few $kcal \cdot mol^{-1}$), an improbable observation as far as there is no any evolutive pressure to optimize the chemical step in enzymes below the rate of the diffusion process.

15.4.2 Slow Protein Motions

In addition to local protein motions occurring within a valley of the multidimensional free energy landscape in an enzymatic reaction, which can be conveniently sampled during simulations and incorporated into the evaluation of the rate constant, more global conformational changes may also occur in proteins. In this case the interconversion rates among the corresponding conformational states can be similar or slower than the catalytic rate constants. In these conditions MD simulations started in a given valley are in practice unable to jump to another valley, because the free energy barrier for the conformational change is too high. Then, while a local equilibrium exists inside each valley, a global equilibrium including all the valleys is very unlikely. The existence of relatively stable reactant valleys in comparison with the time scale of the enzymatic reactions is the required molecular basis to understand the dynamic disorder observed in single molecular studies of some enzyme mechanisms [14, 102–105]. From a theoretical point of view, the study of this conformational diversity require the identification of different conformers and the evaluation of the rate constant for each of them [16].

A particular interesting case of protein slow motions with a possible impact in the rate constant of the chemical step could happen when such a conformational motion is required to activate the substrate. If this motion is a requisite for the reaction and is slower or with similar characteristic times than the chemical step it may be responsible of a reduction in the observed reaction rate [106]. This could be the case of loop motions closing the active site to provide an adequate environment, lowering the activation free energy along the chemical coordinate. EcDHFR has been postulated as a possible example of this interrelationship between conformational changes and chemical reaction, where there is a loop closing the active site, the M20 loop, that has been found in different conformations along the catalytic cycle (see Fig. 15.6) [107, 108].

It was recently suggested that the impact of the motions of this loop in the rate constant of the catalytic step could be observed introducing mutations that reduced the mobility of the loop [109]. The N23PP-S148A DHFR mutant presents reduced dynamics in the millisecond time scale when compared to the wild type enzyme and the loop is unable to undergo the closed to occluded transition observed in Fig. 15.6. In addition this mutant shows a diminution in the catalytic rate constant of about 16 times, which was then attributed to the restriction of active site fluctuations that sample higher-energy substrates that are conductive to formation of the transition state [109]. However, two experimental observations contradict this interpretation: (i) the absence of any dependence of the rate constant with solvent viscosity, which seems to exclude the participation of large conformational changes in this step [110] and (ii) the fact that the mutant and wild type enzymes show the same temperature dependence of KIEs, indicating that both versions of the enzyme

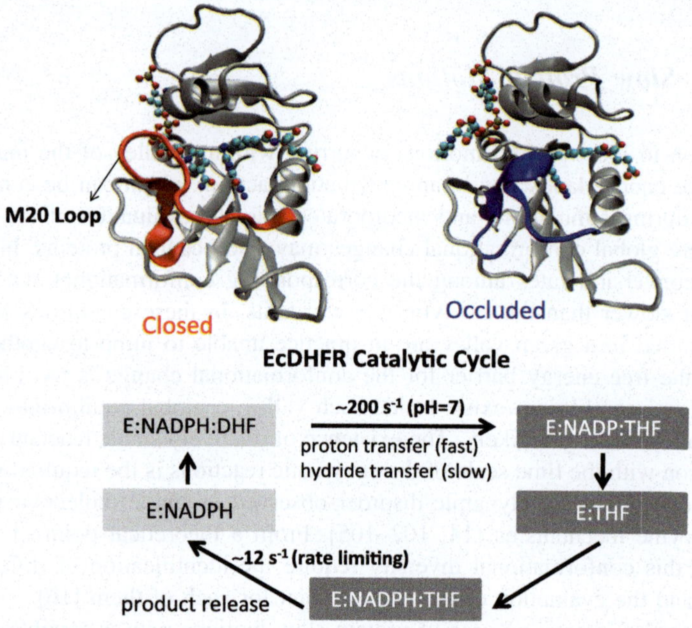

Fig. 15.6 Representation of EcDHFR with two different conformations of the M20 loop and the complete catalytic cycle where the M20 conformation is indicated by means of a color code (light grey = closed; dark grey = occluded)

explore similar conformations at the TS [111]. In a QM/MM study conducted on the N23PP-S148A mutant we showed that the observed reduction in the rate constant can be attributed to an increase in the activation free energy of the chemical step, from 14.6 kcal·mol^{-1} in the wild type to 16.4 kcal·mol^{-1} in the mutant [92, 112]. This is, the main difference between the catalytic efficiency of the mutant and wild type DHFRs seems to be due to an equilibrium and not to a dynamical property as concluded also by Warshel and coworkers [113]. An analysis of the averaged TS structures showed that the mutations introduced in the mutant impair some protein-substrate and protein-cofactor interactions that are responsible of the differential TS stabilization in the active site. We also characterized the frequency of the environmental coordinate in the mutant and the average value obtained was 230 cm^{-1}, almost identical to the value obtained in the wild type. The recrossing transmission coefficient in the mutant was shown to be slightly smaller than in the wild type, 0.53 versus 0.57, which is the expected behavior considering that the mutations reduce the preorganization of the enzyme [92, 112]. These observations seems to exclude the participation of slow protein motions in the conversion from the Michaelis complex to the TS and show that the observed reduction in the rate constant of the mutant can be simply explained considering that mutations change the structural and electrostatic characteristics of the active site.

15.5 Quantum Tunnelling Effects

In H-transfer reactions the motion of the transferred light particle (proton, hydrogen atom or hydride) between the donor and the acceptor atoms must treated quantum mechanically. The description of this motion varies between two limiting cases [114]. The first one corresponds to a quantized vibration of the light particle with a zero point energy above the potential energy barrier found when the light particle is transferred from the donor to the acceptor atoms. This case is usually known as the adiabatic or strong-coupling limit and can be adequately considered in QM/MM simulations including vibrational corrections to the classical free energy barrier (see Eq. 15.5). The second situation corresponds to a vibrational motion in which the zero point energy is below the potential energy barrier and the hydrogen atom is transferred by non-classical penetration of the barrier, or tunnelling. This is the nonadiabatic or weak-coupling limit. In this case the rate constant must be evaluated including not only vibrational corrections to the classical free energy barrier but also the tunnelling probability. In the Ensemble-Averaged Variational Transition State Theory (EA-VTST) [115] tunnelling is incorporated as a prefactor that corrects the classical or adiabatic rate constant (see Eq. 15.1). This prefactor is calculated using an ensemble of reaction paths traced from different TS configurations and for each of the paths the tunnelling transmission coefficient is obtained as the ratio between the quantum and classical probability for transmission through or above the energy barrier, evaluated for all the possible energy values. The final value is averaged over a representative ensemble of reaction paths.

The nature of the H-transfer mechanisms strongly depends on the donor acceptor distance, DAD. At large DAD values, the energy barrier for the motion of the H-atom between the donor and the acceptor atoms will be high and then the transfer essentially occurs via quantum tunnelling. When the distance is reduced the barrier becomes lower and thinner, enhancing tunneling. Finally, at smaller values of donor-acceptor distances the barrier can be lowered below the energy of the vibrational ground state, reaching the adiabatic limit. In connection with the discussion presented in the previous section, some studies on enzymatic H-transfer reactions have emphasized the role of protein motions reducing the donor-acceptor distance. In this sense, a change from temperature independent KIEs in a wild type enzyme to strong temperature dependence of KIEs in a mutant [116] could be the indication of a change in the H-transfer regime, from mainly adiabatic to nonadiabatic, due to a mismatch in the mutant ability to sample short donor-acceptor distances at the TS. According to some interpretations this observation could be the signature of the participation of protein motions in the chemical step. In such a case even remote mutations could affect the rate constant by disrupting a dynamic network of concerted protein motions [117].

We investigated the differences in the tunnelling coefficients between the wild-type and the 'dynamical knockout' N23PP-S148A DHFR mutant [92]. At 300 K we obtained averaged tunneling coefficients of 2.61 ± 0.49 and 2.25 ± 0.45, for the wild type and the mutant, respectively [92, 112]. The difference is below the statistical uncertainties of the two magnitudes, indicating that the change in the tunneling coefficient, if any, must be very small and, in any case, negligible compared to the change due to the modification in the activation free energy. These small differences can be rationalized comparing the equilibrium donor-acceptor distances which has been observed at the TS configurations obtained for the wild type and N23PP-S148A EcDHFR (see Fig. 15.7). Mutations provoke a slight shift to larger values in the donor-acceptor distances distribution, leading to a small decrease of the tunneling coefficient.

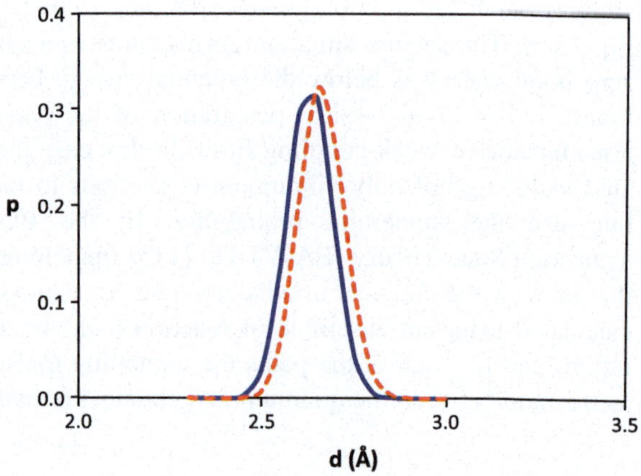

Fig. 15.7 Probability distributions of donor-acceptor distances at the TS of the hydride transfer reaction in wild type EcDHFR (*continuous line*) and N23PP-S148A mutant (*dashed line*)

15.6 Testing Models: Enzyme Kinetic Isotope Effects

In the last years, the use of isotopically substituted enzymes has been revealed as a useful tool to quantify the role of protein motions in the chemical step of enzyme-catalyzed reactions [112]. Isotopic substitutions of atoms involved in the chemical step are traditionally used as a potent tool to know about the characteristics of the TS and, consequently, to study reaction mechanisms [118]. Similar strategies have been applied to the full enzyme from the early works reported in 1969 on Alkaline Phosphatase, where all non-exchangeable hydrogens were substituted by deuterium [119]. The difference on the rate constant between the deuterated ("heavy enzyme") and the natural ("light enzyme") enzyme was attributed to the diminution of the vibrational energy available in the active site. The strategy was later improved replacing not only the non-exchangeable hydrogens by ^{2}H, but also all carbon and nitrogen atoms by ^{13}C and ^{15}N, respectively, on a version of human Purine Nucleoide Phosphorylase [120]. A diminution in the rate constant of the chemical step, relative to the wild or "light enzyme", was interpreted as a dynamical link between mass-dependent bond vibrations of the protein and events in the reaction coordinate [120].

In recent years, we have demonstrated that this effect can be also estimated within the theoretical framework of TST [86]. Combining experimental kinetic studies and QM/MM simulations, the observed differences in reactivity between a heavy EcDHFR (containing ^{13}C, ^{15}N and ^{2}H) and the natural light EcDHFR were analyzed using Eq. (15.1). Values of enzyme KIE (the ratio between the rate constant of the light and heavy versions), showed good agreement between experiments and simulations, 1.10 ± 0.04 and 1.16 ± 0.04, respectively, and were attributed to a lower value of the recrossing transmission coefficient in the heavy enzyme than in the light version. Heavy enzymes present a larger number of recrossings because protein motions coupled to the reaction coordinate are slower. Another important result was the fact that the free energy barrier and the contribution of quantum tunneling were not affected after isotopic labelling, indicating no significant role for "promoting motions" in driving tunneling or modulating the barrier.

The study of enzyme KIEs on EcDHFR was also performed on the catalytically compromised variant, EcDHFR-N23PP/S148A, commented in the previous section [92]. As in the study of wild type EcDHFR, a heavy EcDHFR-N23PP/S148A was prepared and, again, normal enzyme KIE was measured to be 1.37 ± 0.02 at pH 7 while the computationally predicted value was 1.26 ± 0.04 [92]. Again, the change in the rate constant after isotopic substitution was completely attributed to differences in the transmission coefficient. As commented above, the impairing of some protein-substrate interactions in the wild type with respect to the mutant enzyme leads to a larger participation of protein motions in the reaction coordinate. This is translated into smaller values of the recrossing transmission coefficients and a larger value of the enzyme KIEs.

The conclusions of the dynamic role of protein motions in EcDHFR have been later supported by further isotope substitution studies on a moderately thermophilic bacterium *Geobacillus stearothermophilus* DHFR (BsDHFR) [121]. Both, experimental and QM/MM studies, showed enzyme KIEs for hydride transfer close to unity at physiological temperatures and increasing at lower temperatures. This behavior was opposite to that observed for EcDHFR [92], where the enzyme KIEs increased slightly with increasing temperature (see Fig. 15.8).

Our simulations indicate that BsDHFR has greater flexibility than EcDHFR on the ps–ns time scale, which affects the coupling of the environmental motions of the protein to the chemical coordinate and consequently to the recrossing trajectories on the reaction barrier. Then, larger enzyme KIEs in BsDHFR than in EcDHFR at low temperatures can be understood as a consequence of the greater flexibility of the former. At high temperatures, although mass-dependent vibrational frequencies are lowered by enzyme isotope substitution, electrostatic preorganization in BsDHFR remains optimal and, consequently, the probability of dynamic recrossing and enzymes KIEs are small. Instead, at low temperatures, differences due to isotope-dependent protein motions become more evident between the 'light' and 'heavy' enzymes and enzyme KIEs in BsDHFR increase. The motion along the reaction coordinate is reduced at low temperatures and then the hydride transfer step experiences more friction due to protein motions that reorganize the active site. EcDHFR has not evolve to work at high temperatures. Then, the structure becomes more disorder and then the preorganization of the protein is partially lost. This increases the participation of protein motions in the reaction coordinate slightly increasing the enzyme KIEs. This analysis demonstrates that protein motions have distinct effects at different stages of catalysis: enzyme flexibility is critical during the process of preorganization, yet too much flexibility could perturb the stability of the transition state.

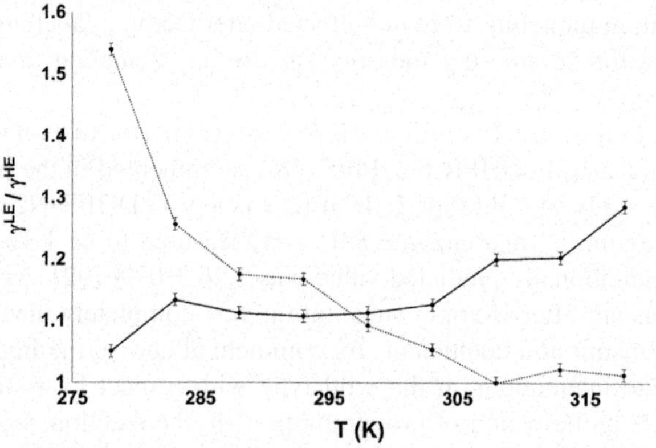

Fig. 15.8 Temperature dependence of the ratio of transmission coefficients of light and heavy (γ^{LE}/γ^{HE}) enzymes computed from QM/MM simulations. EcDHFR is displayed in *solid line* and BsDHFR in *dotted line*

Our theoretical treatment demonstrates that these effects can be satisfactorily reproduced including a transmission coefficient in the rate constant calculation (Eq. 15.1), whose dependence on temperature is affected by the protein flexibility. Protein dynamics would have a small, but measurable, effect on the chemical reaction rate. These studies on DHFRs demonstrate that TST framework, corrected for dynamic recrossings, can satisfactorily be used to characterize the enzyme transition state and to reproduce and rationalize small effects, such as the enzyme KIEs and their temperature dependence.

15.7 Conclusions and Perspectives

The conversion of DHF to THF catalysed by DHFR has been used in this chapter as a reaction model to present the evolution and difficulties on computational methods to describe chemical reactions in enzymes, since the early calculations based at semiempirical level carried out in gas phase to the recent sophisticated simulations based on hybrid Quantum Mechanical/Molecular Mechanics Dynamics schemes.

A lot of controversy about the role of the different aminoacids of the active site, tunnelling contribution to catalysis and problems related with the dynamic and flexibility of the protein has raised for this and other systems during the past decades. Enzyme reactions are electronically and dynamically complex processes because many atoms can be involved. As a consequence of this complexity the analysis of experimental and computational data is not straightforward and different interpretations can be derived from the same set of data. As it has been shown in this chapter, computational simulations of enzymatic processes must be carried out incorporating a realistic description of the environment, both from the energetic and dynamic points of view. Our discussion shows that the comparison between simulations and kinetic experiments can be safely carried out within the framework of Transition State Theory when protein dynamics, tunnelling contributions and conformational changes are introduced in the calculations. In any case, further developments are still needed to improve the definition of the reaction coordinate and to incorporate on an equal-foot treatment the conformational diversity observed in enzymatic systems. Because of their size and the range of timescales available for protein motions we are still far from a complete understanding of their behaviour during the full catalytic process.

The insights obtained from the theoretical studies carried out on the hydride transfer catalysed by different DHFRs clearly show that we are just starting to understand the whole complexity and richness of the dynamic/kinetic behaviour of enzymes. Clearly, further developments of not only computational tools but also of adequate theoretical frameworks are needed to achieve this goal.

Acknowledgements This work was supported by the Spanish Ministerio de Economía y Competitividad for project CTQ2012-36253-C03, Universitat Jaume I (project P1 1B2014-26) and Generalitat Valenciana (PROMETEOII/2014/022).

References

1. Kaplan J, DeGrado WF (2004) De novo design of catalytic proteins. Proc Natl Acad Sci USA 101(32):11566–11570
2. Rothlisberger D, Khersonsky O, Wollacott AM, Jiang L, DeChancie J, Betker J, Gallaher JL, Althoff EA, Zanghellini A, Dym O, Albeck S, Houk KN, Tawfik DS, Baker D (2008) Kemp elimination catalysts by computational enzyme design. Nature 453(7192):190–194
3. Frushicheva MP, Cao J, Chu ZT, Warshel A (2010) Exploring challenges in rational enzyme design by simulating the catalysis in artificial kemp eliminase. Proc Natl Acad Sci USA 107 (39):16869–16874
4. Nanda V, Koder RL (2010) Designing artificial enzymes by intuition and computation. Nature chemistry 2(1):15–24
5. Benkovic SJ, Hammes-Schiffer S (2003) A perspective on enzyme catalysis. Science 301 (5637):1196–1202
6. García-Viloca M, Gao J, Karplus M, Truhlar DG (2004) How enzymes work: analysis by modern rate theory and computer simulations. Science 303(5655):186–195
7. Martí S, Roca M, Andrés J, Moliner V, Silla E, Tuñón I, Bertran J (2004) Theoretical insights in enzyme catalysis. Chem Soc Rev 33(2):98–107
8. Warshel A, Sharma PK, Kato M, Xiang Y, Liu HB, Olsson MHM (2006) Electrostatic basis for enzyme catalysis. Chem Rev 106(8):3210–3235
9. Pauling L (1946) Molecular architecture and biological reactions. Chem Eng News Arch 24 (10):1375–1377
10. Henzler-Wildman KA, Lei M, Thai V, Kerns SJ, Karplus M, Kern D (2007) A hierarchy of timescales in protein dynamics is linked to enzyme catalysis. Nature 450(7171):913–916
11. Hammes GG, Benkovic SJ, Hammes-Schiffer S (2011) Flexibility, diversity, and cooperativity: pillars of enzyme catalysis. Biochemistry 50(48):10422–10430
12. Glowacki DR, Harvey JN, Mulholland AJ (2012) Taking Ockham's razor to enzyme dynamics and catalysis. Nature chemistry 4(3):169–176
13. Francis K, Kohen A (2014) Protein motions and the activation of the C-H bond catalyzed by dihydrofolate reductase. Curr Opin Chem Biol 21:19–24
14. Lu HP (2014) Sizing up single-molecule enzymatic conformational dynamics. Chem Soc Rev 43(4):1118–1143
15. Turner AJ, Moliner V, Williams IH (1999) Transition-state structural refinement with GRACE and CHARMM: flexible QM/MM modelling for lactate dehydrogenase. Phys Chem Chem Phys 1(6):1323–1331
16. Ferrer S, Tuñón I, Martí S, Moliner V, García-Viloca M, González-Lafont A, Lluch JM (2006) A theoretical analysis of rate constants and kinetic isotope effects corresponding to different reactant valleys in lactate dehydrogenase. J Am Chem Soc 128(51):16851–16863
17. Pu JZ, Gao JL, Truhlar DG (2006) Multidimensional tunneling, recrossing, and the transmission coefficient for enzymatic reactions. Chem Rev 106(8):3140–3169
18. Pisliakov AV, Cao J, Kamerlin SCL, Warshel A (2009) Enzyme millisecond conformational dynamics do not catalyze the chemical step. Proc Natl Acad Sci USA 106(41):17359–17364
19. Kanaan N, Ferrer S, Martí S, García-Viloca M, Kohen A, Moliner V (2011) Temperature dependence of the kinetic isotope effects in thymidylate synthase. A theoretical study. J Am Chem Soc 133(17):6692–6702
20. Swiderek K, Tuñón I, Moliner V (2014) Predicting enzymatic reactivity: from theory to design. Wires Comput Mol Sci 4(5):407–421
21. Warshel A, Levitt M (1976) Theoretical studies of enzymic reactions—dielectric, electrostatic and steric stabilization of carbonium-ion in reaction of lysozyme. J Mol Biol 103(2):227–249
22. Field MJ, Bash PA, Karplus M (1990) A combined quantum-mechanical and molecular mechanical potential for molecular-dynamics simulations. J Comput Chem 11(6):700–733
23. Blakley RL, Benkovic SJ, Whitehead VM (1984) Folates and pterins. Wiley, New York

24. Fierke CA, Johnson KA, Benkovic SJ (1987) Construction and evaluation of the kinetic scheme associated with dihydrofolate-reductase from Escherichia-Coli. Biochemistry 26 (13):4085–4092
25. Jurnak FA, McPherson A (1984) Biological macromolecules and assemblies. Wiley, New York
26. Cummins PL, Ramnarayan K, Singh UC, Gready JE (1991) Molecular-dynamics free-energy perturbation study on the relative affinities of the binding of reduced and oxidized nadp to dihydrofolate-reductase. J Am Chem Soc 113(22):8247–8256
27. Warshel A, Weiss RM (1980) An empirical valence bond approach for comparing reactions in solutions and in enzymes. J Am Chem Soc 102(20):6218–6226
28. Aqvist J, Warshel A (1993) Simulation of enzyme-reactions using valence-bond force-fields and other hybrid quantum-classical approaches. Chem Rev 93(7):2523–2544
29. van Duin ACT, Dasgupta S, Lorant F, Goddard WA (2001) ReaxFF: a reactive force field for hydrocarbons. J Phys Chem A 105(41):9396–9409
30. Andrés J, Safont VS, Martins JBL, Beltran A, Moliner V (1995) AM1 and PM3 transition structure for the hydride transfer—a model of reaction catalyzed by dihydrofolate-reductase. Theochem-J Mol Struc 330:411–416
31. Andrés J, Moliner V, Safont VS, Domingo LR, Picher MT, Krechl J (1996) On transition structures for hydride transfer step: a theoretical study of the reaction catalyzed by dihydrofolate reductase enzyme. Bioorg Chem 24(1):10–18
32. Wu YD, Houk KN (1987) Theoretical transition structures for hydride transfer to methyleniminium ion from methylamine and dihydropyridine—on the nonlinearity of hydride transfers. J Am Chem Soc 109(7):2226–2227
33. Wu YD, Houk KN (1987) Transition structures for hydride transfers. J Am Chem Soc 109 (3):906–908
34. Tapia O, Cárdenas R, Andrés J, Colonnacesari F (1988) Transition structure for hydride transfer to pyridinium cation from methanolate—modeling of ladh catalyzed reaction. J Am Chem Soc 110(12):4046–4047
35. Wu YD, Houk KN (1991) Theoretical evaluation of conformational preferences of Nad+ and Nadh—an approach to understanding the stereospecificity of Nad+Nadh-dependent dehydrogenases. J Am Chem Soc 113(7):2353–2358
36. Almarsson O, Karaman R, Bruice TC (1992) Kinetic importance of conformations of nicotinamide adenine-dinucleotide in the reactions of dehydrogenase enzymes. J Am Chem Soc 114(22):8702–8704
37. Almarsson O, Bruice TC (1993) Evaluation of the factors influencing reactivity and stereospecificity in NADPH dependent dehydrogenase enzymes. J Am Chem Soc 115 (6):2125–2138
38. Houk KN, González J, Li Y (1995) Pericyclic reaction transition-states—passions and punctilios, 1935–1995. Acc Chem Res 28(2):81–90
39. Himo F, Siegbahn PEM (2003) Quantum chemical studies of radical-containing enzymes. Chem Rev 103(6):2421–2456
40. Siegbahn PEM, Himo F (2009) Recent developments of the quantum chemical cluster approach for modeling enzyme reactions. J Biol Inorg Chem 14(5):643–651
41. Siegbahn PEM, Himo F (2011) The quantum chemical cluster approach for modeling enzyme reactions. Wires Comput Mol Sci 1(3):323–336
42. Lind MES, Himo F (2014) Theoretical study of reaction mechanism and stereoselectivity of arylmalonate decarboxylase. Acs Catal 4(11):4153–4160
43. de la Lande A, Gerard H, Moliner V, Izzet G, Reinaud O, Parisel O (2006) Theoretical modelling of tripodal CuN_3 and CuN_4 cuprous complexes interacting with O-2, CO or CH_3CN. J Biol Inorg Chem 11(5):593–608
44. de la Lande A, Parisel O, Gerard H, Moliner V, Reinaud O (2008) Theoretical exploration of the oxidative properties of a [(tren(Me1))CuO_2]+adduct relevant to copper monooxygenase enzymes: insights into competitive dehydrogenation versus hydroxylation reaction pathways. Chem-Eur J 14(21):6465–6473

45. Singh UC, Kollman PA (1986) A combined abinitio quantum-mechanical and molecular mechanical method for carrying out simulations on complex molecular-systems—applications to the Ch3cl+Cl- exchange-reaction and gas-phase protonation of polyethers. J Comput Chem 7(6):718–730
46. Théry V, Rinaldi D, Rivail JL, Maigret B, Ferenczy GG (1994) Quantum mechanical computations on very large molecular systems: the local self-consistent field method. J Comput Chem 15:269–282
47. Monard G, Loos M, Théry V, Baka K, Rivail JL (1996) Hybrid classical quantum force field for modeling very large molecules. Int J Quant Chem 58:153–159
48. Assfeld X, Rivail JL (1996) Quantum chemical computations on parts of large molecules: the ab initio local self consistent field method. Chem Phys Lett 263(34):100–106
49. Antonczak S, Monard G, Ruiz-López MF, Rivail JL (1998) Modeling of peptide hydrolysis by thermolysin. A semiempirical and QM/MM study. J Am Chem Soc 120(34):8825–8833
50. Ferré N, Assfeld X, Rivail JL (2002) Specific force field parameters determination for the hybrid ab initio QM/MM LSCF method. J Comp Chem 58:153–159
51. Monari A, Rivail JL, Assfeld X (2013) Theoretical modeling of large molecular systems. Advances in the local self consistent field method for mixed quantum mechanics/molecular mechanics calculations. Acc Chem Res 46:596–603
52. Gao JL, Amara P, Alhambra C, Field MJ (1998) A generalized hybrid orbital (GHO) method for the treatment of boundary atoms in combined QM/MM calculations. J Phys Chem A 102 (24):4714–4721
53. Maseras F, Morokuma K (1995) Imomm—a new integrated ab-initio plus molecular mechanics geometry optimization scheme of equilibrium structures and transition-states. J Comput Chem 16(9):1170–1179
54. Dapprich S, Komaromi I, Byun KS, Morokuma K, Frisch MJ (1999) A new ONIOM implementation in Gaussian98. Part I. The calculation of energies, gradients, vibrational frequencies and electric field derivatives. J Mol Struc-Theochem 461:1–21
55. Maseras F (2000) The IMOMM method opens the way for the accurate calculation of "real" transition metal complexes. Chem Commun 19:1821–1827
56. Ujaque G, Maseras F (2004) Applications of hybrid DFT/molecular mechanics to homogeneous catalysis. Struct Bond 112:117–149
57. Bo C, Maseras F (2008) QM/MM methods in inorganic chemistry. Dalton T 22:2911–2919
58. Moliner V, Turner AJ, Williams IH (1997) Transition-state structural refinement with GRACE and CHARMM: realistic modelling of lactate dehydrogenase using a combined quantum/classical method. Chem Commun 14:1271–1272
59. Castillo R, Andrés J, Moliner V (1999) Catalytic mechanism of dihydrofolate reductase enzyme. A combined quantum-mechanical/molecular-mechanical characterization of transition state structure for the hydride transfer step. J Am Chem Soc 121(51):12140–12147
60. Ferrer S, Silla E, Tuñón I, Martí S, Moliner V (2003) Catalytic mechanism of dihydrofolate reductase enzyme. A combined quantum-mechanical/molecular-mechanical characterization of the N5 protonation step. J Phys Chem B 107(50):14036–14041
61. Cramer CJ, Truhlar DG (1999) Implicit solvation models: equilibria, structure, spectra, and dynamics. Chem Rev 99(8):2161–2200
62. Gilson MK (1993) Multiple-site titration and molecular modeling—2 rapid methods for computing energies and forces for ionizable groups in proteins. Proteins-Struct Funct Genet 15(3):266–282
63. Antosiewicz J, Mccammon JA, Gilson MK (1994) Prediction of Ph-dependent properties of proteins. J Mol Biol 238(3):415–436
64. Jensen JH, Li H, Robertson AD, Molina PA (2005) Prediction and rationalization of protein pK(a) values using QM and QM/MM methods. J Phys Chem A 109(30):6634–6643
65. Li H, Robertson AD, Jensen JH (2005) Very fast empirical prediction and rationalization of protein pK(a) values. Proteins 61(4):704–721
66. Bas DC, Rogers DM, Jensen JH (2008) Very fast prediction and rationalization of pK(a) values for protein-ligand complexes. Proteins 73(3):765–783

67. Ferrer S, Silla E, Tuñón I, Oliva M, Moliner V, Williams IH (2005) Dependence of enzyme reaction mechanism on protonation state of titratable residues and QM level description: lactate dehydrogenase. Chem Commun 47:5873–5875
68. Martí S, Moliner V, Tuñón I (2005) Improving the QM/MM description of chemical processes: a dual level strategy to explore the potential energy surface in very large systems. J Chem Theory Comput 1(5):1008–1016
69. Doron D, Major DT, Kohen A, Thiel W, Wu X (2011) Hybrid quantum and classical simulations of the dihydrofolate reductase catalyzed hydride transfer reaction on an accurate semi-empirical potential energy surface. J Chem Theory Comput 7(10):3420–3437
70. Field M (2007) A practical introduction to the simulation of molecular systems, 2nd edn. Cambridge University Press, Cambridge
71. Eyring H, Stearn AE (1939) The application of the theory of absolute reaction rates to proteins. Chem Rev 24(2):253–270
72. Truhlar DG, Hase WL, Hynes JT (1983) Current status of transition-state theory. J Phys Chem 87(15):2664–2682
73. Gao JL, Ma SH, Major DT, Nam K, Pu JZ, Truhlar DG (2006) Mechanisms and free energies of enzymatic reactions. Chem Rev 106(8):3188–3209
74. Garrett BCTDG (2005) Theory and applications of computational chemistry: the first forty years. Elsevier, Amsterdam
75. Kamerlin SCL, Warshel A (2010) At the dawn of the 21st century: is dynamics the missing link for understanding enzyme catalysis? Proteins 78(6):1339–1375
76. Olsson MHM, Parson WW, Warshel A (2006) Dynamical contributions to enzyme catalysis: critical tests of a popular hypothesis. Chem Revs 106(5):1737–1756
77. Ruiz-Pernía JJ, Tuñón I, Moliner V, Hynes JT, Roca M (2008) Dynamic effects on reaction rates in a Michael addition catalyzed by chalcone isomerase. Beyond the frozen environment approach. J Am Chem Soc 130(23):7477–7488
78. Kollman PA (1993) Free-energy calculations—applications to chemical and biochemical phenomena. Chem Revs 93(7):2395–2417
79. Kollman PA, Kuhn B, Donini O, Perakyla M, Stanton R, Bakowies D (2001) Elucidating the nature of enzyme catalysis utilizing a new twist on an old methodology: quantum mechanical—free energy calculations on chemical reactions in enzymes and in aqueous solution. Acc Chem Res 34(1):72–79
80. Zhang YK, Liu HY, Yang WT (2000) Free energy calculation on enzyme reactions with an efficient iterative procedure to determine minimum energy paths on a combined ab initio QM/MM potential energy surface. J Chem Phys 112(8):3483–3492
81. Schenter GK, Garrett BC, Truhlar DG (2003) Generalized transition state theory in terms of the potential of mean force. J Chem Phys 119(12):5828–5833
82. Roca M, Moliner V, Ruiz-Pernía JJ, Silla E, Tuñón I (2006) Activation free energy of catechol O-methyltransferase. Corrections to the potential of mean force. J Phys Chem A 110(2):503–509
83. Torrie GM, Valleau JP (1977) Non-physical sampling distributions in Monte-Carlo free-energy estimation—umbrella sampling. J Comput Phys 23(2):187–199
84. Kumar S, Bouzida D, Swendsen RH, Kollman PA, Rosenberg JM (1992) The weighted histogram analysis method for free-energy calculations on biomolecules. 1. The Method. J Comput Chem 13(8):1011–1021
85. Verlet L (1967) Computer experiments on classical fluids. I. Thermodynamical properties of Lennard-Jones molecules. Phys Rev 159(1):98–103
86. Luk LYP, Ruiz-Pernía JJ, Dawson WM, Roca M, Loveridge EJ, Glowacki DR, Harvey JN, Mulholland AJ, Tuñón I, Moliner V, Allemann RK (2013) Unraveling the role of protein dynamics in dihydrofolate reductase catalysis. Proc Natl Acad Sci USA 110(41):16344–16349
87. Benkovic SJ, Hammes GG, Hammes-Schiffer S (2008) Free-energy landscape of enzyme catalysis. Biochemistry 47(11):3317–3321

88. Ansari A, Berendzen J, Bowne SF, Frauenfelder H, Iben IE, Sauke TB, Shyamsunder E, Young RD (1985) Protein states and proteinquakes. Proc Natl Acad Sci USA 82 (15):5000–5004
89. Hay S, Scrutton NS (2012) Good vibrations in enzyme-catalysed reactions. Nature chemistry 4(3):161–168
90. Schwartz SD (2013) Protein dynamics and the enzymatic reaction coordinate. Top Curr Chem 337:189–208
91. Glowacki DR, Harvey JN, Mulholland AJ (2012) Taking Ockham's razor to enzyme dynamics and catalysis. Nature Chem 4(3):169–176
92. Ruiz-Pernía JJ, Luk LYP, García-Meseguer R, Martí S, Loveridge EJ, Tuñón I, Moliner V, Allemann RK (2013) Increased dynamic effects in a catalytically compromised variant of Escherichia coli dihydrofolate reductase. J Am Chem Soc 135(49):18689–18696
93. Ruiz-Pernía JJ, Martí S, Moliner V, Tuñón I (2012) A novel strategy to study electrostatic effects in chemical reactions: differences between the role of solvent and the active site of chalcone isomerase in a Michael addition. J Chem Theory Comput 8(5):1532–1535
94. Bergsma JP, Gertner BJ, Wilson KR, Hynes JT (1987) Molecular-dynamics of a model Sn2 reaction in water. J Chem Phys 86(3):1356–1376
95. Mullen RG, Shea JE, Peters B (2014) Communication: an existence test for dividing surfaces without recrossing. J Chem Phys 140 (4):041104
96. Doron D, Kohen A, Nam K, Major DT (2014) How accurate are transition states from simulations of enzymatic reactions? J Chem Theory Comput 10(5):1863–1871
97. Bolhuis PG, Chandler D, Dellago C, Geissler PL (2002) Transition path sampling: throwing ropes over rough mountain passes, in the dark. Annu Rev Phys Chem 53:291–318
98. Castillo R, Roca M, Soriano A, Moliner V, Tuñón I (2008) Using grote-hynes theory to quantify dynamical effects on the reaction rate of enzymatic processes. The case of methyltransferases. J Phys Chem B 112(2):529–534
99. Roca M, Moliner V, Tuñón I, Hynes JT (2006) Coupling between protein and reaction dynamics in enzymatic processes: application of Grote-Hynes theory to catechol O-methyltransferase. J Am Chem Soc 128(18):6186–6193
100. Grote RF, Hynes JT (1980) The stable states picture of chemical-reactions. 2. Rate constants for condensed and gas-phase reaction models. J Chem Phys 73(6):2715–2732
101. Warshel A (1998) Electrostatic origin of the catalytic power of enzymes and the role of preorganized active sites. J Biol Chem 273(42):27035–27038
102. Xue QF, Yeung ES (1995) Differences in the chemical-reactivity of individual molecules of an enzyme. Nature 373(6516):681–683
103. Lu HP, Xun LY, Xie XS (1998) Single-molecule enzymatic dynamics. Science 282 (5395):1877–1882
104. Kou SC, Cherayil BJ, Min W, English BP, Xie XS (2005) Single-molecule Michaelis-Menten equations. J Phys Chem B 109(41):19068–19081
105. Smiley RD, Hammes GG (2006) Single molecule studies of enzyme mechanisms. Chem Rev 106(8):3080–3094
106. Min W, Xie XS, Bagchi B (2008) Two-dimensional reaction free energy surfaces of catalytic reaction: effects of protein conformational dynamics on enzyme catalysis. J Phys Chem B 112(2):454–466
107. Sawaya MR, Kraut J (1997) Loop and subdomain movements in the mechanism of Escherichia coli dihydrofolate reductase: crystallographic evidence. Biochemistry 36 (3):586–603
108. Boehr DD, McElheny D, Dyson HJ, Wright PE (2006) The dynamic energy landscape of dihydrofolate reductase catalysis. Science 313(5793):1638–1642
109. Bhabha G, Lee J, Ekiert DC, Gam J, Wilson IA, Dyson HJ, Benkovic SJ, Wright PE (2011) A dynamic knockout reveals that conformational fluctuations influence the chemical step of enzyme catalysis. Science 332(6026):234–238

110. Loveridge EJ, Tey LH, Behiry EM, Dawson WM, Evans RM, Whittaker SBM, Gunther UL, Williams C, Crump MP, Allemann RK (2011) The role of large-scale motions in catalysis by dihydrofolate reductase. J Am Chem Soc 133(50):20561–20570
111. Loveridge EJ, Behiry EM, Guo JN, Allemann RK (2012) Evidence that a 'dynamic knockout' in Escherichia coli dihydrofolate reductase does not affect the chemical step of catalysis. Nature chemistry 4(4):292–297
112. Swiderek K, Ruiz-Pernía JJ, Moliner V, Tuñón I (2014) Heavy enzymes—experimental and computational insights in enzyme dynamics. Curr Opin Chem Biol 21:11–18
113. Adamczyk AJ, Cao J, Kamerlin SCL, Warshel A (2011) Catalysis by dihydrofolate reductase and other enzymes arises from electrostatic preorganization, not conformational motions. Proc Natl Acad Sci USA 108(34):14115–14120
114. Borgis D, Hynes JT (1996) Curve crossing formulation for proton transfer reactions in solution. J Phys Chem 100(4):1118–1128
115. Truhlar DG, Gao JL, García-Viloca M, Alhambra C, Corchado J, Sanchez ML, Poulsen TD (2004) Ensemble-averaged variational transition state theory with optimized multidimensional tunneling for enzyme kinetics and other condensed-phase reactions. Int J Quantum Chem 100(6):1136–1152
116. Singh P, Sen A, Francis K, Kohen A (2014) Extension and limits of the network of coupled motions correlated to hydride transfer in dihydrofolate reductase. J Am Chem Soc 136 (6):2575–2582
117. Wang Z, Abeysinghe T, Finer-Moore JS, Stroud RM, Kohen A (2012) A remote mutation affects the hydride transfer by disrupting concerted protein motions in thymidylate synthase. J Am Chem Soc 134(42):17722–17730
118. Kohen A, Limbach H-H (eds) (2006) Isotope effects in chemistry and biology. CRC Press, Taylor & Francis Group, Boca Raton
119. Rokop S, Gajda L, Parmerte S, Crespi HL, Katz JJ (1969) Purification and characterization of fully deuterated enzymes. Biochim Biophys Acta 191(3):707–715
120. Silva RG, Murkin AS, Schramm VL (2011) Femtosecond dynamics coupled to chemical barrier crossing in a Born-Oppenheimer enzyme. Proc Natl Acad Sci USA 108(46):18661–18665
121. Luk LYP, Ruiz-Pernía JJ, Dawson WM, Loveridge EJ, Tuñón I, Moliner V, Allemann RK (2014) Protein isotope effects in dihydrofolate reductase from Geobacillus stearothermophilus show entropic-enthalpic compensatory effects on the rate constant. J Am Chem Soc 136(49):17317–17323

Chapter 16
Multiscale Modelling of In Situ Oil Sands Upgrading with Molybdenum Carbide Nanoparticles

Xingchen Liu, Baojing Zhou, Farouq Ahmed, Alexander Tkalych, Akira Miyamoto and Dennis R. Salahub

Abstract This chapter presents multi-scale models of the reactions that occur in the in situ oil sands upgrading process. Its focus is on the various modelling tools and their applications to the benzene hydrogenation reactions catalyzed by molybdenum carbide nanoparticles. As the reaction mechanism of benzene hydrogenation on molybdenum carbide is not clear, we start with density functional theory (DFT) studies to elucidate the reaction mechanism, using both periodic and cluster models. Benzene hydrogenation on molybdenum carbide follows the Langmuir-Hinshelwood mechanism, with the six-member ring tilting up gradually. A tight-binding quantum chemical molecular dynamics (TB-QCMD) method is used to track the physical motion of the atoms in the reaction processes of C_6H_6 on a Mo-terminated α-Mo_2C (0001) surface. The approximate DFT method, density functional tight-binding (DFTB), was parameterized to allow the quantum mechanical treatment of nanoscale systems. With the nudged elastic band method, the potential energy profiles of benzene hydrogenation on molybdenum carbide

X. Liu · D.R. Salahub (✉)
Department of Chemistry, Institute for Quantum Science and Technology, and Centre for Molecular Simulation, University of Calgary, 2500 University Drive NW, T2N 1N4 Calgary, AB, Canada
e-mail: dennis.salahub@ucalgary.ca

B. Zhou
Department of Chemistry, School of Chemical Engineering, Nanjing University of Science and Technology, 200 Xiao Ling Wei Street, Nanjing 210094, China

F. Ahmed
Chemical and Petroleum Engineering Department, Schulich School of Engineering, University of Calgary, 2500 University Drive NW, T2N 1N4 Calgary, AB, Canada

A. Tkalych
Department of Chemistry, Princeton University, Princeton, NJ 08544-5263, USA

A. Miyamoto
New Industry Creation Hatchery Center, Tohoku University, 403, 6-6-10, Aoba, Aramaki, 980-8579 Aoba, Sendai, Japan

J.-L. Rivail et al. (eds.), *Quantum Modeling of Complex Molecular Systems*, Challenges and Advances in Computational Chemistry and Physics 21, DOI 10.1007/978-3-319-21626-3_16 nanoparticles have been obtained. Finally a force field was brought in to describe the solvent environment in the system, leading to a multiscale quantum mechanical/molecular mechanical (QM/MM) model. This study suggests that entropy and the environment play important roles in heterogeneous reactions catalyzed by molybdenum carbide nanoparticles.

List of Acronyms (alphabetical order)

ADFT	Auxiliary density functional theory
DFT	Density functional theory
DFTB	Density functional tight-binding
HDN	Hydrodenitrogenation
HDS	Hydrodesulphurization
HRTEM	High-resolution transmission electron microscopy
KS	Kohn-Sham
L-H	Langmuir-Hinshelwood
LCGTO	Linear combination of Gaussian type orbitals
MCNPs	Molybdenum carbide nanoparticles
MD	Molecular dynamics
MEP	Minimum energy path
PAH	Polyaromatic hydrocarbons
PAW	Projector augmented wave
PMF	Potential of mean force
QM/MM	Quantum mechanics/molecular mechanics
SAGD	Steam-assisted gravity drainage
SCC	Self-consistent-charge
TB	Tight-binding
TB-QC	Tight-binding quantum chemistry
TB-QCMD	Tight-binding quantum chemical molecular dynamics
TPR-MS	Temperature-programmed reaction-mass spectrometry
TS	Transition state
UD	Ultra-dispersed
WHAM	Weighted-histogram analysis method

16.1 In Situ Oil Sands Upgrading, an Introduction

After over a century of exploitation, the world's oil quality is declining, simply because the best quality oil was found and used first. Conventional oil only constitutes 30 % of the world oil reserves [1] (Fig. 16.1), and is approaching depletion very quickly. Unconventional crudes such as heavy oil, extra-heavy oil and oil sands, which, in the past, were not economical for upgrading, are now playing a more and more important role in the world energy market.

Conventional oils (light & medium) have API (American Petroleum Institute) gravities higher than 20°, while heavy crude oils have API gravity values between 10° to 20°. Oil sands are a mixture of 4–6 % water, 83–85 % sediment such as sands, clay, and other minerals, and 10–12 % bitumen. Extra heavy oil and bitumen have API values between 5° and 10°. Generally, heavy crude oil has high in situ viscosities under reservoir conditions and high density. Oil sands are so viscous that they are not mobile in the reservoir. At room temperature, the crude bitumen is in an almost solid state, and must be converted to upgraded crude. High aromatic content (poly aromatic hydrocarbons, or PAH) in the heavy oil has been recognized as one of the major reasons for the increase in density and viscosity of the fuel [2].

One of the most popular in situ recovery methods is steam-assisted gravity drainage (SAGD). In fact, much of the expected future growth of production in the Canadian oil sands is predicted to be from SAGD [3]. In SAGD, two permeable horizontal wells are drilled into the oil sands reservoir, one injector and one producer. The producer well is located on the bottom of the reservoir, and the injector well is higher up. Steam is injected through the injector well for months to soften the oil sands into emulsions, which then flow into the production well by gravity, are pumped to the surface, and sent to upgrading plants and then to a refinery.

The idea of in situ upgrading [4] (Fig. 16.2) is closely related to SAGD, but involves an additional "underground refinery" process. Instead of injecting steam, heavy oil is pre-injected into the oil sand reservoir through the injector well with ultra-dispersed (UD) catalysts and H_2. The UD catalysts are usually on the nano-scale such that they could penetrate inside the porous medium in the reservoir

Fig. 16.1 World oil reserves and percentages [1]

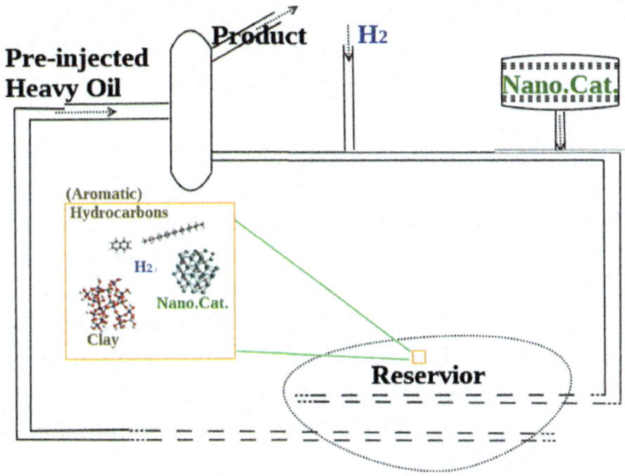

Fig. 16.2 Schematic of in situ oil sands upgrading with nanocatalysts [4, 7]

and react with the bitumen. The best nanocatalysts for this purpose are molybdenum carbide nanoparticles (MCNPs), because of their high activity [5], high abundance [6], and low price. The hydrotreating reactions, including: hydrogenation, hydrodesulphurization (HDS), hydrodenitrogenation (HDN), and hydrocracking, happen underground to upgrade the oil sands into higher quality crude oil. Similarly to SAGD, the product fluid flows into the production with gravity, and is pumped to the surface under pressure. A separator is used in which lighter components are collected through the upper pipe as products, and heavier components continue the cycle through the lower pipe. The temperature in the oil reservoir during the in situ process is usually 200–400 °C and 400 psi. [4] The in situ upgrading takes advantage of the exothermicity of hydrogenation reactions to reduce the viscosity of the oil sands, and improves the quality of the product by hydrogenation. It is also more environmentally friendly as it can be operated at lower temperature and pressure and avoids steam injection, saving energy and water [4].

The main chemistry of the in situ heavy oil upgrading process is the catalytic hydrogenation and hydrocracking (and possibly HDS and HDN as well) of aromatic molecules. However, this is far beyond simple organic reactions because of several factors. First, the interaction between the aromatic molecules with the molybdenum carbide catalyst may strongly change the electronic structure of the molecules on the catalyst surfaces. Second, the catalyst sizes are on the nano-scale, and the direct modelling of the nanocatalyst is still very challenging in computational chemistry. Third, the reactions happen in an emulsion-like aromatic environment at high temperature, which implies possibly strong entropic effects because of the dynamics of the reactants and the nanoparticles.

It is known that poly aromatic hydrocarbons (PAH) hydrogenate via successive steps, each of which is reversible [8]. Hydrogenation of the first ring is usually the least thermodynamically favorable step under typical hydrotreating conditions. We use the simplest mono-aromatic molecule, benzene, as our model molecule in all the related studies.

We will provide in this chapter a brief review of our published work on benzene hydrogenation on the MCNPs, starting from simple surface models and leading towards a multiscale model. Extraordinary features of the MCNPs are found from our suggests that the high flexibility, which suggests that the high flexibility of the MCNPs due to the entropic contributions of the MCNPs and the solvent could greatly affect the free energy profiles of these reactions.

As the reaction mechanism of benzene hydrogenation on MCNPs is unknown in the literature, we start with quantum mechanical density functional theory (DFT) studies to elucidate the reaction mechanism, using both periodic and cluster models. A tight-binding quantum chemical molecular dynamics (TB-QCMD) is used to follow the physical motion of the atoms and molecules in the reaction processes. The approximate DFT method, density functional tight-binding (DFTB), was parameterized to allow quantum mechanical treatment of nanoscale systems. Finally a force field was brought in to describe the solvent environment in the system, leading to a multiscale quantum mechanical/molecular mechanical (QM/MM) model.

16.2 Modelling Tools

16.2.1 Density Functional Theory (DFT)

Density functional theory is a quantum mechanical modelling method based on the Hohenberg-Kohn Theorem [9] and the Kohn-Sham (KS) method [10]. deMon2k [11] (density of Montreal), and VASP [12, 13] codes were used to solve the Kohn-Sham equations numerically. deMon2 k, initially developed in the Salahub group, is a fast molecular DFT code and rich in functionality based on auxiliary density functional theory (ADFT) in the framework of linear combination of Gaussian type orbitals (LCGTO) expansions. VASP (Vienna Ab initio simulation package) [13] is a periodic DFT code based on plane wave basis sets and pseudo-potentials or projector augmented waves (PAW) [12].

The Hohenberg-Kohn theorem states that the total energy of the system is a functional of the electron density ρ:

$$E = E[\rho] = E_{NN} + \mathrm{T}[\rho] + \int \rho(\boldsymbol{r})v(\boldsymbol{r})d\boldsymbol{r} + V_{ee}[\rho]. \tag{16.1}$$

Here, E_{NN} is the nuclear-nuclear repulsion energy, $\rho(\boldsymbol{r})$ is the electron density, $\mathrm{T}[\rho]$ is the kinetic energy of the electrons, $v(\boldsymbol{r})$ is the external/nuclear potential, and $V_{ee}[\rho]$ is the electron-electron repulsion energy.

Instead of calculating the electron density from the wavefunction of the interacting system, in the Kohn-Sham method, it is determined using a non-interacting reference system, which has the same electron density as the interacting system:

$$\rho(\boldsymbol{r}) = \sum_{\mathrm{i}=1}^{\mathrm{M}} \psi_{\mathrm{i}}^{*}(\mathbf{r})\psi_{\mathrm{i}}(\mathbf{r}), \tag{16.2}$$

where $\psi_i(\boldsymbol{r})$ is a set of non-interacting orbitals, or Kohn-Sham orbitals. The Kohn-Sham total energy is:

$$E[\rho] = E_{NN} + T_{KS}[\rho] + \int \rho(\boldsymbol{r})v(\boldsymbol{r})d\boldsymbol{r} + J[\rho] + E_{xc}[\rho], \tag{16.3}$$

Here, $T_{KS}[\rho]$ is the Kohn-Sham kinetic energy, and $J[\rho]$ is the classical coulomb energy with the following expression:

$$J[\rho] = \frac{1}{2}\int\int \frac{\rho(\boldsymbol{r})\rho(\boldsymbol{r}')}{|\boldsymbol{r}-\boldsymbol{r}'|}d\boldsymbol{r}d\boldsymbol{r}'. \tag{16.4}$$

The exchange-correlation energy, for which the analytical functional form is unknown, collects the difference of the kinetic energies of the interacting system and non-interacting system, and the difference of classical and non-classical electron-electron interactions:

$$E_{xc}[\rho] = (T[\rho] - T_{KS}[\rho]) + (V_{ee}[\rho] - J[\rho]). \tag{16.5}$$

The Kohn-Sham total energy can be more explicitly written as

$$\begin{aligned} E[\rho] = \sum_{i}^{M} n_i \langle \psi_i | -\frac{1}{2}\nabla^2 + v(\boldsymbol{r}) + \frac{1}{2}\int \frac{\rho(\boldsymbol{r}')}{|\boldsymbol{r}-\boldsymbol{r}'|} d\boldsymbol{r}' | \psi_i \rangle + E_{xc}[\rho] \\ + \frac{1}{2}\sum_{\beta}^{N}\sum_{\alpha \neq \beta}^{N} \frac{Z_\alpha Z_\beta}{|\boldsymbol{A}-\boldsymbol{B}|}, \end{aligned} \tag{16.6}$$

where n_i is the number of electrons in orbital i, Z_α and Z_β are the nuclear charges of atoms α and β. Treating this equation with the variational method, the Kohn-Sham equations are obtained:

$$\left(-\frac{1}{2}\nabla^2 + v(\boldsymbol{r}) + \frac{1}{2}\int \frac{\rho(\boldsymbol{r}')}{|\boldsymbol{r}-\boldsymbol{r}'|} d\boldsymbol{r}' + V_{xc}[\rho]\right)\psi_i(\boldsymbol{r}) = \varepsilon_i \psi_i(\boldsymbol{r}). \tag{16.7}$$

Since the electron density depends on the Kohn-Sham orbitals, this set of equations is solved iteratively until convergence is reached.

In auxiliary density functional theory (ADFT) [14], the KS orbitals are expanded as linear combinations of Gaussian type orbitals (LCGTO):

$$\psi_i(\boldsymbol{r}) = \sum_{\mu} c_{\mu i}\mu(\boldsymbol{r}), \; and \quad \mu(\boldsymbol{r}) = \sum_{k=1}^{K} d_{\mu k} g_k(\boldsymbol{r}). \tag{16.8}$$

where $\mu(\boldsymbol{r})$ is an atomic orbital, and $g(\boldsymbol{r})$ is an atom-centered Gaussian type orbital. The Coulomb repulsion energy [15] and the exchange-correlation energy [16] are fitted with auxiliary functions of the primitive Hermite Gaussian [17] form:

$$\bar{k}(\boldsymbol{r}) = \left(\frac{\partial}{\partial A_x}\right)^{\bar{k}_x}\left(\frac{\partial}{\partial A_y}\right)^{\bar{k}_y}\left(\frac{\partial}{\partial A_z}\right)^{\bar{k}_z} e^{-\zeta_k(\boldsymbol{r}-\boldsymbol{A})^2} \tag{16.9}$$

which gives a fitted approximate density as:

$$\tilde{\rho}(\boldsymbol{r}) = \sum_{\bar{k}} x_{\bar{k}} \bar{k}(\boldsymbol{r}) \tag{16.10}$$

With this fitted density, the total energy in ADFT is formally [14]:

$$E = \sum_{\mu,\nu} P_{\mu\nu} H_{\mu\nu} + \frac{1}{2} \sum_{\mu\nu} \sum_{\sigma\tau} P_{\mu\nu} P_{\sigma\tau} \langle \mu\nu || \bar{k} x_{\bar{k}} \rangle - \frac{1}{2} \sum_{\bar{k},\bar{l}} x_{\bar{k}} x_{\bar{l}} \langle \bar{k} || \bar{l} \rangle + E_{xc}[\tilde{\rho}] + E_{NN} \tag{16.11}$$

In the SCF procedure of ADFT, the molecular orbital coefficients are changed by the variation of the fitting coefficients [18].

In the periodic DFT code VASP, plane wave basis sets are used to expand the KS orbitals of the form:

$$u_n(\boldsymbol{r}) = \frac{1}{\Omega} \sum_{\boldsymbol{G}} c_n(\boldsymbol{G}) e^{i\boldsymbol{G}\boldsymbol{r}} \tag{16.12}$$

Where Ω is the volume of the unit cell, $\boldsymbol{G}$ is the reciprocal lattice such that the plane-wave $e^{i\boldsymbol{G}\boldsymbol{r}}$ has the periodicity of the real space lattice, and $c_n(\boldsymbol{G})$ are the Fourier coefficients stored on the grid of $\boldsymbol{G}$.

16.2.2 Density Functional Tight-Binding (DFTB) Method

DFTB [19] is a tight-binding approximation of the DFT method. In DFTB, the electron density is written as a reference density and a density fluctuation.

$$\rho(\boldsymbol{r}) = \rho_0(\boldsymbol{r}) + \delta\rho(\boldsymbol{r}). \tag{16.13}$$

The total energy of DFTB is the second-order expansion of the DFT total energy (Eq. 16.6) around the reference density [20]:

$$\begin{aligned} \mathrm{E}[\rho] &= \sum_{i}^{M} \langle \psi_i | \hat{H}_0 | \psi_i \rangle \\ &- \frac{1}{2} \int \int \frac{\rho_0 \rho_0'}{|\boldsymbol{r} - \boldsymbol{r}'|} d\boldsymbol{r} d\mathbf{r}' + E_{xc}[\rho_0] - \int v_{xc}[\rho_0] \rho_0 d\boldsymbol{r} + E_{NN} \\ &+ \frac{1}{2} \int \int \left(\frac{1}{|\boldsymbol{r} - \boldsymbol{r}'|} + \left. \frac{d^2 E_{xc}}{d\rho d\rho'} \right|_{\rho_0} \right) d\rho d\rho', \end{aligned} \tag{16.14}$$

where the reference Hamiltonian $\hat{H}_0$ is defined as:

$$\hat{H}_0 = -\frac{1}{2} \nabla^2 + v(\boldsymbol{r}) + \int \frac{\rho_0'}{|\boldsymbol{r} - \boldsymbol{r}'|} d\boldsymbol{r}' + v_{xc}[\rho_0]. \tag{16.15}$$

The first line in Eq. 16.14 is called the binding energy E_{bnd}, the second line the repulsive energy E_{rep}, and the third line the second order energy E_{2nd}. The E_{2nd} terms is only considered in the self-consistent-charge (SCC) correction introduced by Elstner [21].

These total energy expressions, after a variational treatment, lead to DFTB Kohn-Sham like equations of the form:

$$\sum_{\mu}^{M} c_{\mu i}\left(H_{\mu\nu}^{(SCC)DFTB} - \varepsilon_i S_{\mu\nu}\right) = 0, \quad \forall \nu, i \tag{16.16}$$

The Hamiltonian integrals and the overlap integrals are calculated with atom-centered localized atomic orbitals under a two-center approximation. These atom-centered orbitals are constructed by solving modified KS equations of spherical pseudoatoms with confinement potentials:

$$\left(\hat{T} - \frac{Z_\alpha}{r} + \int \frac{\rho_0^\alpha(\boldsymbol{r}')}{|\boldsymbol{r}-\boldsymbol{r}'|} dr' + v_{\mathrm{xc}}\left[\rho_0^\alpha\right] + V_{\mathrm{conf}}\right)\phi_\mu = \varepsilon_\mu \phi_\mu. \tag{16.17}$$

Once these orbitals are constructed, the Hamiltonian integrals and the overlap integrals in Eq. 16.15 can be calculated for diatomic distances with grids and stored in files (Slater-Koster files), avoiding the most time-consuming SCF procedure in DFT. For SCC-DFTB, the second order term is considered, and the Hamiltonian integral has the form:

$$\begin{aligned} H_{\mu\nu}^{SCC-DFTB} &= \langle \phi_\mu | \hat{H}_0 | \phi_\nu \rangle + \frac{1}{2} S_{\mu\nu} \sum_{\xi}^{N} (\gamma_{\alpha\xi} + \gamma_{\beta\xi}) \Delta q_\xi \\ &= H_{\mu\nu}^{DFTB} + H_{\mu\nu}^{SCC}, \mu \in \{\alpha\}, \nu \in \{\beta\}, \end{aligned} \tag{16.18}$$

where γ is a complicated term that is related to the density fluctuation in an atom, and is usually approximated by the atomic hardness. The values of the integrals, together with the SCC related parameters, are called the electronic part of DFTB parameters.

The short-range repulsive potentials E_{rep} can be represented as a polynomial or a spline in SK files. They are determined as the difference of total energy resulting from a DFT calculation and a DFTB calculation with only the electronic part

$$E_{rep}(\boldsymbol{R}) = E_{DFT}(\boldsymbol{R}) - E_{elec}(\boldsymbol{R}). \tag{16.19}$$

In addition, a force-field-like van der Waals term [22] is usually added in DFTB to account for the dispersion interactions at long range:

$$U_{\alpha\beta}(R) = d_{\alpha\beta}\left[\left(\frac{R_{\alpha\beta}}{R}\right)^{12} - 2\left(\frac{R_{\alpha\beta}}{R}\right)^{6}\right], \tag{16.20}$$

while at short-range, a polynomial potential is taken to avoid the conflicts of the r^{-12} term with the electron density overlapping in DFTB:

$$U_{\alpha\beta}^{(short-range)}(\mathrm{r}) = U_0 - U_1 R^n - U_2 R^{2n}, \tag{16.21}$$

where n = 5, and U_0, U_1, and U_2 are all empirical parameters.

16.2.3 Hybrid QM(DFTB)/MM Method

16.2.3.1 QM/MM Scheme

QM/MM is a multi-scale modelling method that couples quantum mechanical (QM) methods with molecular mechanical (MM) methods using a "hand-shake" scheme. "Hand-shake" multi-scale modelling requires that the program runs concurrently over different spatial regions, with each region treated with different equations of motion. We use the QM/MM scheme, with DFTB as the QM method, implemented in the CHARMM program by Cui et al. [23]. DFTB/MM uses the additive QM/MM scheme first introduced by Warshel [24] and Levitt for treating biological systems:

$$E^{tot} = E^{QM} + E^{MM} + E^{QM/MM}. \tag{16.22}$$

Here the first term is the QM energy of the QM atoms; the second term is the empirical energy of the MM atoms; the last term is the QM/MM coupling term that includes electrostatic and van der Waals interactions:

$$E^{QM/MM} = E_{esd}^{QM/MM} + E_{vdw}^{QM/MM}, \tag{16.23}$$

where $E_{esd}^{QM/MM}$ and $E_{vdw}^{QM/MM}$ represent the electrostatic and van der Waals interactions between QM atoms and MM atoms, respectively. Using the electrostatic embedding approximation, the MM charges are used to polarize the QM electrons, yielding:

$$E^{tot} = \langle \Psi | \hat{H}^{QM} + \hat{H}_{esd}^{QM/MM} | \Psi \rangle + E_{vdw}^{QM/MM} + E^{MM} \tag{16.24}$$

In the above total energy expression, $\hat{H}_{esd}^{QM/MM}$ is the Hamiltonian for the electrostatic interaction between the MM atoms and the QM atoms using the one-electron operator:

$$\hat{H}_{esd}^{QM/MM} = \sum_{QM}\sum_{MM} \frac{Z_{QM} Q_{MM}}{|\boldsymbol{R}_{QM} - \boldsymbol{R}_{MM}|} - \sum_{i=1}^{M}\sum_{MM} \frac{Q_{MM}}{|\boldsymbol{r}_i - \boldsymbol{R}_{MM}|}. \tag{16.25}$$

Z_{QM} is the nuclear charge of a QM atom; Q_{MM} is the partial charge of an MM atom; and M is the total number of QM electrons. To avoid evaluation of two and three-center integrals in $\langle\Psi|\hat{H}_{esd}^{QM/MM}|\Psi\rangle$ in the total energy expression, thc QM/MM electrostatic interaction is approximated by the Coulombic interaction between the Mulliken charges Δq_{QM} of the QM atoms and the MM partial charges Q_{MM} :

$$\hat{H}_{esd}^{QM/MM} \approx \sum_{QM}\sum_{MM} \frac{\Delta q_{QM} Q_{MM}}{|\boldsymbol{R}_{QM} - \boldsymbol{R}_{MM}|}. \tag{16.26}$$

The total energy expression for MM embedding SCC-DFTB is then:

$$\begin{aligned} E_{SCC-DFTB/MM} = & \sum_{i}^{M} \langle\psi_i|\hat{H}_0 + \sum_{QM}\sum_{MM} \frac{\Delta q_{QM} Q_{MM}}{|\boldsymbol{R}_{QM} - \boldsymbol{R}_{MM}|}|\psi_i\rangle + E_{rep} \\ & + \frac{1}{2}\sum_{\alpha,\beta}^{N} \Delta q_\alpha \Delta q_\beta \gamma_{\alpha\beta}, \end{aligned} \tag{16.27}$$

where α and β are both QM atoms. Applying linear variational theory with atom-centered localized atomic orbitals $\phi_\mu(\boldsymbol{r} - \boldsymbol{R}_\alpha)$, similarly a set of secular equations is obtained:

$$\sum_{\mu}^{M} c_{\mu i}\left(H_{\mu\nu} - \varepsilon_i S_{\mu\nu}\right) = 0, \quad \forall \nu, i, \tag{16.28}$$

with the new Hamiltonian matrix elements including the contributions from MM partial charges:

$$\begin{aligned} & H_{\mu\nu}^{SCC-DFTB/MM} \\ & = \langle\phi_\mu|\hat{H}_0|\phi_\nu\rangle + \frac{1}{2} S_{\mu\nu} \sum_{\xi}^{N} \left(\gamma_{\alpha\xi} + \gamma_{\beta\xi}\right) \Delta q_\xi \\ & + \frac{1}{2} S_{\mu\nu} \sum_{MM} \left(\frac{\Delta q_\alpha Q_{MM}}{|\boldsymbol{R}_\alpha - \boldsymbol{R}_{MM}|} + \frac{\Delta q_\beta Q_{MM}}{|\boldsymbol{R}_\beta - \boldsymbol{R}_{MM}|} \right) \\ & = H_{\mu\nu}^{DFTB} + H_{\mu\nu}^{SCC} + H_{\mu\nu}^{esd}, \quad \mu \in \{\alpha\}, \ \nu \in \{\beta\}. \end{aligned} \tag{16.29}$$

16.2.3.2 The Force Field in DFTB/MM

The force field in DFTB/MM is the CHARMM general force field (CGenFF), which was developed [25] to improve the coverage of organic molecules such as drug molecules. The functional form of CGenFF is

$$
\begin{aligned}
E(R) = & \sum_{bonds} K_b(b - b_0)^2 + \sum_{angles} K_\theta(\theta - \theta_0)^2 \\
& + \sum_{dihedrals} K_\phi(1 + \cos(n\phi - \delta)) \\
& + \sum_{\substack{improper \\ dihedrals}} K_\varphi(\varphi - \varphi_0)^2 + \sum_{\substack{Urey- \\ Bradley}} K_{UB}\left(r_{1,3} - r_{1,3:0}\right)^2 \\
& + \sum_{nonbonded} \left\{ \frac{q_\alpha q_\beta}{4\pi D R_{\alpha\beta}} + \varepsilon_{\alpha\beta}\left[\left(\frac{R_{\alpha\beta}}{R}\right)^{12} - 2\left(\frac{R_{\alpha\beta}}{R}\right)^{6}\right] \right\}.
\end{aligned}
\tag{16.30}
$$

The first five terms are bonding terms for bonds, valence angles, dihedral angles, improper dihedral angles, and the Urey-Bradley (i.e. 1,3 interaction) term, respectively. The last term is the non-bonding term, which accounts for electrostatic interactions and van der Waals interactions. Here, $b_0, \theta_0 \varphi_0$ *and* $r_{1,3:0}$ are the equilibrium bond distances, angles, improper dihedrals and Urey-Bradley interaction, respectively; K terms are the corresponding force constants; n and δ are the dihedral multiplicity; $q_i$andq_j are atomic point charges; D is the dielectric constant; R is the diatomic distance; $R_{\alpha\beta}$ is the van der Waals distance; and $\varepsilon_{\alpha\beta}$ is the well depth.

16.2.4 Umbrella Sampling and the Weighted-Histogram Analysis Method (WHAM)

In statistical mechanics, for a system at constant volume and temperature, the (Helmholtz) free energy can be calculated from MD simulations through:

$$
\mathrm{A} = -\frac{1}{\beta} ln Q_{NVT}, \tag{16.31}
$$

where Q_{NVT} is the canonical partition function that can be obtained by integrating over all coordinate space:

$$
Q_{NVT} = \int e^{-\beta V(\boldsymbol{R})} d\boldsymbol{R}. \tag{16.32}
$$

Here, the integration over the momentum space generates a constant and is ignored because the potential energy function $V(\boldsymbol{R})$ is independent of the momentum. For chemical reactions, we are usually interested in three states: the reactant state, the product state, and the transition state (TS). A parameter (usually in one-dimension, but it can also be multi-dimensional) can be defined to describe the continuous change of the system, which is called the reaction coordinate (z). Then the normalized probability distribution of the reaction coordinate $P(z)$ can be calculated by integrating out all other degrees of freedom but z:

$$P(z) = \frac{\int e^{-\beta V(\boldsymbol{R})}\delta[z'(\boldsymbol{R}) - z]d\boldsymbol{R}}{\int e^{-\beta V(\boldsymbol{R})}d\boldsymbol{R}} \tag{16.33}$$

Here, the denominator is essentially Q_{NVT}, and is the normalization factor; the integration goes over the entire configuration space; $\delta[z'(\boldsymbol{R}) - z]$ means all degrees of freedom but z. Ideally this will give $P(z)$ and therefore the free energy along the reaction coordinate A(z) by:

$$A(z) = -\frac{1}{\beta} lnP(z). \tag{16.34}$$

The free energy A(z) with the degree of freedom z extracted is also called the potential of mean force (PMF). However, simulations can only run for a finite time, and problems arise in the sampling part. The TS are often rare events compared to the entire configuration space because they are high in energy and low in probability. Conventional MD spends most of the limited computational time staying in the low energy states during the simulations, and cannot access the states of high energy.

Umbrella sampling is a biased-sampling technique developed by Torrie and Valleau [26] that enhances the sampling by modifying the potential energy function, forcing the system to stay in the desired states. The basic idea of umbrella sampling is presented below; for further details, one can refer to the recent review paper by Kästner [27].

Adding a biasing potential $\omega_i(z)$ (i is the window number) to the original unbiased potential energy function $V_u(\boldsymbol{R})$ with z being the reaction coordinate, we have a new biased potential energy function:

$$V_b(\boldsymbol{R}) = V_u(\boldsymbol{R}) + \omega_i(z) \tag{16.35}$$

where "b" means biased and "u" means unbiased. For each window, the biased potential $\omega_i(z)$ are chosen properly to force the system to stay at the desired point on the reaction coordinate. The simplest potential form is a harmonic potential centered at the desired value of z. In this way, the high-energy and low-probability regions of the phase space can be sampled better. However, one can only directly get a biased probability distribution $P^b(z)$ with the biased potential energy function $V_u(\boldsymbol{R}) + \omega_i(z)$ from these simulations:

$$P_i^b(z) = \frac{\int e^{-\beta[V_u(\boldsymbol{R})+\omega_i(z)]}\delta[z'(\boldsymbol{R}) - z]d\boldsymbol{R}}{\int e^{-\beta[V_u(\boldsymbol{R})+\omega_i(z)]}d\boldsymbol{R}} \tag{16.36}$$

Notice that $\omega_i(z)$ is independent of the integration in the numerator and, using Eq. (16.33), the relationship between $P_i^u(z)$ and $P_i^b(z)$ can be found as:

$$P_i^u(z) = P_i^b(z)e^{-\beta\omega_i(z)} \frac{\int e^{-\beta[V_u(\boldsymbol{R})+\omega_i(z)]}d\boldsymbol{R}}{\int e^{-\beta V_u(\boldsymbol{R})}d\boldsymbol{R}}. \tag{16.37}$$

Here, the fraction is nothing but the ratio of the partition function for the biased potential energy function and the unbiased one. Taking the natural logarithm on both sides of Eq. (16.37) and multiplying with a constant $-\frac{1}{\beta}$, the free energy along the reaction coordinate z can be calculated as:

$$\mathrm{A}_i(z) = -\frac{1}{\beta} lnP_i^u(z) = -\frac{1}{\beta} lnP_i^b(z) + \omega_i(z) + F_i, \tag{16.38}$$

where $P^b(z)$ can be obtained from biased MD simulations, $\omega_i(z)$ is given analytically (usually being a harmonic potential). F_i is a free energy constant independent of z for umbrella sampling calculation of a single window:

$$F_i = -\frac{1}{\beta} ln \frac{\int e^{-\beta[V_u(\boldsymbol{R})+\omega_i(z)]}d\boldsymbol{R}}{\int e^{-\beta V_u(\boldsymbol{R})}d\boldsymbol{R}} = -\frac{1}{\beta} ln\langle e^{-\beta\omega_i(z)}\rangle, \tag{16.39}$$

where the angular brackets denote an ensemble average. However, multiple windows are usually needed to depict the whole course of the reaction coordinate with overlapping probability distributions (Fig. 16.3). To combine these $\mathrm{A}_i(z)$ vs. z curves into one diagram, in other words, to find the global distribution $P^u(z)$ from the distribution of each window $P_i^u(z)$, the weighted-histogram analysis method (WHAM) [28, 29] is used to evaluate F_i to shift the free energies of the windows.

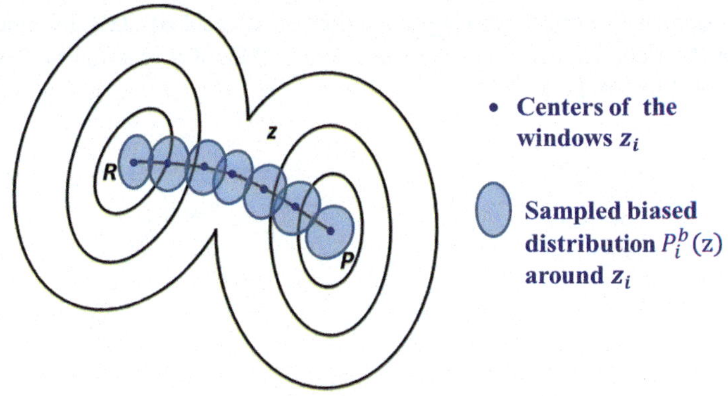

Fig. 16.3 The free energy surface in umbrella sampling: dividing the entire reaction coordinate z into several windows each centered at z_i. With biased forces properly added, umbrella sampling can cover the entire reaction coordinate in the configuration space. R means reactant state and P means product state

In WHAM, the global unbiased distribution $P^u(z)$ is the weighted sum of $P_i^u(z)$:

$$P^u(z) = \sum_{i=1}^{W} p_i P_i^u(z), \tag{16.40}$$

where p_i is the weight factor for the ith window, and W is the total number of windows. To combine the histograms, the probability distribution of each neighbouring window has to be aligned with good overlap. In WHAM, this is done by minimizing an error function $\Delta(p_i)$ related to $P^u(z)$, subject to the constraint that the total weight factor of the windows must be 1:

$$\sum_{i=1}^{W} p_i = 1. \tag{16.41}$$

Once the expression [29] for p_i is assumed, substituting Eq. (16.33) into Eq. (16.36), (the exact derivation can be found in the literature [29]) one obtains:

$$P^u(z) = \frac{\sum_i^W N_i e^{-\beta\omega_i(z)}}{\sum_j^W N_j e^{[-\beta\omega_i(z)+\beta F_i]}} \tag{16.42}$$

where both i and j are window indices, and N_i is the number of counts in the histogram of window i, associated with a particular value of z. Also, F_i can be calculated by:

$$e^{-\beta F_i} = \int P^u(z) e^{-\beta\omega_i(z)} dz \tag{16.43}$$

Because $P^u(z)$ and F_i are interdependent, as Eqs. (16.42) and (16.43) show, the minimization process to reduce the noise has to be done numerically in a self-consistent way until a convergence criterion is reached. The alignment of the individual free energy curves to get the overall PMF of a reaction is shown schematically in Fig. 16.4.

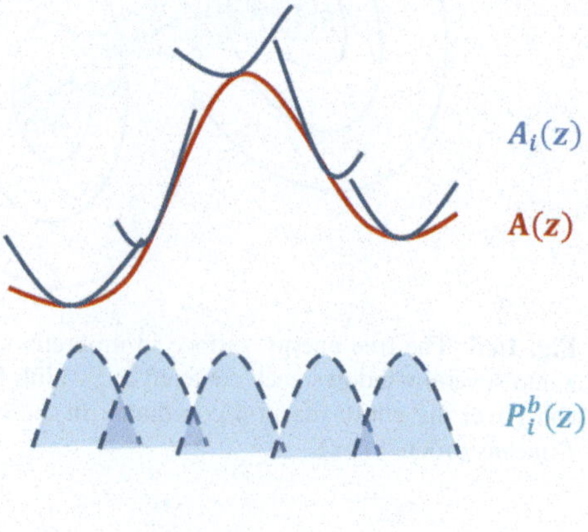

Fig. 16.4 The alignment of free energy curves from all the windows to get the global free energy of a reaction by WHAM

16.2.5 *Tight-Binding Quantum Chemical Molecular Dynamics (TB-QCMD) Method*

The TB-QCMD method, developed by Miyamoto et al. [30], is also based on tight binding (TB) theory, but is at the extended Hückel level of approximation. It is a multi-scale method composed of both quantum mechanics and molecular mechanics. It takes advantage of the fact that most of the configuration changes of the system do not involve chemical reactions, and can be modeled with a force field. In TB-QCMD, the tight-binding quantum chemistry (TB-QC) is first parameterized with first-principle methods. Then the force field is parameterized with TB-QC calculations. The force field is then used to project the system in the phase space with classical molecular dynamics (MD) calculations using Newton's equations of motion. Only for those steps that involve bond breaking/forming, TB-QC calculations shortly take over. The TB-QC is implemented in the "New Colors" program and MD in the "New-Ryudo" program. The TB-QCMD method is valuable in that it can provide us a trajectory of a hydrocarbon molecule reacting on the catalyst, which is currently very hard to get with other computational methods.

16.2.5.1 TB-QC Formalism

The TB-QC method is constructed using a single-ζ Slater-type basis set for each type of atom. The total energy of TB-QC is the sum of the kinetic energy of nuclei, the occupied molecular orbital energies, the coulomb energy, and a repulsion term:

$$E = \sum_{i}^{N} \frac{1}{2} m_i v_i^2 + \sum_{k=1}^{occ} n_k \varepsilon_k + \sum_{i}^{N} \sum_{j \neq i}^{N} \frac{Z_i Z_j e^2}{r_{ij}} + \sum_{i}^{N} \sum_{j \neq i}^{N} b_{ij} exp\left(\frac{a_{ij} - r_{ij}}{b_{ij}}\right), \quad (16.44)$$

where m_i and v_i are the mass and velocity of atom i; n_k and ε_k are the occupation and the energy of orbital k; Z_i and Z_j are the atomic charge of atom i and j with r_{ij} being the distance; a_{ij} and b_{ij} are constants for each atomic pair, which defines the repulsive term E_{rep}. The second term, also called the molecular orbital term E_{MO}, involves both diagonal (from mono-atomic contribution) and off-diagonal elements (from di-atomic contribution) of the Hamiltonian matrix:

$$E_{MO} = \sum_{k=1}^{occ} n_k \varepsilon_k = \sum_{k=1}^{occ} \sum_{r} n_k (C_{kr})^2 H_{rr} + \sum_{k=1}^{occ} \sum_{r} \sum_{s} n_k C_{kr} C_{ks} H_{rs}, \quad (16.45)$$

where C_{kr} is the coefficient of the r-th AO in the k-th MO; H_{rr} and H_{rs} are the diagonal and off-diagonal elements of the Hamiltonian matrix. The diagonal

Hamiltonian element H_{rr} equals the negative of the ionization potential of each atomic orbital:

$$H_{rr} = -I_r^i. \tag{16.46}$$

The off-diagonal elements H_{rs} is determined using the Wolfsberg-Helmholz formula from the extended Hückel method as:

$$H_{rs} = K_{rs}S_{rs}\frac{(H_{rr}+H_{ss})}{2}, \tag{16.47}$$

where S_{rs} is the overlap integral element, and K_{rs} is the Wolfsberg-Helmholtz constant calculated as:

$$K_{rs} = \{1 + \kappa_{rs}(1-\Delta^4) + \Delta^2\} \times \exp[-\delta_{rs}(r_{rs} - d_{rs})]. \tag{16.48}$$

Δ is defined as:

$$\Delta = \frac{H_{rr} - H_{ss}}{H_{rr} + H_{ss}}. \tag{16.49}$$

The parameterization of TB-QC involves the determination of two things: exponents of a Slater-type atomic orbital (AO) denoted as ζ_r for calculating the overlap integral S, and valence state ionization potentials (VSIPs) for calculating the Hamiltonian matrix elements H_{rr} (Eq. 16.44). Both of them are represented as polynomials of atomic charges:

$$\zeta_r = a_0 + \sum_{k=1}^{5} a_k(Z_i)^k \tag{16.50}$$

$$H_{rr} = -I_r^i = b_0 + \sum_{k=1}^{5} b_k(Z_i)^k. \tag{16.51}$$

Here, a_0-a_5 and b_0-b_5 are all free parameters which are fitted to reproduce the electronic structures of the reference systems. Besides, the a_{ij} and b_{ij} parameters in the repulsive term in Eq. 16.42 are fitted to reproduce DFT binding energies of reference systems.

Comparing TB-QC with DFTB, it is clear that the TB-QC method depends more on empirical parameters than DFTB. In DFTB, both the Hamiltonian **H** and overlap **S** matrix of the tight-binding atom-centered orbitals are calculated directly from DFT with only a universal confinement potential for each atom type, while in TB-QC, the Hamiltonian matrix **H** is completely fitted, and the overlap matrix **S** is also fitted by adjusting the exponents of the atomic orbitals. Therefore, DFTB has more transferability than TB-QC.

16.2.5.2 The Force Field in TB-QCMD

The force field in TB-QCMD has a total energy expression as:

$$\begin{aligned} U = & \sum_i \sum_{j \neq i} \left[\frac{Z_i Z_j e^2}{r_{ij}} + f_0 (b_i + b_j) exp\left(\frac{a_i + a_j - r_{ij}}{b_i + b_j} \right) \right] \\ & + \sum_i \sum_{j \neq i} D_{ij} \{ exp[\beta_{ij}(r_{ij} - r_0)] - 2exp[-\beta_{ij}(r_{ij} - r_0)] \} \end{aligned} \quad (16.52)$$

The first term corresponds to the Coulomb potential, and the second term corresponds to the repulsive potential (f_0 is a constant for unit adjustment, a_i is the size, and b_i is the stiffness of atom i), which gives a good account of the repulsive interactions arising from the overlap of electronic clouds. The third term in Eq. 16.52 corresponds to the Morse potential which represents covalent interactions, where D_{ij} is the bond energy (determined from the second term of Eq. 16.42), β_{ij} is the form factor, and r_0 is the bond length at minimum energy. All these parameters are determined from TB-QC calculations of the same atomic configuration.

16.3 Benzene Hydrogenation on a Planar (0001) Surface of MCNPs

16.3.1 Benzene Adsorption on Mo_2C from DFT with a Vacuum Slab Model

Adsorption is commonly known as the first and an essential step in heterogeneous catalysis. To understand the binding of benzene with Mo_2C, we used the periodic DFT code VASP to study benzene adsorption on the Mo-terminated (0001) surface of Mo_2C, which is the most active surface of Mo_2C that resembles the Pt (111) surface [31], with the projector augmented wave (PAW) [12] method and a plane wave basis set with cutoff energy of 500 eV. The DFT-D2 [32] type correction was added to account for the dispersion interactions. We modeled the (0001) surface of Mo_2C with a periodic slab consisting of two Mo layers and two C layers (Fig. 16.5).

Benzene prefers to adsorb on the three-fold hollow site with a vacancy underneath (the Vc site), with an adsorption energy of −3.51 eV. Dispersion interaction is crucial in the benzene adsorption phenomenon, which contributes −1.31 eV (37.3 %) to the total adsorption energy. Re-hybridization of the benzene carbon atoms from sp^2 to sp^3 was observed during the adsorption. Upon adsorption, the C *2p* states delocalize within the Mo *d*-band, and the C *2s* states interact with the 2*s* states of the subsurface C atom and the *4d* states of the Mo atoms.

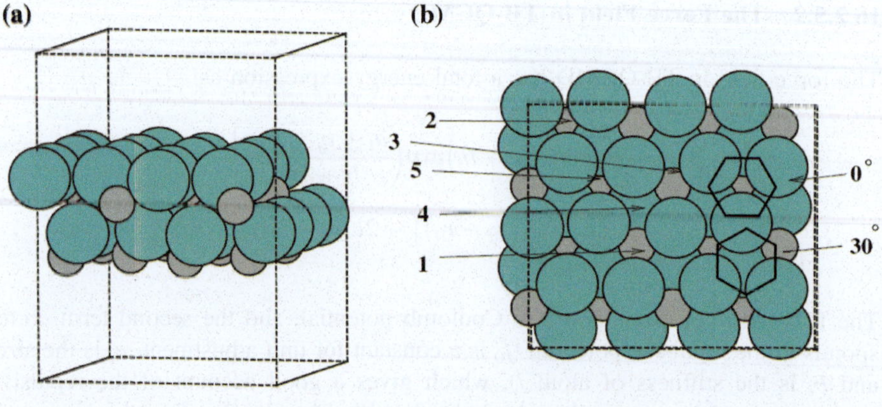

Fig. 16.5 **a** Side view and **b** top view of a Mo_2C surface slab. The large balls are Mo, and the smaller balls are C. The adsorption sites are 1: Hc, 2: Vc, 3: bridge Hc/Hm, 4: Hm, 5: bridge Vc/Hm. Benzene has two possible rotational configurations: 0° and 30°. Reprinted from Structural Chemistry, 2012, 23, 1459–1466, Copyright (2015) with permission from Springer Publishing Company

16.3.2 Benzene Hydrogenation Mechanism from DFT with a Cluster Model

There has been a long debate on the mechanism of benzene hydrogenation on Mo_2C. The debate [33–35] emerged over the reason for the deactivation, which can be attributed to lack of knowledge of the reaction mechanism of Mo_2C-catalyzed hydrogenation. Up to now, it is still not clear about the unique deactivation of Mo_2C-catalyzed hydrogenation of benzene, compared to that of other aromatic molecules such as toluene [36, 37] and naphthalene [38, 39]. Rocha et al. [33, 40] argue that benzene hydrogenation follows the Eley-Rideal mechanism, and the deactivation was because of strong benzene adsorption that blocks the active sites. However, our investigation using the DFT method combined with other experimental evidences strongly suggests that the Eley-Rideal mechanism proposed by Rocha et al. is not correct, and the reaction should follow a Langmuir-Hinshelwood (L-H) mechanism.

In order to understand the mechanism and find the main reaction path for benzene hydrogenation on Mo_2C, the interactions of the reactant, the product, and all the different isomers of partially hydrogenated benzene molecules/biradicals on the α-Mo_2C (0001) surface have been studied with a $Mo_{38}C_{19}$ cluster [41]. All calculations were performed with the linear combination of Gaussian-type orbitals Kohn-Sham density functional theory (LCGTO-KS-DFT) package deMon2 k [11], at the GGA-PBE [42] /DZVP [43] / GEN-A2 [16] level of theory with dispersion corrections.

We defined two properties to characterize the adsorption of the species on the $Mo_{38}C_{19}$ cluster. One is the stabilization energy E_{stab}, which measures the stability of the species on the surface:

$$E_{stab} = E_{C_6H_x(cluster)} - E_{cluster} - E_{C_6H_6} - \frac{x-6}{2}E_{H_2} \tag{16.53}$$

The other one is the adsorption energy E_{ad}, which measures the binding affinity of the species on the surface:

$$E_{ad} = E_{C_6H_x(cluster)} - E_{cluster} - E_{C_6H_x} \tag{16.54}$$

We found that whether dispersion corrections are present or not, C_6H_8 is the least stable species on the surface. However, the adsorption energy shows that C_6H_8 binds the most strongly with the surface before and after dispersion correction (Fig. 16.6). These findings suggest that the intermediate C_6H_8 is very likely to dissociate on Mo_2C during the hydrogenation, in good agreement with the missing of C_6H_8 signal from the post-catalytic Mo_2C from temperature-programmed reduction-mass spectrometry (TPR-MS) experiments [35]. Based on these thermodynamics studies, we propose that the benzene hydrogenation reaction on molybdenum carbide most likely proceeds through the intermediates of 12-C_6H_8 and 1234-C_6H_{10}. With benzene being hydrogenated gradually, the C6 ring becomes

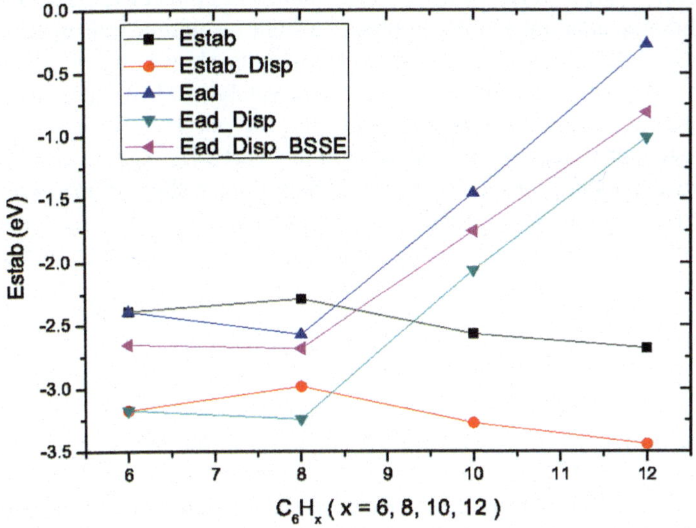

Fig. 16.6 E_{stab} and E_{ad} of the most favorable isomer of adsorbed C_6H_x (x = 6, 8, 10, 12) Those with "_Disp" are dispersion corrected, and those with "_Disp_BSSE" are both dispersion and BSSE corrected. Reprinted from J. Phys. Chem. C, 2013, 117(14), Copyright (2015) with permission from American Chemical Society

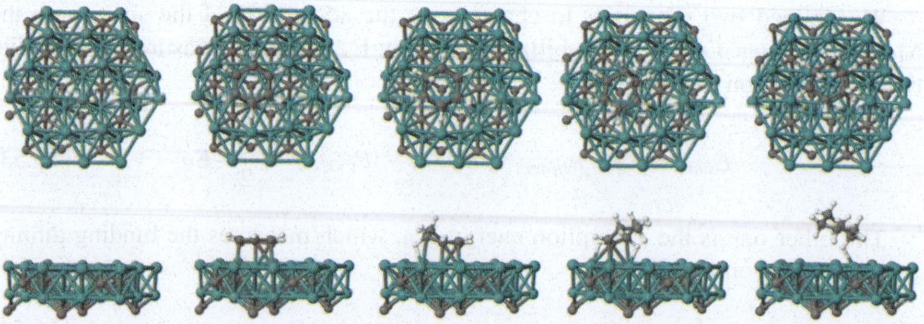

Fig. 16.7 The $Mo_{38}C_{19}$ cluster and the most stable adsorption configurations of the adsorbates (from left to right): C_6H_6, 12-C_6H_8, 1234-C_6H_{10}, and C_6H_{12}. The large balls are Mo; the medium balls are C, and the smallest balls are H

more and more tilted up, and leads to the formation of cyclohexane that interacts with the surface by weak vdW interactions (Fig. 16.7).

The Eley-Rideal mechanism proposed by Rocha et al. [33, 40] may be wrong for several reasons. First, they ruled out the possibility of surface decomposition or the formation of coke as the surface site blocker simply by reasoning that the temperature was not high enough for it to happen at 363 K. This assumption, however, is not valid, because 1,3-cyclohexadiene (12-C_6H_8 in our notation) was found to decompose to form phenyl even at 80 K on carbon-modified Mo surfaces, and it could even dehydrogenate to form benzene at 313 K. [44] Therefore, the formation of coke from C_6H_8 is very possible; so far, there is no direct experimental evidence against this explanation. On the contrary, a high-resolution transmission electron microscopy (HRTEM) experiment [45] has clearly shown that large amounts of amorphous deposits were present on the post-catalytic Mo_2C, indicating the formation of coke during the reaction. Secondly, the observation of a much smaller cyclohexane signal compared to benzene in the TPR-MS experiment [33] does not necessarily mean that benzene is the site blocker. Rather, the carbonaceous layer that may integrate with the Mo_2C surface at high temperature without desorption should not be neglected. Thirdly, if the Eley-Rideal mechanism were true, and the surface underwent gradual benzene accumulation, then when 0 % conversion rate is achieved, the TPR-MS experiment should show the benzene signal as the dominant one, or at least show much less H_2 signal than benzene signal. However, on the contrary, the post-catalytic H_2 signal in their spectrum is much higher than the benzene signal. Fourthly, the benzene desorption temperature on Mo_2C (377–392 K, depending on the synthesis conditions) is similar to the benzene desorption temperature on Ru (384 K) which means that the adsorption strength of benzene over Mo_2C is comparable with that of a Ru catalyst, [46] on which no deactivation of benzene hydrogenation was observed [35, 45]. This means that the deactivation of the catalyst is not necessarily related to benzene adsorption. In fact, comparing the HRTEM of post-catalytic Mo_2C and Ru, no carbon deposits could be seen on the Ru particles, but were found on Mo_2C, [45] strongly supporting the formation of coke as the reason for the deactivation. Lastly, it is impossible to attribute the deactivation of benzene hydrogenation on Mo_2C at 423 K [47] to the adsorption of benzene, which desorbs from Mo_2C readily at 377–392 K [46]. All of these observations, however, can be explained within the L-H mechanism: the sites around benzene molecules are occupied by the carbonaceous species or coke from partially hydrogenated species such as C_6H_8; for the remaining benzene molecules, the attacking H atoms (including those inside the bulk) although of significant amount, are not able to get close enough to benzene for the hydrogenation reaction to happen.

16.3.3 The Dynamics of Benzene Hydrogenation on the Mo_2C (0001) Surface by TB-QCMD

To understand the dynamics of the hydrogenation process, and validate our proposed L-H mechanism from previous work, the TB-QCMD method was applied to the hydrogenation of C_6H_6 on a Mo-terminated Mo_2C (0001) surface. By introducing three consecutive hydrogen molecules from the gas phase, we were able to obtain the information on the whole process of the hydrogenation directly [48].

The C_6H_6 hydrogenation reactions were simulated by applying the TB-QCMD method to a model of the $C_6H_6/\alpha\text{-}Mo_{72}C_{36}$ (0001) system shown in Fig. 16.8 as top and side views, respectively. The $\alpha\text{-}Mo_2C$ (0001) surface was represented by slabs generated from a super cell made of 3 × 3 unit cells, four layers thick. In this model, the $\alpha\text{-}Mo_2C$ (0001) consists of 72 Mo and 36 C atoms in the unit cell. The total

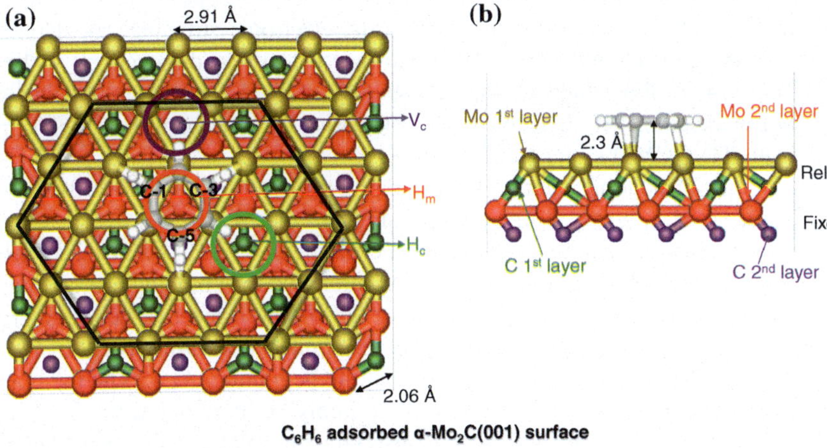

Fig. 16.8 C_6H_6 adsorbed $\alpha\text{-}Mo_2C$ (0001) surface. **a**, **b** shows the *top* and *side* view of the $Mo_{72}C_{36}$ cluster. The *large balls* are Mo, the medium balls are C, and *the small balls* are H. The upper two layers of atoms are relaxed and the lower two layers were fixed during the simulation

number of atoms in our system (including benzene and H_2) is 122. We have symbolized the two hydrogen atoms of the hydrogen molecule as H_1 and H_2 (indices 121 and 122) and we denoted six carbon of C_6H_6 as C_1, C_2, C_3, C_4, C_5 and C_6.

Three-dimensional periodic boundary conditions were applied in the TB-QCMD simulation. The lengths of the unit cell shown in Fig. 16.8 were fixed to $a = 18.03$ Å, $b = 15.63$ Å, $c = 27.83$ Å, and $\alpha = \beta = \gamma = 90°$ during the simulation. As a first step we have stabilized the model by carrying out an MD simulation up to 50000 steps. A total of 35640 steps of MD simulations were performed.

The temperature in the simulation was controlled by scaling the atomic velocities to maintain a substrate temperature of 473 K, with the exception of the emitted molecules and fixed atoms. For the specific case the initial velocity of the H_2 towards $C_6H_6/Mo_{72}C_{36}$ was set to 3730 m/s, chosen from the Maxwell-Boltzmann distribution at 473 K.

In the first hydrogenation pre-adsorbed benzene (C_6H_6) was transformed to cyclohexadiene (C_6H_8). Similarly, after the second and third hydrogenation steps, cyclohexene (C_6H_{10}) and cyclohexane (C_6H_{12}) were formed, respectively. The C_6H_{12} was then found to be desorbed from the surface during the simulation. As an example, the dynamic process of the first hydrogenation reaction ($C_6H_6 + H_2 = C_6H_8$) is shown in Fig. 16.9.

H_2 molecules were placed in the vacuum region of the unit cell above the C_1 atom depicted in Fig. 16.9. An initial distance between the H_2 in the gas phase and the C_6H_6 molecule was approximately 5.6 Å. H_2 was placed in the gas phase in such a way that it can impinge on the vicinity of the C_1 atom of benzene. A snapshot at 90 fs shows that the H_2 molecule approaches the C_6H_6 adsorbed on the $Mo_{72}C_{36}$ cluster whereas a snapshot at 160 fs shows the H_2 interacting with C_6H_6. At 450 fs H_2 interacts with the C_6H_6 molecule and finally it reaches the α-Mo_2C surface. In this stage we have observed the chemisorption of H_2 on the α-Mo_2C surface, which was also confirmed by the electronic structure calculations. As soon as H_2 interacts with the α-Mo_2C surface it starts to dissociate at 712 fs. Finally at 1274 fs dissociated H interacts with C of C_6H_6 and hydrogenation of C_6H_6 take place and C_6H_8 is formed. Our proposed L-H mechanism is well confirmed from TB-QCMD simulations.

The changes in bonding energy, bond length and bond population of the H-H, C_1-H_{122}, and C_2-H_{121} bonds during the MD simulation are examined in Fig. 16.10. Changes in the above parameters confirm the dissociation of H_2 and the hydrogenation of C_6H_6. At the beginning of the simulation, the H-H binding energy in the gas phase was about −101.21 kcal/mol and at 450 fs the respective bonding term reduces to −83.70 kcal/mol and finally becomes about 0.0 kcal/mol (Fig. 16.10a). In addition the H-H bond length was 0.76 Å at the beginning. At 450 fs, the bond length increases to 0.88 Å and finally to 1.54 Å, implying its complete dissociation, which could be further consolidated by observing the plots of the bond length (Fig. 16.10b) and the bond population (Fig. 16.10c) (the latter changing from 0.63 to −0.013 atomic units). The hydrogenation of C_6H_6 could also be elucidated with the same quantities where the C_1-H_{122} and C_2-H_{122} bonds also strengthened considerably. At the starting position, the C_1-H_{122} and C_2-H_{121} bonding energy

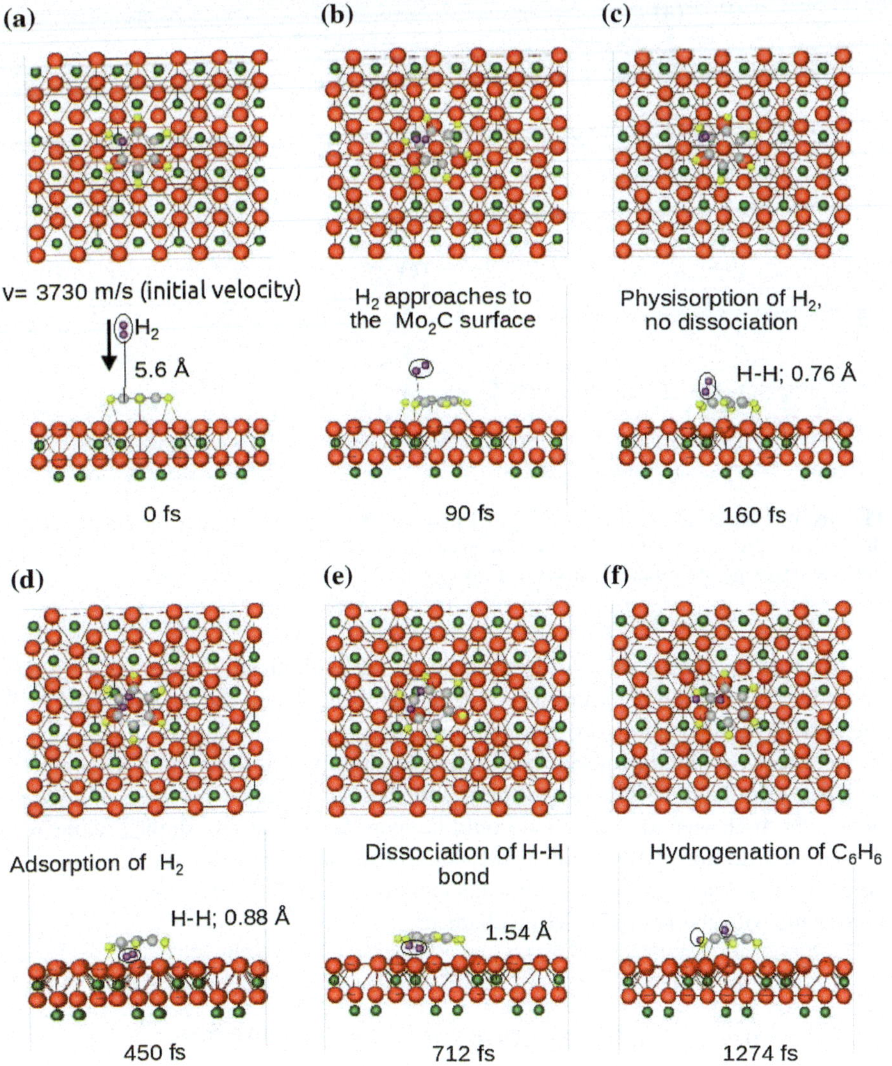

Fig. 16.9 Hydrogenation of C_6H_6 from TB-QCMD simulations. The *large balls* are Mo; the *medium balls* are C, and the *small balls* are H. The H atoms belonging to the incoming H_2 molecule are *circled* to be distinguished from other hydrogen atoms

(Fig. 16.10a) was about 0.0 kcal/mol, whereas at the final stage it becomes −75.34 and −73.23 kcal/mol, respectively. Bond length (Fig. 16.10b) and bond population (Fig. 16.10c) also confirm the similar changes. The mechanism of dissociative adsorption of H_2 followed by the hydrogenation of pre-adsorbed benzene on the supported Mo_2C (0001) surface is therefore confirmed.

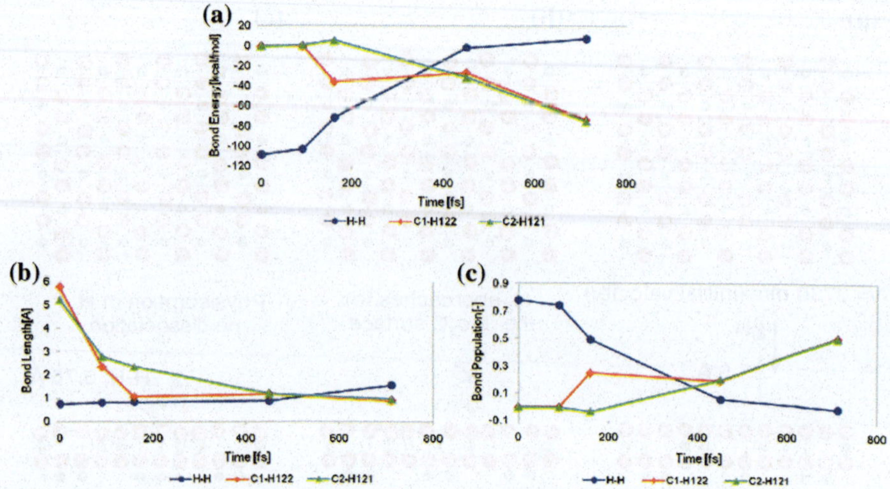

Fig. 16.10 Change of bond energy, bond length and bond population during simulation. The dissociation of H_2 created two H atoms, which are called H121 and H122. The target C atoms that they are attacking are called C1 and C2, respectively

To avoid the bias in the starting position of the H_2 molecule, a similar study was carried out where the incoming H_2 molecule is located atop the α-Mo_2C surface in the same layer adjacent to the benzene molecule. Despite the change in the reaction barriers between the trajectories of different H_2 starting positions, both reactant (H_2, and benzene) have to be co-adsorbed, and the H_2 molecule must be dissociated before the hydrogenation reaction could happen. So, while our studies so far with TB-QCMD do not involve statistically significant sampling, they do shed light on some of the possible intricate details that can occur during the reactions and they are consistent with the overall L-H mechanism.

16.4 Benzene Hydrogenation on Spherical MCNPs in the Gas Phase

Although the Mo-terminated (0001) surface model provides important insights into the benzene hydrogenation on Mo_2C, it may not correctly reflect the whole story of the reaction. The MCNPs, as opposed to the bulk form, are usually roughly spherical amorphous particles [49]. Since the interaction of the surface species with the catalyst (or adsorption) is known as a local phenomenon, the topology of the active site on MCNPs may be essential for the reactions. Also, the quantum size effect of the MCNPs should not be neglected.

To account for these problems, we used the semi-empirical DFTB method, which is fast enough to deal with hundreds of atoms for the reaction path studies.

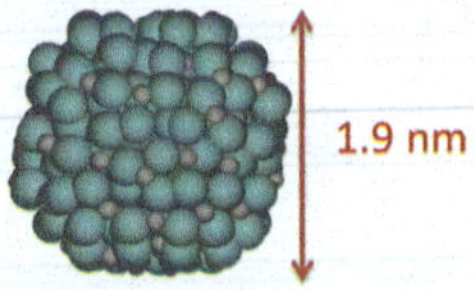

Fig. 16.11 A 1.9 nm MCNP $Mo_{235}C_{115}$ obtained from simulated annealing at DFTB level of theory. The *large balls* are Mo, and the *smaller balls* are C

An example is shown in Fig. 16.11 with the structure of a MCNP $Mo_{235}C_{115}$ with diameter of 1.9 nm, which was obtained by simulated annealing with Born–Oppenheimer molecular dynamics (BOMD) and local geometry optimization. Benzene molecules and the dissociated H atoms are then placed on the surface sites of the MCNP.

With the DFTB parameters that we recently developed, direct calculations on nano-scale models can be driven with great efficiency. The minimum energy paths (MEP) of the benzene hydrogenation reactions can be calculated by the nudged elastic band method. As an example, the minimum energy path of the elementary reaction ($C_6H_6 + H = C_6H_7$) on a two-fold site of the 1.9 nm MCNP is shown in Fig. 16.12. There are 19 intermediate images on the chain-of-states of the reaction. The corresponding reactant state, transition state (TS), and the product state are presented in Fig. 16.13.

In the reactant state, the benzene molecule is adsorbed on a Mo-Mo two-fold site; one of the H atoms is adsorbed also on a two-fold site, and the other on a Mo

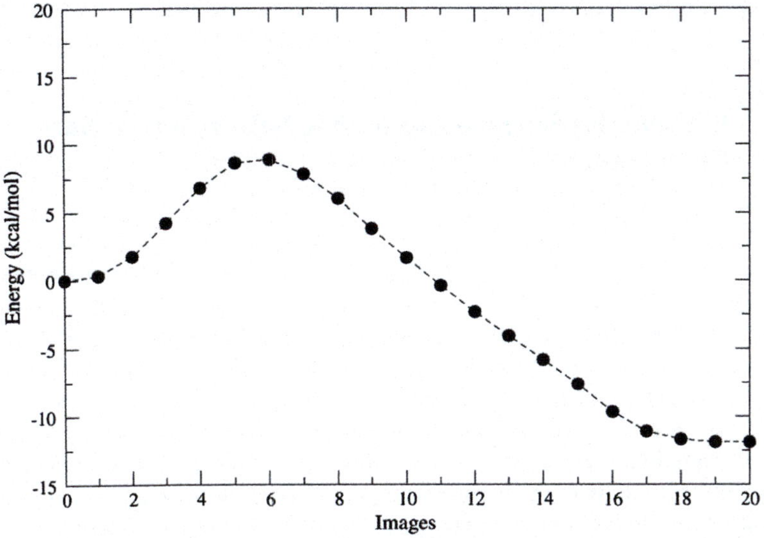

Fig. 16.12 The minimum energy path of the addition of the first hydrogen to benzene on a two-fold site of a 1.9 nm MCNP from NEB calculation

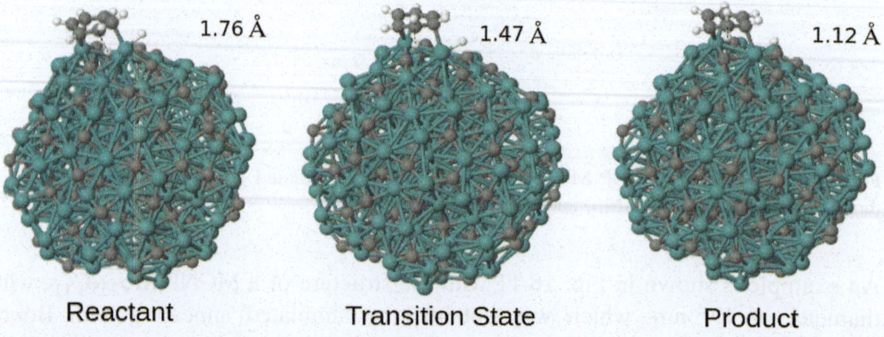

Fig. 16.13 The reactant, TS, and product of the first hydrogen reaction of benzene on a 1.9 nm MCNP from NEB calculation; the *large balls* are Mo; the *medium balls* are C, and the *small balls* are H. All atoms are relaxed

top site. The distance between the attacking H and the corresponding C on the benzene ring is initially 1.76 Å. After going through a transition state with C—H distance of 1.47 Å, a typical C-H bond is formed. During the reaction, the C atom was lifted up, and the original C-H group in benzene turned upwards, in order to accommodate the new H atom. The importance of a full-size model of the nanoparticle is manifested by the fact that the elementary hydrogenation reaction on this particular two-fold site is exothermic, while all others calculated on flat catalyst surfaces, such as Mo_2C (0001) (results not shown), Pt(111) [50], and Ru (0001) [51] are endothermic. This is due to the breaking of the surface symmetry, which causes some active sites to be in better positions than flat surface sites for hydrogenation, while some are worse.

16.5 Benzene Hydrogenation on MCNPs in the In Situ Environment

To simulate the complex in situ catalysis system in a more realistic manner, and to include the solvent and other effects such as temperature and pressure, a multiscale QM/MM model was built. The reacting benzene molecule, the dissociatively adsorbed H atoms for attacking the C atoms in benzene and the MCNP catalyst are included in the QM region. The MM region is made up of a model aromatic solvent with 100 benzene molecules (Fig. 16.14).

The free energy barriers of the hydrogenation reactions are calculated with the umbrella sampling method. Unlike the traditional free energy barriers obtained from thermal corrections with the harmonic approximation on the reactant and transition state structures on the minimum energy path (MEP), the umbrella sampling method includes the anharmonic vibrations. From umbrella sampling, the degrees of freedom related to the free energy (potential of mean force, or PMF) curves can be

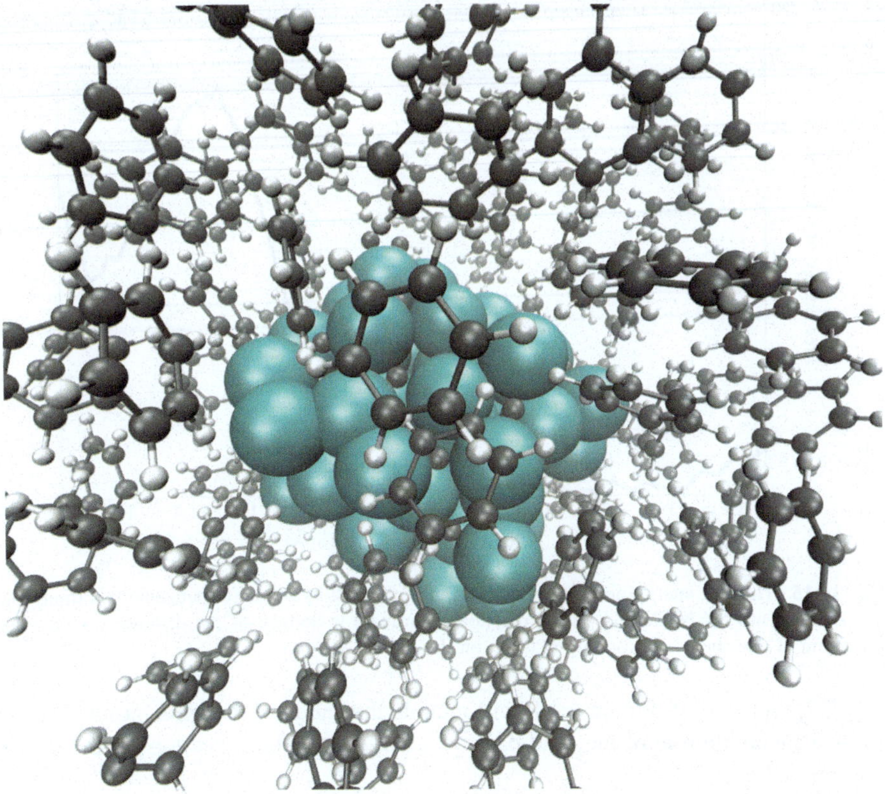

Fig. 16.14 The QM/MM model of 1.2 nm MCNP $Mo_{66}C_{35}$ embedded in a model aromatic solvent of 100 benzene molecules. The *large balls* are Mo; the *medium balls* are C, and the *small balls* are H

obtained (Fig. 16.15). The comparison of the shape, curvature, and the saddle point positions on the PMF curves to the traditional MEP curve can give valuable information on the features of these nanoscale heterogeneous catalysis reactions. Among them are the entropic contributions (including the anharmonic vibrational entropy) of the nanoparticles and the solvent to the reactions in their working conditions. Our calculations show that they could change the shape of the reaction paths, modify the free energy barriers, and affect the initial state and the transition state structures of the hydrogenation reactions on the MCNPs. For this reason, entropic effects must be properly taken into consideration in order to understand nanoscale heterogeneous catalysis reactions with transition metal-containing nanoparticles. Because of the contribution of the anharmonic vibrational entropy of the MCNPs, the thermal corrections based on the harmonic approximation, which is usually used to correct for the entropic contributions, is not sufficient for understanding the reactions on MCNPs. The full details of the modelling of benzene hydrogenation on MCNPs in the in situ environment are in the thesis of X.

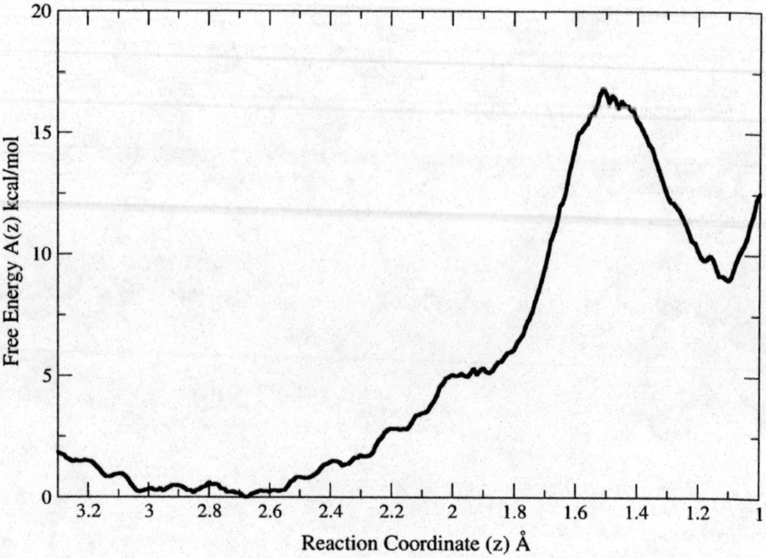

Fig. 16.15 The free energy profile of the addition of the first hydrogen to benzene on a three-fold site of a 1.2 nm MCNP $Mo_{66}C_{35}$; the reaction coordinate is defined as the distance between the incoming H and the corresponding C atom on the benzene ring

Liu [52] and in [53]. Future work will include the sand components into the model to investigate the role of the sand particles in hydrogenation reactions in the in situ environment.

16.6 Summary

The quantum modelling of complex systems is a great challenge to computational chemistry. The in situ oil sands upgrading process in the reservoir is one such complex system. In this chapter, a brief review of our work on the oil sands upgrading with MCNPs is presented, focusing on the computational tools that are used to deal with these systems. We start from periodic and cluster DFT electronic calculations to understand the adsorption of the aromatic hydrocarbons on the MCNPs. Further thermodynamic study with DFT on the reaction intermediates provided important insights into the mechanism of the reactions. The TB-QCMD method reveals the real-time motion of the atoms, and the breaking and forming of chemical bonds during the reactions. The DFTB method, which is an approximate DFT method, is significantly faster than DFT. With proper DFTB parameters, direct calculations on the nanoscale systems become possible. From these calculations, the electronic structures of the MCNPs and the potential energy profiles of the benzene hydrogenation reactions on the MCNPs are obtained. The multiscale QM/MM method brings in the environmental effects, including the temperature (entropy), pressure, and the aromatic solvent, and is the method of choice for modelling these complex reactions in in situ oil sands upgrading. Full details of the work discussed in this chapter may be found in the following references (ref [54] for 3.1, ref [41] for 3.2, ref [48] for 3.3, and ref [52, 53] for 4 and 5).

References

1. Alboudwarej H, Felix J, Taylor S, Badry R, Bremner C, Brough B, Skeates C, Baker A, Palmer D, Pattison K, Beshry M, Krawchuk P, Brown G, Calvo R, Triana JAC, Hathcock R, Koerner K, Hughes T, Kundu D, Cardenas JL, West C (2006) Highlighting heavy oil. Oilfield Rev. Schlumberger. https://www.slb.com/~/media/Files/resources/oilfield_review/ors06/sum06/heavy_oil.pdf
2. Sanati M, Harrysson B, Faghihi M, Gevert B, Jaras S (2002) Catalytic hydrodearomatization. In: Spivey JJ (ed) Catalysis: Volume 16, vol 16. The Royal Society of Chemistry, pp 1–42. doi:10.1039/9781847553287-00001
3. Glassman D, Wucker M, Isaacman T, Champilou C (2011) The water-energy nexus, adding water to the energy sagenda. http://www.worldpolicy.org/sites/default/files/policy_papers/THE%20WATER-ENERGY%20NEXUS_0.pdf
4. Pereira-Almao P (2012) In-situ upgrading of bitumen and heavy oils via nanocatalysis. Can J Chem Eng 90(2):320–329. doi:10.1002/cjce.21646
5. Frauwallner ML, Lopez-Linares F, Lara-Romero J, Scott CE, Ali V, Hernandez E, Pereira-Almao P (2011) Toluene hydrogenation at low temperature using a molybdenum carbide catalyst. Appl Catal A-Gen 394(1–2):62–70. doi:10.1016/j.apcata.2010.12.024
6. Survery USG (2012) U.S. Mineral commodity summaries 2012.107 http://minerals.usgs.gov/minerals/pubs/mcs/2012/mcs2012.pdf
7. Scott C (2013) In-situ oil sand upgrading with nanocatalysts. Pers Commun
8. Stanislaus A, Cooper BH (1994) Aromatic hydrogenation catalysis: a review. Catal Rev 36 (1):75–123. doi:10.1080/01614949408013921
9. Hohenberg P, Kohn W (1964) Inhomogeneous electron gas. Phys Rev 136 (3B):B864–B871. doi:http://dx.doi.org/10.1103/PhysRev.136.B864
10. Kohn W, Sham LJ (1965) Self-consistent equations including exchange and correlation effects. Phys Rev 140(4A):A1133–A1138. doi:10.1103/PhysRev.140.A1133
11. Köster AM, Geudtner G, Calaminici P, Casida ME, Dominguez VD, Flores-Moreno R, Goursot A, Heine T, Ipatov A, Janetzko F, Del Campo JM, Revelevs JU, Vela A, Zuniga B, Salahub DR (2011) deMon2 k. The deMon developers, CINVESTAV, Mexico. www.demon-software.com
12. Blöchl PE (1994) Projector augmented-wave method. Phys Rev B 50(24):17953–17979. doi:10.1103/PhysRevB.50.17953
13. Kresse G, Furthmüller J (1996) Efficient iterative schemes for *ab initio* total-energy calculations using a plane-wave basis set. Phys Rev B 54(16):11169–11186. doi:10.1103/PhysRevB.54.11169
14. Calaminici P, Domínguez-Soria V-D, Flores-Moreno R, Gamboa-Martínez G, Geudtner G, Goursot A, Salahub D, Köster A (2012) Auxiliary density functional theory: from molecules to nanostructures. In: Leszczynski J (ed) Handbook of computational chemistry. Springer Netherlands, pp 573–610. doi:10.1007/978-94-007-0711-5_16
15. Dunlap BI, Connolly JWD, Sabin JR (1979) On first-row diatomic molecules and local density models. J Chem Phys 71(12):4993–4999. doi:10.1063/1.438313
16. Köster AM, Reveles JU, del Campo JM (2004) Calculation of exchange-correlation potentials with auxiliary function densities. J Chem Phys 121(8):3417–3424. doi:10.1063/1.1771638

17. Köster AM (2003) Hermite Gaussian auxiliary functions for the variational fitting of the Coulomb potential in density functional methods. J Chem Phys 118(22):9943–9951. doi:10.1063/1.1571519
18. Köster AM, del Campo JM, Janetzko F, Zuniga-Gutierrez B (2009) A MinMax self-consistent-field approach for auxiliary density functional theory. J Chem Phys 130 (11):114106. doi:10.1063/1.3080618
19. Porezag D, Frauenheim T, Köhler T, Seifert G, Kaschner R (1995) Construction of tight-binding-like potentials on the basis of density-functional theory: Application to carbon. Phys Rev B 51(19):12947–12957. doi:10.1103/PhysRevB.51.12947
20. Oliveira AF, Seifert G, Heine T, Duarte HA (2009) Density-functional based tight-binding: an approximate DFT method. J Braz Chem Soc 20:1193–1205. doi:10.1590/S0103-50532009000700002
21. Elstner M, Porezag D, Jungnickel G, Elsner J, Haugk M, Frauenheim T, Suhai S, Seifert G (1998) Self-consistent-charge density-functional tight-binding method for simulations of complex materials properties. Phys Rev B 58(11):7260–7268. doi:10.1103/PhysRevB.58.7260
22. Zhechkov L, Heine T, Patchkovskii S, Seifert G, Duarte HA (2005) An efficient a posteriori treatment for dispersion interaction in density-functional-based tight binding. J Chem Theory Comput 1(5):841–847. doi:10.1021/ct050065y
23. Cui Q, Elstner M, Kaxiras E, Frauenheim T, Karplus M (2000) A QM/MM implementation of the self-consistent charge density functional tight binding (SCC-DFTB) method. J Phys Chem B 105(2):569–585. doi:10.1021/jp0029109
24. Warshel A, Levitt M (1976) Theoretical studies of enzymic reactions: Dielectric, electrostatic and steric stabilization of the carbonium ion in the reaction of lysozyme. J Mol Biol 103 (2):227–249. doi:10.1016/0022-2836(76)90311-9
25. Vanommeslaeghe K, Hatcher E, Acharya C, Kundu S, Zhong S, Shim J, Darian E, Guvench O, Lopes P, Vorobyov I, Mackerell AD (2010) CHARMM general force field: A force field for drug-like molecules compatible with the CHARMM all-atom additive biological force fields. J Comput Chem 31(4):671–690. doi:10.1002/jcc.21367
26. Torrie GM, Valleau JP (1974) Monte Carlo free energy estimates using non-Boltzmann sampling: Application to the sub-critical Lennard-Jones fluid. Chem Phys Lett 28(4):578–581. doi:10.1016/0009-2614(74)80109-0
27. Kästner J (2011) Umbrella sampling. Wiley Interdiscip Rev: Comput Mol Sci 1(6):932–942. doi:10.1002/wcms.66
28. Roux B (1995) The calculation of the potential of mean force using computer simulations. Comput Phys Commun 91(1–3):275–282. doi:10.1016/0010-4655(95)00053-I
29. Kumar S, Rosenberg JM, Bouzida D, Swendsen RH, Kollman PA (1992) The weighted histogram analysis method for free-energy calculations on biomolecules. I. The method. J Comput Chem 13(8):1011–1021. doi:10.1002/jcc.540130812
30. Selvam P, Tsuboi H, Koyama M, Kubo M, Miyamoto A (2005) Tight-binding quantum chemical molecular dynamics method: a novel approach to the understanding and design of new materials and catalysts. Catal Today 100(1–2):11–25. doi:10.1016/j.cattod.2004.12.020
31. St. Clair TP, Oyama ST, Cox DF (2002) Adsorption and reaction of thiophene on α-Mo_2C (0001). Surf Sci 511(1–3):294–302. doi:10.1016/s0039-6028(02)01505-4
32. Grimme S (2006) Semiempirical GGA-type density functional constructed with a long-range dispersion correction. J Comput Chem 27(15):1787–1799. doi:10.1002/jcc.20495
33. Rocha AS, Rocha AB, da Silva VT (2010) Benzene adsorption on Mo2C: a theoretical and experimental study. Appl Catal A-Gen 379(1–2):54–60. doi:10.1016/j.apcata.2010.02.032
34. Rocha ÂS, Silva VLTd, Leitão AA, Herbst MH, Faro-Jr AC (2004) Low temperature low pressure benzene hydrogenation on Y zeolite-supported carbided molybdenum. Catal Today 98(1–2):281–288. doi:10.1016/j.cattod.2004.07.041
35. Lee JS, Yeom MH, Park KY, Nam I-S, Chung JS, Kim YG, Moon SH (1991) Preparation and benzene hydrogenation activity of supported molybdenum carbide catalysts. J Catal 128 (1):126–136. doi:10.1016/0021-9517(91)90072-c

36. Frauwallner ML (2009) Catalyst development for hydrogenation at low temperature, pressure, and hydrocarbon space velocity. M.Sc. thesis, University of Calgary, Calgary
37. Guzmán HJ, Xu W, Stacchiola D, Vitale G, Scott CE, Rodríguez JA, Pereira-Almao P (2013) In situ time-resolved X-ray diffraction study of the synthesis of Mo_2C with different carburization agents. Can J Chem 91(7):573–582. doi:10.1139/cjc-2012-0516
38. Ardakani SJ, Liu XB, Smith KJ (2007) Hydrogenation and ring opening of naphthalene on bulk and supported Mo_2C catalysts. Appl Catal A-Gen 324:9–19. doi:10.1016/j.apcata.2007.02.048
39. Pang M, Liu CY, Xia W, Muhler M, Liang CH (2012) Activated carbon supported molybdenum carbides as cheap and highly efficient catalyst in the selective hydrogenation of naphthalene to tetralin. Green Chem 14(5):1272–1276. doi:10.1039/c2gc35177c
40. Oliveira RR Jr, Rocha AS, Teixeira da Silva V, Rocha AB (2014) Investigation of hydrogen occlusion by molybdenum carbide. Appl Catal A 469:139–145. doi:10.1016/j.apcata.2013.09.031
41. Liu X, Tkalych A, Zhou B, Köster AM, Salahub DR (2013) Adsorption of Hexacyclic C_6H_6, C_6H_8, C_6H_{10}, and C_6H_{12} on a Mo-Terminated α-Mo_2C (0001) surface. J Phys Chem C117 (14):7069–7080. doi:10.1021/jp312204u
42. Perdew JP, Burke K, Ernzerhof M (1996) Generalized gradient approximation made simple. Phys Rev Lett 77(18):3865–3868. doi:10.1103/PhysRevLett.77.3865
43. Calaminici P, Janetzko F, Koster AM, Mejia-Olvera R, Zuniga-Gutierrez B (2007) Density functional theory optimized basis sets for gradient corrected functionals: 3d transition metal systems. J Chem Phys 126 (4). doi:10.1063/1.2431643
44. Eng J, Bent BE, Frühberger B, Chen JG (1998) Modifying surface reactivities by a carbide overlayer: a vibrational study of the reaction mechanisms of cyclohexene and 1,3-Cyclohexadiene on Mo(110) and (4 × 4)-C/Mo(110) Surfaces. Langmuir 14(6):1301–1311. doi:10.1021/la970709s
45. MarquezAlvarez C, Claridge JB, York APE, Sloan J, Green MLH (1997) Benzene hydrogenation over transition metal carbides, in Studies in surface science and catalysis, vol 106. Hydrotreatment and hydrocracking of oil fractions
46. Choi J-S, Krafft J-M, Krzton A, Djéga-Mariadassou G (2002) Study of residual oxygen species over molybdenum carbide prepared during in situ DRIFTS experiments. Catal Lett 81 (3):175–180. doi:10.1023/a:1016568804706
47. Zhang J, Wu WC (2008) Hydrogenation of benzene over Mo_2C/Al_2O_3 catalyst. China Petrol Process Petrochem Technol 4:41–43
48. Ahmed F, Liu X, Miyamoto A, Salahub DR (2015) Adsorption and hydrogenation of C_6H_6 on α-Mo_2C (0001): a quantum chemical molecular dynamics study. unpublished
49. Hyeon TH, Fang MM, Suslick KS (1996) Nanostructured molybdenum carbide: Sonochemical synthesis and catalytic properties. J Am Chem Soc 118(23):5492–5493. doi:10.1021/ja9538187
50. Saeys M, Reyniers MF, Neurock M, Marin GB (2005) Ab initio reaction path analysis of benzene hydrogenation to cyclohexane on Pt(111). J Phys Chem B 109(6):2064–2073. doi:10.1021/jp049421j
51. Fan C, Zhu Y-A, Zhou X-G, Liu Z-P (2011) Catalytic hydrogenation of benzene to cyclohexene on Ru(0001) from density functional theory investigations. Catal Today 160 (1):234–241. doi:10.1016/j.cattod.2010.03.075
52. Liu X (2015) In-situ heavy oil upgrading with molybdenum carbide nanoparticles: a multiscale modeling approach. Ph.D. thesis, University of Calgary, Calgary
53. Liu X, Salahub DR (2015) Molybdenum carbide nanocatalysts at work in the in-situ environment: a DFTB and QM(DFTB)/MM Study. J Am Chem Soc 137(12):4249–4259. doi:10.1021/jacs.5b01494
54. Zhou B, Liu X, Cuervo J, Salahub DR (2012) Density functional study of benzene adsorption on the α-Mo_2C(0001) surface. Struct Chem 23(5):1459–1466. doi:10.1007/s11224-012-0064-5

Chapter 17
Computational Spectroscopy in Solution: Methods and Models for Investigating Complex Systems

Vincenzo Barone, Enrico Benassi and Ivan Carnimeo

Abstract In this contribution, some issues related to the interpretation, simulation and modelling of solvent effects on the absorption and emission spectra of organic dyes are presented and discussed. First, a brief analysis of the physical basis of solvent effects on the electronic transitions is reported, in order to introduce the most important phenomena and quantities tuning the so-called solvatochromic shifts. This is followed by a general discussion of the most common models employed for the interpretation, simulation and prediction of such effects. A general and effective multilayer scheme is analyzed in some detail, which has been developed in the past years and is known to provide—in most cases—quantitative predictions of the spectral features of solvated molecules. Afterwards, starting from this general model, some approximations are introduced, leading to simplified and cost effective analytical schemes. In order to sketch a more complete perspective of the models still used by spectroscopists, phenomenological methods are critically discussed. Finally, broadening of spectral lines by both symmetric (solvent relaxation) and possibly asymmetric (vibronic) contributions is shortly analysed. In all cases, the theoretical bases of the methods, as well as practical applications and test cases are given, in order to clarify the most interesting aspects of all the discussed models.

Acronyms

AO	Atomic Orbital
CC	Coupled Cluster
Cou	Coulomb
C-PCM	Conductor-like Polarizable Continuum Model
DFT	Density Functional Theory
DFTB	Density Functional Tight Binding

V. Barone (✉) · E. Benassi · I. Carnimeo
Scuola Normale Superiore, Piazza Dei Cavalieri 7, 56126 Pisa, Italy
e-mail: vincenzo.barone@sns.it

I. Carnimeo
Istituto Italiano di Tecnologia (IIT), Via Morego 30, 16163 Genoa, Italy

J.-L. Rivail et al. (eds.), *Quantum Modeling of Complex Molecular Systems*,
Challenges and Advances in Computational Chemistry and Physics 21,
DOI 10.1007/978-3-319-21626-3_17

D-PCM	Dielectric-like Polarizable Continuum Model
EE	Electrostatic Embedding
Eq	Equilibrium solvation regime
FC	Frank-Condon
FX	Fixed Charges
FQ	Fluctuating Charge
FWHM	Full Width at Half Maximum
HOMO	Highest Occupied Molecular Orbital
IEF-PCM	Integral Equation Formalism Polarizable Continuum Model
LUMO	Lowest Unoccupied Molecular Orbital
MD	Molecular Dynamics
ME	Mechanical Embedding
MF	Mean Field
MM	Molecular Mechanics
MO	Molecular Orbital
MRCI	Multi Reference Configuration Interaction
NPBC	Non-Periodic Boundary Conditions
NEQ	Non equilibrium solvation regime
OO	Occupied-occupied orbitals
OPA	One Photon Absorption
OPE	One Photon Emission
OV	Occupied-virtual orbitals
PBC	Periodic Boundary Conditions
PCM	Polarizable Continuum Model
PE	Polarizable Embedding
PES	Potential Energy Surface
SCF	Self-Consistent Field
QM	Quantum Mechanics
QM/MM	Quantum Mechanics/Molecular Mechanics
SS	State Specific
TD-DFT	Time Dependent Density Functional Theory
UV-Vis	Ultra Violet—Visible
vdW	Van der Waals
VV	Virtual-virtual orbitals

17.1 Introduction

This chapter is devoted to computational spectroscopy approaches for the study of one-photon absorption and emission (OPA and OPE, respectively) spectra in solution by means of focused models [1–4] coupling a quantum mechanical (QM) description of the solute (mostly by methods rooted into the density functional theory [5–9]) and a continuum [10–26] or discrete-continuum [27–35] description of the solvent. For interpretative purposes those numerical approaches can be profitably integrated by analytical [36–101] and/or phenomenological [102–142] approaches provided that the physical background and mathematical developments leading to the different simplifications are well defined.

Spectroscopic approaches have become the methods of choice for analysing structural, dynamical, and environmental effects of molecular systems of increasing complexity in a very effective yet non-invasive way. Integrated models, in which different spectroscopic techniques are combined together to offer a comprehensive picture of complex systems and/or phenomena, are finding widespread use. However, the relationship between the spectroscopic outcome and the underlying physical-chemical properties is often indirect, and the disentanglement of the different factors playing a role in determining the overall result can be strongly facilitated by QM computations, provided that the latter are able to couple reliability and feasibility [143]. Medium-size semi-rigid systems in the gas-phase can be nowadays studied by QM methods with an accuracy rivaling that of the most sophisticate experimental techniques. However, the situation is more involved for the large, flexible systems of current technological and/or biological interest. In this connection, methods rooted into the density functional theory (DFT) and its time-dependent (TD-DFT) extension have revolutionised the field allowing effective computations of several properties with sufficient accuracy. When solvated systems are under study, an additional source of complexity arises, since the spectroscopic processes primarily induced by a well-localised chromophore are also significantly affected by the presence of more distant atoms, the "environment". For this reason, many theoretical models have been developed in order to properly take into account the effect of the solvent molecules on the desired spectroscopic property.

Generally speaking, different strategies can be exploited to model environmental effects on structural, thermodynamic, and spectroscopic properties of molecular systems. One common approach relies on (classical) Molecular Dynamics (MD) simulations in order to first sample the configuration space of the solute-solvent system, generating an adequate number of snapshots, and subsequently evaluating the property at the QM level as an average value over all the snapshots. However, proper treatment of boundaries can play a significant role in determining the reliability of the results. Although periodic boundary conditions (PBC) are usually employed in this connection, non-periodic boundary conditions (NPBC) [21–24] show a number of significant advantages and are a more natural choice for the localised basis functions (normally Gaussians) used in computational chemistry approaches. On these grounds, a general multi-layer approach can be introduced for describing localised phenomena, in which the central part of the system is described by a QM approach, the intermediate region by an atomistic description, and the boundary in terms of an effective 'steric' potential together with the reaction field introduced by the (dielectric or conductor) continuum describing bulk solvent. Such a scheme was proposed some years ago [21–24] with encouraging results. With the recent implementation of a QM/MM polarisable embedding based on the fluctuating charge (FQ) approach [29–35], a new feature can be included in such a model, since the water molecules in the nearby of the solute can be polarised by the QM charge density.

Complete neglect of the atomistic part leads naturally to the last-generation continuum models, in which the solute is embedded in a molecularly-shaped cavity (with proper boundary conditions), which follows the motion of the solute [1–4]. This approach allows a significant decrease in the computational cost, leading in state-of-the-art implementations to negligible overheads with respect to a corresponding computation *in vacuo*. For the treatment of the chromophore a wide variety of QM methods are nowadays available, and the most common DFT functionals, such as hybrid [5, 6] and double hybrid [7–9] ones, have shown a high accuracy in the evaluation of a wide number of spectroscopic and thermodynamic quantities. Generally speaking, the implicit models [10–26] are especially suitable for the treatment of bulk effects, where the effect of solvent molecules is included in an averaged way through the definition of an effective cavity. Among the implicit solvent models, the Polarizable Continuum Model (PCM) in all its different variants ("Conductor-like" C-PCM [14, 20], "Dielectric" D-PCM [10], and "Integral Equation Formalism" IEF-PCM [12, 13, 15]) is one of the most reliable methods, and it has been successfully applied for the simulation of many different properties of molecules in condensed phase.

On the other side, experimentally-oriented chemists are still employing phenomenological approaches [102–142, 144–148] based on different macroscopic quantities or over-simplified analytical models. Analytical models [36–101] have some obvious advantages: principally, they enable exact (and fast) solutions and an immediate interpretation of the phenomenon in terms of a simple picture (related to the multipolar expansion of the electronic molecular distribution). However, being able to account for only some molecular properties, and to treat only high-symmetry systems, analytical models cannot offer a fully realistic representation, especially for complex systems. On these grounds, in the second part of this chapter we present a hierarchic set of simplifications leading to qualitative descriptions able, however, to describe general trends in a reasonable way. Following this route, we will analyse strengths, limits, and some incongruences of some widely-used phenomenological approaches. In the following sections, another issue related to solvent effects, i.e. the proper description of spectra broadening, is discussed in terms of solvent relaxation and intrinsic vibronic contributions. Next, a test case is analysed in some detail, in order to compare the performances of the different methods introduced in the previous sections. Some general considerations and perspectives of the different approaches conclude the chapter.

17.2 The Physical Model

Within the Born-Oppenheimer approximation, the electronic motion is assumed to be very fast with respect to the nuclear one, due to the difference between the respective masses. As a consequence, the electrons are usually considered to move in the field created by fixed *nuclei*, while the *nuclei* experience an instantaneous averaged potential induced by the electrons. For this reason, whenever an electronic excitation occurs, in absence of strong non-adiabatic couplings (e.g. near conical intersections), the electronic transitions (fast) can be considered as completely separated from the nuclear rearrangement (slow). Then, in a heuristic simplified picture (Fig. 17.1, left panel), when a molecule M in its gas phase ground state absorbs a photon of proper energy $\hbar\omega_a$, an electronic excited state is populated. During this transition the nuclear configuration of M has not time to relax, thus retaining the same geometry of the ground state, but a different electronic distribution. Depending on the nature of the excited state potential energy surface (PES) in the neighbourhood of the ground state geometry (the so called Franck-Condon (FC) region), M may either decay to its ground state by emitting a photon $\hbar\omega_e$, of the same energy as the absorbed one, or rearrange its geometry, relaxing to a minimum of the excited state PES. Then, M may vertically decay to a non-equilibrium region of the ground state PES, by emitting a photon $\hbar\omega_e'$ of different energy of that involved in the absorption, subsequently relaxing to the ground state equilibrium geometry. The energy difference between the emitted and the absorbed photon is called *Stokes' shift*.

When the molecule is embedded in a solvent S (Fig. 17.1, right panel), the picture is more complex, since the environment may or may not be in equilibrium with M in the aforementioned photophysical processes, from the electronic and nuclear point of view. First of all, the presence of the environment determines a change in the equilibrium geometry of the ground electronic state. The energy difference between the gas phase molecule ($M_{g(eq)}$) and the same molecule in equilibrium with the solvent ($M_{g(eq)}$–$S_{(eq)}$) is called *solvation energy* of the ground electronic state. When one photon is absorbed, a rapid change in the electronic distribution of M occurs, so that also the solvent molecules are in non-equilibrium

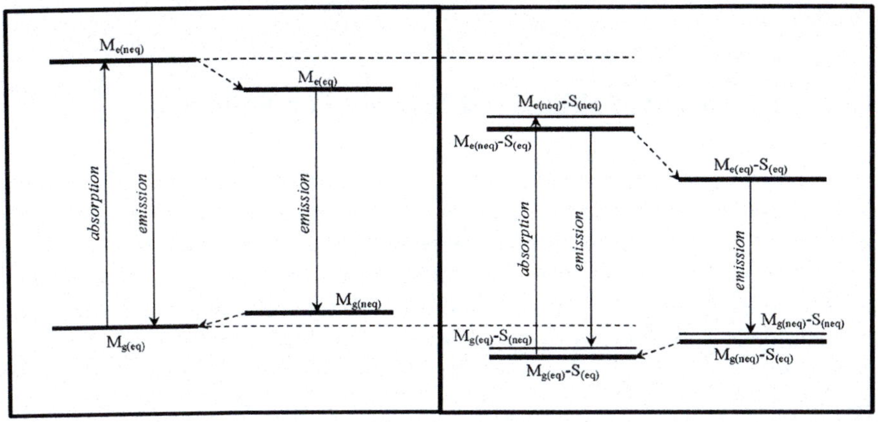

Fig. 17.1 Energy diagram for the ground (g) and excited (e) states of molecule M, *in vacuo* (*left panel*) and in solvent S (*right panel*). "eq" and "neq" refer to equilibrium and non-equilibrium conditions, respectively

conditions ($M_{e(neq)}$–$S_{(neq)}$). Then, the electronic distribution of the molecules of S swiftly rearranges, finding a new equilibrium regime with respect to the FC electronic distribution of M, and this process is generally faster than the relaxation of the nuclear configuration of M toward its excited-state minimum. The energy difference between $M_{e(neq)}$–$S_{(neq)}$ and $M_{e(neq)}$–$S_{(eq)}$ is called *solvation energy* of the FC excited electronic state. Depending on the topology of the FC region and the interaction with S, M may either decay to the ground state or relax to a minimum of the excited state. In the former case, unlike what observed in the gas phase, the photon emitted from the FC region has a different energy with respect to the absorbed one, due to the effect of the relaxation of the solvent molecules. In the latter case, the equilibration of both M and S occurs, towards a minimum of the excited state PES. From such a minimum, the electronic distribution of M decays to the ground state, where neither M nor S is in equilibrium regime. Subsequently, first the fast rearrangement of S occurs, and then the relaxation of the nuclear configurations of M and S follows, towards the ground state minimum.

The difference between non-equilibrium and equilibrium energies is known as *reorganisation energy*, which is one of the factors responsible for the symmetric broadening of the absorption / emission bands (*vide infra*). The difference between the solvation energies of the final and initial states for absorption (emission) processes in various solvents with respect to the isolated phase is the so-called *solvatochromic (fluorosolvatochromic) shift*.

It has long been known that UV-Vis absorption spectra may be influenced by the phase (gas or liquid) and that the solvent can bring about a change in the position, intensity, and shape of the spectral band. The term *solvatochromism* was introduced for the first time by Hantzsch in 1922 [144], and it has never lost its importance and charm since then. In particular, during the last 20 years, non-negligible efforts have been spent in order to investigate molecular systems which can find applications in technology, innovative materials, medicine, biology, etc. thanks to their spectroscopic and photophysical properties.

17.3 A General Multilayered Scheme For Solvation

In the previous section, the effects of the solvent on the absorption and emission spectra have been introduced, in terms of the response of the electronic distribution and nuclear geometry of both the chromophore and environmental molecules to an external applied field (a photon). Such effects, far from being of small magnitude and sporadic occurrence, are usually shuffled in all the spectroscopic experiments in solution, daily performed in every laboratory. For this reason, it is of paramount importance to have reliable tools suitable to assist the interpretation of the measurements.

In this section, we will present a general focused model, in which different regions are treated with levels of sophistication decreasing with the distance from the region where the essential of the studied phenomenon is localised, and proper non-periodic boundary conditions are enforced. Such approaches become mandatory whenever solvent effects cannot be expressed in terms of simple empirical parameters, but are strongly related to both specific and long-range solute-solvent interactions. In Fig. 17.2 a representation at the atomistic level of the solvation environment of the glycine molecule in water is sketched. Starting from the solute placed at the centre, there is a first shell of few water molecules directly (and specifically) interacting with it (the so-called cybotactic region). At larger distances, the water molecules are not involved in direct interactions with the solute, although they still have a significant effect on the solute properties through electrostatic and van der Waals interactions. All the water molecules introduced so far represent the *explicit* solvent. At very large distances, the effect of the bulk solvent molecules can be represented by a regular surface (in the present case a sphere) on which suitable electrostatic and van der Waals potentials are defined in order to represent the average effect of the bulk on the solute. This can be called the *implicit* part of the model.

This general, multi-layer model will be our reference in the following paragraphs devoted to the numerical methods based on QM and classical MM theories, suitable for the quantitative calculation of solvent effects on the electronic distribution of the chromophore, which determines, in turn, the solvatochromic effects.

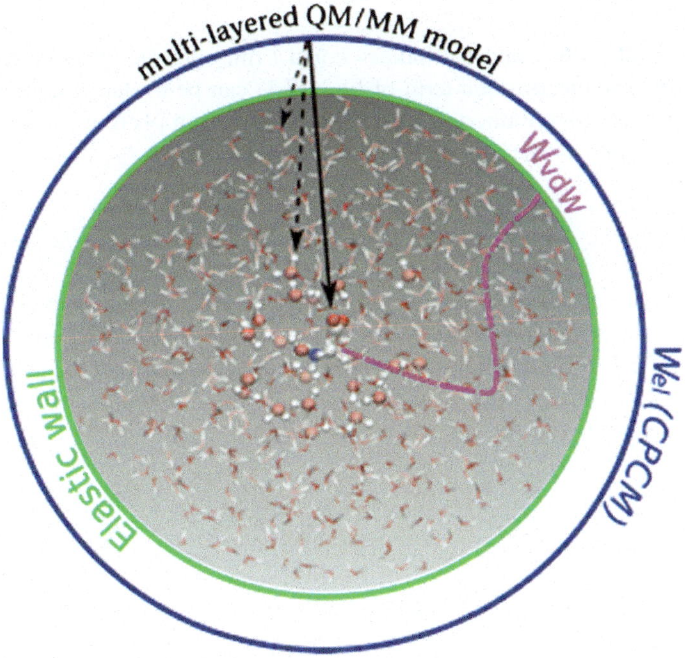

Fig. 17.2 Sketch of the multi-layered model for solvation

17.3.1 The Explicit (Atomistic) Region

For very small molecular systems both the solute and the solvent molecules can be treated using accurate levels of theory. However, when the solvation effects are studied, the number of molecules is usually large (refer to the *explicit* solvent molecules in Fig. 17.2), and the employment of approximate methods becomes mandatory. A very cost-effective and reliable approach is to treat the solute at a high level of theory, usually based on QM methods, and the solvent molecules at a lower level of theory, usually based on MM methods. Since only non-covalent bonds couple the different layers, the interaction energy between the solute (QM subsystem) and the solvent (MM subsystem) contains only Coulomb (Cou) and van der Waals (vdW) contributions:

$$E^{\text{int}} = E^{\text{Cou}} + E^{\text{vdW}}. \tag{17.1}$$

Then, for a molecular system composed by N atoms, with coordinates defined by the vector R, the van der Waals contribution can be written in the usual Lennard-Jones functional form:

$$E^{\text{vdW}} = \sum_{\substack{k \in \text{QM} \\ l \in \text{MM}}} \left(\frac{A_{kl}}{R_{kl}^{12}} + \frac{B_{kl}}{R_{kl}^{6}} \right), \tag{17.2}$$

where $R_{kl} \equiv |\mathbf{R}_k - \mathbf{R}_l|$, and the indices k and l run over the QM and MM atoms, respectively. The electrostatic term of Eq. (17.1) can be evaluated by representing the solvent charge distribution as a point-like distribution $\rho(\mathbf{r})$, of N charges q_k (k = 1, 2,… N) located at $\mathbf{R}_k$:

$$\rho(\mathbf{r}) = \sum_{k \in \text{MM}} q_k \delta(\mathbf{r} - \mathbf{R}_k). \tag{17.3}$$

On the basis of the way of evaluating the solute electrostatic potential, different routes for the calculation of such an energy contribution can be employed. In a simple Mechanical Embedding (ME) scheme (see for example [145] and references therein), a set of charges ($\mathbf{q}'$) is assigned also to the QM atoms, and the Coulomb interaction energy with the solvent is evaluated as:

$$E^{\text{ME}} = \sum_{\substack{k \in \text{QM} \\ l \in \text{MM}}} \frac{q'_k q_l}{R_{kl}}. \tag{17.4}$$

When an electronic embedding scheme (EE) is used ([145] and references therein), the Coulomb interaction energy is evaluated as the electrostatic interaction between $\rho(\mathbf{r})$ and the solute charge density $n(\mathbf{r})$:

$$E^{\mathrm{EE}} = \sum_{l \in \mathrm{MM}} \int \mathrm{d}\mathbf{r} \frac{n(\mathbf{r}) q_l}{|\mathbf{r} - \mathbf{R}_l|}. \tag{17.5}$$

A different (more refined) way to evaluate E^{Cou} includes a two-way polarisation through a polarisable embedding scheme (PE), so that both the QM density matrix is polarised by the MM charge distribution, and the MM charges are polarised by the QM density, by means of a variational formulation. Among different models, in our approach the polarisable embedding scheme for the treatment of the electrostatics is based on the fluctuating charges (FQ) model [29–35].

Then, the expression of the interaction energy for the PE [32, 33] is analogous to Eq. (17.5), except that the FQ classical charge distribution $\tilde{\rho}(\mathbf{r})$ is now obtained from the set of polarisable charges, $\tilde{q}(n(\mathrm{r}))$:

$$\tilde{\rho}(\mathbf{r}) = \sum_{k \in \mathrm{FQ}} \tilde{q}_k \delta(\mathbf{r} - \mathbf{R}_k). \tag{17.6}$$

and the electrostatic interaction with the QM density reads:

$$E^{\mathrm{PE}} = \sum_{l \in \mathrm{FQ}} \int \mathrm{d}\mathbf{r} \frac{n(\mathbf{r}) \tilde{q}_l}{|\mathbf{r} - \mathbf{R}_l|}. \tag{17.7}$$

The determination of $\tilde{\mathbf{q}}$ is done variationally by minimising the total energy ε of the FQ subsystem in presence of the QM electrostatic potential ($\mathbf{V}[n(\mathbf{r})]$):

$$\varepsilon = \frac{1}{2} \tilde{\mathbf{q}} \mathbf{J} \tilde{\mathbf{q}} + \tilde{\mathbf{q}} \boldsymbol{\chi} + \tilde{\mathbf{q}} \boldsymbol{\lambda} + \mathbf{V} \tilde{\mathbf{q}}, \tag{17.8}$$

with respect to the charges, by using the electronegativity equalisation principle [29, 30]. In Eq. (17.8), $\boldsymbol{\chi}$ is an array containing the electronegativities of the MM atoms, $\boldsymbol{\lambda}$ contains Lagrange multipliers for the charge conservation constraints, and J is the interaction matrix between the charges, which in our implementation is based on Ohno functional form [28]. The optimal set of charges which minimises such an energy functional is then found by inverting the J matrix [32, 33]:

$$\tilde{\mathbf{q}} = \mathbf{J}^{-1}(-\boldsymbol{\chi} - \mathbf{V}). \tag{17.9}$$

17.3.2 Non-periodic Boundary Conditions for Electrostatics

Given a molecular or supra-molecular system embedded in a solvent charge distribution, the solute-solvent interaction can be modelled using a mean field (MF) approach [21–24], which treats the solvent as a continuum fully defined by a dielectric constant (ε) and by a shape function, uniquely identifying the space regions where the solute and the solvent are placed. The boundary between the two domains is a compact cavity Σ, which in Fig. 17.2 has been represented as a spherical boundary surface including the explicit molecules. Then, the major issue related to such a scheme is how to model the interactions between the continuum and the explicit molecules placed inside the cavity. In a static picture, such interactions are of two types, electrostatic and non-electrostatic, whereas when such a model is used in MD simulations, an additional potential is included, in order to ensure that all the molecules remain confined inside the cavity during the simulation with a correct density up to the boundary [21].

Regarding the electrostatic interactions, the most common approaches are based on the exact solution of the Classical Electrostatic Laplace's equation for the electric (scalar) potential, Φ,

$$\nabla^2\Phi = 0. \tag{17.10}$$

In particular, let us consider all the molecules inside the cavity as a generic charge distribution $n(\mathbf{r})$, *in vacuo*, and bounded within the closed cavity of surface Σ. The whole space outside Σ is occupied by an homogeneous, isotropic, continuum, dielectric medium, characterised by a dielectric constant, ε_B (where B stands for "bulk"). Such a medium can represent an electrostatic picture of a solvent (or environment), interacting with the inner charge distribution. Once Laplace's equation is solved with the proper boundary conditions, we are able to determine the potentials within the cavity and in the bulk. In other words, the molecular charge distribution within the cavity induces a polarisation in the dielectric, giving rise to an apparent charge distribution σ, which acts back on the original charge distribution, and the effect of the solvation can be expressed by the electrostatic interaction between the two charge distributions.

On the basis of the symmetry properties of Σ and on how complex is the charge distribution source of the potential, different methods can be used in practice for the evaluation of such an interaction. For simple charge distributions (please note that a complex charge distribution can in principle be approximated through a proper multipolar expansion), and rigid cavities described by regular surfaces, exact analytical solutions can be found.

17.3.3 *The Analytical Solution of Laplace's Equation*

These models are based on the exact solution of the Classical Electrostatic Laplace's equation (17.10) for the electric potential, Φ, which can be obtained analytically within particular symmetry conditions of the boundary problem. Let us consider a set of N electric charges $\{q_k\}_{k=1,\ldots N}$, located in a region of a vacuum three-dimensional Cartesian space. We can identify the position of the k-the charge with respect to a prefixed Cartesian system of coordinates with its position

$$\mathbf{r}_k \equiv (\, x_k \quad y_k \quad z_k \,), \tag{17.11}$$

defined with respect to the origin of the Cartesian system of coordinates. This can be a simplified but reasonable electrostatic picture of a molecule in *vacuo*. The set of electric charges can be now supposed to be bounded within a closed cavity of surface Σ, obtained within a homogeneous, isotropic, continuum, dielectric medium, characterised by a dielectric constant, ε_B (where B stands for "bulk"). Such a medium can represent an electrostatic picture of a solvent (or environment), interacting with a solute molecule. Let us account for the general case for which the cavity is filled by a homogeneous, isotropic, continuum, dielectric medium, characterised by a dielectric constant, ε_C (where C stands for "cavity"). Commonly, the cavity is considered evacuated, so that $\varepsilon_C = \varepsilon_0$ is usually set. The boundary conditions needed to solve the problem are:

(a) continuity of the electric potential across the boundary surface Σ, and
(b) continuity of the normal component of the dielectric displacement vector across the boundary surface Σ.

Once solved Laplace's equation for the bounded problem, we are able to determine the potentials within the cavity, Φ_C, and in the bulk, Φ_B. The charges within the cavity induce a polarisation in the dielectric, giving rise to a reaction potential, $\Phi_R(\mathbf{r}_k)$, which acts back on the dissolved charges. Once determined the analytical expression for $\Phi_R(\mathbf{r}_k)$, Helmholtz's free energy of this interaction is just the difference between the reversible work of assembling the charge distribution in the presence of the dielectric and under *vacuum*, and simply reads:

$$A = \frac{1}{2}\sum_k q_k(\mathbf{r}_k)\Phi_R(\mathbf{r}_k). \tag{17.12}$$

For a spherical cavity the problem was solved for the first time by Kirkwood, in 1934 [38] and next generalised by several authors to multipoles of different order and to polarisable solutes [e.g. 36, 37, 40, 46, 68]. Solutions for more or less general multipolar developments in spheroidal cavities have been reported by several authors [45, 47, 48, 79, 95] and this is also the case for general ellipsoidal cavities starting from the work of Westheimer and Kirkwood, in 1938 [41] and proceeding in several steps [96–99] up to the general algorithm for arbitrary multipole orders developed in Ref. [100]. The essential features of these models have been reviewed quite recently [146–148] together with solutes described at a quantum mechanical level in regular cavities, which can be treated through a multipolar expansion of the solute electron density using again spherical [149], spheroidal [150], or ellipsoidal [151] cavities.

17.3.4 The Discretisation Approach for a General Solution of the Electrostatic Problem

Although even more general situations can be treated by analytical approach (for instance the case of two concentric cavities is discussed in Appendices A, B, C), this route, though rigorous and effective, is not general enough to be applied to every molecular system. A more general approach for the solution of Eq. (17.10) can be based on discretisation schemes of the partial derivatives equation. Two main routes have been followed in this connection with special emphasis on a quantum mechanical description of the solute.

The first approach is based on distributed multipolar expansions in general cavities [152] and has been recently reviewed [153].

The second route [145], considered in more detail in the following, is based, instead, on the direct evaluation of the solute electrostatic potential and electric field on the cavity. The cavity Σ is discretised in n_g tiles (usually called *tesserae*), so that the solvent charge distribution σ can be represented in terms of contributions coming from each *tessera*. With such approximations, the total free energy of the solute in presence of the solvent can be expressed as:

$$G = E^0[n] + V_{NN} + \frac{1}{2}E^{\text{Cou}}, \tag{17.13}$$

where $E^0[n]$ is the energy functional associated to the solute charge distribution n, V_{NN} is the nuclear repulsion energy independent of the electronic density, and the solute-solvent interaction energy is:

$$E^{\text{Cou}} = \int_\Sigma \mathrm{d}\mathbf{s} \int \mathrm{d}\mathbf{r}\frac{n(\mathbf{r})\sigma(\mathbf{s})}{|\mathbf{r}-\mathbf{s}|} = \int_\Sigma \mathrm{d}\mathbf{s} V(\mathbf{s})\sigma(\mathbf{s}), \tag{17.14}$$

or equivalently, in the matrix form:

$$E^{\text{Cou}} = \mathbf{V}^\dagger\mathbf{q} = \sum_i V_i\sigma_i; \tag{17.15}$$

where $\mathbf{V}$ is the electrostatic potential generated by the solute charge distribution on the tesserae. At each cycle of the self-consistent field (SCF) procedure, $\mathbf{V}$ is computed from the current density matrix, and σ is obtained by solving the equations in the form:

$$\boldsymbol{\sigma} = \mathbf{Q}^{-1}\mathbf{V}, \tag{17.16}$$

where **Q** is the interaction matrix and depends on the geometry of the solute, the details of the discretisation of the cavity and the particular choice of the model. In our case, the C-PCM [14, 19, 20] model is used, so that the Q_{ij} matrix elements are [19, 21]:

$$\begin{aligned} Q_{ii} &= \frac{1}{a_i}\left[4\pi\frac{\varepsilon}{\varepsilon-1} - \sum_{j\neq i} Q_{ji}a_j\right]. \\ Q_{ij} &= \frac{(\mathbf{s}_i-\mathbf{s}_j)\cdot\mathbf{n}_i}{|\mathbf{s}_i-\mathbf{s}_j|^3} \end{aligned} \tag{17.17}$$

Thus, being the charges obtained from the solute charge density, Eq. (17.15) is equivalent to the electrostatic contribution of Eq. (17.7).

In the framework of the general model for solvation depicted in Fig. 17.2, the PCM accounts for the long-range electrostatic interactions, whereas the short-range dispersion-repulsion can be described by some effective potential able to minimize spurious boundary effects. Further details about this aspect can be found in Ref. [21].

17.3.5 A Multilayered Scheme For Solvation

One approach which has been successfully employed in the past years [21–24], is to apply the EE based on the fixed charges (FX) for the treatment of those solvent molecules close to the solute, involved in local specific interactions with the QM charge density, while treating the rest of the solvent charge distribution in an averaged way through an implicit model (see Fig. 17.2). One limitation of this approach is that the electrostatic model employed for the water molecules is usually designed for the ground state properties, while when excited state properties are under investigation a new set of charges must be used.

In this sense, FQ and FX charges can be simultaneously employed to describe different physical-chemical models of the water molecules, where while the FX charges are used for the treatment of the molecules far from the solute, the FQ charges are used to improve the description of the solvent near the QM/MM interface, taking into account the charge polarisation induced by the solute charge distribution. For this reason both the EE and PE approaches can be simultaneously used in calculations where the solvent molecules far from the solute and belonging to the bulk induce a polarisation on the chromophore, but are not affected by the solute charge density, while the solvent molecules near the solute are affected by the solute charge density.

Eventually, an implicit solvation model can be used for the external outer layer, either for an average treatment of the bulk solvent, in order to sensibly reduce the number of the FX molecules explicitly considered, or for the inclusion of non-periodic boundary conditions [21–24]. In the following we will treat the implicit layer within the MF formalism, and all the relevant aspect emphasised with this model can be easily generalised to the case of a general potential defined on a regular surface.

Hereafter, in order to clearly distinguish between fixed and polarisable charges, we will always use the labels FQ and FX, respectively. The overall energy functional of such a multi-layered solvation model thus becomes:

$$\begin{aligned} E_0 = {} & E^{\mathrm{QM}} + E^{\mathrm{FQ}} + E^{\mathrm{FX}} + E^{\mathrm{MF}} \\ & + E^{\mathrm{int}}_{\mathrm{QM/FQ}} + E^{\mathrm{int}}_{\mathrm{QM/FX}} + E^{\mathrm{int}}_{\mathrm{QM/MF}} + E^{\mathrm{int}}_{\mathrm{FQ/FX}} + E^{\mathrm{int}}_{\mathrm{FQ/MF}} + E^{\mathrm{int}}_{\mathrm{FX/MF}}. \end{aligned} \tag{17.18}$$

The interaction energies between explicit layers ($E^{\mathrm{int}}_{\mathrm{QM/FQ}}$, $E^{\mathrm{int}}_{\mathrm{QM/FX}}$ and $E^{\mathrm{int}}_{\mathrm{FQ/FX}}$) obey Eq. (17.1), containing both van der Waals and Coulomb contributions (see Equations (17.5) and (17.7) for the QM/FX and QM/FQ contributions, respectively, while the FQ/FX contribution obeys Eq. (17.4)). The interaction between the MF and the explicit layers has a purely electrostatic contribution (Eq. (17.16)) and a non-electrostatic short-range dispersion-repulsion contribution. Finally, the other two interaction contributions are the charge-charge interactions (described by Eq. (17.4)) where the PCM charges interact with the two sets of FQ and FX charges. The remaining energy terms are related to the isolated subsystems, which are treated at their own levels of theory.

17.3.5.1 Some Useful Quantities

When the multi-layered scheme is used, all the contributions depending on the QM density in Eq. (17.18) can be expressed in terms of a reference effective ground state Hamiltonian:

$$\langle \Psi | \mathrm{H}_{eff} | \Psi \rangle = E^{\mathrm{QM}} + E^{\mathrm{Cou}}_{\mathrm{QM/FQ}} + E^{\mathrm{Cou}}_{\mathrm{QM/FX}} + E^{\mathrm{Cou}}_{\mathrm{QM/MF}}. \tag{17.19}$$

The generic Fock matrix element in Molecular Orbital (MO) basis can be written as:

$$F_{pq} = h_{pq} + \sum_{i}^{\mathrm{occ}} [(pq|ii) - c_x(pi|iq)] + v^{\mathrm{xc}}_{pq} + v^{\mathrm{QM/FQ}}_{pq} + v^{\mathrm{QM/FX}}_{pq} + v^{\mathrm{QM/MF}}_{pq} \tag{17.20}$$

where the index i runs over the occupied orbitals, c_x is the mixing parameter introduced by Becke [5], which allows a smooth transition from conventional Kohn-Sham DFT ($c_x = 0$) to the Hartree-Fock model ($c_x = 1$, $E_{\mathrm{xc}} = 0$), through the so called hybrid functionals. Assuming real spin-orbitals, the PE Fock operator can be derived from Eq. (17.7), by defining the matrix element $V_{pq,k}$ associated to the electrostatic potential given by the $\phi_p(\mathbf{r}), \phi_q(\mathbf{r})$ charge distributions at $\mathbf{R}_k$:

$$V_{pq,k} = \int d\mathbf{r} \frac{\phi_p(\mathbf{r})\phi_q(\mathbf{r})}{|\mathbf{r} - \mathbf{R}_k|}. \tag{17.21}$$

Then, $v_{pq}^{\mathrm{QM/FQ}}$ is the derivative of $E_{\mathrm{QM/FQ}}^{\mathrm{int}}$ with respect to the generic P_{pq} density matrix element in MO basis:

$$v_{pq}^{\mathrm{QM/FQ}} = \frac{\partial E_{\mathrm{QM/FQ}}^{\mathrm{int}}}{\partial \mathrm{P}_{pq}} = \sum_{k \in \mathrm{FQ}} V_{pq,k}\tilde{q}_k = \mathbf{V}_{pq} \cdot \tilde{\mathbf{q}}. \tag{17.22}$$

The remaining two contributions in Eq. (17.18), $E_{\mathrm{QM/FX}}^{\mathrm{int}}$ and $E_{\mathrm{QM/MF}}^{\mathrm{int}}$, are also described by Eq. (17.22), except that the set of charges and the coordinates are those of the FX atoms and PCM tesserae, respectively.

Another useful quantity is the coupling matrix (**K**), related to the Fock matrix element of Eq. (17.20):

$$\begin{aligned} K_{pq,rs} &= \frac{\partial F_{pq}}{\partial \mathrm{P}_{rs}} \\ &= (pq|rs) - c_x(pr|sq) + f_{pq,rs}^{\mathrm{xc}} + f_{pq,rs}^{\mathrm{QM/FQ}} + f_{pq,rs}^{\mathrm{QM/MF}}, \end{aligned} \tag{17.23}$$

which is used in linear response approaches [154, 155]. Since in the FQ and MF models the charges are obtained from the QM electrostatic potential, explicit FQ and MF terms appear in K, as the derivatives of Eq. (17.23) with respect to another MO density matrix element:

$$f_{pq,rs}^{\mathrm{QM/FQ}} = \frac{\partial v_{pq}^{\mathrm{QM/FQ}}}{\partial \mathrm{P}_{rs}} = \sum_{k,l \in \mathrm{FQ}} \int d\mathbf{r}d\mathbf{r}' \frac{\phi_p(\mathbf{r})\phi_q(\mathbf{r})}{|\mathbf{r} - \mathbf{R}_k|} J_{kl}^{-1} \frac{\phi_r(\mathbf{r}')\phi_s(\mathbf{r}')}{|\mathbf{r}' - \mathbf{R}_l|}. \tag{17.24}$$

Since the double integral is separable, the previous equation can be easily written in a more compact matrix formalism:

$$f_{pq,rs}^{\mathrm{QM/FQ}} = \frac{\partial v_{pq}^{\mathrm{QM/FQ}}}{\partial \mathrm{P}_{rs}} = \sum_{k,l \in \mathrm{FQ}} V_{pq,k} J_{kl}^{-1} V_{rs,l} = \mathbf{V}_{pq}\mathbf{J}^{-1}\mathbf{V}_{rs}, \tag{17.25}$$

which is analogous to the form obtained by means of the time dependent perturbation theory [33].

17.3.5.2 Linear Response Equations

When a time-dependent perturbation is applied to the total multilayered system, the response of the charge distribution to such a perturbation can be calculated through linear response approaches [154–158]. In principle the QM, FQ, and MF charge distributions should respond to the perturbation. However, since our general purpose is the calculation of the spectroscopic properties on the chromophore, which is assumed to be fully included in the model system, we apply the linear response theory only to the QM subsystem, including the effect of the FQ and PCM charges as an environmental effect, referring to the ground state effective Hamiltonian of Eq. (17.20).

In the Casida's formalism [157, 158], the time-dependent equations are:

$$\begin{pmatrix} \mathbf{A} & \mathbf{B} \\ \mathbf{B} & \mathbf{A} \end{pmatrix} \begin{pmatrix} \mathbf{X} \\ \mathbf{Y} \end{pmatrix} = \Omega \begin{pmatrix} 1 & 0 \\ 0 & -1 \end{pmatrix} \begin{pmatrix} \mathbf{X} \\ \mathbf{Y} \end{pmatrix}. \tag{17.26}$$

The energies Ω are the vertical excitation energies, **X** and **Y** are the amplitudes for the single particle excitation and de-excitation, respectively, and the response matrices **A** and **B** are defined in terms of derivatives of the occupied-virtual (ov) blocks of the Fock matrix in the MO basis:

$$\begin{aligned} A_{ai,bj} &= (\varepsilon_a - \varepsilon_i)\delta_{ab}\delta_{ij} + \mathbf{K}_{ai,bj} \\ B_{ai,bj} &= \mathbf{K}_{ai,jb} \end{aligned} \tag{17.27}$$

where ε_a and ε_i are the energies of the a-th and i-th orbital, respectively. In practical implementations [156, 159–162], linear combinations of A and B matrices are used:

$$\begin{aligned} (\mathbf{A}+\mathbf{B})_{ia,jb} &= (\varepsilon_a - \varepsilon_i)\delta_{ab}\delta_{ij} + 2(ia|jb) + 2f^{xc}_{ia,jb} - c_x[(ja|ib) + (ab|ij)] \\ &\quad + 2f^{QM/FQ}_{ia,jb} + 2f^{QM/MF}_{ia,jb} \\ (\mathbf{A}-\mathbf{B})_{ia,jb} &= (\varepsilon_a - \varepsilon_i)\delta_{ab}\delta_{ij} + c_x(ja|ib) - (ab|ij) \end{aligned} \tag{17.28}$$

and the TD-DFT equations are solved in the AO basis.

17.3.6 Test Cases

In order to illustrate how the general model presented in the previous paragraphs can be applied to the study of real molecular systems, two test cases will be presented focusing on the different aspects of the solvation effects on electronic transitions.

17.3.6.1 Local Versus Bulk Solvation Effects on the Electronic Excitation Energies of Uracil in Aqueous Solution

The study of the vertical transition energies of Uracil in aqueous solution is very instructive in this context, since it helps to understand why both explicit and implicit layers must be simultaneously included in the solvation models.

The UV-Vis spectrum of Uracil in the gas-phase is characterized by two main peaks, at about 4.80 and 5.26 eV, corresponding to the n/ π* and π/π* transitions, respectively [160 and references therein]. When Uracil in aqueous solution is considered, the lowest energy transition is blue-shifted with respect to the gas-phase by about 0.5 eV, while the other one is red-shifted by 0.2 eV, resulting in an inversion of stability of the lowest lying excited states [160 and references therein].

It has been shown that while the red-shift of the π/π* transition can be quantitatively predicted by modelling the solvation effects using only implicit models, this is not true for the blue-shift of the n/π* transition. Table 17.1 shows that inclusion of bulk solvent effects alone by PCM leads to an underestimation of the solvent shift on the n/π* transition by about 0.2-0.3 eV, irrespective of the underlying electronic model.

As a matter of fact, the four water molecules forming specific hydrogen bonds with the carbonyl and nitrogen moieties of uracil (see Fig. 17.3) introduce an additional contribution to the total solvent shift. Therefore, a quantitative reproduction of the total solvent shift is achieved only using an integrated discrete/continuum model, which can be viewed as an application of Eq. 17.19 in which the QM level of theory is used for the treatment of both the solute and the first solvation shell, while the MF layer is used for the bulk solvent effects.

It is worth noting that when advanced QM methods cannot be used for the calculations, as in the case of large chromophores interacting with many water molecules, one possible approach is to use semiempirical models. For example, we showed that reliable solvatochromic shifts can be obtained using the Density Functional Tight Binding (DFTB) method [164, 165] in conjunction with PCM (see Table 17.1).

Table 17.1 Vertical excitation energies (eV) of the n/π* transition of Uracil in water predicted with different computational models

Method	Gas phase	Uracil +PCM	Uracil +4H_2O	Uracil +4H_2O + PCM
PBE0/TZP	4.80	+0.29	+0.17	+0.48
PBE/aug-cc-pVTZ	3.85	+0.25	+0.26	+0.45
DFTB	3.64	+0.26	+0.15	+0.33

See Refs. [163, 164] for further details about the original calculations

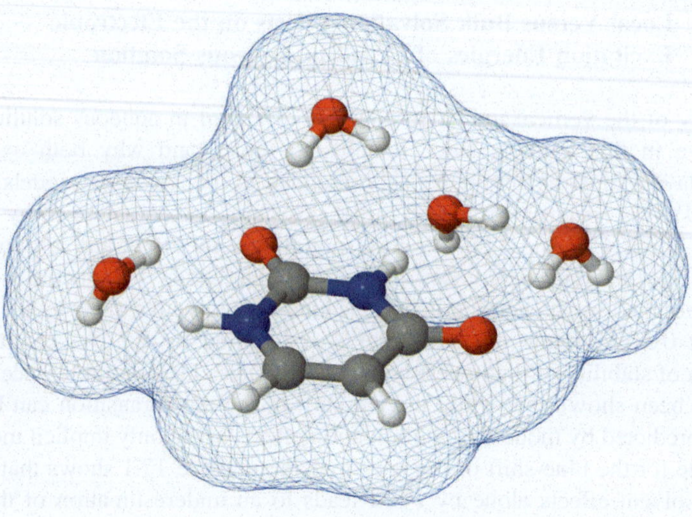

Fig. 17.3 Graphical representation of the geometry of Uracil with the four explicit water molecules of the first solvation shell, embedded in the PCM cavity (see Refs. [164, 165])

17.3.6.2 Fluorescence Spectrum of NfO-TEMPO Using the General Multi-layered Approach

In this paragraph, we report the results obtained through the application of the general multi-layer approach discussed above for the calculation of the fluorescence spectrum of a medium-sized dye, 4-naphthoyloxy-1-methoxy-2,2,6,6-tetramethylpiperidine (NfO-TEMPO). Such a molecule is interesting due to its capability to work as molecular probe for polymers, leading to many potential applications in the design of smart materials [166–171]. Thus, the correct interpretation of luminescence properties of such a molecule in different environments is a major issue when the spectroscopic characterisation of the polymers is done.

A generalised solvent model has been thus successfully applied for the simulation of the fluorescence spectrum [166]. First, the MM force field has been parametrised using excited state TD-DFT (PBE0 [6]/DZP) calculations as reference, and then the MD simulations have been performed with a 0.5 fs time-step for a total period of 3ns, at constant T=300 K and P=1 atm in order to sample the excited state PES. Eventually, 200 configurations have been extracted from the overall trajectory, and were used for running the TD-DFT/MM/PCM (CAM-B3LYP/cc-pVDZ) calculations. The final fluorescence spectrum of the chromophore has been obtained by convoluting with Gaussian functions (FWHM=0.05 eV) the energies and oscillator strengths of each single spectrum, and subsequently averaging over all the configurations. The final spectrum obtained with such a procedure is shown in Fig. 17.4, together with the experimental one [167], and from the comparison it is evident the high accuracy that can be achieved

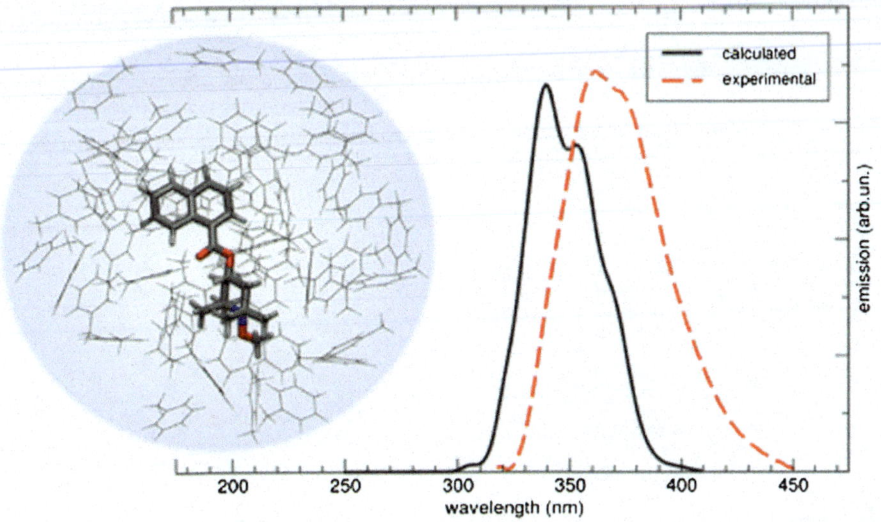

Fig. 17.4 Graphical representation of the NfO-TEMPO molecule, together with the computed [167] and experimental [168] fluorescence spectrum

using this type of computational approaches. It is noteworthy that also the asymmetric broadening observed in the experimental spectra is correctly reproduced by the simulations based on a properly selected set of configurations.

17.4 Fully Implicit Models

17.4.1 The Polarizable Continuum Model

As mentioned in the introduction, elimination of the explicit (atomistic) part of the solvent leads to the conventional polarisable continuum model, in which the solute creates a cavity in the bulk (continuous) solvent, whose shape follows the movement of solute atoms and whose reaction field responds to the electrostatic potential created by the solute wave-function. Note that under such circumstances the solute electron density penetrates the solvent boundary originating the so-called escaped charge effects, whose treatment requires more sophisticate descriptions of electrostatic contributions (e.g. the so-called integral equations formalism, IEF [12, 13, 15]).

Within the approaches based on the discretised finite-element solution of Laplace's equation (Sect. 17.4), the shape of the cavity is not constrained and can be in principle of any morphology. For this reason, the cavity is usually modelled on the basis of the molecular geometry. In particular, given a specific nuclear

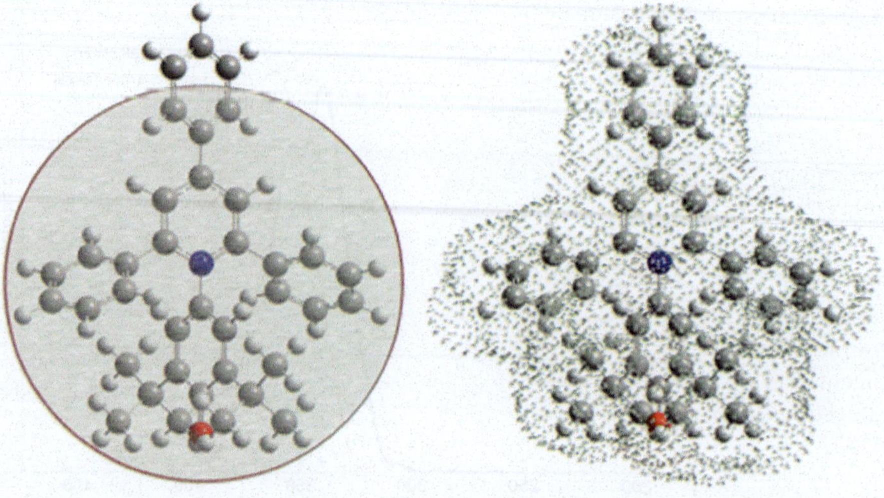

Fig. 17.5 Graphical representations of a spherical versus PCM(UFF) cavity shapes

configuration for the molecule under study, a sphere is centred on each atomic position and the overall cavity is the union of all the spheres. Dedicated smoothing procedures [3], are used in order to ensure that the final molecular cavity shows the proper continuity properties (see Fig. 17.5 right image, for an example of a typical PCM cavity obtained with this approach).

As a consequence, this procedure relies on a reduced number of empirical parameters, including, in particular, the sphere *radii,* which depend, at least, on the atomic number of the corresponding atom. Optimization of these parameters can be based on experimental solvation energies [172–174] or on the solvent shifts of some electronic properties [175], the second approach often providing better results for computational spectroscopy studies [176].

17.4.2 Analytical Models with Regular Cavities

Regular surfaces can be used to define the cavity occupied by the solute leading to analytical solutions of the corresponding Laplace's equation (see Sect. 17.3) and thus avoiding any QM computation for a rough estimate of general trends or for interpretative purposes. The only empirical parameter of such models is the cavity *radius* a_0 (for spherical cavities) or the principal axes (for spheroidal or ellipsoidal cavities) ([177–179] This quantity can be measured as $a_0 = r_{\mathrm{slt}} + r_{\mathrm{slv}}/2$, which is the radius of an equivalent spherical volume for the solute molecules with $r_{\mathrm{slt}} = [3M_{\mathrm{slt}}/(4\pi N_A \rho_{\mathrm{slt}})]^{1/3}$ and $r_{\mathrm{slv}} = [3M_{\mathrm{slv}}/(4\pi N_A \rho_{\mathrm{slv}})]^{1/3}$, for the solute and solvent molecules, respectively (M = molar mass in amu, N_A = Avogadro constant, and ρ = density in g/cm^3). Sometime, the approximation $a_0 \cong r_{\mathrm{slt}}$ is assumed.

The expression is in agreement with that derived by Suppan, starting from Mossotti-Clausius relation [(a)–(c) here below]. There has been set up an indirect measurement technique of a_0 using the solvatochromic data both in absorption and in emission: It consists into performing two regressions, by employing one of the relations (5.76.10)–(5.76.13) and one of the relations (5.76.14)–(5.76.16). (It must be remembered that those relations can be used when the dielectric orientation relaxation time of organic solvent molecules is negligible with respect to the mean lifetime of the solute.) In such a way, we can obtain the coefficients $m_1 = \left|a_m^{(5)}\right|$ and $m_1 = \left|a_m^{(2)}\right|$. If $\boldsymbol{\mu}_e || \boldsymbol{\mu}_g$, then $\mu_e = -\mu_g \frac{m_1+m_2}{m_1-m_2} (m_1 < m_2)$. Knowing the value of μ_g, we calculate μ_e, and then from m_1 or m_2 we obtain a_0). If a_0 is set equal to the *radius R* of the smallest sphere encompassing the molecule, then the energy of interaction will be of course underestimated. Improved accuracy is obtained using an estimated or explicitly determined molecular volume [180, 181], or following [182, 183] the original proposal of Lippert [184, 185], which takes 40 % of the long-molecular axis including Van der Waals radii for elongated molecules. Lippert finds upon estimating the two above mentioned effects that the choice of $a_0 = 0.8$ R should lead to the best value for the difference in dipole moments. Srividya et al. [186] estimate the cavity radius by assuming the molecule to be spherical, and then evaluating the radius from the total surface area obtained by the PCMODEL software [187]. The sphere obtained with such an approach is consistent with the one obtained by defining the diameter from the distance between the coordinates of the farthest atoms of the molecule, obtained from experimental crystal structures. Maus [188] calculates a_0 of some donor-acceptor biphenyls by different methods using PCMODEL and considering an intermediate value, which equals the half-length of the long-molecular axis without Van der Waals *radii* [189, 190]. In a recent work by Carlotti et al. [181], the cavity radius is estimated as $a_0 \cong 0.6\, d_{\mathrm{CT}}$, where d_{CT} is the calculated diameter along the Charge Transfer (CT) direction (CT diameter) of the optimised structure. This procedure is chosen on the basis of the QM results reported in a previous paper [192] where a_0 was calculated by integration of the solvent accessible surface using both the Hartree-Fock and Density Functional Theory optimised geometries and was found to be 60 % of the CT diameter.

The basic problem of these models is, of course, related to the oversimplified cavity shape, as evidenced in Fig. 17.5, which compares spherical and van der Waals cavities for a typical dye. A quite successful empirical recipe was suggested by Kawski et al. who showed that if the condition $\alpha / a_0^3 = 1/2$ is verified (α being the isotropic polarisability of the solute), the specific choice of the solute cavity shape (sphere or ellipsoid of revolution) affects negligibly the value of the excited state dipole moment. It is worth noting that such a condition occurs quite commonly [193–197] and for instance for phtalimide and stilbene derivatives the experimental values found for the radius and the polarisability are in a range from 0.41 to 0.62 [193], and from 0.53 to 0.68 [198, 199], respectively.

17.4.3 *Analytical Treatment of (Fluoro) Solvatochromism in Fully Implicit Models*

Together with the formal limitations discussed above, another problem of current analytical models is that different *formulae* are used for describing the effect of the solvent on the absorption energy [49–56, 59, 60, 74, 76, 77, 80, 84, 85, 88–91], on the emission energy [60, 66, 69, 70, 78, 81, 86, 98, 193], and on the Stokes' shift [57, 61–63, 65, 67, 82, 92, 93].

Three types of electric interactions among a solute and the solvent can be heuristically defined: (a) dispersion interactions, (b) interactions due to the transition multipole moments (normally truncated to the transition dipole contribution), and (c) multipolar interactions. Usually, dispersion contributions are simply neglected, whereas the following equations are employed for the evaluation of the shift of absorption energies in terms of polarisability (solute)-dispersion (solvent) (Stark effect), and the polarisability (solute)-polarisability (solvent) interactions, respectively

$$\Delta\Delta E^{a}_{i-\text{ref.}}(\boldsymbol{\alpha}) = -\frac{\alpha_e - \alpha_g}{a_0^3} L\Delta\left[f(\varepsilon) - f\left(n^2\right)\right]_{i-\text{ref.}}, \tag{17.29}$$

$$\Delta\Delta E^{a}_{i-\text{ref.}}(\boldsymbol{\alpha}) = -\frac{\alpha_e - \alpha_g}{a_0^3} C\Delta\left[f\left(n^2\right)\right]_{i-\text{ref.}}. \tag{17.30}$$

where the left-hand side is the shift of the absorption energy of the i-th solvent with respect to the reference one (ref), $f(x)$ is the polarisation function, ε is the relative dielectric constant of the medium, n is the refractive index of the medium, and a_0 is the *radius* of the spherical cavity (according to spherical model). The value of the polarisation function for the i-th solvent is subtracted from the value of the same function for the reference solvent. L and C are parameters usually referred to as the fluctuation factors. Concerning the Stark's effect, it has been given various mathematical treatments which do not seem to be in very good agreement with experiment, and it has been shown that it is a rather small effect which can be neglected in most cases [200]. Within an intuitive picture, the effect of an electric field on a non-polar polarisable molecule is defined as induction polarisation (*i*), whereas the corresponding effect on a rigid dipole is defined as orientation polarisation (*o*). The induction effects typically occur in shorter times with respect to the orientation effects. If two solvents are considered (namely 1 and 2) the contribution of the terms (b) and (c) to the energy variation due to the solvent change can be described according to the models deriving from Onsager's treatment of the molecule inside a spherical cavity. The expressions for the induction and orientation contributions due to the transition dipole moment M are, respectively

$$\Delta\Delta E^a_{i-\text{ref.}}(\mathbf{M}) = -\frac{M^2}{2a_0^3}\Delta\left[f\left(n^2\right)\right]_{i-\text{ref.}}, (i) \quad (17.31)$$

$$\Delta\Delta E^a_{i-\text{ref.}}(\mathbf{M}) = -\frac{M^2}{2a_0^3}\Delta\left[f(\varepsilon) - f\left(n^2\right)\right]_{i-\text{ref.}} \cdot (o) \quad (17.32)$$

Since the transition dipole moment interaction takes place in very short (instantaneous) time, the orientation contribution is usually neglected. In the case of the multipolar interactions, the solvent is usually still treated as a dipole, whereas for our specific case of solute, the multipolar terms need to be considered up to the quadrupole. Therefore, the following contributions are usually taken into account.

In case of charged species, the total charge (Q)-dipole contributions, according to the Born model [36] are

$$\Delta\Delta E^a_{i-\text{ref.}}(Q) = -\frac{Q^2}{2a_0}\Delta\left[\varphi\left(n^2\right)\right]_{i-\text{ref.}}, (i) \quad (17.33)$$

$$\Delta\Delta E^a_{i-\text{ref.}}(Q) = -\frac{Q^2}{2a_0}\Delta\left[\varphi(\varepsilon) - \varphi\left(n^2\right)\right]_{i-\text{ref.}} \cdot (o) \quad (17.34)$$

The permanent electric dipole moment ($\boldsymbol{\mu}$) -dipole contributions according to the Lippert-Mataga [49–51] (or McRae [52, 54] or other models, *vide infra*) are

$$\Delta\Delta E^a_{i-\text{ref.}}(\boldsymbol{\mu}) = -\frac{\mu_e^2 - \mu_g^2}{2a_0^3}\Delta\left[f\left(n^2\right)\right]_{i-\text{ref.}}, (i) \quad (17.35)$$

$$\Delta\Delta E^a_{i-\text{ref.}}(\boldsymbol{\mu}) = -\frac{\boldsymbol{\mu}_g \cdot \left(\boldsymbol{\mu}_e - \boldsymbol{\mu}_g\right)}{a_0^3}\Delta\left[f(\varepsilon) - f\left(n^2\right)\right]_{i-\text{ref.}} \cdot (o) \quad (17.36)$$

The permanent electric quadrupole moment ($\boldsymbol{\Theta}$)-dipole contributions according to the Suppan-Ghoneim model [101] are

$$\Delta\Delta E^a_{i-\text{ref.}}(\boldsymbol{\Theta}) = -\frac{\Theta_e^2 - \Theta_g^2}{a_0^5}\Delta\left[f\left(n^2\right)\right]_{i-\text{ref.}}, (i) \quad (17.37)$$

$$\Delta\Delta E^a_{i-\text{ref.}}(\boldsymbol{\Theta}) = -\frac{\Theta_g\left(\Theta_e - \Theta_g\right)}{2a_0^5}\Delta\left[f(\varepsilon) - f\left(n^2\right)\right]_{i-\text{ref.}}, (o) \quad (17.38)$$

where $\varphi(x) \equiv 1 - \frac{1}{x}$, ε and n are the dielectric constant and the refractive index of the medium, respectively, and the polarisation function $f(x)$ is $f(x) \equiv 2\frac{x-1}{2x+1}$, when a non-polarisable solvent dipole is considered (such as in Lippert-Mataga model [49–51]), and $f(x) \equiv 2\frac{x-1}{x+2}$, when a polarisable solvent dipole is considered, such as in the McRae model [52, 54]).

Concerning the interpretation of Stokes' shifts as a function of the solvent, the contributions due to the electric dipole moment change have been the most investigated. For a compound m, the interactions with the solvent s are generally expressed as linear combinations of contributions, and in particular the spectral maximum shift $y_{m,s}$ with respect to the gas-phase $y_{m,g}^{(0)}$ is given by:

$$y_{m,s} = y_{m,g}^{(0)} + \sum_{i=1}^{N} a_m^{(i)} x_s^{(i)}, \tag{17.39}$$

where:

$$a_m^{(i)} = a^{(i)}\left(a_0, \boldsymbol{\mu}_g, \boldsymbol{\mu}_e, \boldsymbol{\alpha}_g, \boldsymbol{\alpha}_e, \ldots\right), \tag{17.40}$$

and

$$x_s^{(i)} = x^{(i)}(\varepsilon, n). \tag{17.41}$$

In each term of Eq. (17.39) the $a_m^{(i)}$ factors depend only on the properties of the solute m (the Onsager spherical cavity radius a_0, the electric dipole moments in the ground and excited state $\boldsymbol{\mu}_g, \boldsymbol{\mu}_e$, the electric polarisabilities in the ground and excited state $\boldsymbol{\alpha}_g, \boldsymbol{\alpha}_e$, etc.) whereas $x_s^{(i)}$ factors depend only on the macroscopic properties of the solvent (the dielectric constant ε and refractive index n). The subscript i runs over all the type of the solute-solvent interaction, i.e. (1) dipole-dipole, (2) dipole-induced dipole, (3) dispersion contributions, (4) higher-order interactions, (5) specific interactions, etc.. Since there is currently no general analytical model, which includes all the types of solute-solvent interactions, the only way to describing all the contributions is through a QM treatment. Eq. (17.29) can be, however, assumed to be generally valid in the sense that it involves both the extreme approaches toward solution of the problem of the effect of solvent on the physico-chemical properties of molecules:

(i) the empirical approach, which separates the interactions into specific interactions, expressed by means of basicity and acidity solvent parameters, and nonspecific interactions, expressed by means of the so called polar $x_{polar} = (\varepsilon - 1)/(\varepsilon + 2)$ and polarisability $x_{polarisability} = (n^2 - 1)/(n^2 + 2)$ solvent functions [114]. Of course, this approach does not attribute any physical meaning to the $a_m^{(i)}$ coefficients;
(ii) the approaches, based on the original model of Onsager, in which the solvent is assumed to be a homogeneous continuum and which involves only non-specific multipolar interactions. The coefficients $a_m^{(i)}$ have here an exact physical meaning. Hereafter we shall consider just the theoretical expressions derived for treating the dipole-dipole interactions. In these cases, the sum in Eq. (17.39) is usually reduced to one or two addends, whose coefficients can

be obtained by different expressions, some of which are summarized in Appendix D.

A comprehensive comparison among different expressions was done by Koutek, in a remarkable paper of the late 1970s [201] on the basis of the data available for 9 compounds, *viz.* carbazole, indole, 1,3-diphenyl-2-pyrazoline, (hexamethylbenzene + tetrachlorophthalic anhydride) CT-complex, 4-amino-4′-nitrostilbene, 4-dimethylamino-benzonitrile, 4-dimethylamino-4′-nitrostilbene, 4-dimethylamino-ω-nitrostyrene, and 4-dimethylamino-benzylidene-4′-nitroaniline. Unfortunately, the analysis was limited to neutral molecules characterised by the presence of a symmetry axis, or at least by expected co-directional electric dipole moment vectors.

The most important sources of errors (differences between the particular relationships) in the determination of $\boldsymbol{\mu}_e$ through the use of those equations are: (1) the approximate nature of the solvent-shift theories; (2) problematic approximations in further simplification of the model; (3) assumption of a spherical molecule and estimation of the *radius* of Onsager's spherical cavity; (4) assumption that $\boldsymbol{\mu}_e || \boldsymbol{\mu}_g$; (5) experimental error in the wavenumber (errors in ε and n are not significant). In the previously mentioned analysis, since all the equations have been applied to the same experimental data, the last three factors can be assumed to introduce approximately the same error, which cannot be evaluated. Thus, the differences can be ascribed to the factors (1) and (2), and therefore in the final effect to the variation in the solvent functions. The main conclusions about the analytical models are the followings:

(1) The dispersion can be important for all methods using only absorption data; on the contrary, for methods using the difference $\tilde{\nu}^a - \tilde{\nu}^f$, the second regression term is unimportant.
(2) All equations using uniquely the absorption data, which from the practical point of view could have a much wider scope, exhibit high mean relative error. This could be due to the fact that those relations neglect some interaction terms of the general Eq. (17.14).
(3) Better results are obtained by equations involving also emission data, i.e. for well-fluorescing compounds.

The recommended equations are therefore (D.1.8), (D.1.12) and (D.1.16), based on the theory of Bilot-Kawski [72, 73, 83], or the almost equivalent equations (D.1.13) and (D.1.15), derived by Bakhshiev [62, 65, 67, 71].

As discussed above the most arbitrary quantity in this analysis is surely the cavity radius a_0 and this uncertainty excludes attempts to improve the theory by any of the following refinements: (a) Allowance for the fact that upon absorption the dipole moment of the excited state in the equilibrium configuration of the ground state is involved, whereas upon emission it is the dipole moment in the equilibrium configuration of the excited state. (b) Consideration of the possibility that the dipole moments of the ground and excited states may be not parallel. (c) Choice of a cavity better fitted to the molecular form, an ellipsoid for example. (d) Consideration of the relatively small contribution due to interaction of higher multipoles with the solvent. (e) Consideration of the especially the electric dipole moments, but also the cavity radius can undergo to remarkable changes due to the solvent effects. In other words, both the ground and the excited dipole moment as well as a_0 are differently affected by different solvents, and this important aspect is not described by any analytical model.

17.5 Phenomenological Approaches

The simplest and easiest-to-use tools for the interpretation of solvent effects are the phenomenological models, which rely on a very simple description of the solvent in terms of one (or more) empirical parameter, so that the solvation effects are interpreted in terms of the "strength" of the solvents on the basis on the values of such parameters. As a consequence, the different solvents are classified in scales, on the basis of the value of the chosen parameters, and the experiments are interpreted consequently. One very popular approach tries to rationalise the solvent effects in terms of the solvent *polarity*, empirically defined as the overall solvation power, which can be estimated with the dielectric function or using other more refined and more case-specific approaches.

Among the approaches proposed so far, we recall here single-parameter models [102–111, 115, 118–120, 122, 123, 125, 126, 129], and multi-parametric correlation equations (either based on the combination of two or more existing scales or on the use of specific parameters to account for distinct types of effects) [112, 113, 116, 117, 121, 124]. Additional popular models are the Abraham's scales of solute hydrogen-bond acidity and solute hydrogen-bond basicity [127, 128], and the Catalán et al. solvatochromic scales [130, 132, 133]. Methods based on quantitative structure-property relationships (QSPR) with solvent descriptors derived from the molecular structure [131, 134], and on principal component analysis (PCA) [135, 136] have been also proposed. An exhaustive review concerning the quantification of the solvent polarity has been recently published [138–140], including a detailed list of solvent scales, interrelations between parameters and statistical approaches.

In a very recent study, Jacquemin and co-workers [142] have proposed a new parameterisation of a linear solvation energy relationship (LSER), which seems able to successfully predict many types of solvent-dependent physicochemical properties. The expression of this multi-parameter empirical scale is:

$$A = A_0 + di\,\mathrm{DI} + e\,\mathrm{ES} + a\,\alpha_1 + b\,\beta_1 \qquad (17.42)$$

The parameters are DI (the Dispersion-Induction Parameter), ES (Electrostatic, Non-Hydrogen Bond Donating Solvents), α_1 (Lewis Acidity) and β_1 (Hydrogen-Bond Basicity), while *di*, *e*, *a*, and *b* are the regression coefficients.

As a last noteworthy example, we cite the Kosower approach [102–104, 106, 110], which expresses the solvent polarity (*Z*) in terms of the solvatochromic shift

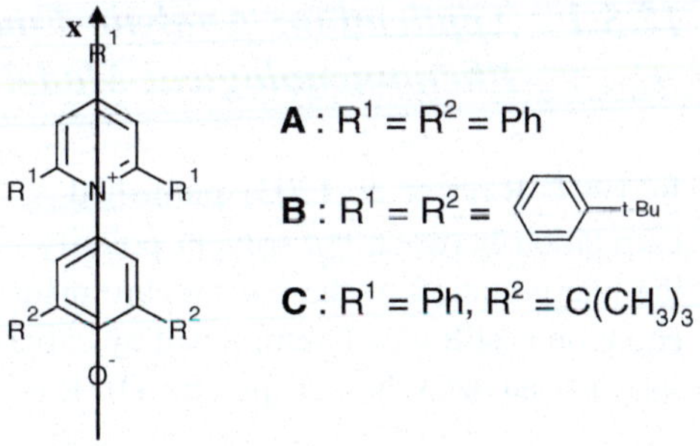

Fig. 17.6 Sketch of different betaines

induced by a given solvent on the maximum absorption wave-length of a reference molecule. Starting from the Kosower Z values, in the early 1960s, Reichard et al. [108] applied an analogous approach using a series of Betaines as reference molecules, and in particular the Betaine-30 (Fig. 17.6, molecule A). Then, the Kosower scale is expressed as:

$$Z \equiv hcN_A\tilde{\nu}_{\max} = 2.8591 \times 10^{-3}\tilde{\nu}_{\max} = 28591/\lambda_{\max} = E_T(30), \qquad (17.43)$$

where $[Z] = \mathrm{kcal} \cdot \mathrm{mol}^{-1}$ and $E_T(30)$ are the two different notation for the solvent polarity, while $[\lambda_{\max}] = \mathrm{nm}([\tilde{\nu}_{\max}] = \mathrm{cm}^{-1})$ is the maximum of the largest wave-length (wavenumber) of the intramolecular charge-transfer $\pi - \pi*$ absorption band of dye A. This parameter was also normalised in order to obtain values belonging to the range [0, 1], for the different (polar) solvents:

$$E_T^N \equiv \frac{E_T(30)^{\mathrm{slv}} - E_T(30)^{\mathrm{TMS}}}{E_T(30)^{\mathrm{W}} - E_T(30)^{\mathrm{TMS}}} = \frac{E_T(30)^{\mathrm{slv}} - 30.7}{32.4}, \qquad (17.44)$$

where TMS = tetramethylsilane ($E_T^N = 0$), W = water ($E_T^N = 1$). Because of solubility issues of the dye A in non-polar solvents, in 1983 Reichardt et al. employed the dye B, since it is sufficiently soluble for solvatochromic measurements in non-polar solvents [122]. Due to the good linearity of both A and B trends (by employing the data published for the dye A (see Ref. [108]) and for the dye B (see Ref. [122]) referring to a set of 17 common solvents, we obtain a linear regression $y = a_0 + a_1 x$ (where $y = E_T(A)$ and $x = E_T(B)$), with: $a_0 = (-3.4 \pm 0.8)$ kcal / mol, $a_1 = (1.09 \pm 0.02)$, and $R^2 = 0.99561$), in this case the scale is simply shifted with respect to the previous one, and thus becomes:

$$E_T(30) = \frac{28591/\lambda'_{\max} - 1.808}{0.9424}. \qquad (17.45)$$

Betaines solvatochromism has been recently object of a refined investigation for the determination of a new acidity scale [202].

17.5.1 Application: Analytic Foundation of Ravi's Phenomenological Model

In 1994, Ravi et al. [203] published a work in which they attempted to find a correlation between the solvent polarity function E_T^N and the analytical models for the interpretation of the solvatochromism of the Stokes shifts. This semi-empirical equation is still widely employed by experimentalists [1, 204–220], in spite of some formal weakness (see Appendix D). Ravi's equation reads:

$$Y(m,s) \equiv 11307.6\left\{(\Delta\mu/\Delta\mu_D)^2(a_D/a)^3\right\}E_T^N + Y^*(m,g), \qquad (17.46)$$

where $Y(m,s) \equiv \tilde{\nu}_a - \tilde{\nu}_f$ is the Stokes' shift for the molecule m in the solvent s, and $Y^*(m,g)$ is related to Stokes' shift for the molecule m in gas phase g (see Appendix D). D is a reference dye, whose electric dipole moment change ($\Delta\mu_D$) and Onsager's radius (a_D) are known. Proper use of this equation requires, in our opinion, a critical examination of the physical and mathematical background together with a clear definition the values of a, a_D, and $(\Delta\mu_D)^2$, where, in principle, D should be the dye A (and not C, see Fig. 5.6). If all of these conditions are verified, then from the solvatochromic shifts we can obtain $(\Delta\boldsymbol{\mu})^2$, and thus we cannot access to μ_e and μ_g (or $\mu_e - \mu_g$) without additional information (measurements or calculations). In order to estimate the value of a_D, the Authors performed least-squares fits of the experimental values of $\left(\tilde{\nu}_a^{m,s} - \tilde{\nu}_f^{m,s}\right)$, a, $\Delta\mu$ of a set of eight molecular systems, reported in the Literature. The experimental $\Delta\mu$ for such systems derive from electro-optical measurements (*eo*, for three of them), where available, and previous solvatochromic data (*sc*, for five of them). The molecules are: 4-dimethylamino-4′-nitrostilbene (*eo*), hexamethylbenzene-tetrachloro-phthalic anhydride (CT complex, *eo*), 4-amino-4′-nitrostilbene (*eo*), 4-aminophthalimide (*sc*), 1-dicyanovinylnaphthalene (*sc*), 9-dicyanovinylphenanthrene (*sc*), 9-dicyanovinylanthracene (*sc*), 1-dicyanovinylpyrene (*sc*). Through this statistical analysis, they estimate a cavity radius of 6.2 Å, with a satisfactory correlation ($R^2 =$ 0.977). A quite accurate QM calculation (Level of theory: PBE0/SNSD//IEF-PCM. Monte-Carlo estimation of the molecular volume using SCF density, based on 0.001 e/bohr^3 density envelope. The spherical cavity radius can be estimated also from the PCM cavity surface ($a_0^{(PCM,S))}$ or volume ($a_0^{(PCM,V))}$. The average of these two radii is close to the value obtained via the Monte-Carlo approach ($a_0^{(MC))}$, i.e. 6.37 and 6.38 Å for dyes A and C, respectively) of the radius of Onsager's model cavity has given $a_0 = 6.46$ and 6.47 Å, for dye A and C of Fig. 17.6 in 1,4-dioxane, respectively. Therefore, the value obtained through the regression represents a reasonable estimation of a_D, for both the dyes.

In order to verify the reliability of their model, the Authors made two tests: on one hand, they compare the above mentioned experimentally determined values of $\Delta\mu$ with those predicted through the usage of Eq. (17.4), with $a_D = 6.2$Å and

$\Delta\mu_D = 9.$D, finding a good agreement; on the other hand, they use another molecule previously characterised in the Literature (*viz.* 4-dimethylamino-benzonitrile), which was not included in the fitting procedure, in order to check the prediction capability of the model. However the apparent reliability of the model is related, in our opinion, to some bias in its validation. As a matter of fact, the first check is somehow tautological (the data taken from the Literature are good, and thus using them in order to build a linear fitting through a regression make the "prediction" of the same data coincident—or at least close—to the them), whereas the second check is not unambiguous, since, as stated by the Authors themselves, only some of the solvatochromic data reported in the Literature were selected.

From Eq. (D.7) of Appendix D, we should have that

$$X(s) \equiv F_3(\varepsilon, n^2) = \left(\frac{\varepsilon - 1}{\varepsilon + 2} - \frac{n^2 - 1}{n^2 + 2}\right)\frac{2n^2 + 1}{n^2 + 2}.$$

and E_T^N are linearly dependent. However, the E_T^N scale does not correlate with $F_3(\varepsilon, n^2)$ (Fig. 17.7a), nor with any other polarity function (see Fig. 17.7b–d). This is true also for the original work by Reichardt [108]. Mente and Maroncelli actually state [214] that a meaningful correlation between the E_T^N scale values and those of

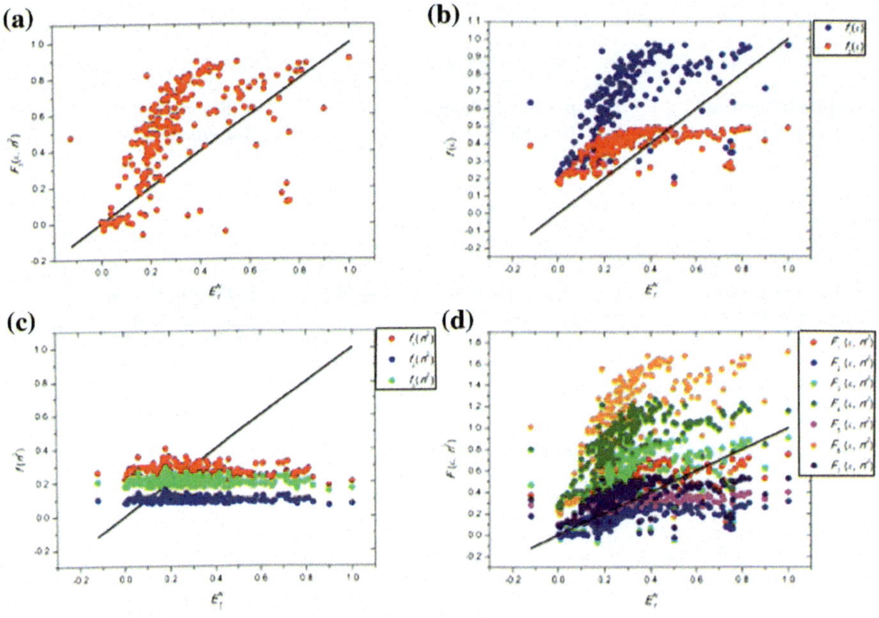

Fig. 17.7 Plot of the values of the polarity functions *versus* those of the E_T^N scale for 204 different solvents. The analytical polarity functions are **a** $F_3(\varepsilon, n^2) = \left(\frac{\varepsilon-1}{\varepsilon+2} - \frac{n^2-1}{n^2+2}\right)\frac{2n^2+1}{n^2+2}$, **b** $f_n(\varepsilon)$, **c** $f_n(n^2)$, and **d** $F_n(\varepsilon, n^2)$

$F_1(\varepsilon, n^2) \equiv \frac{\varepsilon-1}{\varepsilon+2} - \frac{n^2-1}{n^2+2}$ is observable only with the polar aprotic solvents. Anyway, it is evident that the method could empirically work for a sub-set of wisely selected solvents; such a choice can be done once all the quantities referred to the dye Λ that are present in Eq. (D.7) of Appendix D $\Delta\mu_A$, a_0^A, $\tilde{\nu}_a^{A,g}$, $\tilde{\nu}_f^{A,g}$, and $\tilde{\nu}_f^{A,s}$ will have been known with high accuracy. If the fluorescence transition energy is nearly constant ($\tilde{\nu}_f^{A,s}$ = constant for any solvent s), in principle we could estimate this value for a particular solvent s^* (for instance, 1,4-dioxane) and individuate those solvents which satisfy (within a certain range of variability) Eq. (D.7) of Appendix D. As hinted above, we have performed TD-DFT PBE0 [6]/SNSD[221] calculations for the dye A, which show that both the absorption and the emission transition energies are affected by the solvent; for instance, in 1,4-dioxane, $\Delta E_a^{A,\mathrm{dioxane}} = 1.56$ eV (exp.1.560 eV) [108] and $\Delta E_f^{A,\mathrm{dioxane}} = 0.29$ eV; in acetonitrile, $\Delta E_a^{A,\mathrm{acetonitrile}} = 1.97$ eV (exp.1.994 eV) [108] and $\Delta E_f^{A,\mathrm{acetonitrile}} = 1.15$ eV (Fig. 17.8).

In order to check possible limitations of the phenomenological approach, we have collected and analysed experimental data from the Literature. The following requirements have driven our choice: simple, small, neutral organic molecules; predominance of the dipolar character; $\boldsymbol{\mu}_e || \boldsymbol{\mu}_g$; availability of direct measures of $\boldsymbol{\mu}_e$ and $\boldsymbol{\mu}_g$; availability of Literature solvatochromic data, both for absorption and

p–dimethylamino–benzonitrile
I

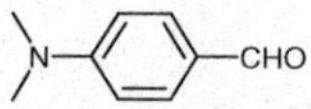

p–dimethylamino–benzaldehyde
II

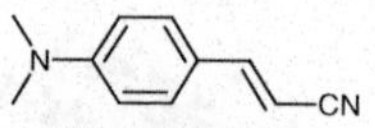

p–dimethylamino–cinnamonitrile
III

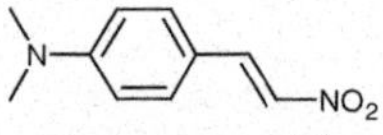

p–dimethylamino–cinnamonitrate
IV

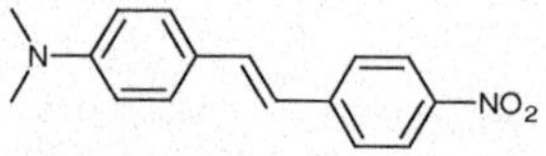

4–dimethylamino–4'–nitro–stilbene
V

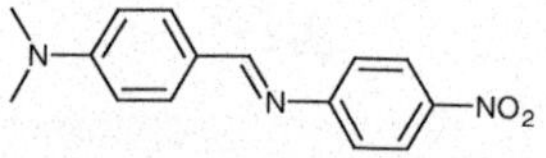

4–dimethylamino–benzylidene–4'–nitroaniline
VI

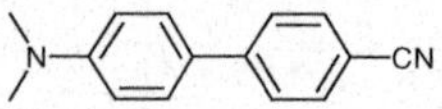

4–dimethylamino–4'–cyano–biphenyl
VII

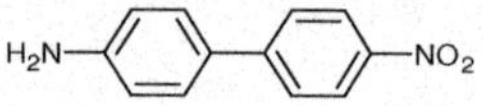

4–amino–4'–nitro–biphenyl
VIII

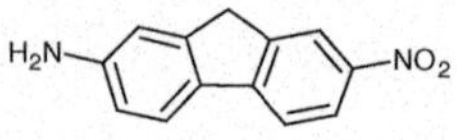

2–amino–7–nitrofluorene
IX

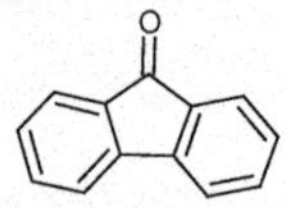

Fluorenone
X

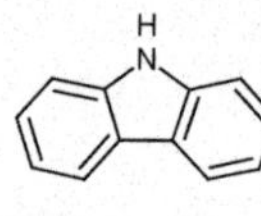

Carbazole
XI

Fig. 17.8 Structures of the 11 molecules used for testing the Ravi equation (see text for details) emission, recorded considering a sufficiently high number of solvents with different characteristics. (The condition of parallelism and same directionality of $\boldsymbol{\mu}_e$ and $\boldsymbol{\mu}_g$ has been checked through DFT calculations.) We have considered eleven different molecules (Fig. 17.8) whose Literature data concerning the dipole moments have been collected in Table 17.2. Please, note that we have also included 4-dimethylamino 4′-nitro-stilbene (molecule V), which was used for building Ravi's model, and the 4-dimethylamino-benzonitrile (molecule I), which was used for the validation of the prediction capability of Eq. (D.8) of Appendix D. Thanks to the values of $\tilde{\nu}_a$ and $\tilde{\nu}_b$ for different solvents, reported in the original works, we have been able to fit them according to Ravi equation (in its original form), and then calculate $\Delta\mu$, once known a_0. It is quite evident that for these simple systems the electrostatic analytical models provide dipole changes in good agreement with the data obtained through analytical absorption and fluorescence measurements. This is straightforwardly explained on the basis of the *criteria* of selection for the molecular candidates for this exploration, which are very close to those required for the correct application of those theories. On the contrary, the phenomenological model leads to unreliable results. If we compare the two solvatochromic approaches, we can observe that for all the molecules but XI, V,VII Ravi's equation underestimates the dipole change. Molecule V gives good results, but this is not surprising since it belongs to the set of molecular systems used for building the statistical model (Fig. 17.9).

17.6 Broadening

Most of the available computational studies focus on excitation energies and oscillator strengths, which correspond to peak positions and intensities, employing a large panel of electronic structure methods, ranging from Multi-Reference Configuration Interaction (MRCI) [222, 223] to Time-Dependent extensions of Coupled Cluster (CC) [224, 225] and Density Functional Theory (TD-DFT) [226–228]. However, even if a high accuracy in the computed energies and intensities is achieved, direct comparison with experiment is not straightforward due to band broadening. Although such an effect is often described in terms of a phenomenological parameter (the Full Width at Half Maximum (FWHM) of the band) a proper modelling requires explicit consideration of the different effects tuning the band-shape. Two main sources of broadening can be identified, originating from intrinsic and environmental effects, respectively.

Regarding the intrinsic contribution, besides non-specific factors (e.g. natural line broadening, Doppler broadening, etc.), one important source of broadening for systems in the gas phase is connected to the vibrational modulation of electronic spectra (the so-called vibronic contributions) resulting from the coupling of the ground and excited vibrational manifolds, which may grant the band its shape and

Table 17.2 Experimental dipole moments of the ground (μ_g) and the excited state (μ_e) of the molecule pictured in Fig. 17.8, in Debye units (1 D ~ 1/$c \times 10^{-21}$ C m = 3.33564 × 10^{-30} C m)

		Electrochromic absorption and fluorescence measurements					Solvatochromic absorption and fluorescence measurements		
		μ_g /D		μ_e/D		Δμ/D	a_0/Å	Δμ/D	Δμ/D
	(a)	(b)	(c)	(c)	(d)			(e)	(f)
I	D	6.6 [40]		12.5 [40]	13 [40]	5.9; 6.4	5.00 [45]	5.5 [45]	3.96
II	H	5.6 [41]		13.3 [41]		7.7	4.28 [46]	6.2 [46]	2.89
III	H	6.9 [40]		14 [40]		7.1	(4.77) [47]	7.1 [48]	2.08
IV	B	7.7 [40]		18 [40]	17.9 [40]	10.3; 10.2	7.7 [54]	9.5 [54]	N.C.
V	B	7.1 [40]		26.5 [40]	25.1 [40]	19.4; 18.0	8.0 [45]	17.9 [55]	16.84
VI	B	8.2 [40]		23 [40]		14.8	8.0 [45]	14.5 [45]	2.72
VII	B	6.0 [40, 44]		16.5 [40]	15.3 [40]	10.5; 9.3	(6.00) [53]	10.6 [53]	5.95
VIII	B	6.0 [40]		23 [40]	22.4 [40]	17; 16.4	7.0 [45]	12.0 [45]	5.76
IX	B	5.8 [40]		23 [40]	19 [40]	17.2; 13.2	(4.93) [47]	17.7 [49]	4.25
X	D	3.3 [42]		5.5 [42]		2.2	(4.62) [47]	2.7 [50]	1.04
XI	D	1.7 [43]	1.6 [43]; 1.9 [43]	3.1 [43]		1.4; 1.5; 1.2	(4.61) [52]	1.4 [51]	4.23

Comparison between the electrochromic and the solvatochromic (absorption and fluorescence) measures
(a) solvents: D = 1,4-dioxane; H = hexane; B = benzene
(b) from dielectric measurements
(c) from electrochromic absorption measurements
(d) from electrochromic fluorescence measurements
(e) from solvatochromic absorption and fluorescence measurements, through analytical models
(f) from solvatochromic absorption and fluorescence measurements, through phenomenological Ravi's model
NC not computable (the linear regression gives a negative slope)

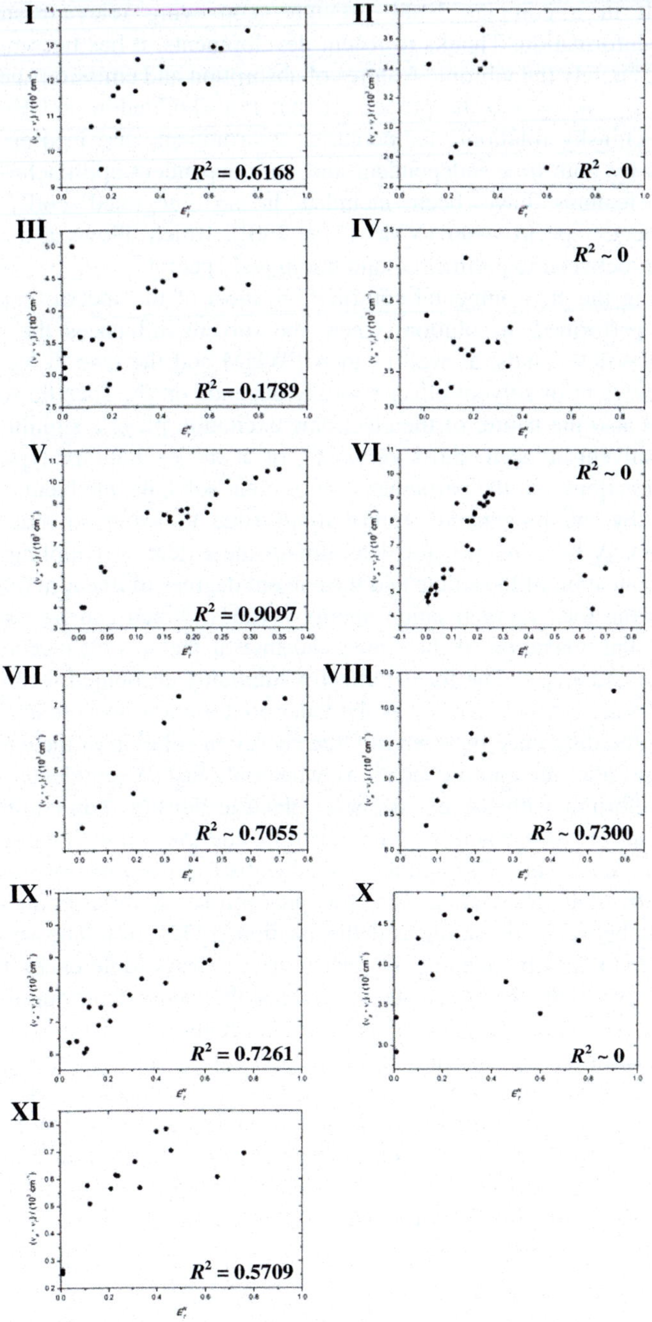

Fig. 17.9 Stokes shifts as a function of the E_T^N scale for the molecules sketched in Fig. 17.8, according to the phenomenological model

even result in a complex band structure, conveying relevant structural and dynamical information. Thanks to recent developments, it has become possible to describe accurately the vibronic features of absorption and emission spectra of fairly large systems taking into the proper account Franck-Condon and Hezberg-Teller terms, Duschinsky rotations, temperature, environment, and leading anharmonic corrections by both time independent and time dependent approaches [226, 229]. All these features have been included in an integrated code, the virtual multi-frequency spectrometer (VMS [229, 230]), which allows a direct vis-à-vis comparison between experimental and computed spectra.

Regarding the environmental contribution, most of the spectroscopic measurements are performed in solution, where the solvent influences the position and intensity of solute bands, as well as their FWHM and the overall band shape. Of course, the extent of solvent effect strongly depends on the specific solute-solvent couple, but also the nature of the electronic excitation plays a significant role (for instance, solvent-induced shifts on n–π^* or π–π^* excitations typically go on opposite directions). In the particular case of polar solvents, the band broadening is dominated by inhomogeneous contributions (due to solute-solvent electrostatic interactions). A key quantity for determining the extent of coupling between an electronic transition of the solute and the solvent degrees of freedom (electronic and nuclear) is the total reorganisation energy λ [231], which can be formally partitioned into the relaxation of the solute and that of the solvent (according to Ref. [231], λ_{inner} and λ_{outer}). The former effect is implicitly accounted for by optimising the excited state geometry, whereas the latter contribution is explicitly obtained by calculating the difference between the free energy in solution of the excited state at $t = 0$ (i.e. just after the photo-excitation), when only fast solvent degrees of freedom are in equilibrium with the excited state electron density (non-equilibrium, neq, level), and at $t \rightarrow +\infty$, when the solvent is fully equilibrated with the solute density (equilibrium, eq, level). Through a classical statistical treatment of solvent nuclear degrees of freedom, Marcus derived an expression for the broadening due to polar interactions between the solute and the medium [232–237]. The so-called inner contribution to this broadening is related to geometry differences between the ground and excited electronic state, whereas the outer contribution is directly related to solvent reorganisation. This latter contribution is, of course, connected to different solvation regimes (equilibrium vs. non-equilibrium) [226, 238], which are well described by the so-called state-specific (SS) methods [239], substantially improving the agreement between experiments and calculations. On these grounds, Ferrer et al. [240] have set up an efficient procedure (employed also in the present work) for the computational evaluation of the inhomogeneous broadening in terms of the solvent reorganisation energy [232–236].

17.6.1 *Solvent Reorganisation Energy Effects and Intra-molecular Charge-Transfer Character*

The spectral shift due to the solvent is obtained by averaging over all the possible solvent configurations whereas band broadening arises from the fluctuations in the energy difference between the initial and final electronic states involved in the transition. Specifically, according to this model the variance of the electronic transition energy is:

$$\sigma^2 = 2\lambda k_B T, \tag{17.47}$$

where T is the temperature, k_B is Boltzmann's constant and $\lambda \equiv F_f^{\mathrm{neq}} - F_f^{\mathrm{eq}}$ is the polar reorganisation energy, i.e. the polar contribution to the difference between non-equilibrium (neq) and equilibrium (eq) Helmholtz's free energies in the final electronic state f at the Franck–Condon (FC) solute geometry. Non-polar contributions are neglected. Interestingly, an analogous expression can be obtained in a fully QM approach describing the solvent as a set of normal oscillators displaced by the electronic transition [226]. Since the width of the experimental spectra is usually given in terms of the FWHM, we use the same measure (W_{pol}) also for the convoluting Gaussian line shape that simulates inhomogeneous broadening. For such a line-shape, W_{pol} and σ^2 are proportional:

$$W_{pol} \equiv FWHM = 2\sqrt{2\ln 2\sigma^2} = 2^2\sqrt{\ln 2\lambda k_B T}. \tag{17.48}$$

Within the framework of the PCM and of other continuum models, the neq and eq regimes are simply defined by two different dielectric constants [2]. In the neq case, the reaction field due to the fast solvent degrees of freedom depends on the dielectric constant at optical frequency ($\varepsilon_{\mathrm{opt}}$), usually related to the square of the solvent refractive index n. Equilibrium solvation is instead determined by the static dielectric constant (ε) [3]. It has been shown [241] that an effective SS implementation of PCM/TD-DFT calculations, where the solvent apparent charges are equilibrated with the true excited state charge density and not with the transition charge density as it happens in a linear-response regime, provides a reliable description of the differences between eq and neq regimes (dynamical solvent effects) [239, 242–248]. When dealing with the absorption process, λ_a can be simply computed as:

$$\lambda_a = E_{S_1}^{\mathrm{neq}} - E_{S_1}^{\mathrm{eq}}. \tag{17.49}$$

Once λ_a is obtained, it is possible to compute W_{pol} by using Equations (17.47) and (17.48). If the experimental band is symmetric and can be described by a Gaussian function, it is possible to estimate the reorganisation energy from the experimental spectrum, via Eq. (17.48), as $\lambda_a = \frac{\mathrm{FWHM}^2}{\ln 2^{2^{2^2}} k_B T} \cong \frac{\mathrm{FWHM}^2}{0.2849\ \mathrm{eV}}$ (at T = 298.15 K).

17.6.2 Asymmetric Vibronic Broadening

Let us consider now the intrinsic broadening due to vibronic effects, which can lead, in several cases, to an asymmetric line-shape. As an example, let us consider the bis (terpyridine)ruthenium(II) complex [249], whose electronic ground state has been optimized at the DFT level, using the B3PW91 [250] functional in conjunction with the LANL2DZ basis set [251], and including bulk solvent effects (CH_3CN) with the PCM. Then, the TD-DFT method was used for the investigation of the excited states and the vibronic contributions were included at the Franck-Condon level with the so-called adiabatic shift approach [252].

In Fig. 17.10 the computed phosphorescence spectrum of the complex (both stick and convoluted) is compared to its experimental counterpart. From Fig. 17.10 it is apparent that the asymmetric character of the experimental line-shape can be accurately reproduced by vibronic computations. On the other side, such a complex line-shape cannot be reproduced by a simple Gaussian function.

In such cases, even a phenomenological description requires a more general function, for instance a Gaussian-exponential hybrid function. The Gaussian (f_g) and exponential (f_e) functions, *viz.*:

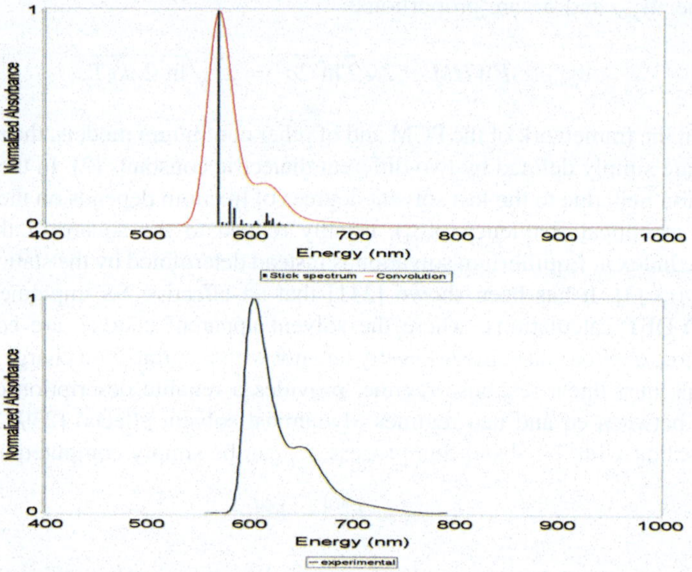

Fig. 17.10 Phosphorescence spectrum of the bis(terpyridine)ruthenium(II) complex. The upper spectrum is computed at the B3PW91/LANL2DZ level and is represented together with the stick spectrum coming from the vibronic structure, while the lower one is the experimental line-shape. See the original work for further details [250]

$$f_g(x) = f_{g0}\exp\left[-\frac{(x-x_0)^2}{2\sigma^2}\right], \tag{17.50a}$$

$$f_e(x) = f_{e0}\exp\left[-\frac{x-x_0}{\alpha}\right], \tag{17.50b}$$

can be combined in a hybrid function:

$$f(x) = f_0\exp\left[-\frac{(x-x_0)^2}{2\sigma^2+\alpha(x-x_0)}\right], \tag{17.51}$$

where f_0 can be taken equal to 1. It is easy to prove that:

$$\lim_{\sigma^2\to 0^+} f(x) = f_e(x), \tag{17.52a}$$

$$\lim_{\alpha\to 0} f(x) = f_g(x), \tag{17.52b}$$

Depending on the values of α and σ^2, the band shape is tuned. The parameters α and σ^2 can be calculated through standard non-linear fitting procedures performed on the experimental spectra, where the initial guess values can be evaluated from the experimental data. Actually, α and σ^2 are related to R_ξ and L_ξ according to:

$$\alpha = -\frac{R_\xi - L_\xi}{\ln\xi}, \tag{17.53a}$$

$$\sigma^2 = -\frac{R_\xi L_\xi}{2\ln\xi}, \tag{17.53b}$$

where R_ξ and L_ξ are the distances from the straight line of equation $x = x_0$ of the right and the left intersection, respectively, of a straight line of equation $y = \xi\,(0<\xi<f_0)$ with the curve. The parameter α can be positive ($R_\xi > L_\xi$, the decay from the maximum to zero is larger for $x < x_0$ than for $x > x_0$) or negative ($R_\xi < L_\xi$, the decay from the maximum to zero larger for $x > x_0$ than for $x < x_0$).

Consequently, we have (with $f_0 = 1$):

$$\Delta_\xi \equiv R_\xi + L_\xi = \sqrt{\alpha^2\ln^2\xi - 8\sigma^2\ln\xi}, \tag{17.54}$$

$$\mathrm{FWHM} \equiv \Delta_{1/2} = \sqrt{\alpha^2\ln^2 2 + 8\sigma^2\ln 2}. \tag{17.55}$$

Additionally we can introduce a parameter for evaluating the asymmetry of the band:

$$\delta_\xi \equiv R_\xi - L_\xi = -\alpha \ln \zeta, \tag{17.56a}$$

$$\delta_{1/2} = \alpha \ln 2. \tag{17.56b}$$

The magnitude $\delta_{1/2}$ indicates the entity of the asymmetry.

17.7 Test Case: Comparison Between Different Approaches for the Simulation of the Spectroscopic Properties of Betaine-26

The experimental absorption spectrum of Betaine-26 (Fig. 17.6, dye C) shows an intense (π, π*) HOMO-LUMO band with intramolecular charge transfer character; the transition maximum is located at $\lambda = 836$ nm ($\Delta E = 1.483$ eV) in *p*-dioxane, and at $\lambda = 563$ nm ($\Delta E = 2.202$ eV) in methanol. Betaine-26 is a neutral, non-fluorescent chromophore, is a quite rigid system due to the extensive conjugation, and therefore the broadening can be totally ascribed to the solvent reorganisation energy and the vibronic asymmetric broadening. For the specific case of Betaine-26, within a multipolar expansion approach, the contributions to the solvatochromic shift are mainly due to the transition dipole moment M, to the electrical dipole moment μ, and to the electrical quadrupole moment Θ.

All the solvatochromic contributions have been then evaluated following the analytical approaches presented in Sect. 4.1. The values of these quantities, as well as the estimation of the spherical cavity radius have been obtained from QM computations, and are reported in Table 17.3 and Fig. 17.11. In particular, Table 17.3 collects the numerical values of the different contributions, while in Fig. 17.11 the computed induction contributions are plotted together with the experimental data. It is apparent that both the contributions due to the transition dipole moment and to the electrical quadrupole moment are one order of magnitude less important than the contributions due to the electrical dipole moment (we remind that within Suppan-Ghoneim model the quadrupole moment [101] is treated as a scalar quantity; this is true just in case of linear and centro-symmetrical quadrupoles. In the other cases, $\Theta = \mathrm{Tr}(\Theta)/3$.

In Fig. 17.12 the solvatochromic data are reported with respect to the polarity functions, in comparison with the experimental data. Regarding the contributions due to the electrical dipole moment, as shown in Fig. 17.12a, b, if we plot the absorption transition energies as functions of one of the polarity functions, we do not obtain a picture that can be reasonably fitted with a linear regression. We have also tried to interpret the solvatochromic shift of the absorption transition energy, when increasing the solvent polarity, invoking the variation of the absorption

Table 17.3 Quantitative evaluation of the multipole contributions for Betaine-26

	a_0/Å				ΔΔE(M)/eV			ΔΔE(μ)/eV
	A	B	C	D	I	O	$i+o$	I
BEN	6.53	7.11	5.63	6.37	−1.33E–03	1.00E–03	−3.23E–04	−4.58E–03
TOL	6.44	7.11	5.63	6.37	−1.48E–03	4.78E–04	−1.00E–03	−4.49E–03
CHCl3	6.49	7.12	5.63	6.38	−3.59E–04	−5.80E–03	−6.16E–03	−2.40E–03
PhCl	6.62	7.12	5.63	6.38	−1.81E–03	−7.13E–03	−8.94E–03	−1.08E–02
ANIL	6.45	7.13	5.63	6.38	−2.47E–03	−6.44E–03	−8.91E–03	−1.79E–02
THF	6.44	7.13	5.63	6.38	3.73E–04	−1.19E–02	−1.15E–02	2.23E–03
DCM	6.49	7.12	5.63	6.38	−2.58E–05	−8.47E–03	−8.49E–03	−2.40E–04
PYR	6.53	7.13	5.63	6.38	−1.23E–03	−8.72E–03	−9.95E–03	−1.18E–02
BuOH	6.47	7.13	5.63	6.38	3.45E–04	−1.06E–02	−1.03E–02	3.59E–03
Ac	6.41	7.13	5.63	6.38	9.76E–04	−1.15E–02	−1.05E–02	1.04E–02
2PrOH	6.62	7.13	5.63	6.38	7.09E–04	−1.40E–02	−1.32E–02	6.01E–03
EtOH	6.52	7.13	5.63	6.38	9.79E–04	−1.22E–02	−1.12E–02	1.02E–02
MeOH	6.53	7.13	5.63	6.38	1.64E–03	−1.39E–02	−1.23E–02	1.64E–02
MeCN	6.52	7.13	5.63	6.38	1.46E–03	−1.49E–02	−1.34E–02	1.36E–02
DMSO	6.77	7.13	5.63	6.38	1.14E–04	−1.65E–02	−1.64E–02	9.02E–04

		ΔΔE(Θ)/eV			ΔΔE(TOT)/eV		
O	$i+o$	I	O	$i+o$	i	o	$i+o$
2.51E–03	−2.07E–03	−1.75E–03	3.28E–04	−1.42E–03	−7.65E–03	3.84E–03	−3.81E–03
1.04E–03	−3.45E–03	−1.67E–03	1.34E–04	−1.54E–03	−7.64E–03	1.65E–03	−5.99E–03
−2.23E–02	−2.47E–02	−6.71E–04	−2.70E–03	−3.37E–03	−3.43E–03	−3.08E–02	−3.42E–02
−2.25E–02	−3.33E–02	−2.95E–03	−2.88E–03	−5.82E–03	−1.55E–02	−3.25E–02	−4.81E–02
−2.31E–02	−4.10E–02	−4.70E–03	−3.04E–03	−7.74E–03	−2.51E–02	−3.26E–02	−5.77E–02
−3.43E–02	−3.21E–02	5.79E–04	−4.57E–03	−3.99E–03	3.18E–03	−5.08E–02	−4.76E–02

(continued)

Table 17.3 (continued)

		$\Delta\Delta E(\Theta)$/eV			$\Delta\Delta E$(TOT)/eV		
O	*i* + *o*	*I*	*O*	*i* + *o*	*i*	*o*	*i* + *o*
–3.58E–02	–3.60E–02	–6.05E–05	–4.92E–03	–4.98E–03	–3.26E–04	–4.92E–02	–4.95E–02
–3.48E–02	–4.66E–02	–2.73E–03	–4.90E–03	–7.68E–03	–1.59E–02	–4.84E–02	–6.43E–02
–4.21E–02	–3.85E–02	8.49E–04	–6.47E–03	–5.62E–03	4.78E–03	–5.92E–02	–5.44E–02
–4.51E–02	–3.47E–02	2.44E–03	–7.11E–03	–4.67E–03	1.39E–02	–6.37E–02	–4.98E–02
–4.35E–02	–3.75E–02	1.41E–03	–6.86E–03	–5.45E–03	8.13E–03	–6.43E–02	–5.62E–02
–4.60E–02	–3.58E–02	2.31E–03	–7.13E–03	–4.82E–03	1.35E–02	–6.53E–02	–5.18E–02
–4.77E–02	–3.12E–02	3.77E–03	–7.97E–03	–4.20E–03	2.19E–02	–6.96E–02	–4.77E–02
–4.75E–02	–3.39E–02	3.04E–03	–7.67E–03	–4.63E–03	1.81E–02	–7.00E–02	–5.19E–02
–4.33E–02	–4.24E–02	2.00E–04	–7.17E–03	–6.97E–03	1.22E–03	–6.70E–02	–6.58E–02

Reference solvent: 1,4-dioxane
[a]Monte-Carlo method
[b]from PCM surface
[c]from PCM volume
[d]mean of *b* and c
[i]induction polarisation
[o]orientation polarisation

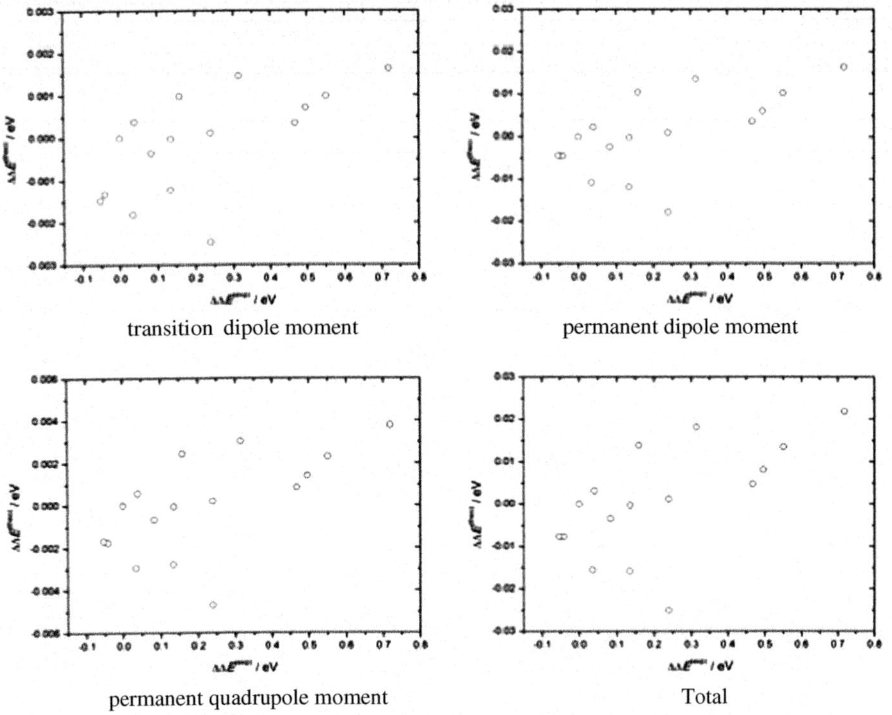

Fig. 17.11 Plot of the computed transition dipole moment, permanent dipole moment, permanent quadrupole moment, and total induction contributions for Betaine-26 in comparison with the experimental measurements. Reference solvent: 1,4–dioxane

transition energy as predicted by Eqs. (4.14.6) and (4.14.7) with respect to a reference solvent (1,4-dioxane):

$$\Delta\Delta E^{a}_{i-\mathrm{ref.}} = -\frac{2\left(\mu_e^2 - \mu_g^2\right)}{a_0^3}\Delta\left[\frac{\varepsilon - 1}{2\varepsilon + 1}\right]_{i-\mathrm{ref.}}, \tag{17.57}$$

$$\Delta\Delta E^{a}_{i-\mathrm{ref.}} = -\frac{\mu_e^2 - \mu_g^2}{a_0^3}\Delta\left[2\frac{\varepsilon - 1}{\varepsilon + 2} - \frac{n^2 - 1}{n^2 + 2}\right]_{i-\mathrm{ref.}}, \tag{17.58}$$

where the electric dipole moments are those measured experimentally. In Fig. 17.12c the results of this analysis are plotted and it is clear that only for few solvents (4–6 on a total of 16, viz. ethanol, 1-propanol, 1-buthanol, aniline, benzene, toluene) the simple analytical model provides reasonable results (absolute value of the discrepancy $\left|\Delta\Delta E_i^{a(\exp)} - \Delta\Delta E_i^{a(\mathrm{theo})}\right| < 0.05{-}0.06$ eV).

Let us now apply a QM approach based on DFT and TD-DFT computations. First, the geometry of the Betaine-26 in the ground state (S_0) has been optimised, both *in vacuo* and in solution (*vide infra*) using the PBE0 functional [6] including

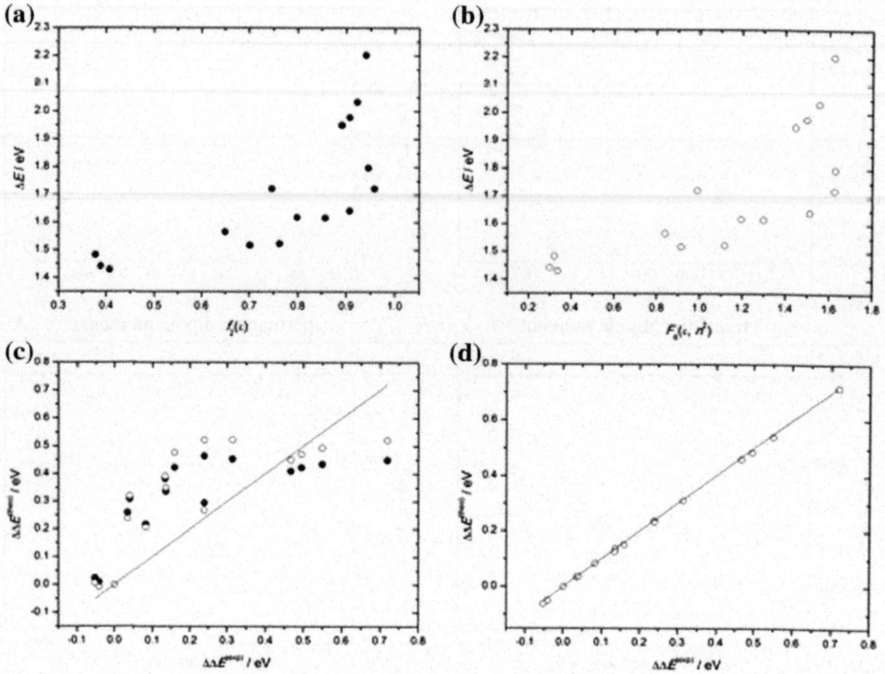

Fig. 17.12 Failure of the analytical models when employed to interpret the experimental solvatochromic data of Betaine-26. **a** Plot of the absorption transition energy as a function of the polarity function $f_2(\varepsilon)$ (according to Eq. (4.14.6)). **b** (a) Plot of the absorption transition energy as a function of the polarity function $F_6(\varepsilon, n^2)$ (according to Eq. (4.14.7)). **c** Comparison between computed *versus* experimental shift of the absorption transition energy, with respect to a reference solvent (1,4-dioxane, $\Delta\Delta E^{a(\text{exp})}_{\text{dioxane}} \equiv \Delta\Delta E^{a(\text{theo})}_{\text{dioxane}} \equiv 0$); filled and unfilled circles refer to Eqs. (17.57) and (17.58) model, respectively. **d** DFT/PCM results: Comparison between computed *versus* experimental shifts of the absorption transition energy, with respect to a reference solvent (1,4–dioxane, $\Delta\Delta E^{a(\text{exp})}_{\text{dioxane}} \equiv \Delta\Delta E^{a(\text{theo})}_{\text{dioxane}} \equiv 0$)

D3 empirical dispersion [253], coupled to the SNSD [221] basis set. Solvent effects have been taken into account by means of the IEF-PCM [28, 254] with scaled UFF atomic radii [167] for building the cavity. The solvents are the same employed in the experimental determinations (*vide infra*) covering a wide range of polarity and chemical features (chlorinated solvents, protic solvents, *etc.*). Harmonic vibrational frequencies have been then computed at the same level of theory employed for the geometry optimisation. Vertical excitation energies have been computed at the TD-DFT level using again the PBE0 functional, but the larger SNST basis set [221]. Both non-equilibrium (neq) and equilibrium (eq) regimes [227, 239] have been enforced in the treatment of solvent effects by IEF-PCM within the State-Specific (SS) [237] formulation. Finally, the excited states have been also optimised in order to compute the vibronic spectra. The vibronic spectra [94–100] of the $S_1 \leftarrow S_0$ electronic transition were simulated including Duschinsky and Herzberg-Teller effects, in 1,4-dioxane and in methanol. All computations have been performed with the GAUSSIAN09 package [255].

In Table 17.4 the computed vertical transition energies of Betaine-26 in different solvents, neglecting vibronic effects, are compared with the experimental results.

The remarkable agreement between computed and experimental results points out the good predictive power of the proposed QM approach. The discrepancies are always below 6 meV, and the largest differences are obtained for 1,4-dioxane, chloroform and dichloromethane (it is worth noting that the 1,4-dioxane is known for its anomalous and challenging behaviour [256]). A linear fit of the data (Fig. 17.12d) leads to $\Delta E^{a(\text{theo})} = 0.9955\ \Delta E^{a(\text{exp})} + 0.0082$ ($R^2 = 0.9997$). Also with respect to a reference solvent (1,4-dioxane), the comparison between predicted ($\Delta\Delta E^{a(\text{theo})}$) *versus* experimental shifts ($\Delta\Delta E^{a(\text{exp})}$) of the absorption transition energies gives good results, having the two $\Delta\Delta E^{a}$ the same trend.

Finally, we have simulated the vibronic spectra, including Duschinsky and Herzberg-Teller effects, for the Betaine-26 in 1,4-dioxane and in methanol. In Fig. 17.13 the computed vibronic spectra (convolution of all vibronic transitions by Gaussian functions with FWHM = 0.1 eV by means of the VMS-Draw toolkit [230]) are plotted together with the experimental ones. The predicted position of the maxima (1.48 eV and 2.20 eV in 1,4-dioxane and in methanol, respectively) and the band shapes around them are extremely accurate, in particular at lower energies. The vibronic analysis points out the dominant role of the 0–0 transition and the

Table 17.4 Comparison between the experimental and computed transition energies of Betaine-26 in different polarity solvents

Solvent[a]	ε_r	n_D	$\lambda_a^{(\text{exp})}$/nm	$\Delta E^{a(\text{exp})}$/eV	$\Delta E^{a(\text{theo})}$/eV
DIOX	2.2099	1.4224	836	1.483	1.489
BENZ	2.2706	1.5011	860	1.442	1.438
TOL	2.3741	1.4961	867	1.430	1.427
CHLR	4.7113	1.4459	792	1.566	1.572
CHLB	5.6968	1.5241	817	1.518	1.523
ANIL	6.8882	1.5863	720	1.722	1.725
THF	7.4257	1.4050	814	1.523	1.526
DCM	8.9300	1.4242	766	1.619	1.625
PYR	12.9780	1.5095	767	1.617	1.614
BUOH	17.3320	1.3993	636	1.950	1.948
AC	20.4930	1.3588	755	1.642	1.639
PROH	20.5240	1.3850	627	1.978	1.973
ETOH	24.8520	1.3611	610	2.033	2.028
MEOH	32.6130	1.3288	563	2.202	2.205
MECN	35.6880	1.3442	690	1.797	1.799
DMSO	46.8260	1.4170	720	1.722	1.721

[a]Legend of the solvents: *DIOX* 1,4-dioxane, *BENZ* benzene, *TOL* toluene, *CHLR* chloroform, *CHLB* chlorobenzene, *ANIL* aniline, *THF* tetrahydrofuran, *DCM* dichloromethane, *PYR* pyridine, *BUOH* 1-butanol, *AC* acetone, *PROH* 1-propanol, *ETOH* ethanol, *MEOH* methanol, *MECN* acetonitrile, *DMSO* dimethylsulfoxide

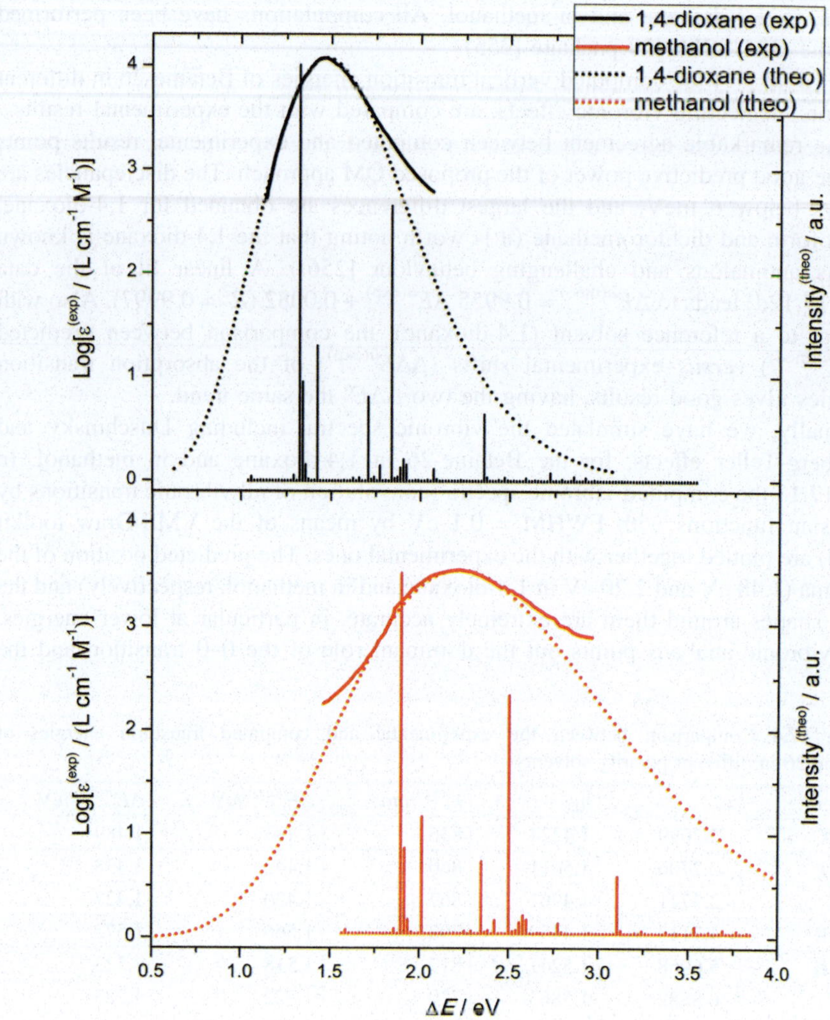

Fig. 17.13 Comparison between the experimental absorption spectra of Betaine-26 and those issuing from vibronic computations (stick and convoluted) for different solvents (1,4–dioxane and methanol). Experimental spectra were digitalised from the original paper by Dimroth et al. [108]

asymmetry of the convoluted spectrum. Therefore, in these cases, the reorganisation energy contribution to the inhomogeneous broadening (which generates a Gaussian bell centred on the absorption transition energy) is not sufficient for a correct description of the band shape. The asymmetric band can be fitted using the hybrid exponential-Gaussian function, obtaining the parameters $\alpha = 0.195$ eV, $\sigma^2 = 0.184$ eV, FWHM = 1.019 eV, and $\delta = 0.135$ eV for 1,4-dioxane [253], and $\alpha = 0.361$ eV, $\sigma^2 = 0.550$ eV, FWHM = 1.764 eV, and $\delta = 0.250$ eV for methanol.

17.8 Concluding Remarks

Thanks to on-going developments in hardware and software, in recent years computational spectroscopy has evolved from a highly specialized field to a tool of general use also for non-specialists. In this framework, processes occurring in solution play a central role from both biological and technological perspectives. Several models with different degrees of approximation are currently employed for interpreting and analysing such kinds of processes. In this chapter, we have tried to describe some of these models and to clarify the relationships among them together with their strengths and weaknesses. Our reference is a general multilayer model, in which different regions are described at different levels of approximation and proper non periodic boundary conditions are enforced for both electrostatic and non-electrostatic forces. Starting from this level, fully implicit models can be derived neglecting the dynamics and atomistic nature of the solvent. Both numerical and analytical approaches can be introduced at this point with different degrees of approximation and reliability. In particular, current analytical models can be slightly improved in order to take into account regular, but non-spherical cavities and higher multipolar developments. Solvatochromic effects on UV spectra can be described quite well at this level at least concerning general trends and role of different effects. However, it should be pointed out that broadening effects are often non-symmetric and thus require proper consideration of vibronic effects, which are well beyond analytical treatments. Also in this case effective numerical models have been introduced in recent years, which can be used also by non-specialists. Finally, phenomenological models have also a role for qualitative considerations provided that their limits are well understood and critical points in their derivation analysed and, in some cases, amended. A specific test case has been analysed in detail to better illustrate all the above points. In summary, even pending further developments in the reliability of different parts of the models (e.g. improved density functionals for describing charge transfer excitations or better description of Pauli repulsion between solute and first-shell solvent) the description of equilibrium or steady-state spectroscopies by computational methods appears well sound and quite reliable. The situation is, in our opinion, quite different for time-resolved spectra and, in general, for fluctuation effects on the studied phenomena. Here, proper integration between more refined QM and statistical mechanical models becomes mandatory and remains a challenging research field for the next years. Analogous considerations apply, of course, to the description of rare events and/or phenomena characterised by long relaxation times and, at the other extreme, to the analysis of ultra-fast events. Judging from the impressive developments of general solvent models in the last decades, we can envisage terrific progresses in those and related fields in the next few years.

Appendix A: Spherical Cavity Embedded By One or More Concentric Dielectric Shells Surrounded By the Bulk

In the following, we show a natural extension of the spherical model for concentric dielectric continua, and in particular the case of a cavity embedded by one concentric dielectric shell surrounded by the bulk (see Fig. A.1a), whose solution is due to Beveridge and Schnuelle [87], and the case of two or more concentric dielectric shells surrounded by the bulk (see Fig. A.1b, c), which we propose here.

A.1 Spherical Cavity Embedded by One Concentric Dielectric Shell Surrounded by the Bulk

We consider now the charge distribution inside a spherical cavity of radius $r = a_C =$ constant and dielectric constant ε_C, with a first shell of polarisable dielectric, extending from $r = a_C$ to $r = a_1 =$ constant, characterised by dielectric constant ε_1. Beyond this shell, there is the bulk region, from $r = a_1$ to infinity, with dielectric constant ε_B. The solution of Laplace's Eq. (3.10) is the same of the single spherical cavity case, but it is needed to introduce coefficients so that the expressions for the potential in the three regions (*viz.* cavity, shell, and bulk) are:

$$\Phi_C = \varepsilon_C^{-1} \sum_n \sum_m \left[\left(A_{nm}^C + A_{nm}^{(1)} \right) r^n + B_{nm}^C r^{-(n+1)} \right] P_n^m(\cos\theta) e^{im\phi}, \quad r < a_C, \tag{A.1}$$

$$\Phi_1 = \varepsilon_1^{-1} \sum_n \sum_m \left[A_{nm}^{(1)} r^n + B_{nm}^{(1)} r^{-(n+1)} \right] P_n^m(\cos\theta) e^{im\phi}, \quad a_C < r < a_1, \tag{A.2}$$

$$\Phi_B = \varepsilon_B^{-1} \sum_n \sum_m B_{nm}^B r^{-(n+1)} P_n^m(\cos\theta) e^{im\phi}, \quad r > a_1. \tag{A.3}$$

The Reaction Potential,

$$\Phi_R = \varepsilon_C^{-1} \sum_n \sum_m \left(A_{nm}^C + A_{nm}^{(1)} \right) r^n P_n^m(\cos\theta) e^{im\phi}, \quad r < a_C, \tag{A.4}$$

contains now two contributions, due to the polarisation of the local dielectric due to the cavity (A_{nm}^C) and the polarisation of the bulk dielectric due to the shell ($A_{nm}^{(1)}$). Concerning the boundary conditions, we have two per each surface, i.e.:

$$(\Phi_C)_{r=a_C^-} = (\Phi_1)_{r=a_C^+}, \tag{A.5a}$$

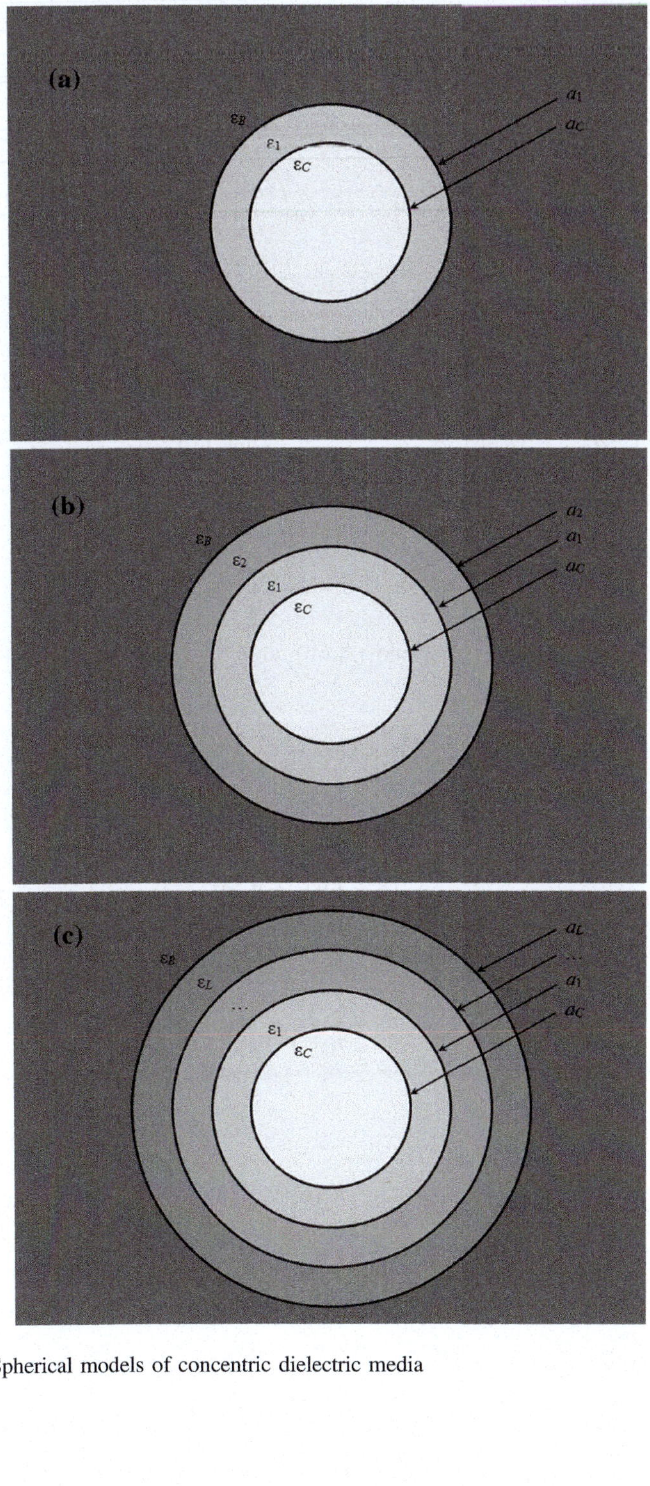

Fig. A.1 Spherical models of concentric dielectric media

$$\varepsilon_C\left(\frac{\partial\Phi_C}{\partial r}\right)_{r=a_C^-}=\varepsilon_1\left(\frac{\partial\Phi_1}{\partial r}\right)_{r=a_C^+}, \tag{A.5b}$$

$$(\Phi_1)_{r=a_1^-}=(\Phi_B)_{r=a_1^+}, \tag{A.5c}$$

$$\varepsilon_1\left(\frac{\partial\Phi_1}{\partial r}\right)_{r=a_1^-}=\varepsilon_B\left(\frac{\partial\Phi_B}{\partial r}\right)_{r=a_1^+}, \tag{A.5d}$$

or, explicitly:

$$\varepsilon_C^{-1}\left[\left(A_{nm}^C+A_{nm}^{(1)}\right)a_C^n+B_{nm}^C a_C^{-(n+1)}\right]=\varepsilon_1^{-1}\left[A_{nm}^{(1)}a_C^n+B_{nm}^{(1)}a_C^{-(n+1)}\right], \tag{A.6a}$$

$$nA_{nm}^C a_C^{n-1}-(n+1)B_{nm}^C a_C^{-(n+2)}=-(n+1)B_{nm}^{(1)}a_C^{-(n+2)}, \tag{A.6b}$$

$$\varepsilon_1^{-1}\left[A_{nm}^{(1)}a_1^n+B_{nm}^{(1)}a_1^{-(n+1)}\right]=\varepsilon_B^{-1}B_{nm}^B a_1^{-(n+1)}, \tag{A.6c}$$

$$nA_{nm}^{(1)}a_1^{n-1}-(n+1)B_{nm}^{(1)}a_1^{-(n+2)}=-(n+1)B_{nm}^B a_1^{-(n+2)}. \tag{A.6d}$$

Eliminating $B_{nm}^{(1)}$ from (A.6a) and (A.6b), and B_{nm}^B from (A.6c) and (A.6d), we obtain:

$$A_{nm}^C=\frac{(n+1)\left(1-\varepsilon'_{C^*}\right)}{(n+1)\varepsilon'_{C^*}+n}\frac{B_{nm}^C}{a_C^{2n+1}}, \tag{A.7}$$

$$A_{nm}^{(1)}=\frac{(n+1)(1-\varepsilon_{1^*})}{(n+1)\varepsilon_{1^*}+n}\frac{B_{nm}^{(1)}}{a_1^{2n+1}}, \tag{A.8}$$

$$\begin{aligned}B_{nm}^{(1)}&=\varepsilon'_{C^*}\left[A_{nm}^{(1)}a_C^{2n+1}+B_{nm}^C\right]\\&=\left[1-\frac{n}{n+1}g'_n\left(\varepsilon'_{C^*}\right)\right]B_{nm}^C,\end{aligned} \tag{A.9}$$

where:

$$\varepsilon'_{C^*}=\varepsilon_{C^*}\left[1+(1-\varepsilon_{C^*})g'_n(\varepsilon_{1^*})\left(\frac{a_C}{a_1}\right)^{2n+1}\right]^{-1}, \tag{A.10}$$

$$g'_n(\varepsilon)\equiv\frac{1-\varepsilon}{\frac{n}{n+1}+\varepsilon}, \tag{A.11}$$

and $\varepsilon_{C^*} \equiv \varepsilon_1/\varepsilon_C$, $\varepsilon_{1^*} \equiv \varepsilon_B/\varepsilon_1$. Finally, Helmholtz's free energy reads:

$$A = \frac{1}{2\varepsilon_C} \sum_n \frac{Q'_n}{a_C^{2n+1}}, \tag{A.12}$$

$$Q'_n \equiv \left\{ g'_n(\varepsilon'_{C^*}) + \left[1 - \frac{n}{n+1} g'_n(\varepsilon'_{C^*})\right] g'_n(\varepsilon_{1^*}) \left(\frac{a_C}{a_1}\right)^{2n+1} \right\} Q_n. \tag{A.13}$$

A.2 Spherical Cavity Embedded by Two Concentric Dielectric Shells Surrounded by the Bulk

If we consider two spherical concentric dielectric shells, defined by the conditions:

$$r = a_C = \text{constant}, \tag{A.14a}$$

$$r = a_1 = \text{constant} > a_C, \tag{A.14b}$$

$$r = a_2 = \text{constant} > a_1, \tag{A.14c}$$

we have now four different dielectric regions, *viz.* the cavity (with dielectric constant ε_C), the inner shell (with dielectric constant ε_1), the outer shell (with dielectric constant ε_2), and the bulk (with dielectric constant ε_B). The electric potential in the four regions is expressed by:

$$\Phi_C = \varepsilon_C^{-1} \sum_n \sum_m \left[\left(A_{nm}^C + A_{nm}^{(1)} + A_{nm}^{(2)} \right) r^n + B_{nm}^C r^{-(n+1)} \right] P_n^m(\cos\theta) \mathrm{e}^{\mathrm{i}m\phi}, \quad r < a_C, \tag{A.15}$$

$$\Phi_1 = \varepsilon_1^{-1} \sum_n \sum_m \left[\left(A_{nm}^{(1)} + A_{nm}^{(2)} \right) r^n + B_{nm}^{(1)} r^{-(n+1)} \right] P_n^m(\cos\theta) \mathrm{e}^{\mathrm{i}m\phi}, \quad a_C < r < a_1, \tag{A.16}$$

$$\Phi_2 = \varepsilon_2^{-1} \sum_n \sum_m \left[A_{nm}^{(2)} r^n + B_{nm}^{(2)} r^{-(n+1)} \right] P_n^m(\cos\theta) \mathrm{e}^{\mathrm{i}m\phi}, \quad a_1 < r < a_2, \tag{A.17}$$

$$\Phi_B = \varepsilon_B^{-1} \sum_n \sum_m B_{nm}^B r^{-(n+1)} P_n^m(\cos\theta) \mathrm{e}^{\mathrm{i}m\phi}. \quad r > a_2. \tag{A.18}$$

The reaction potential is given by:

$$\Phi_R = \varepsilon_C^{-1} \sum_n \sum_m \left(A_{nm}^C + A_{nm}^{(1)} + A_{nm}^{(2)} \right) r^n P_n^m(\cos\theta) e^{im\phi}, \quad r < a_C, \tag{A.19}$$

and the boundary conditions are imposed for each surface, i.e.:

$$(\Phi_C)_{r=a_C^-} = (\Phi_1)_{r=a_C^+}, \tag{A.20a}$$

$$\varepsilon_C \left(\frac{\partial \Phi_C}{\partial r} \right)_{r=a_C^-} = \varepsilon_1 \left(\frac{\partial \Phi_1}{\partial r} \right)_{r=a_C^+}, \tag{A.20b}$$

$$(\Phi_1)_{r=a_1^-} = (\Phi_2)_{r=a_1^+}, \tag{A.20c}$$

$$\varepsilon_1 \left(\frac{\partial \Phi_1}{\partial r} \right)_{r=a_1^-} = \varepsilon_2 \left(\frac{\partial \Phi_2}{\partial r} \right)_{r=a_1^+}, \tag{A.20d}$$

$$(\Phi_2)_{r=a_2^-} = (\Phi_B)_{r=a_2^+}, \tag{A.20e}$$

$$\varepsilon_2 \left(\frac{\partial \Phi_2}{\partial r} \right)_{r=a_2^-} = \varepsilon_B \left(\frac{\partial \Phi_B}{\partial r} \right)_{r=a_2^+}, \tag{A.20f}$$

which enable to reach the final *formulae* for Helmholtz's free energy:

$$A = \frac{1}{2\varepsilon_C} \sum_n g_n(\varepsilon''_{C^*}) \frac{Q''_n}{a_C^{2n+1}}, \tag{A.21}$$

where:

$$Q''_n \equiv \sum_k \sum_l q_k q_l r_k^n r_l^n \sum_m g''_{nm} \frac{(n-m)!}{(n+m)!} P_n^m(\cos\theta_k) P_n^m(\cos\theta_l) e^{im(\phi_l - \phi_k)}, \tag{A.22}$$

$$\begin{cases} \varepsilon_{C^*} \equiv \varepsilon_1/\varepsilon_C \\ \varepsilon_{1^*} \equiv \varepsilon_2/\varepsilon_1, \\ \varepsilon_{2^*} \equiv \varepsilon_B/\varepsilon_2 \end{cases} \tag{A.23}$$

$$\begin{cases} \varepsilon'_{C^*} \equiv \varepsilon_{C^*} \left[1 + g_n(\varepsilon'_{1^*}) \dfrac{1 - \varepsilon_{C^*}}{(n+m)!} \left(\dfrac{a_C}{a_1} \right)^{2n+1} \right]^{-1} \\ \varepsilon'_{1^*} \equiv \varepsilon_{1^*} \left[1 + g_n(\varepsilon_{2^*}) \dfrac{1 - \varepsilon_{1^*}}{(n+m)!} \left(\dfrac{a_1}{a_2} \right)^{2n+1} \right]^{-1}, \end{cases} \tag{A.24}$$

$$\varepsilon''_{C^*} \equiv \varepsilon'_{C^*}\left\{1-\frac{\varepsilon'_{C^*}}{\varepsilon_{C^*}}\left[(n+m)!-\left(1+\frac{\varepsilon'_{1^*}}{n(1+\varepsilon'_{1^*})}\right)^{-1}\right]g_n(\varepsilon_{2^*})\frac{1-\varepsilon_{C^*}}{[(n+m)!]^2}\left(\frac{a_C}{a_2}\right)^{2n+1}\right\}^{-1}, \tag{A.25}$$

$$g''_{nm} \equiv \frac{1}{(n+m)!}\left\{\begin{aligned}&1+\frac{g_n(\varepsilon'_{1^*})}{g_n(\varepsilon''_{C^*})}\left(\frac{a_1}{a_C}\right)^{2n+1}\left[1-g_n(\varepsilon''_{C^*})\frac{\frac{n}{n+1}}{(n+m)!}\left(\frac{a_1}{a_C}\right)^{2n+1}\right]\\&+\frac{g_n(\varepsilon_{2^*})}{g_n(\varepsilon''_{C^*})}\left(\frac{a_2}{a_C}\right)^{2n+1}\left[1-g_n(\varepsilon'_{1^*})\frac{\frac{n}{n+1}}{(n+m)!}\right]\left[1-g_n(\varepsilon''_{C^*})\frac{\frac{n}{n+1}}{(n+m)!}\right]\end{aligned}\right\}, \tag{A.26}$$

$$g_n(\varepsilon) \equiv \frac{1+\varepsilon}{\frac{n}{n+1}+\varepsilon}. \tag{A.27}$$

Appendix B Spheroidal Cavity Embedded By One or More Concentric Dielectric Shells Surrounded By the Bulk

B.1 Spheroidal Cavity Embedded by One Concentric Dielectric Shell Surrounded by the Bulk

As we have seen for the spherical model, also the spheroidal model can be extended, considering concentring shells (Fig. B.1).

We can investigate first the charge distribution inside a prolate spheroidal cavity, defined by the condition $\lambda = \lambda_C = a_C/d =$ constant and dielectric constant ε_C, with a first shell of polarisable dielectric, extending from $\lambda = \lambda_C$ to $\lambda = \lambda_1 = a_1/d =$ constant $> \lambda_C$, characterised by dielectric constant ε_1. Beyond this shell, there is the bulk region, from $\lambda = \lambda_1$ to infinity, with dielectric constant ε_B. The solution of Laplace's Equation (3.10) is the same of the single spheroidal cavity case, but it is needed to introduce coefficients so that the expressions for the potential in the three regions (*viz.* cavity, shell, and bulk) are:

$$\Phi_C = \varepsilon_C^{-1}\sum_n\sum_m\left[\left(\mathrm{A}^C_{nm}+\mathrm{A}^{(1)}_{nm}\right)d^nP^m_n(\lambda)+\mathrm{B}^C_{nm}d^{-(n+1)}Q^m_n(\lambda)\right]P^m_n(\mu)\mathrm{e}^{im\phi}, \quad \lambda<\lambda_C, \tag{B.1}$$

$$\Phi_1 = \varepsilon_1^{-1}\sum_n\sum_m\left[\mathrm{A}^{(1)}_{nm}d^nP^m_n(\lambda)+\mathrm{B}^{(1)}_{nm}d^{-(n+1)}Q^m_n(\lambda)\right]P^m_n(\mu)\mathrm{e}^{im\phi}, \quad \lambda_C<\lambda<\lambda_1, \tag{B.2}$$

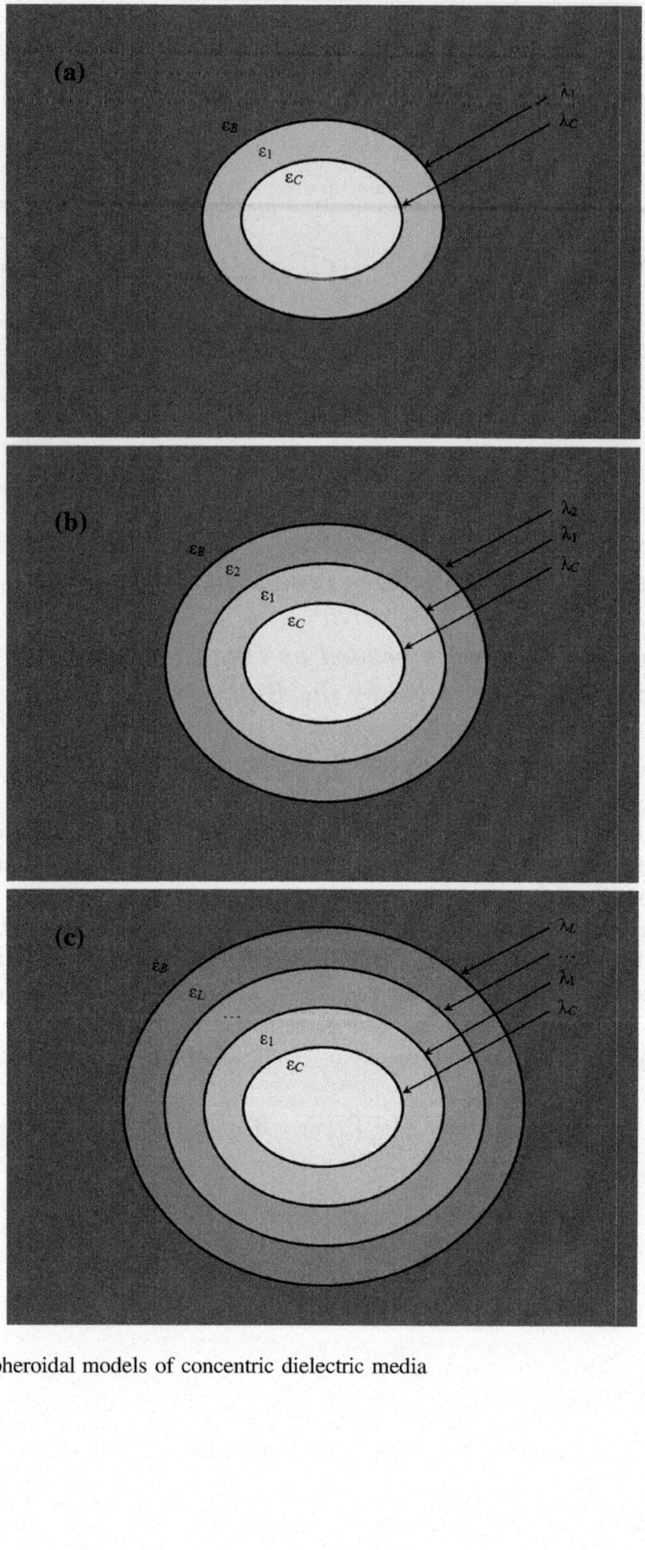

Fig. B.1 Spheroidal models of concentric dielectric media

$$\Phi_B = \varepsilon_B^{-1} \sum_n \sum_m \mathrm{B}_{nm}^B d^{-(n+1)} Q_n^m(\lambda) P_n^m(\mu) \mathrm{e}^{\mathrm{i}m\phi}, \quad \lambda > \lambda_1. \tag{B.3}$$

The reaction potential is given by:

$$\Phi_\mathrm{R} = \varepsilon_C^{-1} \sum_n \sum_m \left(\mathrm{A}_{nm}^C + \mathrm{A}_{nm}^{(1)} \right) d^n P_n^m(\lambda) P_n^m(\mu) \mathrm{e}^{\mathrm{i}m\phi}, \quad \lambda < \lambda_C, \tag{B.4}$$

and the boundary conditions are imposed for each surface, i.e.:

$$(\Phi_C)_{\lambda=\lambda_C^-} = (\Phi_1)_{\lambda=\lambda_C^+}, \tag{B.5a}$$

$$\varepsilon_C \left(\frac{\partial \Phi_C}{\partial \lambda} \right)_{\lambda=\lambda_C^-} = \varepsilon_1 \left(\frac{\partial \Phi_1}{\partial \lambda} \right)_{\lambda=\lambda_C^+}, \tag{B.5b}$$

$$(\Phi_1)_{\lambda=\lambda_1^-} = (\Phi_B)_{\lambda=\lambda_1^+}, \tag{B.5c}$$

$$\varepsilon_1 \left(\frac{\partial \Phi_1}{\partial \lambda} \right)_{\lambda=\lambda_1^-} = \varepsilon_B \left(\frac{\partial \Phi_B}{\partial \lambda} \right)_{\lambda=\lambda_1^+}, \tag{B.5d}$$

or, explicitly:

$$\begin{aligned} &\varepsilon_C^{-1} \left[\left(\mathrm{A}_{nm}^C + \mathrm{A}_{nm}^{(1)} \right) d^n P_n^m(\lambda_C) + \mathrm{B}_{nm}^C d^{-(n+1)} Q_n^m(\lambda_C) \right] \\ &\quad = \varepsilon_1^{-1} \left[\mathrm{A}_{nm}^{(1)} d^n P_n^m(\lambda_C) + \mathrm{B}_{nm}^{(1)} d^{-(n+1)} Q_n^m(\lambda_C) \right] \end{aligned} \tag{B.6a}$$

$$\boldsymbol{A}_{nm}^C d^n \dot{P}_n^m(\lambda_C) + \boldsymbol{B}_{nm}^C d^{-(n+1)} \dot{Q}_n^m(\lambda_C) = \boldsymbol{B}_{nm}^{(1)} d^{-(n+1)} \dot{Q}_n^m(\lambda_C), \tag{B.6b}$$

$$\varepsilon_1^{-1} \left[\boldsymbol{A}_{nm}^{(1)} d^n P_n^m(\lambda_1) + \boldsymbol{B}_{nm}^{(1)} d^{-(n+1)} Q_n^m(\lambda_1) \right] = \varepsilon_B^{-1} \boldsymbol{B}_{nm}^B d^{-(n+1)} Q_n^m(\lambda_1), \tag{B.6c}$$

$$\boldsymbol{A}_{nm}^{(1)} d^n \dot{P}_n^m(\lambda_1) + \boldsymbol{B}_{nm}^{(1)} d^{-(n+1)} \dot{Q}_n^m(\lambda_1) = \boldsymbol{B}_{nm}^B d^{-(n+1)} \dot{Q}_n^m(\lambda_1). \tag{B.6d}$$

Eliminating $\mathrm{B}_{nm}^{(1)}$ from (B.6a) and (B.6b), and B_{nm}^B from (B.6c) and (B.6d), we obtain:

$$\mathrm{A}_{nm}^C = \gamma_{nm}\left(\varepsilon'_{C^*}, \lambda_C\right) \frac{\mathrm{B}_{nm}^C}{d^{2n+1}}, \tag{B.7}$$

$$\begin{aligned} \boldsymbol{A}_{nm}^{(1)} &= \gamma_{nm}(\varepsilon_{1^*}, \lambda_1) \frac{\boldsymbol{B}_{nm}^{(1)}}{d^{2n+1}} \\ &= \left[1 + \gamma_{nm}\left(\varepsilon'_{C^*}, \lambda_C\right) \frac{\dot{P}_n^m(\lambda_C)}{\dot{Q}_n^m(\lambda_C)} \right] \gamma_{nm}(\varepsilon_{1^*}, \lambda_1) \frac{\boldsymbol{B}_{nm}^C}{d^{2n+1}}, \end{aligned} \tag{B.8}$$

where:

$$B_{nm}^{C} = (-1)^m (2n+1) \left[\frac{(n-m)!}{(n+m)!}\right]^2 d^n \sum_k q_k P_n^m(\lambda_k) P_n^m(\mu_k) e^{-im\phi_k} \tag{B.9}$$

$$\varepsilon'_{C^*} = \varepsilon_{C^*} \left[1 + (1 - \varepsilon_{C^*}) \gamma_{nm}(\varepsilon_{1^*}, \lambda_1) \frac{P_n^m(\lambda_C)}{Q_n^m(\lambda_C)}\right]^{-1}, \tag{B.10}$$

and $\varepsilon_{C^*} \equiv \varepsilon_1/\varepsilon_C$, $\varepsilon_{1^*} \equiv \varepsilon_B/\varepsilon_1$. Finally, Helmholtz's free energy reads:

$$A = \frac{1}{2\varepsilon_C} \sum_n \frac{2n+1}{d} \sum_k \sum_l q_k q_l \sum_m (-1)^m \gamma'_{nm}(\varepsilon'_{C^*}, \varepsilon_{1^*}, \lambda_C, \lambda_1) \left[\frac{(n-m)!}{(n+m)!}\right]^2$$
$$\times P_n^m(\lambda_k) P_n^m(\lambda_l) P_n^m(\mu_k) P_n^m(\mu_l) e^{im(\phi_l - \phi_k)}, \tag{B.11}$$

where:

$$\gamma'_{nm}(\varepsilon'_{C^*}, \varepsilon_{1^*}, \lambda_C, \lambda_1) \equiv \gamma_{nm}(\varepsilon'_{C^*}, \lambda_C) + \gamma_{nm}(\varepsilon_{1^*}, \lambda_1) \left[1 + \gamma_{nm}(\varepsilon'_{C^*}, \lambda_C) \frac{\dot{P}_n^m(\lambda_C)}{\dot{Q}_n^m(\lambda_C)}\right]. \tag{B.12}$$

The final expressions for the oblate spheroid problem can be straightforwardly obtained by those of the prolate case, by making the substitutions (4.47).

B.2 Spheroidal Cavity Embedded by Two or More Concentric Dielectric Shells Surrounded by the Bulk

If we consider two prolate spheroidal concentric dielectric shells, defined by the conditions:

$$\lambda = \lambda_C = a_C/d = \text{constant}, \tag{B.13a}$$

$$\lambda = \lambda_1 = a_1/d = \text{constant} > \lambda_C, \tag{B.13b}$$

$$\lambda = \lambda_2 = a_2/d = \text{constant} > \lambda_1, \tag{B.13c}$$

we have now four different dielectric regions, *viz.* the cavity (characterised by dielectric constant ε_C), the inner shell (characterised by dielectric constant ε_1), the outer shell (characterised by dielectric constant ε_2), and the bulk (characterised by dielectric constant ε_B). The electric potential in the four regions are expressed by:

$$\Phi_C = \varepsilon_C^{-1} \sum_n \sum_m \left[\left(\mathrm{A}_{nm}^{C} + \mathrm{A}_{nm}^{(1)} + \mathrm{A}_{nm}^{(2)} \right) d^n P_n^m(\lambda) + \mathrm{B}_{nm}^{C} d^{-(n+1)} Q_n^m(\lambda) \right] P_n^m(\mu) \mathrm{e}^{\mathrm{i}m\phi}, \quad \lambda < \lambda_C, \tag{B.14}$$

$$\Phi_1 = \varepsilon_1^{-1} \sum_n \sum_m \left[\left(\mathrm{A}_{nm}^{(1)} + \mathrm{A}_{nm}^{(2)} \right) d^n P_n^m(\lambda) + \mathrm{B}_{nm}^{(1)} d^{-(n+1)} Q_n^m(\lambda) \right] P_n^m(\mu) \mathrm{e}^{\mathrm{i}m\phi}, \quad \lambda_C < \lambda < \lambda_1, \tag{B.15}$$

$$\Phi_2 = \varepsilon_2^{-1} \sum_n \sum_m \left[\mathrm{A}_{nm}^{(2)} d^n P_n^m(\lambda) + \mathrm{B}_{nm}^{(2)} d^{-(n+1)} Q_n^m(\lambda) \right] P_n^m(\mu) \mathrm{e}^{\mathrm{i}m\phi}, \quad \lambda_1 < \lambda < \lambda_2, \tag{B.16}$$

$$\Phi_B = \varepsilon_B^{-1} \sum_n \sum_m \mathrm{B}_{nm}^{B} d^{-(n+1)} Q_n^m(\lambda) P_n^m(\mu) \mathrm{e}^{\mathrm{i}m\phi}. \quad \lambda > \lambda_2. \tag{B.17}$$

The reaction potential is given by:

$$\Phi_\mathrm{R} = \varepsilon_C^{-1} \sum_n \sum_m \left(\mathrm{A}_{nm}^{C} + \mathrm{A}_{nm}^{(1)} + \mathrm{A}_{nm}^{(2)} \right) d^n P_n^m(\lambda) P_n^m(\mu) \mathrm{e}^{\mathrm{i}m\phi}, \quad \lambda < \lambda_C, \tag{B.18}$$

and the boundary conditions are imposed for each surface, i.e.:

$$(\Phi_C)_{\lambda=\lambda_C^-} = (\Phi_1)_{\lambda=\lambda_C^+}, \tag{B.19a}$$

$$\varepsilon_C \left(\frac{\partial \Phi_C}{\partial \lambda} \right)_{\lambda=\lambda_C^-} = \varepsilon_1 \left(\frac{\partial \Phi_1}{\partial \lambda} \right)_{\lambda=\lambda_C^+}, \tag{B.19b}$$

$$(\Phi_1)_{\lambda=\lambda_1^-} = (\Phi_2)_{\lambda=\lambda_1^+}, \tag{B.19c}$$

$$\varepsilon_1 \left(\frac{\partial \Phi_1}{\partial \lambda} \right)_{\lambda=\lambda_1^-} = \varepsilon_2 \left(\frac{\partial \Phi_2}{\partial \lambda} \right)_{\lambda=\lambda_1^+}, \tag{B.19d}$$

$$(\Phi_2)_{\lambda=\lambda_2^-} = (\Phi_B)_{\lambda=\lambda_2^+}, \tag{B.19e}$$

$$\varepsilon_2 \left(\frac{\partial \Phi_2}{\partial \lambda} \right)_{\lambda=\lambda_2^-} = \varepsilon_B \left(\frac{\partial \Phi_B}{\partial \lambda} \right)_{\lambda=\lambda_2^+}, \tag{B.19f}$$

which enable to reach the final *formulae* for the reaction potential:

$$\Phi_{\mathrm{R}} = \varepsilon_C^{-1} \sum_n \sum_m \gamma''_{nm}(\varepsilon''_{C^*}, \varepsilon'_{1^*}, \varepsilon_{2^*}, \lambda_C, \lambda_1, \lambda_2) \mathrm{B}^{(1)}_{nm} d^{-(n+1)} P_n^m(\lambda) P_n^m(\mu) \mathrm{e}^{\mathrm{i}m\phi}, \quad \text{(B.20)}$$

and Helmholtz's free energy:

$$A = \frac{1}{2\varepsilon_C} \sum_n \frac{2n+1}{d} \sum_k \sum_l q_k q_l \sum_m (-1)^m \gamma''_{nm}(\varepsilon''_{C^*}, \varepsilon'_{1^*}, \varepsilon_{2^*}, \lambda_C, \lambda_1, \lambda_2) \left[\frac{(n-m)!}{(n+m)!}\right]^2$$
$$\times P_n^m(\lambda_k) P_n^m(\lambda_l) P_n^m(\mu_k) P_n^m(\mu_l) \mathrm{e}^{\mathrm{i}m(\phi_l - \phi_k)}, \quad \text{(B.21)}$$

where:

$$\begin{cases} \varepsilon_{C^*} \equiv \varepsilon_1/\varepsilon_C \\ \varepsilon_{1^*} \equiv \varepsilon_2/\varepsilon_1 \,, \\ \varepsilon_{2^*} \equiv \varepsilon_B/\varepsilon_2 \end{cases} \quad \text{(B.22)}$$

$$\begin{cases} \varepsilon'_{C^*} \equiv \varepsilon_{C^*} \left[1 + (1 - \varepsilon_{C^*}) \gamma_{nm}(\varepsilon'_{1^*}, \lambda_1) \dfrac{P_n^m(\lambda_C)}{Q_n^m(\lambda_C)}\right]^{-1} \\ \varepsilon'_{1^*} \equiv \varepsilon_{1^*} \left[1 + (1 - \varepsilon_{1^*}) \gamma_{nm}(\varepsilon_{2^*}, \lambda_2) \dfrac{P_n^m(\lambda_1)}{Q_n^m(\lambda_1)}\right]^{-1} \end{cases}, \quad \text{(B.23)}$$

$$\varepsilon''_{C^*} \equiv \varepsilon'_{C^*} \left\{1 + \varepsilon'_{C^*} \left(1 - \frac{1}{\varepsilon_{C^*}}\right) \left[1 + \gamma_{nm}(\varepsilon'_{1^*}, \lambda_1) \frac{\dot{P}_n^m(\lambda_1)}{\dot{Q}_n^m(\lambda_1)}\right] \gamma_{nm}(\varepsilon_{2^*}, \lambda_2) \frac{P_n^m(\lambda_C)}{Q_n^m(\lambda_C)}\right\}^{-1}, \quad \text{(B.24)}$$

$$\begin{aligned} \gamma''_{nm}(\varepsilon''_{C^*}, \varepsilon'_{1^*}, \varepsilon_{2^*}, \lambda_C, \lambda_1, \lambda_2) &\equiv \gamma_{nm}(\varepsilon''_{C^*}, \lambda_C) \\ &+ \gamma_{nm}(\varepsilon'_{1^*}, \lambda_1) \left[1 + \gamma_{nm}(\varepsilon''_{C^*}, \lambda_C) \frac{\dot{P}_n^m(\lambda_1)}{\dot{Q}_n^m(\lambda_1)}\right] \\ &+ \gamma_{nm}(\varepsilon_{2^*}, \lambda_2) \left[1 + \gamma_{nm}(\varepsilon'_{1^*}, \lambda_1) \frac{\dot{P}_n^m(\lambda_1)}{\dot{Q}_n^m(\lambda_1)}\right] \left[1 + \gamma_{nm}(\varepsilon''_{C^*}, \lambda_C) \frac{\dot{P}_n^m(\lambda_C)}{\dot{Q}_n^m(\lambda_C)}\right]. \end{aligned} \quad \text{(B.25)}$$

The expressions for the oblate spheroidal cavity can be obtained with the substitutions (4.47).

This approach can be generalised for a system accounting for L prolate spheroidal concentric dielectric shells, where the S-th shell is defined by the condition $\lambda = \lambda_S = a_S/d =$ constant, and characterised by a dielectric constant ε_S. The potentials can be expressed as:

$$\Phi_S = \varepsilon_S^{-1} \sum_n \sum_m \left[\left(\sum_{\sigma=S}^{L} \mathrm{A}_{nm}^{(\sigma)} \right) d^n P_n^m(\lambda) + \mathrm{B}_{nm}^{(S)} d^{-(n+1)} Q_n^m(\lambda) \right] P_n^m(\mu) \mathrm{e}^{\mathrm{i}m\phi}, \quad S = 0, 1, \ldots L, \tag{B.26}$$

$$\Phi_B = \varepsilon_B^{-1} \sum_n \sum_m \mathrm{B}_{nm}^{B} d^{-(n+1)} Q_n^m(\lambda) P_n^m(\mu) \mathrm{e}^{\mathrm{i}m\phi}, \quad \lambda > \lambda_L. \tag{B.27}$$

where $S = 0 = C$ correspond to the inner cavity. The reaction potential is given by:

$$\Phi_{\mathrm{R}} = \varepsilon_C^{-1} \sum_n \sum_m \left(\sum_{\sigma=0}^{L} \mathrm{A}_{nm}^{(\sigma)} \right) d^n P_n^m(\lambda) P_n^m(\mu) \mathrm{e}^{\mathrm{i}m\phi}, \quad \lambda < \lambda_C, \tag{B.28}$$

and the boundary conditions are imposed for each surface, i.e.:

$$(\Phi_S)_{\lambda=\lambda_S^-} = (\Phi_{S+1})_{\lambda=\lambda_S^+}, \tag{B.29a}$$

$$\varepsilon_S \left(\frac{\partial \Phi_S}{\partial \lambda} \right)_{\lambda=\lambda_S^-} = \varepsilon_{S+1} \left(\frac{\partial \Phi_{S+1}}{\partial \lambda} \right)_{\lambda=\lambda_S^+}, \tag{B.29b}$$

where S = 0, 1,… L (S = 0 = C corresponds to the inner cavity, S = L + 1 = B corresponds to the outer bulk). which enable to reach the final *formulae* for the reaction potential:

$$\Phi_{\mathrm{R}} = \varepsilon_C^{-1} \sum_n \sum_m \gamma_{nm}^{(L)} \left(\left\{ \varepsilon_{\sigma^*}^{(L-\sigma)}, \lambda_\sigma \right\}_{\sigma=0,1,\ldots L} \right) \mathrm{B}_{nm}^{(1)} d^{-(n+1)} P_n^m(\lambda) P_n^m(\mu) \mathrm{e}^{\mathrm{i}m\phi}, \tag{B.30}$$

and Helmholtz's free energy:

$$A = \frac{1}{2\varepsilon_C} \sum_n \frac{2n+1}{d} \sum_k \sum_l q_k q_l \sum_m (-1)^m \gamma_{nm}^{(L)} \left(\left\{ \varepsilon_{\sigma^*}^{(L-\sigma)}, \lambda_\sigma \right\}_{\sigma=0,1,\ldots L} \right) \left[\frac{(n-m)!}{(n+m)!} \right]^2 \times P_n^m(\lambda_k) P_n^m(\lambda_l) P_n^m(\mu_k) P_n^m(\mu_l) \mathrm{e}^{\mathrm{i}m(\phi_l - \phi_k)}, \tag{B.31}$$

where:

$$\begin{cases} \varepsilon_{\sigma^*} \equiv \varepsilon_{\sigma+1}/\varepsilon_\sigma \\ \varepsilon'_{\sigma^*} \equiv \varepsilon_{\sigma^*} \left[1 + (1 - \varepsilon_{\sigma^*}) \gamma_{nm} \left(\varepsilon'_{(\sigma+1)^*}, \lambda_{\sigma+1} \right) \frac{P_n^m(\lambda_\sigma)}{Q_n^m(\lambda_\sigma)} \right]^{-1} \\ \varepsilon''_{\sigma^*} \equiv \varepsilon'_{\sigma^*} \left\{ 1 + \varepsilon'_{\sigma^*} \left(1 - \frac{1}{\varepsilon_{\sigma^*}} \right) \left[1 + \gamma_{nm} \left(\varepsilon'_{(\sigma+1)^*}, \lambda_{\sigma+1} \right) \frac{\dot{P}_n^m(\lambda_{\sigma+1})}{\dot{Q}_n^m(\lambda_{\sigma+1})} \right] \gamma_{nm} \left(\varepsilon_{(\sigma+2)^*}, \lambda_{\sigma+2} \right) \frac{P_n^m(\lambda_\sigma)}{Q_n^m(\lambda_\sigma)} \right\}^{-1}, \\ \vdots \end{cases} \tag{B.32}$$

$$\gamma_{nm}^{(L)}\left(\left\{\varepsilon_{\sigma^*}^{(L-\sigma)},\lambda_\sigma\right\}_{\sigma=0,1,\ldots L}\right)=\left\{\sum_{\sigma=0}^{L}\gamma_{nm}\left(\varepsilon_{\sigma^*}^{(L-\sigma)},\lambda_\sigma\right)\prod_{\tau=0}^{\sigma}\left[1+\gamma_{nm}\left(\varepsilon_{\sigma^*}^{(L-\sigma)},\lambda_\sigma\right)\frac{\dot{P}_n^m(\lambda_\sigma)}{\dot{Q}_n^m(\lambda_\sigma)}\right]\right\}. \tag{B.33}$$

Appendix C Ellipsoidal Cavity Embedded By One or More Concentric Dielectric Shells Surrounded By the Bulk

C.1 Ellipsoidal Cavity Imbedded by One Concentric Dielectric Shell Surrounded by the Bulk

The formalism described above lends itself to be easily extended to the case of cavity imbedded by a concentric dielectric shell surrounded by the bulk. We aim here to investigate the charge distribution inside an ellipsoidal cavity, defined by the condition $\rho = \rho_C =$ constant and dielectric constant ε_C, with a first shell of polarisable dielectric, extending from $\rho = \rho_C$ to $\rho = \rho_1 =$ constant $> \rho_C$, characterised by dielectric constant ε_1. Beyond this shell, there is the bulk region, from $\rho = \rho_1$ to infinity, with dielectric constant ε_B. The solution of Laplace's Equation (3.10) is the same of the single ellipsoidal cavity case, but it is needed to introduce coefficients so that the expressions for the potential in the three regions (*viz.* cavity, shell, and bulk) are:

$$\Phi_C=\varepsilon_C^{-1}\sum_n\sum_m\left[\left(\boldsymbol{A}_{nm}^C+\boldsymbol{A}_{nm}^{(1)}\right)E_n^m(\rho)+\boldsymbol{B}_{nm}^C F_n^m(\rho)\right]E_n^m(\mu)E_n^m(\nu),\quad \rho<\rho_C, \tag{C.1}$$

$$\Phi_1=\varepsilon_1^{-1}\sum_n\sum_m\left[\mathrm{A}_{nm}^{(1)}E_n^m(\rho)+\mathrm{B}_{nm}^{(1)}F_n^m(\rho)\right]E_n^m(\mu)E_n^m(\nu),\quad \rho_C<\rho<\rho_1, \tag{C.2}$$

$$\Phi_B=\varepsilon_B^{-1}\sum_n\sum_m\mathrm{B}_{nm}^B F_n^m(\rho)E_n^m(\mu)E_n^m(\nu),\quad \rho>\rho_1. \tag{C.3}$$

The reaction potential is given by:

$$\Phi_\mathrm{R}=\varepsilon_C^{-1}\sum_n\sum_m\left(\mathrm{A}_{nm}^C+\mathrm{A}_{nm}^{(1)}\right)E_n^m(\rho)E_n^m(\mu)E_n^m(\nu),\quad \rho<\rho_C, \tag{C.4}$$

and the boundary conditions are imposed for each surface, i.e.:

$$(\Phi_C)_{\rho=\rho_C^-}=(\Phi_1)_{\rho=\rho_C^+}, \tag{C.5a}$$

$$\varepsilon_C\left(\frac{\partial \Phi_C}{\partial \rho}\right)_{\rho=\rho_C^-} = \varepsilon_1\left(\frac{\partial \Phi_1}{\partial \rho}\right)_{\rho=\rho_C^+}, \tag{C.5b}$$

$$(\Phi_1)_{\rho=\rho_1^-} = (\Phi_B)_{\rho=\rho_1^+}, \tag{C.5c}$$

$$\varepsilon_1\left(\frac{\partial \Phi_1}{\partial \rho}\right)_{\rho=\rho_1^-} = \varepsilon_B\left(\frac{\partial \Phi_B}{\partial \rho}\right)_{\rho=\rho_1^+}, \tag{C.5d}$$

or, explicitly:

$$\varepsilon_C^{-1}\left[\left(\mathrm{A}_{nm}^C + \mathrm{A}_{nm}^{(1)}\right)E_n^m(\rho_C) + \mathrm{B}_{nm}^C F_n^m(\rho_C)\right] = \varepsilon_1^{-1}\left[\mathrm{A}_{nm}^{(1)}E_n^m(\rho_C) + \mathrm{B}_{nm}^{(1)}F_n^m(\rho_C)\right], \tag{C.6a}$$

$$\boldsymbol{A}_{nm}^C \dot{E}_n^m(\rho_C) + \boldsymbol{B}_{nm}^C \dot{F}_n^m(\rho_C) = \boldsymbol{B}_{nm}^{(1)} \dot{F}_n^m(\rho_C), \tag{C.6b}$$

$$\varepsilon_1^{-1}\left[\boldsymbol{A}_{nm}^{(1)}E_n^m(\rho_1) + \boldsymbol{B}_{nm}^{(1)}F_n^m(\rho_1)\right] = \varepsilon_B^{-1}\boldsymbol{B}_{nm}^B F_n^m(\rho_1), \tag{C.6c}$$

$$\boldsymbol{A}_{nm}^{(1)}\dot{E}_n^m(\rho_1) + \boldsymbol{B}_{nm}^{(1)}\dot{F}_n^m(\rho_1) = \boldsymbol{B}_{nm}^B \dot{F}_n^m(\rho_1). \tag{C.6d}$$

Eliminating $\mathrm{B}_{nm}^{(1)}$ from (C.6a) and (C.6b), and B_{nm}^B from (C.6c) and (C.6d), we obtain:

$$\mathrm{A}_{nm}^C = \gamma_{nm}(\varepsilon'_{C^*}, \rho_C)\mathrm{B}_{nm}^C, \tag{C.7}$$

$$\begin{aligned}\boldsymbol{A}_{nm}^{(1)} &= \gamma_{nm}(\varepsilon_{1^*}, \rho_1)\boldsymbol{B}_{nm}^{(1)} \\ &= \left[1 + \gamma_{nm}(\varepsilon'_{C^*}, \rho_C)\frac{\dot{E}_n^m(\rho_C)}{\dot{F}_n^m(\rho_C)}\right]\gamma_{nm}(\varepsilon_{1^*}, \rho_1)\boldsymbol{B}_{nm}^C,\end{aligned} \tag{C.8}$$

where:

$$\mathrm{B}_{nm}^C = N_{nm}Q_{nm}, \tag{C.9}$$

$$\varepsilon'_{C^*} = \varepsilon_{C^*}\left[1 + (1 - \varepsilon_{C^*})\gamma_{nm}(\varepsilon_{1^*}, r_1)\frac{E_n^m(\rho_C)}{F_n^m(\rho_C)}\right]^{-1}, \tag{C.10}$$

and $\varepsilon_{C^*} \equiv \varepsilon_1/\varepsilon_C$, $\varepsilon_{1^*} \equiv \varepsilon_B/\varepsilon_1$. Finally, Helmholtz's free energy reads:

$$\begin{aligned}A = \frac{1}{2\varepsilon_C}\sum_n\sum_m \gamma'_{nm}(\varepsilon'_{C^*}, \varepsilon_{1^*}, \rho_C, \rho_1)N_{nm}\sum_k\sum_l q_k q_l \\ \times E_n^m(\rho_k)E_n^m(\rho_l)E_n^m(\mu_k)E_n^m(\mu_l)E_n^m(\nu_k)E_n^m(\nu_l),\end{aligned} \tag{C.11}$$

where:

$$\gamma'_{nm}\left(\varepsilon'_{C^*},\varepsilon_{1^*},\rho_C,\rho_1\right)=\gamma_{nm}\left(\varepsilon'_{C^*},\rho_C\right)+\gamma_{nm}(\varepsilon_{1^*},\rho_1)\left[1+\gamma_{nm}\left(\varepsilon'_{C^*},\rho_C\right)\frac{\dot{E}_n^m(\rho_C)}{\dot{F}_n^m(\rho_C)}\right]. \tag{C.12}$$

C.2 Ellipsoidal Cavity Embedded by Two Concentric Dielectric Shells Surrounded by the Bulk

If we consider two ellipsoidal concentric dielectric shells, defined by the conditions:

$$\rho=\rho_C=\text{constant}, \tag{C.13a}$$

$$\rho=\rho_1=\text{constant}>\rho_C, \tag{C.13b}$$

$$\rho=\rho_2=\text{constant}>\rho_1, \tag{C.13c}$$

we have now four different dielectric regions, *viz.* the cavity (characterised by dielectric constant ε_C), the inner shell (characterised by dielectric constant ε_1), the outer shell (characterised by dielectric constant ε_2), and the bulk (characterised by dielectric constant ε_B). The electric potential in the four regions are expressed by:

$$\Phi_C=\varepsilon_C^{-1}\sum_n\sum_m\left[\left(\mathrm{A}^C_{nm}+\mathrm{A}^{(1)}_{nm}+\mathrm{A}^{(2)}_{nm}\right)E_n^m(\rho)+\mathrm{B}^C_{nm}F_n^m(\rho)\right]E_n^m(\mu)E_n^m(\nu), \quad \rho<\rho_C, \tag{C.14}$$

$$\Phi_1=\varepsilon_1^{-1}\sum_n\sum_m\left[\left(\mathrm{A}^{(1)}_{nm}+\mathrm{A}^{(2)}_{nm}\right)E_n^m(\rho)+\mathrm{B}^{(1)}_{nm}F_n^m(\rho)\right]E_n^m(\mu)E_n^m(\nu), \quad \rho_C<\rho<\rho_1, \tag{C.15}$$

$$\Phi_2=\varepsilon_2^{-1}\sum_n\sum_m\left[\mathrm{A}^{(2)}_{nm}E_n^m(\rho)+\mathrm{B}^{(2)}_{nm}F_n^m(\rho)\right]E_n^m(\mu)E_n^m(\nu), \quad \rho_1<\rho<\rho_2, \tag{C.16}$$

$$\Phi_B=\varepsilon_B^{-1}\sum_n\sum_m\mathrm{B}^B_{nm}F_n^m(\rho)E_n^m(\mu)E_n^m(\nu). \quad \rho>\rho_2. \tag{C.17}$$

The reaction potential is given by:

$$\Phi_\mathrm{R}=\varepsilon_C^{-1}\sum_n\sum_m\left(\mathrm{A}^C_{nm}+\mathrm{A}^{(1)}_{nm}+\mathrm{A}^{(2)}_{nm}\right)E_n^m(\rho)E_n^m(\mu)E_n^m(\nu), \quad \rho<\rho_C, \tag{C.18}$$

and the boundary conditions are imposed for each surface, i.e.:

$$(\Phi_C)_{\rho=\rho_C^-} = (\Phi_1)_{\rho=\rho_C^+}, \tag{C.19a}$$

$$\varepsilon_C\left(\frac{\partial\Phi_C}{\partial\rho}\right)_{\rho=\rho_C^-} = \varepsilon_1\left(\frac{\partial\Phi_1}{\partial\rho}\right)_{\rho=\rho_C^+}, \tag{C.19b}$$

$$(\Phi_1)_{\rho=\rho_1^-} = (\Phi_2)_{\rho=\rho_1^+}, \tag{C.19c}$$

$$\varepsilon_1\left(\frac{\partial\Phi_1}{\partial\rho}\right)_{\rho=\rho_1^-} = \varepsilon_2\left(\frac{\partial\Phi_2}{\partial\rho}\right)_{\rho=\rho_1^+}, \tag{C.19d}$$

$$(\Phi_2)_{\rho=\rho_2^-} = (\Phi_B)_{\rho=\rho_2^+}, \tag{C.19e}$$

$$\varepsilon_2\left(\frac{\partial\Phi_2}{\partial\rho}\right)_{\rho=\rho_2^-} = \varepsilon_B\left(\frac{\partial\Phi_B}{\partial\rho}\right)_{\rho=\rho_2^+}, \tag{C.19f}$$

which enable to reach the final *formulae* for the reaction potential:

$$\Phi_R = \varepsilon_C^{-1}\sum_n\sum_m \gamma''_{nm}(\varepsilon''_{C^*}, \varepsilon'_{1^*}, \varepsilon_{2^*}, \rho_C, \rho_1, \rho_2) N_{nm} Q_{nm} E_n^m(\rho) E_n^m(\mu) E_n^m(\nu), \tag{C.20}$$

and Helmholtz's free energy:

$$\begin{aligned} A = \frac{1}{2\varepsilon_C}\sum_n\sum_m \gamma''_{nm}(\varepsilon''_{C^*}, \varepsilon'_{1^*}, \varepsilon_{2^*}, \rho_C, \rho_1, \rho_2) N_{nm} \sum_k\sum_l q_k q_l \\ \times E_n^m(\rho_k) E_n^m(\rho_l) E_n^m(\mu_k) E_n^m(\mu_l) E_n^m(\nu_k) E_n^m(\nu_l), \end{aligned} \tag{C.21}$$

where:

$$\begin{cases} \varepsilon_{C^*} \equiv \varepsilon_1/\varepsilon_C \\ \varepsilon_{1^*} \equiv \varepsilon_2/\varepsilon_1, \\ \varepsilon_{2^*} \equiv \varepsilon_B/\varepsilon_2 \end{cases} \tag{C.22}$$

$$\begin{cases} \varepsilon'_{C^*} \equiv \varepsilon_{C^*}\left[1 + (1-\varepsilon_{C^*})\gamma_{nm}(\varepsilon'_{1^*}, \rho_1)\dfrac{E_n^m(\rho_C)}{F_n^m(\rho_C)}\right]^{-1} \\ \varepsilon'_{1^*} \equiv \varepsilon_{1^*}\left[1 + (1-\varepsilon_{1^*})\gamma_{nm}(\varepsilon_{2^*}, \rho_2)\dfrac{E_n^m(\rho_1)}{F_n^m(\rho_1)}\right]^{-1}, \end{cases} \tag{C.23}$$

$$\varepsilon''_{C^*} \equiv \varepsilon'_{C^*} \left\{ 1 + \varepsilon'_{C^*} \left(1 - \frac{1}{\varepsilon_{C^*}} \right) \left[1 + \gamma_{nm}(\varepsilon'_{1^*}, \rho_1) \frac{\dot{E}_n^m(\rho_1)}{\dot{F}_n^m(\rho_1)} \right] \gamma_{nm}(\varepsilon_{2^*}, \rho_2) \frac{E_n^m(\rho_C)}{F_n^m(\rho_C)} \right\}^{-1}, \tag{C.24}$$

$$\begin{aligned} \gamma''_{nm}(\varepsilon''_{C^*}, \varepsilon'_{1^*}, \varepsilon_{2^*}, \rho_C, \rho_1, \rho_2) &\equiv \gamma_{nm}(\varepsilon''_{C^*}, \rho_C) \\ &+ \gamma_{nm}(\varepsilon'_{1^*}, \rho_1) \left[1 + \gamma_{nm}(\varepsilon''_{C^*}, \rho_C) \frac{\dot{E}_n^m(\rho_1)}{\dot{F}_n^m(\rho_1)} \right] \\ &+ \gamma_{nm}(\varepsilon_{2^*}, \rho_2) \left[1 + \gamma_{nm}(\varepsilon'_{1^*}, \rho_1) \frac{\dot{E}_n^m(\rho_1)}{\dot{F}_n^m(\rho_1)} \right] \left[1 + \gamma_{nm}(\varepsilon''_{C^*}, \rho_C) \frac{\dot{E}_n^m(\rho_C)}{\dot{F}_n^m(\rho_C)} \right]. \end{aligned} \tag{C.25}$$

Appendix D: Different equations for Stokes shifts

$$y_{m,s} = y_{m,g}^{(0)} + a_m \left[x_s^{(1)} + x_s^{(2)} \right], \tag{D.1}$$

Equation	Refs.	$y_{m,s}$	$a_m^{(i)}$	$x_s^{(1)}$	$x_s^{(2)}$
(D.1.1)	[52, 54–56, 59, 74]	$\tilde{\nu}^a$	$a_m^{(1)}$	$F_1(\varepsilon, n)$	$f_1(n^2)$
(D.1.2)	[49–51]		$a_m^{(1)}$	$F_2(\varepsilon, n)$	$f_2(n^2)$
(D.1.3)	[64, 76, 84]		$a_m^{(1)}$	$F_3(\varepsilon, n)$	$f_1(n^2)$
(D.1.4)	[60]		$a_m^{(1)}$	$F_3(\varepsilon, n)$	$f_3(n^2)$
(D.1.5)	[77, 88–90]		$a_m^{(1)}$	$f_2(\varepsilon)$	$f_2(n^2)$
(D.1.6)	[80, 85, 91]		$2a_m^{(2)}$	$f_2(\varepsilon)$	
(D.1.7)	[80, 85, 91]		$a_m^{(2)}$	$F_6(\varepsilon, n)$	
(D.1.8)	[60, 168]	$\tilde{\nu}^f$	$a_m^{(3)}$	$F_3(\varepsilon, n)$	$f_3(n^2)$
(D.1.9)	[66, 69, 70, 78, 81, 86, 98]		$a_m^{(4)}$	$F_5(\varepsilon, n)$	
(D.1.10)	[92]	$\tilde{\nu}^a - \tilde{\nu}^f$	$a_m^{(5)}$	$F_1(\varepsilon, n)$	
(D.1.11)	[57, 61]		$a_m^{(5)}$	$F_2(\varepsilon, n)$	
(D.1.12)	[63, 75, 82, 93]		$a_m^{(5)}$	$F_3(\varepsilon, n)$	
(D.1.13)	[62, 65, 67]		$a_m^{(5)}$	$F_3(\varepsilon, n)$	$f_1(n^2)$
(D.1.14)	[58]	$\tilde{\nu}^a + \tilde{\nu}^f$	$2a_m^{(2)}$	$F_2(\varepsilon, n)$	$f_2(n^2)$
(D.1.15)	[62, 65, 67, 71]		$2a_m^{(2)}$	$F_4(\varepsilon, n)$	$f_1(n^2)$
(D.1.16)	[72, 73, 83]		$4a_m^{(2)}$	$F_7(\varepsilon, n)$	

where:

$$f_1(\xi) = \frac{\xi - 1}{\xi + 2}, \quad f_2(\xi) = \frac{\xi - 1}{2\xi + 1}, \quad f_3(\xi) = \frac{3}{2}\frac{\xi^2 - 1}{(\xi + 2)^2},$$

$$F_1(\varepsilon, n) = f_1(\varepsilon) - f_1(n^2), \quad F_2(\varepsilon, n) = f_2(\varepsilon) - f_2(n^2),$$

$$F_3(\varepsilon, n) = F_1(\varepsilon, n)\frac{f_1(n^2)}{f_2(n^2)}, \quad F_4(\varepsilon, n) = f_1(\varepsilon)\frac{f_1(n^2)}{f_2(n^2)},$$

$$F_5(\varepsilon, n) = f_2(\varepsilon) - \frac{1}{2}f_2(n^2), \quad F_6(\varepsilon, n) = 2f_1(\varepsilon) - f_1(n^2),$$

$$F_7(\varepsilon, n) = \frac{1}{2}F_3(\varepsilon, n) + f_3(n^2),$$

$$a_m^{(1)} = -\frac{2}{hca_0^3}\boldsymbol{\mu}_g \cdot (\boldsymbol{\mu}_e - \boldsymbol{\mu}_g), \quad a_m^{(2)} = -\frac{1}{hca_0^3}\left(\mu_e^2 - \mu_g^2\right),$$

$$a_m^{(3)} = -\frac{2}{hca_0^3}\boldsymbol{\mu}_e \cdot (\boldsymbol{\mu}_e - \boldsymbol{\mu}_g), a_m^{(4)} = -\frac{2}{hca_0^3}\mu_e^2, \quad a_m^{(5)} = +\frac{2}{hca_0^3}(\boldsymbol{\mu}_e - \boldsymbol{\mu}_g)^2.$$

Appendix E: Derivation of Ravi's Equation

The starting point for the derivation of Ravi's *formula* is the following equation [193, 211, 255–258],

$$\tilde{\nu}^a - \tilde{\nu}^f = \frac{2}{hca_0^3}(\boldsymbol{\mu}_e - \boldsymbol{\mu}_g)^2\left[f_1(\varepsilon) - f_1(n^2)\right], \tag{E.1}$$

which is the best analytical model for the description of the solvatochromism (see Sect. 4.1); the equation is written—according to the Authors' notation—as follows:

$$Y(m, s) = B(m)X(s) + Y^\circ(m, g), \tag{E.2}$$

where:

$$\begin{cases} Y(m, s) \equiv \tilde{\nu}_a - \tilde{\nu}_f \\ B(m) \equiv \frac{2(\Delta\boldsymbol{\mu})^2}{hca^3} = \frac{2(\boldsymbol{\mu}_e - \boldsymbol{\mu}_g)^2}{hca^3} \\ X(s) \equiv \left(\frac{\varepsilon - 1}{\varepsilon + 2} - \frac{n^2 - 1}{n^2 + 2}\right)\frac{2n^2 + 1}{n^2 + 2} \equiv F_3(\varepsilon, n^2) \end{cases}, \tag{E.3}$$

and $Y^\circ(m, g)$ is a constant, independent by the solvent, which represents Stokes shift of the molecule m in gas phase. The Authors actually do not use the squared difference of the two vectors when expressing $B(m)$ (as it should be), but the squared difference of their *moduli*, *viz.* $(\Delta\mu)^2 = (\mu_e - \mu_g)^2$: in other words, they implicitly assume that $\boldsymbol{\mu}_e$ and $\boldsymbol{\mu}_g$ must have (always) the same direction and *versus*.

(Since this is not explicitly stated in the paper, their equation, that will be shown soon, has been often improperly evoked in the experimental Literature.)

They apply Eq. (E.2) to the special case of the so-called Reichardt's betaine–30, $m = D$, which the Eq. (17.1) was built on (Chart 5.1, molecule **A**). Also at this point there arise some issues. First of all, they use (approximated) experimental values of the dipole moments which actually refer to another betaine (i.e. Chart 5.1, molecule **C**), *viz.* $\mu_e = 6$ D and $\mu_g = 15$ D, and thus $\Delta\mu_{\mathrm{C}} = 9$D, although this is not explicitly stated. In order to evaluate whether the two molecules have different values of electric dipole moments, we have calculated the values of μ_g and μ_e (in the Franck-Condon S_1) for the dyes **A** and **C** at PBE0 [6]/SNSD [219] level of theory, both *in vacuo* and in the reference solvent (i.e. 1,4-dioxane); in the latter, $\Delta\mu_{\mathrm{A}} = 6.120$ D and $\Delta\mu_{\mathrm{C}} = 9.018$ D. These QM calculations confirm that $\boldsymbol{\mu}_e$ and $\boldsymbol{\mu}_g$ are parallel and directed along the **x**-axis [259], and seems to suggest that actually the two dyes show a different dipole moment change when passing from S_0 to S_1. So, according to our calculations, $\Delta\mu_{\mathrm{A}}$ is smaller than $\Delta\mu_{\mathrm{C}}$. Therefore, by employing the improper value of $\Delta\mu_{\mathrm{C}}$ leads to an under-estimation of the electric dipole moment change of the system in study.

For the reference compound, they assume $Y(D,s) \equiv \tilde{\nu}_a^D$ – constant, since the dye is written not to fluoresce at all and hence one can only follow the shifts of absorption maximum with change in solvent polarity. This is actually another contradictory point. As explained in Sect. 4.1, Eq. (D.1) gives good results, but to be employed from the experimental point of view the knowledge of the emission data is required and thus the compound must be well-fluorescing. They put $\tilde{\nu}_f^D \equiv$ constant, since they consider $\mu_e^D = \mu_g^D$, and then they feel confident of making the "safe" assumption that the solvatochromic shift of the florescence peak, if any, will be negligible compared to the shift of the absorption peak, i.e. $\tilde{\nu}_f^D(s) =$ constant for any solvent *s*, but this is not necessarily true. (In particular in the case of the dye **A**, according to our PBE0 [6]/SNSD [219] calculations, the energy of the relaxed excited state is affected by the polarity of the *medium*; *vide infra*) Anyway, they write Eq. (D.2) for the dye $m = D$, as:

$$\begin{aligned}
Y(D,s) &= B(D)X(s) + Y^{\circ}(D,g)\\
\tilde{\nu}_a^{D,s} - \text{constant} &= \frac{2(\Delta\mu_D)^2}{hca_D^3}X(s) + Y^{\circ}(D,g)\\
\tilde{\nu}_a^{D,s} &= \frac{2(\Delta\mu_D)^2}{hca_D^3}X(s) + Y^{\circ\prime},
\end{aligned}$$

where $Y^{\circ\prime} \equiv Y^{\circ}(D,g) + \text{constant} = \left[\tilde{\nu}_a^{D,g} - \tilde{\nu}_f^{D,g}\right] + \tilde{\nu}_f^{D,s}$, and thus:

$$X(s) = \frac{hca_D^3}{2(\Delta\mu_D)^2}\tilde{\nu}_a^{D,s} - Y^{\circ\prime\prime}, \tag{E.4}$$

where $Y^{\circ\prime} \equiv \frac{hca_D^3}{2(\Delta\mu_D)^2} Y^{\circ\prime\prime}$ [259]. (The Authors state that the nature of the function $X(s)$ can be left unspecified, but the mathematical expression for the polarity function is well-defined once the theoretical framework is chosen.) From Eq. (E.1), we have:

$$\tilde{\nu}_a^{D,s} = E_T(30)/2.8591 \times 10^{-3} = 349.7604 E_T(30), \quad \text{(E.5)}$$

whereas from Eq. (D.9), we have:

$$E_T(30) = 32.4 E_T^N + 30.7, \quad \text{(E.6)}$$

so that Eq. (D.6) becomes:

$$\tilde{\nu}_a^{D,s} = 349.7604\left(32.4 E_T^N + 30.7\right) = 11332.2 E_T^N + 10737.6. \quad \text{(E.6')}$$

(Please, note that the Authors actually report slightly different values of the numerical constants, in particular: 11307.6 instead of 11332.2. In the algebraic derivation, we shall continue in using the value of the numerical constants that we have found, but during the numerical evaluations, we shall adopt their equation as is, *vide infra*.) Substituting into Eq. (E.4), we have:

$$\begin{aligned} X(s) &= \frac{hca_D^3}{2(\Delta\mu_D)^2}\left(11332.2 E_T^N + 10737.6\right) - Y^{\circ\prime} \\ &= 11332.2 \frac{hca_D^3}{2(\Delta\mu_D)^2} E_T^N + Y^{\circ\prime\prime}, \end{aligned} \quad \text{(E.7)}$$

where
$Y^{\circ\prime} \equiv \left(10737.6 - Y^{\circ\prime}\right)\frac{hca_D^3}{2(\Delta\mu_D)^2} = \left\{10737.6 - \left[\left(\tilde{\nu}_a^{D,g} - \tilde{\nu}_f^{D,g}\right) + \tilde{\nu}_f^{D,s}\right]\right\}\frac{hca_D^3}{2(\Delta\mu_D)^2}$.
Finally, substituting Eq. (D.7) into Eq. (D.2) instead of the correct $X(s)$ function, we have:

$$\tilde{\nu}_a^{m,s} - \tilde{\nu}_f^{m,s} = 11332.2 \frac{(\Delta\mathbf{\mu})^2/a^3}{(\Delta\mu_D)^2/a_D^3} E_T^N + Y * (m, g), \quad \text{(E.8)}$$

where $Y * (m, g) \equiv Y^{\circ}(m, g) + \frac{2(\Delta\mathbf{\mu})^2}{hca^3} Y^{\circ\prime\prime}$.

References and Notes

1. Pandey N, Gahlaut R, Arora P, Joshi NK, Joshi HC, Pant S (2014) J Mol Struct 1061:175
2. Tomasi J, Mennucci B, Cammi R (2005) Chem Rev 105:2999
3. Scalmani G, Frisch MJ (2010) J Chem Phys 132:114110

4. Cossi M, Rega N, Scalmani G, Barone V (2003) J Comp Chem 24:669
5. Becke AD (1993) J Chem Phys 98:5648
6. Adamo C, Barone V (1999) J Chem Phys 110:6158
7. Neese F, Schwabe T, Grimme S (2007) J Chem Phys 126:124115
8. Kozuch S, Gruzman D, Martin J (2010) J Phys Chem C 114:20801
9. Biczysko M, Panek P, Scalmani G, Bloino J, Barone V (2010) J Chem Theory Comput 6:2115
10. Cossi M, Barone V, Cammi R, Tomasi J (1996) J Chem Phys Lett 255:327
11. Barone V, Cossi M, Tomasi J (1997) J Chem Phys 107:3210
12. Cances E, Mennucci B, Tomasi J (1997) J Chem Phys 107:3032
13. Mennucci B, Cances E, Tomasi J (1997) J Phys Chem B 101:10506
14. Barone V, Cossi M (1998) J Phys Chem A 102:1995
15. Cances E, Mennucci B (1998) J Math Chem 23:309
16. Mennucci B, Cammi R, Tomasi J (1998) J Chem Phys 109:2798
17. Cappelli C, Corni S, Cammi R, Mennucci B, Tomasi J (2000) J Chem Phys 113:11270
18. Cossi M, Barone V (2001) J Chem Phys 115:4708
19. Cossi M, Scalmani G, Rega N, Barone V (2002) J Comp Chem 117:43
20. Cossi M, Rega N, Scalmani G, Barone V (2003) J Comp Chem 24:669
21. Brancato G, Di Nola A, Barone V, Amadei A (2005) J Chem Phys 122:154109
22. Rega N, Brancato G, Barone V (2006) Chem Phys Lett 422:367
23. Brancato G, Rega N, Barone V (2006) J Chem Phys 124:214505
24. Brancato G, Rega N, Barone V (2008) J Chem Phys 128:144501
25. Becke AD (1993) J Chem Phys 98:5648
26. Cappelli C, Lipparini F, Bloino J, Barone V (2011) J Chem Phys 135:104505
27. Scalmani G, Frisch MJ (2010) J Chem Phys 132:114110
28. Ohno K (1964) Theor Chem Accounts 2:219
29. Mortier WJ, van Genechten K, Gasteiger J (1985) J Am Chem Soc 107:829
30. Rappe A, Goddard W (1991) J Phys Chem 95:3358
31. Rick SW, Stuart SJ, Berne BJ (1994) J Chem Phys 101:6141
32. Lipparini F, Barone V (2011) J Chem Theory Comput 7:3711
33. Lipparini F, Cappelli C, Barone V (2012) J Chem Theory Comput 8:4153
34. Lipparini F, Cappelli C, Scalmani G, Mitri ND, Barone V (2012) J Chem Theory Comput 8:4270
35. Lipparini F, Cappelli C, Barone V (2013) J Chem Phys 138:234108
36. Born M (1920) Z. Phys. 1:45
37. Bell RP (1931) Trans Faraday Soc 27:797
38. Kirkwood JG (1934) J Chem Phys 2:351
39. Kirkwood JG, Westheimer FH (1936) J Chem Phys 6:506
40. Onsager L (1936) J Am Chem Soc 58:1486
41. Westheimer FH, Kirkwood JG (1938) J Chem Phys 6:513
42. Kirkwood JG (1939) J Chem Phys 7:911
43. Böttcher F (1942) Physica (Utrecht) 9:937
44. Böttcher F (1942) Physica (Utrecht) 9:945
45. Scholte ThG (1949) Physica 15:437
46. Bonnor WB (1951) Trans Faraday Soc 47:1143
47. Böttcher F (1952, 1973) Theory of electric polarization. Elsevier, New York
48. Buckingham AD (1953) Trans Faraday Soc 49:881
49. Lippert E (1955) Z Naturforsch 10A:541
50. Mataga N, Kaifu Y, Koizumi M (1955) Bull Chem Soc Jpn 28:690
51. Mataga N, Kaifu Y, Koizumi M (1956) Bull Chem Soc Jpn 29:115
52. McRae EG (1957) J Phys Chem 61:562
53. Lippert E (1957) Ber Bunsenges Phys Chem 61:962
54. McRae EG (1958) Spectrochim. Acta A 12:192
55. Ito M, Inozuka K, Imanishi S (1960) J Am Chem Soc 82:1317

56. Robertson W, King A, Weigang O (1961) J Chem Phys 35:464
57. Czekalla J, Meyer KO (1961) Z. Phys. Chem. (Frankfurt am Main) 27:185
58. Lippert E, Lüder W, Moll F, Nagele W, Boos H, Prigge H, Seibold-Blankenstein I (1961) Angew Chem 73:695
59. Kubota T, Yarnakawa M (1962) Bull Chem Soc Jpn 35:555
60. Bilot L, Kawski A (1962) Z Naturforsch A 17A:621
61. Lippert E, Lüder W, Boos H (1962) Fluoreszenzspektrum und Franck–Condon–Prinzip in Lösungen Aromatischer Verbindgungen. In: Magnini A (ed) Advances molecular spectroscopy, vol I. Pergamon Press, London, p 443
62. Bakhshiev NG (1962) Opt Spektrosk 13:192
63. Bakhshiev NG (1964) Opt Spektrosk 16:821
64. Bakhshiev NG (1965) Opt Spektrosk 19:196
65. Bakhshiev NG, Saidov GV (1966) Dokl Akad Nauk SSSR 168:120
66. Beens H, Knibbe N, Weller A (1967) J Chem Phys 47:1183
67. Bakhshiev NG, Saidov GV (1967) Dokl Akad Nauk SSSR 175:1090
68. Abraham RJ, Cooper MA (1967) J Chem Soc B 202
69. Beens H, Weller A (1968) Acta Phys Pol 34:593
70. Weller A (1968) Pure Appl Chem 16:115
71. Bakshiev NG, Knyazhanskii MI, Osipov OA, Minkin VI, Saidov GV (1969) Usp Khim 38:1644
72. Chamma A, Viallet P (1970) CR Acad Sci C 270:1901
73. Abribat G, Viallet P (1970) CR Acad Sci C 271:1029
74. Figueras J (1971) J Am Chem Soc 93:3255
75. Bartoszewicz B, Kawski A (1971) Bull Acad Pol Sci Ser Sci Math Astronom Phys 19:249
76. Bakhshiev NG (1972) Spektroskopiya Mezhdumolekulyarnykh Vzaimodeistvii. Nauka, Leningrad, p 165
77. Varma CAGO, Groenen EJJ (1972) Rec Trav Chim Pays-Bas 91:296
78. Okada T, Fujita T, Kubota M, Masaki S, Mataga N, Ide R, Sakata Y, Misumi S (1972) Chem Phys Lett 14:563
79. Thiebaut JM, Rivail JL, Barriol J (1972) J Chem Soc, Faraday Trans 2(68):1253
80. Amos AT, Burrows BL (1973) Adv Quantum Chem 7:303
81. Dekkers AWJD, Verhoeven JW, Speckarnp WN (1973) Tetrahedron 29:1691
82. Marty A, Viallet P (1973) J Photochem 1:443
83. Sun M (1974) Thesis. University, Lubbock, Texas Tech
84. Karelov AA, Chmutova GA, Vtyurina NN, Kataev EG (1975) Zh Fiz Khim 49:1901
85. Grasso D, Bellio E (1975) Chem Phys Lett 30:421
86. Rao KG, Bhujle VV, Rao CNR (1975) Spectrochim Acta A 31:885
87. Beveridge DL, Schnuelle GW (1975) J Phys Chem 79:2562
88. Sahini VE, Telea L (1976) Rev Roum Chim 21:321
89. Sahini VE, Weinberg J, Vasilescu M (1976) Rev Roum Chim 21:1121
90. Koutek B, Pisova M, Soucek M, Exner O (1976) Coll Czech Chem Commun 41:1676
91. Grasso D, Fasone S, Gandolfo C, Buemi G (1976) Tetrahedron 32:2105
92. Yamaguchi H, Ikeda T, Mamatsuka H (1976) Bull Chem Soc Jpn 49:1762
93. Deurnie M, Viallet P, Jacquignon P, Croisy-Delcey MC (1976) CR Acad Sci C 283:57
94. Strahle H, Seitz W, Glisten H (1976) Ber Bunsenges Phys Chem 80:288
95. Harrison SW, Nolte HJ, Beveridge DL (1976) J Phys Chem 80:2580
96. Ehrenson S (1976) J Am Chem Soc 98:7510
97. Perram JW, Stiles PJ (1976) Proc R Soc Lond A 349:125
98. Masuhara H, Mataga N, Yoshida M, Taternitsu H, Sakata Y, Misumi S (1977) J Phys Chem 81:879
99. Revail JL, Terryn B (1982) J Chim Phys 79:2
100. Rinaldi D (1982) Comp Chem 6:155
101. Ghoneim N, Suppan P (1995) Spectrochim Acta A 51A:1043
102. Kosower EM (1958) J Am Chem Soc 80:3253

103. Kosower EM (1958) J Am Chem Soc 80:3261
104. Kosower EM (1958) J Am Chem Soc 80:3267
105. Brownstein S (1960) Can J Chem 38:1590
106. Kosower EM, Skorcz JA, Schwarz WM, Patton JW (1960) J Am Chem Soc 82:2188
107. Allerhand A, Schleyer P (1963) J Am Chem Soc 85:371
108. Dimroth K, Reichardt C, Siepmann T, Bohlmann F (1963) Liebigs Ann Chem 661:1
109. Brooker LGS, Craig AC, Heseltine DW, Jenkins PW, Lincoln LL (1965) J Am Chem Soc 87:2443
110. Kosower EM (1968) An introduction to physical organic chemistry. Wiley, New York
111. Dubois JE, Bienvenûe A (1968) J Chim Phys 65:1259
112. Koppel IA, Palm VA (1971) Reakts. Sposob. Org. Soedin 8:291; Engl. Transl. In: Chapman NB, Shorter J (eds) Advances in linear free–energy relationships. Plenum Press, London, p 203
113. Fowler FW, Katritzky AR, Rutherford RD (1971) J Chem Soc B 460
114. Koppel IA, Palm VA (1972) The Influence of the solvent on organic reactivity. In: Chapman NB, Shorter J (eds) Advances in linear free energy relationships, Chapter 5. Plenum Press, London
115. Walter DJ (1974) Prakt Chem 316:604
116. Krygowsky TM, Fawcett WR (1975) J Am Chem Soc 97:2143
117. Dougherty RC (1975) Tetrahedron Lett 8:385
118. Knauer BR, Napier JJ (1976) J Am Chem Soc 98:4395
119. Gutmann V (1977) Chemtech 7:255; and references therein
120. Kamlet MJ, Abboud JL, Taft RW (1977) J Am Chem Soc 99:6027
121. Jones ME, Taft RW, Kamlet MJ (1977) J Am Chem Soc 99:8452
122. Reichardt C, Harbusch-Görnert E (1983) Liebigs Ann Chem 1983:721
123. Dong DC, Winnik MA (1984) Can J Chem 62:2560
124. Swain CG, Powell AL, Alunni SG (1984) J Am Chem Soc 105:502
125. Buncel E, Rajagopal S (1989) J Org Chem 54:798
126. Drago RS (1992) J Chem Soc, Perkin Trans 2:1827
127. Abraham MH (1993) J Phys Org Chem 6:660
128. Abraham MH (1993) Pure Appl Chem 65:2503
129. Reichardt C (1994) Chem Rev 94:2319
130. Catalán J, López V, Pérez P, Martín-Villamil R, Rodríguez JG (1995) Liebigs Ann 1995:241
131. Katritzky AR, Lobanov VS, Karelson M (1995) Chem Soc Rev 24:279
132. Catalán J, Díaz C, López V, Pérez P, Paz JLG, Rodríguez JG (1996) Liebigs Ann 1996:1785
133. Catalán J, Díaz C (1997) Liebigs Ann 1997:1941
134. Katritzky AR, Karelson M, Lobanov VS (1997) Pure Appl Chem 245
135. Katritzky AR, Tamm T, Wang Y, Sild S, Karelson M (1999) J Chem Inf Comput Sci 39:684
136. Katritzky AR, Tamm T, Wang Y, Sild S, Karelson M (1999) J Chem Inf Comput Sci 39:692
137. Catalán J (2001) Solvent effects based on pure solvent scales, in Handbook of solvents (ed Wypych G), Section 10.3. ChemTec Publishing, Toronto
138. Katritzky AR, Fara DC, Hong Y, Tamm K, Tamm T, Karelson M (2004) Chem Rev 104:175
139. Katritzky AR, Fara DC, Kuanar M, Hur E, Karelson M (2005) J Phys Chem A 109:10323
140. Reichardt C, Welton T (2010) Solvents and solvent effects in organic chemistry. Wiley–VCH Verlag
141. Laurence C, Legros J, Chantzis A, Planchat A, Jacquemin D (2015) J Phys Chem B 119:3174
142. Barone V (ed) (2011) Computational strategies for spectroscopy: from small molecules to nanosystems. Wiley, Chichester
143. Hentzsch A (1992) Ber Deutsch Chem Ges 55:953
144. Vreven T, Byun KS, Komaromi I, Dapprich JAM Jr, Morokuma K, Frisch MJ (2006) J Chem Theory Comput 2:815
145. Tomasi J, Persico M (1994) Chem Rev 94:2027
146. Rivail JL, Rinaldi D (1976) Chem Phys 18:233

147. Rinaldi D, Rivail JL (1973) Theor Chim Acta 32:57
148. Rinaldi D, Ruiz-Lopez MF, Rivail JL (1983) J Chem Phys 78:834
149. Dillet V, Rinaldi D, Rivail JL (1994) J Phys Chem 98:5034
150. Ruiz-Lopez MF (2008) The multipole moment expansion solvent continuum model: a brief review. In: S.Canuto (ed) Solvation effects on molecules and biomolecules: computational methods and applications, Springer Series: Challenges and advances in computational chemistry and physics, Chapter 2. Springer
151. Casida M (ed) (1997) Recent advances in density functional methods (Part I). Chapter 5: time-dependent density functional response theory for molecules.; World Scientific Publishing Co. Pte. Ltd., Singapore
152. Jamorski C, Casida ME, Salahub DR (1996) J Chem Phys 104:5134
153. Stratmann RE, Scuseria GE, Frisch MJ (1998) J Chem Phys 109:8218
154. Furche F (2001) J Chem Phys 114:5982
155. Furche F, Ahlrichs R (2002) J Chem Phys 117:7433
156. Scalmani G, Frisch MJ, Mennucci B, Barone V (2006) J Chem Phys 124:094107
157. Jørgensen P, Simons J (eds) (1981) Second quantization based methods in quantum chemistry. Academic, New York
158. Weiss H, Ahlrichs R, Haser M (1993) J Chem Phys 99:1262
159. Davidson ER (1975) J Comput Phys 17:87
160. Improta R, Barone V (2004) J Am Chem Soc 126:14320
161. Barone V, Carnimeo I, Scalmani G (2013) J Chem Theory Comput 9:2052
162. Elstner M, Porezag D, Jungnickel G, Elstner J, Haugk M, Frauenheim T, Suhai S, Seifert G (1998) Phys Rev B 58:7260
163. De Mitri N, Monti S, Prampolini G, Barone V (2013) J Chem Theory Comput 9:4507
164. Cicogna F, Coiai S, Pinzino C, Ciardelli F, Passaglia E (2012) Reac Funct Polym 72:695
165. Egidi F, Russo R, Carnimeo I, DÚrso A, Mancini G, Cappelli C (2015) J Phys Chem A doi:10.1021/jp510542x
166. Cicogna F, Coiai S, Passaglia E, Tucci I, Ricci I, Ciardelli F, Batistini A (2011) J Polym Sci A 49:781
167. Monti S, Cicogna F, Passaglia E, Prampolini G, Barone V (2011) Phys Chem Chem Phys 13:21471
168. De Mitri N, Prampolini G, Monti S, Barone V (2014) Phys Chem Chem Phys 16:16573
169. Marenich AV, Cramer CJ, Truhlar DG (2009) J Phys Chem B 113:6378
170. Rappé AK, Casewit CJ, Colwell KS, Goddard WA, Skiff WM (1992) J Am Chem Soc 114:10024
171. Benassi E, Egidi F, Barone V (2015) J Phys Chem B 119:3155
172. Suppan P (1983) Chem Phys Lett 94:272
173. Mossotti OF (1850) Memorie di matematica e fisica in Modena 24 II:49
174. Clausius R (1879) Die mechanische U'grmetheorie 2:62
175. Karelson MM, Zerner MC (1992) J Phys Chem 96:6949
176. Létard JF, Lapouyade R, Rettig W (1993) J Am Chem Soc 115:2441
177. Van Damme M, Hofkens J, De Schryver FC (1989) Tetrahedron 45:4693
178. Klock AM, Rettig W, Hofkens J, van Damme M, De Schryver FC (1995) J Photochem Photobiol, A 85:11
179. Lippert E (1957) Z. elektrochemie 61:962
180. Lipper E (1955) Z Naturforsch 10A:541
181. Srividya N, Ramasamy P, Radhakrishanan VT (1997) Spectrochim Acta A 53:1743
182. PCMODEL, Molecular modeling software. Serena Software, Bloomington
183. Maus M (1998) Ph.D. Thesis, Humboldt Universität zu Berlin
184. Lahmani F, Breheret E, Zehnacker-Rentien A, Amatore C, Jutand A (1993) Photochem Photobiol A 70:39
185. Lahmani F, Breheret E, Benoist d'Azy O, Zehnacker-Rentien A, Delouis JF (1995) Photochem. Photobiol A 89:191
186. Carlotti B, Flamini R, Kikaš I, Mazzuccato U, Spalletti A (2012) Chem Phys 407:9

187. Baraldi I, Benassi E, Ciorba S, Šindler-Kulyk M, Škorić I, Spalletti A (2009) Chem Phys 361:61
188. Kawski A, Gryczyński I, Jung C, Heckner KH (1977) Z Naturforsch A 32A:420
189. Bakhshiev NG (1961) Opt Spcktrosk 10:379
190. Bilot L, Kawski A (1963) Z Naturforsch A 18A:10
191. Bilot L, Kawski A (1963) Z Naturforsch A 18A:256
192. Bakhshiev NG (1964) Opt Spektrosk 16:821
193. Bakhshiev NG (1962) Opt Spectrosk 13:24
194. Kawski A (2002) Z Naturforsch A 57A:255
195. Kawski A, Kuklinski B, Bojarski P (2002) Z Naturforsch A 57A:716
196. Suppan P, Ghoneim N (1997) Solvatochromism. The Royal Society of Chemistry, London
197. Koutek B (1978) Collect Czech Chem Commun 43:2368. Onsager cavity radius $a_0 = 4.41$ Å, from DFT Cam-B3LYP/6-31 + G(d)//IEF-PCM(UFF) calculations (present work)
198. Cerón-Carrasco JP, Jacquemin D, Laurence C, Planchat A, Reichardt C, Sraïdi K (2014) J Phys Chem B 118:4605
199. Ravi M, Samanta A, Radhakrishnan TP (1994) J Phys Chem 98:9133
200. Patil SK, Wari MN, Yohannan Panicker C, Inamdar SR (2014) Spectrochim Acta A 123:117
201. Manohara SR, Udaya Kumar V, Shivakumaraiah, Gerward L (2013) J Mol Liq 181:97
202. Homocianu M, Airinei A, Dorohoi DOJ (2011) Adv Res Phys 2:011105
203. Butler RS, Cohn P, Tenzel P, Abboud KA, Castellano RK (2009) J Am Chem Soc 131:623
204. Choi JY, Park EJ, Chang SH, Kang TJ (2009) J Bull Korean Chem Soc 30:1452
205. Patra A, Hebalkar N, Sreedhar B, Radhakrishnan TP (2007) J Phys Chem C 111:16184
206. Vázquez ME, Blanco JB, Imperiali B (2005) J Am Chem Soc 127:1300
207. Webb MA, Morris BC, Edwards WD, Blumenfeld A, Zhao X, McHale JL (2004) J Phys Chem A 108:1515
208. Saha S, Samanta A (2002) J Phys Chem A 106:4763
209. Bruni S, Cariati E, Cariati F, Porta FA, Quinci S, Roberto D (2001) Spectr Acta A 57:1417
210. Zhao X, Burt JA, Knorr FJ, McHale JL (2001) J Phys Chem A 105:11110
211. Samanta A, Saha S, Fessenden RW (2001) J Phys Chem A 105:5438
212. Samanta A, Fessenden RW (2000) J Phys Chem A 104:8577
213. Bardeen CJ, Rosenthal SJ, Shank CV (1999) J Phys Chem A 103:10506
214. Mente SR, Maroncelli M (1999) J Phys Chem B 103:7704
215. Saroja G, Ramachandram B, Saha S, Samanta A (1999) J Phys Chem B 103:2906
216. Saha S, Samanta A (1998) J Phys Chem A 102:7903
217. Matyushov DV, Schmid R, Ladanyi BM (1997) J Phys Chem B 101:1035
218. The N07 and SNS basis sets are available in the download section of http://dreams.sns.it
219. Shamasundar KR, Knizia G, Werner HJ (2011) J Chem Phys 135:054101
220. Neese F (2003) J Chem Phys 119:9428
221. Watson TJ, Lotrich VF, Szalay PG, Perera A, Bartlett RJ (2013) J Phys Chem A 117:2569
222. Caricato M, Trucks GW, Frisch MJ, Wiberg KB (2010) J Chem Theory Comput 6:370
223. Biczysko M, Bloino J, Santoro F, Barone V (2011) Computational strategies for spectroscopy: from small molecules to nanosystems (ed Barone V), Chapter 8. Wiley, Chichester
224. Jacquemin D, Wathelet V, Perpète EA, Adamo C (2009) J Chem Theory Comput 5:2420
225. Vlček A Jr, Záliš S (2007) Coord Chem Rev 251:258
226. Barone V, Baiardi A, Biczysko M, Bloino J, Cappelli C, Lipparini F (2012) Phys Chem Chem Phys 14:12404
227. Licari D, Baiardi A, Biczysko M, Egidi F, Latouche C, Barone V (2015) J Comput Chem 36:321
228. Matyushov DV, Newton MD (2001) J Phys Chem A 105:8516
229. Marcus RA (1963) J Chem Phys 38:335
230. Marcus RA (1963) J Chem Phys 39:460
231. Marcus RA (1963) J Chem Phys 39:1734
232. Marcus RA (1965) J Chem Phys 43:1261

233. Marcus RA (1989) J Phys Chem 93:3078
234. Ferrer FJA, Cerezo J, Soto J, Improta R, Santoro F (2014) Comp Theor Chem 1040–1041:328
235. Mennucci B, Cappelli C, Guido CA, Cammi R, Tomasi J (2009) J Phys Chem A 113:3009
236. Improta R, Barone V, Scalmani G, Frisch MJ (2006) J Chem Phys 125:54103
237. Ferrer FJA, Improta R, Santoro F, Barone V (2011) Phys Chem Chem Phys 13:17007
238. Improta R, Scalmani G, Frisch MJ, Barone V (2007) J Chem Phys 127:074504
239. Improta R, Barone V, Santoro F (2007) Angew Chem Int Ed 46:405
240. Improta R, Barone V, Santoro F (2007) J Phys Chem B 111:14080
241. Latouche C, Baiardi A, Barone V (2014) J Phys Chem B (in press)
242. Perdew JP (1986) Phys Rev B 33:8822
243. Perdew JP, Burke K, Wang Y (1996) Phys Rev B 54:16533
244. Becke AD (1993) J Chem Phys 98:5648
245. Jr Dunning T H, Hay PJ (1977) Mod Theor Chem 3:1
246. Wadt WR, Hay PJ (1985) J Chem Phys 82:284
247. Hay PJ, wadt WR (1985) J Chem Phys 82:299
248. Hay PJ, Wadt WR (1985) J Chem Phys 82:270
249. Bloino J, Biczysko M, Santoro F, Barone V (2010) J Chem Theory Comput 6:1256
250. Grimme S, Antony J, Ehrlich S, Krieg H (2010) J Chem Phys 132:154104
251. Tomasi J, Mennucci B, Cancès E (1999) J Mol Struct: Theochem 464:211
252. *GAUSSIAN 09*, Revision D.01, Frisch MJ, Trucks GW, Schlegel HB, Scuseria GE, Robb MA, Cheeseman JR, Scalmani G, Barone V, Mennucci B, Petersson GA, Nakatsuji H, Caricato M, Li X, Hratchian HP, Izmaylov AF, Bloino J, Zheng G, Sonnenberg JL, Hada H, Ehara M, Toyota K, Fukuda R, Hasegawa J, Ishida M, Nakajima T, Honda Y, Kitao O, Nakai H, Vreven T, Montgomery JA, Peralta JE Jr, Ogliaro F, Bearpark M, Heyd JJ, Brothers E, Kudin KN, Staroverov VN, Kobayashi R, Normand J, Raghavachari K, Rendell A, Burant JC, Iyengar SS, Tomasi J, Cossi M, Rega N, Millam JM, Klene M, Knox JE, Cross JB, Bakken V, Adamo C, Jaramillo J, Gomperts R, Stratmann RE, Yazyev O, Austin AJ, Cammi R, Pomelli C, Ochterski JW, Martin RL, Morokuma K, Zakrzewski VG, Voth GA, Salvador P, Dannenberg JJ, Dapprich S, Daniels AD, Farkas O, Foresman JB, Ortiz JV, Cioslowski J, Fox DJ, Gaussian, Inc., Wallingford CT (2009)
253. Barata-Morgado R, Sánchez ML, Galván IF, Corchado JC, Martín ME, Muñoz-Losa A, Aguilar MA (2013) Theor Chem Acc 132:1390
254. Benassi E, Cappelli C, Carlotti B, Barone V (2014) Phys Chem Chem Phys 16:26963
255. Bartosewicz B, Kawski A (1971) Bull Acad Pol Sci Ser Sci Math Astronom Phys 19:249
256. Deurnie M, Viallet P, Jacquignon P, Croisy-Delcey MC (1976) R Acad Sci C 283:57
257. Strahle H, Seitz W, Glisten H (1976) Ber Bunsenges Phys Chem 80:288
258. For dye C, the angle formed by g and the transition moment has been measured experimentally, viz. ≈ 30°, see Ref. 242
259. Liptay W, Weisenberger H, Tiemann F, Eberlein W, Konopka G (1968) Z Naturf 23A:377

Index

J.-L. Rivail et al. (eds.), *Quantum Modeling of Complex Molecular Systems*, Challenges and Advances in Computational Chemistry and Physics 21, DOI 10.1007/978-3-319-21626-3

The manufacturer's authorised representative in the EU is Springer Nature Customer Service Centre GmbH, Europaplatz 3, 69115 Heidelberg, Germany. If you have any concerns regarding our products, please contact ProductSafety@springernature.com

Printed and bound by CPI Group (UK) Ltd, Croydon, CR0 4YY
01/10/2026
02225872-0018